AF344467

Biomaterials Science and Tissue Engineering

In the last couple of decades, the field of biomaterials science and tissue engineering has arrived at the frontier of research and innovation, considering the number of scientific discoveries and their potential impact in treating human diseases. This topical area is the focus of this texbook. The textbook has been divided into four sections. Section I provides an overview of the subject area with a focus on the application in human healthcare. Many important terms related to the field of Biomaterials Science and Tissue Engineering are also defined. Section II primarily focuses on discussing fundamental topics of Materials Science, including manufacturing of biomaterials, probing material structures at multiple length scales as well as mechanical properties. Section III comprehensively covers topics such as structure and properties of cell and proteins, cell-material interaction and biocompatibility, probing cell response, *in vitro*; bacterial growth and biofilm formation and probing tissue response, *in vivo*. The contents of Section III certainly make the book a unique pedagogical asset to readers without any formal knowledge in biological sciences. Section IV discusses several case studies, including corrosion/wear of Ti-based alloys and calcium phosphate-based multifunctional composites for bone replacement applications. This section also closes with author's perspectives on future growth of the field. The book offers extensive pedagogical features including multiple choice questions, fill in the blanks, review questions, numerical problems and solutions to selected problems.

Bikramjit Basu is Professor at Materials Research Center and Associate faculty at Center for BioSystems Science and Engineering, Indian Institute of Science, Bangalore. His current research integrates biomaterials and bioengineering approaches to develop new generation biomaterials and to address clinically relevant research problems for human diseases. Professor Basu is the recipient of the prestigious Shanti Swarup Bhatnagar Prize (2013) by the Government of India and Robert L. Coble Award for Young Scholars (2008) by the American Ceramic Society. He is an elected Fellow of the American Institute for Medical and Biological Engineering (2017), Indian National Academy of Engineering (2015) and National Academy of Sciences, India (2013).

CAMBRIDGE–IISc SERIES

Cambridge–IISc Series aims to publish the best research and scholarly work on different areas of science and technology with emphasis on cutting-edge research.

The books will be aimed at a wide audience including students, researchers, academicians and professionals and will be published under three categories: research monographs, centenary lectures and lecture notes.

The editorial board has been constituted with experts from a range of disciplines in diverse fields of engineering, science and technology from the Indian Institute of Science, Bangalore.

IISc Press Editorial Board:

G. K. Ananthasuresh, *Professor, Department of Mechanical Engineering*

K. Kesava Rao, *Professor, Department of Chemical Engineering*

Gadadhar Misra, *Professor, Department of Mathematics*

T. A. Abinandanan, *Professor, Department of Materials Engineering*

Diptiman Sen, *Professor, Centre for High Energy Physics*

Titles in print in this series:

- *Continuum Mechanics: Foundations and Applications of Mechanics* by C. S. Jog
- *Fluid Mechanics: Foundations and Applications of Mechanics* by C. S. Jog
- *Noncommutative Mathematics for Quantum Systems* by Uwe Franz and Adam Skalski
- *Mechanics, Waves and Thermodynamics* by Sudhir Ranjan Jain
- *Finite Elements: Theory and Algorithms* by Sashikumaar Ganesan and Lutz Tobiska
- *Ordinary Differential Equations: Principles and Applications* by A. K. Nandakumaran, P. S. Datti and Raju K. George

Cambridge–IISc Series

Biomaterials Science and Tissue Engineering

Principles and Methods

Bikramjit Basu

CAMBRIDGE
UNIVERSITY PRESS

CAMBRIDGE
UNIVERSITY PRESS

University Printing House, Cambridge CB2 8BS, United Kingdom

One Liberty Plaza, 20th Floor, New York, NY 10006, USA

477 Williamstown Road, Port Melbourne, vic 3207, Australia

4843/24, 2nd Floor, Ansari Road, Daryaganj, Delhi – 110002, India

79 Anson Road, #06–04/06, Singapore 079906

Cambridge University Press is part of the University of Cambridge.

It furthers the University's mission by disseminating knowledge in the pursuit of education, learning and research at the highest international levels of excellence.

www.cambridge.org
Information on this title: www.cambridge.org/9781108415156

© Bikramjit Basu 2017

This publication is in copyright. Subject to statutory exception and to the provisions of relevant collective licensing agreements, no reproduction of any part may take place without the written permission of Cambridge University Press.

First published 2017

Printed in India by Nutech Print Services, New Delhi

A catalogue record for this publication is available from the British Library

ISBN 978-1-108-41515-6 Hardback

Additional resources for this publication at www.cambridge.org/9781108415156

Cambridge University Press has no responsibility for the persistence or accuracy of URLs for external or third-party internet websites referred to in this publication, and does not guarantee that any content on such websites is, or will remain, accurate or appropriate.

Contents

Section II　Fundamentals – Materials Science

Section III   Fundamentals – Biological Science

Section IV Illustrative Examples of Biomaterials Development

Note:
- A number of statements appear within boxes in various chapters. The statements within shaded box refer to definition-like statements, while the statements within unshaded boxes reflect the author's perspective / important concepts.
- Coloured figures are placed at the end of the book and cross-referred in the main text at their respective appearances.

Foreword I

Over the last few decades, biomaterials science and biomedical engineering have been perceived as being among the fastest growing areas of research and innovation within the engineering science community when considering the number of scientific discoveries and their societal impact. To substantiate the relevance for human healthcare, degenerative and inflammatory problems of bone and joints affect millions of people worldwide. Due to the growing geriatric population, these problems account for half of the chronic diseases in people over 50. Osteoporosis, for example, is the most prevalent bone degenerative disease, particularly among the middle-aged population and post-menopausal women worldwide. The upsurge in the clinical demand for reconstructive joint replacements requires new implants with better biocompatibility properties, outweighing existing biomaterial solutions. In spite of remarkable advances in pharmacological, interventional, and surgical therapies, neurodegenerative and stroke disorders remain among the leading causes of mortality and lifelong impairments in humans.

In order to address biomedically relevant challenges in orthopedics as well as neural and cardiovascular diseases, researchers must blend the fundamental concepts of engineering sciences (materials science and electrical engineering), basic sciences (chemistry and physics), and biological sciences (cell and molecular biology) to engineer synthetic tissue replacements and develop novel healing strategies. Such an interdisciplinary research approach requires understanding across the boundary of remotely linked scientific disciplines. Researchers can develop innovative ideas, as well as understand the language of this important research area of societal relevance.

It has been noted globally that many accomplished researchers, as well as young researchers, pursuing the field of biomaterials and biomedical engineering are not formally trained in biology and medical sciences. Nevertheless, they are attempting to think laterally, blending sufficient knowledge of biological systems with engineering sciences to develop biomedical materials, ultimately impacting the field of Biomedical Engineering. Lately, unprecedented growth in the fields of biomaterials and biomedical engineering has revolutionized personalized healthcare.

This book emphasizes the fundamentals of both Materials and Biological Sciences. On the Materials science front, it contains chapters which provide non-specialists with a fundamental understanding on the conventional and advanced manufacturing techniques as well as mechanical properties. Clearly, the strength of this textbook lies in the clear description of the *in vitro* and *in vivo* biocompatibility assessment protocols, an asset for non-biologists. The conclusion presents a number of chapters describing case studies, primarily from the author's own research. The number of problem sets and assignments are also important attributes.

I find this much-needed textbook timely and valuable for the biomaterials community.

Cato T. Laurencin, MD, PhD
University Professor,
Albert and Wilda Van Dusen Distinguished Professor of Orthopedic Surgery,
Professor of Chemical and Biomolecular Engineering,
Professor of Materials Science and Engineering,
Professor of Biomedical Engineering,
Director, The Institute for Regenerative Engineering,
Chief Executive Officer, Connecticut Institute for Clinical and Translational Science
The University of Connecticut, USA

Foreword II

Professor Bikramjit Basu's new book "Biomaterials Science and Tissue Engineering" is ideal for biologists who wish to understand, in more depth, the biology-materials connection. On the other hand, materials scientists and materials engineers will find a wealth of information on biological concepts needed to fully exploit applications of materials in biology. The book is extremely well structured and every chapter is critical for anyone planning to design medical devices or implants. For example, the inclusion of a chapter on biofilms is a wonderful addition, which you will not normally find in a biomaterials text. Many implants suffer from infection-related failures.

The basic materials science chapters deal with a variety of biomaterials, principles governing the properties of materials, including materials processing and manufacturing. Then the book develops the principles of biocompatibility, materials-tissue and materials-cell interactions. It impressively integrates materials and biology concepts needed for selecting materials for human application as well as engineering them to optimize their properties, biocompatibility and bio-functionality.

Another attractive aspect of the book, especially for experimentalists and students, is the inclusion of testing and characterization techniques, both materials and biological test methodologies, and the designing and planning of animal experiments, and associated ethical issues to be considered.

The book could be an excellent textbook for a course in biomaterials and a great reference for those working in the field. For the last thirty years I have been teaching graduate level materials science both in the US and India. This would be a book I would use as a mandatory course reference, because I want to include more biological considerations into the standard materials science course.

Shantikumar Nair, PhD
Dean of Research
Director, Center for Nanosciences,
Amrita Vishwa Vidyapeetham University
Kochi, India

Formerly Professor, University of Massachusetts, Amherst, MA, USA.

Preface

Biomaterials, recognized as a new class of materials in the Materials Science community, are being widely developed in last few decades. This specific class of materials has received significant attention because of their potential applications to repair and regenerate tissues in human musculoskeletal system, and thereby augment disease treatment modalities. The field of biomaterials science and tissue engineering has therefore large relevance for human healthcare. The design and development of biomaterials requires the integration of the concepts and expertise from two disconnected disciplines, i.e. Materials Science and Biological Science. While such integration is not an easy task by any means, researchers have put in extensive efforts in this direction. The importance of the field of biomaterials is increasingly being noticed in the Materials community; a compulsory course on this subject is being taught at undergraduate and graduate levels in most top universities around the world. It has been widely perceived that the education and training of next generation researchers can be accomplished more effectively with the availability of a textbook on the subject, which should cater to the requirements of the readers from both materials and biological sciences disciplines.

In the above backdrop, this book "Biomaterials Science and Tissue Engineering: Principles and Methods" begins with an overview of biomaterials and tissue engineering scaffolds (section I). This is followed by three well-structured sections, with section II discussing the fundamentals of Materials Science, as relevant for Biomaterials Science. Considering the significant breadth of the field of biological sciences, Section III of this book describes only the most necessary concepts and techniques of cell and molecular biology with a focus on the application of such knowledge in evaluating the biocompatibility property in a broad sense. The last section essentially illustrates various aspects of biomaterials development, primarily from author's own research. This book is meant for readers, who will be introduced to the broad area of biomaterials. Sections II and III set the floor for the readers to understand necessary fundamentals related to Materials and Biological sciences, as applied to Biomaterials science. Therefore, this textbook will be extremely useful to those readers, who want to pursue research in the field of Biomaterials Science without a formal background either in Materials or Biological science. While conceiving this textbook, the author wanted to motivate young researchers as well as to provide experts in the area with a healthy balance of topics for teaching/academic purposes. It is expected that the book will benefit senior undergraduate as well as graduate students.

In particular, this textbook has the following distinguishing features:

(a) Integration of the concepts of Materials Science and Biological Science, facilitating the use of this book as a textbook for teaching as well as for research purposes.

(b) Coverage of the necessary fundamentals of cell / molecular biology, which is often difficult to extract to an appropriate extent from various available textbooks of biological sciences in qualitative and quantitative manner [structure and properties of cells, tissues, bones, collagen, proteins, cell fate process (migration, differentiation, apoptosis, division) as well as cellular signaling processes].

(c) Detailed discussion on the processing, structure and properties of materials for biomedical applications (metals, ceramics, polymers and their composites), together with techniques and guidelines.

(d) Coverage of *in vitro* and *in vivo* biocompatibility property evaluation of materials for bone, neural as well as cardiovascular tissue engineering applications, together with protocols.

Altogether the book contains 18 chapters, with eleven chapters in the fundamentals section and the rest of the chapters being illustrative examples of biomaterials development. In chapter 1, the field of biomaterials is introduced with a special emphasis to distinguish biomaterials as a special class of functional materials and portray how they are different from other material classes. This introductory chapter also outlines the motivation for the development of new biomaterials to mimic the natural tissue properties. Various important terms are defined in this chapter. In chapter 2, the use of different primary material classes (metals, ceramics and polymers) for biomedical applications is discussed, which is followed by the classification of biomaterials on the basis of their biocompatibility. A detailed explanation on the various processing aspects of the tissue engineering scaffolds, with a special emphasis on the surface modification to enhance the biocompatibility properties, has been discussed in chapter 3. The processing of scaffolds and implants are markedly different as the latter involves the conventional processing approaches. To this end, chapter 4 introduces a number of fabrication techniques to prepare metals, ceramics and polymeric biomaterials with an emphasis on the processing science aspects. Chapter 4 also discusses the bulk deformation processes as well as machining and joining techniques, as applicable to metallic implants. After discussing the conventional processing methods, the additive manufacturing techniques are discussed with emphasis on powder based 3D printing technique.

Many researchers, without any formal background in material science, utilize a number of materials characterization techniques to characterise the structure or to measure the physical properties. The fundamental aspects of many of these techniques is discussed in chapter 5 with necessary theoretical background. For bone-tissue engineering applications, the implantable biomaterials need to have a set of desired mechanical properties. While the mechanical reliability of metallic implants is well-established, such reliability for ceramic implants is a matter of major concern for many clinicians. To this end, the fundamental aspects of deformation and fracture of ceramics and polymers are largely discussed in chapter 6. Apart from providing theoretical foundation, the experimental techniques to measure various mechanical properties are also mentioned. An important aspect of this section is the science-based discussion on the origin of brittle fracture and strength variability of ceramics. The concept of fracture toughness, measurement of various mechanical properties as well as brief discussion on toughening mechanisms are also presented in chapter 6. Five chapters in the fundamental section discuss the necessary biological foundation of this book. In particular, chapter

7 discusses the structure and properties of cells, proteins and bacteria. Various cellular adaptation processes as well as cell fate processes are also discussed. In chapter 8, the biocompatibility concept is introduced and the implication of biocompatibility in the context of cell-material interaction is critically discussed. The mechanistic description of cell-material interaction is also discussed. In chapter 9, various *in vitro* biochemical assays for cytocompatibility of biomaterials are extensively discussed. A large number of complimentary assays are mentioned to quantify the cell viability and proliferation. More importantly, many advanced cell biological techniques, like flow cytometry are also critically discussed. The ethical issues related to stem cell study are also mentioned in chapter 9. Bacterial growth and biofilm formation is addressed in chapter 10, which mainly includes bacterial classification, bacteria-material interaction and experiment assessment. Chapter 11 includes tissue compatibility assessment with a particular focus on pre-clinical testing in different animal models as well as ethical issues related to such studies. A few examples on the protocols to be followed in conducting pre-clinical studies.

In section IV, six chapters contain illustrative examples of biomaterials development. In the first of such chapters (chapter 12), the corrosion properties of some selected new titanium based alloy are presented. In chapter 13, the processing of calcium phosphate-mullite composites and their cytocompatibility, genotoxicity and *in vivo* biocompatibility are discussed. The next chapter i.e. Chapter 14 presents the case study on HDPE based hybrid composites using compression molding route and their biocompatibility properties are also summarized. One of the major issues in the development of HA based materials is the bactericidal property without compromising cytocompatibility properties. These aspects have been illustrated while discussing the development of HA-Ag composites in chapter 15. Next, chapter 16 discusses the processing related challenges as well as good toughness and biocompatibility properties together with desired functional properties in HA-based electroconductive composites with $CaTiO_3$ second phase. The next chapter discusses the compatibility of neuronal and cardiac cells on patterned carbon substrates. The proliferation of cardiac tissue-specific cells on PLGA-carbon nanofiber substrates is majorly discussed in chapter 17. At the end, chapter 18 closes with author's perspective on the subject. For the benefit of the students and college teachers, this book also contains an Appendix section with an array of questions of various formats for the self-assessment of the readers and for the examination purpose. The solution of some of the selected problems is also provided.

This book is an outcome of the several years of teaching undergraduate and postgraduate level courses on Biomaterials, being offered to students of Indian Institute of Technology Kanpur, India and Indian Institute of Science, Bangalore. Several chapters also reflect the extensive research by the author's research group, both at IIT Kanpur and IISc, Bangalore during the last two decades, which has been supported by the Council of Scientific and Industrial Research (CSIR), Department of Biotechnology (DBT), Department of Science & Technology (DST), Indo-US Science and Technology Forum (IUSSTF), and the UK-India Education Research Initiative (UKIERI). The author would also like to mention the recent multi-institutional research program 'Translational Centre on Biomaterials for Orthopedic and Dental Applications', supported by Department of Biotechnology, Government of India. Similarly, the funding from Science and Engineering Research Board of DST under the umbrella of 'National Network for Mathematical and Computational Biology' supported recent research in the author's group. The author also acknowledges the ongoing collaboration and interaction under the umbrella of this centre with several colleagues, including

Drs H. K. Varma, Vamsi Krishna Balla, Amit Roy Chowdhury, Debasish Sarkar, R. Joseph Ben Singh, Biswanath Kundu, Manoj Komath, Sivaranjani Gali, Vibha Shetty, A. Sabareeswaran, Aroop Kumar Dutta, Ranjana C. Dutta, Tony McNally, B. Ravi, Michael Gelinsky, Jonathan Knowles, Tom Joyce and B. Vaidhyanathan.

Some present and past group members, who deserve special mention, include Greeshma T., B. Sunil Kumar, Ravi Kumar K., Yashoda Chandorkar, Anupam Purwar, Ragini Mukherjee, Atasi Dan, B. V. Manoj Kumar, Amartya Mukhopadhyay, G. B. Raju, Indu Bajpai, Shekhar Nath, Subhadip Bodhak, Atiar R. Molla, Naresh Saha, Shouriya Dutta Gupta, Garima Tripathi, Alok Kumar, Shilpee Jain, Ashutosh K. Dubey, Shibayan Roy, Ravi Kumar, Prafulla Mallik, R. Tripathy, Brajendra Singh, U. Raghunandan, Divya Jain, Nitish Kumar, and Sushma Kalmodia. The author also acknowledges the past and present research collaboration with a number of researchers and academicians, including Drs Omer Van Der Biest, Jozef Vleugels, K. Lambrinou, M. Chandrasekaran, R. K. Bordia, B. V. S. Murthy, Dileep Singh, M. Singh, T. Goto, Suk-Joong L. Kang, F. Wakai, Subhash Risbud, William Fahrenholtz, Bill Lee, Jon Binner, David Green, F. Sanjay Mathur, T. J. Webster, Amar S. Bhalla, Ruyan Guo, Mauli Agrawal, Artemis Stamboulis, G. Sundararajan, Ananya Barui, Pallab Datta, S. Kanagaraj, M. Ravishankar, D. Mazumdar, Srikumar Banerjee, K. S. Ghosh, V. Verma, Rajeev Gupta, Mira Mohanty, P. V. Mohanan, Ender Suvaci, Hasan Mondal, Ferhat Kara, Nurcan Kalis Ackibas, S. J. Cho, Doh - Yeon Kim, J. H. Lee, Alok Pandey, Arvind Sinha, and Animesh Bose.

The author is grateful for the suggestions from several colleagues, including Professors P. Balaram, Ashutosh Sharma, David Kaplan, B.D. Malhotra, S. C. Koria, David Williams, G. Padmanaban, Indranil Manna, Kamanio Chattopadhyay, Sandya Visweswariah, Annapoorni Rangarajan, Polani Seshagiri, Saumitra Das, K. K. Nanda, N. Ravishankar, Abhishek Singh, Prabeer Barpanda, A. M. Umarji, Vikram Jayaram, Dipankar Banerjee and N. K. Mukhopadhayay. I also acknowledge the long term association with a few of my colleagues, including Professors Seeram Ramakrishna, Thomas Webster, Mauli Agrawal, Arvind Sinha, Suprabha Nayar, Rinti Banerjee, Raman Singh, Anish Upadhyaya, Kantesh Balani, Malay Banerjee, Krishanu Biswas, Santanu Dhara, Satyam Suwas, Debrupa Lahiri and Sourabh Ghosh, who have provided several suggestions during the course of writing this book. A few chapters of this book are critically reviewed by Professors M. S. Valiathan, H. S. Maiti, Abhay Pandit, Brian Derby, C.P. Sharma, Dieter Scharnweber, Alok Dhawan, Aditya Murty, K. Chatterjee, S. Basu and S. Bose. I am particularly grateful to them. The author also acknowledges support from the colleagues of SCTIMST, Trivandrum, including Drs C. P. Sharma, H. K. Varma, C. V. Muraleedharan, A. Sabareeswaran and Sahin J. Shenoy during the course of writing this book particularly on the biomedical application perspective of biomaterials. The clinical and commercial perspective on biomaterials development, as summarized briefly in this book is largely credited to the author's continuous discussion with his colleagues, Dr D. C. Sundaresh (Honorary Director, Sri Sathya Sai Institute of Higher Medical Sciences, Bangalore), Mr. Ravi Sarangapani (Vice President, Smith & Nephew, Pune), Dr Tanvir Momen (Orthopedic surgeon, Woodlands hospital, Kolkata), Dr K. H. Sancheti (Sancheti hospital, Pune) and Dr T. R. Rajesh (Cardiothoracic surgeon, Sparsh hospital, Bangalore). I profusely thank Smith & Nephew, Inc. for providing me several biomedical device images and for according necessary approval to use them in this book. Some chapters of this book, particularly those of section III, were exclusively reviewed by my colleagues at IISc (Profs. Deepak Saini, Ramray Bhat and Ravi Sundareshan) and I am grateful to them. Similarly, I appreciate significant help rendered by my group members

(Manoj Kumar, Ashutosh, Amartya, Greeshma, Sunil, Ravi, Nitu, Sharmistha, Madhuri Dey) in scientifically reviewing and/or extensively editing a few chapters of this book.

I would also like to thank the Centre for Continuing Education, IISc, Bangalore for extending financial support during the writing of this book. The author likes to express his gratitude to his long-time friend and collaborator, Dr Jaydeep Sarkar for his constant inspiration during the writing of this book. Last but not the least, the author is extremely grateful to Dr Baldev Raj, Director, National Institute of Advanced Studies, Bangalore for his constant inspiration to motivate me to take up this important and satisfying assignment. The author also expresses sense of gratitude for the help rendered by the office of IISc Press (Profs. G. K. Ananthasuresh, T.A. Abinandanan, G. Mishra, A. Chakrabarti, Mrs. Kavitha Harish and Mr. Sreenivasa Rao). The great help rendered by Nitu Bhaskar, Prerana S., Shubham Jain, Asish Kumar Panda, Vignesh, Sherine Alex, Srimanta Barui, Sourav Mandal, Subhadip Basu, Sharmishtha Naskar, Subhomoy Chatterjee, Sivaranjani Gali, Rahul Kumar Upadhyay, Ranjith Kumar P, Shardul Bhusari, Subhendu Pandit, Karunakaran K, Swati Sharma, Vidushi and Shruti Nair during the writing of this book is gratefully acknowledged. I express my sincere thanks to Gopinath N. K. for his painstaking efforts during the manuscript preparation and proof corrections of this book. I profusely thank Professor Cato Laurencin and Professor Shantikumar Nair for writing the foreword for this book. Finally, I am grateful to my parents (Manoj Mohan Basu and Chitra Basu), my wife (Pritha Basu) and son (Prithvijit Basu) for their constant moral support and encouragement. I received significant support and constant inspiration in my academic life from the members of extended family, uncles (particularly, Late Mihir Mohan Basu, Late Ranjit Majumdar, Late Sadhan Karmakar, Dr. J. N. Hazra, Dr. Amitava Maiti, Dr. S. K. Pal), aunts (particularly, Mrs. Maya Mukherjee, Mina Majumdar, Manju Hazra, Sipra Karmakar, Pika Maiti), cousins (particulalry, Mr. Jayanta Bose, Diganta Bose, Prasanta Bose, Simanta Basu, Dr. Navojit Basu, Dr. Smita Ganguly, Mrs. Simi Gupta, Mrs. Ipsita Banerjee) and my in-laws family (particularly, Sri Parimal Kumar Choudhury, Late Nirmal Kumar Choudhury, Late Sima Choudhury, Dr. Rita Paul Choudhury, Mrs. Mousumi Mitra). I am grateful to all of them. It is heartening to mention that both my father, at the age of 75, and wife have put in extensive efforts to critically proofread significant portions of this manuscript!

SECTION I

Overview

Introduction

Biomaterials Science, a multidisciplinary subject primarily at the intersection of materials and biological sciences, will be discussed in this chapter together with key concepts. The chapter will also introduce the readers to the major developments in the field of biomaterials over the last few decades with the recent paradigm shift towards more biologically active materials. The relevance of this important interdisciplinary field of science is further highlighted by referring to biomaterials based solutions for human diseases. Summarising, this chapter serves as a platform to understand the evolution of biomaterials as a distinct class of materials, while emphasizing the important role played by this class of materials in human healthcare.

1.1 | Background

Although biomaterials as a field formally came into existence only after the Second World War, the development of this important class of materials has been well documented throughout history[1, 2, 3]. Ancient civilizations used ants to naturally stitch a wound. In those days, the two ends of the wound were held together and the ant bites were used to close the wounds[4]. As mentioned in Fig. 1.1, Sushruta, the ancient Indian surgeon reported the use of sutures made from naturally occurring materials such as human hair, flax and hemp to close skin incisions[5,6]. It is worthwhile to mention here that an Indian orthopedic company (Sushrut Surgicals Pvt. Ltd.) is named after this ancient Indian surgeon (Sushruta) and it was established in 1973. This specific company was known in the field of orthopedic implants and instrumentation for traumatology, spine and reconstructive surgery. This company, which is until recently known as Adler Mediequip Pvt. Ltd. to orthopedic surgeons, is lately acquired by Smith & Nephew.

As far as early evolution of biomaterials is concerned, the Greek physician, Galen of Pergamum has also described catgut sutures to close large wounds[1,7]. Nacre from sea shells was used as a substitute for teeth, while metals, such as gold and wrought iron were used as dental implants in the early ages[1]. Such examples highlight the use of an implant to compensate or retrieve the loss of an anatomical function. However, it is important to note that these materials were used without an assessment of the host response, as will be scientifically explained later in this book. Their long term success itself was assumed as a proof of their efficacy.

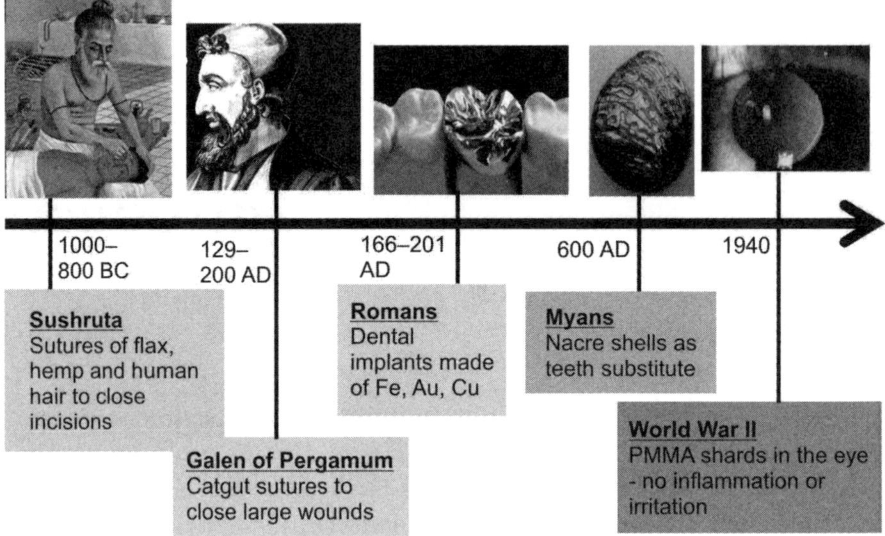

Fig. 1.1 Historical evolution of the biomaterials over last several decades[35].

The field of biomaterials science blossomed into a multidisciplinary subject mainly after the Second World War. Polymethyl methacrylate (PMMA) shards that were accidentally lodged in the eyes of aviators did not create inflammation or an adverse tissue reaction. This inspired the British Ophthalmologist Sir Harold Ridley to create early intra-ocular lenses made from poly methyl methacrylate. Today, they have been replaced by polydimethyl siloxane (PDMS) lenses and cataract associated blindness can be avoided with the use of PDMS lenses. Some recent projections state that around 14 million cataract surgeries will be needed every year in 2016–2020 in India alone[8], and more than 20 million surgeries were performed worldwide in 2015[9].

Another fine example of biomaterials in the context of musculoskeletal diseases is that of the total hip replacement prosthesis, which has evolved to become one of the most successful surgeries[2,3]. A retrospection of the history of development of this prosthesis is a proof of how prudent material selection can increase success rates. While early attempts using metals such as stainless steel and acrylics failed due to material design related issues, the use of a teflon acetabular cup by Dr. Charnley marked the onset of the modern age orthopedics[1]. High density polyethylene and stainless steel are used for total hip replacement and other orthopedic applications. Total hip replacement has emerged as one of the most successful surgeries, and over 300,000 surgeries are performed annually worldwide[10].

Other examples of successful translations of biomaterials research include dental implants, the artificial kidney, breast implants, vascular grafts and stents, pacemakers, heart valves, etc. The sheer range of applications of these biomedical devices and their widespread use are indicated by their ever-increasing numbers. These examples highlight how biomaterials have evolved and become an integral part of our life today. In most of these cases, the materials used initially were off-the-shelf materials, without substantial redesign, and were developed primarily for non-biological applications. Their interaction with biological systems was not sufficiently understood until three decades ago.

To this end, researchers have made significant efforts in last few decades to create new materials with specific requirements for particular biological applications. In this context, a material science tetrahedron as shown in Fig. 1.2 can be adapted to describe the concept of biomaterials science. The traditional materials science tetrahedron emphasizes the interdisciplinary research on materials processing, structure, properties and performance.

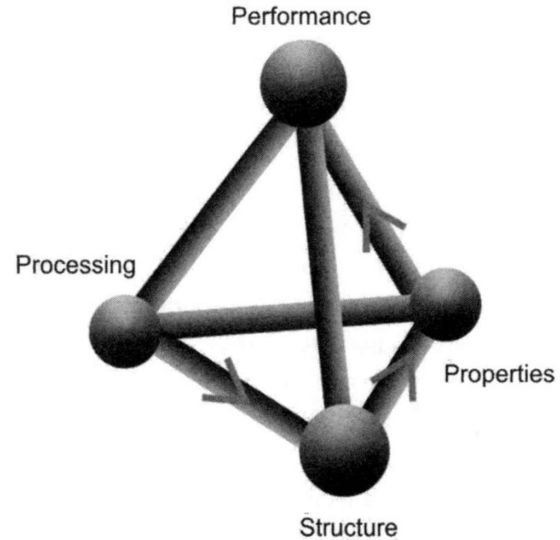

Fig. 1.2 The materials science tetrahedron [Adapted from Ref. 36].

In the context of biomaterials, the conventional materials manufacturing technique can be adopted. For example, solidification of melt followed by different bulk deformation processes (forging, rolling, extrusion, wire drawing, etc.) with machining as the final stage can be used in particular for metallic biomaterials. As will be mentioned later in this book, different coating deposition techniques, like plasma spray, sol-gel, and physical vapour deposition can be adopted to produce biocompatible coatings on an otherwise bioinert bulk implants. In case of ceramics, the conventional sintering or slip casting technique can be used. For polymers, set of different processing approaches, like extrusion, compression/injection moulding, etc., can be utilized to consolidate thermoplastic polymers into useful shapes. Apart from the conventional processing approaches, two distinct manufacturing techniques are particularly relevant for biomaterials and those include: a) additive manufacturing (e.g., 3D printing, 3D plotting, etc.) and b) porous scaffold manufacturing techniques (e.g., salt leaching, gel casting, etc.). While additive manufacturing can be used to fabricate patient-specific implants, the salt-leaching or similar techniques are more suitable for porous scaffolds but without much control over size/distribution of porosity. The requirement of having porosity in materials for specific biomedical applications (e.g., tissue engineering) necessitates one to use specific set of processing routes and this will be discussed in details in a later chapter.

Concerning structure, the bulk microstructure and more importantly, surface is extremely significant in the context of interaction with biological systems. In particular, grain size or phase assemblage in bulk determines many macroscale properties, like strength, toughness, etc. The surface topography, i.e., roughness or presence of specific patterns, guides the way the cells will function on a material surface. As part of the bulk structure, 3D porous architecture is particularly important in the context of bone tissue engineering. The fine scale characterization in terms of pore distribution, interconnectivity in 3D requires one to use specific characterization tool, e.g., micro-computed tomography, which is not widely used for various structural or functional materials. Various surface and bulk characterization techniques are discussed in sufficient details in a separate chapter of this book. In the context of properties, the bulk strength or toughness and more importantly elastic modulus play an important role in the biomechanical compatibility of a material in osseous system. While these properties can be determined using standard array of materials characterization tools, the surface properties, in terms of wettability require the use of contact angle goniometer. Apart from these, the key requirement of biocompatibility in the context of biomedical applications is to be sufficiently and ethically characterized using a host of biochemical assays, molecular biology techniques, etc. This makes the field of biomaterials distinct from other fields of materials science. This last aspect requires appropriate level of understanding of biological systems (cell, protein, bacteria, blood, etc.) and also to adopt techniques of remotely linked disciplines of biological sciences. In view of its paramount importance, a host of widely used biocompatibility characterization techniques are discussed in a separate chapter. The performance of any biomaterial is to be tested in pre-clinical studies involving experimental animals. The selection of animal model, defect type as well as study duration or follow up evaluation depend on the specific biomedical application. These aspects are to be approved by ethical committees at appropriate level and also involve expertise from veterinary surgeons, clinicians, histopathologists, etc. The result of such study provides scientific understanding of tissue–material interaction. Once a material is proven to be compatible in animal studies, the ultimate level of performance can be assessed in human clinical trials, which require even stricter approval from institutional and national ethical committee. Once the outcome measures based on such study involving voluntary human patients are found to be clinically acceptable, a biomaterial can be commercialized. The above discussion certainly indicates an interactive and longer product development cycle for biomaterials, when compared to other non-biomedical materials. It also emphasizes the need to adopt a different skillset or approach to establish processing-structure-property-performance correlation in case of biomaterials, which is unique in nature. An understanding of materials science is necessary, but not sufficient. This makes the field challenging and the interaction among several remotely linked expertise also makes the field interesting.

1.2 | Defining Key Elements of Biomaterials Science

This section introduces the subject of biomaterials science and related concepts.

> The study of materials with a special reference to their interaction with biological system is called as *biomaterials science.*

As explained in Fig. 1.3, an understanding of biomaterials science requires multidisciplinary approach of drawing expertise from multiple disciplines of engineering, biological and medical sciences[11]. The field of biomaterials science is a relatively young field with a history of 70–80 years, but it does have a significant impact on human society, since the use of PMMA as the first Intraocular lens (IOL) by Sir Harold Ridley in 1949[12], and the discovery of Bioglass® by Late Professor Larry Hench and his colleagues during the late 1960s[13]. The research and development in the biomedical materials sector has grown exponentially in the last few decades due to the advent of new materials, technological expertise and novel applications. According to a report by global market research companies, the biomedical materials industry is estimated to be worth $130 billion worldwide by the year 2020 with an emphasis on growth and development in the Asia–Pacific region, especially in India and China[14].

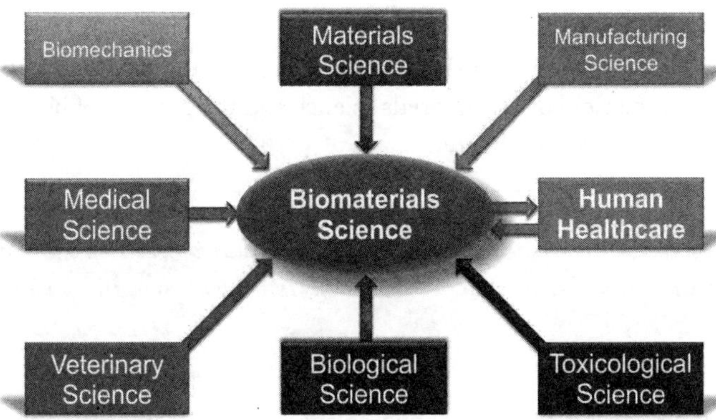

Fig. 1.3 Schematic illustration of the integration of concepts drawn from various fields of science and engineering to create multidisciplinary research theme in biomaterials science.

The materials used in the biomedical industry can be classified into metallic materials (stainless steels, titanium and its alloys, cobalt–chrome alloys, etc.), ceramic materials (calcium phosphates, alumina, zirconia, glass ceramics, etc.) and polymeric materials (polyethylenes, PMMA, nylon, etc.). Also, some of the naturally occurring materials, such as, fibrin, collagen, hyaluronic acid, and silk are also used for biomedical applications. Before delving further into the matter, it is therefore essential to understand what makes a material appropriate for biomedical applications.

A biomedical material, more commonly known as **Biomaterial,** possesses many unique properties that allow it to interact with living system comprising of proteins, cells, bacteria, etc. Many definitions and descriptions have been proposed by researchers for the terms 'biomaterial' and 'biocompatibility', which have changed over the years with respect to progress in biomaterials science. At the infancy of this field, biomaterials were defined as materials of natural or manmade origin to interface with biological systems to evaluate, treat, augment, or replace any tissue, organ, or function of the body. When these materials display universally 'good' or harmonious behaviour in contact with tissue and body function and evoke a minimal biological response, they are considered to have good biocompatibility property [15]. According to another definition, biomaterial is a term used to indicate materials that constitute parts of medical implants, extracorporeal devices, and disposables that have been utilized in medicine, surgery, dentistry, and veterinary medicine as well as in every aspect of patient healthcare. William[16] has provided a more acceptable definition of Biomaterial, which is as follows:

> Biomaterial is defined as *'Any substance that has been engineered to take a form which, alone or as a part of a complex system, is used to direct, by control interactions with components of living systems, the course of any therapeutic or diagnostic procedure, in human or veterinary medicine.'*

It is important to mention here that all the above definitions too have evolved with time as our understanding has improved. In a set of recently published articles in Nature journal, the following definition is proposed:

'Biomaterials are those materials—be it natural or synthetic, alive or lifeless, and usually made of multiple components—that interact with biological systems'. Biomaterials are often used in medical applications to augment or replace a natural function[17].

At this juncture, it is instructive to define tissue engineering and scaffolds as both of these concepts are relevant in the field of biomaterials science and the concept of biocompatibility will be introduced in a later section of this chapter.

> *'Tissue engineering is the creation of new tissue for the therapeutic reconstruction of the human body, by the deliberate and controlled stimulation of selected target cells through a systematic combination of molecular and mechanical signals'.*

Tissue engineering can be conceptualized as the means of orchestrating cells, engineering materials and suitable biological factors to enable relevant biological functions[2,3].

In the field of tissue engineering, the challenge lies in designing a suitable scaffold together with growth factors for actuating the regeneration process of the human body, culminating in full recovery of the structure and function of the damaged tissues[18]. The repair process is initiated *in vitro* by culturing cells onto the scaffolds, to be implanted in the host[19]. Usually this technique requires porous scaffolds that are biocompatible and bioresorbable. Tissue growth factors are incorporated into this material in order to promote the cellular functionality.

Scaffolds are defined as physical structure of synthetic/biological materials or both, wherein the biological cells are often implanted or 'seeded' with an aim to facilitate three-dimensional tissue formation *in vitro*. These structures, typically called scaffolds, are often critical, both *ex vivo* as well as *in vivo*, for recapitulating the *in vivo* milieu and allowing cells to influence their own microenvironments. Scaffolds usually a) allow cell attachment and migration and b) exert certain mechanical and biological influences to modify the cell behaviour.

It is conceptually difficult to mark a line of distinction between tissue engineering and biomaterials science. It has been widely perceived that tissue engineering deals with porous scaffolds, which can be made of ceramics, metals and polymers as well as biological materials, like collagen or other proteinaceous materials. Nonporous solid implants that are used for load-bearing articulating joints are commonly defined as **implants**. The popular examples are ceramic femoral ball heads, stainless steel/titanium based stems in total hip joint replacement applications. The implants therefore should not be confused with scaffolds. The use of implants in any biomedical applications cannot be considered as 'tissue engineering' approach and can be related to biomedical engineering. However, the interaction of scaffolds or implants with biological systems comes under the purview of biomaterials science.

> An **implant** is to be conceptually distinguished from a **scaffold.**
>
> - An implant is perceived as a synthetic non-porous biomaterial, which would remain in a living system for a longer timescale, unless it fails under biomechanical stresses.
> - A scaffold is generally considered as a porous implantable biomaterial platform, which would support cell functionality.
> - An implant provides biomechanical support to the surrounding osseous system.
> - A scaffold is often perceived as non-permanent and degradable entity in a biological system. Depending on porous architecture, a scaffold has generally weaker mechanical properties, but has better biocompatibility than an implant.

Today's scaffold-based tissue engineering, pivots on an efficient combination of viable cells, biomolecules and scaffold to promote repair and/or regeneration of tissues. One of the most taxing applications would be the repair and creation of musculoskeletal tissues, mostly bone, where a scaffold with high elastic modulus is sacrosanct for catering temporary mechanical support without symptoms of fatigue or failure. The scaffold or neo-tissue must be retained in the space they were designated for and conjointly, adequate space must be provided for simultaneous growth and development. Although bones have inherent regeneration ability, large bone defects due to trauma or tumours are still challenging for surgeons, as no spontaneous healing occurs in such cases. Bone tissue engineering presents a new approach to these sets of problems, whereby a porous scaffold is used, onto which tissue-specific cells can be seeded. Furthermore, they can also be coated with proteins and other biologically active species. Another alternative approach can be to use implant materials with appropriate surface modifications either in terms of graded surface porosity or in terms of porous bioactive surface coatings or in terms of chemical surface treatment with an endpoint objective to promote osseointegration. Some of these approaches will be discussed in the subsequent chapters of this book.

1.3 | Interdisciplinary Nature of Biomaterials Science

From the preceding discussion, it should be clear by now that, Biomaterials Science is a field of interdisciplinary research, which is discussed in this section. As illustrated in Fig. 1.3, the biomaterials science essentially integrates the concepts and expertise drawn from multiple engineering and science disciplines. The fabrication of material for biomedical applications should be driven by clinical needs, which are to be defined by clinicians. For specific biomedical application, one has to consider various material options primarily from biocompatibility perspective. The design of biomaterial should also critically consider the biomechanics aspect. For example, femoral ball head in total hip joint replacement (THR) application needs to sustain complex biomechanical stresses, which are to be evaluated using burst strength measurement and hip joint simulator experiments. Similarly, the heart valve prototype has to sustain a large number of fatigue cycles. The biomechanical considerations are equally important in dental restorative applications as well. Next step should be to identify the manufacturing options to develop device prototypes or to obtain lab-scale samples with optimal combination of physical and biocompatibility properties. The *in vitro* biocompatibility assessment requires expertise drawn from molecular biology, microbiology and other branches of biological sciences. The safety evaluation of biomaterials requires toxicity study, both at cell and gene level. This requires the involvement of toxicologists. Finally, the *in vivo* biocompatibility study requires the participation of veterinary surgeon to assist in experiments with small to medium to large sized animals.

> The effective integration of ideas and expertise from remotely linked disciplines (Materials, Biological sciences, etc.) will have a larger impact to human healthcare, once the clinical trials (after regulatory approval) on the *in vivo* biocompatible materials provide satisfactory outcome and the technology is commercialized in the market.

As will be discussed in this book, the approvals from ethical committee at various stages of research in the biomaterials science make this field very challenging. Nevertheless, the endpoint objectives towards human healthcare also equally make the researchers motivated to actively contribute to this socially relevant field of science and technology.

A student (from a non-biological science background) wishing to pursue research on biomaterials science, always seeks to answer himself/herself as to how far or to what extent one has to learn or understand the core of biological sciences discipline. Keeping this in mind, the author has made an attempt to discuss the salient concept or ideas or phenomenon of this remotely linked disciplines in a number of subsequent chapters. In order to illustrate this aspect in this introductory chapter, Fig. 1.4 provides the idea of marriage between engineering and biological sciences. While the fundamentals of engineering ideas can be translated to fabricate and characterize synthetic materials, the understanding of biological sciences is necessary to analyse how proteins, cells and bacteria can interact with synthetic biomaterials. Such response can be studied both *in vitro* (in glassware or in culture/growth medium at lab scale experiments) or *in vivo* (experiments conducted in animal models). The integrated understanding of *in vitro* and *in vivo*, cells/tissue interaction with biomaterial constitutes the central theme of the field of biomaterials science. The above discussion is schematically summarized in the form of a concept triangle in Fig. 1.4. Also, the interaction between two remotely linked disciplines make biomaterials science a truly interdisciplinary field of research.

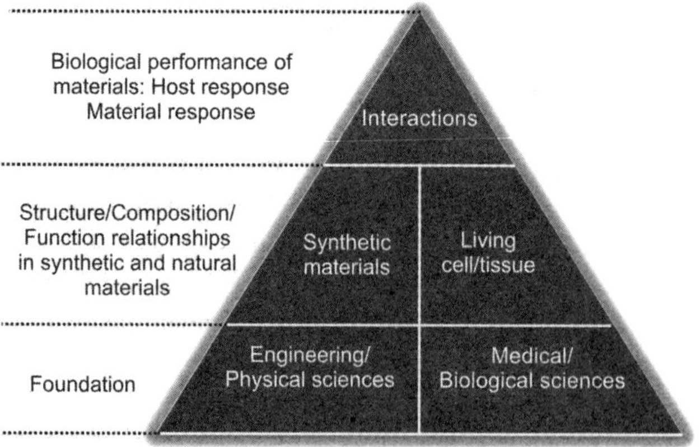

Fig. 1.4 Concept triangle to describe the field of biomaterials science.

Extending the discussion further, Fig. 1.5 describes some key aspects of cell–material interaction, which is considered as the core or central concept in the field of biomaterials science, as well as other aspects of biological sciences. For example, a biological cell can function in a different manner on a material substrate, when compared to that in isolation, in a culture medium. Although an understanding of cell fate processes from biological science point of view is necessary, but this needs to be adopted considering a cell has to interact with a synthetic material substrate. As will be discussed in a subsequent chapter, such interaction is mediated via adsorbed protein on a material substrate. Similarly, the growth kinetics of bacteria in culture (in isolation) can be different as in how they multiply in a microenvironment around a biomaterial. This also has significant implication towards biofilm formation, leading to prosthetic infection. Another important idea from biological sciences is how to qualitatively and quantitatively analyse the cell functionality modulation on a biomaterial. This aspect is often supported by how specific genes are upregulated/downregulated, which requires the quantification of mRNA expression after mRNA being extracted from cells grown on a material. Once a biomaterial is screened to have acceptable or desired cytocompatibility, the next important step is to evaluate the tissue level compatibility of a biomaterial, when implanted in an animal model. A large number of biological assays are to be used for the above discussed biological study, which are discussed in chapters 8-11 of this book.

As the subject of biomaterials science grows, deeper understanding into how cellular signalling processes are involved in the functionality of the cell population on material substrate is being more appreciated in the community.

Clearly, the knowledge of the above aspects requires more in-depth understanding of biomaterials science. In the early stage of evolution of biomaterials science, the first generation of researchers were excited to see different cell types growing on materials with different composition. With the advancement of the field in terms of gene-level understanding, the field of biomaterials science is getting more and more integrated firmly with biological science discipline. On the basis of the above discussion, the key concepts of biological sciences are summarized in Fig. 1.5. It needs to be emphasized once again that the core concept of biomaterials science intelligently uses some elements of the biological science discipline at an interface with a synthetic material without necessarily delving into a greater detail of this remotely linked discipline.

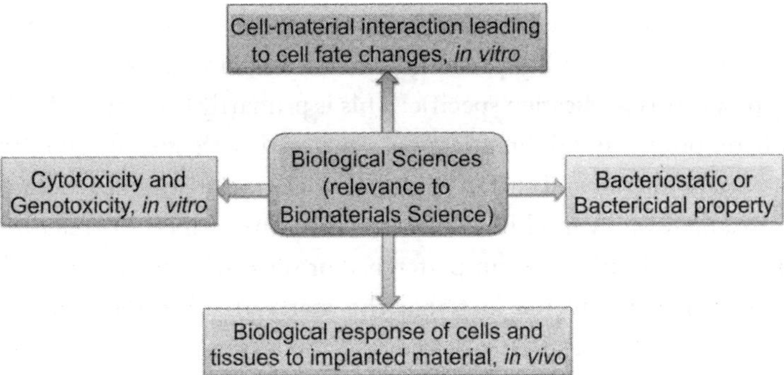

Fig. 1.5 Sketch to illustrate the fact that biomaterials science requires biological science-based specific approaches to probe into the influence of materials on biological cells and tissues.

1.4 | **Defining Biocompatibility and Related Concepts**

One of the pre-requisites a successful biomaterial ought to fulfil is that it should be accepted by the body, i.e., it should be compatible with the body. This performance of implants, along with their interaction and co-existence with biological tissues is studied in the realm of the broad and elusive subject of biocompatibility. While it is relatively common to describe a successful biomaterial as biocompatible, the exact nature of biocompatibility is still uncertain[20]. On the lines of the first principle of Hippocrates that a doctor should do no harm, the main requirement for biocompatibility of a material, whatever be its end application, is that the material should do no harm. Ratner et al.[11] have defined the term 'biocompatibility' of a medical device in terms of the success of that device in fulfilling its intended function. The common denominator in all the definitions that have been proposed for 'biomaterials' is the undisputed recognition that biomaterials are distinct from other classes of materials because of the special biocompatibility criteria that these materials must satisfy[21]. A more recent definition of biocompatibility is well accepted in the biomaterials community and is as follows[20],

*'**Biocompatibility** refers to the ability of a biomaterial to perform its desired function with respect to a medical therapy, without eliciting any undesirable local or systemic effects in the recipient or beneficiary of that therapy, but generating the most appropriate beneficial cellular or tissue response in that specific situation, and optimising the clinically relevant performance of that therapy.'*

The word 'biocompatibility', as defined above, is a broad term and needs to be carefully used. If an *in vitro* assay is performed to confirm the cell viability, adhesion, proliferation, it is more appropriate to use the word 'cytocompatibility'. However, biocompatibility describes the compatibility of material, both *in vitro* and *in vivo*.[17]

The most essential and universal characteristic of the aforementioned materials that sets them apart from other materials allowing for their use in conjunction with living tissue is their biocompatibility. It is that property of the material which allows it to perform the desired function in a specific application without eliciting adverse effects.

In this respect, a biomaterial itself is relegated to play subservient role, with central importance being attached to its biocompatibility property. At this juncture, it must be mentioned that the biocompatibility property is application specific. This is primarily because of the fact that each cell type likes to adhere, proliferate/differentiate on a material with specific substrate composition, elastic stiffness, surface energy and surface wettability characteristics. To be more precise, some cell types prefer to adhere on hydrophilic surfaces, while other cell types prefer to adhere more on hydrophobic surfaces. Such difference in preference or sensitivity for various cell types towards different substrates substantiate the fact that the material substrate composition or surface properties are to be tailored for a specific biomedical application. A multitude of issues, e.g., lack of desired physical properties (strength/toughness), piezoelectric property/bactericidal property have engendered significant research activities to develop HA-based biocomposites with bone-mimicking properties[22,23,24,25,26,27]. While changes in surface chemistry or surface properties (e.g., wettability) can influence cell fate processes, recent studies have also shown that application of external field (electrical/magnetic) within a narrow window can enhance cell proliferation or elicit bactericidal response[28,29].

> ***Host response*** *is the reaction of a living system to the presence of a synthetic material, in vivo.*

The formation of structural and biological bond with the host tissue is a key aspect of host response[30]. It can be envisioned that upon biomaterial implantation/injury to the tissue concerned, a cascade of inflammatory and wound healing responses are elicited. The inflammatory response consists of two phases: an initial acute phase and subsequent chronic phase. The acute inflammatory response phase may last from a few hours to days. This phase, characterized by fluid and protein secretion and neutrophilic reaction, induces wound healing and neo-tissue formation. On the other hand, chronic inflammatory response depends on the nature and extent of injury, and histologically, the formation of vascular, fibrous capsule, 50–200 μm thick, encapsulating the implant marks off the last stage of host response[30]. In the following section, a few concepts related to the host response (particularly for bone tissue engineering applications) are defined.

Osteoconduction is defined as a phenomenon regularly encountered in bone implants and this signifies bone growth on a biomaterial surface. Implant materials of low biocompatibility such as copper, silver and bone cement show little or no osteoconduction[31].

Osteoinduction is the process to induce *osteogenesis*. It is a phenomenon regularly seen in any type of bone healing process. Osteoinduction indicates recruitment of immature cells and their simultaneous stimulation into preosteoblasts. Bone healing processes, as in fractures, are primarily dependent on osteoinduction.

Osseointegration is defined as the stable anchorage formation around an implant by direct bone-to-implant contact. In craniofacial implantology, this is the only mode of anchorage for which high success rates have been reported. Osseointegration is possible in other parts of the body, but its importance for the anchorage of major arthroplasties is under debate. However, bone in-growth in a porous, coated prosthesis may or may not represent osseointegration.

For a complex biological and sensitive system (e.g. human body), integrating biocompatibility and tissue-mimicking structural properties, obligate the fabrication of a suitable scaffold and this is in itself an immensely challenging task. For bones, the materials should be preferably *osteoconductive*, i.e., capable of promoting differentiation of progenitor cells down an osteoblastic lineage, *osteoinductive*, i.e., support bone growth and facilitate in-growth of surrounding bones particularly for critical size defects and capable of *osseointegration*, i.e., integrate into the surrounding bone.

In the context of prosthetic infection, the **bactericidal** property refers to the lack of viability of bacterial cell population on a biomaterial substrate. In contrast, **bacteriostatic** property implies the deactivation of bacterial growth i.e. bacterial colonies may survive, but will not grow with time on a given biomaterial substrate.

1.5 | Implication of Biomaterials Science in Human Healthcare

In the last few decades, the human diseases as well as human healthcare has stimulated significant research activities in the field of biomaterials science and biomaterials engineering. Commensurate with the improvement in medical science, considerable research has been pursued in the area of biological, chemical and physical sciences. In this perspective, the researchers put considerable efforts in understanding the molecular biology aspects of some of the life threatening diseases and the engineers have attempted to develop various diagnostic tools as well as synthetic implant materials to assess the disease state or to repair the damaged tissues. The quantification aspect in the disease diagnosis has enabled better healthcare or more timely treatment. The above research perspective has been reflected in more 'biology' related research in multiple engineering disciplines (materials science, chemical engineering, mechanical engineering, etc.), chemistry and physics.

Among various diseases impacting human life, cardiac and cancer are the life threatening diseases, while a large number of patients also suffer from orthopedic, i.e., bone/joint related diseases. In most cases, complicated revision surgery is needed after a few years of implantation. The selection and design of a suitable material for hard tissue replacement with long term durability without significant degradation in physical and mechanical properties are the major area of research in orthopedics. Significant evolution in materials science and engineering in terms of developing the potential scaffold for healthcare has been recorded during the last few decades. Depending upon the anatomical location in the body, metals, ceramics, polymers and their composites with acceptable biocompatible property are being used as artificial implants.

In order to illustrate the need for biomedical research, it may be noted that there are more than 300,000 cases every year of Total Knee Replacements (TKR) and more than 100,000 cases every year of Total Hip Arthroplasty (THA) in India. The available articulating joint-implants generally offer a trouble free life for about 10–15 years, which is inadequate considering the increased lifespan for humans in many developing nations. Therefore, a search for ideal prosthetic materials together with treatment methods, constructive solutions and surface designs is currently being pursued in the field of orthopedic biomaterials[2,3].

> A specific biomaterial can exhibit good biocompatibility property in reference to the bone replacement applications, but the same material may not be biocompatible for cardiovascular applications.

Cardiovascular diseases are considered to reduce mortality worldwide. It has also been recognized that the scarred cardiac muscle results in heart failure for millions of heart attack survivors worldwide. In 2009, an estimated 785, 000 Americans had a new coronary attack and about 470, 000 had a recurrent heart attack, leading to a coronary event. Despite significant progress in the field of medical sciences, heart attack is still considered as a serious killer till today. This is primarily because of the fact that it is difficult to regenerate the diseased cardiac tissues. In case of myocardial infarction, i.e., 'heart attack', the cardiac tissues are largely damaged. The functionality of cardiomyocytes (a specialized contractile muscle cell that generates part of the myocardium tissue of the heart) and neurons (an electrically excitable cell that processes and transmits certain information by electrical and chemical signalling in the heart) depends on continuous conductivity property. However, such conductivity may break down during heart disease or malfunction. For instance, a myocardial infarction usually occurs because a major blood vessel supplying the heart's left ventricle is suddenly blocked by an obstruction, such as a blood clot. During myocardial infarction, part of the cardiac muscle, or myocardium, is deprived of blood and therefore oxygen, which destroys cardiomyocytes and neurons leaving dead tissue as well as denervation of the myocardium. In particular, nerve damage to cardiac tissue can result in nerve sprouting in the left ventricle and development of arrhythmias. It is in this context, various bioengineering approaches for heart regeneration as well as challenges related to the *in vitro* and *in vivo* biocompatibility assessment of what they call as a 'Perfect Tissue Engineered Graft' are being investigated, rather to a limited extent. Apart from the biocompatibility properties, it has been mentioned that an ideal graft should have sufficient electrical and micro-mechanical properties to ensure the natural contractile nature of myocardium. Recently, the development of Polylactic-polyglycolic acid-Carbon nanofiber (PLGA–CNF) biocomposites has received wider media attention for its potential to be used as a new generation cardiac patch materials.[32]

The development of synthetic materials for cardiovascular applications is therefore important to treat life-critical diseases, like myocardial infarction. The recent advances in anti-thrombogenic approaches for surface modification have been utilized to develop better biomaterials for cardiovascular tissue engineering applications. Amongst various surface modification approaches, the use of anti-coagulants like heparin, hirudin appears to be very effective for various polymeric substrates. The restriction in platelet adhesion and activation is critical to produce non thrombogenic implants. The adsorption of nitric oxide or polyethylene glycol (PEG) or albumin is reported to be very effective. Amongst cell based approaches, *in vitro* or *in situ* endothelialisation or seeding of bone marrow mesenchymal stem cells can be used to mediate thrombosis.

1.6 | Relevance of Biomaterials Science to Biomedical Device Development

In the preceding section, the various important concepts of biomaterials science were introduced. The relevance of the field towards human healthcare is mentioned and this aspect has motivated researchers to develop biomedical devices, which requires translation of biomaterials science to implantable products.

In the context of biomaterials science and biomedical engineering, the 'science' aspects provide a base to understand the cell/tissue interaction with a biomaterial, but the 'engineering' aspect involves the design and manufacturing of a biomedical device for the targeted clinical application.

In this section, some illustrative examples are provided to substantiate this point. In Fig. 1.6, various anatomical sites of a human osseous system are shown to indicate potential areas of tissue/ bone replacement with synthetic biomaterials. Some illustrative examples of orthopedic biomedical devices are shown in Fig. 1.7. In Fig. 1.7 (a, b), the commercially available femoral hip stems are shown. While bone cement (e.g., PMMA) is used for cemented stem, this is not the case for

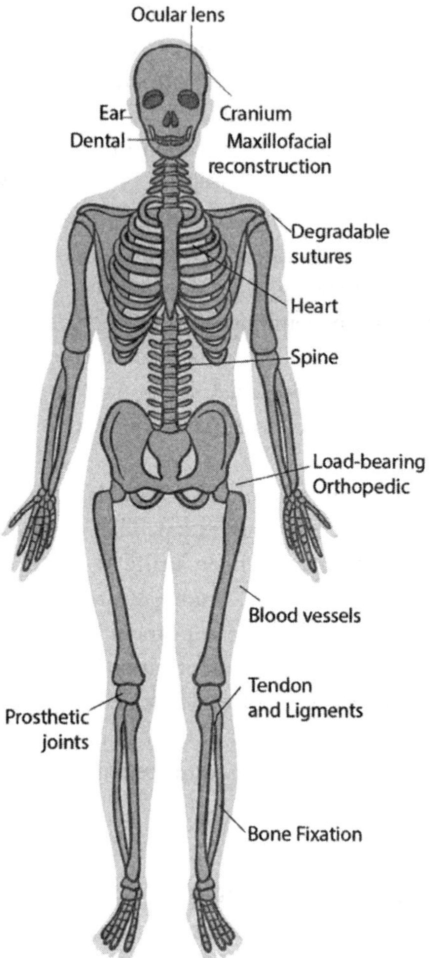

Fig. 1.6 Schematic showing the application of synthetic biomaterials in human body.

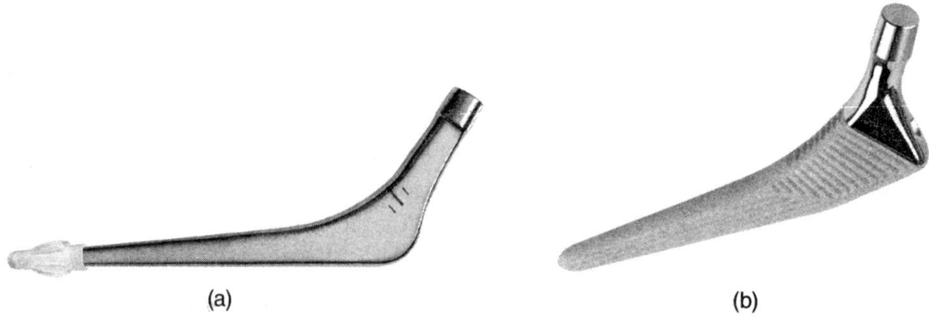

Fig. 1.7 Illustrative examples of femoral component of total hip joint (THR) based orthopedic devices: (a) Cemented femoral hip stem, typically forged from high nitrogen stainless steel or chrome-cobalt alloy and (b) Hydroxyapatite coated forged titanium alloy femoral hip stem for cementless device. (Courtesy of Smith & Nephew, Inc.).

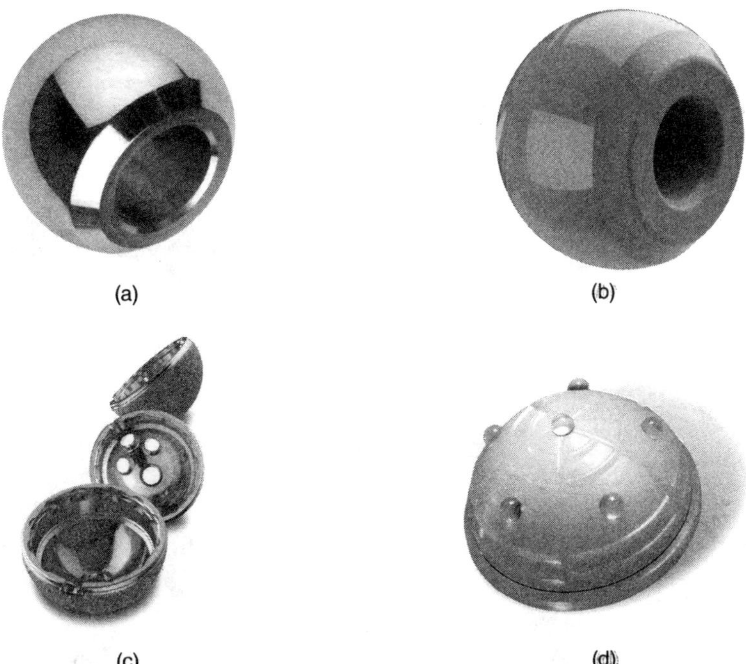

(a)

(b)

(c)

(d)

Fig. 1.8 Commercially available articulating components of total hip joint replacement: (a) Metallic modular ball-head, typically taper-locked into the femoral hip stem, (b) Modular ceramic ball-head, typically taper-locked into the femoral hip stem, (c) Hydroxyapatite or porous-coated acetabular shell, typically used as the acetabular component in a cementless total hip replacement and (d) UHMWPE acetabular cup (Courtesy of Smith & Nephew, Inc.).

cementless stem, wherein bioactive HA is coated onto stem for better osseointegration. The material for femoral stem can be typically stainless steel or Ti-alloys. In Fig. 1.8, various device components for articulating part of THR assembly are shown. The ball-head combination can be ceramic–polymer or metal–polymer or metal–metal. Among these options, ceramic–polymer mating combinations are by far the most widely accepted clinically because of its better biocompatibility property. As far as THR assembly integration is concerned, the 'press fit' concept as well as cementless or cemented hip systems are the options. In term of patient specificity, different size of femoral ball head and respective socket combinations are commercially available. In case of knee replacement surgery, the device components have different sizes and shapes than hip replacement as shown in Fig. 1.9. In particular, knee femoral component and tibial trays are the major components of total knee joint replacement, TKR. The commercial device products, as shown in Figs. 1.7–1.9, also illustrate the need to adopt appropriate fabrication technique to shape the best biocompatible material for a given biomedical application.

Although not shown in Figs. 1.7–1.9, biomaterials play an important role in neural, dental, opthalmological and cardiovascular applications. For example, in cardiovascular applications, NiTi-based coronary stents are used to remove plaques attached to blood vessel walls. Polytetra fluoroethylene (PTFE) or Ti-based heart valves are another important application area. The present technology is to use diamond-like carbon (DLC) or TiN coated Ti6Al4V based heart

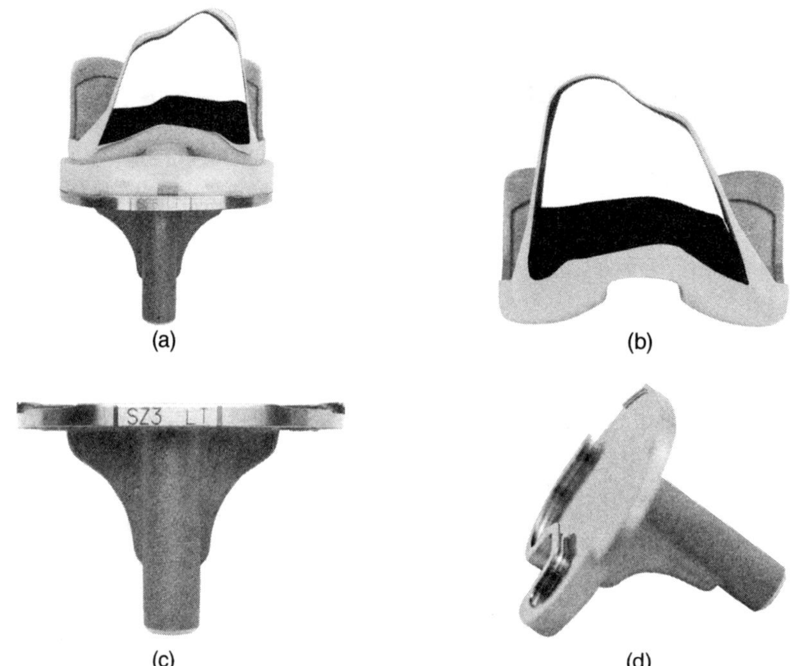

Fig. 1.9 Different device components of total knee replacement system: (a) total knee joint replacement assembly, (b) knee femoral, and (c, d) tibial tray (Courtesy of Smith & Nephew, Inc.).

valves. Another application area is related to the use of PMMA based intraocular lenses in opthalmological applications.

1.7 | Closure

This chapter has introduced some important concepts and various elements of biomaterials science and tissue engineering. It is expected that all the concepts, discussed in this chapter should be used as guidelines to design biomaterials. Various biomedical applications as well as relevance for human healthcare are also discussed. The discussion in this chapter also establishes a base for many of the subsequent chapters, which discuss the specific aspects of materials and biological sciences as relevant for biomaterials science.

Biomaterials science is, now more than ever, a converging field of various science and engineering disciplines[33]. The field of biomaterials has witnessed an emergence of advanced multi-functional materials for various applications, such as bone replacement and healing, biodegradable scaffolds based drug delivery devices, etc.[34]. It should be emphasized that, while cell/tissue interaction with a biomaterial substrate constitutes the central theme of 'science' aspects of the field, the design and fabrications of the material with the acceptable biocompatibility property to a desired shaped product/ device for a given biomedical application is the 'engineering' aspect of this important discipline[2,3].

Materials for Biomedical Applications

In line with the discussion in the preceding chapter, this chapter will discuss the generic classification of materials to illustrate the conceptual evolution of biomaterials. This will be followed by the more classical distinction among biomaterials, based on biocompatibility and host response. A major emphasis is placed on the use of three primary material classes i.e., metals, ceramics and polymers for clinical applications with some representative examples. The advantages as well as the performance-limiting properties of these three material classes will lead to a discussion on the biocomposites. Overall, this chapter will provide readers with knowledge of the multitude of biomedical application of implantable biomaterials.

2.1 | Conceptual Evolution of Biomaterials

In the early stage of development in the field of biomaterials, researchers used to make an attempt to tailor the use of engineering materials which are traditionally used in non-biological applications. As the understanding of the cell/tissue interaction with biomaterials matures, scientists are now trying to mimic the structure and properties of natural bone, teeth or other tissues with innovative designs and combination of different materials.

The evolution of biomaterials science can thus be traced to three distinct generations of biomaterials (see Fig 2.1)[37].

> The **first generation of biomaterials** were used to achieve a suitable combination of physical properties to match those of the tissue being replaced with minimal host response. Examples include high strength materials such as titanium and stainless steel, as bone implants.

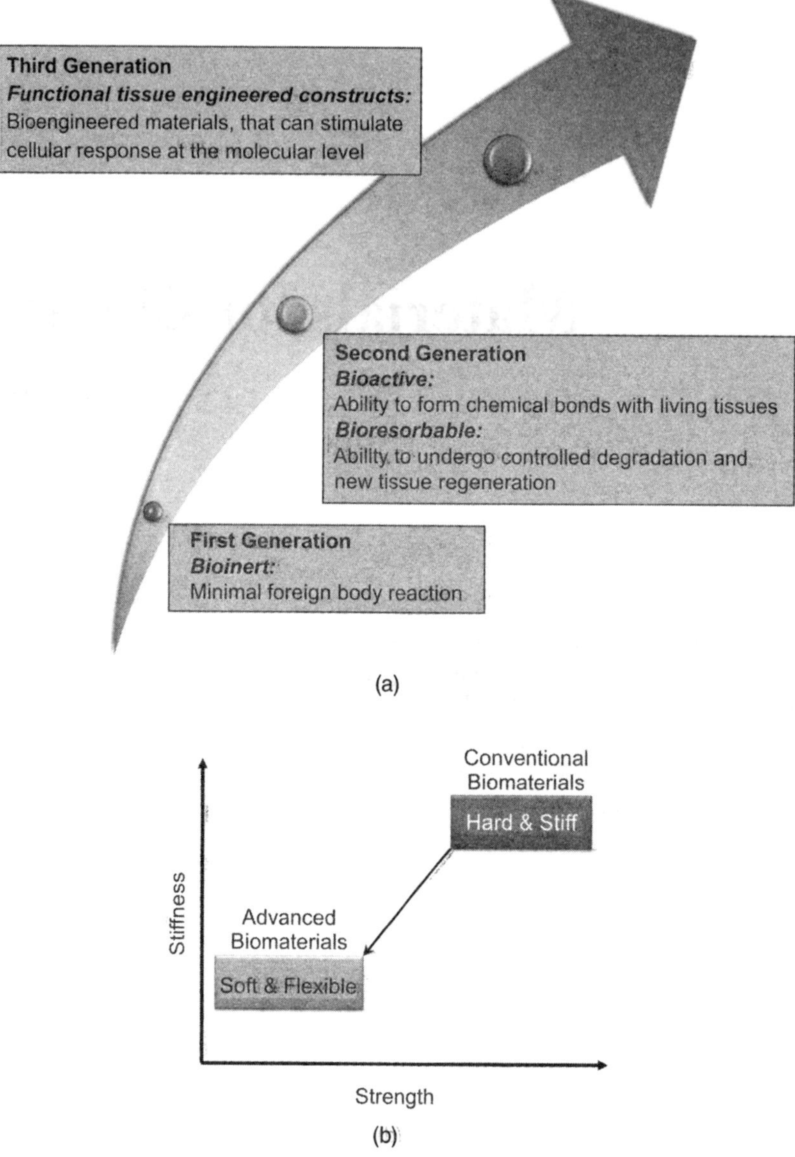

Fig. 2.1 (a) Illustration of the conceptual progress in biomaterials development and (b) paradigm shift towards material property requirement in line with the change in such conceptual development[44].

The first generation biomaterials were known for their bioinertness and their development was focussed on reducing the foreign body immune response. Common examples of the first generation biomaterials include 'off-the-shelf' industrial materials not specifically meant for clinical use, such as silicone rubber and pyrolytic carbon for inert coating on artificial heart valves[38,39]. Early heart valves (made of teflon), cardiovascular stents, dental implants and total hip replacement prostheses also belong to the first generation biomaterials. Here, the focus was on material properties and the bioinert nature of the materials[40,41,42].

While developing the **second generation biomaterials**, importance was given to materials that could elicit a desired response in a controlled physiological environment, while simultaneously minimizing the immune response.

The second generation biomaterials were bioactive with the ability to form chemical bonds with interacting living tissue[43]. Some classic examples are bioactive glasses, which allow the nucleation and growth of a hydroxy-carbonate apatite layer (biomineralization), the mineral component of the host bone. Such a layer is reported to facilitate a strong interface between living tissue and implant[44]. Apart from glasses, synthetic hydroxyapatite [$Ca_{10}(PO_4)_6(OH)_2$], is also osteoconductive (supports bone ingrowth) and has been in clinical use since the 1980s as orthopedic and dental implants in various forms, such as porous implants, powder and coatings on Ti and stainless steel implant prostheses[45,46,47,48,49,50,51]. Another important development about the second generation biomaterials was the use of bioresorbable materials with a controlled chemical breakdown (degradation) and resorption. In these materials, the interface issue is irrelevant as the implant material is replaced by the host tissue with time. Biodegradable sutures, scaffolds and drug carriers made of Poly(lactic-co-glycolic) acid (PLGA), nylon, polyesters, and polyanhydrides are well-known examples of biodegradable polymers, which also fall into the category of second generation biomaterials[52,53,54].

The successes of materials which are bioinert, bioactive and bioresorbable has no doubt helped improve the quality of life for millions around the world. However, it has been observed that approximately 50% of prostheses/artificial implants fail in the long term (10–20 years), necessitating painful and expensive revision surgery. The failure of an implant may be due to various reasons such as microbial infection, stress shielding, aseptic loosening and corrosion followed by subsequent mechanical failure[55,56,57,58]. Though strategies to counteract implant failures in the first and second generations of biomaterials are available, it will only delay the inevitability of failure due to the fundamental reason that man-made materials, unlike living tissues, cannot adapt to the dynamic changes in the physiological environment.

There has been an increasing need to develop biomaterials with multifunctional properties, that can facilitate a more natural way for the repair and regeneration of host tissues. This has given way to the development of **third generation biomaterials**, designed to elicit specific reactions at cellular and sub-cellular levels to control cell fate processes, such as proliferation, differentiation, etc.

The available techniques for tissue repair now have led to the new concept, **tissue engineering** for *in situ* tissue regeneration. While the underlying principles of biomaterials, biocompatibility and host tissue response remain unchanged, the current trend is to realize alternative and equally (or more) efficient routes for the repair and regeneration of tissues. The three major components of tissue engineering are cells, scaffolds and growth factors. For a given biomedical application, the tissue-specific cells are seeded together with growth factors into a three dimensional scaffold. The scaffold can be made of synthetic materials (metals, ceramics, polymers) or biological materials (collagen, etc). The growth factors stimulate specific functionality of the targeted tissue-specific cells. This innovative approach has shown significant promise to develop artificial organs and many advanced manufacturing approaches are currently being investigated in this direction, as will be discussed later in this book.

2.2 | Classification of Biomaterials Based on Biocompatibility and Host Response

Depending on biocompatibility and host response, the biomaterials can be broadly categorized into three primary classes[59]:

(a) *Bioinert* materials, subsuming stainless steel, zirconia, Ti-based alloys, etc., which, although is biocompatible, but fail to induce an interfacial biological bond between the host bone and the implant.

> Bioinert materials do not exhibit any biological bond with the host bone, *in vivo*. However, bioinert materials do exhibit cell adhesion and proliferation, although to a much lower extent than bioactive materials. Therefore, these materials are essentially cytocompatible.

Most metallic materials without surface modifications are examples of bioinert materials. A discussion on the application of bioinert materials is provided in a later section. These bioinert materials are generically used in load-bearing articulating joints.

(b) *Bioactive* materials, which are biocompatible and attach easily to body tissues forming chemical and biological bonds at the early stage of the post-implantation period. Synthetic hydroxyapatite [$Ca_{10}(PO_4)_6(OH)_2$], glass–ceramic A–W and Bioglass® are archetypal to this class of biomaterials.

(c) *Biodegradable* materials, which are gradually degraded and are replaced by new tissues, *in vivo*. Common examples include tri-calcium phosphate [$Ca_3(PO_4)_2$], poly(lactic-co-glycolic)acid (PLGA) copolymers, etc. Broadly, biodegradable polymers can be categorized into natural and synthetic polymers. Natural polymers include polysaccharides (e.g., starch, alginate, chitin/ chitosan, and hyaluronic acid derivatives) or proteins (e.g., collagen, fibril gels, and silk) and a variety of lignocellulosic biofibres as reinforcement. PLGA is an important example of synthetic biodegradable polymer.

It is clear from the preceding discussion that bioinert, bioactive and bioresorbable materials alone cannot cater to all the needs of the growing population of the world. However, instead of looking for new materials, another strategy adopted by scientists is to impart additional properties to well-known bioinert/bioactive materials. In some of the most interesting cases, the materials are engineered so that they possess both bioactive and bioresorbable properties. We can find this is, for example, in autogenous bone grafts loaded with growth factors that speed up the process of healing fractures[60]. Another example is the bioactive glasses used to treat bone defects, where chemicals from bioglass stimulates cells in the vicinity of the implant site, initiating a cascade of signalling processes over multiple generations of cells inducing the formation of tissues[61]. Similar examples exist where the concept of *in situ* tissue generation is used in skeletal muscle grafts, skin grafts, and nerve grafts in treating various disorders[62,63,64]. The other alternative is the well-known method of tissue engineering. For example, resorbable scaffolds are differentiated and then implanted in the body resulting in host tissue repair over time with the resorption of scaffolds[65,66]. For example, Polyethylene glycol (PEG) is known for its low cell adhesion and low protein adsorption properties. Polysaccharides used in knee cartilage repair also exhibit low cell adhesion and low protein adsorption. Biodegradable polymers (e.g., PLGA) are used in scaffold designing for temporary structures[37]. It is important to mention here that the rate of degradation of the scaffold influences the inflammatory response. Also, the mechanism of degradation such as surface erosion or bulk erosion affects the mechanical properties. Collagen and chitosan are natural biodegradable polymers. Non-degradable i.e., stable polymers (e.g., acrylonitrile-co-vinyl chloride) are used in encapsulated cell therapy.

In the following section, the characteristics and materials aspects of some of the aforementioned classes of biomaterials are described.

2.2.1 | Biodegradable polymer scaffolds

The most popular choice for biodegradable synthetic polymers is the saturated poly-α-hydroxy esters, namely, polylactic acid (PLA) and polyglycolic acid (PGA), poly(lactic-co-glycolic) acid (PLGA) copolymers[67]. PLA and PGA are easy to process with desirable mechanical and structural properties. Also, degradation rates can be attuned by using different molecular weights and copolymers. However, the rapid degradation of such polymers indicates the premature failure of the scaffold. Besides, the acidic end products released on degradation may elicit strong inflammatory responses[68]. Biodegradable polymer degradation is usually mediated by the uptake of water and subsequent hydrolysis. The rate governing factors are many fold and these include chemical composition and configurational structure, processing history, molar mass, polydispersity, environmental conditions, stress and strain, crystallinity, device size, morophology (e.g., porosity) and chain orientation, distribution of chemically reactive compounds within a matrix, additives[69], types of original monomers and overall hydrophilicity. The blends containing higher proportions of PLGA polymers are known to degrade faster. Contrarily, poly-ε-caprolactone (PCL) takes several years to degrade *in vivo*.[70]

Poly (lactic-co-glycolic) acid (PLGA) is a commonly used biopolymer. PLGA is commercially available as a biodegradable polymer in the form of various devices such as Polysorb[TM] sutures and vicryl meshes[71]. On degradation, PLGA and its homopolymers polylactic acid (PLA) and polyglycolic acid (PGA) release lactic acid and glycolic acid, respectively, which are not benign and can invoke an inflammatory response[72]. Low inflammation levels caused by PLA, PGA and their copolymer PLGA may be acceptable due to the lack of a large amount of the implanted polymer (e.g., surgical sutures). It is however, not acceptable in tissue engineering and other applications such as bone screws that involve larger implants[73]. An important example is the late degradation tissue response to poly-L-lactic acid (PLLA) implants for zygomatic bone repair. PLLA is widely used as surgical sutures and as implants for the fixing of bone fractures due to its mechanical strength and biocompatibility[74]. Pain, inflammation and swelling were observed in patients after as long as 3 years. A revision surgery followed by an investigation showed that that the implants had degraded and were surrounded by a thick fibrous capsule. This was due to the bulk erosion of PLLA that led to a release of noxious, acidic degradation products, which was further accelerated by autocatalysis[75]. Further, the rapid depletion of an externally loaded drug from PLGA occurs even before the device is completely degraded. This is due to the bulk erosion of PLGA, which also makes the continuous release of drugs from PLGA difficult.

To overcome the problem of the creation of a local acidic tissue micro-environment due to the bulk erosion of PLGA, certain strategies aimed at controlling the characteristic sudden erosion of PLGA have been adopted. Cameron and co-workers have incorporated α-tricalcium phosphate nanoparticles and microparticles in PLGA matrices. α-tricalcium phosphate also acts as a buffering agent as it is a mild alkaline salt[76]. Jansen and co-workers incorporated nanoapatite in PLGA-based electrospun nanofibrous scaffolds and showed that the tissue response was better when compared to pristine PLA[77]. However, these are mitigating strategies which aim to control degradation and delay the release of degradation products, which still remain undesirable. While PLGA has served as an excellent starting point and has helped millions of patients, the necessity of next generation biodegradable polymer materials which have more acceptable degradation products is easy to recognize.

A second limitation of biodegradable polymers is the lack of independently tunable mechanical properties. The influence of substrate mechanical properties on cell fate processes has been very well established for stem cells, which differentiate into different lineages based on the substrate modulus. Also, different tissues have different elastic moduli. A mismatch between the mechanical properties of the scaffolds and the tissue, that are meant to regenerate is believed to compromise their performance. Also, the time scales for the growth and regeneration of different tissues are different, indicating that the degradation time of the scaffolds on which this tissue is grown *in vitro* should also be tailorable.

To a limited extent, this can be achieved by varying the molecular weight of the polymers and by changing the co-polymer ratios. The degradation rate and hydrophilicity of PLGA can be varied by changing the ratio of lactide and glycolide. While a high glycolide ratio increases the rate of degradation, it also affects the mechanical strength[78]. Cell fate is greatly influenced by the mechanical properties of substrates. Hence, it is important to create a polymer platform for tissue engineering, where mechanical and degradation properties can be independently tuned. The major challenge is to achieve this without sacrificing other essential properties like biodegradability and cytocompatibility.

Thick samples of such biodegradable polymers are susceptible to heterogeneous degradation, faster inside than on the surface. This can be attributed to two factors: (a) easier diffusion of soluble oligomers from the surface into the external medium compared to the interiors, and (b) neutralization of the carboxylic acid end groups located at the surface by an external buffer solution, both *in vivo* and *in vitro*. The hydrolytic degradation rate is determined by the diffusion coefficient of chain fragments within the polymer matrix, temperature, buffering capacity, pH, ionic strength, additions in the matrix and medium and the initial degree of alignment in the polyesters and processing history. Some groups have incorporated basic compounds, like bioactive glasses and calcium phosphates to attune the background pH[79,80]. However, adverse tissue reactions from the degradation of polyester matrices must be averted.

Poly-DL-lactic acid (PDLLA) is another widely explored biomedical coating to orthopedic material. Such extensive use is due to its excellent conformatory features with respect to implants, and its high mechanical stability[81,82]. Growth factors, antibiotics or thrombin inhibitors can be easily entrapped into the PDLLA matrix scaffold, to engineer a locally-acting drug-delivery system[83,84]. PCL, another aliphatic polyester can efficiently encapsulate antibiotics and serve as a drug-delivery system. Thereby, bone in-growth and regeneration can be beefed up[85]. Akin to PLA, PCL degrades via random hydrolytic ester cleavage and weight loss through the diffusion of oligomeric species from the bulk. However, the degradation of PCL with a high molecular weight ($M_n \approx 50,000$) is much slower and may take as long as 3 years to complete[70].

Another biodegradable polymer is Polypropylene fumarate (PPF), which is an unsaturated linear polyester and can be moulded into a hard composite. It is important to reiterate that the mechanical properties and degradation times can be tailored depending on the molecular weight of the polymer. The degradation products of PPF, propylene glycol and fumaric acid are known to be biocompatible and readily removable from the body[86]. In view of the desired property combination, PPF is investigated as a tissue engineering scaffold as well as osteoblast growth promoting substrate[87].

Another well-investigated biodegradable polymer is polyhydroxyalkanoates (PHA), which is aliphatic polyester and can be synthesized in the laboratory from microbes in unstable growth conditions[88,89]. They are endowed with properties like biocompatibility and thermoprocessibilty, which render them attractive for medical devices and tissue engineering applications. They can be blended with other polymers, enzymes, inorganic materials and among themselves as well to augment their mechanical properties. Surprisingly, substantial changes in mechanical properties and biocompatibility can be observed, when PHAs are intermixed[89,90].

Poly-3-hydroxybutyrate (PHB) has been upheld to be consistent and favourable as bone tissue material. No unwanted chronic inflammatory response has been reported after implantation periods of up to 12 months[91]. Bone forms in close apposition with the implant surface (around 80%) and subsequently attains a good deal of structural organization. More importantly, no indication of loss of any post-implantation structural stability is reported[91]. In spite of this, the complicated processing methodology of PHB from micro-organisms impedes its usage in scaffold engineering.

Polymers like polyanhydrides, polyorthoesters and polyphosphazenes undergo heterogeneous hydrolysis predominantly at the polymer–water interface[92]. Such polymers are commonly categorized under the family of surface-eroding polymers. This surface-eroding characteristic holds out three key advantages over bulk degradation, subject to their usage as scaffold material: (a) retention of mechanical integrity over the degradative lifetime of the device by virtue of a constant mass to volume ratio; (b) insignificant toxic effects (i.e., local acidity), subject to lower solubility and concentration of degradation products, and (c) fleshed out bone ingrowth into the porous scaffolds with a gradual increment in pore size as erosion commences[92]. This class of polymers can also be metamorphosed into corresponding bulk degradable forms by tailoring the bulk to surface ratio of the scaffold.

> The degradation kinetics of biodegradable polymers is dependent on molecular weight as well as polymer chemistry. The growing thrust is therefore to develop polymers with tunable chemistry.

2.2.2 | Bioactive glasses and ceramics

The applications of bioactive ceramics in tissue engineering can be traced back to Hench's discovery that certain glass compositions had excellent biocompatibility and dexterity towards bone bonding[93,94]. A noted feature of bioactive glasses and ceramics is the post-implantation kinetic modification of the surface with time. Through interfacial and cell-mediated reactions, the surface conceives a biologically active hydroxycarbonate apatite (HCA) layer, which administers bonding with the tissues. This bone-binding behaviour is commonly referred to as bioactivity[93,94]. The HCA layer that forms on the bioactive implants is chemically and structurally concurrent with the mineral phase in bone and hence caters interfacial bonding[95]. This form is reproducible *in vivo* on the surface of the bioactive ceramic in a protein-free and acellular simulated body fluid (SBF), with ion concentrations nearly equal to that of human blood plasma. In fact, it can be authenticated that the formation of a biologically potent HCA layer is essential for bone-scaffold bonding.

In Chapter 3 of this book, the influence of surface patterning on the cell functionality in the context of bone tissue engineering is summarized[96]. The formation of a biological interface with the native surrounding tissues is sacrosanct for averting chances of scaffold loosening. Bioactive glasses are also known to support enzyme activity[97], vascularization[98], stimulate osteoblast adhesion, growth, differentiation and induce the differentiation of mesenchymal cells to osteoblast[99]. In addition, the dissolution byproducts from bioactive glasses, notably 45S5 bioglass, upregulate gene expression, that controls osteogenesis and the production of growth factors[100]. Such facets further substantiate the great importance of such a group of materials in tissue engineering.

Among various glass ceramics, 45S5 Bioglass has already been used successfully in clinical treatments of periodontal disease (Perioglass), bone filler material (Novabone) and also in replacing damaged middle ears. Recently, melt and sol-gel derived glasses have also found potential applications as promising scaffold materials as fillers or coatings of polymer surfaces[101]. The principal constituents of 45S5 bioactive glasses are SiO_2, Na_2O, CaO and P_2O_5. 45S5 Bioglass has also been shown to induce a greater secretion of the vascular endothelial growth factor (VEGF) *in vitro* and accelerate vascularization *in vivo*. It can, hence, be envisaged that controlled concentrations of bioactive glass in scaffold materials dynamizes neo-vascularization, advantageous for large tissue engineered constructs[102]. Conjointly, chemical properties, rates of bioresorption can be quadrated, specific to bone tissue engineering application by modifying the structure and chemistry of glasses at different compositions, and processing routes. However, the crystallization of bioactive glasses can render them bioinert[103,104]. This hinders the extensive use of bioglass as a scaffold material. Prolonged sintering is essential for the densification of the struts of a scaffold to impart the necessary mechanical strength, but this may also lead to the full crystallization of glass[105].

Another crucial defect of bioactive glass is its low mechanical strength, particularly among porous structures[106]. In fact, it is well documented that porous scaffolds needed for tissue engineering applications exhibit low mechanical strength compared to cancellous and cortical bone. Notwithstanding this, bioactive glasses are mechanically superior to amorphous glasses and calcium phosphate ceramics. However, high bending strength can still be achieved as in apatite–wollastonite (A/W) glass–ceramic. Here, the precipitates of wollastonite as well as the apatite phases impart high fracture toughness[107].

Calcium phosphates have a significant chemical and crystal similitude to the bone mineral[108] and are biocompatible, osteoconductive and osteoinductive. Numerous *in vivo* and *in vitro* assessments have corroborated the fact that this group of compounds supports adhesion, differentiation and proliferation of relevant cell types (osteoblasts and mesenchymal cells), irrespective of their form (bulk, coating, powder or porous) and phase (amorphous or crystalline). Hydroxyapatite (HA) is the most popular scaffolding material in this group[109]. However, their relatively poor biodegradability and mechanical strength as well as toughness impede their application in the engineering of new bone tissue, particularly at load-bearing anatomical sites.

On a similar note, there has been a recent interest on the development of biphasic calcium phosphate (BCP), which is a composite of biostable HA and bioresorbable tricalcium phosphate (TCP). It is important to note that the dissolution rate of synthetic calcium phosphate compounds largely depend on the Ca/P ratio, crystallinity of CaP phases, and the dissolving media (pH, solution saturation, etc.). In addition, the amount and characteristics of the micro and macro-porosities also determine the dissolution behaviour[108]. In fact, HA has the slowest degradation kinetics among the calcium phosphates of biological relevance.

> The biological and biomechanical environments, to an extent, regulate the crystallinity, porosity and composition of bone so that the mechanical properties of natural bone are synchronized with their anatomical location.

The properties of synthetic calcium phosphates can be remodelled and reformed by a modification of their crystallinity, grain size, porosity and composition (e.g., calcium deficiency). An increase in the amorphous phase, microporosity and grain size orchestrates deterioration in mechanical strength, whereas high crystallinity, low porosity and finer grain size are conducive to higher stiffness, compressive and tensile strengths and greater fracture toughness. It has been enunciated that wet conditions induce greater flexural strength and fracture toughness of dense HA in contrast to dry conditions[110].

If we collate the properties of cortical bone with those of HA and related calcium phosphates, we may arrive at the fact that HA is superior to bone in terms of compressive strength, although the latter markedly surpasses its fracture toughness. It may be worthwhile to note that the combination of high tensile strength and fracture toughness of natural bone can be attributed to the unique distribution of reinforcing HA nanocrystals in the collagenous fibrous matrix. The crystalline HA itself has poor fracture resistance properties. While mimicking the natural bone microstructure, efforts are therefore invested in making attempts to develop synergistically toughened HA composites for load-bearing scaffold applications with uncompromised biocompatibility and osteoconductivity.

Some studies have shown that bioactive ceramics need not be only bioglass, HA or glass ceramics containing HA or its components like CaO and P_2O_5; rather, some materials can be rendered bioactive after simple heat treatment as well[111]. For example, the chemical treatment of metals and ceramics e.g., by alkali and heat treatments of Ti/Ti-alloys and Ta or the H_3PO_4 treatment of tetragonal zirconia culminated into the functional gradation of surface and such approach encouraged the formation of CaP phases in simulated body fluid (SBF). As far as the underlying mechanisms are concerned, the presence of Ti-OH, Zr-OH, Nb-OH and Ta-OH surface groups in anatase perhaps promote the epitaxial nucleation of apatite crystals. In parallel, the adsorption of Ca^{2+} ions from a solution is facilitated by negatively charged surfaces. Over and above that, chemical treatment leads to Na^+ ion incorporation on metal oxide surfaces, whereby the crystalline structure metamorphoses into an amorphous phase inducing greater ion solubility[112]. However, such models need further investigation to better establish the mechanisms.

Summarizing, the bioactivity of synthetic biomaterials can be induced either by selecting appropriate materials or by the heat/chemical treatment of bioinert materials.

2.3 | Generic Classification of Biomaterials

In this section, the classification of biomaterials based on three primary material classes, namely, metals, ceramics and polymers is discussed in reference to their characteristics as well as their application areas. Such a broad and non-biocompatibility specific classification is useful for general readers to know the wide spectrum of the usage of synthetic materials in biomedical applications.

2.3.1 | Metallic biomaterials

Metallic biomaterials are by far the most widely used biomaterials in various load-bearing articulating joints as well as in other biomedical applications. These materials have a characteristic metallic bonding with metal ions floating in a cloud of non-localized electrons. Such a bonding characteristic is also responsible for isotropic transport (electrical/thermal conductivity) as well as elastic stiffness/ strength. The most advantageous properties of metals include high fracture toughness as well as greater malleability or ductility, enabling metals to be shaped into various complicated shapes. Also, metals have good weldability characteristics and machinability properties. This allows metals to adapt to various bone plates, screws and nails and metallic plates can be tightened against the host bone through the use of screw tightening. Table 2.1 summarizes the advantages as well as the clinical applications of the three widely-used metallic materials.

Table 2.1 Orthopedic materials and their characteristics[59,155].

	Stainless steels	*CoCr alloys*	*Ti and Ti–based alloys*
Advantages	Cost effective, easily available, easy processing	Wear resistance, corrosion resistance, high fatigue strength	Biocompatible, corrosion resistance, low elastic modulus, high fatigue strength
Disadvantages	Long term adverse effects, high elastic modulus	High elastic modulus	Poor wear resistance
Applications	Fracture plate, screws, hip nails, femoral stem in THR (total hip replacement)	Prosthesis stem, load-bearing components in TJR (total joint replacement), knee joint replacement	Femoral stem in THR, pacemaker, screws in dental implants for prosthodontics, heart valves

Although metals have the advantage of high resistance to fracture as well as a higher elastic modulus (not good for load-bearing orthopedic applications), the leaching of metal ions due to corrosion/dissolution and wear at articulating surfaces in body fluids are reported to cause the irritation, inflammation and loosening of implants *in vivo*. Another factor that limits the long term performance of metallic implants is the corrosion in a physiological environment. It may be noted here that the physiological environment has a dynamic change in pH and such change can have a significant influence on the corrosion of implants. For example, blood contains Na^+, K^+, Mg^{2+}, Ca^{2+}, Cl^-, HCO_3^-, HPO_4^{2-} and SO_4^{2-} ions. The blood pH drops to 5.2 in the hard tissue during implantation and it takes at least two weeks to reach 7.4 (pH of blood). Therefore, an implant should be highly corrosion-resistant to such an acidic environment.

Although stainless steel has been used in orthopedic applications for long time, the release of ions such as Fe, Cr and Ni results in toxicity. The release of toxic ions into body fluids leads to the failure of an implant due to the reduced biocompatibility of stainless steel. Lanthanum was added to improve the corrosion resistance of 316L stainless steel. The *in vitro* tests carried out in a simulated body fluid (SBF) at 37°C showed the suppression in pitting corrosion with 0.01% lanthanum addition. In a different work, nickel-free stainless steel was designed to resolve the issue of nickel sensitization.

Along with good osseointegration, its lower density (4.5 g/cm^3) and good mechanical properties compared to 316L stainless steel (7.9 g/cm^3) makes Ti a suitable material for orthopedic applications. Titanium and its alloys are also found to be better than stainless steel and CoCr alloys due to their superior corrosion resistance property. The passive layer on Ti/Ti alloys, formed during corrosion, can be broken due to wear and repassivation. Such a process can be affected by the presence of proteins and a local electrochemical environment.

> The wider clinical acceptability of many metallic biomaterials can be attributed to their reliable property combinations as well as the ease of manufacturing to obtain various sizes and shapes.

Although stainless steel, Co–Cr alloys and Ti-alloys are being used as implant materials, the Ti alloys show superior biocompatibility as compared to stainless steel and Co–Cr alloys. It is to be noted that the loosening of the implant due to stress shielding, caused by a mismatch in the elastic modulus of implant and bone, is a serious issue. As compared to bone, the elastic moduli of Co-Cr (240 GPa), and 316L stainless steel (210 GPa), and commercially pure Ti (110 GPa) are very high. Therefore, they may lead to the failure of the implant. However, Ti has the lowest elastic modulus among all aforementioned materials, and therefore the extent of stress shielding will be at lower level.

A relatively new class of metallic biomaterial is Mg-alloys, which have a unique biodegradable property[113]. However, a poor corrosion resistance property does not allow its immediate use as a biomaterial. Other disadvantages of the use of Mg-alloys are the processing/fabrication related challenges as well as their ceramic-like brittle fracture behaviour.

The bioinert nature of metals results in weak bioactivity properties. Therefore, a protective ceramic (e.g., HA) coating over metallic implants is being used to solve these problems. A list of coating techniques is given in Table 2.2. Each coating technique has its own advantages and disadvantages. Also, depending on the coating technique, the thickness can vary from submicron thickness to 1–2 nm. Among all the deposition techniques, thermal spraying and in particular, plasma spraying is by far the most widely used because of its high deposition rate. Also, thicker coatings of 30–200 μm can be obtained on metallic substrates (see Table 2.2). Prior to coating deposition, it is recommended that one use sand blasting or a similar technique to roughen the substrate for better coating adhesion. Similarly, composites of HA and Ti6Al4V are of greater interest as coatings and the major role is occupied by processing techniques for the preparation of coatings such as plasma spraying, electrodeposition and micro arc-oxidation[128]. A HA coating on implant materials (Ti6Al4V, Ti, stainless steel) is necessary to provide an appropriate bioactive interface and also to limit the corrosion and wear of implants. Microstructural characterization and the bond strength of the coating-implant interface are of major interest in such studies. It is instructive to note that coating on implantable biomaterials is a large research area. Apart from CaP-based coatings, metallic implants are often coated with other ceramics, like TiN or diamond-like carbon. For example, diamond-like carbon (DLC)-coated Ti-screws are expected to reduce friction with the host bone during implantation and, therefore, will reduce the patient's pain.

Table 2.2 Summary of processing approaches used for calcium phosphate coating on metal implants[59,155].

Technique	*Coating thickness*	*Advantages*	*Disadvantages*
Thermal spraying	30–200 µm	High deposition rate, low cost	High temperatures/rapid cooling often induce thermal decomposition/amorphous coating
Sputter coating	0.5–3 µm	Dense and uniform coating thickness on flat surface	Expensive and time consuming
Pulse laser deposition	0.05–5 µm	Coating by crystalline and amorphous phases, both porous and dense coating	Expensive and time consuming
Electrophoretic deposition	0.1–2 mm	Uniform coating thickness, rapid, can coat complex shapes	Difficult to produce a uncracked coating
Hot isostatic pressing	0.2–2 µm	Dense coating	High temperature required and difficult to coat complex shapes
Electrochemical deposition	0.05–0.5 mm	Uniform coating thickness, rapid, can coat complex shapes, low cost	Poor bonding strength between coating and substrate
Biomimetic coating	< 30 µm	Low temperature processing technique, bone like apatite can form, can coat complex shapes	Time consuming, requires close control of pH

In cardiovascular applications, Ti6Al4V based heart valves are coated either with DLC or TiN to enhance the haemocompatibility (reduced platelet adhesion). While the DLC coating is deposited using plasma-enhanced chemical vapour deposition (PECVD), TiN or TiAlN based coating deposition is pursued using physical vapour deposition by vaporizing a ceramic target of similar composition in an inert chamber. TiN coatings are also deposited on NiTi-based coronary stents in angioplasty applications. In all such applications, the coating thickness and adhesion remain equally significant, like CaP-based coatings.

Stress shielding, poor osteoinductivity, corrosion and wear remain major concerns with metallic biomaterials.

2.3.2 | **Bioceramics**

Ceramics are defined as a class of inorganic solids with ionic/covalent bonding, which are typically processed at high temperature. In simplistic terms, the ceramics are oxides/carbides/nitrides of metals.

> The term 'bioceramics' essentially refers to only those ceramics, which are used for biological applications. The synthetic ceramics, like alumina, silica, zirconia have long been investigated for biomedical application, whereas hydroxyapatite or calcium phosphates are the most well-known examples of natural bioceramics, which are found in human musculoskeletal systems.

While searching for implantable biomaterials, it is believed that ceramic-based articulating surfaces in a joint would give superior longevity as compared to available metal or polymer joints. Some examples of commercially available biomedical device components of a THR assembly are shown in Fig. 2.2.

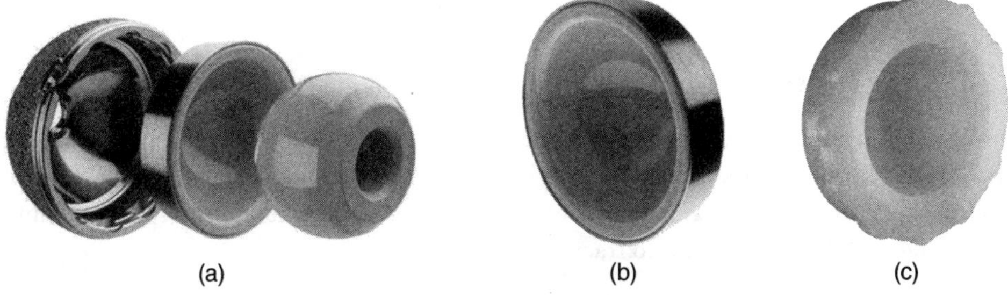

Fig. 2.2 Biomaterial components of total hip joint replacement, THR: (a) ceramic liner and head, (b) ceramic liner and (c) UHMWPE liner typically taper locked into an acetabular shell in a cementless total hip replacement (Courtesy of Smith & Nephew, Inc.).

Oxide ceramics, such as alumina have been extensively studied for load-bearing bio-implant applications. Because of the inherent brittleness of alumina ceramics (K_{IC} = 4 MPa.m$^{1/2}$), zirconia ceramics are considered for load-bearing implants, due to higher fracture toughness (7–10 MPa.m$^{1/2}$). But the problem of aging with zirconia ceramics causes the failure of implants. In contrast, non-oxide ceramic materials such as dense nitrides, sintered by hot isostatic pressing or hot pressing shows better hardness and strength than Al_2O_3. It has been reported in the literature that *in situ* toughened nitrides have high flexural strength (~1 GPa) and fracture toughness (10–12 MPa.m$^{1/2}$).

SiAlON ceramics with good biocompatibility have also been investigated due to their chemical stability and high fracture toughness. For load-bearing biomedical applications, it is important to mention that SiAlON are structural ceramics and can be subcategorized into α-SiAlON and β-SiAlON, which are solid solutions based on α-Si_3N_4 and β-Si_3N_4, respectively. In the SiAlON phase, Si and N are substituted with Al and O to incorporate better properties than in the Si_3N_4 phase.

Another important class of structural ceramics is SiC. Dense SiC, produced by hot pressing or hot isostatic pressing, possesses a better strength and hardness than Al_2O_3, with comparable fracture toughness. Nowadays, it is possible to prepare tougher SiC with a flexural strength of ~650 MPa and a fracture toughness of 9–10 MPa.m$^{1/2}$. Naji et al. reported the non-cytotoxicity of SiC to human osteoblast and fibroblast cells. In a different work, Wan et al. showed the non-cytotoxic behaviour of SiC using an MTT assay. A recent paper from the author's research group reports the compatibility of these hot pressed SiAlON-SiC composites with connective tissue cells (L929) and bone cells (Saos-2)[114].

One of the advantages of ceramics is the large compressive strength and better wear resistance, due to a high hardness and elastic stiffness. While the compressive strength is almost identical to tensile strength in magnitude for metals, the compressive strength of ceramics can be eight times larger than the tensile strength. In many biomedical applications, such advantageous properties of ceramics are judiciously utilized.

> The brittleness of ceramics remain a key issue towards their reliability in clinical applications.

Therefore, significant research efforts have been invested in the last few decades to design microstructures to enhance the fracture toughness property. This will be discussed in detail in a subsequent section on biocomposites.

2.3.3 | Biopolymers

Polymers are organic compounds, characterized by the large molecular structure with generally C, H, O, and N, forming the backbone of a long chain. For biomedical applications, polymeric biomaterials remain attractive candidates, because of their lightweight and their ability to undergo large viscoelastic deformation prior to fracture.

> A distinction is to be made between synthetic polymers and biopolymers. While the latter refers to the polymers (e.g. collagen) in natural living systems of mammals, the former is those polymers (e.g. PHA, UHMWPE, etc.), which are synthesized in the laboratory and subsequently tested to have desired biocompatibility in reference to targeted biomedical applications.

In this book, the formal distinction is not being truly followed and in general, synthetic polymers to be used for biomedical applications are referred to as biopolymers. In the early decades of the use of synthetic polymers for biological applications, the sutures made of nylon were reported in early 1940s and other polymers, like polymethyl methacrylate (PMMA), Dacron polyester, and polyvinyl chloride attracted attention in mid 1940s. Subsequently, other polymers (Teflon, high-density polypropylene, polyurethanes, etc.) were also investigated for biomedical applications. Here, it is not out of place to mention that many of these polymers, like many metallic biomaterials, were initially used for non-biomedical applications, before they received wide attention in the biomedical community. Some of these polymers are currently used as permanent prosthetic devices, including hip implants, artificial lenses, vascular grafts, catheters, etc., after they were proved to have acceptable biocompatibility (long term stability and appropriate host response), *in vivo*. The applications in the field of tissue engineering, where polymers are used to assist the regeneration of three-dimensional tissue and organ structures, are more nascent and are more integrated with biological demands. In spite of their good biocompatibility, polymeric materials are mainly used for soft tissue replacement (like skin, blood vessel, cartilage, ligament replacement, etc.).

> Low stiffness, high wear rate and flexibility limit the use of polymers alone as load-bearing biomaterial for hard tissue applications (like bones, teeth, joints, etc.).

As biomaterials, polymers have several applications, e.g., bone cement (polymethyl methacrylate), contact lenses (polyhydroxyethyl methacrylate), degradable components (polyglycolic acid) and bone screws (polyglycolic-co-lactic acid). However, the degradation products of some polymers, like polyglycolic acid (PGA) and polyglycolic-co-lactic acid (PGLA) in acidic environments lead to toxicity. Also, the poor load-bearing capability restricts their wide application. Considering their mechanical and structural properties, polymers are most suited for soft tissue applications (Table 2.3).

Table 2.3 Mechanical properties (nominal values) of polymeric biomaterials and soft tissues[59,155].

Polymer/biological tissue	Elastic modulus (GPa)	Tensile strength (MPa)
Polyethylene (PE)	0.88	35
Polyurethane (PU)	0.02	35
Polytetrafluoroethylene (PTFE)	0.5	27.5
Polyacetal (PA)	2.1	67
Polymethyl methacrylate (PMMA)	2.55	59
Polyethylene terepthalate (PET)	2.85	61
Polyetheretherketone (PEEK)	8.3	139
Silicone rubber (SR)	0.008	7.6
Polysulfone (PS)	2.65	75
Articular cartilage	10.5	27.5
Ligament	303.5	29.5
Tendon	401.5	46.5
Skin	0.1–0.2	7.6
Intraocular lens	5.6	2.3

Polymeric biomaterials can be sorted into two distinct categories- (i) Biodegradable polymers such as polyesters, polyamides, polyanhydrides and polycarbonates are among the most commonly used polymers in biomedical applications and (ii) Non-biodegradable polymers; a few well known examples being polyethylene, polystyrene, polytetrafluoroethylene, polymethyl methacrylate, polyaniline, polypyrrole. In particular, non-biodegradable polymers also account for a significant fraction of use in implantable devices. Some of the examples that have already been mentioned are PMMA used for intraocular lens and bone cements. Similarly, HDPE and UHMWPE are well known for their use in orthopedic devices (see Fig. 2.2). Hydrogels, which can be either chemically stable or biodegradable, have been used as tissue engineering matrices[115]. Also, a review of polymeric materials for biomedical applications is incomplete without the mention of polydimethyl siloxane (PDMS), a bioinert polymer mainly used in soft lithography and microfluidic devices but also as an inert coating on implant materials and biomedical micro/nano systems[116].

> The ability of tailoring the chemistry makes polymeric materials attractive for biomedical applications.

The major advantages of polymers are their attractive properties and availability in a wide variety of compositions, and forms (solid, fibres, fabrics, films and gels). Polyethylene (PE), polyurethane (PU), polytetrafluoroethylene (PTFE), polymethyl methacrylate (PMMA), polylactic acid (PLA), polyether etherketone (PEEK), polyethylene terephthalate (PET), polyglycolic acid (PGA), poly-L-lactic acid (PLLA), Poly(D,L-lactic acid) (PDLLA), etc., are a few examples of the biopolymer family.

Electroactive polymers have been gaining increased attention over the past decade or so, due to developments in the characterization and processing of such materials. Conducting polymers exhibit electrical properties similar to that of metals/semiconductors, but also exhibit the ease of synthesis and fabrication common to conventional polymers. These electroactive polymers are already well known in the field of biomedical applications and the effect of electrical conductivity and external electric stimuli have been sufficiently demonstrated in modulating cell behaviour such as cell adhesion, migration, proliferation and even DNA synthesis and protein secretion. Apart from biocompatibility, conducting polymers have other unique advantages, which allow the alteration of electrical, physical, chemical and mechanical properties to suit the needs of the application. These characteristics have made the use of conducting polymers popular in biomedical applications. To this end, a system that is relatively less common in the biomedical field is the polyvinylidene difluoride (PVDF). PVDF is an inert polymer with a good resistance to corrosive environments. PVDF is generally used in filtering membranes and surgical sutures due to its inertness. Many studies with conducting polymers involve bone, muscle and nerve tissue/cells that can be excited/stimulated through electrical impulses. It was initially shown that in electrically conducting polypyrrole substrates, the shape and growth of mammalian cells can be controlled using an electrical stimulus. Further studies on the same material, polypyrrole, revealed that the electrical stimulus increased neurite outgrowth, indicating its potential application in nerve regeneration. Since then, conducting polymers have been used as tissue engineering constructs, neural probes, drug delivery devices, biosensors and bio-actuators. Conducting polymers, used as tissue engineering scaffolds are usually hydrophobic and non-biodegradable. This has been demonstrated by the use of polypyrrole–hyaluronic acid composite materials with the films of such composites retaining hyaluronic acid on the surface for several days *in vitro* which was found to promote vascularization *in vivo*, hinting at the promise of such materials in wound healing and tissue engineering applications.

Recent studies have shown that it is possible to eliminate the disadvantages of 'conventional' conducting polymers by using polypyrrole nano-particles in a blend with poly(D,L-lactic acid) (PDLLA), which is also biodegradable and hydrophilic. Another excellent example is that of poly(3,4-ethylenedioxythiophene) (PEDOT), which is used in neural probes, and drug delivery devices, and there are even cases where the drug release profile can be controlled through an electric stimulus (such as in Biotin doped polypyrrole), thereby altering the drug dosage as and when required depending on the physiological state. Conducting polymer substrates such as polyaniline, polypyrrole and carbon nanotubes are also used as biosensors and actuators and as substrates to alter/direct the stem cell fate processes to direct the differentiation towards a particular lineage. It has been reported that different electrical stimuli (pulse width, frequency, amplitude, etc.) given to human mesenchymal stem cells (hMSCs) on HCl doped polyaniline substrates can direct stem cell differentiation towards neurogenic and cardiomyogenic lineages[34].

2.3.4 | Biocomposites

A composite material is a material that has a chemically and/or physically distinct second phase within a continuous matrix phase. The composite generally has properties/characteristics better than or different from those of either component. The matrix phase is the continuous phase, while the distributed phase, commonly called as the reinforcement phase, can be in the form of particles, whiskers or short fibres, continuous fibres or sheets.

Similar to bioceramics or biopolymers, the term 'biocomposite', in its strictest sense, refers to a material comprising two different phases of biological origin, with one of them being embedded in another matrix. The most relevant example is the human bone, which is characterised by dispersion of a bioceramic phase, calcium phosphate nanoplatelets and biomolecules, like non-collagenous proteins (e.g. osteopontin) in a collagenous (biopolymer) fibrous matrix. In a similar sense, another popular example of a biocomposite material is the natural teeth. The unique hierarchical structure of human bone in terms of the size and distribution of different phases leading to a gradient in properties has not yet been mimicked in any of the synthetic composites so far!

In most of the published literature as well as in this book, the term biocomposite however will be referred to a synthetic composite having metals/ceramics/polymers with each of the phases having the acceptable biocompatibility property for a specific biomedical application. It is important to bear in mind that all prosthetic biomaterials are used to maintain a balance between the mechanical properties of the replaced tissues and the biochemical effects of the material on the tissue. Both areas are of great importance as far as the clinical success of materials is concerned. However, in most (if not all) biological systems, a range of properties are required e.g., biological activity, mechanical strength, chemical durability, etc. Therefore, often a clinical need can only be fulfilled by a material which exhibits a complex combination of properties. The use of composite technology has also enabled materials researchers to develop a wide range of new, tailored materials, 'biocomposites', which offer the promise of greatly improving the quality of life of many people. The aim of making the composite is to strengthen the materials, without affecting the biocompatibility.

The demands made on materials for better performance are so rigorous and diverse that no material is able to satisfy them. This has naturally led to a resurgence of the ancient concept of combining different materials to a composite material for satisfying user requirements. Such composite material systems result in a performance unattainable by the individual constituents. According to Chawla, "composites have introduced an extraordinary fluidity to design engineering, in effect forcing the designer-analyst to create a different material for each application as he pursues savings in weight and cost"[117].

Therefore, a new approach to the design of implantable biomaterials with tailor-made properties has been attempted for various polymer biocomposite materials. Polymer biocomposite materials consist of one or more reinforcing phases (ceramics, metals and polymers) in a polymer matrix. These composites are considered an alternative choice to overcome the various shortcomings of homogeneous polymers. Besides significant improvements in creep, fatigue resistance and other mechanical properties, one major advantage of polymer biocomposites is their radio transparency and non-magnetic features, which allow them to be scanned by various modern imaging and modern diagnostic methods, like computed tomography (CT) and magnetic resonance imaging (MRI). Such advantages can be exploited in their applications in cranioplasty.

The specific examples related to lab-scale development of biocomposites will now be briefly mentioned. While the citation to some of the studies is not provided below, a reader can refer to the author's recent books[1,2,3] as well as other references provided under 'for further reading' at the end of reference of Chapter 2.

In the last decade, researchers have attempted to develop biocomposites by combining bioactive HA with various reinforcements, like HA whiskers and other bioinert ceramic phases, like Al_2O_3 or ZrO_2. Overall, the mechanical properties, in terms of the hardness and toughness of HA-based composites remain poor, except for the FHA–ZrO_2 composite (H_v ~ 10 GPa, K_{IC} ~ 2.5 MPa.m$^{1/2}$). Although biocomposites in other ceramic systems like Si_3N_4-bioglass composites exhibit better toughness (K_{IC} ~ 4.4 MPa.m$^{1/2}$), the biocompatibility property remains to be investigated. In the following, some discussion has been made to illustrate potential issues in the processing/densification of bulk bioceramic composites.

Calcium phosphate based composites offer great potential in many applications, requiring bonding with the bone. In view of good toughness and strength properties of ZrO_2 (in particular, Y-TZP), a few attempts were made to develop HA-ZrO_2 composites. However, improvements in toughness and strength, achieved for the HA–ZrO_2 composites were not accompanied by enhanced strength reliability, in terms of the Weibull modulus. Bioverit®III glass–ceramic matrix/Y–PSZ particulate composites, successfully prepared by pressureless sintering, confirmed the toughening effect of the Y–PSZ particles. In another interesting work, HA/Al_2O_3 composites of functionally graded structure were fabricated by the underwater-shock compaction technique. The composite showed continuous compositional variation after heat treatment up to 1200°C. A highly dense (>98% of relative density) Si_3N_4–bioglass composite is being developed with the aim of taking potential advantage of each constituent, i.e., the high fracture toughness of the Si_3N_4 with the bioactivity of a bioglass. The most significant feature concerning the mechanical properties of this new biomaterial composite is the improvement in fracture toughness (4.4 MPa.m$^{1/2}$) and bending strength (383±47 MPa) with respect to currently used bioceramics glasses and glass ceramics for load-bearing applications. However, detailed biocompatibility study on Si_3N_4-bioglass composite is yet to be carried out. It can be commented that any composite, which shows superior mechanical and physical properties and is also biodegradable, would be the material of choice in the near future.

2.3.4.1 | Biocomposites with antimicrobial properties

Silver (Ag) has been a very popular antibacterial element for its toxicity towards the bacterial viability and has been widely used whenever antimicrobial action is needed. Ag^+ containing HA (Ag/HA) and Cu^{2+} containing HA (Cu/HA) have been made and confirmed to have good antibacterial ability.

For example, Nath et al. prepared HA–Ag biocomposites with the addition of 10 wt.% of silver in an HA matrix[118]. Antibacterial experiments involving gram-negative bacteria and Escherichia coli confirmed the excellent bactericidal property of HA-10 wt.% Ag composites.

> One of the important aspects to note is that the compositional design of a biocomposite to promote antibacterial property should not compromise the cell functionality property.

In view of the above assertion, the cell and bacterial study are both conducted in the development of antibacterial biocomposites. For example, *in vitro* cell functionality on HA-10 wt.% Ag composites was assessed using mouse fibroblast (L929) and human osteosarcoma (MG63) cell lines[118]. The reported results, qualitatively and quantitatively, indicated comparable or even better cellular viability, osteogenic differentiation and bone mineralization of HA-10 wt.% Ag than HA.

Similar to Ag, other antimicrobial agents, like ZnO is also used to develop HA-based antimicrobial composites. However, the mechanism for the antibacterial property of ZnO is still not clear. According to several studies, the reaction of zinc oxide with water produces a hydroxyl radical (OH^-), which provides antimicrobial properties by disrupting microbial cell membranes. In contrast, another report mentions that hydrogen peroxide generated from the surface of ZnO is instrumental in the inhibition of bacteria growth while they deny the significant effect of Zn^{2+} towards bacterial growth. Although ZnO is considered a good antimicrobial agent, limited efforts have been put to comprehensively understand and develop HA–ZnO composites[119]. The main advantage of ZnO, except producing H_2O_2, is that it does not increase the pH of the environment, which may be detrimental to the osseous tissues.

2.3.4.2 | Biocomposites with better toughness properties

The major shortcoming of ceramic biomaterials is their low fracture toughness value that limits their use as load-bearing implant materials. In this backdrop, low modulus materials, such as polymers, could be combined with ceramic to obtain a combination of desired material properties. In several years of research, HA/HDPE, SiO_2/silicone rubber, HA/EVA, BCP/PMMA, HA/PLA, HA/PEEK, bioactive glass/PMMA, nano-HA/Polyhexamethylene adipamide composite materials have been developed. Polymer–ceramic biocomposite materials are effectively used in different applications such as hard tissue replacement (e.g., load-bearing implant applications like joint replacement, bone replacement materials, dental implants, bone fracture repairing, etc.) and soft tissue replacement (e.g., artificial tendons and ligaments, maxillofacial reconstruction, etc.).

One of the major shortcomings of HA as a load-bearing implant material is its lack of fracture toughness (0.7 $MPa.m^{1/2}$)[120]. Since metals usually possess high fracture toughness due to their inherent plasticity, it is a good strategy to use metallic additives to increase the fracture toughness of HA. However, only a few metals/alloys can be used with HA due to relative differences in density (which causes an aggregation of the metallic phase), the reactivity of the metal and the poor interface bonding of HA with the metallic phase. Also, a higher amount of the metallic phase is known to induce the formation of β-TCP and α-TCP, which reduce the mechanical properties of HA further. The ideal choices for additives are lightweight metals such as Ti, Al and Mg. In fact, a number of HA–Ti[121,122,123,124,125], HA–Ti6Al4V[126,127,128] and a few HA–Mg[129,130] composites have been studied in the literature. It appears that, based on the depth of the literature, HA–metal composites are more efficient as coatings rather than as sintered objects or scaffolds. Nevertheless, due to its corrosion resistance and favourable elastic modulus, Ti and its alloys have been studied extensively both as coatings and as sintered composites. In particular, the HA–Ti system is interesting, mainly because HA coatings on Ti implants are generally used in bone replacement applications. Due to the reactivity of Ti, difficulties in sintering are common and Yang et al.[131] reported that the sintering of HA-10,20 wt.% Ti at 1100°C in vacuum led to the formation of sintering reaction products, such as β-TCP, α-TCP and $CaTiO_3$ with little retention of metallic Ti. In order to reduce the sintering reactions (due to the dissociation of HA) and retain metallic Ti, the effect of a change in the sintering atmosphere was demonstrated in a study by Nath et al.[132], where an argon atmosphere was used in the sintering process in the temperature range of 1000–1400°C. Functionally graded HA–Ti composites were prepared by the powder metallurgy route and their effectiveness in enhancing the mechanical properties was demonstrated[122].

Recently, advanced sintering techniques such as Spark Plasma Sintering (SPS) have been adopted to prevent/reduce the formation of sintering reaction products. These methods are usually characterized by high heating rates (50–200°C/min) and short sintering duration (30–45 min). The major advantage in such methods is the limited time available for grain growth and sintering reactions along with attaining densities close to that of the theoretical density of the composites. In the study by Kumar et al., the sintering of HA–Ti (limited to 20 wt.% Ti) in an argon atmosphere through SPS to obtain dense composites was demonstrated[125]. The evaluation of mechanical properties of SPSed HA–Ti composites showed improvement in the toughness and flexural strength (3-point bending). The improvement in toughness (4.7 MPa.m$^{1/2}$) measured by the single edge V-notch bend test is of enormous importance. Such higher toughness is mainly due to the ductile Ti phase, which reduces the crack-tip driving force and hence increases the resistance to crack propagation. The relevant mechanical properties of HA–Ti composites are summarized in Table 2.4. *in vitro* studies of HA–Ti/Ti alloys coatings in simulated body fluid (SBF) have shown good mineralization upto 8 weeks[126]. Saos-2 cell (osteoblast like cells) studies with HA–Ti composites show excellent cell adhesion and proliferation in addition to the formation of an apatite layer when immersed in SBF. Similar to Ti, mullite ($3Al_2O_3.2SiO_2$) is also used as a sinter-additive with HA and after SPS, HA-mullite powders with 30 wt.% mullite indeed showed better mechanical properties with an increase in fracture toughness of 1.5 MPa.m$^{1/2}$, flexural strength of 70–80 MPa and a hardness of 4–5 GPa. Such composites also support the formation of a porous apatite layer in SBF and *in vivo* studies showed bone integration with HA–mullite implants in a rabbit model, establishing uncompromised biocompatibility[133,134].

Table 2.4 Mechanical properties of SPSed HA–Ti composites [Adapted from Ref. 95].

Composite (in wt.%)	*Elastic modulus (GPa)*	*Flexural strength (MPa)*	*Fracture toughness (MPa.m$^{1/2}$)*
HA-5% Ti	58.1±0.2	79.0±4.8	3.6±0.3
HA-10% Ti	54.5±0.3	89.7±8.4	4.7±0.1
HA-20% Ti	50.2±0.8	99.7±1.2	4.3±0.3

In the case of HA–ZrO_2 composites, sintering reaction products are observed as ZrO_2 needs a temperature of 1200–1400°C for efficient consolidation, whereas HA is unstable at such temperatures. Secondary phases, such as $CaZrO_3$ and β-TCP are usually observed when the ZrO_2 content is increased beyond 15–20 wt.%[135]. In another study, CaF_2 was added to the HA–ZrO_2 mixture to aid in the densification of HA–ZrO_2 when sintered conventionally in air at 1350°C to attain a 98% of theoretical density. The increase in flexural strength and fracture toughness was highest for HA-20 wt.% ZrO_2-5 mol.% CaF_2 with 125 MPa and 1.5 MPa.m$^{1/2}$, respectively[136]. *in vitro* and *in vivo* studies using L929 mouse fibroblasts and rabbit animal models have shown that it supports cell growth and osseointegration (in rabbit femur)[137].

In HA–Al_2O_3 composites, 40–50 vol.% of Al_2O_3 has been known to increase flexural strength (250 MPa) and fracture toughness(2.5 MPa.m$^{1/2}$), but larger amounts of Al_2O_3 cause secondary reactions during sintering[138]. Instead of adding more Al_2O_3, studies have shown that the addition of a third component such as ZrO_2, HDPE, CNTs or SiC might give way to better mechanical properties with a lesser number of sintering reaction products[139,140,141]. *in vitro* studies for such composites show the usual biocompatible behaviour, with better hardness (in the case of SiC), toughness (ZrO_2) and good scratch resistance (with CNTs in coatings), but the fracture toughness still remains modest (1.5 MPa.m$^{1/2}$).

2.3.4.3 | HA–Polymer composites with better mechanical properties

HA–polymer composites have been developed with the idea that the viscoelastic behaviour of polymers under mechanical loads can impart additional toughness to HA making it more reliable as an implant material. In this regard, the processing, mechanical and biological response of HA–HDPE (high density polyethylene)[142,143], HA–UHMWPE (ultra-high molecular weight polyethylene)[144], HA–PMMA[145], and HA–PVA[146] have been well studied. Some composites such as HA–PVDF and HA–CNT not only increase the toughness but also make the composite a high dielectric (in the case of PVDF) and high strength (with CNTs) material, which according to the discussion so far are very attractive properties for a potential bone implant material[147,148].

HA–HDPE composites, prepared by blending HA (10–50 vol.%) particles and HDPE at temperatures in the range of 200–250°C were observed to exhibit an elastic modulus in the range of 1–6 GPa, similar to that of cortical bone (7–30 GPa). Also, the fracture strain for the composites with a lower HA content was naturally higher, but the transition from ductile to brittle behaviour was noticed at 30 vol.% HA due to the debonding of HA particles from the HDPE matrix. Novel strategies have been used to improve the bonding of HA particles to further improve the properties. In particular, chemically treated HA (a silane coupling agent) along with grafted polytheneacrylic acid has yielded promising results. The mechanical properties were found to increase with the silanization of HA and the acrylic grafting of polyethylene, especially the tensile strength. The fracture strain was also found to be higher compared to the previous case (ungrafted and untreated HA), whereas there was a slight decrease in Young's modulus at a higher vol.%(40%) of HA[149].

A similar study was conducted by Fang et al.[144], but with UHMWPE. It was observed that even at 23 vol.% of HA, there was an order of magnitude improvement in the elastic modulus of UHMWPE and there was a retention in ductile (or viscoeleastic to be more precise) behaviour with fracture strains going upto 300%, something not seen in other biocomposites. HA–HDPE composites support the proliferation of osteoblasts and HA–UHMWPE composites have been shown to be bioactive and aid the formation of an apatite layer, when immersed in SBF.[150,151] While some

HA–polymer composites are used as coatings or melt–mixed substrates, HA–PMMA bone cements have found applications in rapid prototyping techniques such as 3D printing and 3D plotting[151]. It has been observed that a larger amount of HA in the PMMA matrix enhances the osteoblast response on HA–PMMA composites, indicating their applicability to bone healing applications[152].

2.4 | Closure

This chapter discussed the various modes of classification of biomaterials. One such classification clearly shows the paradigm shift in the development of three generations of biomaterials, while another classification is based on the biocompatibility and host response. A major emphasis has been placed on discussing the characteristics of biodegradable polymers and bioactive ceramics/glasses with illustrative examples. A substantial part of the discussion in this chapter can be traced to three generic material classes, i.e., metals, ceramics and polymers for biomedical applications. The recent approach in coating on metallic implants is also briefly mentioned. The toughness enhancement to ceramics has led to an extended discussion in attempts to develop ceramic biocomposites. Some mention has been made on the recent advances in electroactive polymers, apart from a detailed discussion on various biodegradable polymers[153]. Among biocomposites, a specific discussion is made on the development of HA or bioglass composites with antimicrobial properties as well as on HA-based composites with better mechanical reliability. Some attempts to develop HA–polymer composites are also highlighted. Overall, the discussion in this chapter establishes the growing need to develop futuristic biomaterials to address various issues, including prosthetic infection, multifunctional properties, better strength/toughness properties, better corrosion and wear resistance properties together with uncompromised compatibility with biological cells and tissues[154,155].

Tissue Engineering Scaffolds: Principles and Properties

The fabrication of scaffolds for bone tissue engineering demands an amalgamation of a multitude of attributes including a suitable porosity to encourage vascular invasion, desired surface chemistry for controlled deposition of calcium phosphate-based mineral as well as ability to support attachment, proliferation, and differentiation of lineage specific progenitor cells. In this perspective, this chapter summarizes the important natural and synthetic scaffolds as well as the surface treatment techniques to manoeuver the structure–property (physico-chemical and biological properties) relationships of engineered scaffolds. Currently employed fabrication strategies, like the peptide self-assembly, electrospinning for the fabrication of extracellular matrix-mimicking polymeric/ ceramic/hybrid fibrous mesh, etc., have been discussed. In particular, surface functionalization and patterning techniques used for architecturing the surface topography for the enhancement of the biological and mechano-chemical properties of the scaffolds are also discussed. Apart from reviewing a host of experimental studies reporting the functionality of osteoblast-like (bone) cells and stem cells on surface modified or textured bioceramic/biopolymer scaffolds, some attempts to provide theoretical insights to predict cell behaviour are also mentioned.

3.1 | Introduction

Tissue engineering is an emerging multidisciplinary research area at the interface of biology, engineering as well as medicine with immense potential to recast our means of healthcare by offering potential solutions in terms of regenerative medicine for maintenance, restoration and amelioration of damaged tissues and hence, organ functions[156].

Tissue engineering ventures primarily on site-specific tissue architecture by manipulation of cells, scaffolds and biological cues.

Underlying this concept of tissue engineering for both therapeutic and diagnostic applications, is the ability to exploit the functionality of living cells in diverse ways on engineered scaffolds. The ability to develop materials, which concert with tissues structurally, mechanically and biofunctionally, i.e., biocompatibility is of prime importance in regenerative medicine and all such materials, with appropriate biocompatibility property, are widely known as biomaterials. One prime approach for developing artificial constructs for tissue regeneration is autologous transplantation, i.e., direct transplant from the host only[157,158]. Although this method offers the 'gold standard' and promises the best clinical results by virtue of inherent reliability with the host, a lack of immune and disease related complications possible from external transplantation sources, it severely restricts large-scale application of such an approach. Efficacious design of a tissue-engineered scaffold requires sufficient understanding of the interplay among the cells, tissues and the extracellular matrix (ECM) at the molecular level.

> The scaffold should ideally provide an environment for tissue regeneration, evading hostile immunogenic responses, while degrading in a sustained manner, *in vivo.*

The tissue interactions at molecular level involve predominantly weak interactions: (hydrogen bonding, electrostatic interactions and solvation effects), manifested in hydrophobicity[159]. Such lower energy couplings are responsive and adaptive, facilitating rapid assembly and disassembly as well as changes in protein conformation. The self-assembly of collagen, a primary component of the ECM protein of hard tissues occurs sequentially from the expression of pro-collagen proteins to their eventual assembly into triple-helical fibres[159]. The entire phenomenon is strongly dependent on the hydrophobic interactions within the physiological environment. The weak forces aid the dynamicity of tissues and the shock absorbing nature of cartilage can be attributed to the movement of water from a nexus of hydrophilic proteoglycan molecular chains[159]. On the same note, the nucleation of hydroxyapatite into the nano-dimensional gaps between collagen molecules tones up a bone[159]. The proteins, like collagen and elastin are stabilized by the enzymatic cross-linking of lysine and the phenomenon brought about by stronger covalent bonds between tissues is known as covalent capture[160]. However, apart from conglomerating the biological components into a structure, the ECM must essentially provide necessary instructive cues for sustenance of cell phenotype behaviour[161].

Another objective of this chapter is to highlight the plethora of biomaterials developed for tissue engineering with particular focus on osteoblast functionality. A comprehensive study of various natural and synthetic, polymeric and ceramic biomaterials and their composites has been presented here along with the brief descriptions of the most widely used techniques like electrospinning, self-assembly, etc. Subsequently, the techniques for enhancing biocompatibility of the scaffold are also discussed. Bioactivity of a substrate can be augmented either by direct topographical patterning or by surface treatment techniques, whereby surface nanofeatures can be introduced. It has already been widely appreciated that the bone cells respond to multidimensional cues, including topography. Along with this, an overview of the influence of patterned surfaces on cell responses is also presented, with special focus on the role of protein adsorption in manoeuvring cell–substrate interplay.

Tissue engineering combines the principles and methods of life sciences and engineering sciences to create materials and develop strategies to heal damaged tissues or create tissue replacement substitutes[162]. The three main approaches of tissue engineering include the implantation of (a) engineered scaffolds, (b) only cells and (c) engineered scaffolds seeded and cultured with appropriate cells. The most ideal strategy for regeneration is tissue specific. For instance, cells are not always included in scaffolds for nerve repair as host axons need to connect with the target tissue to restore function[163]. Cells injected directly do not survive and invoke an immune response[164]. The most common approach involves a scaffold with cells. The scaffold is a construct that supports cell growth and proliferation and allows adequate nutrient exchange, and ideally, degrades as it is gradually replaced by a functional tissue. Different biofactors (cells, genes and proteins) may be transplanted along with this scaffold and stem cell and gene therapy approaches may be used to stimulate tissue repair[165]. Depending on the type of tissue, the scaffold may be an elastomer (for soft tissue engineering involving skeletal muscles, skin, etc.), or rigid polymers/metals/ceramics (for hard tissue engineering involving bones, teeth, etc.).

Tissue engineering offers a potential solution to many problems. Tissue and organ shortage is a severe challenge worldwide, and this gap is bound to exacerbate in the wake of an aging population. One of the aims of tissue engineering is also to achieve artificial organs, as the availability of donor organs and tissues is very limited[166]. Tissue engineering offers a possibility of eliminating (or at least reducing) the need for organs by generating organ scaffolds and by regenerating the organ. Products such as tissue-engineered skin, bone, blood vessels and pancreas exist as already approved therapies[167]. Success in the field of tissue engineering can create tremendous financial opportunity. In this regard, the sales of regenerative biomaterials, which exceed USD 240 million per year, also hint at the tremendous potential that the field beholds[168]. However, the complexities and the challenges involved in generating whole organs have largely given way to smaller, attainable goals. For instance, the goal to make an artificial heart has been replaced by a realistic goal to replace coronary arteries, valves and myocardium.

3.2 | Structure and Properties of Bone

The property requirements of tissue engineered scaffolds are application specific. For example, a scaffold for cardiovascular applications requires the material to interact suitably with blood as well as to allow adhesion and functionality cardiac tissue-specific cells (e.g. cardiomyocytes).

> The scaffolds for bone tissue engineering requires the material to mimic the functional properties of natural bone, while allowing proliferation of bone cells or stem cells to undergo osteogenesis.

At this juncture, it is instructive to briefly mention the bone structure. The hierarchical structure of natural bone is described in Fig. 3.1. At the finest structural length scale, the natural bone is composed of HA nanocrystals or nanoplatelets dispersed in collagenous matrix. Such inorganic–organic structural units, at the next level form osteons, the basic structural unit of macrostructure of bone. A number of different canals, like Haversian canals, Volkman's canals, surround the osteon in a bone structure. These canals essentially transport oxygen through blood vessels and also allow waste products to be removed. Various cell types, like osteoblasts (bone forming cells), osteoclasts (bone resorbing cells) or osteocytes (matured osteoblasts) contain the natural bone. Each cell type has distinguishable morphological features, as also shown in Fig. 3.1. Summarizing, the unique structural arrangement of HA and collagen at multiple length scales provide unique set of bone properties. For example, the piezoelectric, electrical conductivity properties of bone are attributed to the presence of collagen, while the elastic stiffness and strength properties are due to the presence of large volume fraction (65%) of HA. The unique viscoelastic properties of natural bone are largely due to the presence of collagenous fibrous matrix, which behaves like viscous polymeric matrix. The above-described unique structural and functional characteristics of natural bone have motivated significant research activities around the world. Despite decades of efforts, the ideal bone replacement scaffolds are still awaited. In this chapter, some of the generic approaches illustrating such efforts are discussed.

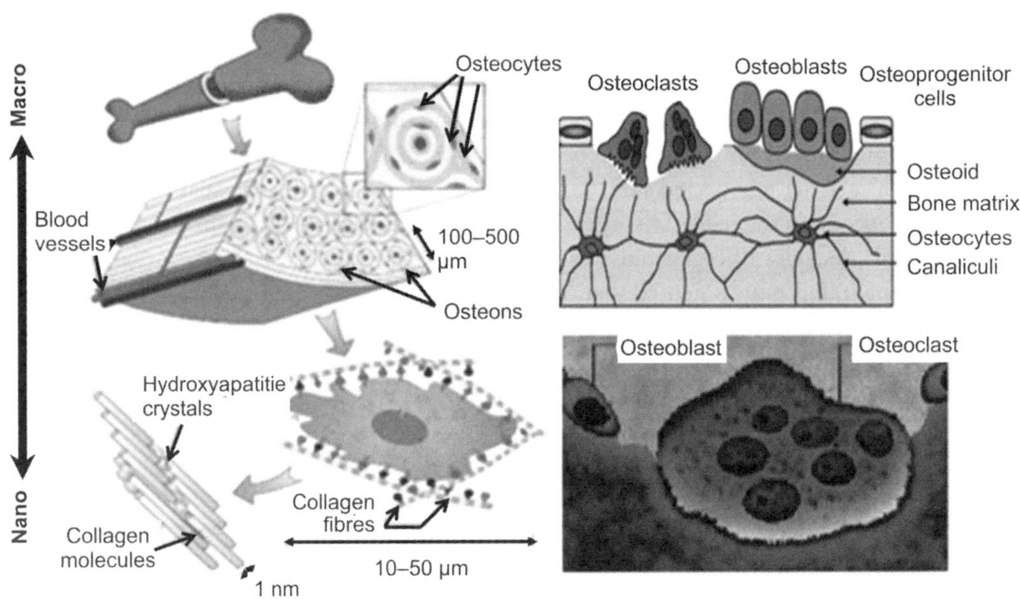

Fig. 3.1 Hierarchical structure of natural bone, including that at cellular level[444] [see Colour Plate].

3.3 | Property Requirements for Bone Tissue Engineering Scaffolds

Bones have the unique capacity to remodel themselves throughout the lifetime of an individual. This highly regenerative capacity of bones particularly in younger people should imply easier recovery from fractures without surgery.

Large bone defects and many of the non-union fractures in orthopedic surgery require orchestrated tissue regeneration at the site of injury through surgical intervention.

In this regard, bone tissue engineering uniquely presents a promising approach for bone repair and reconstruction. Typically, this approach draws together fabrication of suitable scaffolding material in conjunction with cells and biological cues.

It can be argued that the origin of the 'scaffold-based tissue engineering concept'—as we recognize it at present—was in the mid-1980s. The initial approach was to fabricate scaffolds, for seeding cells onto or mixing cells into naturally available matrices. However, the early scaffolds had physical and chemical properties which were difficult to manipulate, thus resulting in wide variations of the results produced *in vitro* and *in vivo*.

It is important to emphasize at the outset that the domain of scaffold-based tissue engineering is still in its formative years and various diverse approaches are under experimental investigations. Thus, it is by no means clear what defines an ideal scaffold even for a specific tissue type. Indeed, while some tissues proffer manifold functional roles, it is improbable that a single scaffold would serve as a universal foundation for the regeneration of even a single tissue. Hence, the considerations for scaffold design are multifaceted. They consist of material composition, porous structural design, structural mechanics, surface properties, degradation properties in tandem with the paraphernalia of any biological component, which may have been added to the scaffold to improve function. In addition, one must also consider the behaviour and the consequences of how all of the aforementioned factors may transmute with time. Material chemistry jointly with processing methodologies determines the functional properties that a scaffold can attain and how cells interact with the scaffold (Fig. 3.2)[169].

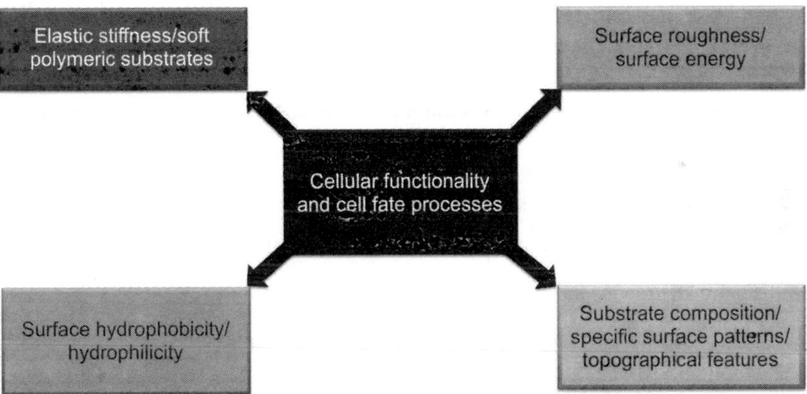

Fig. 3.2 Various factors involved in 'scaffold-based' tissue engineering.

The design of scaffolds to guide cell growth is an important challenge in tissue engineering. The aim is to fabricate porous substrates that support and structure the newly formed tissue from the cultured cells. The development of a porous component starts with the pore structure requirements. Porous materials can be classified into three categories as per ASTM standards. They are as follows:

1. Interconnecting pores (open pores)
2. Closed or non-interconnecting pores
3. Combination of both 1 and 2

Porous solids were classified into two groups, foams and honeycombs. Gibson and Ashby[170] connoted that the mechanical properties of a porous solid depend mainly on its relative density, the properties of the material that make up the pore edges or walls and the anisotropic nature, if any, of the solid.

Of course, the scaffolding material must be biocompatible and cater to advanced levels of therapeutic and surgical assistance. Biocompatibility implies that the material transplanted must not elicit unresolved inflammatory response nor induce immunogenicity and cytotoxicity. It also calls for sufficient mechanical strength so that it should not yield, while handling or during the individual's normal biological activities. Notably for degradable bulk scaffolds coming in contact with human cells and tissues, all surfaces must be sterilized to avert chances of infection[171]. The ideal tissue engineering paradigm is that the biomaterial should be resorbed and replaced over time by the body's own newly regenerated tissue.

> The scaffolds for bone tissue engineering must be endowed with strong bone bonding ability, culminating into osteoconduction and osteoinduction; its mechanical strength should be commensurate with the mechanical constancy required in load bearing.

In addition, the scaffold must possess substantial controllable interconnected porosity to guide the cells to grow into desired physical form and to promote vascularization of the ingrown tissue. A typical porosity of 90% and a minimum pore size of 100 µm (Table 3.1)[172] is known to be *de rigueur* for cell penetration and efficient vascularization of the ingrown tissue (Table 3.1)[173,174]. In fact, the three dimensional scaffolds provide the necessary support for adhesion, growth and differentiation and limit the overall shape of the engineered implant[156].

Table 3.1 Pore size distribution for an ideal scaffold in bone tissue engineering application[172,174,452].

Pore size (µm)	Biological relevance
<1	Protein interaction and imminent bioactivity
1–20	Cell attachment and supervenient directed cell growth
100–1000	Cell growth and concomitant bone growth
>1000	Programming the shape and functionality of implant

3.4 | Overview of Biological and Porous Scaffolds

Tissue engineering annexes diverse disciplines and calls for equal attention in developing suitable scaffolds and amassing appropriate cells and biomolecules; e.g., angiogenic factors, growth and differentiation factors and signalling molecules. In this section, we document the biomaterials potentially used for scaffold engineering. However, it is to be borne in mind that the biomaterials are not mere vehicles for introducing cells to the damaged or diseased tissues; rather they must aggrandize the functions of the endogenous progenitor cells. As mentioned before, scaffolds should essentially be osteoconductive as new bone is formed from creeping substitution of adjacent living bones. In addition, they must be equally proficient in routing cytokines for activating or recruiting precursor cells from the host into osteogenic lineage.

The selection of material is of cardinal importance while engineering tissue regeneration scaffolds. In lieu of the manifold requirements of scaffold material for tissue engineering, today, the popularly used materials are synthetic or natural polymers such as, polysaccharides, poly (α-hydroxy ester), hydrogels or thermoplastic elastomers.[157,175]. Apart from these, bioactive ceramics including bioactive glasses, calcium phosphates and glass-ceramics are widely used as biomaterials (Fig. 3.3)[174,176,177]. As will be discussed later, a current endeavour is to blend the organic–ceramic polymers into a scaffold. In conjunction, the ongoing efforts are to engineer the scaffolds with drug delivery capacity, which would enable *in situ* release of growth factors or antibiotics to augment bone in-growth, nurse bone defects and support wound healing[178].

Fig. 3.3 Overview of different bone graft materials [Adapted from Ref. 445].

A multitude of processing techniques has been developed to fabricate porous scaffolds complying with the property requirements, as outlined above. The most commo among these are electrospinning, gel casting[179], slip casting[180,181], fibre compaction[182], freeze casting[183] and gas foaming[184] phase separation. A schematic overview of the nitty-gritties as well as the diverse processing techniques instrumental in fabrication of tissue engineering scaffolds is provided in Fig. 3.4. SEM images of HA-based macroporous scaffolds and 45S5 bioglass foams (after SBF treatment) are shown in figs 3.4d and 3.5, respectively. Also, some 'intelligent' scaffolds have the unique ability to refashion themselves in response to certain external stimuli[185].

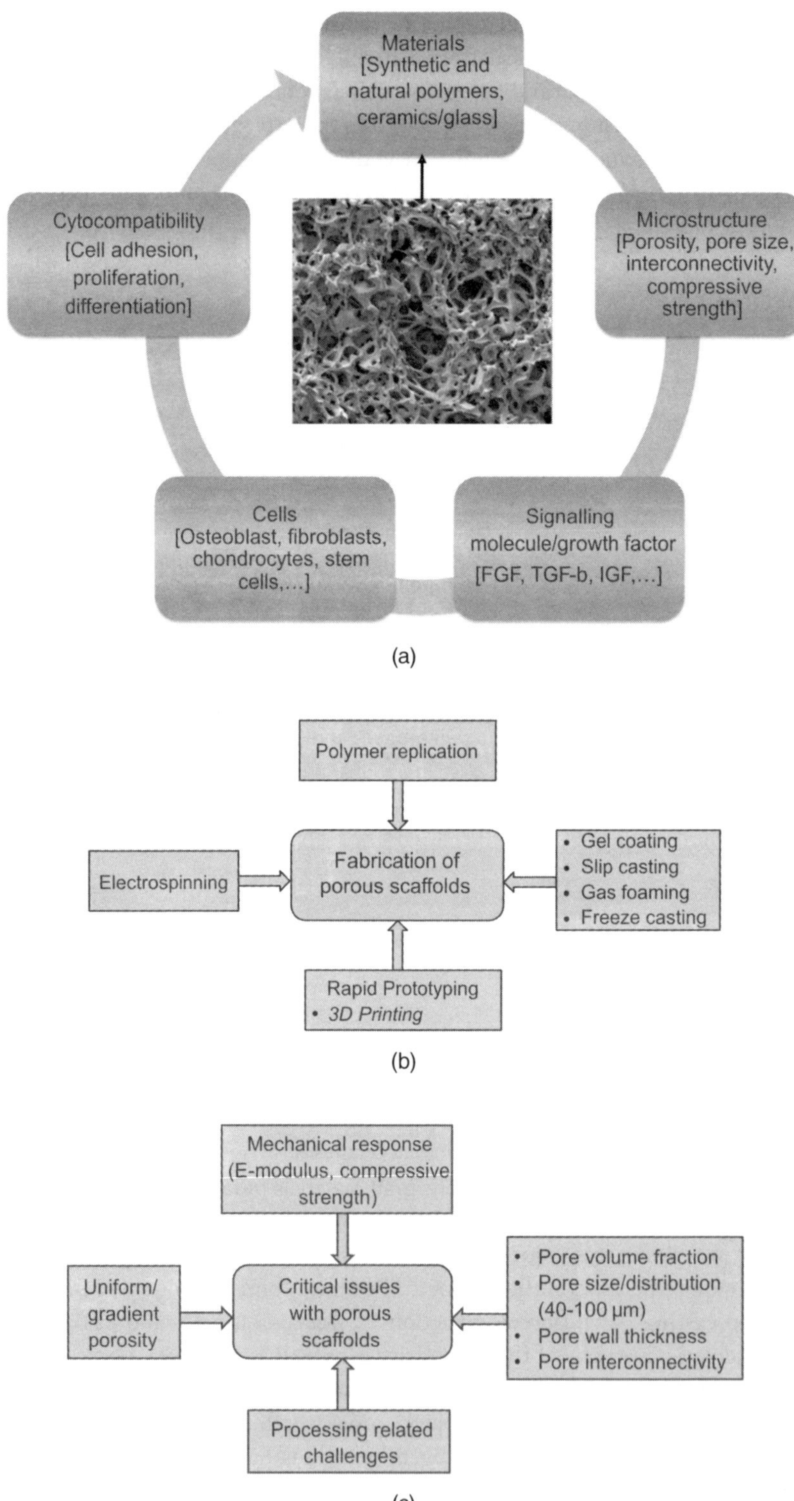

Materials
[Synthetic and natural polymers, ceramics/glass]
Cytocompatibility
[Cell adhesion, proliferation, differentiation]
Microstructure
[Porosity, pore size, interconnectivity, compressive strength]
Cells
[Osteoblast, fibroblasts, chondrocytes, stem cells,...]
Signalling molecule/growth factor
[FGF, TGF-b, IGF,...]
(a)
Polymer replication
Electrospinning
Fabrication of porous scaffolds
Gel coating
Slip casting
Gas foaming
Freeze casting
Rapid Prototyping
3D Printing
(b)
Mechanical response
(E-modulus, compressive strength)
Uniform/ gradient porosity
Critical issues with porous scaffolds
Pore volume fraction
Pore size/distribution (40-100 µm)
Pore wall thickness
Pore interconnectivity
Processing related challenges
(c)

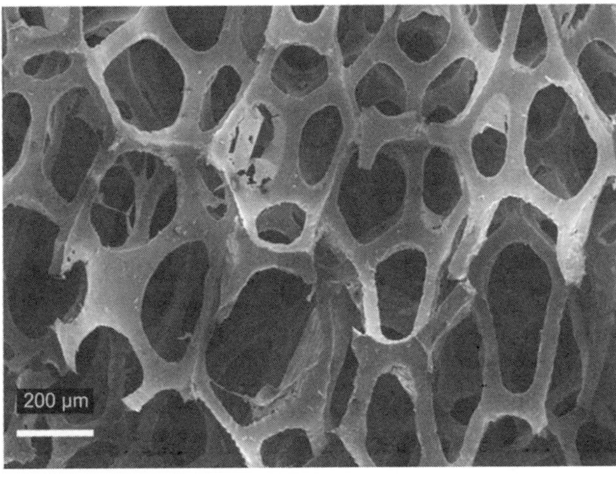

(d)

Fig. 3.4 (a) Schematic illustration of various factors to be considered while developing porous tissue engineering scaffolds, (b) various classes of the fabrication to make porous scaffolds, (c) various issues of relevance to porous scaffolds, and (d) SEM image of HA-based macroporous scaffold revealing pore sizes, distribution and interconnectivity of pores [(d) reproduced from Ref. 446].

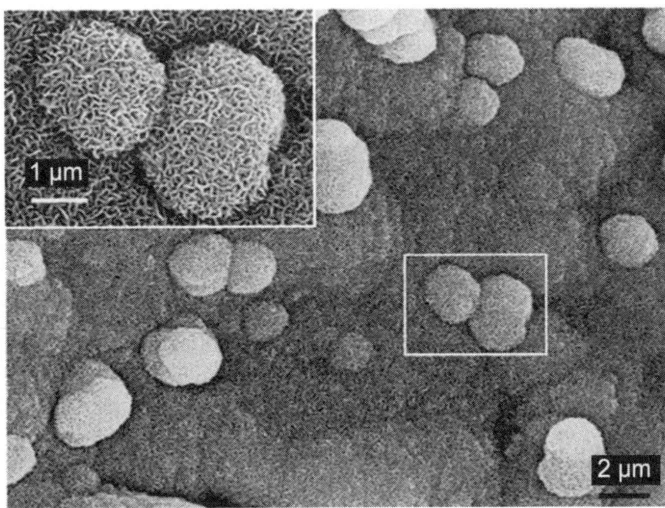

Fig. 3.5 SEM micrograph depicting the cauliflower morphology of hydroxyapatite formed on the surface of a 45S5 Bioglass-based foam, when immersed in simulated body fluid (SBF). Inset shows the magnified view of the framed area revealing rod-shaped crystal morphology of hydroxyapatite [Adapted from Ref. 447].

The scaffolds fabricated for tissue engineering should essentially have interconnected pores, with sizes between 20 and 400 µm.

In the following section, major classes of tissue engineering scaffolds are briefly discussed with reference to the type and material composition as well as their distinguishing properties.

3.4.1 | Protein templates

Proteins being natural biomaterials, can be easily mixed with organic and inorganic components to form nanocomposites with innate biomimetic and excellent physico-chemical characteristics, which stimulate cellular interactions, initiating tissue regeneration. Purified and reconstituted natural proteins, like chitosan, gelatin, and collagen can form hydrogel matrices with unique 3D architecture, and can hence be reinforced into templates. However, limited control over their physical properties and biodegradability impede extensive use of such recombinant proteins. Some commonly used proteins used in tissue engineering are delineated in the following section.

3.4.1.1 | Type I collagen

Type I collagen, the most abundant connective tissue protein, is the key component of ECM. It exists as a monomer in acidic pH, but polymerizes *in vitro* under physiological conditions.

Hydrogels, formed from such proteins, manifest excellent cell adhesive properties, foster growth, differentiation and proliferation of different cell lines. Nevertheless, the lack of sufficient load-bearing ability and mechanical strength during early implantation entails admixing of such hydrogels with natural and synthetic materials. Such additives should not only supplement the fabricated scaffold, but must also furnish relevant cues vital for differentiation of precursor cells into functional osteoblasts. Coating recombinant collagen fibrils with crystalline hydroxyapatite has been shown to be osteoconductive (with substantial mechanical strength) to heal critical size defects in animal models[186]. Recently, a scaffold of collagen foams with apatite crystals and blends of collagen, nanoapatite and polylactic acid has been shown to possess excellent cell attachment properties *in vitro* and osteoinductive properties *in vivo*[187]. Collagen–chitosan blend demonstrated commendable mechanical properties, antimicrobial effects along with cell attachment properties (Figs. 3.6 and 3.7)[179]. Zhang et al.[188] reported osteogenic properties with electrospun collagen/HA/chitosan nanofibres. The incorporation of collagen was found to promote osteogenic ability of the cells with respect to HA/chitosan blends.

Recently, Chen et al.[189] developed a novel scaffold, where porous polycaprolactone (PCL) meshes, fabricated by fused deposition modelling (FDM) were embedded in a matrix of hyaluronic acid, methylated collagen and terpolymer via polyelectrolyte complex coacervation. On being cultured in osteogenic stimulation medium both statically and dynamically, such embedded scaffolds demonstrated higher cell seeding efficiency, homogeneous cell distribution, osteogenically differentiated cells, substantiated by notable gene expression of bone markers like alkaline phosphatase, osteocalcin, bone sialoprotein II, with respect to the virgin PCL scaffolds. Interestingly, dynamic cultures manifested higher amounts of DNA and calcium compared to static culture and, the surface modification synergistically intensified the calcium deposition of hMSCs.

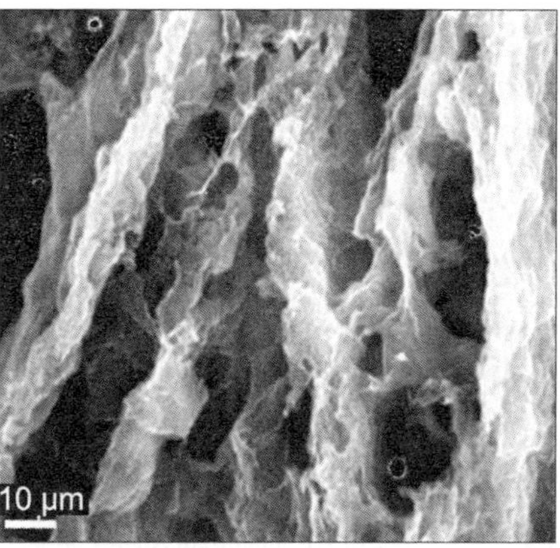

Fig. 3.6 Calcium phosphate coated collagen–chitosan (1:1) scaffold [Adapted from Ref. 448].

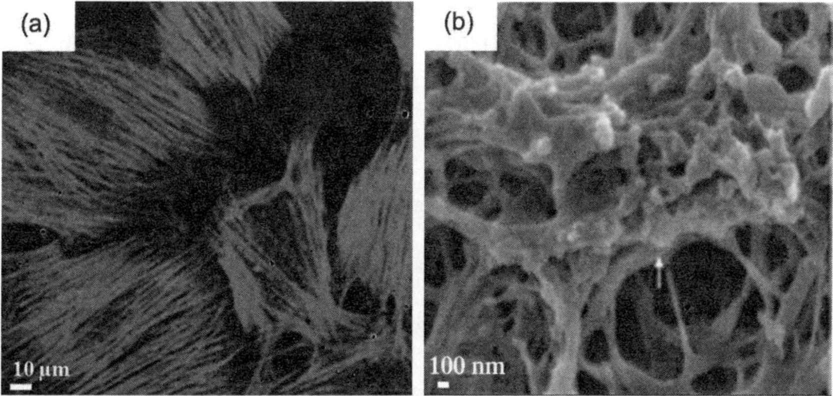

Fig. 3.7 The cellular morphological differences between rat marrow stromal cells (RMSCs) grown on 2D and 3D matrices: (a) confocal image of RMSCs grown on tissue culture plates and stained with phalloidan conjugated to TRITC for labeling actin filaments. (b) SEM image of RMSCs embedded within type-I collagen–chitosan (1:1) biopolymer matrix [Adapted from Ref. 451].

3.4.1.2 | Silk fibroin

Hydrogels, meshes and membranes of silk fibroin have already been used without cells for guided bone regeneration[190]. Osteoconductivity of silk fibroin scaffolds is already widely acknowledged. It has also been demonstrated that the incorporation of nanoscale cell motifs, like RGD peptides accentuated the rate of calcium deposition and differentiation of cells along the scaffold[191,192]. The surface modification can also be effected by reinforcing silk fibroins with calcium phosphate; e.g., poly-aspartic acid residues can be synergized with silk fibroins for efficient calcium binding and simultaneous improvement in mechanical properties[193]. On incorporation of BMP2, electrospun silk scaffolds also exhibited enhanced osteogenic differentiation of mesenchymal stem cells[194]. On similar lines, Kim et al.[195] reported increased osteoconductivity with silk fibroin protein scaffolds, pre-mineralized with $CaCl_2$ and Na_2HPO_4. In a recent study, composite matrices reinforced with micron sized silk fibres (10–600 µm) were shown to promote mesenchymal stem cell differentiation into bone-like tissue[196]. Regardless of its immense potential in bone tissue engineering, the effects of degradation products of silk fibroins are yet to be studied. A recent study reported significant increase in levels of osteoblastic regeneration genes and alkaline phosphatase activity, when MG63 osteoblasts were treated with low molecular weight silk fibroins, similar to degradation products of silk fibroins. In another recent study, Hota et al.[197] reported potential usage of silk-fibroin protein in bio-memristor devices for applications in advanced bio-inspired VLSI (very large scale integration) circuit design.

3.4.1.3 | Extracellular matrix

Extracellular matrix (ECM) gels fundamentally serve as dynamic tissue specific matrices, wherein different types of cells are embedded[198,199]. Imperatively, *in vivo* cellular response can be induced only with biomimicking microenvironments. 3D-natural extracellular matrix materials, being replete with bioactive molecules, provide physiological environment conducive for cell growth and proliferation[179]. Ultrafine continuous fibres obtained from proteins like collagen, laminin, fibronectin and proteoglycans have high surface-to-volume ratio, high porosity and variable pore size.

> Ideally, a tissue engineered scaffold should be characterized by highly interconnected porous structures to facilitate cellular migration and transport of nutrients as well as metabolic wastes, while promoting the neotissue formation.

The structure of the ECM, like other structural constituents of living system, is complex in nature. So far, there is no report as how to derive ECM from a cell line such that the readily synthesized ECM can provide structure and properties of natural ECM in a biological scaffold. Among the published studies, Narayanan et al.[200] reported that hMSCs grown on 'reconstituted' ECM scaffolds can differentiate into bone cells with an endpoint result being the cell-scaffold construct with a bone phenotype. ECM derived from osteogenic cell line has also been evidenced to promote osteogenesis in embryonic stem cells significantly, when juxtaposed with ECM derived from other cell lines or type I collagen alone[201].

Although the exact interplay of the different constituents and factors in the natural environment are yet to be elucidated, the observations clearly emphasize advantages of using the ECM as a whole, instead of isolated ECM components in a scaffold. However, the effect of ECM proteins on cell adhesion is lineage-specific; e.g., although fibronectin accentuates osteogenesis in osteoblasts, its effect on mesenchymal cells (MSCs) is negligible. Nonetheless, multiple ECM proteins can be combined in a scaffold, each serving different purposes like, enhancing cell adhesion or differentiation, so as to develop functional osteo-constructs, which are better adept in clinical skeletal regeneration[202].

Biomimetics is one emerging discipline that consolidates nature's best ideas into novel designs and principles. In line with such principles, it has been possible to synthesize peptide based 'designer scaffolds' with biologically active peptide sequences, which can be tailored into 'instructive extracellular matrices'.

Bone tissue engineering necessitates an effective integration of optimal properties within the scaffolds, like strength, toughness, porosity, controlled degradation rate, minimal inflammatory response, moldability, osteoconductivity and osteoinductivity.

3.4.1.4 | Biomaterials with cell instructibility

The current endeavour is to come up with a new class of biomaterials conforming to the cellular and extracellular protein domains so that signals can be transmitted in a controlled spatiotemporal manner. For example, RGD and BMP peptides, when grafted on to the PLEOF hydrogel (poly-lactide-co-ethylene oxide-co-fumarate) substrate were found to influence differentiation and mineralization of bone marrow stromal cells[203]. It was postulated that RGD peptides provided sites for cell attachment to the substrate and BMP peptides interacted with the type I and type II transmembrane receptors and synergistically accelerated osteogenic differentiation, when the respective concentrations were 1.62 and 5.2 pmol/cm^2, respectively.

3.4.1.4.1 | *Biomaterials based on self-assembly*

Nature is abounding in sophisticated functional materials, engendered by hierarchical self-assembly of simple nanoscale motifs[204]. Such self-assembled biomolecular structures are endowed with versatile chemistry, manifested in their molecular recognition properties and biocompatibility[205,206]. These natural systems provide stimulus towards engineering novel assemblies through *de novo* design as in bottom-up approach for fabricating nanostructures for tissue engineering. Self-assembling peptides can be tailored into biocompatible nanoscale and molecular structures so that the differentiation of precursor cells can be invigorated. The versatility of such motifs is also exuded in novel supramolecular architectures and composites[207].

Self-assembly can be envisaged as the ramification of non-covalent interactions like, hydrogen bonds, ionic bonds, electrostatic and van der Waals interactions under thermodynamic equilibrium that lead to spontaneous organization of molecules into structurally stable arrangements[208]. Zhang et al.[209] recently demonstrated an increased osteoblast adhesion and mineralization on incorporation of tailorable amino acid and peptide side chains (e.g., lysine, RGD and peptide Lys-Arginine-Serine-Arginine). RAD 16-I is another self-assembling peptide that can be moulded into nano-fibres. The biochemical properties as well as nanoporosity of such fibrous scaffolds have been extensively exploited for cell culture experiments. RAD 16-I was also found to accelerate bone regeneration, when administered to small bone defects (3 mm) in mice calvaria[210].

Until now, tissue engineering templating applications have been extensively studied with three primary protein-folding motifs: α-helices[211], β-pleated structures[212] and collagen triple helices[213]. One of the most widely investigated motifs is leucine-zipper coiled-coil motif. A tri-block artificial protein with an alanylglycine-rich coil between two-terminal leucine zipper motifs, conceived by Petka et al.[214] was found to form a stable and reversible hydrogel under physiological condition. This strategy was utilized to incorporate multiple bioactive motifs into the construct without altering the properties of leucine-zipper self-assembly (Fig. 3.8a). Besides, the recombinant DNA technology was leveraged into the engineering of functional tissues with leucine zipper motifs[215].

Leucine-zipper polypeptide (LZ) has been modelled to assimilate proteins from the hydroxyapatite nucleating domain and cell adhesive motifs of dentin matrix protein 1 (DMP1)[216]. Cells seeded on LZ-DMP1 coated substrate facilitated integrin clustering, which in turn piloted the actin filaments, leading to directed focal adhesion complex formation (Fig. 3.8b–3.8d). Moreover, this self-assembled biomaterial also has the potential to actuate nanoscale mineral formation.

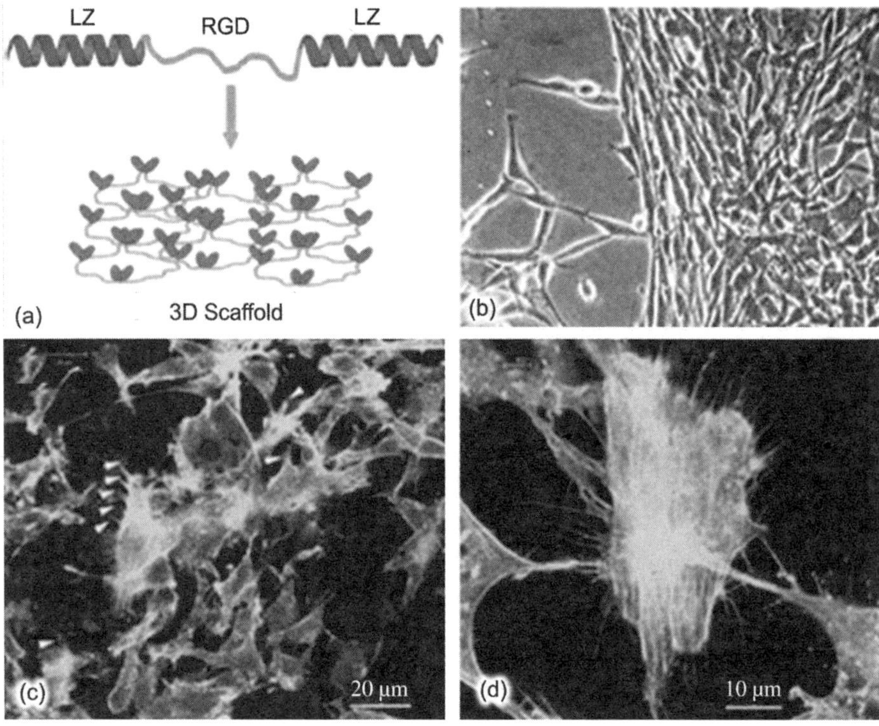

Fig. 3.8 Leucine zipper-DMP1 hydrogel facilitates cell–matrix interplay: (a) schematic representation of the leucine zipper-DMP1 chimeric protein. (b) a comparative study of cell attachment on a LZ-DMP1 substrate and glass palte (control) by light microscopy. (c) and (d) actin stress fibres and focal adhesion complexes promoting cell—cell communications, when coated on LZ-DMP1 substrate[291] [see Colour Plate].

Chemically synthesized artificial self-assembling peptides, called 'molecular Lego' have also been widely studied for tissue engineering applications. In a recent study, Brun et al.[217] intimated improved osteoblast functionality on PEO scaffolds, enriched with β-hairpin or β-sheet peptides, like RGD–EAK. In a nutshell, emulating nature can allow us to synthesize self-assembled biological and inorganic components that exhibit order from molecular to macro level.

An important example of self-assembled biomaterials is 3D hydrophilic polymeric networks, formed due to chemical or physical cross-linking.

3.4.1.4.2 | *Hydrogels*

Hydrogels have also been used to imitate the biochemical ambience of an extracellular matrix. In fact, the properties of hydrogels can be tailored to influence cell fate and bone formation. Porous alginate scaffolds modified with RGD when tested with animal models evidenced minimal immune response, little encapsulation and significant tissue growth. Also, ectopic bone formation was reported in PLA-DX-PEG and PLA–PEG scaffolds with BMP-2, photo-cross-linked individually in each of them[218]. Burdick et al.[219] documented increased osteoblast attachment and spreading on non-adhesive PEG-hydrogels, modified with RGD-peptide sequences. A new domain of hydrogels is being processed simultaneously, known as 'smart hydrogels' that apes the spatial and temporal variability of the natural cellular environment and hence, reaches out to the surrounding tissues via biomolecular recognition. Biomimetic gels housing building blocks, which are sensitive to cell-secreted enzymes[220], are appropriately used for bone tissue engineering. Polymer gels, as well as PEG-based hydrogels are degradable through hydrolytic breakdown of ester bonds. Also, the cell-modulated breakdown of the matrix capacitates cell migration and proliferation[221].

It is reported that peptide amphiphiles (PA) can be used for bone and cartilage regeneration[222]. A peptide molecule consists of peptide segment, covalently bonded to hydrophobic segment, such as an alkyl tail. These charged amphiphiles self-assemble in the presence of ions, say in aqueous medium, to form nanofibers with high aspect ratio[223]. Such self-assembly is mechanized by hydrogen bonding among peptide segments into β-sheets and hydrophobic collapse of their alkyl segments. Advantageously, the PAs may bear terminal bio-signalling peptides, which upon self-assembly, crowd on the surface of the nanofibers. These nanofibers direct hydroxyapatite mineralization, such that crystallographic c-axes of hydroxyapatite are aligned with the long axes of the fibres[215].

Nowak et al.[224] synthesized di-block co-polypeptide amphiphiles, containing charged and hydrophobic segments, with high thermal and mechanical stability upto temperatures of about 90°C. It was also established that amphiphilic nature of polypeptides is not the only factor instrumental to gelation, rather chain conformations (α-helix, β-strand or random coil) exert significant influence on this phenomenon. In another study, it was demonstrated that basic fibroblast growth factor (bFGF)-incorporated collagen sponge reinforced with self-assembled PA nanofibers provided a dynamic, hybrid 3D tissue culture scaffold conducive to bone regeneration. All these findings testify the potential of self-assembly as a novel technique towards biomimicking 3D scaffold for hard tissue repair or regeneration[225,226].

Dinca et al.[206] reported a unique synthesis approach, wherein controlled self-assembly of peptides was produced in laboratory using biotin–avidin and thiol chemistry on ORMOCER substrate. The advantage of this technique rests on the easy availability of thiol functionalized biomolecules commercially as well as in the lab scale. It was also proposed that using the same technique, larger scaffolds with biodegradable peptide structures can be conjoined into 'scaffold on scaffold' architecture to allow directed growth of a myriad of cells into ordered arrays of functional biological units. Taking into account the ambidexterity of protein templates, they can be rightly crowned as the 'Prometheus' of hard tissue engineering applications.

3.4.2 | **Electrospun scaffolds for bone regeneration**

Tissue engineering primarily ventures to bridge the gap between isolated cells and the functioning tissue through the engineered scaffold. This also pertains to an instructive extracellular microenvironment to sacrosanct for spatio-temporal guidance of the cells towards neo-tissue formation[227]. A wide variety of synthetic and biological materials can be reinforced into scaffold structures by electrospinning technique[228]. Electrospun scaffolds befit the basic requirements of microenvironment in view of close resemblance to the native extracellular environment and capacity for controlled release over time[227,229]. All these advantages have been exploited widely in designing scaffolds specifically to foster the formation and regeneration of a wide range of different tissue types[229,230]. The current research focuses on designing nanofibers endowed with biofunctional surface, because the cellular activities are primarily modulated in terms of the scaffold surface initially presented to the cells. One way of modifying the surface for bone regeneration is to remodel the surface with materials propitious for active bone growth.

> Electrospinning affords easy functionalization of fibres by surface conjugation or by direct incorporation of proteins and genes and thereby, boosts sustained delivery of bioactive molecules (cytokines, proteins, drugs, etc.) to the cells over prolonged duration.

However, maximum pore size in an electrospun scaffold being around several micrometres, introduction of larger sized pores within nano-fibrous network is coveted for improved 3D scaffolding role of the matrix.

3.4.2.1 | Polymeric nanofibres

The superiority of polymeric nanofibers pre-eminently rests on their easy processibility, flexibility and shape-availability. Among all polymeric materials, a group of poly-α-hydroxyl acid, such as polylactic acid (PLA), polyglycolic acid (PGA), poly-ε-caprolactone (PCL) and their copolymers have been exclusively examined for regeneration of tissues, including bones[229,231]. Bioactivity of PCL was substantiated by the superior adhesion of rat bone marrow stromal cells (rBMSCs) onto PCL scaffolds[232]. Additionally, the formation of collagen I and mineralization analogous to bone ECM on cell-nanofiber construct implanted in rat omenta for 4 weeks evidenced the efficacy of PCL scaffolds in bone tissue engineering[233].

However, the MC3T3-E1 cell response towards PLA electrospun nanofibers was found to be markedly fibre size dependent[234]. Interestingly, higher cell density was measured on PLA nano-fibrous scaffolds than on flat PLA scaffolds, when osteogenic culture medium was used. However, the absence of osteogenic media could not induce significant differences. This may be attributed to the use of osteogenic factors, which aids in osteoblastic responses to nano-fibrous topology. Similarly, polyhydroxybutyrate (PHB) and polyhydroxybutyrate-co-hydroxyvalerate (PHVB) microfibers exhibited a better cell growth of Saos-2 cells w.r.t. flat films without compromising the osteoblastic phenotype[235].

Nonetheless, the inherent hydrophobicity of synthetic polymers impairs initial cell adhesion. In this respect, blending with natural polymers serves to ameliorate cytocompatibility[232,236]. Some studies revealed a greater penetration of BMSCs within PCL–gelatin (1:1) matrix, compared to pure PCL nanofibers[235]. On similar lines, gelatin blended with PLA improved cell viability (MC3T3-E1), compared to pure PLA nanofibers[237]. Blending with natural polymers not only rendered these hydrophobic substrates hydrophilic, but also evinced significantly higher levels of gene expression. Human MSCs, pre-committed to an osteogenic lineage, were reported to secrete bone matrix on heparin sulphate containing PCL nano-fibrous scaffold. Bone formation was also observed under subcutaneous model in nude mice using the same scaffold[237]. Conjointly, synthetic nanofibers can be encrusted with natural polymers, like collagen and gelatin[238], alginate, hyaluronic acid, starch[229] to improve initial cell adhesion and growth characteristics of cells, including osteoblasts.

Natural polymers are immensely attractive subject to their innate biocompatibility and biofucntionality[229]. Among natural polymers, type I collagen has been extensively studied for scaffolding applications[239] as mentioned previously. Nanofibers of collagen type I are reported to have bolstered adhesion and proliferation of BMSCs, irrespective of fibre diameters.

> Despite the fact that nano-fibrous morphology of electrospun collagen emulates the native ECM, the researchers are still puzzled with much controversy on the question whether native structure and biological characteristics are compromised to some extent in electrospun collagen.

In this context, some postulated that typical biological properties of collagen educed from the triple helical structure were concealed, when collagen was electrospun out of fluroalcohols,[239]; others reported the presence of native periodic bands in electrospun collagen[239]. Nevertheless, cross-linked electrospun collagen appears to have high potential as nano-fibrous scaffold for cells to adhere, populate, and induce osteogenic development and mineral deposition. Apart from collagen, fibronectin, chitosan and elastin have been electrospun for biomedical applications. DNA nanofibers have also been fabricated by electrospinning[240]. However electrospinning of natural polymers is somewhat cumbersome, as a suitable solvent which conserves the integrity of the polymer must be found out for this purpose.

3.4.2.2 | Inorganic nanofibers

Apart from the biodegradable natural or synthetic polymer fibres, bone-active inorganic fibres, including calcium phosphates and bioactive glasses/glass ceramics serve as excellent scaffolding materials catering germane microenvironments, as mentioned previously. Among many studies, Kim et al. reported the processing of a nano-fibrous scaffold by blending silica-based-sol-gel glass ($70SiO_2.25CaP.5P_2O_5$) with a polymer binder.[241] The heat treatment of the electrospun mats yielded fibres with widely varying sizes depending on the initial concentration (Fig. 3.9). Moreover, significantly higher levels of osteogenic differentiation of rat BMSCs on such nano-fibrous scaffolds, compared to dense sintered bioactive glass or PCL polymer, ratified the morphological and biological boons associated with the usage of bioactive glass nanofibers.

In parallel, fabrication of a range of inorganic nanofibers, including hydroxyapatite[242], fluorohydroxyapatite[242a], and silica nanofibers[243] have been accomplished by mixing the sol-gel solution with a polymeric binder [e.g. Poly (vinyl pyrrolidone) and poly(vinyl butyral)], followed by heat treatment. Recently, cell-fibronectin was efficiently integrated onto the surface of bioactive glass nanofiber and subsequently, there was a notable augmentation in initial osteoblast adhesion and spreading[244].

In spite of their elegant bone-bioactivity, the application of electrospun inorganic nanofibers, including calcium phosphates and bioactive glasses in tissue regeneration is handicapped on account of their brittleness. In fact, heat treatment sometimes enervates their drug delivery potential as well. The current research in this field centres on production of nano-fibrous scaffolds with inorganic nano-fillers and degradable polymers.[245]

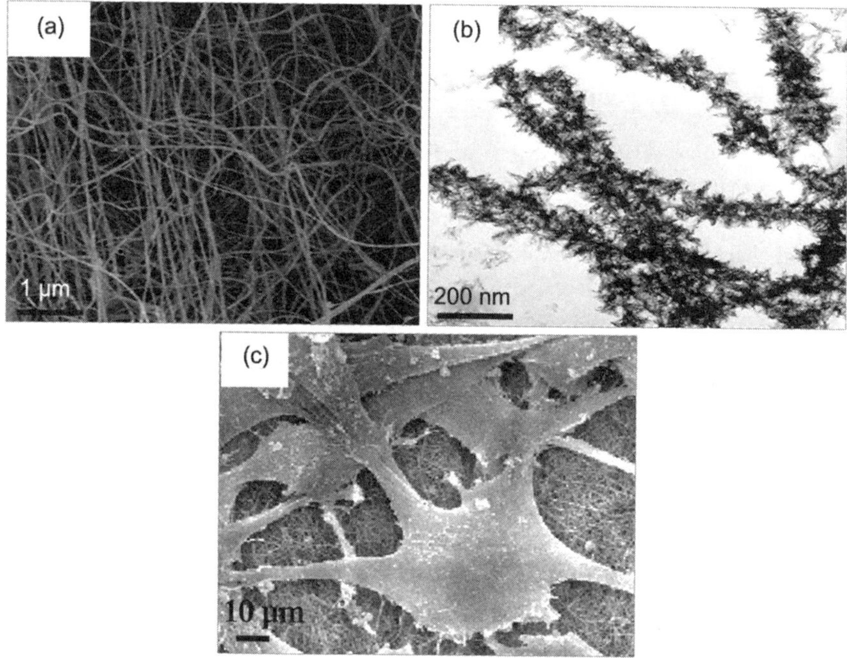

Fig. 3.9 (a) Electrospun inorganic nanofibers (sol-gel glass $70SiO_2.25CaP.5P_2O_5$). (b) TEM image revealing the formation of bone-mineral like apatite. (c) morphological features of rat BMSCs grown on bioactive glass nanofibers for 7 days [Adapted from Ref. 241].

3.4.2.3 | Polymeric–inorganic composite/hybridized nanofibers

Inorganic–organic composites, targeted to emulating the natural environment of the ECM, lace in the compressive strength of inorganic phase and toughness of polymeric phase to generate bioactive materials with upgraded mechanical properties and degradation profile. The incorporation of polymeric phase adds to mechanical flexibility of pure inorganic fibres. On the other hand, the latter not only amends the biological properties such as cell compatibility and bone formation of the polymeric phase, but also, for some polymer blends (e.g., with PLA), the autocatalytic degradation of polymer is neutralized by the alkalinity of the inorganic filler[246,247]. It has also been adduced that a nano-sized inorganic component is more bioactive than a micro-sized one. Also, the fact that thermal treatment of inorganic fibres is rendered redundant on account of the binding with polymer matrix, and hence they can be efficiently employed for drug delivery purposes. Eventually, the fact that bone ECM is a composite, constituted mainly of collagenous fibres and embedded with hydroxyapatite nano-crystallites, entails the need for development of nanocomposites to simulate bone structure[248].

Inorganic phase and polymers have already been assimilated into porous scaffolds and membranes[249]. The same concept can be applied for developing nano-fibrous scaffolds too. However, one major hurdle in this respect is the blending of solutions. Gelatin–hydroxyapatite composite fibres were fabricated by dissolving both the components in an organic solvent and simultaneous electrospinning[250]. A uniform distribution of hydroxyapatite nano-crystallites was evidenced in the gelatin matrix within the nano-fibrous milieu. In contrast, a different processing approach involves the mixing of HA nano-powders with a gelatin solution, followed by electrospinning. Such scheme often results in beaded nanofibers. Collagen–hydroxyapatite system[251] was generated on similar lines along with chitosan–hydroxyapatite[252] and similar systems to better mimic the bone ECM.

The hydrophobicity of polymers, such as, PLA, PCL and PHBV precludes easy blending of such systems with inorganic phases. For example, PCL–$CaCO_3$ composite fibres could be developed by implanting ultrafine $CaCO_3$ (~40 nm in size) particles[253]. On optimization of process parameters, such blends demonstrated improved water affinity and tensile properties, and also routed osteoblastic adhesion and growth, substantiating its use for guided bone regeneration. Nevertheless, the propensity of inorganic nanoparticles to agglomerate easily is the major impediment in homogenizing the same with synthetic polymer solution. Imperatively, beaded structures are commonly formed in such cases. Recently, Kim et al.[254] fabricated PLA nanofibers containing ultrafine hydroxyapatite nano-crystallites stabilized by a surfactant, 12-hydroxysteric acid. Probably, the amphiphilic nature of the surfactant stabilized the interplay at the interface of the hydroxyapatite nano-crystallites and PLA–organic solvent. Bead free microfibers could be electrospun with hydroxyapatite nano-crystallites well distributed within the PLA framework. This composite fibre exhibited considerably enhanced growth of osteoblastic cells and their phenotype expression compared to pure PLA fibres. The current endeavour is the development of inorganic nanoparticle incorporated within polymeric matrix with uniform fibrous morphology. This has been achieved to an extent through addition of ultrafine particles or by controlling the level of homogenization.

> Using the precursors containing both the organic and inorganic phases, degradable and bioactive nanofibers can also be produced by sol-gel process.

In one of the studies, an aqueous solution of gelatin was mixed in different ratio with polysilane [3-(glycidopropyl) trimethoxysilane] and the solution was hydrolysed before electrospun into nanofibers[255]. Besides, enhanced osteoblastic differentiation on such hybrid scaffolds upheld their potential application for bone regeneration (Fig. 3.10).

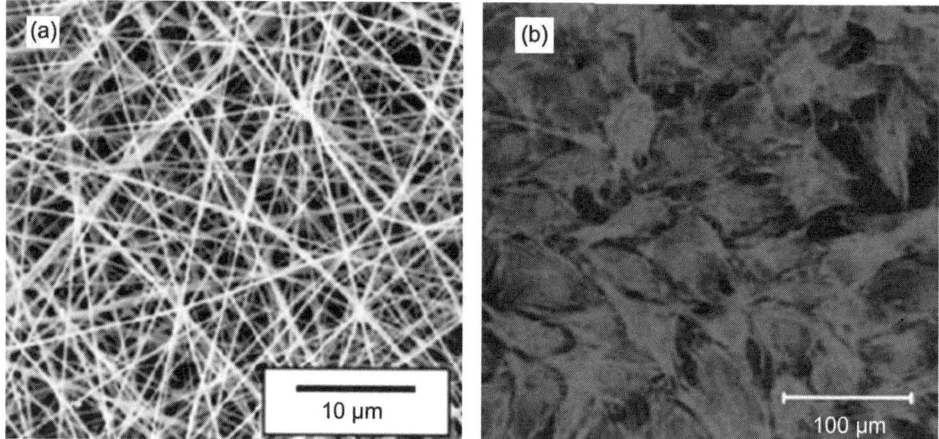

Fig. 3.10 (a) Gelatin–siloxane nanofibers (gelatin:polysilane = 1:1, CaCl$_2$ addition = 2.5 wt %). (b) MC3T3-E1 growth on the hybridized nanofiber after 3 days (confocal laser scanning microscopy). [Adapted from Ref. 255].

Whited et al. [256] developed an apatite-coated electrospun PLLA scaffold with variable pore size and porosity by utilizing a three step water-soluble PEO fibre inclusion, dissolution and mineralization process. The scaffolds with a greater initial PEO content were found to accelerate infiltration of MC3T3-E1 pre-osteoblast cells compared to scaffolds with lower PEO concentrations. Besides, apatite-based scaffolds significantly upregulated MC3T3-E1 alkaline phosphate activity, osteocalcin content along with cell-mediated mineralization with respect to PLLA scaffolds alone.

Francis et al.[257] recently demonstrated fabrication of gelatin/HA nano-fibrous scaffolds by simultaneous electrospraying and electrospinning. The spin-spray synergistic method facilitated complete exposure of HA on nano-fibrous scaffolds, thereby presenting osteophilic environment with favourable surface topography, invigorating cell attachment, proliferation and mineralization of hFoB cells (Fig. 3.11). Gel/HA scaffolds, obtained via spin-spray method, exhibited greater cell proliferation through enhanced cell–scaffold interplay and subsequently, increased osteoconductive and osteoinductive effects, inducing increased ALP activity and hFoB mineralization.

> The efficacy of engineered scaffolds is manifested in furnishing germane osteoconductive cues to shepherd the host's own stem cells towards osteogenesis and tissue regeneration.

Imperatively, the scaffolding material must bear complex information encrypted in their physical or chemical structure(s) to regulate cell behaviour. This can be accomplished by biomimicking, i.e., recreating the complexity of natural tissue *ex vivo*. Another emerging technology, in this field, correlates fabrication of synthetic scaffolds, endowed with the potential to communicate with the cells in a way so as to unleash their inborn regenerative and self-healing capacity. The surface morphology, microstructure, composition and properties of the implant material are therefore of cardinal importance in tissue engineering. In fact, all such factors synergistically modulate protein **adsorption, which mediates adhesion of desirable cell types**[258].

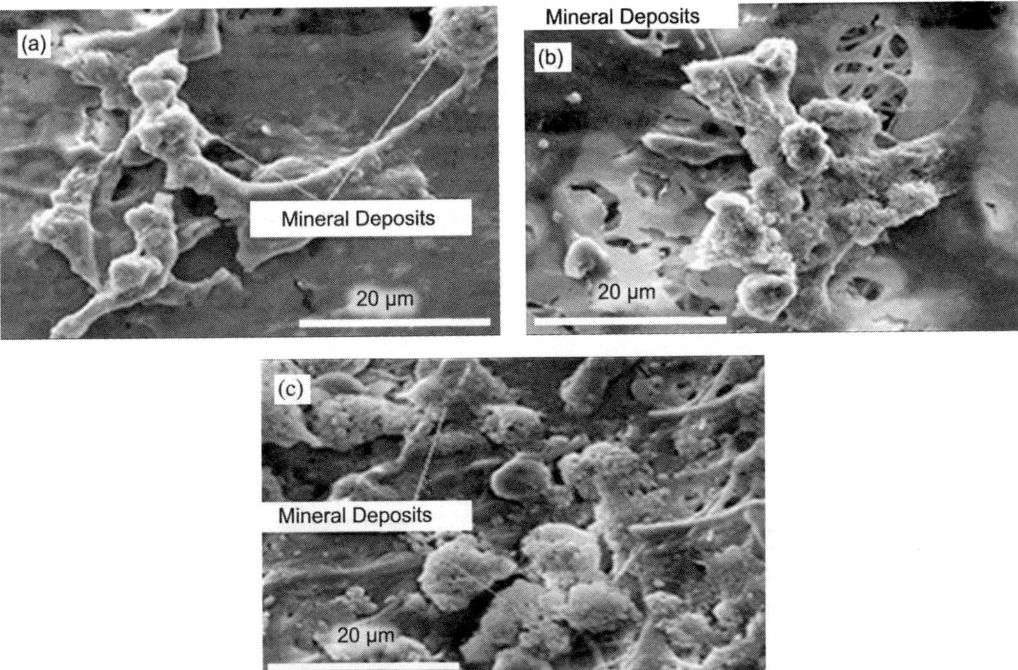

Fig. 3.11 FESEM micrographs of hFoB cultured on nano-fibrous scaffolds after 15 days on (a) TCP, (b) gelatin, (c) Gel/HA (blend 4:1)[444].

3.5 | Some Routes to Enhance Biocompatibility

Cells are innately receptive of their surroundings, typically to a broad spectrum of feature sizes from the macro down to molecular level, between 100 µm and 10 nm. Typically, the outer cell membrane is blanketed by specific carbohydrate structures and receptor systems, which can be stimulated by interactions with the adjacent cells, ligands in the surrounding ECM, and secreted signalling molecules[259]. The activation of such cell receptors is translated into a plethora of responses including, cell migration in the early embryo, coordinated organogenesis, and wound healing. In the ECM, a multitude of cryptic, latent information is embodied in the form of a multidimensional map, which is used by the cells to route their activities and perpetuate differentiation within tissues.

Each cell type fashions its own nexus of precisely encoded ECM proteins in line with its innate structural properties and information content. It is through an elaborate and dynamic feedback mechanism of signal transfer between the ECM and the cells, that the behaviour of the latter is coordinated into complex functional tissues[260]. Hence, engineering these dynamic ECM mechanisms into biomaterials is the key to control the cell behaviour.

Literature is replete with evidences on topographic sensitivity of cells (i.e., interplay of cells with surface features) to nanoscale as well as micrometer-range features, like grooves, ridges and wells[261]. Nanoscale alterations in topography evoke multifaceted cell responses, including changes in cell adhesion, cell orientation, cell motility, surface antigen display, cytoskeletal condensation, activation of tyrosine kinases, and modulation of intracellular signalling pathways, which in turn coordinate transcriptional activity and gene expression[159,262]. Notwithstanding the feature size, cell behaviour is also administered by the nature of ordered topography (e.g., ridges, grooves, steps, pits, pillars, and channels) and their symmetry (e.g., orthogonal or hexagonal packing) (Fig. 3.12)[263,264,265].

Importantly, the sensitivity towards nanoscale topography varies from one cell phenotype to the other; e.g. osteoblasts have been found to adhere preferentially to carbon nanofiber compacts in competition with chondrocytes, fibroblasts and smooth muscle cells.

Nanoscale features, in general, accentuate cell attachment and proliferation; even ridges as thin as 70 nm are also known to influence cytoskeletal assembly[263,264], as will be illustrated in the subsequent sections.

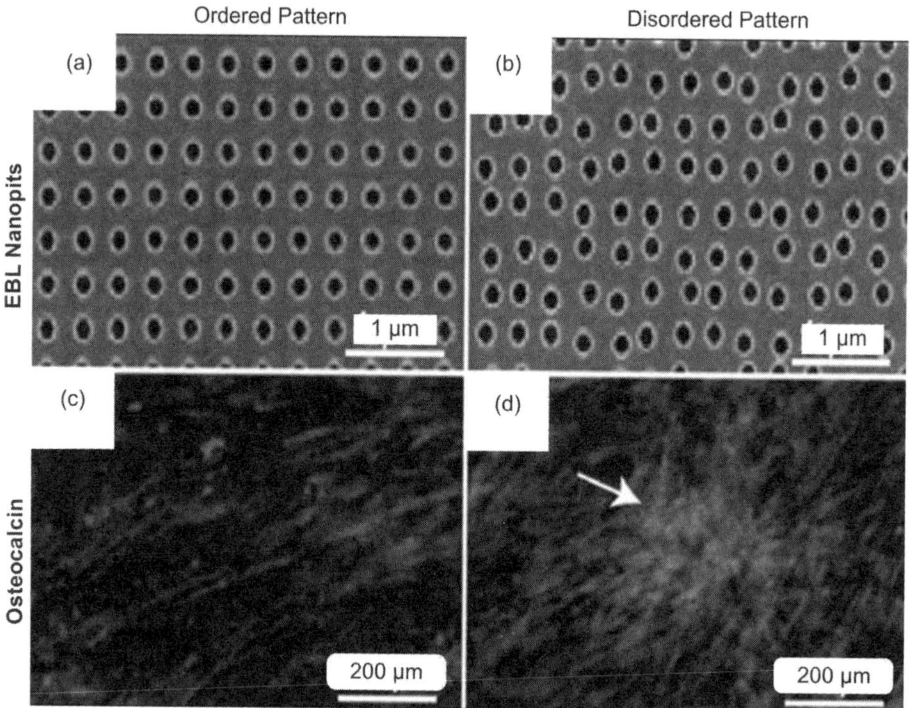

Fig. 3.12 Nanotopography and cell differentiation. (a,b) The nanopits (120 nm in diameter and 100 nm deep) were generated by electron beam lithography (EBL) (a) in a square arrangement and (b) with increasing disorder (displaced square ±50 nm from true center). The disordered topography is more proficient in stimulating human mesenchymal stem cells to increase the expression of bone-specific ECM protein osteopontin (arrow, (d)), compared to that in case of the orderly square arrangement (c) [Adapted from Ref. 263] [see Colour Plate].

In all probability, nanoscale topology orchestrates the interfacial forces that induce cytoskeleton formation and membrane receptor organization in the cell, culminating into necessary modifications of intracellular signals[262,264]. Besides, nanoscale features also monitor adsorption and conformation of integrin binding proteins, changing availability of binding sites and modifying integrin signaling[264]. In fact, some studies have predicted similar biological reactions towards nanofeatures of similar geometry and size, irrespective of underlying materials chemistry[264].

In all, nanoscale material structuring leverages control over cell behaviour, which has manifold implications while designing novel materials for tissue engineering. During tissue regeneration, engineered scaffolds host cells derived from natural tissue. Hence, incorporation of complexity and nanoscale details observed in real organs at the level of cell–matrix interaction upholds potential usage of such engineered scaffolds (Fig. 3.13)[161]. Even, an increase in surface roughness of pore walls enhances cell attachment, proliferation, and expression of ECM components[266].

One way of introducing nanostructured topology to enhance biocompatibility of conventional biomaterials is afforded by surface modification. Optimization of bulk properties of a scaffolding material[267], followed by surface modification often segregates conflicting requirements and thereby contributes to tailoring a wide range of properties. As envisaged from the foregoing discussion, extensive studies have been made on effects of geometrical confinements on cell spreading; lately, the researchers are also focusing on surfaces characterized by gradients of chemical compositions or protein densities[268] to investigate cellular responses on asymmetric surfaces and in pseudo-3D microwell geometries[269].

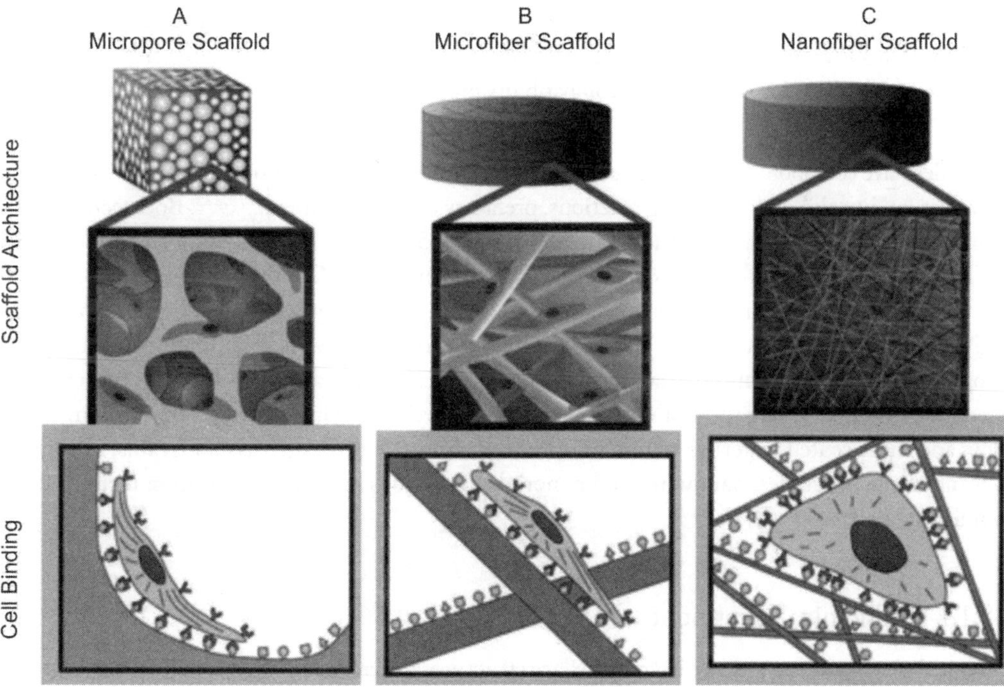

Fig. 3.13 Effect of scaffold architecture on cell binding and spreading. (A and B) Cells binding to scaffolds with microstructures flatten and spread as if cultured on flat surfaces. (C) Scaffolds with nanostructures have larger surface areas to adsorb proteins, catering additional binding sites to cell membrane receptors[447] [Adapted from Ref. 161].

Depending on the modifications reinforced, nano-functionalization techniques can be broadly classified under two heads: nano-coating and film deposition, and *in situ* surface nano-functionalization. Furthermore, these two techniques can be coupled to produce a variegated surface topology of coating/film and nanostructured zone. Popular coating and film deposition techniques include plasma spraying, plasma immersion ion implantation and deposition (PIII&D), sol-gel, chemical vapour deposition (CVD), physical vapour deposition (PVD), cold spraying, self-assembly, etc. On the other hand, *in situ* surface modification techniques subsume laser etching, shot blasting, acid and alkali treatments, anodic oxidation, micro-arc oxidation, ion implantation, etc.[270]. In the following section, some of the surface treatment techniques are briefly discussed.

3.5.1 | Surface functionalization of bioceramics

The biocompatibility of ceramics largely depends on the microstructure, particularly the grain sizes. Although the biocompatibility of conventional ceramic materials such as alumina, zirconia, titania, hydroxyapatite (HA) and calcium silicate is well established, their clinical applications are constrained. The ability of nanophase alumina and titania is reported to enhance protein interactions[271].

Various carbonaceous materials (carbon nanotubes, graphene and graphene oxide, etc.) are being investigated for biomedical applications on account of attractive electrical transport and elastic stiffness properties. However, their biocompatibility properties needs to be convincingly established.

Studies have corroborated their propensity not only in stimulating the pool of proteins constituting native tissues, but also in dynamizing reactions, prefatory to cell attachment[272,273]. Boccaccini et al.[274] developed highly porous bioactive and biodegradable 45S5 Bioglass scaffold, uniformly coated with CNT, which on immersion in SBF occasioned orderly formation of CNT/hydroxyapatite composite layer on the surface, attesting the preservation of scaffold bioactivity even in the presence of CNTs.

Similarly, metal surfaces (e.g. Ti, Ti6Al4V and CoCrMo) with micrometer to nanophase topography have been evidenced to augment deposition of calcium and phosphorous and consequently, osteoblast adhesion[275]. Coatings and films have been developed with substrates of titanium and its alloys and those coated materials exhibit multitude of nanostructures such as nanograins, nanopores, nanotubes, nanoparticles, nanowires and nanorods via various surface modification techniques, to be illustrated subsequently.

3.5.1.1 | Chemical treatment

Chemical treatment engenders different surface structures (micro/nano sized topology) on large area surfaces and multi-faceted devices, as well, by alkali, acid or hydrogen peroxide mediated chemical reaction at the substrate/solution interface. Physicochemical characteristics and topographies procreated thereby can be tailored to facilitate adhesion and proliferation of osteogenic cells[276], precipitation of apatite[277] and expression of bone-related genes and proteins[278]. For example, by controlling the exposure time and initial surface topography, rough surfaces with lower cytotoxicity and superior biocompatibility can be generated upon submerging polished Ti surfaces in 0.2% HF in solution[279]. Improved biocompatibility on chemical treatment of such surfaces was attributed to low hydrocarbon content, presence of fluoride, hydride and oxide, and the micro and nano-topography spawned on HF treatment.

Large-scale direct growth of nanostructured bioactive titanates (predominantly, 1D nanobelts and naowires) closely resembling the lowest level of hierarchical organization of collagen and hydroxyapatite on 3D microporous Ti-based metal (NiTi and Ti) scaffolds have been achieved by treatment with 10M NaOH at 60°C inside a Teflon–lined autoclave (Fig. 3.14). The engineered surface demonstrated superhydrophilicity, expedited hydroxyapatite deposition and thereby accentuated cell attachment and proliferation.

Similarly, bioactivity can be imparted to metal surfaces like tantalum, zirconium and their alloys by simple chemical treatments. In fact, such metal surfaces on treatment with NaOH showed accelerated apatite nucleation in stimulated body fluids[280,281,282].

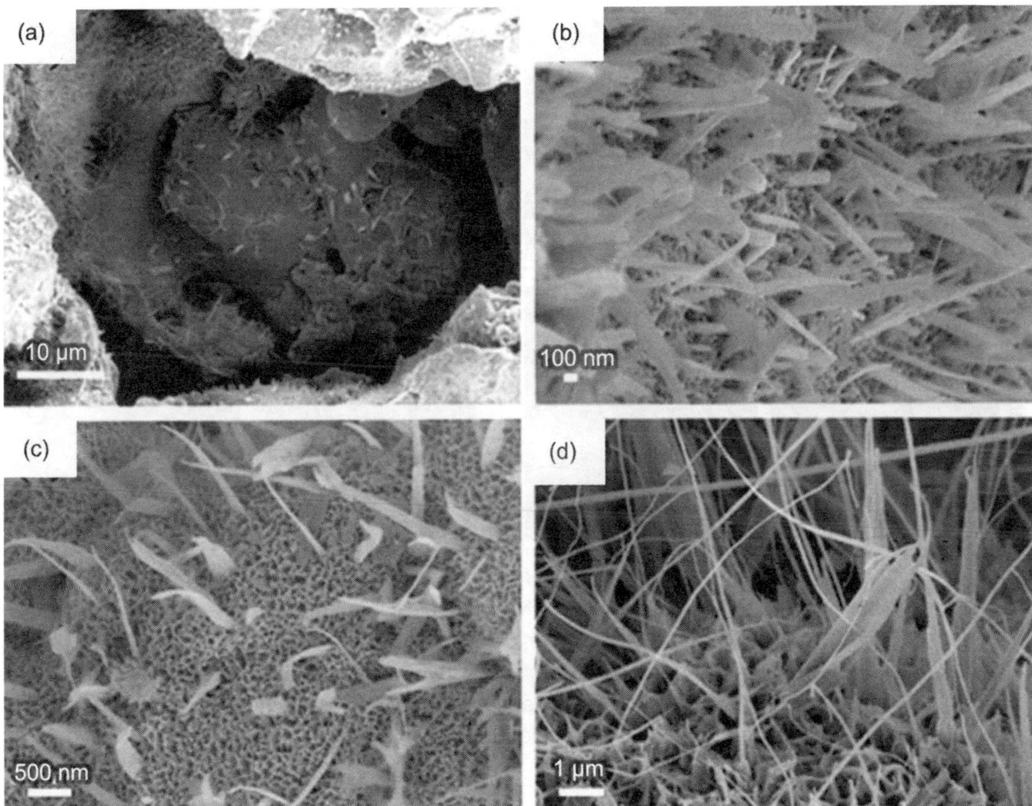

Fig. 3.14 SEM images of hydrothermally treated 3D porous NiTi scaffold. (a) Post-treatment surface morphology of exposed internal pore (b) Morphology of the wall of the exposed pore in (a),(c) Morphology of the bottom of the pore in (a). (d) General structure of nanowires/nanobelts on the exposed surface after treatment at 60°C for 240 h [Adapted from Ref. 449].

3.5.1.2 | Anodic oxidation

Employing electrode reactions in tandem with electric field driven metal and oxygen ion diffusion, the process of anodic oxidation facilitates the formation of an oxide film on the anode surface. This self-ordering electrochemistry approach has also been employed for the development of nanostructures[283]. Interestingly, optimization of a number of parameters, like anodization potential, electrolyte composition, anodization time and temperature and solution properties, including pH, conductivity, viscosity has led to evolution of high aspect ratio[283a], smooth[283b], ultra long[283c], double walled[283d], and bamboo type nanostructures[283e]. A study by Park et al.[284] substantiates the significance of nanostructured topology in cell differentiation, osteoclastic activation and ultimately bone formation. Side by side, a universal geometric constant (15 nm) for surface topography, necessary for promoting cell adhesion and differentiation (Fig. 3.15), has been suggested. Additionally, adhesion/propagation of osteoblasts culminating in the formation of interlocked cell structures improved manifold on surfaces with aligned TiO_2 nanotubes[285].

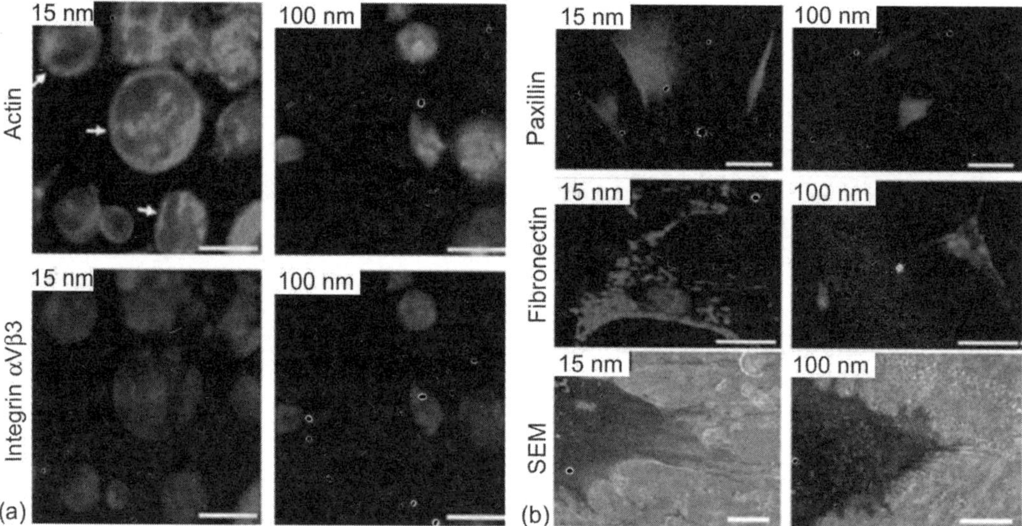

Fig. 3.15 Some illustrative results showing the clear evidence of osteoclast adhesion on 15 nm nanotubes (a). Scale bars: 50, 200, and 200 μm; (b) Morphological features of primary human osteoblasts to TiO_2 nanotubes. SEM images also reveal focal contacts by paxillin staining (upper panel) (scale bars: 100 μm), extracellular deposition of fibronection matrix (middle panel) (scale bars: 50 μm), and filopodia formation (lower panel) are higher on the 15 nm nanotubes (scale bars: 10 μm) [Adapted from Ref. 270] [see Colour Plate].

3.5.1.3 | Micro-arc oxidation

Micro-arc oxidation (MAO), also known as anodic spark oxidation or plasma electrolytic oxidation (PEO), is a complex plasma enhanced physico-chemical process, integrating a number of sub-processes such as, micro-arc discharge, diffusion, plasma chemical reactions, and cataphoretic effects[286]. This technique is employed for depositing ceramic coatings on the surface of metals like

(Al, Ti, Mg, Ta, W, Zn and Zr) and their alloys. The surfaces of such metals are shielded by thin, self-healing, strongly adherent dielectric oxide films in their native state, which refuse passage of current in the anodic direction.

MAO process guarantees firm adhesion of films (nanoporous) to substrates, irrespective of the geometric configuration[287]. The chief attributes of MAO coatings are manifested in superior corrosion resistance[288], thermal stability[289], photocatalytic activity[290], wear resistance[291], and CO sensing[292], all of which can be tailored by deposition parameters, such as electrolyte composition, electrolyte temperature, alloy composition, voltage, current density, time, etc.[270].

The potentiality of MAO coatings for biomedical applications has been demonstrated by many researchers[293,294]. MAO coated Ti surfaces have been shown to intensify osteoblast adhesion, ALP activity with respect to polished and sand blasted surfaces (Fig. 3.16) and osseointegration in animal models[295,296]. Further improvement in bioactivity can be effectuated by coupling this process with other techniques, like UV irradiation, chemical treatment, electron beam evaporation and so on[297]. MAO treated Mg, Zr, Al and their alloys, when employed in biomedical applications, also evinced improved bioactivity[298].

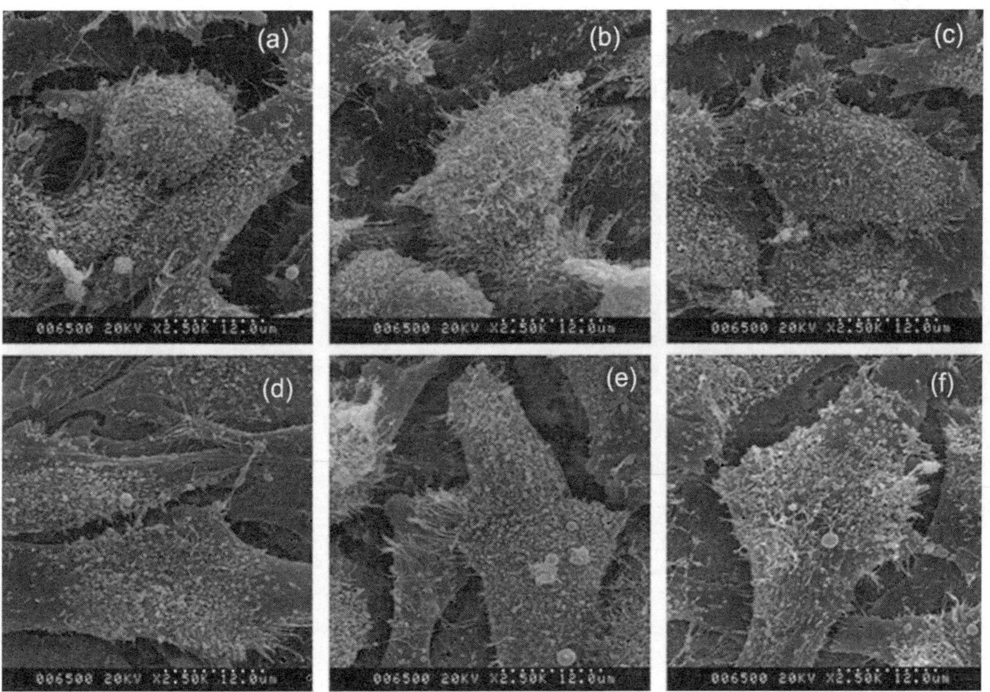

Fig. 3.16 Saos-2 cell morphology after 1 day (a–c) and 3 days(d–f) of incubation on machined (a, d), sand blasted (b, e) and micro-arc oxidised (c, f) Ti surfaces [Adapted from Ref. 450].

3.5.1.4 | Sol-gel

Sol-gel technique for deposition of thin coatings ensures a better control of chemical composition and microstructure of the coating[270]. The process comprises five basic steps: (1) hydrolysis and polycondensation, (2) gelation, (3) aging, (4) drying and (5) sintering[299]. Sol-gel derived coatings are usually obtained by spin-coating or dip-coating. The former, consists of four sequential steps in order, deposition, spin up, spin off and evaporation. On the other hand, dip coating, primarily employed for coating complex shaped substrates involves three steps: dipping, withdrawal and heating[300]. However, in both the processes, the formation of micro or nanostructures in the coated film can be conditioned by the precursor and relative rates of condensation and evaporation during film deposition[301,302,303].

in vitro bioactivity studies on titania and titania–silica composite films, formed by the said process, evidenced better bioactivity of the latter[304]. Even, titania films with smallest/narrow particle size distribution showed an inferior hydroxyapatite precipitation compared to other titania films. Mouse mesenchymal stem cells, cultured on layer-by-layer self-assembled TiO_2 nanoparticle thin films, demonstrated enhanced proliferation and attachment with an increase in the number of layers, i.e., with increase in roughness[305]. In a similar work, encapsulation of TiO_2 nanoparticles in 205 nm thick TiO_2 films on NiTi alloy led to an improvement in biocompatibility[306].

Investigations by Eisenbareth et al.[307] demonstrate a clear coherence of migration, adhesion and collagen type I production of MC373-E1 cells from the surface of Nb_2O_5 coated cp-Ti, concomitantly substantiating a better adhesion and cell migration on smoother surfaces. Intermediately rough surface (R_a = 15 nm) led to slower but stronger cell adhesion and migration (Fig. 3.17); whereas, a rather rough and hilly surface (R_a = 40 nm) impeded cell migration, but stimulated appreciable spreading and cell migration on the surface.

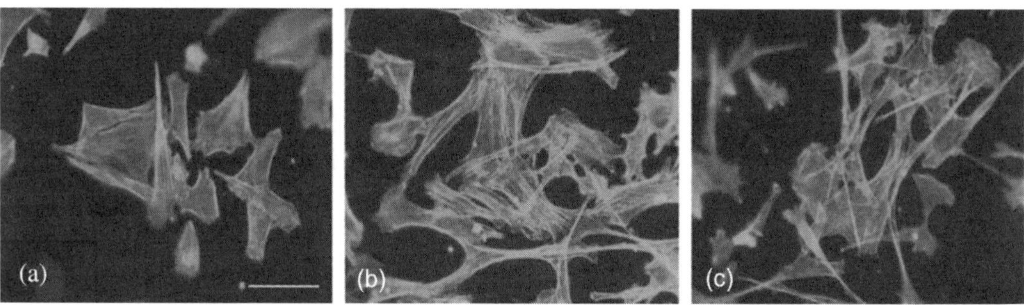

Fig. 3.17 Actin cytoskeleton indicating adhesion strength on Nb_2O_5-coated cp-Ti with varying surface roughness: (a) R_a = 7 nm, (b) R_a = 15 nm, (c) R_a = 40 nm (bar = 100 μm) [Adapted from Ref. 307].

3.5.1.5 | Plasma spraying

Plasma spraying is a rapid heating/cooling solidification process, wherein the desired material is melted at around 12000 K in a plasma jet and propelled towards a substrate. On the substrate, the molten droplets flatten, and rapidly solidify to form a deposit[270].

Plasma sprayed nanostructured zirconia coatings, stabilized by 3 mol% yttria, induced bone-like apatite formation on the surface after immersion in SBF for 28 days. In contrast, no apatite formation could be observed on the surface of polished coating. It can hence be surmised that surface plasma treatment is a possible route for rendering a surface bioactive[308].

3.5.1.6 | Physical vapour deposition

Physical vapour deposition (PVD) involves atomistic deposition of a surface layer, vaporized from a solid/liquid precursors and transported through vapour/plasma in vacuum or low pressure gaseous environment to the substrate of different composition, where they condense to form films of thicknesses ranging from a few to thousands of nanometres[309].

PVD roughly integrates processes, such as evaporation, sputtering, vacuum arc deposition, ion plating, ion implantation and deposition, all of which have been individually used extensively for modification of the substrate's physico-chemical nature, like surface energy, surface electrical charges, etc., with respect to development of tissue-compatible implants[310].

The modification of surface properties of implants by changing surface composition through introduction of functional groups or adding a surface layer, etc. can also be realized by a process called plasma immersion ion implantation (PIII)[311]. In most cases, long term durability and the performance of implant materials are menaced by the presence of bacteria in the vicinity of the implant surface. To overcome such undesired effect, silver and copper ions are engrafted into metallic[312] and non-metallic[313,314] biomaterials.

3.5.1.7 | Chemical vapour deposition

Chemical vapour deposition (CVD) involves deposition of atoms or molecules onto substrates, after reduction or decomposition of chemical vapour precursors at elevated temperatures. CVD constitutes a family of techniques such as vapour phase epitaxy (VPE) for deposition of single crystal films, metalorganic CVD (MOCVD) using precursor gases containing metal–organic species; plasma enhanced CVD (PECVD) effectuating plasma catalyzed decomposition and reaction, and low pressure CVD (LPCVD) for pressures less than the ambient.

Nanocrystalline diamond (NCD) films have been produced by microwave plasma enhanced CVD (MWPECVD)[315] and hot filament CVD[316]. Superior hardness, low friction coefficient, high chemical resistance and expediency in coating a wide variety of such films, have already been exploited for biomedical applications, especially for durable orthopedic implants. Interestingly, nanocrystalline diamond and submicron crystalline diamond (SMCD) films, obtained by MWPECVD and hydrogen plasma treatment were found to exhibit similar surface chemistry and wettability, yet, different topographies (Fig. 3.18). Additionally, significant enhancement in short and long-term functions of osteoblasts on the former with respect to SMCD, could be conceived from parallel extension and radial filopodia protrusion from the cell membrane in NCD in contrast with less protruded cell appendages, converging on specific spots, with SMCD (Fig. 3.19)[451].

Apart from these, the advent of new nanotechniques such as polymer demixing[317], dip-pen lithography[318], photo or electron-beam nanolithography[319] and imprint lithography[209b] have enabled engineering of geometrically defined environment for cells in the micro/nanoscale, as will be illustrated in the subsequent section.

Fig. 3.18 Nano and submicron scale topographies of nanocrystalline diamond (a, b) and submicron crystalline diamond (c, d) revealed by SEM and AFM images respectively [Adapted from Ref. 451].

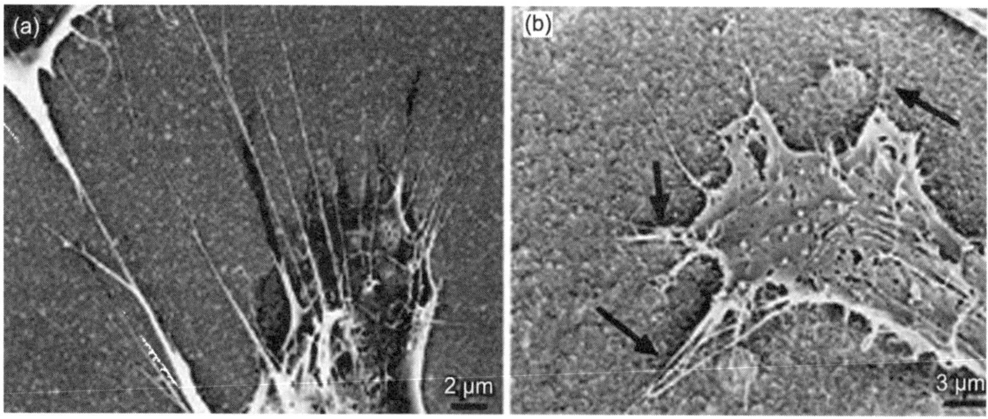

Fig. 3.19 Some representative results revaling cellular features of osteoblasts: (a filopodia extensions on nanocrystalline diamond and (b) submicron crystalline diamond, after 48 h of culture on the diamond surfaces [Adapted from Ref. 451].

3.5.2 | Surface functionalization of biopolymers

In view of the potential applications of biopolymers in prosthetic systems and controlled drug delivery, there has been a significant interest in modulating the biocompatibility property.

Since the chemistry of polymeric materials can be tailored to introduce surface functional groups, numerous attempts are made to enhance biocompatibility through surface functionalization approach.

3.5.2.1 | Deposition of coatings/films

Khang et al.[320] demonstrated the selective alignment of osteoblasts on carbon nanofibers/nanotube arrayed on a variety of polymer matrices. It was postulated that carbon nanotubes/nanofibers promote protein adsorption, which in turn mediate subsequent bone cell adhesion. The same group also documented enhanced chondrocytes (cartilage synthesizing cells) functionality, when carbon nanotubes are highly dispersed in polycarbonate urethane (PCU). All these results are suggestive of the synergism between nanosurface roughness and electrical stimulation towards chondrocyte functionality. In addition, such results also highlight the positive influence of conductive CNT/PCU films in promoting functions of chondrocytes for cartilage regeneration.[238]

The charged metallic ions or plasma guided onto polymer surfaces at ambient temperature orchestrates formation of metallic nanoislands. Ultrahigh molecular weight polyethylene (UHMWPE) and polytetrafluroethylene (PTFE) coated with nanostructured titanium, developed in accordance with the above principle, significantly boosted osteoblast proliferation and calcium deposition in comparison to titanium or uncoated polymer substrates (Fig. 3.20)[321].

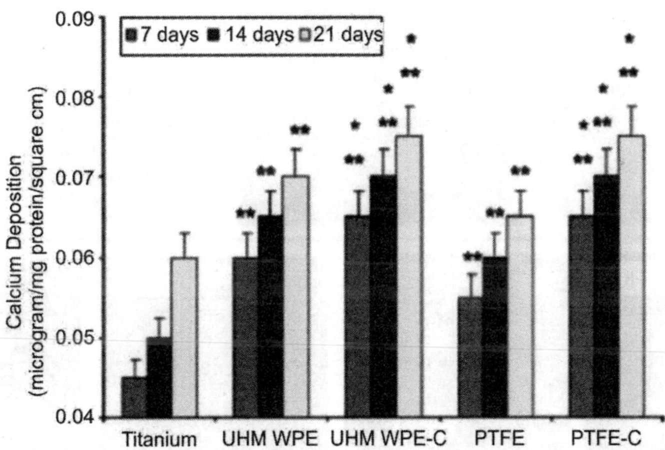

Fig. 3.20 Increased osteoblast calcium mineral deposition on UMMWPE and PTFE coated with nano Ti after 7, 14, and 21 days in culture. Data = mean + SD, n = 3; *p < 0.01 (compared to respective uncoated polymer sample) and **p < 0.01 (compared to currently-used Ti, labeled as Titanium). C, coated with nano Ti using Ionic plasma deposition [Adapted from Ref. 321].

3.5.2.2 | Chemical treatment

As in the case of bioceramics, surface functionalization of biopolymers can be effectuated by chemical treatment methods. Investigations by Park *et al.*[322] evidenced improvement in osteoblast functions on 2D NaOH treated polylactic-co-glycolic acid (PLGA). Also, chemically modified nanopatterns have been shown to reinforce biomolecular activity such as, increased sensitivity in antibody assays[323].

> Importantly, the role of surface topography against chemistry in controlling biological responses is still under controversy. In some cases, both the factors are in concert, whereas in others, one predominates.

A multitude of techniques is being exercised nowadays to generate definite chemical and biological micro- and nanopatterns on surfaces. Such techniques often require expensive equipment, even with low throughput. In light of these drawbacks, novel low cost patterning techniques, adept in fabricating large areas (cm^2), such as, colloidal lithography (CL)[324,325]. elastic contact lithography (ECL)[326], direct binary colloidal crystal patterning[327], shadow nanosphere lithography (SNL)[328], sequential colloidal shadowing[329], binary colloidal crystal based plasma polymer patterning[330] and combinations of evaporation and sputtering[331] are gaining prominence. Masks, such as colloids are used, in some cases, for generation of highly complex patterned surfaces with graded surface chemistry. Also, chemically concurrent nanopatterns can be conceived from colloidal layers by a combination of plasma polymerization, sputtering and electron beam evaporation techniques[332]. As shown in Fig. 3.21, the patterned Au and SiO_2 regions, selectively functionalized with fluorinated thiol and chlorinated silane, respectively, can also be functionalized with protein-resistant molecules to promote selective adsorption of proteins onto unmodified portions of the pattern.

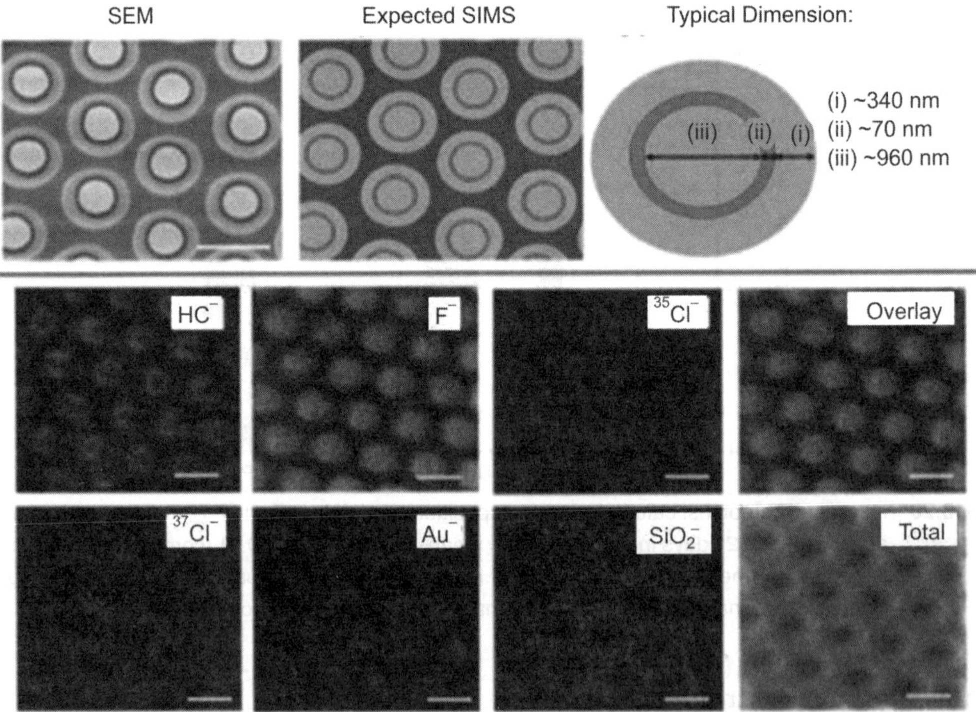

Fig. 3.21 Chemically concurrent nanopatterns obtained by synchronizing a number of techniques: High spatial resolution negative ion ToF-SIMS imags of Au-ppOct-Au-SiO$_2$ surfaces with Au central spots, chemically functionalized with ppOct and different SAMs. ppOct is a hydrophobic polymer generated from 1,7-octadiene. Scale bar is 2 μm [Adapted from Ref. 332].

3.5.2.3 | Plasma treatment

The fourth state of matter, plasma, is composed of highly excited atomic, molecular, ionic, and radical species. Plasma provides a highly unusual and reactive chemical environment for plasma-surface chemical reactions to occur. Contingent to the interplay with materials, plasma modification techniques can be roughly catalogued into three heads: plasma treatment, plasma polymerization and etching. Plasma treatments employ gases such as Ar, N_2, O_2, NH_3, CF_4 for implanting or substituting chemical functionalities onto a substrate or for creating radicals for crosslinking or subsequent surface grafting, subject to the process conditions, process vapors and substrates. Plasma treatment often renders an otherwise chemically non-reactive substrate chemically reactive. The functionalities introduced may be in the form of chemical groups and/or radicals. In case of biopolymeric scaffolds, grafting onto radicals can be accomplished in two different ways: direct irradiation and post irradiation. In post-irradiation grafting, the plasma engenders radicals such as peroxide on the substrate, prior to exposure to a grafting monomer. On the other hand, direct irradiation method (also known as plasma immobilization) irradiates the substrate, creating radicals in the presence of grafting molecules.

Plasma technologies have manifold applications, from inking industries down to improving biocomaibility of biomaterials[333] and also in aggrandizing the antibacterial properties of the same, e.g., by plasma immersion ion implantation[334]. Such a diverse range of applications can be envisaged in a plethora of advantages, which include the ability to create or treat materials of diverse sizes, shapes, geometries, without altering the surface and bulk properties; imparting specific functionality and even sterility to the surface concerned[270]. As a result of the diverse chemical reactions occurring in the plasma, a forest of different functional groups are generated. The plasma induces various homolytic bond fissions, ionization events and secondary collisions. Such a sequence of events materializes into further molecular fragmentations, reactions, ionization process and hence, functionally variegated plasma treated surface. Also, plasma polymers may possess irregular, non-repeat polymer structures with much branching and crosslinks. Such chemical disparities can be circumvented to an extent by techniques, like remote (or downstream) plasma polymerization and pulsed-plasma polymerization[335].

In a recent study with murine dermal fibroblasts and neuroblastoma spinal cord hybrid cells (NSC-34), Teixidor et al.[336] demonstrated the potentiality of plasma treatment in enhancing the biocompatibility of carbon microelectromechanical (C-MEMS) systems. AFM and FTIR studies further evidenced modification of physio-chemical properties of the plasma treated carbon substrates, concomitant to enhancement of adsorption of ECM forming proteins.

Apart from the surface functionalization techniques delineated above, the well established polymer processing techniques like electrospinning[337], thermally induced phase separation[338] and protein self assembly[339], used for generating nano-fibrous matrices[340,341], provide an avenue for procreation of biomimetically enhanced environment[342].

> Surface functionalization annexes different levels of complexity, ranging from changes in hydrophilicity to functionalization with peptides or proteins (biofunctionalization).

The various approaches briefly discussed in the context of surface functionalization are conceptually summarized in Fig. 3.22.

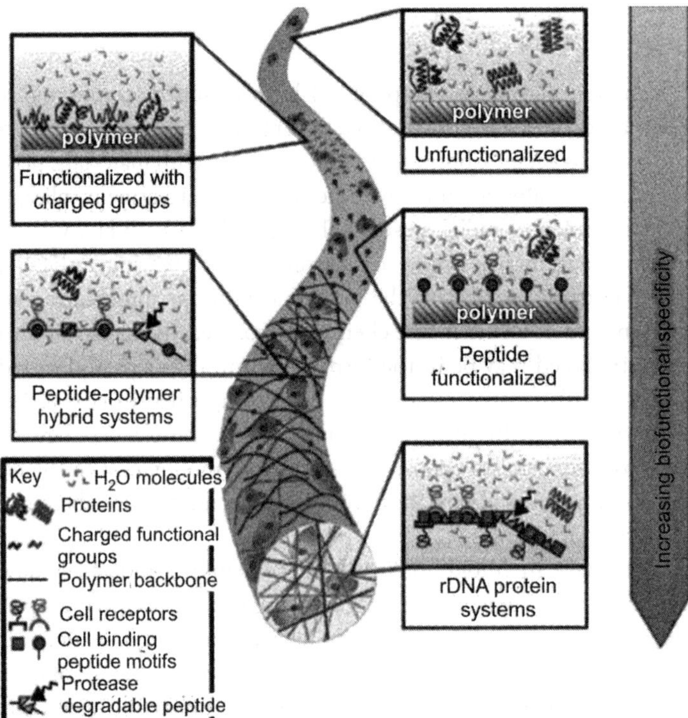

Fig. 3.22 Routes for enhancing material biofunctionality for efficient scaffold material design. Unfunctioanlized polymers nonspecifically adsorb proteins through weak interactions between the protein–water and water–surfaces interfaces. Functionalization with charged end groups (e.g. $-OH^-$, $-COO^-$, $-NH_3^+$) increase electrostatic interactions culminating into stronger protein adsorption and structural metamorphosis, which may unmask cryptic sites for cell attachment. Peptide functionalized and peptide–polymer hybrid systems enhance cell-mediated migration and degradation. Recombinant DNA (rDNA) technology permits incorporation of synthetic artificial proteins structurally and functionally akin to the specific ECM constituents [Adapted from Ref. 342].

3.5.3 | Biofunctionalization

Cellular interplay with material programs further growth and cellular mass, if feasible, migration of cells, gaining of special functions and concomitantly formation of a new tissue. Biofunctionalization involves the known biological principles and relationships in regulating the properties of the material concerned. The cell response (e.g., migratory response) is influenced by the ligand surface density, binding affinity of the ligand concerned and their co-localization with synergistic ligands[343]. The incorporation of bioactive peptide motifs, such as arginine-glycine-aspartic acid (RGD), which can be identified by the cell's transmembrane integrin receptors, or, proteolytically degradable peptide motifs. The surface functionalisation using such peptides are the common strategies for enhancing biofunctionality and likewise imparting biodegradability, as in the latter, for tissue remodelling and regeneration[344,345].

Similarly, other biological components from the ECM, such as, glycosaminoglycans (GAGs) and heparin have been incorporated into engineered scaffolds with the objective of recreating the biological context of native ECM[346]. Heparin, for example, has been incorporated as heparin-functionalized poly-(ethylene glycol), PEG hydrogels[346], by electrospinning of a heparin-PEG star copolymer into PLA fibers[346b] Jiang et al.[347] reported the fabrication of chitosan/poly(lactic acid-co-glycolic acid) (chitosan/PLAGA) sintered microsphere scaffolds, functionalized with immobilized heparin. Such heparinized chitosan/PLAGA scaffolds evidenced superior MC3T3-E1 cell proliferation as well as differentiation, in contrast with non-heparinized scaffolds. Also, heparinized scaffolds, supplemented with vascular endothelial growth factor (VEGF), have been shown to stimulate angiogenesis *in vivo*[348]. Recently, Sun et al.[349] demonstrated electrospinning mediated biofunctionalization, wherein, the electric field used in electrospinning induced surface segregation of a polarizable bio-organic peptide segment on PEO microfibers. In another attempt, it has been used to catalyze the self-assembly of nanostructures from designed peptide amphiphile molecules[346c]. All such approaches related to electrospinning and self-assembly have been elucidated in details in the previous sections.

To sum up, biofunctionalization of tissue engineering scaffolds can be conceptualized by any of the following methodologies:[350]

(a) immobilization (i.e., adsorption or covalent binding) of biologically active molecules (BAMs) on the scaffold surface,
(b) grafting of RGD peptides or similar meterials for accentuating cell adhesion,
(c) introduction of RGD peptides via macromolecular compounds (e.g., PEG) and
(d) creation of special layers on the scaffold surface by targeted delivery of BAMs.

> The surface modification techniques engender alterations in surface properties for better performance in a biological system, while retaining the bulk properties of the scaffolds.

In a nutshell, the techniques can be loosely catalogued under (a) physico-chemical and (b) biological surface modification techniques, as already outlined in this section (Fig. 3.23).

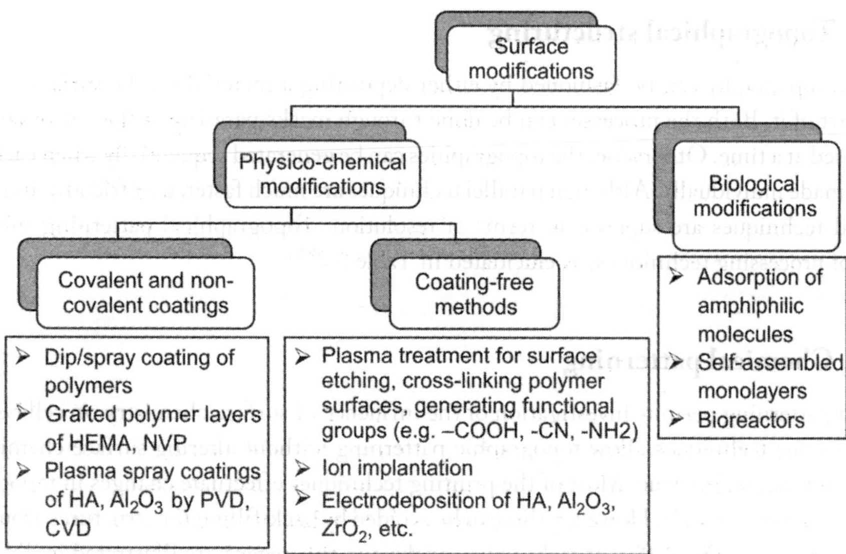

Fig. 3.23 Surface modification techniques for biomaterial scaffolds.

3.6 | **Biocompatibility of Patterned/Textured Biomaterial Surfaces**

The domain of biological micro-electromechanical systems and allied biomedical nanotechnologies annex a multitude of subfields, like biomolecular and cell analysis, microfluidics, biomedical drug delivery and medical prostheses. The pivotal point in all these disciplines lies in apprehending the cell-substratum interactions. A scaffold can be considered at different length scales: macro, micro as well as nanoscales. It is already circumstantiated that on macro and micro scales, cell behaviour is regulated by the chemical composition and surface topography[351]. On the same lines, cells are equally likely to respond to nanotopography, as not only the extracellular matrix housing the cells comprise of a nanoscale collagen fibrils, but the cell surface also is constituted by nanoscale receptors and filopodia. In fact, fabrication of nano and microscale topographies and subsequent tailoring of the surface chemistry and topography to explicate cellular interplay with the substratum constitutes an intriguing field of research[352].

> The surface topography with micrometer and nanometer scale features have marked influence on cell fate processes, such as, migration, adhesion, proliferation and differentiation, *in vitro.*

Additionally, materials organized on multiple length scales have a better conformity to biological matrices than those with single scale features and are hence, more propitious for all kinds of biomedical applications. But, it must be borne in mind that as cells have to compete with bacteria for surfaces in many environments, engineering bacteria repellant topography is also of paramount importance. Two popular surface patterning methodologies involve structural topography and chemical patterning.

3.6.1 | **Topographical structuring**

Structural topography can be fashioned by either depositing a material on the surface or etching away a part of it. Both the processes can be done through masks parallely, if the entire sample can be processed at a time. Otherwise, the topographies can be generated sequentially when each feature has to be made individually. Although parallel techniques are much faster, user friendly and cheaper, sequential techniques are superior in terms of resolution. Topographical patterning subsumes a number of processing techniques, as elucidated in Table 3.2[353].

3.6.2 | **Chemical patterning**

Chemical patterning permits investigation of the influence of surface chemistry on cell behaviour. Although some techniques allow topographic patterning without altering surface chemistry, the reverse is not necessarily true. Most of the printing techniques effectuate changes in topography at least at the nanometer scale. However, this can be avoided by back-filling the structures around some areas in some cases. The different techniques involved in this regard are illustrated in Table 3.2.

Table 3.2 Summary of surface patterning techniques [Adapted from Ref. 201].

	Technique	Physico-chemical modification	Lateral resolution	Advantages	Disadvantages
Topographical patterning	Photolithography	Topographical modification induced without alterations in material chemistry	40 nm (EUV)	Robustness; almost any shape can be engineered	Best resolutions require a synchrotron and are hence expensive
	Transfer from self-assembled polymer film	Modified topography, but chemical modification can be eluded	50 nm	Independent of nature of substrate	Mostly restricted to formation of hexagonal pegs
	Nanoimprint lithography	Only topographical modification	100 nm	Chemically unaltered features	Low resolution
	Colloidal lithography	Can be inflicted simultaneously	50 nm	Even spacing	Predominantly hexagonal structures are formed
	Surface roughening	Modified feature structures without chemical metamorphosis	Not easily controlled (in nm usually)	Uniform surface with controllable chemistry	Lack of control over surface structures
	Anodic oxidation of metals	Reformed topography	15 nm for nanotubes	Easily tailored chemistry and sizes	Limited variation in nature of structures produced
	E-beam lithography	No chemical modification	15 nm	Good resolution	Slow and expensive
Chemical patterning	Micro-contact printing	Chemical remodelling; topographical change can be avoided by backfilling	40 nm linewidth	Facile, variety of inks possible	Alteration in topography cannot be entirely averted
	Nanoparticle arrays	both	Dots with 28 nm spacing	Small features, tunable sizes	Topography induced
	Molecular assisted patterning by lift-off	Changes in topography can be avoided	100 nm linewidth	Plethora of structures obtained	resolution
	Supra-molecular nanostamping	Both	14 nm dots, 77 nm apart	Modifiable chemistry	Aqueous condition
	LB films/spin coated block-copolymers	Both	50 nm	Tunable chemistry	Demanding technique, inflexible in terms of shapes and chemistry of the features engendered
	Dip-pen nanolithography	Both	15 nm	Control over feature size and shape	Slow, as each feature needs to be developed separately

3.6.3 | Influence of surface topography on surface energy

Although nanotopographied surface is chemically congruent with the bulk, the presence of nanostructures can trigger changes in surface chemistry and concomitantly, surface energy. In some instances, this is manifested in an increase in hydrophobicity on microstructured surfaces, commonly envisaged as the 'lotus effect'. This increases the surface contact angle, as described by the Cassie–Baxter equation[354]. Conceivably, the effect is more pronunced at the nanoscale[355]. Additionally, the surface chemistry is also regulated by the density of structures[356].

The difference between the surface and bulk properties can be ascribed to the presence of dangling bonds. However, this effect is predominantly associated with highly clean surfaces, as exposure to fluids promotes adsorption of atoms or molecules, which exhaust the dangling bonds. Imperatively, the surface chemistry, that the cells are exposed to under normal culture conditions is totally conditioned by the adsorbed molecules from solution. Hence, it is equally important to analyze the effect of nanotopography on the adsorption of biomolecules.

Recently, a mathematical model, developed to predict the strength of cellular adhesion to a nanorough surface[357,358], identified three different regimes: (a) for small surface energies, cell adhesion is hindered by an increase in surface roughness; (b) for large surface energies, adhesion is maximum at an optimal roughness; and, (c) for intermediate surface energies, cell adhesion is virtually independent of surface energies. In this study[203] nanopatterns were functionalized with streptavidin, so as to provide an adhesive interface for biotinylated probes. The surface energies were calculated using the Johnson-Kendall-Robert's model for the adhesion-induced deformation of elastic biotinylated agarose beads, ligated to streptavidin, as probe. According to the results of this model, the cell adhesion increases at an optimal roughness with surface energy as per the following expression:

$$a^3 = \frac{6\pi R^2 W}{K} \tag{3.1}$$

where, 'a' is the contact radius for attachment, 'W' is the total surface energy (inclusive of the non-specific interaction), 'K', the elastic modulus of the substrate and 'R', the radius of the bead used in the experiments.

> A larger cell viability or proliferation can be achieved on surfaces with larger surface energy.

This aspect has been confirmed in recent experiments with Neuroblastoma cells (N2a) on amorphous carbon substrates and this study from our research group indicates an increase in N2a cell viability with an increase in roughness at nanoscale[359].

It was also adduced that the strength of adhesion is manifested in the ratio of surface enegies of patterned ($\gamma_{eff,}$ h>0) and flat surfaces (γ_o), with congruent physico-chemical properties[203]:

$$\frac{\gamma_{eff}}{\gamma_o} = \frac{2E\lambda\tilde{F}_{tot}(\tilde{a}, \tilde{h} > 0)}{2E\lambda\tilde{F}_{tot}^{flat}} = -2\frac{\tilde{F}_{tot}(\tilde{a})}{eC_{eq} - \tilde{\gamma}_{ns}} = -\frac{2}{\tilde{\gamma}}\tilde{F}_{tot}(\tilde{a})$$

$$\tilde{h} = \frac{h}{\lambda}; \tilde{a} = \frac{a}{\lambda}; C_{eq} = C - \log\frac{m_l}{m_r}; e = \frac{m_b k_b T}{E\lambda}; \tilde{\gamma}_{ns} = \frac{\gamma_{ns}}{E\lambda} \qquad (3.2)$$

The final expression derived was,

$$\gamma_{eff} = -\frac{2E\lambda\gamma_0 F_{tot}(\tilde{a})}{m_b k_b T(c - \log\frac{m_l}{m_r}) - \gamma_{ns}} \qquad (3.3)$$

Here, 'a' is the contact radius of cell attachment on a surface, 'F_{tot}' represents the total enegy, including the non-specific interaction energy, K_b is the Boltzmannn's constrant, T is the temperature of cell culture medium, h and λ, being the wave amplitude and wavelength of a hypothetical thin bilayer (cell membrane) of elastic modulus ('E') and thickness ('s'), resting on a substrate with sinusoidal wavy profile (Fig. 3.24). The parameters 'm_l' and 'm_r' respectively, quantify the densities of the ligands and receptors on the substrate surface, 'm_b' represents the density of bonds formed and 'C' stands for the binding energy factor. Although E is defined as the elastic stiffness of a curved cell membrane in the above equations (Fig 3.2 and 3.3), it may be more appropriate to consider composite elastic stiffness ($\bar{E}$), assuming the elastic interaction between curved cell membrance and localized element of surface topography of the patterned surface. Also, the dependence of various parameters on the surface energy of patterned surface can be predicted. Higher is the density of bonds formed between patterned surface and cells (m_b) or higher is the binding energy factor (C), the larger will be the γ_{eff} i.e., the interfacial energy of patterned surfaces, when all other parameters remain unchanged. Imperatively, repulsive non-specific interactions, manifested in a positive $\tilde{\gamma}_{ns}$ impede adhesion, in contrast to a negative $\tilde{\gamma}_{ns}$, which suggests relevant non-specific interactions.

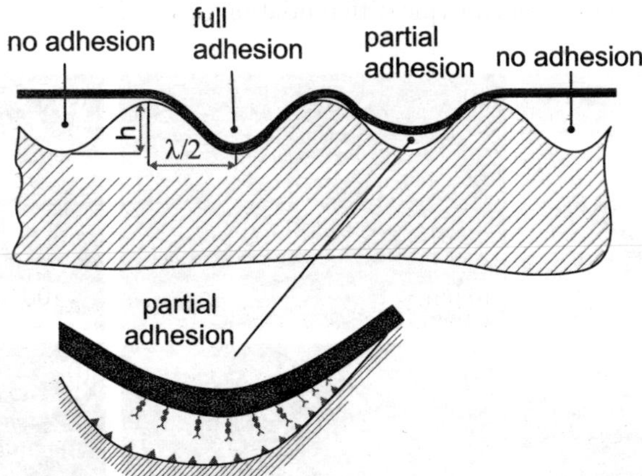

Fig. 3.24 A thin biolayer over a wavy surface, depicting the three adhesion conditions: partial, full and no adhesion [Adapted from Ref. 358].

In the same line, Kemkemer et al.[360] formulated an automatic controller model for elucidating the finer scale details of cell behaviour on a micro-grooved surface. One basic result derived by considering the symmetry of the cell–substrate system, suggested the predominance of square of the product of groove height and spatial frequency to the aspect ratio of the grooves, in eliciting cellular response. In conjunction with the Fokker–Planck equation[361], the equation proposed to describe the time dependence of the orientation of cellular machinery is as follows:

$$\frac{d\psi}{dt} = -k_p g(\psi, \text{signal}) \tag{3.4}$$

$$\frac{d\psi}{dt} = -k_p (h^n L^m) \sin 2\psi + \Gamma(t) \tag{3.5}$$

where, the reaction of the cell is reviewed on the basis of temporal change of angle of orientation, 'ψ' at time 't', with respect to the groove direction. The angle and field dependence of the detection unit of the automatic controller was detailed in the unknown function $g(\psi, \text{signal})$ and the reaction co-efficient, 'k_p'. The cellular reaction is effected by the height, 'h' and spatial frequency, 'L' of the grooves, with positive power exponents (n, m > 0). Additionally, the stochastic signal, corresponding to cellular orientation is quantified by $\Gamma(t)$.

> The quantitative dependence of the cellular response on various topographical parameters of the patterned substrates remain unexplored.

3.6.4 | Cell responses to material surfaces

The basic reactions of osteoblasts on material surfaces are exteriorized in a series of different time-related phenomena: protein adsorption at the material surface, followed by a sequence of rapid short term events constituting the attachment phase and the subsequent adhesion phase, followed by proliferation, migration and phenotypic differentiation (Fig. 3.25)[362].

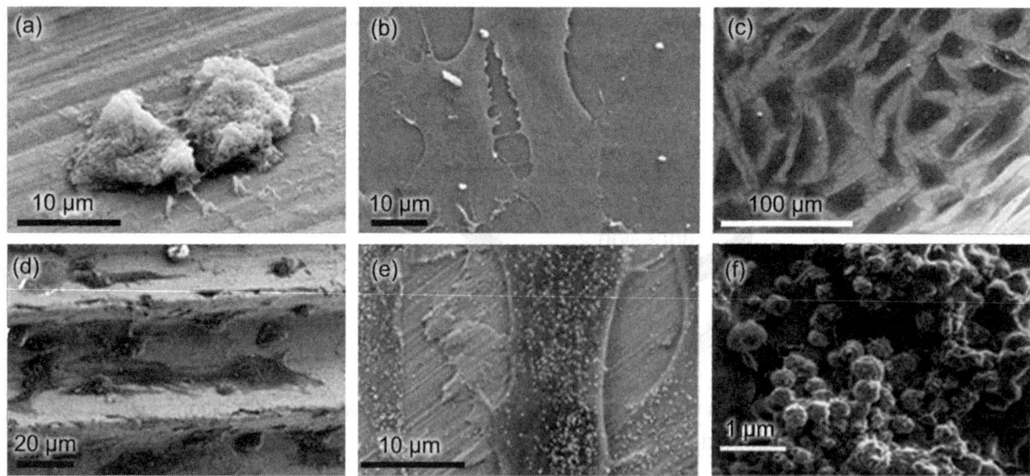

Fig. 3.25 SEM of stages of osteoblast/titanium surface interplay: (a) cell attachment, (b) cell spreading, (c) cell proliferation, (d) cell migration, (e) matrix synthesis (immune staining of osteonectin) and (f) mineral formation (bar 1 µm)[447].

> Osteoblast adhesion and functionality on material surfaces involve interactions among a host of biological molecules, to induce signal transduction and concomitantly, cell response.

However, at each step, the material is redressed by the biological fluid constituents prior to interplay with the cells[363]. Protein adsorption is hypothesized to be the foremost step to mediate cell-material interaction. A gamut of factors, such as, the pH, ionic composition and strength of the solution, temperature, functional group of proteins and the substrates, regulate protein adsorption[364]. Interestingly, some of the bone proteins are known to exhibit chemotactic or adhesive properties. Such proteins are earmarked by the presence Arg-Gly-Asp (RGD) sequence which is specific to the attachment of cell membrane integrin receptors. In fact, the potentiality of isolated RGD-sequence containing peptides e.g., fibronectin, vitronectin, in aiding bone cell adhesion and simultaneously, matrix maturation and mineralization, are well evidenced in literature[365,366].

Cellular interaction with the surface can be either direct or indirect. Oxygen tension and other parameters manoeuver cell behaviour in the culture media in a complex manner[208] Evidently, the surface topography and physico-chemical surface characteristics are of paramount importance in affecting cell-substrate interaction. Nevertheless, the interaction of these parameters in actuating biological responses is still incomprehensible.

The 3D morphology of a surface can be described using its size, shape and surface texture. Numerous studies corroborate the sensitivity of bone cells to patterned surfaces[367]. Notwithstanding, it is difficult to compare osteoblast behaviour on surfaces with different topographies, primarily due to lack of consensus regarding the proper representation of material surface as 'rough' or 'smooth'[368]. Effect of nanotopography on osteoblast functionality is imperative, subject to the nanotopographical features of the *in vivo* adhesion structures like the cell membrane and the basement membrane[369]. Such naturally occurring nanotopographic structures, inherent to the cell's extracellular matrix provide mechano-transductive cues which modulate local migration, cell polarization and cytoskeleton organization. Previously, albeit the general acceptance of the fact that cells like osteoblasts preferentially align themselves along defined surface morphologies by a process called contact guidance, the effect of ordered nano topography on cell behaviour was largely unknown. Recently, investigations on ordered nanostructured surfaces ratified the significance of structured nanophase surfaces on osteoblast orientation and migration[370,371], a prerequisite for cell colonization culminating into directed tissue formation.

3.6.4.1 | Cell responses to nanotopography

The interplay of cells with nanostructures is roughly similar to those at other scales. The cell attachment is mediated by specific molecules, commonly integrins at specific binding sites. Integrins are heterodimeric proteins comprising of two different subunits: α and β. On attachment to a surface, subject to the environment outside the cell, integrin receptors aggregate together and mobilize cytoplasmic proteins into a complex, known as focal contact[372]. Notwithstanding the microscale features of these complexes in contrast to the nanoscale features of integrins (8–12 nm), the latter can be influenced by surface topography on microscale[373]. Upon adhesion to a substrate, the cell initially explores its environment and migrates using nanometer scale processes such as filopodia and lamelli podia. Filopodia probe the environment around the cells and their ends provide points of anchorage for movement. The movement of filopodia is likely to be influenced by nanotopography of the surface[374,375]. Recently, it has been surmised that probing by filopodia was not the critical act in cell alignment on nanogrooves, rather, the quality of focal adhesions formed on the nanogrooves played the most significant role[198]. In fact, focal adhesions, formed and extended along the ridge were more stable, resulting in lesser filopodia retraction than those formed perpendicular to the ridge (Fig. 3.26).

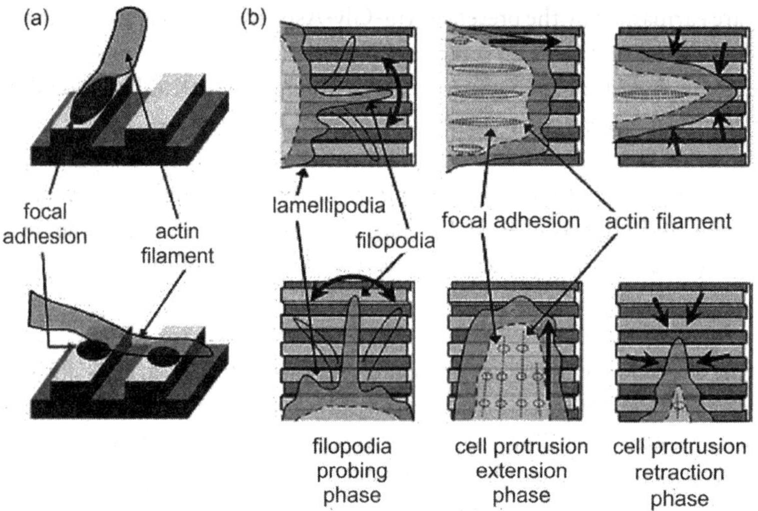

Fig. 3.26 Model for cell alignment on the nanogrooved substrate. (a) Actin filaments parallel to the grooves form wide focal adhesions at filament terminations. On the other hand, termination of perpendicular filaments is fragmented because focal adhesions form only on the ridge. (b) Filopodia movements are isotropic, i.e., no specific direction was observed for their extension and retraction against the nanogrooved structure[447].

Apart from these cytoplasmic processes, cell-substrate interactions can also be coordinated by cell membrane deformation. This, in turn, affects different processes occurring at the cell surface. Notably, it has been established that mechanical stretching of cells opens 'stretching-activated channels (SACs)', leading to ion influx and a series of concomitant downstream signalling events[376].

> The surface topography dependent cell functionality modulation is mediated via signal transduction by various cellular mechanisms.

As in the micrometer scale, the cytoskeleton is directly involved in both sensing and responding mechanisms of the cell. Two common hypotheses have been propounded to explain direct transduction of information from outside the cells to the nucleus via mechanical signals: percolation and tensegrity. The percolation hypothesis involves transmission of intercellular mechanical signals via a nexus of fibres[377]. On the other hand, tensegrity or tensional integrity hypothesis is based on continuous alternating compression and tension of cytoskeleton filaments[201,378]. Based on this hypothesis and accompanied evidences, Dalby et al.[379] proposed a theory of 'self-induced mechano-transduction', according to which, the surface nanotopography induces alterations in the nucleus morphology and interface chromosome positioning, occasioning possible changes of gene transcription.

The other mechanism of cell interaction with surface topography is indirect mechano-transduction, whereby mechanical signals, metamorphosed into chemical signals, are conveyed to the nucleus via a series of integrin mediated signalling cascades[198,380].

3.6.4.2 | Influence of physical structuring of features on cellular response

Nanotopography can be perceived by a cell as an entrenchment hindering movement of processes on the cell surface (receptor clustering, filopodia) or through surface contortion. The chemical topography can similarly be considered as a barrier, where it may hinder cell adhesion.

> A multitude of factors (height, width, shape, organization of nanofeatures, chemistry and rigidity of surfaces) synergistically manoeuver cell response to surface topography.

3.6.4.2.1 | *Influence of pattern size*

One major endeavour in the recent years has been to demarcate the nanosize threshold capable of eliciting cellular response. Large areas of nanotopography prepared by polymer demixing with polystyrene and poly(4-bromostyrene) and subsequent annealing evinced improved cell adhesion and proliferation over 13 nm tall islands[381] than on a flat surface[382]. Contrarily, another study reported an increased filopodia production together with smaller focal adhesions on surfaces with 10 nm tall islands[383]. Nanoscale textured surfaces with 14 nm tall worm like structures, obtained by polystyrene and PLLA polymer demixing, enhanced human foetal osteoblastic cell attachment, spreading, integrin subunit expression and focal adhesion protein synthesis with respect to flat PLLA surfaces[384]. It was enunciated that nanotopography upregulated specific integrin expression, leading to paxillin recruitment and focal adhesion kinase (FAK) activation. Also, murine osteoblasts, when grown on functionally gradient nanocrystalline PLLA, were reported to differentiate in a manner, which depends on root mean square (RMS) roughness between 0.5 nm and 13 nm and the results also revealed the proliferation rate was inversely proportional to surface roughness[385]. To sum up, these studies indicate that size threshold for cell detection is <10 nm. Possibly, at such reduced size scales, the surface topography impediments formation of focal adhesions subject to impaired integrin clustering on the nanostructures.

In comparison to microstructures or conventional materials, surfaces characterized by isotropic nano-structures demonstrated convalescent cell response as manifested by better proliferation and differentiation of osteoblasts[198,386]. Such finding was further substantiated by augmented vitronectin adsorption on nanophase ceramics with higher wettability[387].

However, several studies on the behaviour of cells on diameter controlled TiO_2 nanotube surfaces presented mixed findings[388,389]. Some studies indicated improved cell adhesion and spreading, yet no noticeable cell differentiation on smaller nanotubes (30 and 50 nm), in contrast to substantial elongation (in view of cell differentiation) on larger nanotubes (70 and 100 nm). It was theorized that in stretched cells on larger nanotubes, the cytoskeleton is under tension, which induces elongation of the nucleus, complementary to differentiation[128]. Contradictorily, research by Park et al.[284] ennunciated greater adhesion, proliferation, migration and differentiation on smaller nanotubes. This discrepancy was primarily correlated to the difference in nature of TiO_2 nanotubes manufactured via diverse synthetic routes, in the abovementioned studies.

It can hence be surmised that an increase in feature size has negative ramifications in matters of cell behaviour. Contrarily, Andersson et al.[390] reported enhancement of cell spreading with an increase in feature size. It was argued that larger features presented larger continuous surfaces for attachment, thereby aiding formation of focal adhesions. In contrast, increasing the sizes of nanotubes resulted in diminution of the surface area for attachment. In fact, apart from the size of nanofeature, a wide spectrum of parameters, such as morphology, organization, distance between nanfeatures, synergistically mediate cell response[198,391], as will be discussed in the following sections. Biggs et al.[236] annotated that integrin activation and clustering, phenomena sacrosanct for focal adhesion formation in cellular populations can be modulated by nanotopographical features of critical dimensions and density. Formation of focal adhesions is hindered when feature dimension is less than 70 nm and inter-feature spacing is between 70–300 nm. However, anisotropic elongation of the adhesion plaque is restored as the feature dimensions approach the micron range. Contrarily, a pit of diameter and depth < 70 nm has been found to be favourable for integrin clustering and concomitant focal adhesion reinforcement.

3.6.4.2.2 | *Influence of pattern shape*

Literature is replete with findings connoting alignment of cells with microgrooves[392]. Teixeira et al.[393] were the first to publish findings on epithelial cell alignment on well defined nanogrooved substrates of silicon dioxide patterns. Their studies revealed elongation and alignment of the cells along the patterns, and the response amplified with increasing groove depth. Andersson et al.[394] likewise established improved cell alignment and elongation with increasing depths of TiO_2 grooves, on smooth surfaces only. It was further evinced that the surface protrusions disconcerted the alignment mechanisms of cells. Two possible reasons could be either perturbation in the filopodial direction or size of the protrusions not allowing cell spreading and growth.

However, no significant difference was reported between the cell area and cytoskeleton formation after 4 days and cell differentiation after 21 days of culturing human osteoprogenitor cells on semi-ordered columns and random island nanostructures of PMMA[395]. Nanometric islands of height 21.0±4.7 nm generated by polystyrene/PLLA demixing, on seeding with human foetal osteoblasts demonstrated greater cell adhesion and spreading in islands compared to pits, 28.5±4.3 nm in depth, fabricated by the same process[396]. Such a finding was ascribed to finer topographic features and larger total area of the islands. Similar investigations made with normal adult human osteoblasts and STRO-1 MSCs showed that although focal adhesion formation was reduced on nano-islands compared with planar surfaces, it was still higher with respect to the nanopit substrates. Not withstanding, focal adhesions and complexes were pre-eminently found within the inter pit regions. Immunofluorescence and genomic analyses revealed up-regulation of genes essential for osteogenesis and cell adhesion on the nano-islands and attenuation of the same on nanopits[397,398].

Recently, Biggs et al.[399] reconnoitered the effects of nanoscale pits and grooves on focal adhesion formation in human osteoblasts (HOBs). The experimental observations demonstrated the demented adhesion and cell spreading of HOBs on nanopit arrays: e.g., HOBs cultured on 10 μm wide groove/ridge arrays and planar surfaces demonstrated preferential formation of focal adhesions in the latter and hence, concomitant significant differentiation along osteospecific lineage. In contrast, augmented integrin mediated focal adhesion formation and cellular spreading was observed in 100 μm wide grooved arrays.

3.6.4.2.3 | *Influence of distance between the pattern features*

Previously, efforts have been made in defining the optimum distance between nanofeatures, obligatory for orientation of human osteoblasts on substrates with varied hemispherical protrusion densities[400]. Interestingly, recent studies reported substantially lower proliferation of rat calvarial osteoblasts at sites with higher particle density[401].

With a similar objective, Spatz and co-workers[402,403] prepared RGD peptide alkanethiol functionalized gold nanoparticles (size <8 nm), partitioned by PEG to prevent cell and protein adhesion. The size of each nanoparticle permits binding of only one integrin molecule. Osteoblasts cultured on such surfaces, with gold nanoparticle spacing of 58 nm or less, showed attachment and spreading behaviour analogous to that of cells grown on a monolayer of RGD. However, the cell attachment and spreading were attenuated, when the gold nanoparticle spacing was increased beyond 73 nm. Such observations clearly demonstrated that the minimum spacing required for the clustering of integrins and activation of focal complexes, a prerequisite to cell attachment and spreading, is in the range of 58–73 nm[403].

More recently, the same group also evidenced the ability of cells to sense a difference of 0.9±0.5 nm in the gradient of gold nanoparticles between the front and back of the cell[404]. Thus, the minimum distance for formation of focal adhesion was reported to be around 50 nm, the minimal spatial variation that the cells can sense being <1 nm. Besides, the spacing between nanofeatures is of immense significance, as enough space must be provided for bending of the cell membrane[405].

3.6.4.2.4 | *Influence of the arrangement of the pattern features*

The influence of the organization of microfeatures on cell behaviour has been elaborately chronicled.[265,398,406] Biggs et al.[399] recently reviewed the role of nanoscale topography on early cell-biomaterial interactions, integrin-mediated cellular adhesion and differential function, which in turn modulate cellular behaviour. The effect of such features can be visualized with the help of a schematic, as shown in Fig. 3.27. The influence of nanofeature organization on human mesenchymal stem cells (MSCs) was perused by Dalby et al. using PMMA stamps made by electron beam lithography[406a]. The results suggested the possibility of inducing differentiation of stem cells, using nanoscale organization of the surface concerned.

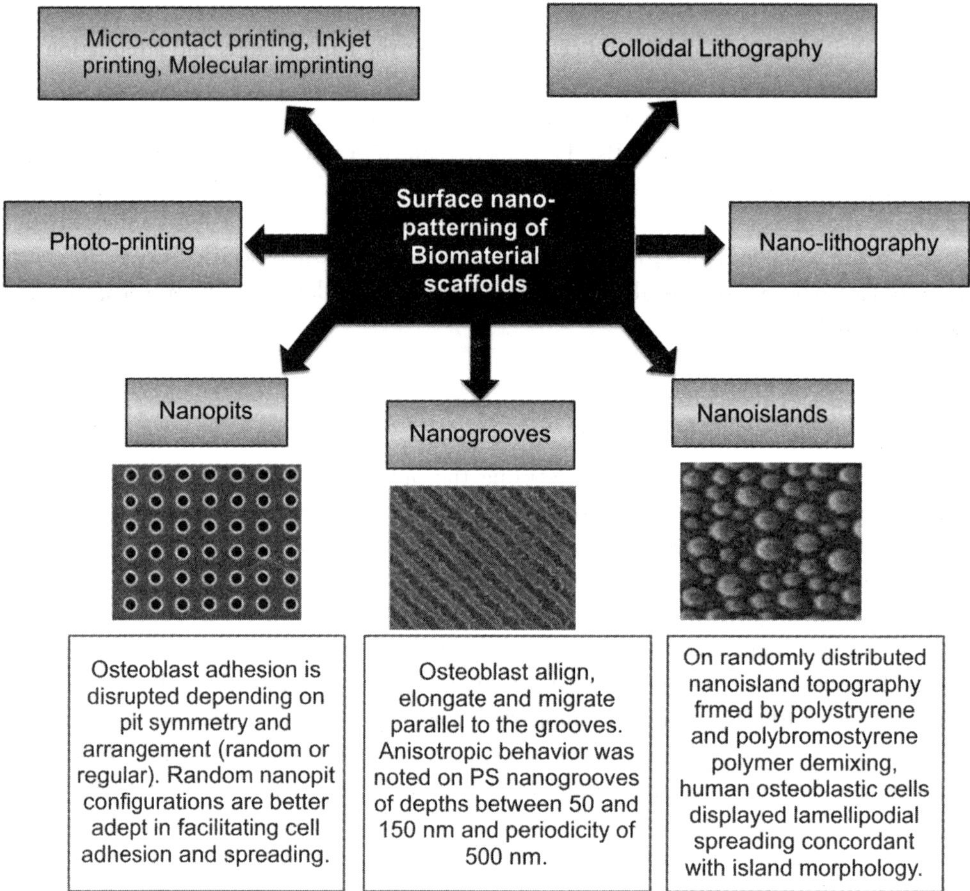

Fig. 3.27 Nano-structuring of biomaterial scaffold surfaces and ramifications on osteoblast functionality [Adapted from Ref. 447].

3.6.4.3 | Cell response to chemical structuring of surface

The influence of complex implant environment, embodied in the form of anisotropic topographies, such as, topographical grooves, chemically patterned stripes, or even the curved contour of fibre, on morphology and physio-chemical activities of a cell, has already been assayed by numerous research groups[198]. Of the physico-chemical characteristics of surfaces, zeta potential and interfacial tension are of primary importance.

Möller et al.[407] studied the influence of surface charge on proliferation of osteoblasts. However, no direct correlation between proliferation of osteoblasts and wettability of the biomaterial surface could be arrived at from their investigations. Later, Redey et al.[408] reported that poor osteoblast cell adhesion and collagen production might be occasioned by low surface wettability. They also theorized that the introduction of polar components on the surface promotes cell attachment and matrix synthesis.

Recently, Ishikawa et al.[409] reported significant spreading and functional expression by human osteoblast-like SaOS2 cells on imogolite (a naturally occurring alumino-silicate) nanotube scaffolds compared to carbon nanotube scaffolds. It was argued that substantial amount of imogolite presented favourable environments in terms of topography, roughness, wettability and protein adsorption for spreading and differentiation of osteoblasts[410,411].

> The quantitative correlation between the physico-chemical characteristics and cell functionality are yet to be conclusively established.

It can still be envisaged that interactions of a cell with the surface amounts to the cellular interactions at the attachment points with the surface. Lehnert et al.[412] analyzed the geometrical limits of ECM binding sites required for cell attachment and spreading. Square patches of ECM proteins, fibronectin and vitronectin were generated using micro-contact printing technique and the regions around the patches were passivated with a non-adherent, protein-resistant alkanethiol. It was evidenced that cells cultured on such substrata adhered and spread on ECM patches as small as $0.1\ \mu m^2$, when the inter patch distance was <5 μm. Also, the cell morphology conformed to that of the ECM patterns, when the distance between the dots was increased to 5–25 μm. However, cell proliferation and migration on dots ($\geq 1\ \mu m^2$) was demented, when the dot separation was ≥ 30 μm. In fact, it was surmised that the number of adherent cells and their area of spread remains unaffected for patch spacing below a threshold value, which in turn is dictated by the patch size (<2 μm for 300 nm patches).

Slater et al.[413] demonstrated that, in contrast with control glass, cells grown on nanopatterns (90–400 nm patches) had smaller focal adhesions, which can scale with the pattern features. In a similar line, the cells grown on patterned surfaces showed much greater proliferation, subsequently ratifying the significance of local fibronectin density at the focal adhesion sites. Initially, although the area of spread between the samples was found to be comparable, the samples with the largest patch size of 400 nm, demonstrated maximum spreading compared with other nanopatterned and even control samples after 3 days of incubation. These findings are clearly prognostic of effects of chemical pattern spacing on cell spreading.

Mitchell et al.[414] studied the ability of a protein containing C-terminal cysteine to influence cell function, on assemblage onto a gold surface, after addition of bone morphogenetic protein-2 (BMP-2) or osteopontin. The observations revealed that osteoblast adhesion and short-term responsiveness to BMP-2 banks on the surface density of a cell adhesive motif derived from osteopontin. Moreover, even in complete absence of osteogenic supplements other than BMP-2, immobilized cell interaction motif derived from the former upheld bone formation *in vitro* over 28 days. Besides, two dimensional patterning of the ligand, engineered by soft lithography technique, eventuated in spatial control of osteogenesis. The results hence highlighted the contributions of immobilized ligands on the control of cell activity.

Further investigations established that chemical patterning at the nanoscale is primarily monitored by the interaction of integrins with the surfaces[244,259,415,416]. However, the spatial organization of integrins and their interplay with different chemical moieties are yet to be detailed[390].

3.6.5 | **Protein adsorption and its role in cell responses**

Apart from the geometrical properties, the chemical properties of the biomaterial surface administer cellular functions (e.g., migration, proliferation, differentiation) and also modulate the transduction pathways of extracellular signals[417]. It has been authenticated that the cells sense a foreign surface via the layer of protein adsorbed onto it. In fact, cell adhesion is predominantly triggered by the adsorbed proteins, to which the cells respond initially[418]. Hence, a clear perception of the influence of structured material surfaces on the structure, activity and stability of conjugated proteins is of cardinal importance for predicting cell response to the structured material supports[419]. It can however be envisioned that dimensions of proteins being in the nanometer range, a surface characterized by nano-topography is more likely to affect protein functionality in size dependent manner. This was attributed to a greater surface curvature of smaller nanoparticles[420].

Denis et al.[421] demonstrated that the collagen molecules preferentially adsorb on hydrophobic surfaces (CH_3 functionalized) compared to hydrophilic ones (OH functionalized). Also, elongated features of collagen, recognizable on smooth surfaces were seemingly absent on rough surfaces, presenting nanoscale protrusions. Evidently, such observation can be correlated to differences in protein motility on smooth and rough surfaces. In fact, nano-protrusions of rough surfaces are adverse to collagen motility and subsequent assembly. The authors propounded that although the amount of protein adsorbed is primarily dictated by the surface chemistry, supra-molecular organization of the adsorbed layer is governed by both the surface chemistry and topography. The research by Webster et al.[387] also established increased adsorption of vitronectin on nanophase materials in contrast to conventional materials.

In this regard, inspite of several contradictions, some generalizations can be made from *in vitro* studies.

Moderately hydrophilic surfaces are seemingly more proficient in inducing cell responses, compared to hydrophobic surfaces, by virtue of accentuated vitronectin and/or fibronectin adsorption and better preservation of bioactivity of the adsorbed proteins.

Besides, net charges on hydrophilic surfaces provide driving forces for protein adsorption and conjointly influence the selectivity and orientation of adsorbed proteins. However, albeit the fact that the ability of both negatively and positively charged surfaces in binding adhesion proteins have been validated, the optimum charge inducing osteoblast adhesion is yet to be documented. Nanotopography of artificial substrates undeniably effects protein adsorption. However, influence of nanostructures towards protein adsorption is protein specific[417]. Besides, in a real time sceanario, a pool of different biomolecules is involved synergistically and hence, the composition of adsorbed protein layers changes with time (Vroman effect). The production of ECM by cells further complicates the situation[417]. Nonetheless, it must be borne in mind that the implant environment *in vivo* significantly differs from the cell-biomaterial interface, *in vitro*. Thus, a key unresolved issue in this regard revolves around the interface composition to be actually encountered by osteoblasts 'within the sequence of inflammation and bone regeneration[422].

3.6.6 | Biophysical constraints of osteoblast and surface interaction

In order to visualize long term interactions between cells and material interfaces, one must also concentrate on the biophysical constraints of the material under function. Such aspects are conspicuous not only in directly loaded implants (dental implants, osteosynthesis plates), but also at the material interface in functionally loaded sites[423]. Biophysical constraints are of paramount importance for synthesis of adhesion proteins, cell growth and cell differentiation[424].

The biophysical mechanisms underlying cell-substrate interaction can be broadly classified as direct and indirect mechanical effects[425], electromechanical effects[426], or mechanisms inducing acceleration of molecular transport[427]. The load transfer between material surfaces and osteoblasts most likely has profound influence on cell behaviour[428]. The forces arising from deformations at the biomaterial surface are apprehended by the osteoblasts via the attachment sites[429]. Also, the mechanical stimulation was found to spawn changes in expressions of alkaline phosphate, osteopontin, and collagen-I, and the synergistic effect was manifested in an enhancement of proliferation activity[430]. Evidently, as mechanical stimulation promotes tissue formation *in vitro*[431], an interesting group of materials, the *mechano-active* biomaterials are currently being developed by several research groups, with optimized innate physical properties[432,433].

Surface deformations not only pose as disfigured microenvironments to a proliferating cell, they can also generate dynamics of electric potentials in flow field[434]. Investigations have already advocated the effects of electric field on bone cell behaviour, *in vitro*[435,436]. In fact, it has also been established that electrical stimulation triggers a cascade of biochemical processes at the substrate surface engendering pristine extracellular matrix, which leads to improved biomineral formation[437].

Static magnetic fields (SMFs) have been found to enhance proliferation and even induce differentiation in murine MC3T3-E1 osteoblasts. It was rationalized that diamagnetic ECM proteins present in the serum accelerated cell signalling in presence of magnetic field[438]. Polte et al.[439] recently fabricated nanostrcutured magnetizable materials, which on combination with cell-adhesion lignad coated magnetic microbeads, promoted cell attachment. Conjointly, cell death response (apoptosis) could be actuated by tailoring the applied magnetic field, and so can be processes like cell retraction, rounding and detachment. In this regard, a mammoth problem commonly encountered in the functioning of bioMEMS is biofouling, brought about by the non-specific adsorption of proteins, or accumulation of cell debris and biomolecules, which ultimately transpires into impaired functioning of the device. Such problems can however be averted by galvanizing the potent parts of the material, coming into direct contact with the biological media, with nanostructured brushes of non-adhesive materials like polyethylene glycol, (PEG) and polyethylene oxide, (PEO)[440]. Nonetheless, unwanted cell adherence can be precluded in some cases by tailoring the surface topography alone, without incorporating additional chemical functionality to the surface[344,441].

Such nanostructured surfaces also have potential applications in cell sorting, cell differentiation and selective cell staining[442]. It may henceforth be deduced that topographical signals elicit a plethora of cell responses, especially differentiation and proliferation[406a]. All such manifestations if integrated into engineered implants, might improve acceptance of the engineered biomaterials in the body.

3.7 | **Closure**

Tissue engineering offers promising results for mitigating bone and cartilage tissue trauma for millions of people all around the world. One of the main principles of the conventional tissue engineering process is the fabrication of a supporting structure in the form of scaffold biomaterials for the control of cell behaviour/functionality, vascularization and the mechanical stability. This novel venture began with the development and application of bioactive glasses and ceramics in view of their excellent bone-bonding properties[349]. Subsequently, biodegradable scaffolds were fabricated, and those scaffolds have the capacity for stimulating cellular responses at the molecular level concomitant to tissue regeneration[443].

This chapter chronicles the mammothly challenging ongoing work, aimed at fabrication of 'smart' biomaterial scaffolds for tissue engineering. A basic tenet in this perspective entails recapitulation of the cellular milieu, or in other words, mimics the native ECM. Hence, architecturing of designer scaffolds with clinically relevant dimensions and homogeneous distribution of cells, reinforcing the functions, characteristic to native ECM, must be addressed for future tissue engineering applications.

> Despite the significant progress made in utilizing different processing approaches as well as chemical treatment to fabricate scaffolds with various surface chemistry/features, few theoretical models to correlate the cellular functionality with surface properties are currently available. Future research should be directed to quantitatively address this aspect.

Although numerous studies have demonstrated the great potential of nano-fibrous scaffolds as well as scaffold surfaces with various designed topographical features/patterns/textures, fewer *in vivo* studies have been carried out. The limited number of *in vivo* studies will impede the development of this field in terms of realizing the clinical potential of several of the engineered scaffolds developed in various research groups.

Some of the models discussed in this chapter correlate the total surface energy with the cell attachment/adhesion. However, none of the theoretical models satisfactorily explain the following aspects: a) influence of nanoscale roughness, imparted by designing various topographical features and polar component of surface energy on time dependent cell growth behaviour and protein adsorption on scaffolds, b) correlation among the substrate properties (elastic stiffness, substrate conductivity) and cell proliferation, and c) influence of porosity/porosity gradient on the cell functionality. Such models, if developed, will provide a better understanding of the complex relationship among surface functionalization/surface texture, surface roughness, surface energy and cell functionality.

The elastic modulus, among various other mechanical properties plays an important role in the cell functionality. For surface treated scaffolds or scaffolds with various textures/patterns, the elasticity at nanoscale can be determined using nanoindentation technique. In fact, such a technique uniquely provides the loading-unloading response of nanoscale region on the scaffold and provides the estimate of the elastic stiffness. The compressive or flexural strength can be measured for scaffolds developed for intended bone replacement applications. Often, a complete spectrum of mechanical properties along with biocompatibility evaluation for a new scaffold material are not presented in a single source, making it difficult for one to assess the suitability for a specific biomedical application. This aspect needs to be considered in future study.

Recent progress in the design and functionalization of scaffolds and corresponding processing technologies enables engineering of tunable physico-chemical characteristics, specific to a given application. A number of natural and synthetic biomaterials have already been employed for engineering scaffolds to support cell proliferation and furnish germane cues that shepherd the cells through the biological milieu, culminating into tissue regeneration. A plethora of new techniques subsuming electrospinning, peptide self assembly, biomineralization, surface patterning are being employed in recent times, to fashion dexterous scaffolds which mimic the extracellular environment found within the tissues, *in vivo*. All such advances *en masse*, have culminated into engineering of highly tailored cellular environments to recast our means of healthcare.

The current trend in the field of tissue engineering involves introduction of stem cells seeded on to suitable scaffolds. Such an approach affords emulation of the complex biological environment, with immense potential for future usage during *in vivo* and *in vitro* growth of organs. A current research focus is to elucidate the mechanisms operative at the cell-scaffold interface, issuing growth factors and communicative signals prior to cell adhesion and proliferation. However, a major challenge still rests with the fabrication of 'scaffolds of choice' with the desired porosity, pore size distribution and yet with targeted application-specific biocompatibility property.

Another novel approach in tissue engineering would be to develop scaffolds with gradient in porosity or surface texture so as to obtain differential cellular response on the same surface. Apart from fundamental scientific interest, the scaffolds with gradient porosity would be suitable for bone replacement applications as biological tissues can favorably grow into various pore assemblies, providing better osseointegration property.

The existing and ever-expanding interdisciplinary field of nanobioscience has leveraged advancements in schemes and synthesis techniques, instrumental in rendering biomaterials more adaptive towards the biological molecules. The 'holy grail' is to have sufficient control over the mechanisms governing the protein-mediated cellular interactions with the bio-scaffolds. With the pool of data unambiguously intimating topographical modification at the cell-substrate interface as a regulator of cell-functionality, it can be envisioned that future research would be directed towards topographic modifications of advanced materials. However, major detriments here are the complicated methodologies and their low reproducibility associated with currently available techniques for the fabrication of complex 3D biomedical devices.

Even then, the degree of improvement over static culture with the best 3D scaffold is limited, primarily in view of inefficient nutrient transfer across the cell–material interface. Clinically effective perfusion bioreactor systems are to be designed to circumvent the impediments related to transport phenomena. Ideally, a bioreactor based bone tissue engineering strategy should start at stem cell seeding on biodegradable, biocompatible, 3D scaffolds in a bioreactor system. After sufficient culture time, the construct should be removed and directly implanted in the patient to actuate osteoconduction and osteoinduction for rapid repair of the defect tissue.

In a nutshell, tissue engineering is one of the most stimulating areas of interdisciplinary and multidisciplinary researchers today. Scaffold materials and fabrication technologies play a crucial role in tissue engineering, and are rapidly evolving. This chapter presents a summary on many exciting materials developed for scaffold application including biodegradable and bioactive polymers, ceramics, glass–ceramics, hydrogels, electrospun nanofibers, etc. As scaffolds play a vital role in tissue engineering, the feasibility of designing nanofeatured scaffolds was also reviewed. Although bone is a functionally diverse tissue, new biomaterials with promise to cure orthopedic problems are yet to be discovered. In conclusion, it may be anticipated that the promising arena of tissue engineering unveiled via the usage of stem cells, in conjunction with the multitude of biomaterials and processing techniques outlined in this chapter, might bear possible solutions to therapeutic repair of bone tissues both in preclinical and clinical human models. It can be envisaged that with relevant animal studies and coherent understanding of the principles governing cell–biomaterial interactions, the tissue engineering doctrine can be efficiently leveraged to revolutionize the current scenario of regenerative medicine, as discussed in a review paper from the author's research group[444].

Section II

Fundamentals – Materials Sciences

Chapter 4

Conventional and Advanced Manufacturing of Biomaterials

In reference to the discussion in some of the preceding chapters, it is instructive to realize that the size and shape of the implantable biomedical device is specific to the anatomical location in a human patient to treat specific diseases. To substantiate this, the device components of hip and knee joints are of different shapes, and can be made from different material classes (metals/ceramics/polymers). Although the lab scale research involves the processing of biomaterials with simplistic shapes, the fabrication of a biomedical device demands manufacturing of design-specific components made from materials with acceptable biocompatibility for a given biomedical application. Such a scheme of research requires an understanding of a range of processing techniques, which can be adapted for a wide spectrum of materials. The fabrication approaches for metallic implants are relatively well established and reproducibility of such approaches to produce implants with reliable properties have resulted in wider clinical acceptance. In the above perspective, this chapter discusses conventional processing approaches for metals, ceramics and polymers. While discussing patient-specific implant fabrication, an emphasis has been laid on the discussion of low temperature additive manufacturing techniques, i.e., 3D printing and 3D plotting. In particular, processing science of these two rapid prototyping techniques is critically analysed. Summarizing, this chapter is expected to provide a platform for the reader to understand and identify specific processing route for a given biomaterial. More discussion on the conventional and advanced implant fabrication processes can be found elsewhere[453]. The processing approaches for tissue engineering scaffolds are already discussed in a preceding chapter.

4.1 | Conventional Manufacturing of Metallic Biomaterials

The manufacturing of metallic implants is relatively well established, and approaches of processing/ fabrication are routinely combined to produce metallic implants of various shapes and sizes. These metallic implants are widely used for orthopedic and dental restorative applications and more specifically for load bearing articulating joints.

> The established manufacturing techniques can produce metallic implants with desired shapes/sizes together with reliable material properties in a reproducible manner. This aspect has made metallic implants more acceptable to clinicians.

In Fig. 4.1, various manufacturing methods, typically used for metallic implants, are summarized with the flow or sequence of different processing approach in shaping or making different products. For ceramics, the starting material is mostly in powder form and powder-based processing approaches are to be used. In case of polymers, the starting material can be either powder or granules. The consolidation of ceramics and polymers takes place in strikingly different routes. For the former, high temperature sintering is commonly used and for the latter, the injection/compression moulding or extrusion is adopted. For either class of material, melting is not involved, although processing temperatures can be close to the melting point in case of polymers.

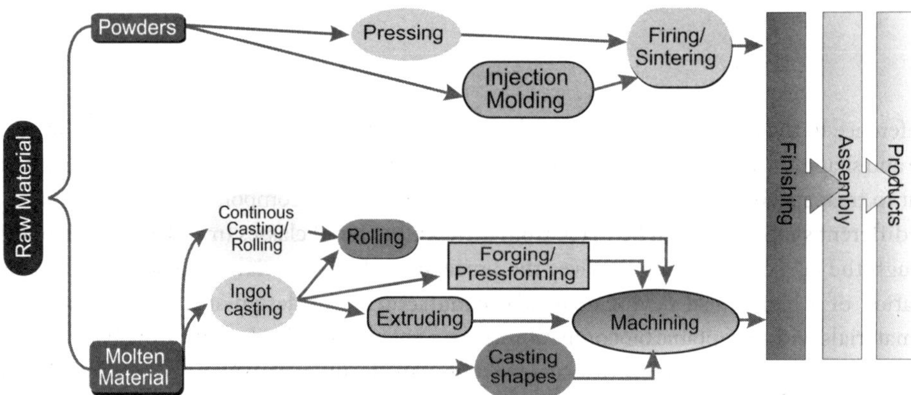

Fig. 4.1 Flow chart showing the sequence of processing and post-processing approaches to obtain metallic/ceramic product with desired shape and size.

4.1.1 | Casting

In general, metallic implants are mostly fabricated by melting–casting route i.e., starting from molten metal. The classification of the casting processes is shown in Fig. 4.2. The generic classes include (a) expendable mould processes (mould is sacrificed to remove part) to obtain more complex shapes and (b) permanent mould processes (metal mould can be used to make many castings) for higher production rates.

In a conventional casting process, molten metal with good fluidity is poured into a mould cavity and after slow cooling with heat being dissipated primarily by conduction through mould material, one can get a cast product (see Fig. 4.3a).

> Casting is defined as a process in which molten metal flows by gravity or other forces into a mould, where it solidifies in the shape of the mould cavity.

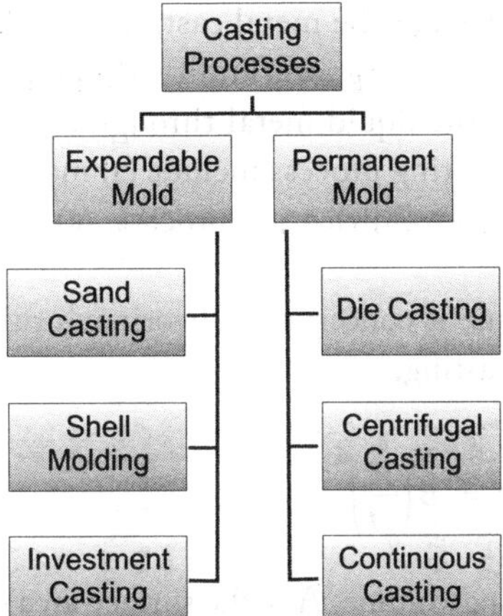

Fig. 4.2 General classification of casting processes.

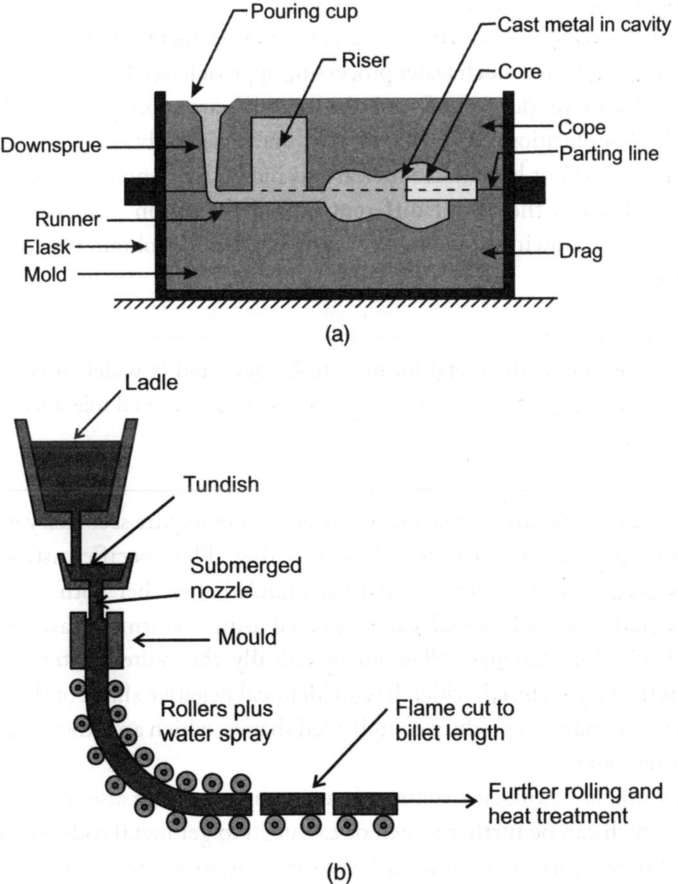

Fig. 4.3 Schematic of (a) conventional sand casting of metals and (b) continuous casting of metal plates.

Various sequential steps involved in the metal casting include (a) melting of the metal above the equilibrium melting point to ensure good fluidity, (b) pouring of the superheated melt into the mould cavity, (c) freezing of the liquid metal through heat dissipation through the mould. Six basic factors involved in the casting process include (a) mould cavity preparation, (b) melting process, (c) pouring technique, (d) solidification process, (e) part removal process and (f) post-processing.

As far as the solidification time is concerned, Chvorinov's rule can be used to predict time of solidification for simple shaped casting,

$$t_s = B\left(\frac{V}{A}\right)^n \tag{4.1}$$

where, n = 1.5–2.0, V is the casting volume, A is the surface area of casting and B is the moulding constant. The value of 'B' depends on metal property (specific heat capacity, latent heat of fusion) and mould material. The above rule implies that under identical casting conditions, the simple shaped cast product with small volume and large surface area would cool more rapidly than a casting with a large volume and a small surface area.

Alternative to the conventional casting process, the molten metal can be directly cast into plates via continuous casting route in a much faster processing approach (see Fig. 4.3b). Continuous casting is used for industrial scale production of steel plates, which are subsequently rolled to desired size for use in biomedical applications. The shapes, that one usually obtain either via conventional or continuous casting, would not be suitable for various biomedical applications. The size reduction and shape changes demand the use of different bulk deformation processes including forging, rolling, extrusion or wire drawing. One of the casting processes to obtain net shaped product is the investment casting.

> Investment casting is one of the metal-forming techniques and is widely used commercially to produce biomedical grade metallic components with good dimensional tolerances and exceptional surface qualities.

Often the as-cast materials, after investment casting, do not require secondary machining process, which are otherwise used for die casting and sand casting. This specific casting route derives its name from a designed pattern being invested (surrounded) together with a refractory material. Many biomedical grade materials, which can be shaped using investment casting, include stainless steel, carbon steel, Ti-alloys and glass. Phenomenologically, the molten material is allowed to pour into a cavity in a refractory material, which has an identical negative shape of the desired part. This specific metal forming route can produce complicated shapes, which are otherwise not possible with conventional casting routes.

Other than investment casting, conventional casting processes are also used to obtain blooms or metal workpiece, which can be further forged or extruded to get metal rods for various orthopedic surgeries. Since all these processes require bulk deformation, some related elements are emphasized now.

4.1.2 | Bulk deformation processes

For a better understanding of the bulk deformation processes, a general understanding of the tensile stress–strain response of a ductile metal is required. As shown in Fig. 4.4a, the point of transition from linear elastic to non-uniform plastic deformation is marked as yield strength (YS) and the point of maximum load or the stress corresponding to the largest load bearing capability is known as ultimate tensile strength (UTS). The total strain to the point of fracture is a measure of ductility. All the three parameters, i.e., YS, UTS and ductility vary with temperature, as shown in Fig. 4.4b. The strength parameters decrease with temperature, while ductility increases in a non-linear manner.

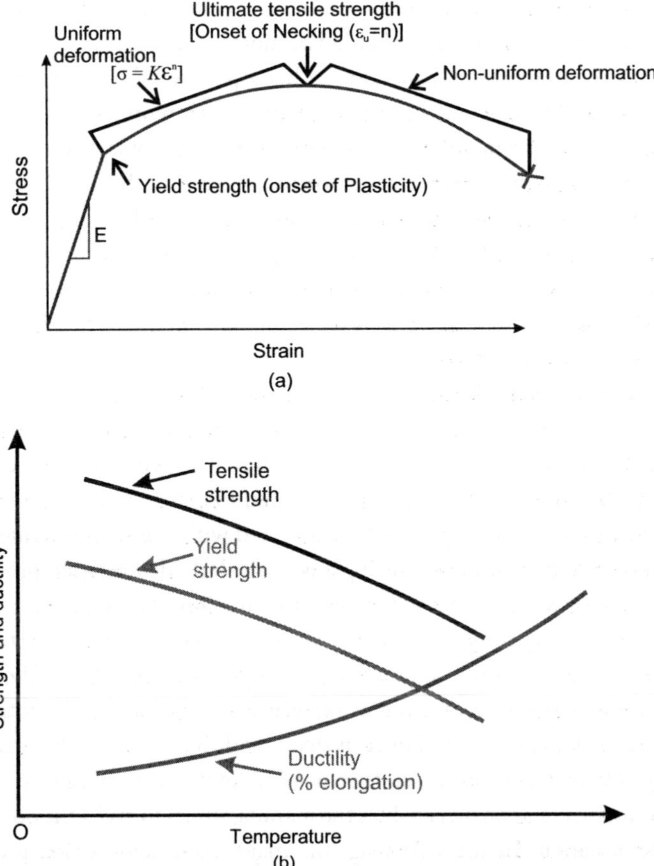

Fig. 4.4 (a) Typical stress–strain response of a metal under tension and (b) variation of strength and ductility with temperature.

At a higher processing temperature, the deformation of a metallic workpiece would take place at a lower stress level and also a metal can be deformed to a larger extent without fracture, when compared to that at ambient temperature.

Since the metallic materials are expected to flow during bulk deformation processes leading to size/shape changes, the metal deformation temperature therefore needs to be suitably selected to facilitate desired metal 'flow' without failure/fracture. Accordingly, the deformation processes are performed at a stress level larger than yield strength (YS), but lower than UTS, so that uniform deformation of a bulk metal can be ensured without the onset of necking.

> Bulk deformation of a metal needs to be conducted in the region between YS and UTS. Above UTS, necking starts and non-uniform deformation leads to shape instability.

Also, at a lower yield strength, the deformation can be initiated at lower stress and larger the difference between strain at yield point and at UTS, more uniform strain can be imparted as metal can flow or deform over a larger strain window. This aspect will be better understood, when the mechanical properties will be discussed in a later chapter with more details.

At this juncture, one can distinguish between hot working and cold working. In materials science/metallurgy, 'working' means the metal being deformed using any of the metal deformation processes. Hot working essentially means deformation at a temperature, $T/T_m > 0.5$ (T_m being the melting point of the metal). As shown in Fig. 4.4(b), with temperature, UTS decreases with an increase in the strain to failure. Clearly, metal should have more ductility i.e. it can withstand larger strain to failure, if deformed at higher temperature. It is therefore well perceived that large scale deformation can be given to a metal during hot working.

However, close dimensional tolerance cannot be achieved, when a metal undergoes significant deformation. This can be sorted out in cold working processes, when a metal is deformed at or near room temperature or at a temperature of much lower than $T/T_m = 0.5$. During cold working, much less deformation can be expected, but close dimensional tolerance can be achieved. Therefore, all the deformation processes are to be necessarily correlated with the shape changes.

An important aspect to be considered during any of the bulk deformation processes is the stress–strain response of specific metal under compression or tension. The maximum deformation to be imparted to a metal in rolling, forging and extrusion should be less than the total deformation that the metal can experience in the uniform or homogenous deformation region. If one wants to impart a large amount of deformation to a metallic workpiece, then one can pursue total deformation in a number of steps. The strain accumulated in each step is additive in nature. For example, if a metallic plate has to undergo 60% total deformation, then one can do it in three stages with 20% deformation in each stage. This intelligent approach enables the manufacturers to avoid the occurrence of cracking or non-uniform deformation. In the following, the major bulk deformation processing routes are discussed briefly. One should remember that the total volume of metallic workpiece during the bulk deformation processes will always remain constant. This means that the volume of the metal before deformation and after deformation is equal.

4.1.2.1 | Rolling

In case of rolling, metal plates with large thickness can be thinned to lower thickness by forcing through the roll gap of two oppositely rotating rolls.

Rolling is defined as a bulk deformation process, in which a workpiece is deformed between two rotating rolls, while being subjected to compressive stresses from the rolls and surface shear stresses from the friction between roll surface and workpiece surface.

For the rolling to take place, the exit speed of the plate should be higher than the roll speed and the entry speed of the work piece into the deformation zone between the two rolls. Depending on the arrangement of rolls, the rolling process can be classified to several classes, e.g., 2-high mill, 4-high mill, etc. The example of the roll stand shown in Fig. 4.5 is the case of 2-high mill. In case of 4-high mill, one roll is placed vertically up on the top roll and one vertically down under the bottom roll of the roll stand. The thickness reduction in rolling is largely dependent on roll pressure, which in turn is influenced by roll speed, flow strength of workpiece metals, etc. In the rolling process, the thickness of the plate is reduced and since the volume of the metal remains constant during plastic deformation zone, the velocity (v_o) of the rolled plate exiting from roll gap can be written as,

$$v_o = v_i \, (h_i / h_o) \tag{4.2}$$

In the above, v_i is the entry speed of the plate entering the roll gap; h_i and h_o are the thickness of the initial plate (prior to rolling) and final thickness of post-rolled plate.

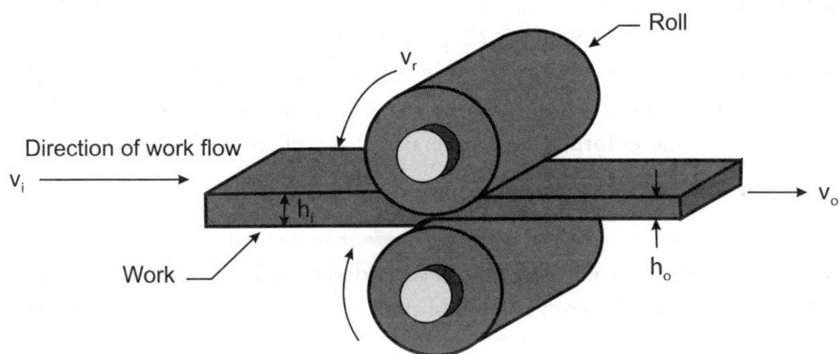

Fig. 4.5 Schematic illustration of the rolling of metals.

From Fig. 4.5, it is clear that the following relationship must be satisfied for the rolling process to occur,

$$v_i < v_r < v \tag{4.3}$$

where, v_r is the equivalent linear speed of the roll.

Also, the deformation of the metallic plate takes place only at the roll contact zone, i.e. wherein the plate is in physical contact with the roll surfaces. By carefully tailoring the roll speed, rolling temperature, roll gap and roll pressure, one can impart specific bulk deformation to a metallic plate being rolled. In practice, multiple pass rolling with a number of rolling station in series with different roll gap or roll pressure are used and this is known as tandem rolling. In order to maintain specific bulk temperature, the rolled plate is intermittently soaked in a high temperature furnace between two roll passes, as and when required. The rolled plates are typically used in many orthopedic fixation processes.

4.1.2.2 | Forging

Similar to rolling, large compressive forces are applied to reduce the thickness of the workpiece in the forging process.

> Forging is defined as the bulk metal deformation process, wherein large compressive stress is applied on a metal, longitudinally constrained between two platens, with or without lateral constraints.

The forging can be primarily classified as open die forging and closed die forging. In view of the relative importance of the forging processes in Ti-based orthopedic alloys, various types of forging techniques are shown in Fig. 4.6 (a–c). The most conventional forging process is open die forging, wherein a cylindrical metallic block is compressed between a moving platen of a die and a stationary base plate. If the friction between the die surface and metal is ignored, then the cylindrical workpiece will be deformed uniformly to be reduced in height with corresponding increase in diameter. In reality, the interfacial friction results in the barrelled shaped final forged product with the change in diameter being commensurate with the decrease in height (see Fig. 4.6a). Two alternative forging techniques are impression die and flashless forging. As shown in Fig. 4.6b, the die cavity has a specific shape in commensurate with the final forged shape in case of impression die forging. The sidewise flash is removed from the forged part by metal cutting. In case of the flashless forging, the die cavity has a more regular geometric shape, allowing the forged part to undergo constrained deformation with the designed die cavity (see Fig. 4.6c). Like in rolling, the extent of deformation during forging depends on flow stress (i.e., the stress at which metal undergoes uniform plastic deformation at a given metal deformation temperature) as well as thickness or height change to be pursued.

> Many biomedical castings are usually forged to obtain a desired shape. For example, forged Ti6Al4V or high nitrogen stainless steel is used as femoral stem in total hip joint replacement surgery.

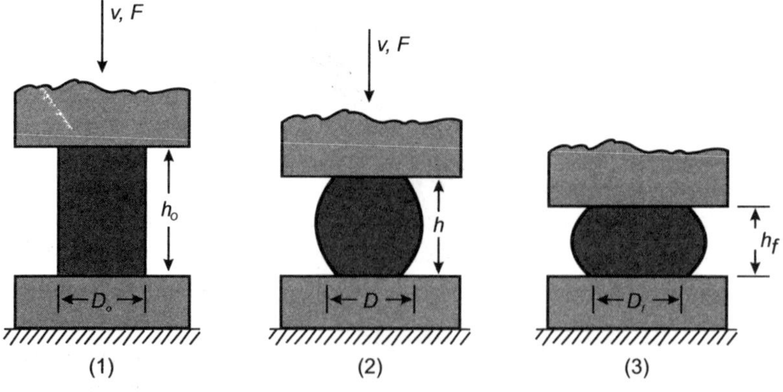

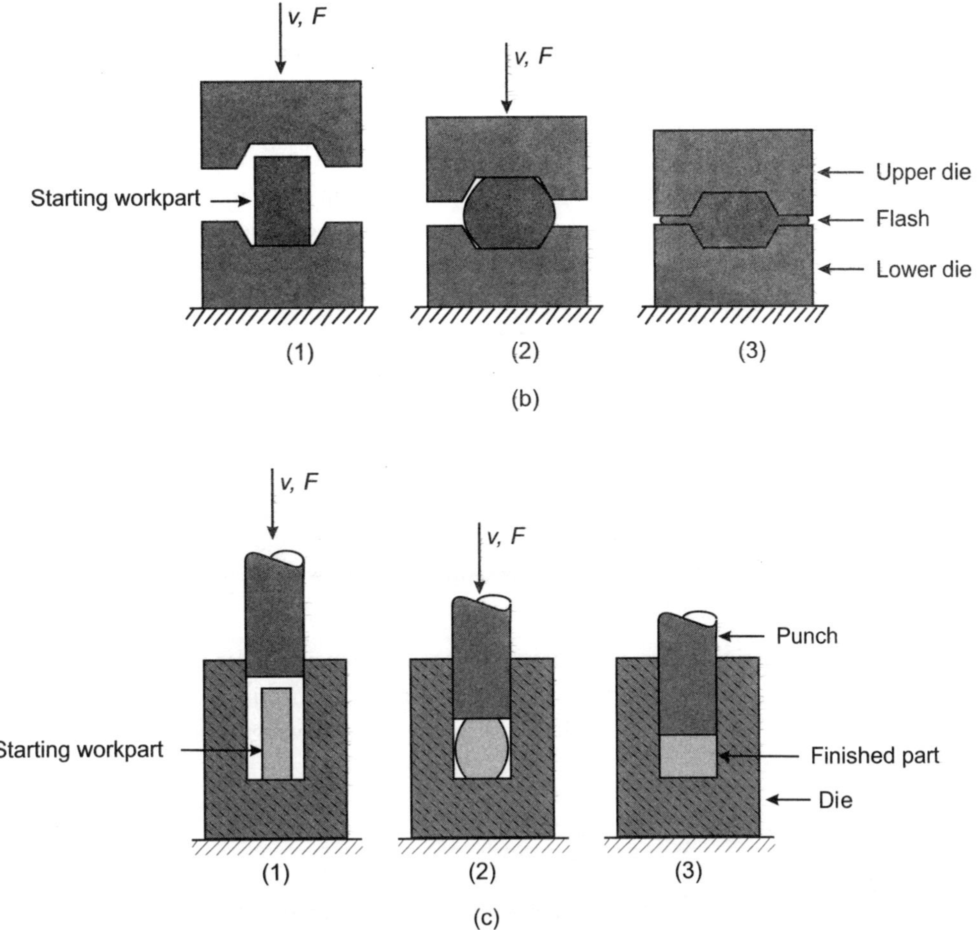

Fig. 4.6 Different forging processes: (a) open die forging with friction at die/metal interface, (b) impression die forging, and (c) flashless forging of metals. While v, f denotes the ram velocity and compressive force respectively; D_0 and h_0 are respectively the original diameter and height of the workpiece. Similarly, h_f and D_f are the final height and final diameter respectively. Also, h, D are the instantaneous height and workpiece diameter, respectively at an intermittent stage of forging.

4.1.2.3 | Extrusion and drawing

Extrusion is a bulk deformation process to obtain rods or similar components with a specific cross-sectional profile, when a material is being pushed with compressive force through a die.

The schematic of the extrusion process is shown in Fig. 4.7a. In the extrusion process, a ram is used to transfer compressive forces to deform a workpiece and the extruded rod with identical diameter of that of the die orifice is obtained at the die exit (see Fig. 4.7a). Here, the maximum deformation is imparted at the die constriction region, when the metal is forced to exit from the die orifice. As far as the extrusion process is concerned, the variables include, (a) die angle, (b) extrusion ratio, (c) temperature of workpiece, (d) ram speed and (e) lubricant. As far as the die angle is concerned, the contact surface area of the metal with die would be large for low die angle, leading to increased friction at die/workpiece interface. This would require large ram force, meaning larger capacity of extrusion press. In contrast, more redundant work or dead zone would develop in case of larger disc angle. The extrusion process can be classified as direct and indirect extrusion. Depending on ram pressure application and the die placement, one can use either of these two extrusion processes. As will be discussed later, the extrusion of polymer depends on different process science aspects. The extruded rods are also used for many orthopedic applications.

In contrast to all the above-described metal working processes, the wire drawing involves the application of tensile force.

> The wire drawing can be defined as a metalworking process to obtain a wire with reduced cross-section by pulling it through a single, or series of, drawing die(s) with a tensile force.

Clearly the cross-sectional area of the drawn wire will be much smaller than the extruded rod. Depending on the die angle (α), the deformation zone length (L_c) in wire drawing process extends to the die orifice (see Fig. 4.7b). In both the extrusion and wire drawing, the extent of deformation is quantified using % area reduction,

$$r = \frac{A_i - A_o}{A_i} \tag{4.4}$$

where, A_i and A_o are the cross-sectional area prior to deformation and after deformation.

As shown in Fig. 4.7b, a tensile force is effectively utilized to pull a metallic workpiece in the form of a thin wire through the orifice of a die in the wire drawing process. Like in rolling process, one can place a series of die with each die having correspondingly lower orifice diameter to impart the amount of deformation in an intelligent manner. Such tandem drawing process can be coupled with intermediate annealing treatment, as and when applicable. Such a process is phenomenologically known as thermomechanical processing, which is defined as a manufacturing process combining bulk deformation processing (e.g., forging, rolling, extrusion, wire drawing) with heat treatment. Depending on the material composition and microstructure, the extent of deformation can be tailored by carefully tailoring the process parameters at each bulk deformation stage and heat treatment conditions.

The fabrication of different biomedical devices, like guide wires, catheters, pacemakers, stents, staples, functional electrical stimulation systems, eyeglass frames and orthodontic braces need wires with diameters in the range of a few millimetre to 10–15 micrometres (see for example, Fig. 4.7c). As far as the performance of such devices is concerned, the fatigue and fracture under physiologically relevant stress cycles are important.

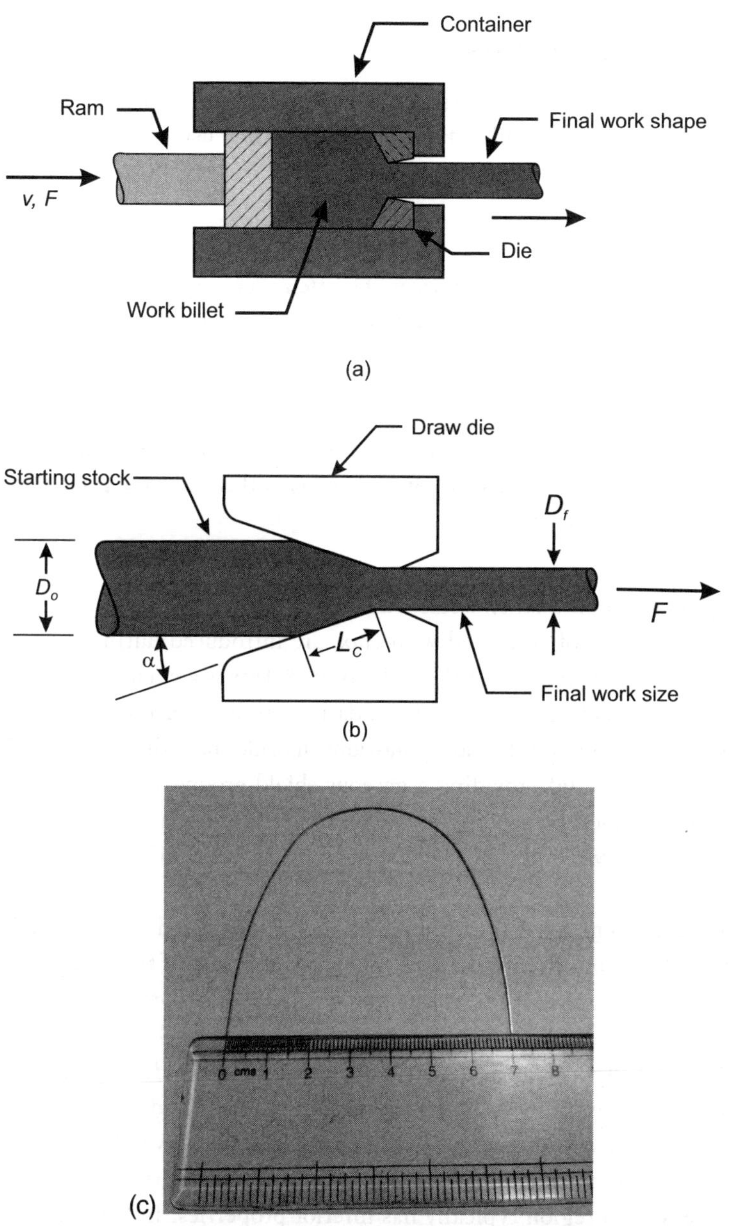

Fig. 4.7 (a) Schematic illustration of the extrusion process to obtain a metal rod, (b) the wire drawing process to produce a metal wire; and (c) an example of NiTi wire for orthodontic application.

An important example of the application of wire drawing is to produce NiTi wire for vascular and peripheral stents as well as to obtain NiTi or stainless steel based wires for orthodontic applications (e.g., dental arch wire).

4.1.3 | Metal joining processes

Once a metallic structure is deformed to a specific shape, it may need to be joined with either a similar or different shaped structure of similar/dissimilar composition using different joining processes.

4.1.3.1 | Welding

In manufacturing of various biomedical devices, it is often necessary to join two materials of identical composition or two dissimilar materials. The engineering techniques, largely adopted for such purpose are welding and brazing.

> Welding is defined as a material joining process, wherein two or multiple solid structures, either of similar or dissimilar composition, can be joined together at their contacting surfaces using suitable application of heat and/or pressure.

In order to produce good quality welds, filler and flux are to be used. Filler metals (similar composition as base metal) often coated with flux, are introduced during the welding operation to increase the volume and strength of the weld joint. A flux is a chemical cleaning agent, which facilitates brazing and welding by removing oxidation products of the metals to be joined. The examples of fluxes include sodium chloride, potassium chloride and sodium fluoride. Mechanistically, a flux retards oxidation by, (a) providing a gaseous shield around weld zone, and (b) forming a 'slag', which dissolves oxides. Depending on the welding technique and the materials to be welded, different fluxes are used, while the filler material typically has similar composition to that of the base metal to be welded.

Typically, a welded joint consists of weld pool and heat affected zone (HAZ) [see Fig. 4.8a]. The composition of the solidified weld pool can be a composite of the base metal, filler and the flux. Depending on the severity of heat flux and the filler/flux composition, the compositional gradation in the weld pool can be noticed. In a cross-sectional metallurgical microscopic structure, one can therefore identify an interface between the base metal, HAZ and the weld pool. The generic classification of welding includes, (a) fusion welding and (b) solid state welding. As shown in Fig. 4.8(b), the microstructure development at a fusion welded joint can be described as the metallurgical bond formation via continuity in grain structure. While weld metal exhibits coarse grains, the welded region typically has inferior properties. In contrast, HAZ extends on both sides of weld metal. Also, HAZ has typically a finer grain structure and thereby better mechanical properties.

Phenomenologically, the fusion welding process involves two sequential stages: (a) melting of localized regions, and (b) solidification of molten pool, while the heat is being extracted from molten pool. In practice, a number of welding techniques are used for various engineering applications, including automotive, shipbuilding, etc. In fact, any engineering application, requiring large material structures, do require plates or other shapes to be joined together. It needs to be ensured that the welded joint has similar, if not better strength so that the load bearing capability of the welded structure is not compromised. As far as the heat source is concerned, three principal sources in case of fusion welding can be, (a) chemical (e.g., gas welding), (b) electrical (e.g., arc welding) and (c) high energy density (e.g., laser or electron beam welding). In contrast to fusion welding, solid state welding is considered as a variant of welding process, in which the joining of two surfaces is achieved by pressure alone or by heat and pressure. In order to ensure good weldability under this process, two surfaces must be brought into close physical contact with each other to allow atomic level bonding.

Various fusion welding processes involving chemical reactions as heat sources include oxyacetylene gas welding (reactions between acetylene, C_2H_2 with air). Similarly, the welding processes based on electrical energy input include shielded metal arc welding, submerged arc welding and gas tungsten arc welding, etc. Depending on the type of electric arc welding and metals, the plates of different thickness or various shapes can be welded. Also, depending on sensitivity of metals to oxidation as well as required tolerances of oxide inclusion, a specific fusion welding process is chosen for targeted engineering application.

The rationale for such choice is the additional requirement of biocompatibility property of the welded structure. While joining two similar or dissimilar materials for a given biomedical application, the requirement of uncompromised biocompatibility needs to be ensured. From that perspective, laser/electron beam welding under protective atmosphere provides controlled atmosphere; thereby the chances of oxide formation can be avoided.

In case of high energy density processes, laser or electron beam is focused into a small area, and depending on the laser/electron beam energy, the metallic plates of a few mm thicknesses can be welded (see Figs. 4.8c and 4.8d). In case of laser beam welding (LBW), no filler/flux is required and low heat input produces low distortion of the metallic workpiece. Therefore, the chances of the formation of oxides or other compounds are minimal in case of LBW. For metallic samples, LBW is carried out under inert atmosphere (see Fig. 4.8d). As shown in Fig. 4.8c, a laser beam is made to scan at the interfacial area between two metallic structures to be welded.

For biomedical application, however, the welding processes involving high energy density, e.g. laser or electron beam are preferred. One of the important requirements of laser beam welding is the acceptable level of laser absorption ability of the material to be welded. In case of poor absorption with laser, the electron beam welding process can be adopted.

4.1.3.2 | Brazing

Another joining process is brazing, which has the ability to join materials with different properties/thickness/cross-sections. Similar to fusion welding, both filler and flux are used in case of brazing. Also, a strong joint is obtained by capillary action between closely fitting components upon cooling and solidification of the filler metal (see Fig. 4.8e).

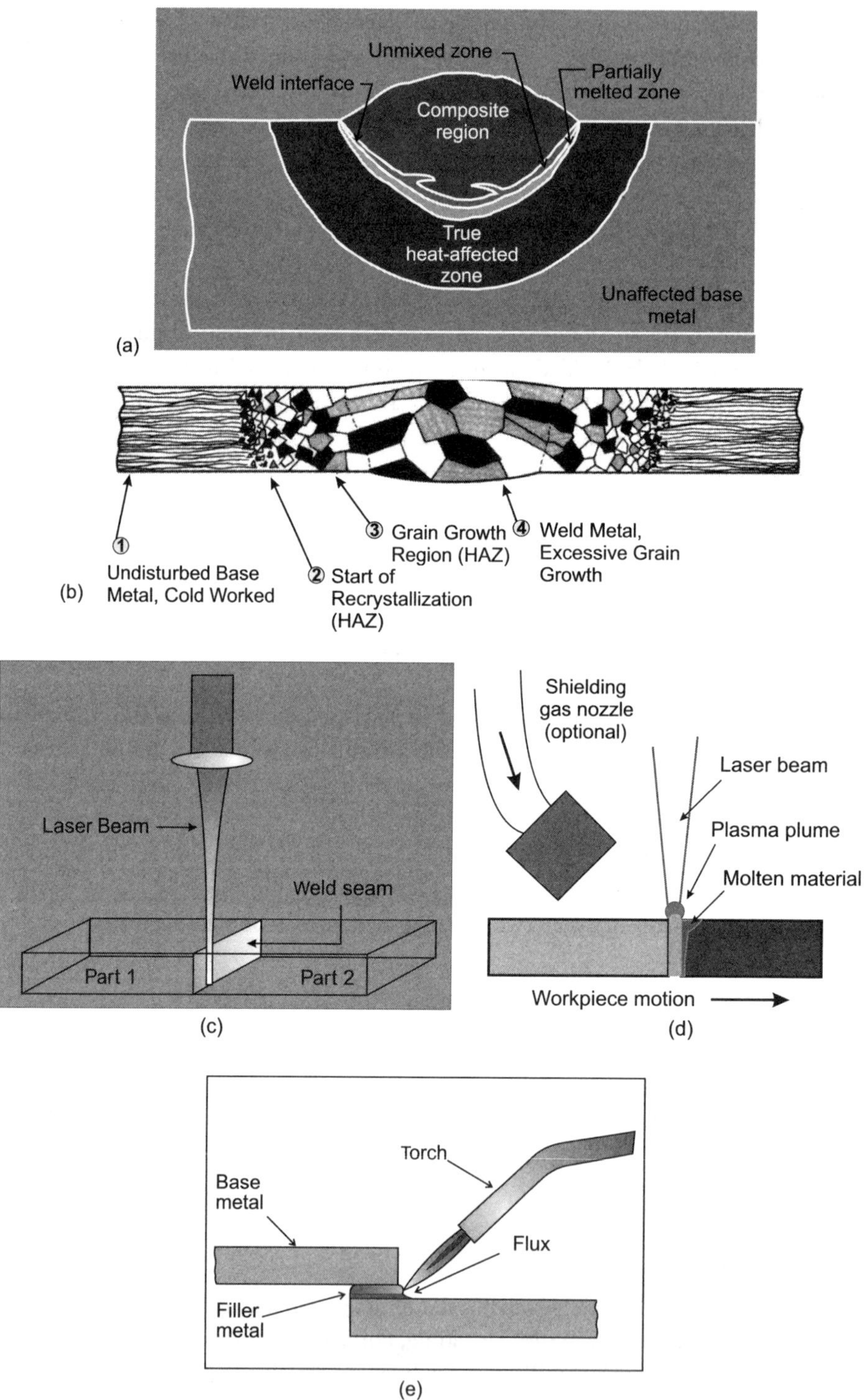

Fig. 4.8 Schematic illustration showing (a) various zones in a fusion welded joint, (b) grain structure across a welded joint, (c, d) Laser Beam Welding (LBW), and (e) brazing process.

Brazing can be defined as a material joining process in which a filler metal is melted and distributed by capillary action between the metal surfaces.

The fundamental difference between fusion welding and brazing is that in case of the latter, the filler material with lower melting point is melted at a lower temperature and the base metal/workpiece is not melted. Also, brazing takes place at a much lower temperature than fusion welding. Again, filler composition should be selected such that the biocompatibility of the brazed joint is not compromised. In addition, the use of flux is important to prevent oxidation and to remove oxide films from workpiece surfaces. In case of brazing, dissimilar metals can also be assembled with good joint strength.

The strength of a brazed joint is generally less than welded joint. The melting of filler along with workpiece metal during brazing must satisfy the following criteria,

(a) high fluidity for penetration into interfaces,
(b) wetting of molten filler material on workpiece
(c) low surface tension and low viscosity of filler (liquid) ensures good wettability
(d) clearance between mating surfaces important (recommended clearance: up to 0.2 mm)
(e) surface roughness and surface cleanliness are also important to ensure capillary action

Another metal joining process is soldering, which takes place still at a temperature lower than brazing. In case of soldering, the filler metal contains lead, but the recent developments enable the use of lead free alloys as filler material for environmental friendly alloys for electronics applications. While lead is certainly not biocompatible, but soldering with Pb-free alloy filler can lead to biocompatible alloy joints. This remains an unexplored area in the area of biomedical device manufacturing.

4.1.4 | Machining processes

In the manufacturing of any metallic biomedical device with a desired size and shape, the finishing and assembly are also equally important.

Among different classes of biomaterials, one of the reasons for the better clinical acceptability of metallic implants is the easier adaptability to various machining techniques.

For example, making a hole on a metallic plate without fracture or significant shape distortion is possible with conventional machining techniques, which can however not be adopted for brittle ceramics. In order to obtain desired shapes, both conventional and non-conventional, like Electro discharge machining can be used for metals. Prior to the biomedical usage, the finished metallic implants are sterilized before packaging.

In manufacturing of biomedical devices, machining processes, particularly when conventional fabrication processes are used, play an important role in removing excess material from starting workpiece to obtain desired geometry. Two generic type of machining processes, i.e., conventional and non-conventional machining processes are used. The conventional machining processes include tuning, drilling, milling, grinding. Some widely used conventional machining processes are shown in Fig. 4.9. In all conventional processes, excess material is removed in the form of a chip. In case of turning a cylindrical workpiece, the diameter is constantly reduced during the progression of material removal (see Fig. 4.9a). For making hole in a metallic plate, the drilling processes can be adopted, as shown in Fig. 4.9b.

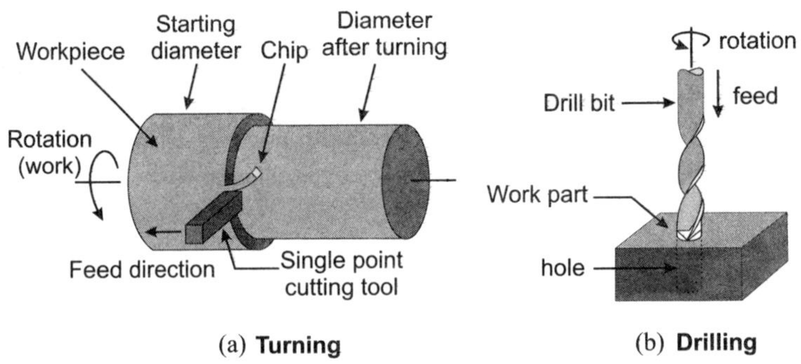

Fig. 4.9 Conventional machining process for metals: (a) turning and (b) drilling.

In the conventional machining processes, highly complex stress conditions with high levels of shear lead to controlled fracture of the material. Electrical discharge machining (EDM) is an important example of non-conventional machining process and EDM is regarded as a potential manufacturing route to fabricate dies and moulds of conductive materials[454,455,456]. In principle, EDM is a thermoelectric process that uses repetitive electrical discharge (sparks) to erode electrically conductive workpiece and electrode materials. Also, the EDM process facilitates to machine hard, brittle, and difficult-to-machine materials, with appreciable electrical transport property.

> The major advantage of EDM is the ability to obtain components with desired shape and closer dimensional tolerance in a shorter time.

Two competing EDM processes are die-sinking and wire-EDM. In case of the former, the die cavity is directly impressed on the workpiece while accomplishing large material removal (see Fig. 4.10a). Typical materials, used as EDM electrodes in die-sinking process, include copper, graphite, tungsten, brass, steel, copper–tungsten, copper–chromium alloy, etc. Materials having good electrical and thermal conductivity with a high melting point are always preferred to minimize electrode wear. The EDM technology has become attractive in precision die making, since no physical contact occurs between the electrode and the workpiece. During die-sinking, the voltage, current and duty cycle are varied to obtain the machined materials good surface roughness properties. The efficacy of the EDM for a given material is also assessed in quantifying the material removal rate and tool wear rate. For wire EDM, highly conductive oxygen free high conductivity copper (OFHC) is widely used as the wire to cut intricate or complex shaped product from a workpiece (see Fig. 4.10b).

Another non-conventional machining process is the laser beam machining, wherein a high power laser source is focused on a flat material and the metal locally melts and vaporizes (see Fig. 4.11). Depending on the traversing path of the laser beam, the specific area of the metal will be machined. Here, the laser power can be varied depending on the material. The laser absorption ability i.e., the ability of the metal to get spontaneously heated upon laser beam interaction is the only property requirement in case of laser beam machining. For metals, the laser beam machining is to be conducted in inert gas atmosphere to prevent oxidation.

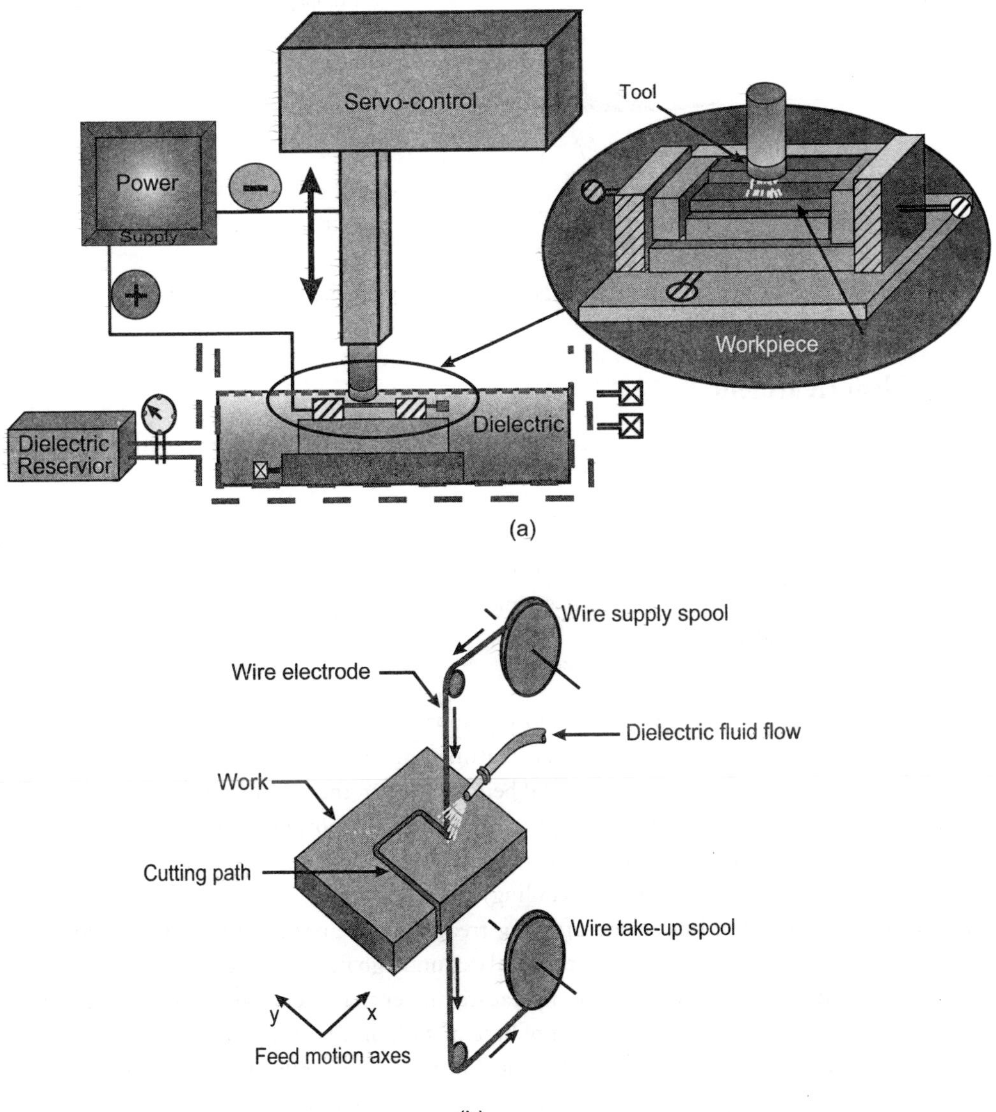

Fig. 4.10 Schematic illustration of the EDM processes: (a) die-sinking and (b) wire-EDM for machining electroconductive materials.

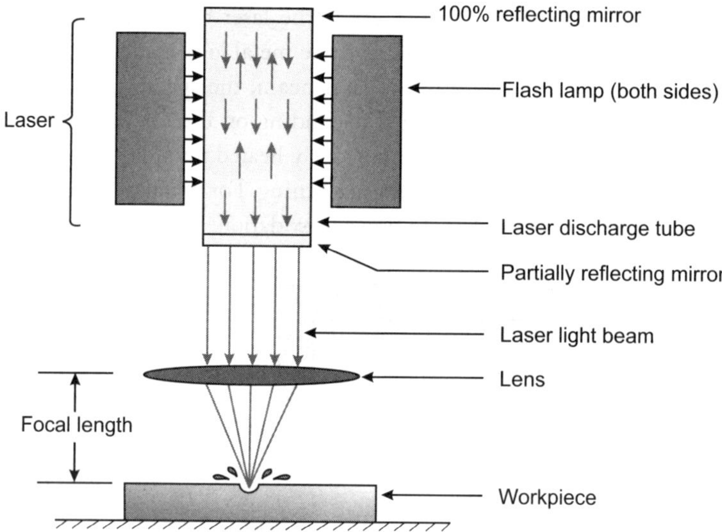

Fig. 4.11   Schematic illustration of the laser beam machining process.

4.1.5 | Heat treatment

This is one of the intermediate stages in an integrated cycle of materials processing of metals or ceramics.

> During heat treatment, a material is heated to a higher temperature, but much below the melting point or sintering temperature of the metal/ceramic. The heat treatment changes the microstructure and properties of a material.

The heating can be done under inert atmosphere or in ambient atmosphere inside a furnace. Depending on the nature of cooling, the heat treatment process can be generically classified into three different classes: (a) annealing: (the metal being cooled in the furnace itself), (b) normalizing (the metal taken out of the furnace and kept in ambient air atmosphere for normal cooling), (c) quenching (the metal taken out of furnace at the heat treatment temperature and placed directly into water bath/salt bath/oil bath for rapid cooling).

Annealing is alternatively known as softening treatment, as annealing increases ductility and reduces the hardness. Therefore, the annealed metal can undergo further bulk deformation processes. The annealing heat treatment is done at a temperature higher than recrystallization temperature, T (where, $T > 0.5 T_m$, where, T_m is melting point of base metal). Similarly, normalizing is alternatively known as air quenching or air hardening treatment. The term 'recrystallization' means the formation of strain-free finer newly crystallized grains due to the heat treatment of the deformed metal. In materials science, the crystallization refers to the formation of a solidified microstructure from the liquid melt via the 'nucleation-growth' mechanism. In contrast, the strain energy in the deformed metal is the driving force for the formation of the recrystallized microstructure.

Depending on the targeted microstructure and properties, a combination of heat treatment temperature, time and nature of cooling are employed for a given material. As far as the mechanical properties are concerned, annealing provides poor hardness due to coarse grain structure, while quenching can potentially provide better properties. The normalizing treatment provides intermediate properties. In case of metals, the mechanical properties, like strength or hardness scales with the microstructure. Therefore, as far as the microstructural length scale (e.g., grain size) is concerned, the microstructure of annealed material is characterized by coarser, normalized material by intermediate and the quenched material by finer grain size. However, due to rapid cooling, surface distortion takes place as well as residual stresses are generated. In order to relieve quenching associated residual stress, tempering is carried out by heating the quenched metal to a temperature 100–150°C above room temperature. The oil or salt bath quenching is preferred over water quenching to reduce severity of quenching processes.

> The distinction between annealing and other heat treatments (tempering/aging) is that the former always involves the heating and holding a metal above the recrystallization temperature, whereas the temperature is much lower than the recrystallization temperature in case of the latter.

The above-described heat treatment techniques are adopted as an intermediate or an intermittent step between successive deformation processes. For example, a continuously cast or conventionally cast product is annealed before being forged or rolled. Similarly, heat treatment processes are adopted in between the multiple passes of rolling or forging or extrusion. The intermediate anneal softens the metal so that hot/cold working can be further performed. Similarly, heat treatment is adopted after/before welding or machining operations to relieve residual stress or to obtain metals with desired microstructure and mechanical properties.

4.2 | Processing of Ceramics

In contrast to metals, the manufacturing of ceramic implants typically starts from fine powders. Various properties of the powders, including size, shape, flowability, are important. In the conventional manufacturing approaches, the powders can be pressed into a green body and subsequently, sintered under specific conditions to obtain a solid compact. For polymers, different moulding techniques like injection moulding or compression moulding are used to obtain various products. Some of the relevant processing approaches of polymers will also be discussed in a later section.

Technologically, the process of sintering can be described as powder metallurgy based technique to produce high density components from metal or ceramic powders by applying mechanical pressure and/or thermal energy.

> Fundamentally, sintering refers to the process of consolidation of a porous powder compact to a dense (nonporous) solid at temperature, $T > 0.5T_m$ ($T_m \sim$ melting point) and is assisted by diffusional mass transport. For biomedical applications, sintering parameters are often adjusted to obtain porous compact with desired pore architecture.

Typically, sintering involves compaction and consolidation of a powder compact (often containing sinter-aids) at high temperature. Alternatively, it can be described as a process of transformation from a porous state to a state of dense material and it must involve the process of neck formation among the powder particles. The relevance of sintering in the context of the consolidation of ceramics can be realized in reference to the following factors: (a) conventional melting–solidification route, as widely used for metals, cannot be applied in case of ceramics. This is in view of the fact that the melting points of ceramics are in the range of 2000–3500 °C and therefore, the use of furnaces with capability of operating at such extreme conditions is required. This is not a cost-effective engineering solution. (b) Ceramics are essentially brittle and lack any room temperature ductility. Therefore, the bulk deformation process, used for the metal forming processes (see Fig. 4.1), e.g., forging, rolling cannot be adopted to produce products with desired shapes in case of ceramics. (c) Also, low thermal conductivities of ceramics (2–50 W/m/K) cause large temperature gradients, resulting in thermal stress and shock during melting/solidification of ceramics.

4.2.1 | **Sintering mechanism**

Sintering can be accomplished using pressure (e.g., hot pressing, hot isostatic pressing) or without pressure (e.g., pressureless sintering in air/protective environment). At this point, it is important to note that sintering route is also adopted to produce porous ceramics for engineering applications, including biomedical applications. In order to fabricate ceramics with desired porosity, two strategies can be adopted: (a) powders can be mixed with volatile material (naphthalene, PMMA, PVA, sugar, etc.) before sintering. The idea is that the evaporation of volatile material will leave behind a porous structure. Thus by selecting the amount and size of the volatile material, a porous ceramic with desired microstructure can be obtained. This strategy enables one to control volume fraction and size of porosity. (b) Sintering at a temperature much lower than that required to obtain dense ceramic can also produce a porous material. The idea here is to prevent the complete removal of open porosity. So far as the distinction between closed and open pores is concerned, it has been widely observed that closed porosity is left in the sintered material, when density is above 95% theoretical density. Below 95% theoretical density, pores are open to free surface and often interconnected in the bulk of the sintered ceramics.

> The conventional processing approaches can only produce porous ceramics with uncontrolled pore size, shape and volume distribution. The additive manufacturing techniques are capable of producing patient-specific scaffolds with close control over pore architecture in 3D space.

From thermodynamic point of view, all the solid/vapour (s/v) interfaces in the porous compact are essentially replaced by solid/solid (s/s) interface (see Fig. 4.12). Total driving force for sintering is therefore the reduction in surface energy of the system. Any irreversible process with reduction in total energy of the system is thermodynamically preferred and this can be expressed as,

$$d(\gamma A) = Ad\gamma + \gamma dA < 0 \qquad (4.5)$$

where, A = surface Area, γ = interface/surface energy.

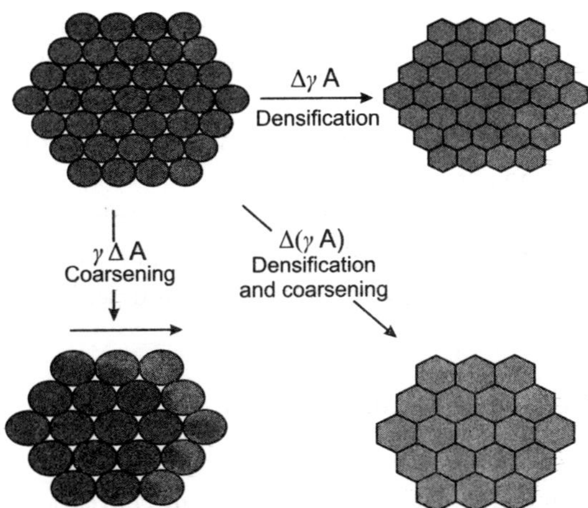

Fig. 4.12 Schematic demonstration of the sintering of a porous powder compact to a dense ceramic with equiaxed grain morphology.

Typically, s/s or s/v surface area is replaced by only s/s during initial/second stage of sintering and therefore, (Adγ) term dominates (see Fig. 4.12). The grain coarsening occurs during the final stage of sintering (if sintering time/holding time not optimized) and therefore, (γdA) term dominates at last stage of the process. Ideally, sintering of powder compact should lead to densification in the closest packed arrangement of hexagonal array of grains, evolving from a state of largely loose packed spherical shaped powder particles.

Three distinguishable stages can be identified in sintering of a powder compact:

(a) Initial stage

Initially, the particles (of diameter D) come in contact and form a weak cohesive bond. Vapour phase transport occurs due to the surface diffusion, leading to neck formation at the interparticle region. In the initial stage of sintering, vapour phase transport of atoms by surface diffusion takes place. It should be noted that at initial sintering stage, large pores are immobile and pin the grain boundaries, maintaining a small grain size. If held for longer time at initial stage, coarsening instead of shrinkage takes place for a powder compact.

(b) Intermediate stage

During continuous heating, neck growth effectively takes place due to grain boundary and lattice diffusion. In parallel to neck growth, the pore coarsening occurs, which results in the evolution of the interconnected pore channel. Towards the end of the intermediate stage, the continuous pore channel breaks and finally the pores are removed. The theoretical densities of around 85–92 % are achieved during this stage.

(c) Final stage

In the intermediate stage, most of the open pores are removed and in the last or final stage of sintering, the closed pores undergo shape changes primarily by grain boundary/lattice diffusion based mass transport mechanism. While being attached to grain boundaries, the isolated pores will be eventually removed. This would result in the attainment of 98-99 % theoretical density. If the sintering temperature and holding time (at final stage) are not appropriately selected, the closed pores become isolated. The improper selection of sintering temperature /time can lead to the simultaneous occurrence of grain growth and pore growth.

As far as sintering atmosphere is concerned, the different atmospheres generally used are hydrogen, nitrogen, argon and natural gas, vacuum and dissociated ammonia. In the following, the typical process methodology of consolidation of a powder compact is described.

4.2.2 | **Conventional processing of ceramics**

In this section, we will discuss the various process steps involved in obtaining dense ceramics using the standard powder metallurgical processes, including conventional and advanced sintering processes. Following the discussion in the preceding section, it is important to mention here again that in order to ensure faster densification, it is important to use agglomerate-free high purity ceramic powders with finer particle size. To enhance sintering, often a desired amount and type of binder or sinter-aid is used. This requires efficient milling, which is otherwise also used for two reasons; (a) to reduce the starting particle size powder feed stock, and (b) to mix second phase of a different composition with matrix phase.

4.2.2.1 | Ball milling

Ball milling is used to mix the elemental/prealloyed powders along with the grinding balls (milling media) to cause milling of powders, as shown in Fig. 4.13. Usually stainless steel, WC or other ceramic (Al_2O_3, ZrO_2, agate) balls and jars are used to mix and homogenize ceramic powders. Normally, the jars are made of agate or steel. The ball milling process leads to a reduction in particle size by comminution due to repeated crushing of powder particles by the balls, moving at high speeds. The physical mechanisms include extensive deformation (for metallic powders) and subsequent fracturing. During the initial stage, the particles break due to impact of the balls as well as due to the friction between the balls and particles. In the second stage, cold-welding occurs, which causes particles to adhere to 'the surface of other particles. Consequently, the particles are reduced in size and become spherical. As the ball milling progresses, the powders repeatedly flatten, cold weld, fracture and re-weld. The flattening of particles primarily occurs for ductile metals, which consequently work harden and become brittle. Thereby, successive impacts break the particles into fragments. Similarly, in case of ceramic powders, the particles undergo repeated fracture.

It must be noted that a critical speed is required for effective milling, which is inversely proportional to the internal diameter of the mill. If the speed of ball mill is lower, then ball-impact to cause milling is minimal. On the other hand, any speed greater than the critical speed causes balls to revolve with the container due to centrifugal force. Therefore, only critical speed causes cascading (frictional milling) and impacting (impact-milling) of the powder media. An ideal movement of balls and powder is shown in Fig. 4.13b. It should also be mentioned that ball to powder ratio (BPR) is critical in achieving the desired results. Ideally, the value of BPR can vary in the range of 4:1 to 10:1 for effective milling. Increasing the ball diameter or the number density of balls also leads to increase in the BPR.

Fig. 4.13 (a) Experimental set up of a planetary ball-mill, (b) schematic illustration showing the movement of balls and powders in a ball mill.

As mentioned earlier, effective ball milling can refine the powder particle size. The milling time needs to be carefully optimized to reduce total production time. Various reports suggest that depending on the type of the material, a steady state particle size is attained at longer duration of milling (24 or 48 hours). Any additional milling often causes agglomeration of fine powders which is not desirable.

To sum up, the ball milling variables include BPR, type of balls (harder than powders), liquid medium (acetone, toluene), or dry medium (air/vacuum/inert), mill speed. The contamination from milling balls as well as from milling vials is also of great concern as balls of lower hardness than powders can cause wear of balls. Dry milling in specific cases is preferred. In case of high speed milling, high temperatures are generated at particle/particle interfaces, which leads to localized temperature rise or oxidation of metallic powders. The milling medium (dry/wet) as well as milling time should be optimized to ensure good homogeneous mixing of binder/sinter additive with powders. There is an enhanced tendency to form agglomerates. Also, handling in vacuum/inert gas conditions and special chemical treatments are generally required. A typical problem experienced while handling nanopowders is the agglomerate formation due to Van Der Waals forces of attraction. Since the intercrystallite pores are of smaller sizes, they can be removed during the sintering process. However, it would be difficult for interagglomerate pores to be completely removed because of their large sizes. Any residual porosity in the sintered ceramic would act as potential sites for stress concentration, leading to cracking and therefore, degradation in mechanical properties.

4.2.2.2 | Compaction

The second step after the ball milling is the compaction, wherein the powders are pressed into the desired shape and size using hydraulic/mechanical press to obtain 'green compact'. Compaction can be carried out at room temperature and therefore is known as cold pressing.

4.2.2.2.1 | *Uniaxial Compaction*

Cold pressing involves compacting the powders in a tool steel die with the application of high uniaxial pressure. Cold pressing involves the pressure-mediated denser packing of loose powders, while being contained in dies/punches. Herein, the rearrangement of powder packaging followed by mechanical deformation results in compaction. Subsequently, the cold-pressed pellets can be densified to dense compacts via high temperature sintering.

A typical cold compaction cycle starts with filling the die with loose powders (see Fig. 4.14). This can be done via automatic feeding, wherein the die-cavity is filled by a hopper prior to the compaction. Then, uniaxial pressure is applied by the punch to compact the powders to obtain a green body. Correspondingly, one of the dies can push out the part for its removal after cold compaction. In order to obtain smooth outer surface of the cold compact, the die wall is often lubricated with steric oil. Also, the compaction pressure can be tailored to obtain a compact with sufficient strength for further handling. Typically, a cold compact has an apparent density of 50–55% theoretical density.

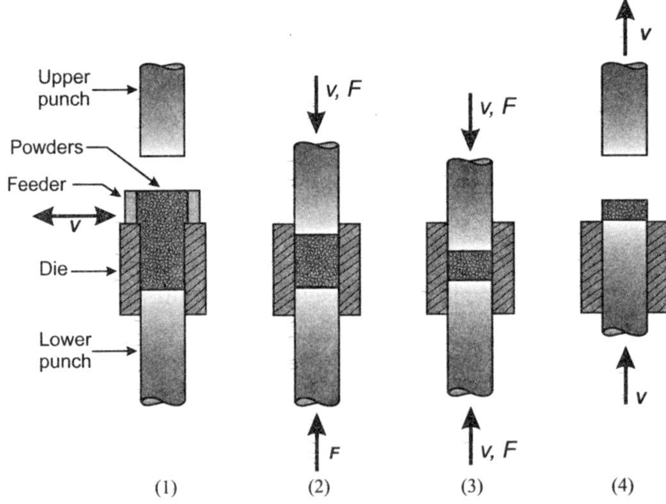

Fig. 4.14 Conventional method of compaction; (1) filling die cavity with powder by automatic feeding system, (2) initial, (3) final positions of upper/lower punches, and (4) ejection of part.

4.2.2.2.2 | *Cold isostatic pressing (CIP)*

The basic problem with the uniaxial compaction is the variation of apparent density across the thickness of the compact. Depending on the loading direction, the compact is denser along the pressure direction than the transverse direction. Therefore, in order to achieve a more uniform compact, cold-isostatic pressing (CIP) can be alternatively adopted. During CIP, a pressure of ~ 300 MPa is applied, while immersing the sealed container of powder/compact in water or oil. CIP is particularly useful for complicated shaped part without density gradient. The ceramic powder is sealed in a flexible bag (such as a metallic container) and submerged in a fluid to achieve uniform hydrostatic pressure in all the directions (Pascal's law).

The CIP process can be classified into two schemes: wet-bag and dry-bag processes. In the wet bag process, the ceramic powder is filled in a rubber mould, and then this bag is immersed in the fluid for compaction. The isostatic pressure around the powder compact is achieved as the pressure is transferred from pressurized fluid to the rubber mould. In case of dry-bag process, only radial pressure (along the horizontal direction) is applied between flexible mould and rigid shell, while the bag can rest on its top/bottom surface. Overall, CIP process is useful to obtain a better quality surface finish for complicated parts.

4.2.2.3 | Pressureless sintering

In pressureless sintering, a powder compact is heated to sintering temperature in a furnace and in the absence of external gas or mechanical pressure. Depending on whether the compact would be densified via solid state or liquid phase sintering, both sintering temperature and time are to be optimized for a given powder compact.

Generally, heat treatment cycle includes a first stage of preheating to allow removal of volatile or organic components entrapped within a powder compact. This stage is followed by sintering, where the ceramic component is held at sintering temperature for a specified dwell time. It is important to note that the selection of temperature and time of sintering, and the sintering environment decides the evolution of microstructure (porosity elimination and grain growth). Also, the cooling at the end of heating cycle induces thermal stress in the final component. High thermal stresses may also lead to cracking in the ceramic component and hence a proper control of cooling cycle becomes essential. Typically, furnace cooling is adopted to minimize the thermal stresses.

In industries, a continuous sintering furnace is adopted to densify a large number of samples in batch process. Here, the samples are fed from one side, and the conveyer belt speed is maintained in such a manner that each component receives identical sintering cycle of heating, holding and cooling down, when the component comes out from the exit side. In certain cases, pressureless sintering does not completely densify the material and in such cases, pressure assisted sintering (e.g. hot isostatic pressing, HIP) becomes essential.

4.2.3 | **Advanced processing of ceramics**

4.2.3.1 | Hot pressing (HP)

Hot pressing is a conventional densification process which utilizes simultaneous application of pressure and high temperatures to powder compact contained within a (graphite) die-punch assembly. High pressure and temperature synergistically allow rearrangement of particles, allow plastic flow and facilitate faster mass transport, leading to enhanced neck growth in vacuum/inert atmosphere. A picture of typical hot pressing unit and a schematic of the process are shown in Fig. 4.15. For HP, the ceramic powders are fed in a die, which is pressed with a pressure of ~ 10–50 MPa at temperatures ranging between 1000–2200°C for a certain duration (from minutes to few hours) to achieve a dense ceramic, often with densities of more than ~ 95% of the theoretical density. Most often graphite is utilized as a die material for containing the powder, as it can withstand high temperature and compact the ceramic powders at high pressures.

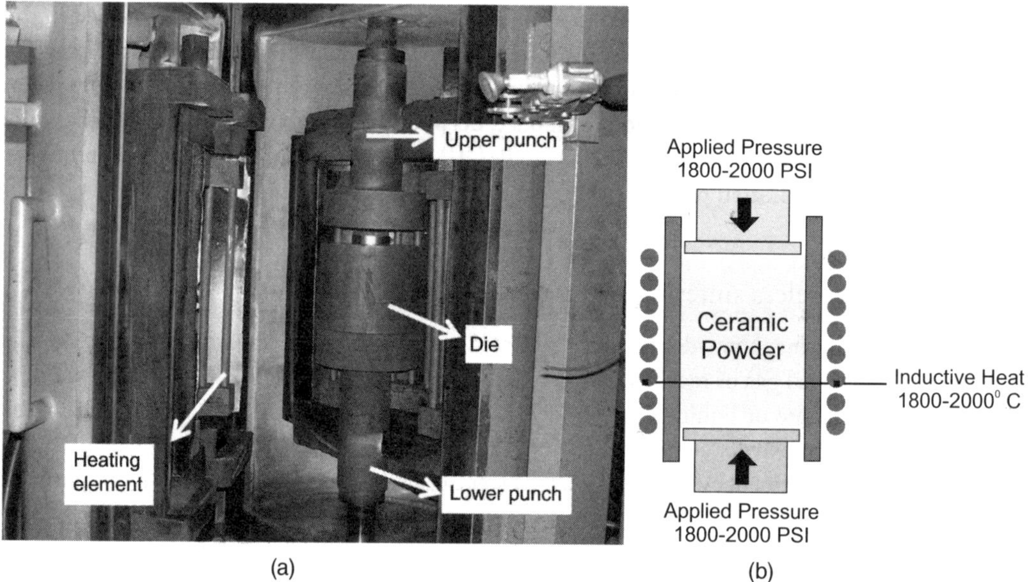

Fig. 4.15 (a) Industrial hot-pressing set-up and (b) schematic showing the heating and pressure application in HP.

4.2.3.2 | Microwave sintering

Microwave (MW) sintering is regarded as an advanced sintering process, which involves rapid heating of a powder compact to full density. In order to substantiate further on the advantages of such technique with respect to conventional pressureless sintering, it is instructive to mention that the heating rate of 100 °C/min or higher can be achieved for MW/SPS, while it is 2–5 °C/min for the pressureless sintering. This allows significant reduction in the overall process cycle time.

Volumetric heating is particularly more feasible in MW sintering route even for complicated shapes. The ceramic powder compact is contained in the microwave cavity for its consequent sintering and densification (Fig. 4.16). Since high frequency microwave causes skin effect by preferentially heating the surface, a combination of low and high frequency during microwave may be required to achieve uniform heating. Otherwise, strong thermal gradients can be generated within the body and cause immediate fracture. Hence, controlling the degree of microwave interaction specific to the absorption of the material should be considered. In addition, the ceramic requires an envelope of non-absorbing material to create insulation and limit the thermal loss. For insulating ceramics, like ZrO_2 or Al_2O_3, hybrid heating using SiC rods enclosing such powder compact in MW cavity can be adopted[457]. SiC can absorb MW radiation very fast and subsequently, transfers the heat to insulating ceramics.

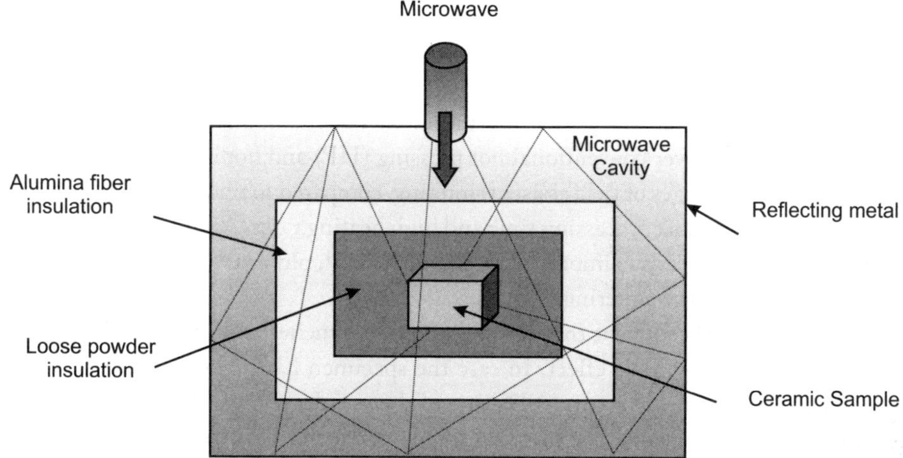

Fig. 4.16   Schematic of microwave sintering set up.

Summarizing, the advantages associated with the microwave sintering (MW) are short heating and cooling cycle i.e. rapid volumetric heating, which restricts grain growth during sintering. Unlike, conventional sintering, uniform heating of the material does not allow generation of steep surface to bulk thermal gradients in case of MW sintering.

Microwave sintering results in uniform volumetric heating of powder compact via oscillation of free electrons and ions via application of microwaves (frequency ~2.5-85 GHz).

4.2.3.3 | Spark plasma sintering

Spark plasma sintering (SPS) has emerged as a superfast consolidation route for metals and ceramics in last few decades. In this book, some case studies also report the use of SPS to process bioceramics. Therefore, this section exclusively discusses the SPS process as well as influence of various sintering variables on densification and properties of some model ceramics. The SPS process is based on the electrical spark discharge phenomenon, involving the use of a high current together with low voltage pulse momentarily generating spark plasma in the interparticle neck regions, resulting in optimum thermal and electrolytic diffusion[458]. Using SPS, the materials can be consolidated to full density at temperatures of 200–500°C lower than the respective conventional sintering temperature[458]. In view of high heating rates (100 °C/min), the entire heating cycle is completed in short periods of approximately 5–20 minutes, including temperature rise and holding times.

Phenomenologically, a large pulsed electric current is applied to heat the graphite mould and the powder compact, which is subjected to a modest pressure, as shown in Fig. 4.17. In view of the fact that plasma formation (local high temperature-state) during SPS heating remains to be conclusively proven/authenticated, other terms are often used in literature for this technique, such as pulse electric current sintering (PECS) or field assisted sintering technique (FAST). FAST process utilizes ON–OFF DC pulse, energizing of the powder compact. The repeated application of an ON–OFF DC pulse voltage to the powder compact leads to spark discharges at the particle/particle interfaces or at neck regions and the Joule heating is accomplished. The repeated spark discharges result in efficient sintering at low power consumption. For many materials, SPS process offers significant improvements over conventional hot pressing (HP) and hot isostatic press sintering[459]. Some technological advantages of field assisted sintering, compared to traditional hot pressing or hot isostatic pressing include short processing time and high heating rates. All these lead to minimizing grain growth, which often leads to improved mechanical[460,461], physical[462], or optical[463] properties, while eliminating the need of sintering aids.

During SPS process, the current through the graphite punches and dies and the pressing tool acts as a heating element by Joule effect. In case the specimen has higher electrical conductivity, the electrical current contributes to direct heating of the powder compact[542]. In contrast, graphite heating elements in HP surround the pressing tool and transfer heat by radiation/convection[464]. It can be reiterated that the above explained spark discharges cause high temperature between the particle contacts and voids.

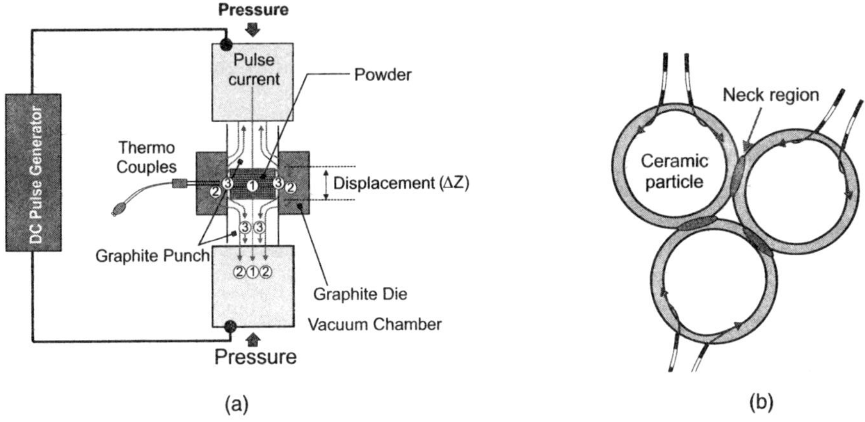

Fig. 4.17 Spark plasma sintering (SPS): (a) schematic showing the DC current flow during the sintering process, and (b) mechanism of temperature rise at the interface of particles due to Joule heating effect [see Colour Plate].

4.3 | **Consolidation and Shaping of Polymers**

In contrast to high temperature sintering process, the polymeric materials are consolidated at much lower temperature and typical processing temperature is most often below 300°C. For example, the thermoplastic polymers, like high density polyethylene (HDPE), can be moulded at 140°C. As far as the process parameters are concerned, both the temperature and pressure are important parameters. Both these parameters also influence the physical properties of the polymers, including shear rate, viscosity, which in turn determines the flowability of the polymer at the moulding/extrusion temperature. With easier flowability, more complex shaped polymeric parts can be fabricated. Depending on the composition or shape of the polymeric product, different processing techniques, like compression moulding, injection moulding or extrusion moulding can be efficiently used. The salient features of these processes are discussed below.

4.3.1 | Extrusion and melt compounding

Extrusion is one of the widely used polymer processing technique which has the capability to blend multiple polymers or to fuse different additives (antioxidants, UV stabilizers, etc.) / inorganic fillers with the base polymer. In this process, the molten polymer is sheared between two screws and is conveyed through the die. The polymer granules are fed into a heated barrel through hopper. During the extrusion process, various ingredients in the material are forced to interact closely, while undergoing size reduction (dispersion), to create homogenous mass as a result of weak adhesive bonds. The melted polymer is then forced through a die to form a continuous structure (see Fig. 4.18). The die design determines the shape of the extruded structure (continuous strand/polymeric thin films).

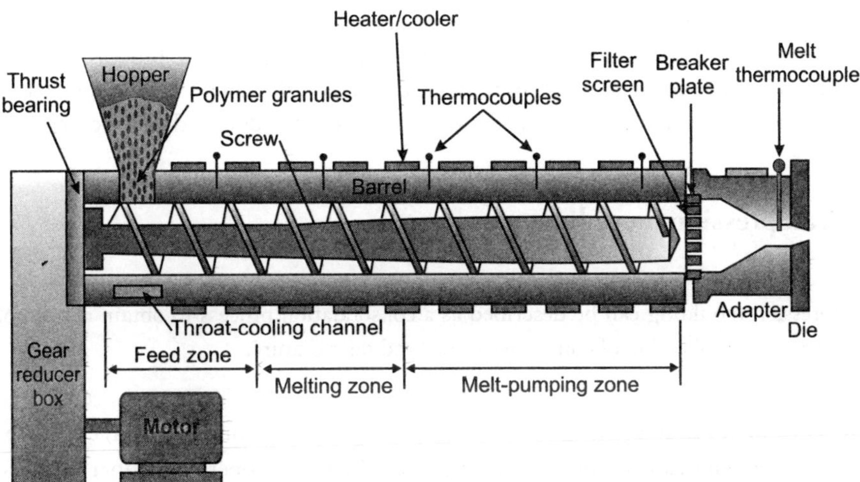

Fig. 4.18 A schematic of extrusion process to extrude polymeric rods [see Colour Plate].

The extrusion process is typically accomplished using single-screw or twin-screw extruder (TSE). The basis for a TSE is two Archimedean screws positioned in a parallel configuration. The use of dual screws as opposed to a single screw enables a degree of positive conveyance as a means for material transport in contrast to a single screw's drag flow mode of conveyance also enabling more uniform distribution of shear in the material. Two distinct design approaches were successfully implemented and characterized by the direction of rotation of the screws in relation to one another, namely co-rotating, where the screws rotate in the same direction and counter-rotating, where the screws rotate in the opposite direction. In a co-rotating TSE system, the progress of material through the extruder is highly controlled and such a system can be used to conduct sophisticated chemical reactions during polymer processing.

Also, the twin screws can be intermeshing or non-intermeshing, depending upon the distance between the shafts. If the distance between the shafts is less than the screw diameter, they are called intermeshing and if the distance between the shafts is equal to the screw diamter, they are called non-intermeshing.

The extrusion set up consists of a motor, an extrusion barrel, rotating screws and an extrusion die. The motor acts as a drive unit of the extrusion system. Various processing parameters like screw speed, temperature and thus the pressure can be controlled from a central electronic unit attached to the melt compounder. The extruder rotates the screws with a predetermined speed, simultaneously compensating the shear and torque generated from the extruded material. The screw speed is set at a sufficient rpm to generate enough power to prevent the extruder from surpassing its torque limit and shutting down.

The earliest generation extruders built in the 1950s had a mean residence time of 300 seconds. During the past 50 years, the residence time has been reduced to half its original value every ten years or so with a steady increase in shear capacity. Since the 1900s, the state-of-the-art extruders have the capability to reduce the residence time to less than 10 seconds and mean shear rate to greater than 1000 s^{-1}. Nowadays, commercially available melt extruders are made up of materials that are resistant to high temperatures, wear of abrasive components and a wide range of pH. They can go upto temperatures as high as 450°C and screw speeds of 250 RPM. The volume of material that can be processed can range from 3 to 15 ml. The prior knowledge of polymer properties, like melting temperature, viscosity and degradation behaviour, is important to set the process variables in the melt extruder. It is noteworthy to mention that one can record the temperature dependent polymer properties like melt torque, melt viscosity, shear stress and shear rate during the melt extrusion process.

4.3.2 | Compression moulding

> The compression moulding can be described as a consolidation process to obtain dense polymer with simultaneous application of compressive load and temperature.

In general, a mould in the compression moulding consists of four components: (a) cavity, (b) core or plunger, (c) retainer plates, and (d) heaters (see Fig. 4.19). Before the operation, inner surface of cavity component area and the outer surface of core component area are cleaned to avoid the component sticking either in the core or cavity. After cleaning, the heater with power supply is connected to a controller. Core retainer plate is clamped on the upper plate of the machine and cavity retainer plate is clamped on the base plate of the machine. Both, the core and cavity are to be closed with proper alignment. The heater is switched on and the desired temperature is set. Depending on the size and shape of the final product, the compression moulding parameters, e.g., temperature, time, pressure and ramp rate have to be optimized. This can be further explained on the basis of recent experimental results. While developing HDPE–20% HA– 20% Al_2O_3 based acetabular sockets for total hip joint replacement applications, the cylindrical shaped 10 mm diameter sample was moulded at 130–140°C in a lab-scale experiment. While fabricating the prototype, a typical patient-specific acetabular socket design with outer diameter, inner diameter and thickness as 44 mm, 28 mm and 8 mm, respectively was used. For moulding such socket prototype of the same polymer composite composition [(HDPE–(20% HA–20% Al_2O_3)], the optimal compression moulding conditions were a temperature of 190°C with mould pressure varying in the range of 3–7 MPa. As far as the total time taken in a compression moulding cycle is concerned, the time taken to reach the desired temperature, e.g., 185–190°C is around 1 hour and total cycle including cooling is around 2.5–3 hours. This specific example signifies the optimization of compression moulding parameters depending on the volume/shape/size of the polymer to be moulded.

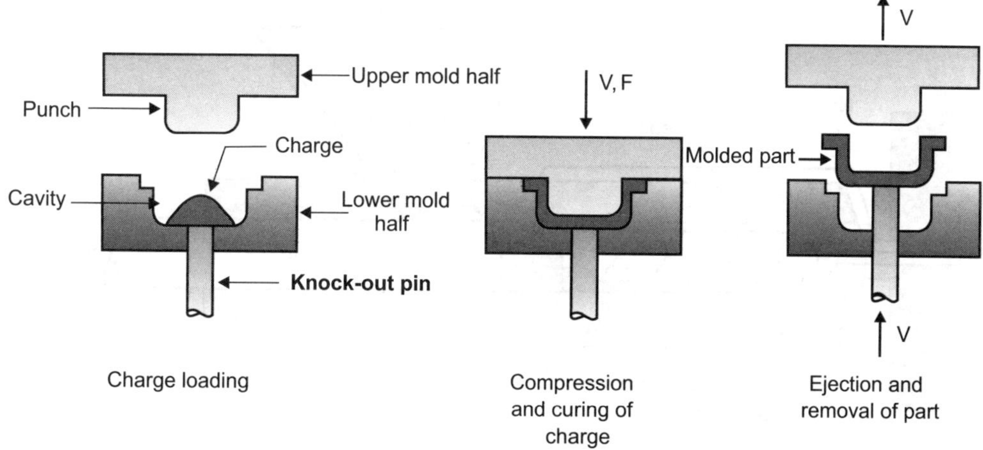

Fig. 4.19 Compression moulding steps for thermosetting plastics: loading of charge followed by compression and curing of charge; then ejection and removal of the part.

4.3.3 | Injection moulding

The process of injection moulding utilizes the shape forming capability of engineered polymers and the similar process is also adapted to fabricate metals, ceramics or composites with desired shape and size.

Another polymer shaping process is injection moulding. The injection moulding is often attached to melt compounding facility, which is used either to mix co-polymer or mix polymer with inorganic fillers. Once a polymer mix is compounded in a melt compounder machine, the product is fed to the injection moulding set up in an integrated processing approach. Alternatively, injection moulding is also used as a stand-alone or independent shaping facility for polymers.

Various advantages of injection moulding include its ability to obtain: (a) small and extremely complex shapes, even from difficult to machine materials, (b) moderate to high volume parts, (c) near net-shaped product. As far as the process methodology is concerned, the polymeric powders are mixed with appropriate binders and they are premixed. Subsequently, they are pelletized and fed into an injection moulding machine. The moulded component undergoes debinding process, which can be either thermal or solvent-based. For ceramics or metallic components, the additional step of sintering is followed. The entire sequence of process methodology is schematically shown in Fig. 4.20.

The injection moulding utilizes the granular polymer (mostly thermoplastics), which is fed by gravity into a heated barrel through hopper. A screw-plunger is used to slowly move forward the granules. As the plunger moves forward, the melted material is pushed through a nozzle and charged into a mould cavity. In contrast to the compression moulding, the mould in injection moulding is kept cool so that the molten polymer solidifies as soon as it fills the cavity. A typical injection moulding machines consists of a hopper, a screw-plunger, and a heating unit.

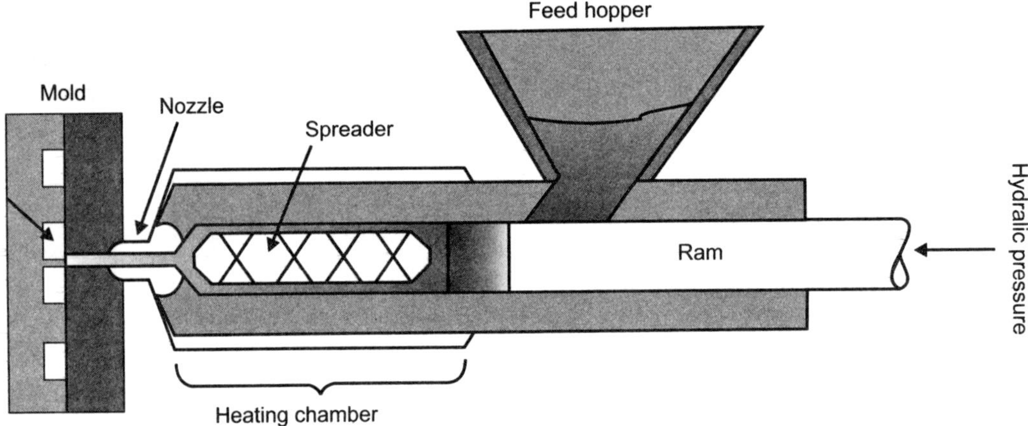

Fig. 4.20 A schematic of injection moulding process.

At closure, more than one single processing approach is often used to fabricate polymers with various shapes and composition. For example, a melt-mixer together with injection moulding is used to process shaped polymers with homogeneous bulk chemistry and properties. In the next section, near-net shape fabrication technique is discussed in particular reference to additive manufacturing, which has lately received global attention. It is also instructive to mention that the conventional processing approaches, as discussed above, cannot be used to fabricate patient-specific implants.

4.4 | Patient-specific Implant/Scaffold Fabrication using Additive Manufacturing

Conceptually, additive manufacturing technology is a layer-by-layer fabrication technology of three-dimensional physical models directly from computer-aided design (CAD)[469]. Additive Manufacturing (AM), also known as 'Rapid Prototyping' is considered as the next global industrial and technological revolution[470]. In most sectors of industry, the utilization of AM is possible and might lead to a radical change of production processes. The outcome of this will be cost reduction and more customer-specific solutions. It has been claimed that additive manufacturing can cut down the product costs by up to 70– 90%.[471,472] In the medical field, AM will be a crucial factor for the implementation of the 'Personalized Medicine' concept, aiming for more individual and therefore patient-specific therapeutic options[473].

As will be discussed in this section, a complex internal pore network can be constructed using the additive manufacturing process[474]. Furthermore, the build-on-demand nature of additive manufacturing provides the flexibility to create a patient-specific implant with custom-made architectures[475,476]. Therefore, in order to reconstruct the damaged tissue, scaffolds fabricated by additive manufacturing methods are reported to be more advantageous over conventional 3D scaffolds. This is in view of the fact that additive manufacturing allows the incorporation of drugs/proteins as well as cells during scaffold manufacturing to produce complex architecture similar to bone[477].

Medical applications of additive manufacturing (AM) are second largest only after automobile sector. This indicates the potentially as well as high commercial feasibility (market readiness) of existing AM technology to fabricate with metallic and ceramic materials as well as synthetic polymers. With these technologies, implants with patient-specific size and shape can be manufactured, based on three-dimensional imaging techniques like computer tomography (CT) and magnetic resonance tomography (MRT). Such implants in most cases aim for a permanent replacement of a missing tissue, like in case of skull implants made of titanium. Concerning the economical aspect, AM represents a potentially growing market in many manufacturing sector with a global market of $1843.2 million in 2012 and expected to grow at a CAGR of 13.5% to reach $3,471.9 million by 2017[478].

Evolved as an alternative to the conventional scaffold fabrication technologies, additive manufacturing has been used in the biomedical engineering field, mainly for producing models and prototypes[479,480]. Depending on the applications, various additive manufacturing technologies are available and each technique has specific advantages[481,482]. On the basis of ASTM standard F2792-12a, additive manufacturing methods are categorized in seven categories: binder jetting, direct energy deposition, material extrusion, material jetting, powder bed fusion, sheet lamination, and vat photo-polymerization (see Table 4.1). In this chapter, we discuss 3D powder printing (3DPP) and 3D plotting (3DP), which fall under binder jetting and material extrusion (see Table 4.1). While additive manufacturing method was primarily used to produce models and prototype parts in civil engineering applications, the advancement of this technology in human health care provided biomedical engineers to create larger and more complex implants/scaffolds to treat skeletal disorders[483,484].

In the context of orthopedic applications, selective laser sintering (SLS) and electron beam melting (EBM) methods are well recognized due to their ability to produce patient-specific metallic implants with high architectural complexity.

SLS/EBM techniques employ high energy to consolidate or melt the material and these methods are only limited to the fabrication of monolithic materials (mostly metals/alloys). Although these designed implantable devices perform seamlessly in reference to bone architecture, it is desirable to produce a device that mimic the bone microstructure and compositional details. Apart from this, the application of methods like EBM and SLS is restricted in biofabrication due to the fact that cells/proteins/growth factors can get denatured at high processing temperature.

In the above broad perspective, additive manufacturing methods, especially, 3D plotting (3DP) and 3D powder printing (3DPP) are considered as more appropriate for the fabrication of implantable 3D porous scaffolds at low temperature. The following discussion revolves around a critical analysis of the physico-chemical aspects of binder-material interaction and related parameters.

It needs to be reiterated here that the fabrication of scaffolds with desired porous architecture requires the adaptability to additive manufacturing processes. It is worthwhile to mention that numerous conventional techniques, including solvent casting with particulate leaching, fibre bonding, membrane lamination, freeze drying, and gas foaming were developed to fabricate porous scaffolds[485]. However, these techniques are incapable of precisely controlling pore size, geometry and interconnectivity between pores, and therefore cannot be adapted to fabricate patient–specific scaffolds[486]. In addition, the size of the scaffold is often limited due to difficulties in removing porogens and some of these conventional techniques use toxic organic solvents that may partially remain in the scaffold post-processing. Most of the above-listed problems can be overcome by additive manufacturing methods[487]. Probably, the most significant advantage of additive manufacturing methods over conventional scaffold fabrication methods is that additive manufacturing methods can be used to produce more intricate, yet selectively designed external and internal geometries[488].

> The patient-specific scaffolds/implants with precise control on the porous architecture (size, shape and interconnectivity) can be manufactured using an additive manufacturing technique.

Table 4.1 summarizes the most frequently used fabrication methods.[489] Stereolithography (SLA), selective laser sintering (SLS), fused deposition modelling (FDM), laminated object manufacturing (LOM) represent some of the promising AM techniques, which can be used in manufacturing of implantable biomaterials. The comprehensive reviews about the principles of these methods and their application in biomedical engineering and tissue regeneration are reported in literature[490,491,492]. Since, these techniques are not discussed in this chapter, a brief overview of the capability of some of these AM techniques to produce implants in various materials systems. Most popular techniques used in the 3D printing of metals and alloys are EBM, SLS, laser engineered net shaping (LENS), and direct metal laser sintering (DMLS). For example, the device fabricated by electron beam melting (EBM) of Co26Cr6Mo0.2C powder was characterized by a hardness of 4.4 GPa, and corresponding yield and ultimate tensile strength of 0.51 and 1.45 GPa, respectively[493]. The Ti-alloys (Ti6Al4V) are in particular extensively used in limb salvage applications due to considerable osseointegration properties and also due to its higher strength to weight ratio as compared to other metals such as stainless steel[494,495]. In a different work, Ti24Nb4Zr7.9Sn solid components fabricated by EBM displayed a hardness of ~2.5 GPa, which was close to the hardness (~2.3 GPa) of the component made by the selective laser sintering (SLS) method[496]. It is however important to reiterate that the above-mentioned methods use the high energy probe to melt the powder particles, followed by joining during 3D printing[497]. This involves a sharp increase in temperature, followed by a fast cooling, which results in inhomogeneous microstructures[498]. Due to involvement of very high temperature, these methods are only limited to the fabrication of scaffolds of metals and alloy and not found suitable for the composite materials (e.g., ceramic–metal and polymer–metal or polymer–ceramic) due to large difference in the melting temperature of the powder components. In this context, low temperature processing methods like 3D powder printing or 3D plotting are found to be more viable to produce the components of composite materials.

In particular, 3D powder printing and 3D plotting methods are extensively used to fabricate the scaffolds at room temperature. Also, post-processing, such as sintering or chemical treatment can be utilized to tailor the properties of the scaffolds.

Table 4.1 Summary of various rapid prototyping techniques as well as scaffold features that can be obtained for designed porous scaffolds[543].

Additive manufacturing categories	Technique	Principle	Parameters	Materials	Advantages	Disadvantage
Binder jetting	3D powder printing	Using binder: The inkjet print head droplets of a binder fluid on a powder bed. This fluid binds the powder and thus builds up part of the solid's cross section. This process is repeated until the 3D structure is printed.	Layer thickness: 20–100 µm Smallest feature: 350–500 µm.	Metals, ceramics, polymers	Speed, low cost	Fragile parts, Post processing required
Direct energy deposition	Electron beam melting	Using Electron beam: Electron beam initiates melting of a thin layer of metal powder that is repeated for each layer.	Layer thickness: ~50 µm Smallest feature: ~100 µm	Metals	High mechanical properties and high accuracy	Very high local temperature
	Selective laser sintering	Using Heat: Laser beam Selectively initiates melting in a thin layer of powdered material. Iterative repetition for every layer. Unmelted powder serves as support structures and remaining powder is removed	Layer thickness: 76–100 µm Smallest feature: 45–100 µm	Metals, ceramics, polymers	High mechanical properties and high accuracy	Very high local temperature
	Laser engineered net shaping	Metal particles are injected into a laser beam, melted and deposited onto a substrate. The trace of the laser beam on the substrate is driven by the CAD models until the desired component is produced.	Ability to tailor deposition parameters to feature size for speed, accuracy, and property control.	Metals	Wide range of alloys	Does not meet exact specifications

Additive manufacturing categories	Technique	Principle	Parameters	Materials	Advantages	Disadvantage
Materials extrusion	3D plotting	Extrusion of pasty biomaterials: Strands of in principal any pasty material are deposited layer by layer on the surface of the working platform. The pastes are extruded either by air-pressure or with the help of a syringe and a step-motor	Layer thickness: 40–1000 µm Smallest features: 100 µm	Blend of metals and/or ceramics with polymer	Wide range of materials can be used	No support for overhanging structures
	Fused deposition modelling	Using Melt: Thermoplastic fibre is heated and selectively extruded via a nozzle layer by layer. Small feature size of scaffolds allows the fibre to bridge across unconnected parts without support structures.	Layer thickness: 250–370 µm smallest feature: rod diameter: 260–700 µm	Polymer	Low cost	Limited to thermoplastic polymers
Sheet lamination	Laminated object manufacturing	Using Sheet and a laser to cut the sheet: Sheet is adhered to a substrate using a heated roller, followed by cutting of sheet using laser based on CAD design	Layer thickness: 50–500 µm	Paper, polymers, composites	High speed process and large parts can be made	Dimensional accuracy is low
	Ultrasonic consolidation	Using ultrasonic energy to join the metal sheets on the basis of CAD design. The excess material is then trimmed out using CNC machine.	Layer thickness: ~100 µm	Metals like aluminium	Low temperature process. Can be used to join dissimilar material, high dimensional accuracy	Narrow range of material can be printed

Additive manufacturing categories	Technique	Principle	Parameters	Materials	Advantages	Disadvantage
Vat photo polymerization	Stereolithography	Using Laser: Laser beam selectively initiates solidification in a thin layer of liquid photopolymer. Iterative repetition for every layer. Requires support structures for unconnected parts.	Layer thickness: minimum feature size of 100 nm is easily producible	Reactive resin	High accuracy	Only reactive resin can be used. Out of these mostly toxic
	Digital light processing	Using Light: This is similar to the stereolithography in which photopolymer used for the printing. The digital light processing method uses conventional light source in place of laser.	Layer thickness: minimum feature size of 100 nm is easily producible	Reactive resin	High accuracy, low running cost, less waste	Only reactive resin can be used. Out of these mostly toxic

4.4.1 | 3D powder printing

The importance of 3D powder printing (3DPP) lies in its ability to fabricate the devices with complex architecture for bone tissue engineering applications at room temperature using a binder[499,500]. 3D powder printing employs inkjet printing technology for processing powder-based materials[501,502]. During fabrication, a print head is used to spray a liquid onto thin layers of powder and such process is intended to follow the contour or design profile generated by the computer aided design (CAD).

> Phenomenologically, any additive manufacturing process involves the creation and slicing of the physical model, followed by the layer-by-layer fabrication process.

Significant research has been pursued in 3DPP to evaluate and optimize the parameters, like powder bed thickness[503,504], binder and powder composition[505,506], surface roughness and flowability of powder[507], interaction of powder and binder, etc.[508].

The description of 3D printing process has been reported in some of the review papers[509]. Based on such reports as well as our understating, the entire 3DPP process can be considered to involve six sequential steps, as discussed below, with the first three being common to any additive manufacturing method and rest are specifically followed for the 3DPP process. The functioning of 3DPP process along with the scientific description of the powder printing is schematically shown in Figure 4.21. In the following, 3DPP is described in terms of five sequential steps.

STEP 1: A CAD model is created digitally or captured from a physical object by digital means, for example, a CT scanner. Afterwards, the region of interest from the CT scan data can be extracted and converted into 3D design by commercially available softwares, like MIMICS and CATIA. For a patient specific scaffold, the CT scan of the defect part is acquired, which is followed by data separation from CT scan of defect part using the MIMICS software. It can be noted here that CT images are a pixel map of the linear X-ray attenuation coefficient of tissue. Depending on the bone density, CT scan data in the form of DICOM consist of two dimensional grey scale images of bone. The Hounsfield Units (HU) corresponding to each element is averaged and is converted into grey values. The pixel values are scaled so that the linear X-ray attenuation coefficient of air equals -1024 and that of water equals 0, on a scale known as the Hounsfield scale. The Hounsfield unit (HU) scale is a linear transformation of the original linear attenuation coefficient and represents a method for defining the radio-density of a scaffold (x). HU is defined as follows:

$$HU = \frac{\mu_x - \mu_{water}}{\mu_{water} - \mu_{air}} \times 100 \tag{4.6}$$

where, μ is the X-ray attenuation coefficient.

STEP 2: The STL file is sliced digitally into a large numbers of cross-sectional layers. Because the CAD model is numerically represented in an STL file, it is possible to slice the model into layers with each representing a cross-sectional contour of the CAD model. The layer thickness is an important parameter in the slicing process, because it affects both model accuracy and build time. The lowering of layer thickness increases the model accuracy, but would increase the build time.

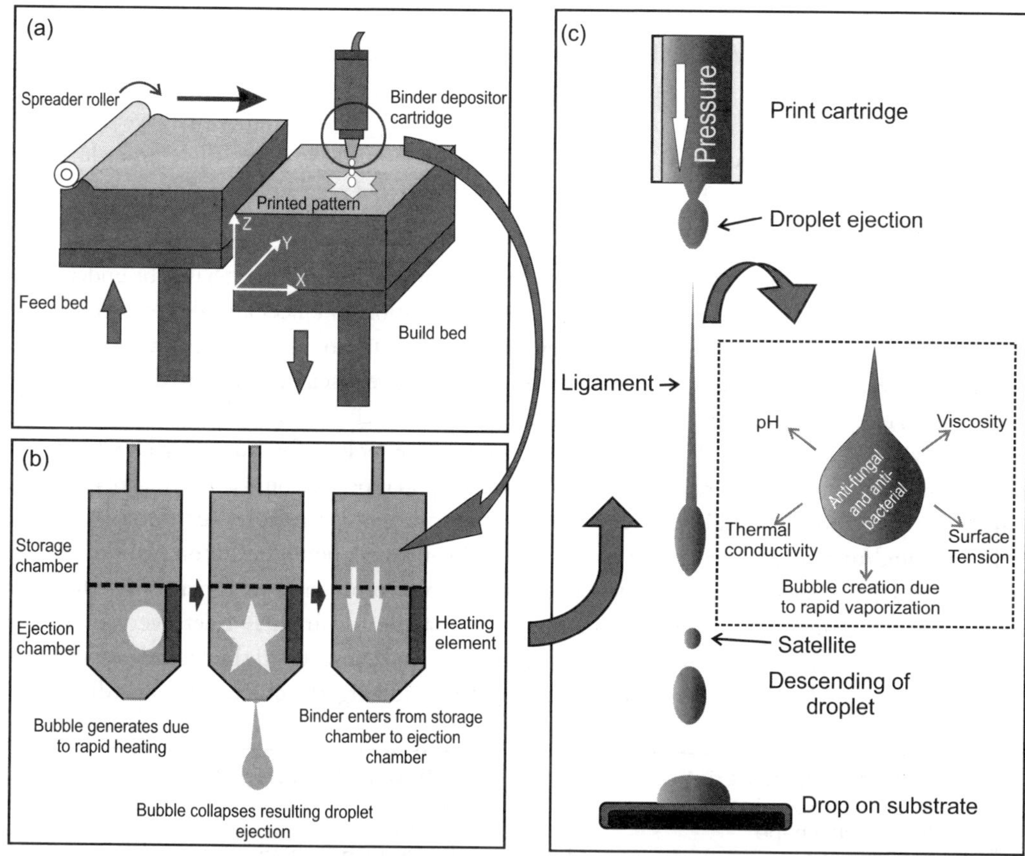

Fig. 4.21 Schematic illustration of 3D printing process: (a) the sequential stages in bubble formation inside printhead, (b) and the binder droplet ejection and subsequent interaction with powder bed, and (c) the various physical parameters of relevance of the binder are shown in inset[543] [see Colour Plate].

STEP 3: After pre-processing, the digital data of the sliced layers are sequentially sent to the 3DPP machine having an interface with a computer. The 3DPP machine drops binder solution ('ink') onto a layer of loose powder to form a solid cross-section (see Figure 4.21). The powder adheres firmly where the binder solution is deposited. The surrounding powder does not bind and serves as self-support. 3D printing has been shown to lead to anisotropic material properties. This is in view of the fact that cross-sections are printed in continuous strips along the x-axis, powder layers are prepared along the y-axis and laminated layers along the z-axis, as shown in Fig. 4.21. The prototype is built as a layer at a time on top of previously laid layer. Once a layer is finished, the build chamber is lowered down by a distance of one powder layer thickness. The process repeats itself until the entire model is completely printed.

STEP 4: Finally, the model is cleaned to remove loose powder. Depending on the design of a CAD model as well as the nature of the 3DPP process, the support material may be required during the fabrication, and the removal of these support materials must be appropriate. This step calls for the most attention and perfection as the interior details should be perfectly revealed to achieve the actual 3D printed features. While cleaning of the outer surface is relatively easy, e.g., by blowing air; the removal of loose powder particles from inner parts (e.g. small pores <500 μm intended for blood vessel ingrowth) is much more demanding. This may require additionally wet methods, such as ultrasonication or microwave assisted boiling of the sample. However, these procedures are only possible in liquids, which do not dissolve the binder to avoid mechanical disintegration of the printed sample.

STEP 5: Depending on the scaffold composition, the post-processing treatment has to be undertaken to obtain scaffolds with desired porous architecture, composition and mechanical properties. Some of the commonly used post-processing treatments include infiltration, high temperature sintering, chemical conversion or a combination of both. This step will be discussed in more details later.

Summarizing, 3D powder printing (3DPP) can fabricate mechanically stable scaffolds of nearly any shape with high resolution. It employs ink-jet printing technology for processing powder-based materials in ambient temperature. During fabrication, a liquid is printed onto thin layers of powders. Subsequently, the powder bed is lowered, allowing the spread of the next powder layer. After the build, unbound/unreacted powders are vacuumed or blown away upon process completion, leaving the finished part. Depending on powders and binders, post processing treatments like chemical infiltration/high temperature heat treatment process may be required. A brief overview of the different factors which influence the structural, mechanical and consequently biological properties of 3DPP/3DPL scaffolds is summarized in Fig. 4.22. In the following, the process is briefly discussed.

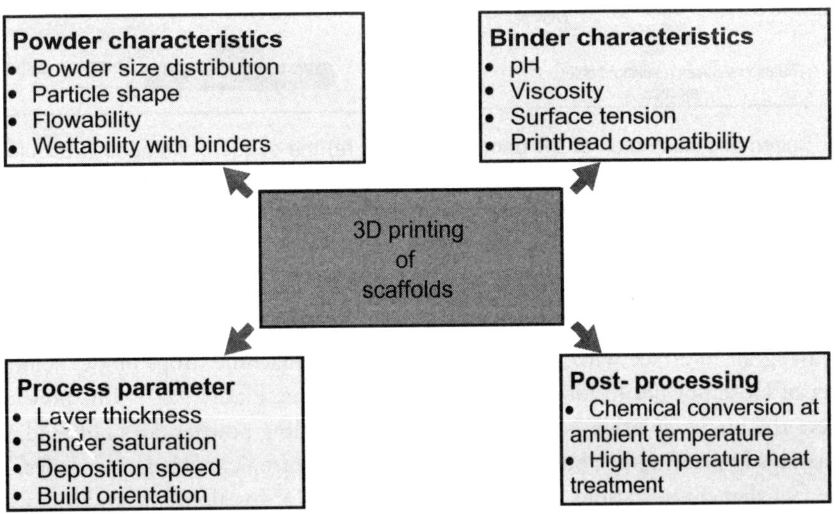

Fig. 4.22 Important parameters to be considered prior to 3D powder-printing of scaffold[543].

4.4.1.1 | Fundamentals of thermal inkjet powder printing

The foremost requirement for inkjet powder printing is the compatibility of the ink (physico-chemical properties) with the printer components[510]. Depending upon the printing technology, the ink/binder ejection system can be divided into two broad categories: continuous inkjet printing (CIJ) and drop on demand (DOD). The CIJ is used in commercial purpose printing, while the DOD is used in small scale printing (e.g. laboratory) due to minimum waste of ink as well as higher resolution than CIJ. A typical DOD based inkjet printer can generate the ink drop in the volume range of 1–100 pl. Depending upon the mechanism involved in the drop ejection, the DOD print head can be further divided into thermal and piezoelectric print head. In both types of print head system, the ink drops are ejected by the pressure generated in the cavity filled with ink just behind the orifice from where ink drops are ejected (Fig. 4.21). In case of thermal inkjet printing, this pressure is generated due to rapid heating at 200–300°C, which is achieved in 2 μs[511,512]. Bubbles form and collapse due to that rapid heating. The temperature due to bursting of bubbles inside the printhead can be estimated from the following expression,

$$E = TC_p V_{drop} \tag{4.7}$$

where, C_p is the heat capacity of the ink, V_{drop} is the volume of the drop, and E is the energy supplied by the heating unit[510,513]. An implication of the above equation is that for a given energy input to the inkjet system the ink drop volume will be smaller if the temperature increases locally and vice versa.

As far as the mechanism of DOD is concerned, the bubble formation pushes the ink outside the cavity, while the collapsed bubble locally evacuates the space which pulls the ink from the ink reservoir into the cavity (see Fig. 4.21). The process of heating the ink, followed by bubble formation and drop ejection is completed within 80–200 μs[514]. The mechanistic description of DOD process can be discussed in reference to the following points[509]:

(a) Electrical signals are used to control the ejection of an individual droplet.
(b) The pressure pulse created causes a small drop of ink to be ejected from the nozzle, followed by capillary-driven refilling of the cavity.
(c) The printing speed is limited by the frequency of drop ejection approximately 5–10 kilohertz, which correspond to a linear speed of 0.1 meter/sec for a drop size of 20 microns. Thus, a 0.5 meter by 0.5 meter part could be printed in 5 seconds/layer.
(d) The printing rate determining step is likely to be a powder or binder drying time. For the realistic printing speed of 7 layers/min, built up rate will be 0.01 m/hour, if layer thickness is 25 μm.
(e) The 3D printing resolution is defined as the minimum drop size, whereas the final resolution of the printed structure is dependent on the way in which the drop spreads on the substrate. As shown in Fig. 4.21, drop size on the other hand is a function of multiple variables, like binder viscosity and surface tension[510].

4.4.1.2 | Binder for 3D powder printing

Most of the commercially available 3D printing units use a thermal inkjet head to print the binder on powder bed and they used mostly water based binder[515]. The binding mechanism can be mainly of two types:

Physical: There is no chemical reaction between the binder and powder as the particles are held together by adhesive action of the binder. Polymer-based materials (cellulose, polyvinyl alcohol, dextrin, etc.), either suspended in a solvent or dispersed in the powder, are used as binder.

After printing, the solvent is evaporated and the particles get bound together by means of the adhesiveness[516].

Chemical: The binder and powders are chosen in such a way that they react as soon as they come into contact. One such example is the use of acidic binder to print CaP based materials. The powder partially dissolves into the binder and recrystallizes to form an entangled crystal network[517,518,519,520]. Another commonly used binder is water for binding POP (Plaster of Paris) by conversion into gypsum[521,522], as water is not harmful for print head components. POP is not initially converted into the dehydrate during printing. This being not fast enough, commercial suppliers always add polymer additives (e.g., gelatin, cellulose) to the POP to achieve binding. To print a new powder, the suitable binder has to be optimized for the printing procedure. Binders can leave a residue that contributes either partially or fully to the final strength of the part.

It is important to mention that there is no 'universal' binder, which can be used for printing all classes of materials (metals/ceramics/polymers). Therefore, significant research efforts must be directed towards the formulation of such binder.

The most important properties controlling the size and shape of binder droplet and jet reliability are the surface tension and rheology of the binder. Practically, four types of liquid have been reported to be used as the binder in the 3DPP system for processing powdered biomaterials, which can be tailored to have a wide range of surface tension and viscosity[516]: (a) water solution suspended with polyvinyl materials; (b) chloroform solvent; (c) acidic solution; and (d) water-based liquid. The suspended solids within the polymeric solution may increase the rate of nozzle wear and the chloroform solvent/acid solutions can cause the degradation of heating elements. The use of these liquids in the thermal print head may reduce the working life of the print head. In practice, it is therefore restricted to water-based media (>98% water), which is also the commercially used binder in the 3DPP system.

4.4.1.3 | Binder formulation and properties

As discussed in this section, the binder composition needs to be tailored while considering a number of factors. In reference to Fig. 4.21, some key requirements for the binder systems are as follows.

(i)   The binder solution must have sufficient concentration, while still having a low viscosity (~1.35 cps), allowing easy ejection from the print cartridge micro-orifices.
(ii) The binder must not dry or cure rapidly before the next layer of powder is added to the build structure.
(iii) pH value should be nearly 7 as strong acidic or basic nature of the binder can cause damage of printhead and associated machine components.
(iv) Surface tension should be low ($\leq$ 72 mN/m, similar to DI water), so that ligament formation is minimal. The surfactants can facilitate to decrease the surface tension.
(v)   Bubbles are usually created due to rapid vaporization at 350–400°C and therefore the localized temperature inside the printhead is considered as a factor.

The binder composition should also be designed to maintain the pH above the predetermined pH value to prevent premature coagulation in some specific powder-binder systems. For example, for colloidal silica, the base should be chosen to maintain the pH between 9 and 10, which is the range of maximum stability of the colloidal silica and is most compatible with the stainless steel and nickel printhead components. The quantity of the base in a binder must be sufficient to allow the binder pH to drop to 7 or lower in the powder bed. The quantity of the base must also be sufficient to overwhelm the effect of impurities that may be picked up during storage. For example, CO_2 in air is mildly acidic and will cause the pH of a very weak solution to change.

In addition, binder viscosity is one of the key properties to influence printing. The binder with low viscosity is preferred to prevent nozzle clogging, often caused by drying of the ink. Also, low-viscosity inks allow for quick refills in situations, where repeated printing is necessary. Soluble low-viscosity surfactants, for instance, Pluronic can be added to a medium to reduce nozzle clogging.[523] If an ink is particle-loaded, the viscosity can be reduced by lowering the solid loading, i.e., by increasing the mean particle size with the same solids loading. For a polymer loaded solution, the viscosity can be lowered by shortening the polymer chain or lowering the polymer loading. More universal approaches to alter the viscosity include changing the carrier fluid and diluting the ink[524]. From the above discussion, it should be clear that, the behaviour of ink drop in the DOD process can be therefore controlled by a number of physical properties, like surface tension, viscosity, gravitational force, etc. The influence of binder concentration and post-processing on compression strength property are shown in Fig. 4.23. As shown in Fig. 4.23, the strength property is not only dependent on percentage of H_3PO_4, but also on post-fabrication sintering temperature. A combination of 10% H_3PO_4 together with sintering at 1300°C results in the best compressive strength of 50 MPa.

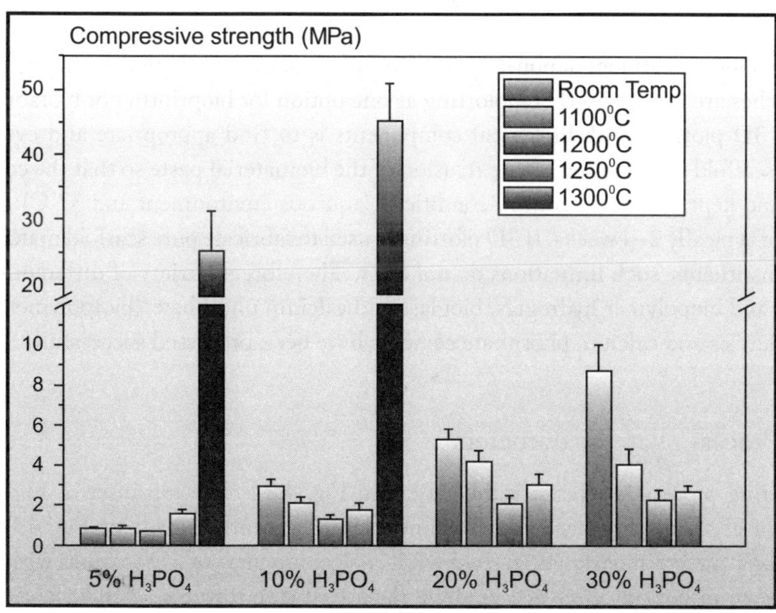

Fig. 4.23 Plot of the influence of post-processing conditions as well as binder concentration on compressive strength properties[543].

4.4.2 | 3D plotting

Among the available AM methods, suitable for the preparation of clinically relevant implants/devices, 3D plotting is one of the widely investigated AM technique that has the flexibility of simultaneous printing of cells during scaffold fabrication. This is possible due to its mild operating conditions at room or physiological temperature. In addition, 3D plotting is considered as a more versatile method than 3D powder printing due to the ease of selection of biomaterial to be processed and possibility to plot multiple layers with different composition[525,526].

In contrast to the other RP methods, 3D plotting was originally developed for biomedical applications and is distinguished by the fact that it was not translated from mechanical engineering discipline.

> Three dimensional plotting (3D plotting) is defined as the manufacturing of 3D scaffolds by dispensing pasty biomaterials at ambient or physiological temperature. The principle of 3D plotting is similar to that of fused deposition modelling (FDM) with the exception that materials, which are pasty at room (or physiological) temperature are extruded in a strand-like fashion.

However, a heating stage can be incorporated in devices for 3D plotting to preheat or even melt a material. Therefore, 3D plotting can be combined with FDM. Other terms, sometimes used for 3D plotting applications are bioplotting[527], direct write assembly[528] or robocasting[529].

In principle, any material with a suitable viscosity at the operating temperature can be processed by 3D plotting. In addition, sensitive components like drugs, growth factors or even living cells easily can be integrated just by mixing with the plottable paste prior to extrusion. This is feasible if organic solvent, heat treatment or non-physiological pH is not used/adopted during manufacturing. Such approaches are known as 3D bioplotting as one option for bioprinting or biofabrication. The challenge in 3D plotting with biological components is to find appropriate and cytocompatible methods for scaffold stabilization after extrusion of the biomaterial paste so that the constructs can be handled and kept under cell culture conditions (aqueous environment and 37°C) for a suitable time period of typically 2–4 weeks. If 3D plotting is used to fabricate pure scaffold materials without biological constituents, such limitations do not exist. Therefore, a variety of different biomaterials like polymer and biopolymer hydrogels, bioglass and calcium phosphate/(bio)polymer composites, bioceramic slurries and calcium phosphate cements have been processed successfully so far.

4.4.2.1 | Process related parameters

The 3D plotting process is schematically shown in Fig. 4.24. The top inset of Fig. 4.24 shows the deposition of slurry/paste using multiple nozzles. The central image of Fig. 4.24 shows the 3D plotting of a three dimensional scaffold while a close-up view of 3D scaffold with rectangular strands is shown in bottom inset of Fig. 4.24. As a first step towards scaffold fabrication in 3D plotting route, the properties of the paste are to be optimized in terms of viscosity and injectibility. The incorporation of various additive needs to be tailored to reduce the viscosity of paste. PVA is commonly used for this purpose. Also, the viscosity needs to be measured at both constant and dynamic shear rate in order to understand the viscosity dependent flowability properties. Scaffold fabrication in 3D plotting route requires a high solid content in a polymeric matrix (to allow extrusion) with desired viscosity (associated with the needle diameter), controlled drying rate (to avoid distortion in shape due to drying), and plottability at room temperature (or below) to manufacture a complex architecture with stable dimensions. Although different HA-based pastes, suitable for 3D plotting, were prepared in the past, the usage of several additives to control the various parameters (e.g., viscosity, injectability, gelification, etc.) makes the ink preparation a very challenging job. Also, such additives may deteriorate the cytocompatibility of non-sintered samples (due to progressive dissolution/degradation of the additives) as well as mechanical properties of sintered samples (due to removal of these additives at high temperature treatment and possible activation of unwanted sintering or side reactions).

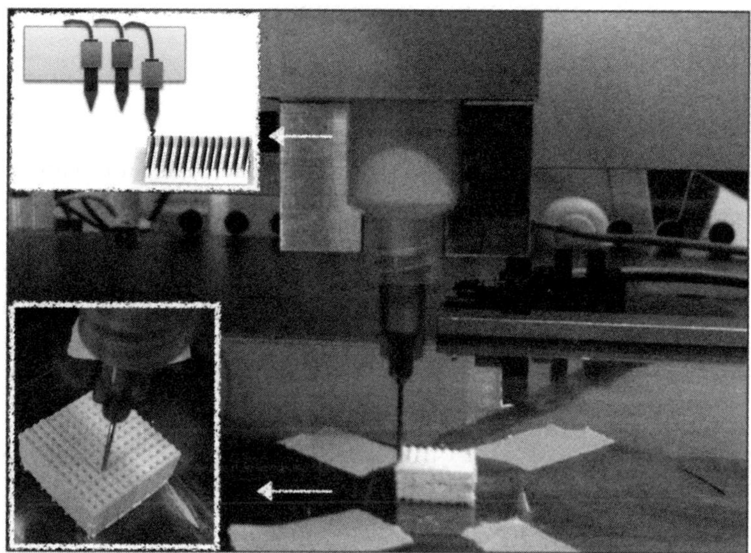

Fig. 4.24 Photographic image showing 3D plotting of a scaffold; the schematic of a multi-nozzle 3D plotter is also shown (top inset) and a close view of the 3D plotting process (bottom inset)[543].

After control of viscosity, the plotting speed needs to be adjusted, followed by optimization of tear off speed (speed of the plotting head required to break the strand after completion of strand deposition) and length (the distance required to break the strand after completion of strand deposition). Briefly, for a given viscosity, the increase in plotting speed leads to the deposition of discontinuous stands. In contrast, the plotting at a slower speed results in over-deposition of the material.

As far as other plotting parameters are concerned, the tear off speed and tear off length determine the plotting of extra length of the strand in addition to pre-programmed strand length. In case of viscoelastic solids, the extra length of plotting material needs to be deposited to avoid an incomplete strand in the plotted scaffold. Typically, the tear off speed and tear off length can be set to 1 mm/s and 1 mm, respectively. In case of plotting materials with high viscosity (faster drying rate), it has been noticed that the plotting starts with a delay, which further results in the formation of shorter strand than the programmed length. In order to further ensure the complete structure construction without any discontinuity in the construct, the start break time (time for which pressure is applied before the plot head moves) and end break time (time for which pressure is applied after the plot head has completed a strand) were set to 0.3 s and 0 s, respectively. In this row, cut off time for the air pressure before strand break was set to 0 s to avoid the over-deposition of plotting material at the end of the strand.

For a fixed needle diameter and air pressure, the strand diameter of a plotted scaffold strongly depends on the plotting speed. Typically, slow plotting speed (< 1 mm/s) resulted in over-deposition of plotting material with no discontinuity in the strand length. This reduced the microporosity in strands and therefore, increased the strength of both green as well as sintered scaffolds. In contrast, higher plotting speeds (> 5 mm/s) result in less deposition of plotting material and reduced strand diameters. Also, a very high plotting speed (25 mm/s) resulted in discontinuous strands at many places.

4.4.3 | Post-processing

Post-processing or post-fabrication treatment is often considered as an essential step after 3D printing/plotting of scaffolds. Based on the chemical nature of the powder, the post-processing step is followed to improve the mechanical strength. High temperature sintering is usually practiced to consolidate metallic and ceramic structures, while chemical treatments (infiltration, solvent leaching, crosslinking, etc.) are used in case of ceramics and polymers depending on their chemistry. The techniques are briefly discussed to explain the effectiveness of this step to enhance the scaffold properties.

4.4.3.1 | De-powdering and/or curing

It can be reiterated here that this step is used to remove the loose powder from the 3D printed part. For a part with no internal feature, this can be performed manually by brushing or gently blowing away the excess powder. For slurry-deposited parts, the entire bed may need to be deposited in a solvent to break up the unbound particles. Complex or internal features can be more difficult to completely depowder and may require additional steps. Dry options include blowing air, and vibration. Wet depowdering is an option as long as the binder and powder is not soluble in the fluid. After depowdering, the part should be dried, prior to any additional processing.

For 3D plotting, a curing step is followed as a post-processing step to enhance binding strength as well as to facilitate polymerization. Some widely reported curing options include the exposure to radiation and heat, reduction with a salt-based binder or conversion of a pre-ceramic polymer[530]. Alternatively, high temperature sintering can be followed for post-processing of a 3D plotted scaffold[475,499].

4.4.3.2 | Sintering and infiltration

The two most common post-processing steps for 3DPP/3DPL scaffolds are sintering and infiltration. These steps can increase the strength of a part considerably, and infiltration can also significantly affect other bulk properties. Prior to sintering, the part can be settered by packing the part in a higher sintering temperature material to give the part support during thermal treatment. The applicability of settering to high shrinkage sintering is limited, since the support material may resist the dimensional changes of the part[531,532].

One of the post-processing steps that has been widely used for ceramics or metal is sintering. It can be reiterated here that sintering refers to the consolidation of a powder compact and its transformation to a non-porous solid, when a powder compact is heated at a temperature, $T/T_m > 0.5$ (T_m is the melting temperature)[533]. The shrinkage of a powder compact is driven by the diffusional mass transport into the pores or void spaces of the green powder compact. In view of this, sintering parameters as a post-processing step for a 3DPP/3DPL scaffold needs to be intelligently tailored in reference to sintering temperature and time. For example, HA-based scaffolds are sintered at 1200–1300°C, while stainless steel or Ti-alloy are heat treated at 1000°C or above in argon atmosphere. The sintering conditions should be close to but less than that nominally used to obtain non-porous solid of similar composition. This will allow one to obtain the desired combination of micro and macro porosity in the 3D scaffold. As far as sintering atmosphere is concerned, metallic scaffolds e.g., Ti6Al4V scaffolds are normally sintered in flowing argon atmosphere, while CaP-based scaffold are sintered in air.

Prior to sintering and/or infiltration, it is possible to improve the surface finish of the part by coating the part with a layer of finer particles. This can be done using a polymer smaller than the bulk powder or by slip casting a thin layer of fine particles (0.1–1 microns) onto the part. The slip can be prevented from penetrating the pores of the part by selecting self-locking slip particles, precoating the construct. The coating does not need to be the same material as the part, but the potential difference in thermal processing behaviors must be considered to reduce the cracking problems from differential thermal expansion[534,535].

The use of an unconventional post-processing, namely microwave-sintering on the 3D printed scaffolds is reported by Tarafder et al.[531]. The conventional sintering, in which heat dissipates from outside to inside of the scaffold, results in non-uniform heating and undesired grain growth. In contrast, microwave-sintering leads to uniform heating throughout the scaffold with a better control over grain growth[536]. The scaffolds with three different macroporosity ranges were fabricated with β-TCP and 3D printed samples were sintered at 1150 and 1250°C in conventional furnace and in a 3 kW microwave furnace for 1 hour. The comparison of compressive strengths of the scaffold after above-mentioned treatments is shown in Fig. 4.25. It is found that the macropore size strongly affects the compressive strength. Although both conventional and microwave sintered scaffolds exhibit similar morphology with micropores, microwave sintered scaffolds exhibited shrinkage, lower porosity, higher density (data not shown here) and higher strength. In particular, cylindrical scaffolds (height 10.5 mm, diameter 7 mm) with 500 μm square shaped macropore achieved the best compressive strength of 11 MPa after 1hour of microwave sintering at 1250°C.

Infiltration is a way to achieve high-density parts without the large shrinkage associated with sintering to full density. Both low and high temperature infiltrations are possible, depending on the part material and binding mechanism. The only constraint on the process relies on the fact that the infiltrate must melt at a temperature below the melting point or solidus temperature of the bulk material, so that the structure should not be affected during infiltration. An exception to this is when the powder is coated with a higher temperature material (for example, carbon coated with a ceramic), so only the coating comes in contact with the infiltrate. The preferred traits for melted infiltrates include sufficient fluidity and viscosity to flow through the part pores, and a low contact angle with the bulk material to make the infiltration more effective[537,538].

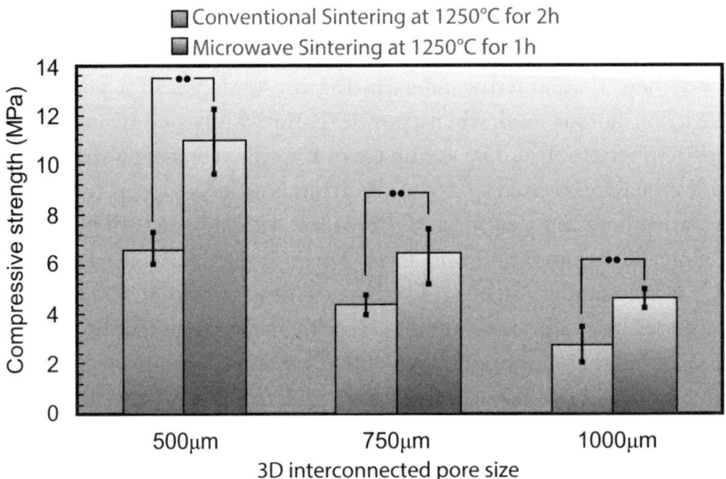

Fig. 4.25 Effect of macro-pore size and sintering methods on the compressive strength of 3D powder printed scaffolds[543].

4.4.3.3 | Chemical conversion/post hardening

In case of sintering and infiltration, one can observe shrinkage and coating respectively with an infiltrate. However, chemical conversion/post-hardening is quite different in terms of processing approach and outcome. As the name suggests, the principle of chemical conversion is based on phase transformation. It is mainly used for reactive cements/ceramics (calcium phosphate, magnesium phosphate, POP, etc.) based scaffolds[539,540]. The 3D printed part (green body) is dipped in a chemical solution which will react with the powder and the outer layer of the scaffold will be converted to another phase. It is advised to carry out this procedure only after de-powdering, otherwise loose powders trapped in small pores will be attached to the scaffold permanently leading to decrease in resolution and porosity. Chemical conversion is particularly becoming the preferred choice, as it has the following advantages compared to other post-processing techniques:

(a) It is carried out at an ambient temperature, opening up the possibilities of temperature sensitive additives like drug, growth factors, etc.

(b) By intelligently choosing the powder and phase-conversion-liquid combination, one can have a scaffold having two different phases at the inner core and outer surface, e.g., a scaffold with resorbable surface and less soluble core phase.

(c) Another important reason to adopt chemical conversion is post-hardening, which can increase the mechanical properties (tensile, compressive, flexural strength) due to phase conversion.

(d) Chemical conversion can be also combined with other post-processing steps, e.g., sintering, as demonstrated by Meininger et al. for 3D printed SrMPC (Strontium substituted magnesium phosphate cement) scaffold[541]. In this study, sintering and then chemical conversion with

Na_2HPO_4 was used in a combination to increase mechanical strength along with maintaining the resorbability of the scaffold.

4.5 | Closure

The fabrication of different materials demand one to adopt specific processing approach(es) and it is extremely important to follow material specific processing parameters to obtain desirable physical, chemical and mechanical properties. In order to substantiate this aspect, the processing approaches for the major classes of biomaterials (metallic biomaterials, bioceramics, polymeric biomaterials) are largely discussed in this chapter. Metallic biomaterials are generally fabricated in various shapes and sizes. They can be drawn into wires for cardiovascular stents or cast and shaped into plates/rods for orthopedic fixation, etc. Thus, the processing parameters should be optimized in such a manner that the bulk deformation processes can be optimized to avoid fracture. Premature fracture of metallic implants will certainly render it unsuitable in clinical settings. This chapter also summarizes various types of densification processes for ceramics, including pressureless and pressure-assisted sintering. Porous bioceramic implants are mostly desirable for favourable cell-biomaterial interaction and osseointegration. Thus, the sintering parameters have to be carefully optimized to obtain clinically relevant porosity, without compromising much on the mechanical properties of the implant. Polymeric biomaterials are also sensitive to processing parameters. Moulding techniques, such as injection moulding and compression moulding have also been briefly discussed in this chapter. An equal emphasis has also been placed to discuss the additive manufacturing of patient-specific scaffolds. In reference to the discussion on low temperature AM techniques, the major challenges involve the selection and formulation of binders as well as optimization of post-processing treatments. Some specific issues in case of 3D printing include the material (to be printed)-specific binder chemistry and property optimization. Also, the interaction of ink with scaffold surface is of important consideration. It has been emphasized that the binder chemistry needs to be closely tailored with respect to some characteristic physical parameters. The currently used post-processing approaches are also briefly discussed and a specific route should be adopted while critically considering the targeted application specific structural and mechanical characteristics. In summary, this chapter provides the breadth and necessary depth of information, which can be used as guidelines to select and design processing methodology for implantable biomaterials.

Probing Structure of Materials at Multiple Length Scales

It is now widely known that the properties of the biomaterial substrate significantly influence the cellular functionality, *in vitro*. Therefore, one of the important aspects of the research on biomaterials is to sufficiently characterize the physical and functional properties, and to establish linkages with the microstructural characteristics. In this perspective, this chapter briefly discusses a number of important materials characterization techniques, which a researcher pursuing the field of biomaterials should be aware of. From the biomaterials science perspective, an important aspect is to establish the correlation between substrate properties and biocompatibility. The purpose of this chapter is to introduce the readers from diverse backgrounds with the fundamental concepts of multiple materials characterization techniques, as well as the related practical guidelines. A reader can refer to some of the standard textbooks to know more details on any of the material characterisation techniques.

5.1 | Introduction

In the preceding chapter, the multitude of manufacturing processes for different classes of materials are discussed. Irrespective of adopted processing approach, any crystalline material is characterized by unique microstructural characteristics. In simplistic terms, a microstructure describes the structural features, when viewed using a microscope. Extending this further, the microstructure essentially reveals how different compositional phases are distributed in 2D or 3D space. The schematic shown in fig. 5.1 shows different such phases in 2D space. Depending on the length scale, shape and composition of microstructural phases, a given material substrate can have a set of physical and functional properties. Therefore, the determination of various phases at multiple length scales, as well as the measurement of bulk functional properties, is equally important in comprehensively analysing the role of substrate properties towards cell-material interaction. This chapter describes various characterization tools, which can be used alone or in a complementary manner to characterize material composition and phases.

Depending on the composition of the substrate, the structural characterization of each material requires a unique set of characterization tools. For example, polymers, which are essentially made of C, H, O, N, etc., require different spectroscopic techniques, like Fourier Transform Infrared Spectroscopy (FT-IR), Nuclear Magnetic Resonance (NMR) to probe into various molecular configurations. The structural characterization of crystalline materials and ceramics mostly start with X-ray diffraction (XRD) for initial phase assemblage identification, which can be followed by electron microscopy techniques to characterize various phases at different length scales. In the following, the commonly used characterization techniques are described, starting with spectroscopic techniques.

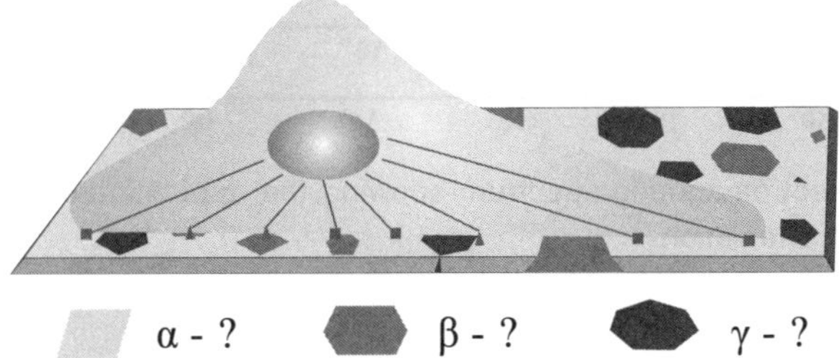

Fig. 5.1 Schematic showing that the cell-material interaction at the biomaterial interface is dependent on composition and functional property. Accordingly, one needs to know the composition and morphology of various phases (α, β, γ ...).

5.2 | Spectroscopic Analysis

Spectroscopy is usually defined as the study of interaction of electromagnetic radiation with matter.

Spectroscopy has become an indispensable tool for analysing the structural and functional properties of any material. The electromagnetic spectrum extends from the high frequency gamma rays to low frequency radio waves, distributed over a range of 10^{21} Hz to about 10^7 Hz (depicted in Fig. 5.2). Further, as shown in Fig. 5.3, the electromagnetic spectrum can be subdivided into various regions corresponding to a unique molecular process, which generates specific spectroscopic information. Spectroscopic data are generated in the form of a spectrum, which is generally a plot of intensity vs wavenumber/frequency/wavelength. Here, it is instructive to mention the difference between a spectrum and a pattern. If the intensity, which is always expressed in arbitrary units (unless otherwise mentioned), is plotted against angle or a parameter other than wavelength, then such plot is to be called as a pattern.

Depending on the energy of the incident wavelength, we observe different transitions, on the basis of which spectroscopy may be classified (see Fig. 5.3). For example, incident energy falling in the UV and visible range gives rise to electronic excitations in molecules, which forms the basis of UV-visible spectroscopic technique. Incident energy falling in the infrared region gives rise to atomic vibrations, which is the governing principle behind IR and Raman spectroscopy techniques. In the following sections, the spectroscopy techniques of special relevance to biomaterial characterization have been described in details.

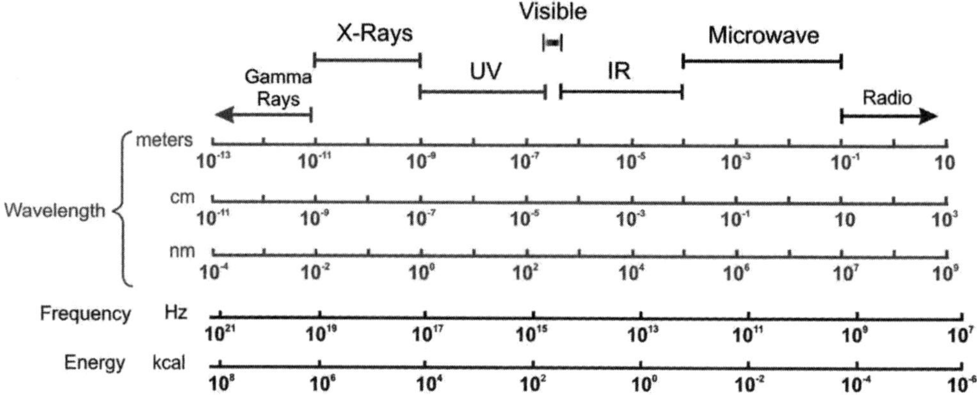

Fig. 5.2 The entire electromagnetic spectrum showing the approximate ranges for each type of electromagnetic radiation.

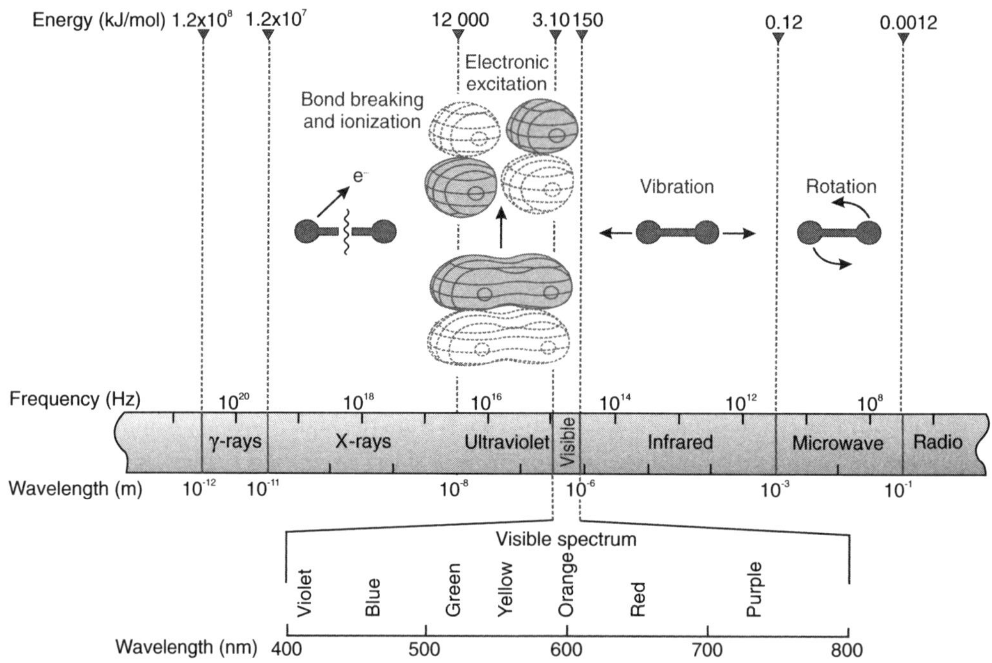

Fig. 5.3 Regions of electromagnetic spectrum and types of energy transitions observed.

5.2.1 | **Infrared spectroscopy**

Infrared spectroscopy is the study of the interaction of infrared radiation with matter. It is extensively used to analyse chemical microenvironment, i.e., the presence of functional groups, while determining bonding characteristics of organic as well as inorganic molecules.

FT-IR stands for Fourier Transform Infrared spectroscopy. Fourier transform, a signal processing technique, is used to transform the signal from time domain to its representation in frequency domain. The basic advantage of using a Fourier transform IR instrument is its speed and sensitivity. Such an instrument is able to acquire multiple interferograms in a very small span of time, thus minimizing the signal to noise ratio.

The IR transitions occur between the different vibrational levels as are shown in Fig. 5.4, governed by quantum mechanical selection rules. When a molecule is irradiated with IR, it gets excited to a higher energy level. As a result, the mean amplitude of vibration increases. The atoms are excited to the next allowed vibration level. This is depicted in Fig. 5.4. The absorption of Infrared radiation causes the bonds to vibrate with a frequency (v), which is related to the bond strength or the force constant (k) of a bond, as per the following equations:

$$v = \frac{1}{2\pi}\left(\sqrt{\frac{k}{\mu}} \right) \tag{5.1}$$

$$\frac{1}{\mu} = \frac{1}{m_1} + \frac{1}{m_2} \tag{5.2}$$

where, v is the bond vibrational frequency, k is the force constant and μ is the reduced mass of the system, m_1, m_2 are the masses of the atoms. Thus, IR radiation excites the analyte to be studied and the changes observed are generally recorded in terms of wavenumber, $\bar{v}$.

$$\tilde{v} = \frac{1}{\lambda} \tag{5.3}$$

Wavenumbers are expressed as cm^{-1} and are related to the frequency (v) as,

$$v = c\bar{v} = \frac{c}{\lambda} \tag{5.4}$$

The range of wavenumbers generally encountered in IR spectroscopy extends from 4000–400 cm^{-1}. So, the force constant of a bond is proportional to the wavenumber, which in turn is directly proportional to the energy (E).

$$E = hv = hc\bar{v} \tag{5.5}$$

Thus, stronger bonds show a peak at a higher wavenumber as compared to weaker bonds.

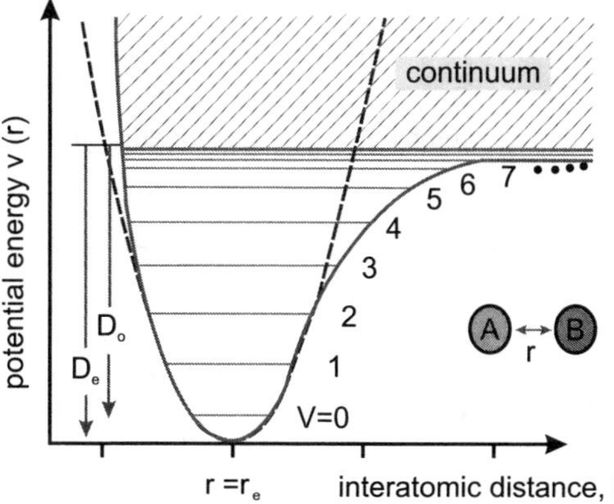

Fig. 5.4 The potential energy diagram for an asymmetric diatomic molecule showing harmonicity.

Like all other absorption processes, the absorption of IR radiation is also quantized. The transitions are governed by quantum mechanical selection rules. Hence, there are particular frequencies which are preferentially absorbed, giving rise to detectable signals. When a molecule is irradiated with IR radiation, only those frequencies of radiation are absorbed which match the natural vibrational frequency of the molecule. This leads to increased amplitude of the vibrational motion of bonds in the molecule. The bonds, which undergo change in dipole moment with time, absorb infrared radiation.

Dipole moment (μ) of a bond is defined as the product of the charge (Q) and the distance between the charges (r), as per the following relation;

$$\mu = Qr$$

The inter-nuclear distance changes as a bond vibrates. This gives rise to a change in the dipole moment μ. Diatomic molecules (such as O_2, H_2, Cl_2, etc.) or molecules with symmetric bonds having identical functional groups on each end will not absorb infrared radiation. Such systems are said to be infrared inactive. Only molecules with a permanent net dipole moment (like HI, HF, CO, etc.) are IR active.

This is because only transitions which are accompanied by a change in the dipole moment are allowed according to quantum mechanical theory. There are 3N-6 normal modes of vibration for a non-linear molecule and 3N-5 for a linear molecule, where N denotes the number of atoms present in the molecule. For example, H_2O has three normal modes of vibration (N = 3, so 3N-6= 3*3-6 = 9-6 = 3). In the expression, 3N denotes the number of degrees of freedom. The simplest modes of vibrational motion which gives rise to IR active modes are mostly stretching and bending motion, as shown in Fig. 5.5. Stretching can be symmetric or asymmetric and causes a change in the bond length due to change in the interatomic distance. In general, symmetric stretching modes appear at a higher wavenumber than antisymmetric modes. It would be worthwhile to remember that bending motion of vibration causes a change in the bond angle or changes in the position of groups of atoms with respect to rest of the molecule.

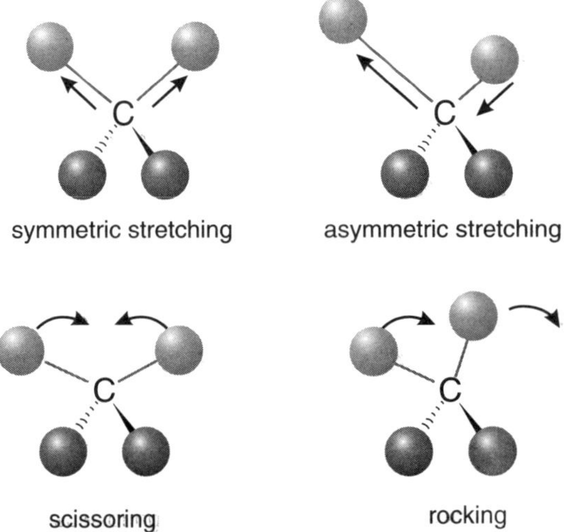

Fig. 5.5 Principal modes of molecular vibration of relevance to infrared spectroscopy.

A typical IR spectrum is a plot of % transmittance vs wavenumber, as shown in Fig. 5.6. Stronger the bond, the higher is its k value, and higher is the wavenumber at which its peak occurs. Figure 5.7 depicts the approximate regions, where the different kinds of bonds absorb IR radiation. As shown, C=O stretching vibration appears around 1700 ± 100 cm^{-1} depending on the chemical environment. Similarly, O-H stretch occurs at 3200–3650 cm^{-1}. A band due to O-H bending motion appears at 1650 cm^{-1}. So both these peaks together can confirm the presence of O-H group in the sample.

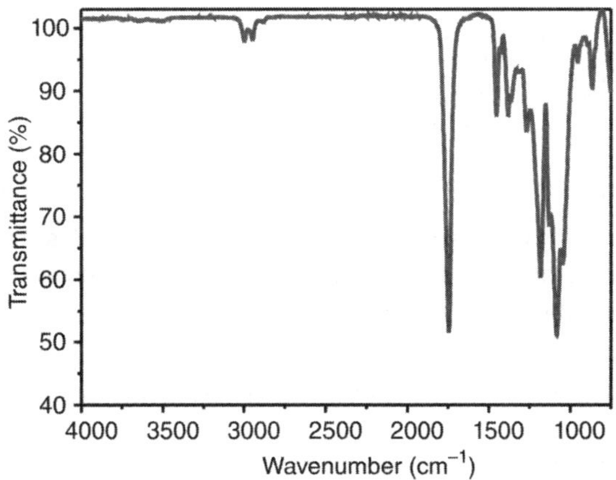

Fig. 5.6 A typical FT-IR spectrum of a biodegradable polymer, PLGA.

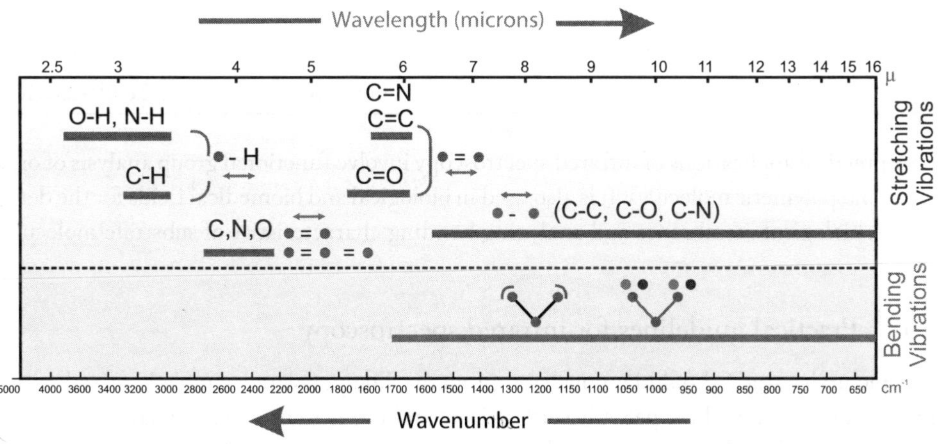

Fig. 5.7 The approximate windows for various common types of bonds.

In attenuated total reflectance IR, a beam is directed at an angle in an optically dense, high refractive index medium (see Fig. 5.8). Attenuation of the evanescent wave occurs only when the sample absorbs in the IR range. The applications of infrared spectroscopy can be split into three main regions, near-infrared (4000–14000 cm^{-1}), mid-infrared (670–4000 cm^{-1}) and far-infrared (less than 670 cm^{-1}) [see Table 5.1]. The fingerprint region appears in the range of 670–1280 cm^{-1} and the group frequency region appears at 1280–5000 cm^{-1}.

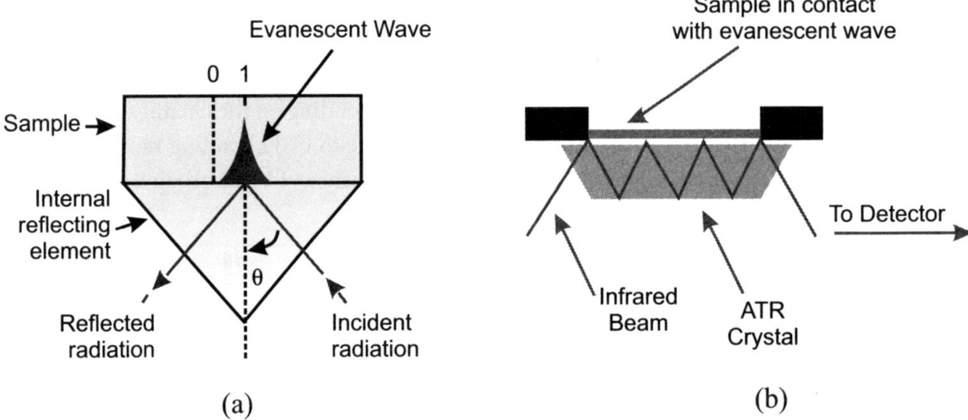

Fig. 5.8 A representation of the UATR accessory showing the sample in close contact with the ATR crystal to facilitate total internal reflection (a); the formation of an evanescent wave upon irradiation with an incident beam which is detected (b).

Table 5.1 Different regions in the Infrared spectrum and their corresponding wavelength, wavenumber and frequency ranges.

Region	Wavelength Range, (μm)	Wavenumber Range, (cm^{-1})	Frequency Range, (Hz)
Near	0.78–2.5	12800–4000	$3.8 \times 10^{14} – 1.2 \times 10^{14}$
Middle	2.5–50	4000–200	$1.2 \times 10^{14} – 6.0 \times 10^{12}$
Far	50–1000	200–10	$6.0 \times 10^{12} – 3.0 \times 10^{11}$
Most used	2.5–15	4000–610	$1.2 \times 10^{14} – 2.0 \times 10^{13}$

Some practical applications of infrared spectroscopy involve functional group analysis of organic, inorganic and polymeric molecules. It is also used in biological and biomedical fields for the detection of water in biological membranes and analyzing bonding characteristics of substrate molecules.

5.2.1.1 | Practical guidelines for infrared spectroscopy

The sample under study should be powdered properly. The powdered sample is mixed with KBr and pressed into a pellet. This pellet is very delicate and needs to be carefully handled. The pellet is then put into an FT-IR instrument and scanned over the range of 4000–400 cm^{-1}.

Although Infrared spectroscopy gives basic information about the functional groups present in a molecule, it is extremely sensitive to the presence of water.

So before recording an IR spectrum one should make sure that the sample under analysis does not adsorb moisture as it will interfere with proper analysis of the sample.

The main advantage of ATR over normal IR lies in the ease of sample preparation. It is not necessary to make the KBr pellet, while using ATR. The powdered sample/neat liquid can be directly placed in contact with the crystal, and the spectrum is recorded. However, the refractive index of the sample should be significantly less than that of the crystal for UATR for total internal reflection to occur.

Before recording the spectra for the sample, the background is recorded, which has to be subtracted from the spectra of the sample. This spectrum obtained has to be analyzed by correlating with standard IR data tables. The peaks obtained may be shifted due to the influence of the chemical environment surrounding the bonds. Resonance, conjugation and the symmetry of the molecule have a great influence on the absorption wavenumber. The position of the peaks is also affected by the presence or absence of hydrogen bonding or other electrostatic interactions with the solvent. IR peaks are not sharp lines, but always have a finite width. This is due to various factors, such as natural line broadening due to Heisenberg's uncertainty principle and Doppler broadening. In addition, hydrogen bonding and rotational fine structure also account for peak broadening.

5.2.2 | Raman spectroscopy

When a beam of light interacts with the substrate molecules, photons may undergo elastic or inelastic collision with the molecules. If the collision is elastic, then the frequency of the emergent photon will be the same as that of the incident photon. Such scattering of photons is called Rayleigh scattering. The energy obtained in Rayleigh scattering is the same as that of the energy of the incident radiation. However, if there is an exchange of energy between the photon and molecules, then the collision is inelastic. Radiation scattered with a frequency lower than the incident radiation is called Stokes scattering. This inelastic scattering results in spectral lines, called Raman lines. Anti-Stokes scattering results in Raman lines from photons scattered with higher energy than the incident radiation. All the probable processes, which may occur during Raman scattering, have been depicted in Fig. 5.9. In case of Stokes scattering, molecules gain energy. On the contrary, in case of Anti-Stokes scattering, molecules lose their energy to the photons. The anti-stokes scattering is only possible if the molecules were initially at an excited vibrational or rotational state. Hence, the intensity of Stokes scattering is generally much higher than that of anti-Stokes scattering.

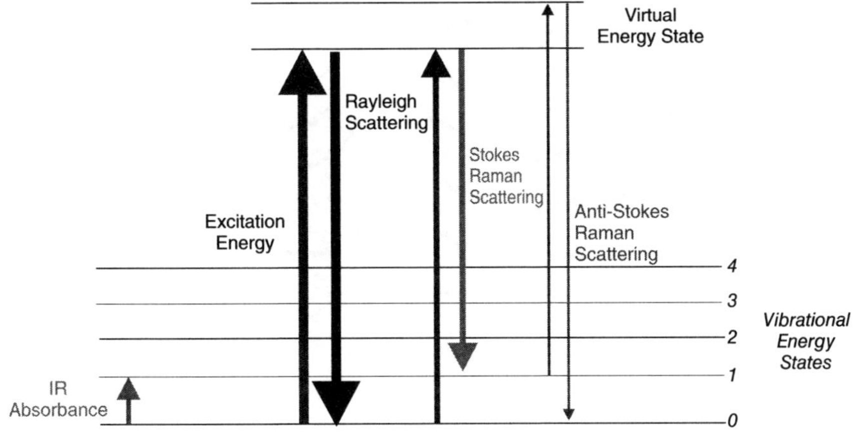

Fig. 5.9 All the probable processes which may occur during Raman scattering.

A basic requirement for a molecule to be Raman active depends on its polarizability. When a molecule is placed in an electric field, it undergoes separation of charges which creates an induced dipole. The molecule is said to be polarized and this induced dipole moment (μ) is directly proportional to the strength of the applied electric field (E). The proportionality constant here is the polarizability (α) of the molecule. The relationship among μ, E and α is given by the following equation:

$$\mu = \alpha E \tag{5.6}$$

Polarizability of a molecule indicates the extent of change in distribution of electron cloud around vibrating atoms. Hence, for a molecule to be Raman active, there should be a change in the polarization of the molecule, as compared to the change in dipole moment in the case of Infrared activity, during any molecular rotation or vibration.

An important point to be noted here is that, for molecules with a centre of symmetry (for example CO_2), Raman active modes are not Infrared active and vice versa. This is known as the rule of **mutual exclusion**. However, this rule may not be true for molecules which do not have a similar centre of symmetry. If the molecule is not centrally symmetric, then some modes may be both Raman as well as Infrared active.

Figure 5.10 shows a Raman spectrum taken from hot pressed zirconia ceramic showing the evolution of tetragonal ZrO_2 and monoclinic ZrO_2 at different spatial locations from the centre of worn surface. Figure 5.11 also depicts a Raman spectrum, obtained for Cholesterol molecule. Raman spectroscopy from the centre of worn surface is advantageous over IR as aqueous solutions can also be analyzed.

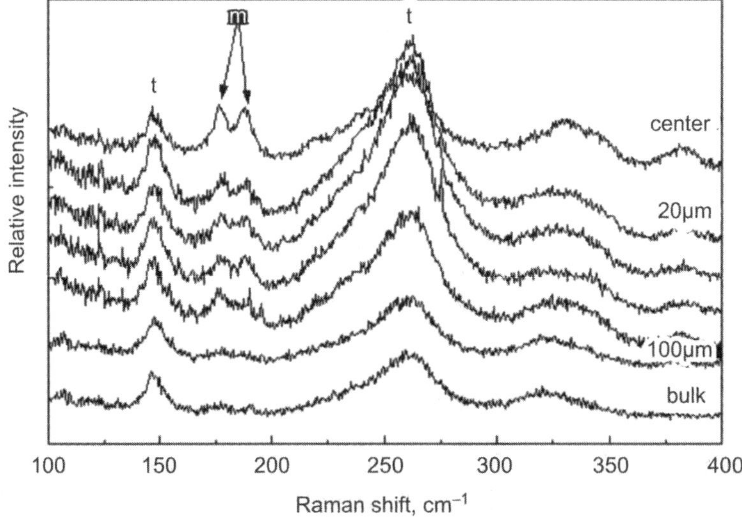

Fig. 5.10 Typical Raman spectra taken from hot pressed zirconia ceramic showing the evolution of tetragonal (t) ZrO_2 and monoclinic (m) ZrO_2 at different spatial locations.

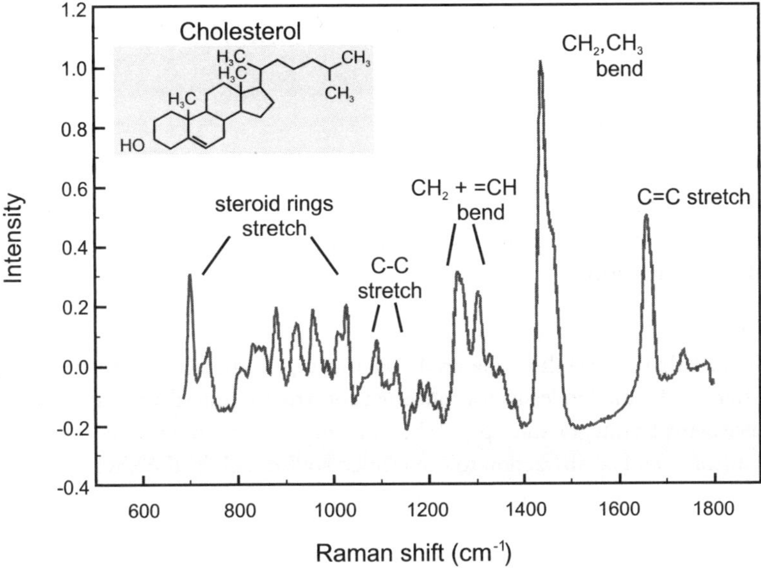

Fig. 5.11 Raman spectrum showing Raman peaks for different molecular vibrations present in cholesterol molecule[553].

5.2.2.1 | Practical Guidelines for Raman spectroscopy

Raman spectroscopy 'fingerprints' molecules by characterizing interactions among photons and molecular vibrations.

An important point of distinction between Raman spectroscopy and infrared spectroscopy is that Raman signals are obtained due to scattering of light by vibrating molecules, while IR peaks are obtained due to absorption of light by molecular vibrations.

Hence, the following points may be considered while performing Raman spectroscopy experiments. Typical lasers used in Raman spectroscopy are UV (244 nm, 257 nm, 325 nm, 364 nm), visible (457 nm, 473 nm, 488 nm, 514 nm) and near IR (785 nm, 830 nm, 980 nm, 1064 nm). The actual choice depends on the availability of lasers as well as the specimen to be examined.

The laser used also influences the data obtained. Raman scattering intensity is directly proportional to $1/\lambda^4$. Therefore, higher λ leads to weaker signal intensity. Hence, if the sample permits, a higher frequency is preferable. Spatial resolution is governed by the diffraction limit.

$$\text{Laser spot diameter} = \frac{1.22\ \lambda}{\text{numeric aperture}}$$

Therefore, lower wavelength (high energy) lasers are preferred for better spatial resolution. Generally, high wavelength, low frequency infrared lasers are preferred for biological applications. Additionally, recent optical fibre probe developments allow accurate real-time analysis *in vivo*.

5.3 | Crystal Structure and Compositional Analysis

5.3.1 | X-ray diffraction

X-ray diffraction (XRD) is a technique generally used to identify the crystal structure of various phases in crystalline materials. When X-rays interact with a material surface, they are scattered by the electron cloud of the atoms of the material without significant loss in the energy. This phenomenon is known as **elastic scattering**.

> Based on the fact that a crystalline material contains a periodic array of scattering centers, the coherent scattering of radiation leads to a concerted constructive interference at specific angles (the path difference being an integral multiple of the wavelength; as per Bragg's law). This phenomenon is known as diffraction. For diffraction to occur in crystalline solids, the wavelength of the source radiation must be of the order of the atomic spacing in a crystal lattice.

The X-rays used to obtain diffraction data are 'Hard X-rays', which are electromagnetic radiation with wavelength between 1–2 Å. X-rays are generated under vacuum, when electrons from a heated tungsten filament (anode) are accelerated under a high potential ($\approx$40 kV) and are suddenly decelerated by a metal target like Cu, Mo (cathode). The rapid deceleration results in the generation of characteristic X-rays, Bremsstrahlung radiation as well as a large amount of energy dissipated as heat. Monochromatic X-rays are produced by using a Ni-filter, which absorb the Cu-Kβ and transmit the Cu-Kα (1.54 Å) radiation.

5.3.1.1 | Principle

For X-ray diffraction to occur in crystalline solids, Bragg's law must be satisfied. From Fig. 5.12, it can be observed that for constructive interference to occur, the path difference between the two adjacent rays must be an integral multiple of the wavelength (nλ; n is non–zero integer).

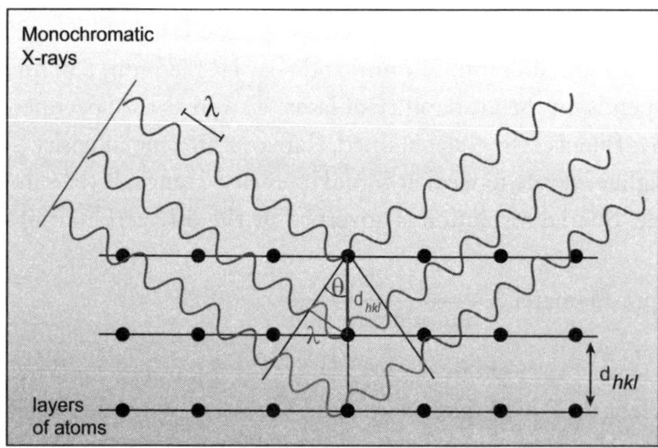

Fig. 5.12 Schematic showing the diffraction of X-rays by lattice planes in a crystalline solid.

Under the above boundary conditions, the Bragg's equation for X-ray diffraction can be represented as:

$$2d_{hkl} \sin\theta = n\lambda \tag{5.7}$$

where,
θ is the Bragg/diffraction angle,
λ is the wavelength of the X-rays (λ = 1.54 Å for Cu-Kα),
d_{hkl} is the spacing of the (hkl) lattice planes,
n is the order of diffraction. For example, the second order diffraction from (100) plane is equivalent to the first order diffraction from (200) plane.

5.3.1.2 | Working with XRD

In a typical XRD unit, the X-ray source and the detector are placed in either of the two specific configurations, as shown schematically in Fig. 5.13.

(i)   Bragg–Brentano geometry (θ–θ configuration): The source and detector move by equal angles about the sample which is stationary in the centre.
(ii) Seeman Bohlin geometry (θ–2θ configuration): The sample station and detector move by θ and 2θ angles, respectively, keeping the source stationary.

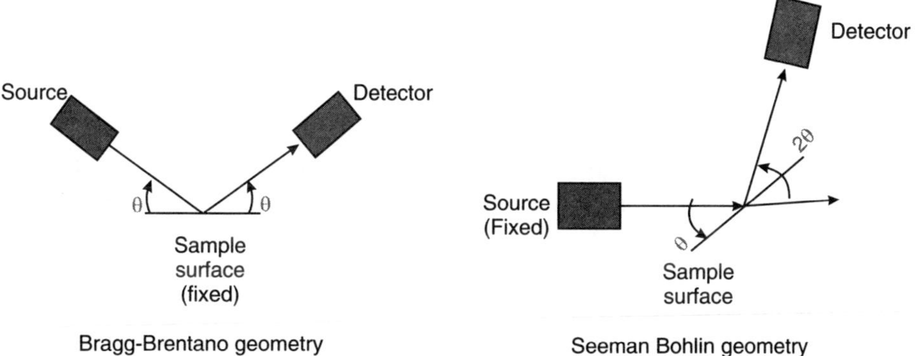

Fig. 5.13   Pictorial representation of the two types of goniometer geometries in diffractometers.

In a typical XRD scan, one can obtain intensity of diffracted beam as a function of angle of diffraction. After the necessary background intensity correction, the characteristic peaks are identified. The peak positions give unit cell dimensions and crystal symmetry (greater symmetry leads to lesser reflections and vice-versa). From the peak positions (2θ), the d_{hkl} spacing can be calculated using Bragg's law. The following are the relationships between d-spacing and the lattice parameters for the most commonly encountered crystal systems.

Cubic

$$\frac{1}{d^2} = \frac{1}{a^2}\left(h^2 + k^2 + l^2\right)$$

Tetragonal

$$\frac{1}{d^2} = \frac{1}{a^2}\left(h^2 + k^2\right) + \frac{1}{c^2}l^2$$

Orthorhombic

$$\frac{1}{d^2} = \frac{1}{a^2} h^2 + \frac{1}{b^2} k^2 + \frac{1}{c^2} l^2$$

Hexagonal

$$\frac{1}{d^2} = \frac{4}{3} \left(\frac{h^2 + hk + k^2}{a^2} \right) + \frac{l^2}{c^2}$$

The diffraction pattern can be used as a fingerprint to identify the material and its crystal structure due to the difference in the intensity and position of characteristic peaks in the pattern. The intensity of the peaks from diffracting planes depends on the structure factor (F_{hkl}).

> The structure factor is defined as a summation of contributions from all the atoms in the crystal structure towards the scattering intensity for a particular crystal plane.

It is mathematically written as follows:

$$F_{hkl} = \sum_j f_j e^{2\pi i (hx_j + ky_j + lz_j)} \tag{5.8}$$

Based on the structure factor calculation, one can predict the intensity (I) of the diffracted X-ray beam from a set of (hkl) planes as per the following formula,

$$I = \left| F_{hkl} \right|^2 \tag{5.9}$$

where,
f_j is the atomic scattering factor,
(hkl) are the indices of the crystal planes,
(x_j, y_j, z_j) are the coordinates of the atoms in the unit cell.

In determining the peak intensity (I), one has to additionally consider the contributions from Lorentz factor, multiplicity factor, polarization factor and temperature related factor. For allowed reflections, F_{hkl} is non-zero and for forbidden reflections, $F_{hkl} = 0$. Based on the relative intensities and systematic absences of peaks, it is possible to determine the symmetry and to identify the phase assemblage of the materials. For a simple cubic lattice, all the reflections have a non-zero structure factor. As the symmetry of crystal increases, some of the reflections (determined by the 'selection rules') disappear. Thus, systematic absences give information about the type of unit cell. Table 5.2 provides the diffraction conditions for the materials with cubic symmetry.

Table 5.2 Types of Bravais lattices (cubic) and selection rules for allowed and forbidden reflections from structure factor calculations.

Bravais lattice	Reflections present	Reflections absent
Simple cubic	All (hkl)	None
Base-centred/C-centred cubic	h and k unmixed (both h & k (2,4,6,8..), (1,3,5,7..) are either even or odd)	h and k mixed
Body-centred cubic	(h+k+l) = 2n or even	(h+k+l) ≠ 2n (i.e., odd)
Face-centred cubic	h, k, l are unmixed (all h, k, l indices are either even/odd)	h, k, l are mixed

Ideally, the diffraction peak should be a δ-function at the angle of diffraction (Bragg's angle), but it usually has a finite peak width due to instrumental limitations, quality and size of the crystal/crystallites. This provides a way to determine the particulate size from powder X-ray patterns by analyzing the peak parameters (FWHM and shape). The observed peak broadening arises due to instrumental errors, crystallite/particulate size and strain in the crystal. The net broadening from these sources can be fit into a linear relation as shown below:

$$B = B_i + B_c + B_s \tag{5.10}$$

where,

B_i is the instrumental broadening,

B_c is the crystallite size broadening,

B_s is the sample strain.

With a standard Si sample, the instrumental broadening can be determined, while the strain component is proportional to $\tan\theta$, the broadening B_c and the corresponding crystallite size (t) can be determined by the Scherrer formula, as in the following;

$$t = 0.9\lambda/B \cos\theta \tag{5.11}$$

where,

B is the peak full width at half maximum (FWHM),

θ is the Bragg angle.

5.3.1.3 | Practical guidelines for XRD

Depending on sample composition Cu, Mo or Cr source is used. For example, Cr source is used for samples containing Cu. A fast scan is done in the first run and then a slow scan (0.02°/minute) is performed. After acquiring the pattern, it is analyzed using Joint Committee on Powder Diffraction Standards files (JCPDS) to match peak positions. For finer characterization of the closely spaced X-ray peaks, a much slower scan of <0.02°/min is used, over a narrow 2θ region.

> The first three strongest peaks are to be matched with the standard JCPDS file to confirm the presence of any phase. In case one of the characteristic peaks is shifted due to sample height error, then systematic shift should appear for other peaks as well.

5.3.2 | X-ray photoelectron spectroscopy (XPS)

XPS belongs to the generic characterization technique, 'Electron Spectroscopy for Chemical Analysis (ESCA)'. Other ESCA techniques are Auger electron spectroscopy (AES) and Electron Energy loss spectroscopy (EELS).

> X-ray photoelectron spectroscopy (XPS) is a surface analytical technique, which is based on Einstein's photoelectric effect.

Each atom on a material surface has core electrons with characteristic binding energy, which is similar to the ionization energy of that electron.

5.3.2.1 | Principle

When an X-ray beam is directed onto a sample surface, the energy of the X-ray photon is absorbed by the core electron of an atom (Fig. 5.14). If the photon energy ($h\upsilon$) is sufficiently large, the core electron will escape the interatomic potential well as a photoelectron from the surface of the material. The binding energy is calculated from the measured kinetic energy of the photoelectron and the incident X-ray photon energy (Al Kα = 1486.6 eV; Mg Kα = 1253.6 eV) using the below equation:

$$KE = h\upsilon - BE - \Phi \tag{5.12}$$

where,
KE is the kinetic energy of the photoelectron,
$h\upsilon$ is the incident X-ray photon energy,
Φ is spectrometer work function (few eV and constant for the instrument),
BE is the binding energy of the core electron.

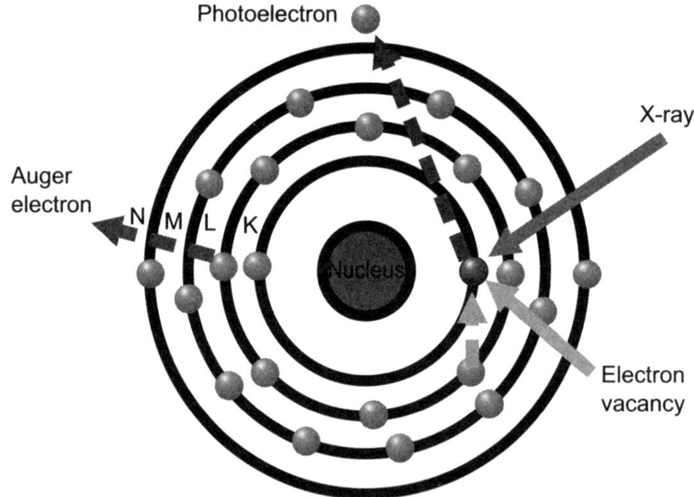

Fig. 5.14 Planetary model of an atom showing the emission of photoelectrons and Auger electrons when bombarded with high energy X-rays.

A related surface-sensitive technique is Auger electron spectroscopy (AES). The sequence of events in AES includes:

(a) Removal of electron from core shell
(b) Vacancy occupied by electron from higher shell leads to emission of photons and
(c) Energy released causes the ejection of third electron – Auger electron, with characteristic energy for different elements.

5.3.2.2 | Working of XPS

The working principle is summarized below:

(a) XPS is surface-sensitive technique. It can detect even slight surface contamination and hence operated in ultrahigh vacuum (UHV $\leq 10^{-9}$ torr). Surface contamination can be removed by Argon-ion etching (3–5eV).

(b) For insulating materials (metal oxides), the removal of core electrons creates a positive surface potential, which can shift the binding energies to higher values. A low-energy electron flood gun is used to deliver the electrons to the sample surface to neutralize the positive charge.

(c) X-rays penetrate the sample to a depth of few μm, while electron signals are recorded from a depth of 1–10 nm.

(d) The C 1s (BE = 285 eV) signal from adventitious carbon is taken as the reference peak. This is to account for peak shift during data analysis.

5.3.2.3 | Information from XPS signals

(i) XPS provides information on elemental composition and surface chemical/electronic states of elements that exist in a material.

(ii) Other useful analysis is possible. For example,
    (a) peak positions can be used to identify elemental composition;
    (b) peak intensities reveal atom percent of different elements based on relative sensitivity factors of the constituent elements;

(iii) high resolution scanned peaks are deconvoluted and fit in order to determine the different valency/oxidation states of elements;

(iv) electrons ejected from non-'s' sub-shells like p, d and f orbitals are observed in pairs with related intensities, as a result of spin-orbit coupling. The doublet states are characterized by $j = l \pm \frac{1}{2}$ quantum number with $2j + 1$ spin multiplicity of states (Table 5.3). The resulting energy states and the peak intensities are presented in the table 5.3;

(v) depth profiling gives information about the sample at lower depths by alternating steps of data acquisition and surface etching.

Table 5.3 Spin-orbit coupling leading to the splitting of the XPS signals into doublets with a specific intensity ratio (see Ref. 2 in the 'for further reading').

Sub-shell 'l'	$2\,(l-1/2) + 1: 2\,(l + \frac{1}{2}) +1$ Doublet intensity ratio	Designation of energy states
p	1:2	$p_{1/2}$, $p_{3/2}$
d	2:3	$d_{3/2}$, $d_{5/2}$
f	3:4	$f_{5/2}$, $f_{7/2}$

A typical XPS spectrum is shown in Fig. 5.15.

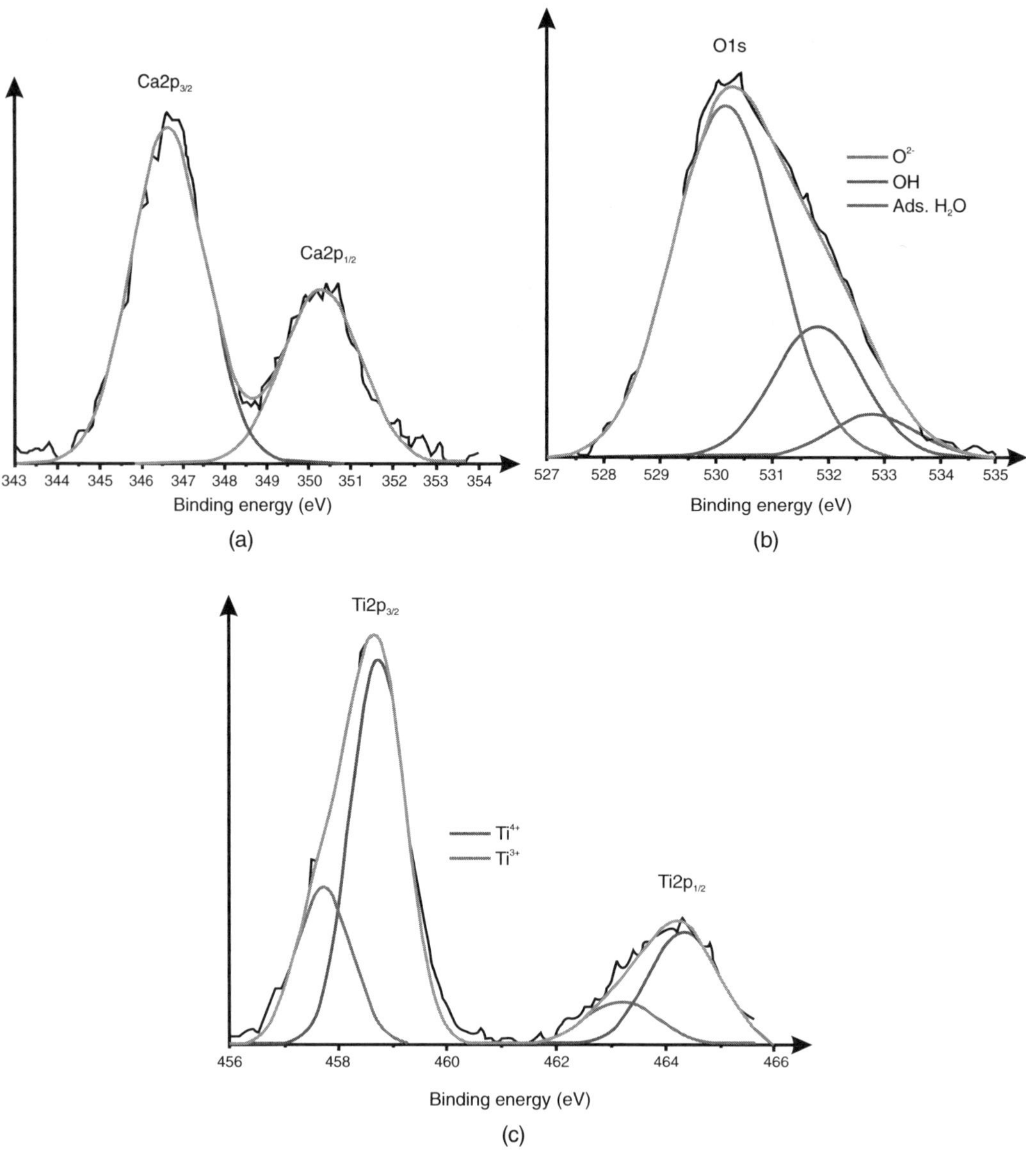

Fig. 5.15 XPS analysis showing deconvoluted peaks obtained from spark plasma sintered $CaTiO_3$, which has proven biocompatibility property: (a) Ca2p, (b) O1s and (c) Ti2p.

5.3.2.4 | Practical guidelines for XPS

Since XPS is mainly a surface analytical method, samples to be analyzed by XPS must be free of contaminants for acquiring a good quality data. One should ensure keeping the equipment clean and use clean tweezers. Powder free nitrile or polyethylene gloves should be used. Proper arrangements should be done for moisture or air sensitive samples like arranging for a glove box, etc.

5.4 | Imaging Techniques for Microstructure Characterization

A host of techniques of microscopy can be used to image cells and tissues. Depending on their resolution, only specific microscopes like fluorescence or confocal microscopy are largely utilized in cell biology related research (see Fig. 5.16). In another chapter in this book, the fluorescence and confocal microscopy are described in details in the context of the qualitative and quantitative understanding of the cell morphological changes. In the following, other microscopy techniques are briefly discussed.

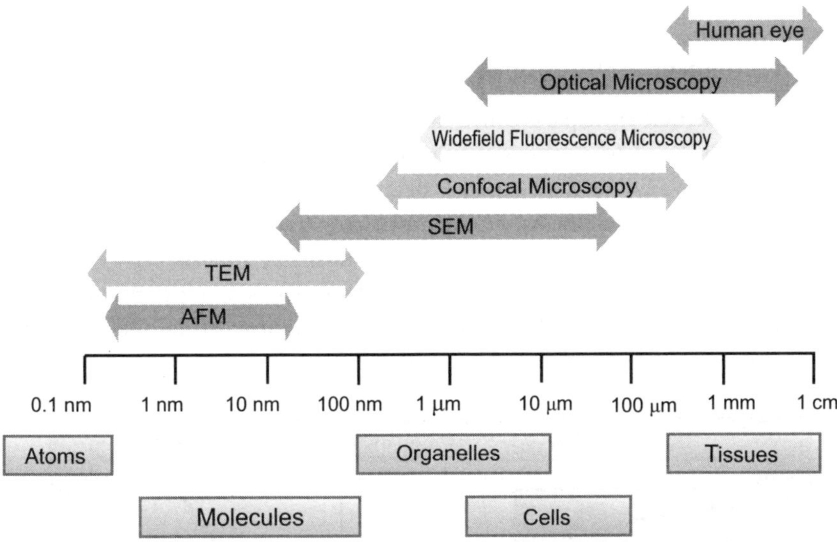

Fig. 5.16 Schematic illustrating the spatial resolution of various imaging techniques to study the morphological characteristics of cellular organelles, cells and tissues.

5.4.1 | Atomic force microscopy (AFM)

Atomic force microscopy analysis generates profile by measuring the forces between a sharp probe tip (radius of curvature<10 nm) and the material surface at small distance (0.2-10 nm).

AFM is exclusively used in biomaterials research. In particular, AFM can be effectively used to scan the surface to reveal the clusters of atoms and molecules, macromolecules like proteins and DNA or cells or bacteria. The probes are typically made of Si_3N_4 or Si. By rastering the tip on the sample surface and recording the change in force as a function of position, AFM provides information on surface topography/roughness, as well as on electrical and magnetic properties (depending on the probe tip). The image resolution obtained in AFM depends on the probe–sample interaction volume and is typically 1 nm in the x–y plane and 0.1 nm in the z-direction. The low scanning rate and the requirement of extremum smooth surface are considered as the notable drawbacks of AFM.

5.4.1.1 | Principle

When the probe tip is in close proximity to the sample surface, the deflection of the cantilever occurs due to a force between the sample and the probe tip. This force (F) is governed by Hooke's law.

$$F = -kx \tag{5.13}$$

where,

x is cantilever deflection,

k stands for spring constant (~0.1–1 N/m) of the cantilever.

As far as the fundamental principle is concerned, the force curve measurements portray the force experienced by the cantilever, when the probe tip approaches and retracts from the sample (Fig. 5 17). Typically, force curve analysis can be used to determine the mechanical hardness and elasticity as well as chemical properties such as adhesion and bond strength. In Fig. 5.17, the slope of the deflection (C) provides information on the sample hardness. The adhesion (D) is a measure of the interaction between the sample and the probe tip, as the probe is being retracted. In the context of biomaterials characterization, AFM can also be used to obtain substrate and cell stiffness. Such measurements can be correlated with cellular functionality.

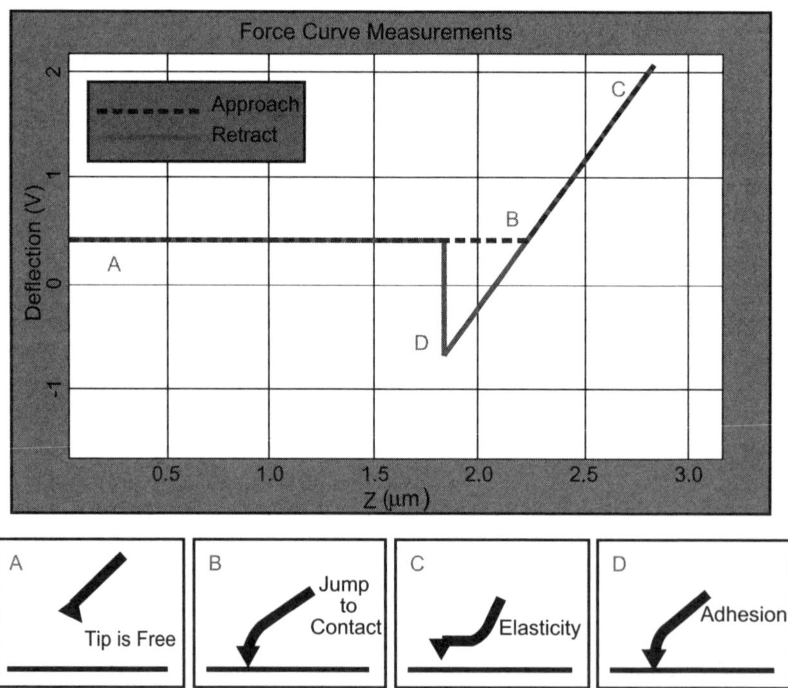

Fig. 5.17 A typical force curve measurement showing different types of probe-sample interaction.

5.4.1.2 | Working of AFM

The operation of AFM involves measurement of probe deflection by the 'beam bounce method'. The laser beam is bounced off the back of the cantilever onto a position sensitive photodetector (Fig. 5.18). The detector measures the bending of the cantilever as the tip is rastered over the sample surface. The measured cantilever deflections are used to generate the surface topography.

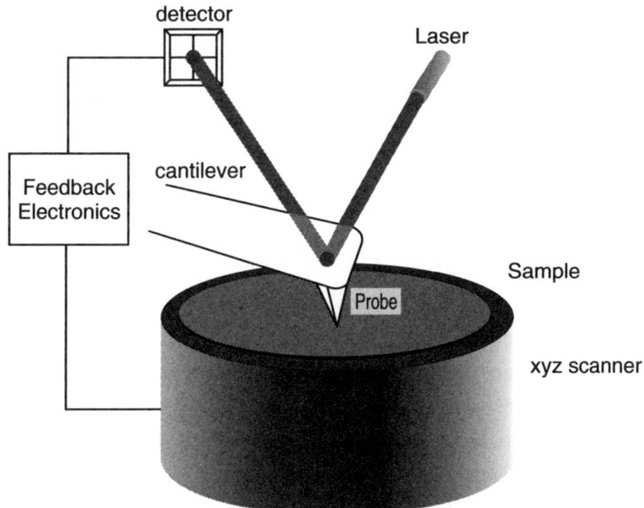

Fig. 5.18 Basic working of an AFM showing how cantilever deflection due to probe-sample interaction is monitored by a feedback mechanism with the help of a diode laser.

5.4.1.3 | Imaging modes in AFM

Three primary imaging modes in AFM are possible and the spatial extent of each of these regions belongs to the Lenard–Jones potential curve, as shown in Figs. 5.19 and 5.20.

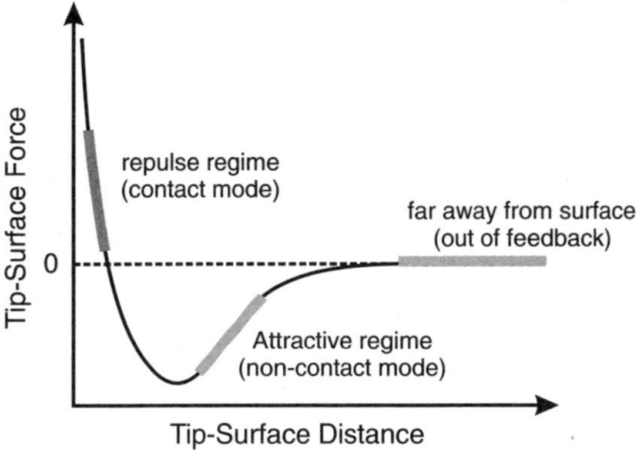

Fig. 5.19 Lennard–Jones potential curve showing the attractive and repulsive force regimes corresponding to non-contact and contact modes of AFM operation.

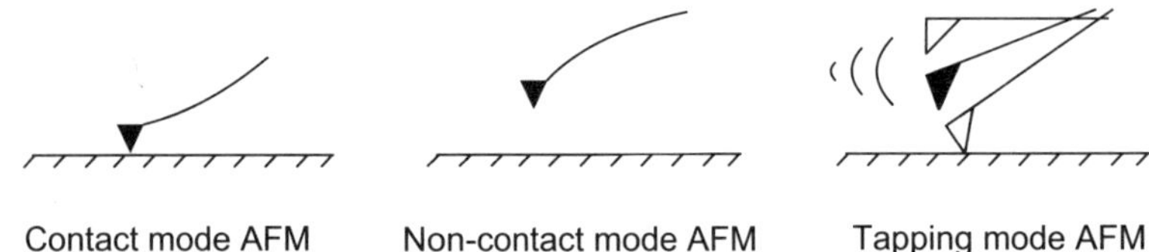

Fig. 5.20 Pictorial representations of the three modes of AFM operation: viz., contact, non-contact and tapping modes.

(i) Contact mode: < 0.5 nm separation between the probe tip and surface; in the strongly repulsive force regime; constant force or distance.

(ii) Non-contact mode: 0.1–10 nm probe tip to surface separation; weak attractive force regime, vibrating probe.

(iii) Tapping mode: 0.5–2 nm probe tip to surface separation; strongly repulsive; vibrating probe.

5.4.1.4 | Applications of AFM

In the field of materials science, AFM is used to investigate the morphology of nanoparticles or to capture the topographic images containing nanoscale features. With AFM, one can also obtain force–displacement plots to estimate the surface stiffness. In biological applications, AFM technique can act as a force sensor to estimate the bond strength between different biological molecules. For example, this system has been successfully applied to probe biotin–streptavidin, cell-adhesion proteoglycans, antigen–antibody, protein dynamics and complementary DNA strands interactions, etc. AFM has also found tremendous applications in implant materials to test material host/cell interactions and in tissue engineering. In particular, AFM can be used to image the cell adhesion as well as cell morphological changes on a material substrate.

5.4.1.5 | Practical guidelines for AFM

An extremely smooth surface with a roughness of less than 0.1 μm should be used. Tapping mode should be used to obtain topography. A lower scanning speed is better to obtain a good resolution.

5.4.2 | **Scanning electron microscopy (SEM)**

Scanning electron microscope uses an electron beam as the probe to scan the conducting surface of a material. The electrons are generated from a heated tungsten or LaB_6 filament (thermal gun); or by the application of a large positive potential at the tip of the filament (field emission gun). The electron beam is guided to the sample surface with the help of condenser/electromagnetic lenses, which works on the principle of Lorentz Force,

$$F = -q(v \times B) \tag{5.14}$$

where,

v = velocity of the electrons,

B = magnetic fields of the lens, which can be varied by changing the current through the coils.

q = charge on the electron

5.4.2.1 | Electron beam–material interaction

As an electron beam strikes a material surface, the interaction between the electron and the sample leads to the generation of different signals (shown in Fig. 5.21), which can be used to obtain various kinds of sample information. The electron beam, travelling through electromagnetic lens guided path inside the column of a SEM, interacts with the conducting surface of a bulk material. For a bulk non-conducting material, the surface is made to be conductive by sputtering a conducting thin layer of gold/silver/carbon onto the surface.

Depending on the accelerating voltage of the electron gun, the interaction volume extends to some limited depth from the surface of the material. Theoretical calculations revealed such interaction zone to be of pear-shaped and all the signals, are generated from the interaction zone.

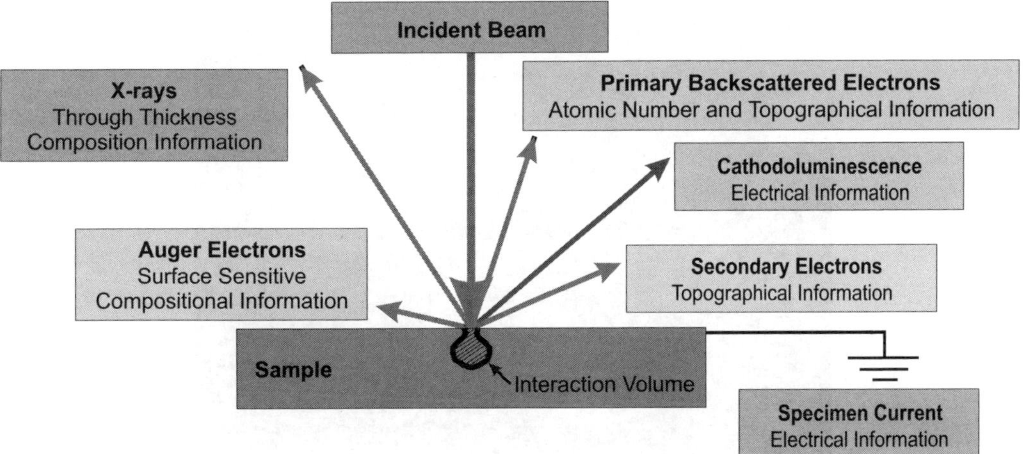

Fig. 5.21 Types of signals generated due to the interaction of an electron beam with a bulk material surface. The interaction volume is also shown (not to scale).

Typically, Auger electrons are released with the topmost region of within 10–20 Å of the surface, followed by around 5–50 nm for secondary electrons. The backscattered electrons are released from the 1 µm depth within the pear-shaped interaction zone, while X-rays are generated from a depth of 2–5 µm. While interpreting the information from SEM analysis, the above information should be duly considered. In particular, the compositional analysis using EDS attached to SEM is typically carried out using X-rays generated due to the above-mentioned interaction. Depending on the grain size of a particular phase, one therefore has to assess the source of X-rays generated during spot analysis using SEM-EDS facility. Also, SE-BSE based morphological analysis using SEM is restricted to only surface region of a material. In case 2D microstructural analysis is not sufficient to correlate specific properties with microstructure, then 3D microstructural analysis is to be conducted using X-ray micro computed tomography (micro-CT; as detailed in section 5.5).

It is recommended to use both SEM and micro-CT to obtain pore morphology in 2D surface and 3D space, respectively with both having its independent implications towards cell adhesion/growth and functionality changes.

It may be worthwhile to mention that SEM, in the context of cell study, can only be used to describe cell morphological changes on a material surface. For identification of cytoskeleton changes or mitochondria or nucleus, one has to use either fluorescence or confocal microscope. A representative SEM image showing the adhesion of fibroblast cells on a bioceramic surface is shown in Fig. 5.22.

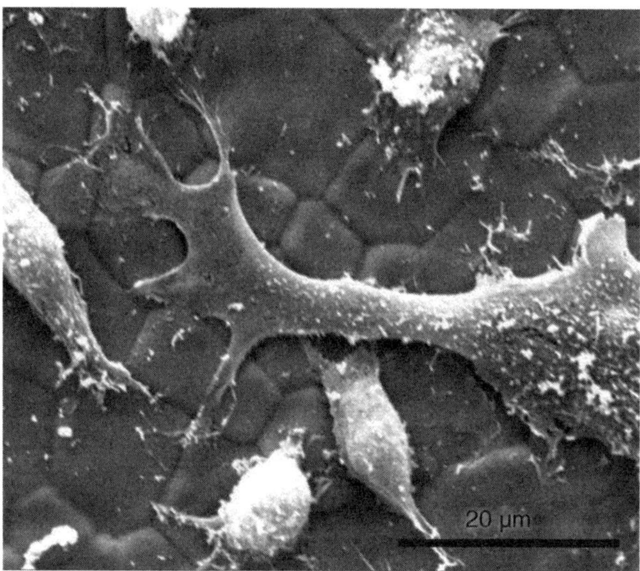

Fig. 5.22 SEM image revealing cell morphology and spreading of L929 mouse fibroblast cells on sintered hydroxyapatite. The grains and grain boundaries are clearly visible on thermally etched biomaterial surface (author's unpublished result).

In the context of SEM, the image formation is possible in the following two modes:

(i) Secondary electrons (SE)- Signals generated by inelastic scattering, from a region within 5λ (λ is the mean free path) of the sample surface. SE provides surface **topography contrast**. They possess a very low energy < 50eV and are detected by Everhart Thornley (ET) detector.

(ii) Back scattered electrons (BSE)- These signals are generated by elastic scattering with a positive bias (+200V) and have an escape depth of much greater than SEs. BSEs possess energy slightly lesser than the incident electron beam. The BSEs generated are sensitive to the atomic number of the sample (**Z-contrast/compositional contrast**). Hence, heavier elements appear brighter than elements with low atomic number in BSE images.

The image formation in SEM depends on the following key parameters:

(a) Probe size- Smaller the probe diameter (d), greater is the resolution with a corresponding decrease in brightness.

(b) Probe current- Larger the probe current, greater is the brightness/signal to noise ratio. However, to enable larger probe current, larger convergence angle (α) of the beam is necessary which can lead to spherical aberration. The probe current is determined by a number of parameters as per the following equation,

$$I = \frac{\beta \pi^2 d^2 \alpha^2}{4} \tag{5.15}$$

where,

I stands for the probe current,

β represents brightness,

d represents probe diameter,

α stands for convergence angle.

(c) Convergence angle (α)- Smaller convergence angle increases the depth of focus while simultaneously decreasing the probe diameter and hence the brightness. Smaller α requires lesser aperture width, increasing diffraction effects.

(d) Depth of focus (DOF)- It is the region above and below the plane of optimum focus where the specimen appears to be in focus. Smaller convergence angle and larger working distance give better DOF.

In addition to obtaining images, compositional analysis also can be conducted using Energy Dispersive X-ray Spectroscopy (EDS), wherein inelastic collision of the high energy electron beam can lead to the removal of the core level electrons from the atoms. This results in the generation of characteristic X-rays. Thus, EDS provides information on the sample composition.

5.4.2.2 | Sample preparation and operation of SEM

The following precautions must be taken while using SEM:

(a) The samples must be dried thoroughly and desiccated overnight in order to ensure the complete removal of water/solvent.

(b) Insulating samples must be sputtered with a thin layer of Au, Au/Pd or carbon (especially for imaging at higher resolution). This will prevent charging of the samples due to the removal of surface electrons.

(c) The specimen chamber of the SEM is maintained at a pressure of $\sim 10^{-7}$ Pa while the operating acceleration voltages range from 0–30 kV.

(d) The maximum resolution that can be obtained in a SEM, as of date, is ~ 2 nm (with field-emission gun). However, such resolution can only be obtained with ultra-high vacuum level in the sample chamber.

(e) Environmental scanning electron microscope (ESEM) is used to image biological specimens in their native state without dehydration of the sample. The sample chamber is saturated with water vapour to maintain the chamber humidity. However, such environmental chamber leads to compromise with the resolution that can be achieved.

5.4.2.3 | Practical guidelines for SEM

One can use both bulk as well as powder samples for SEM analysis. In case of non-conducting samples, a thin layer of gold or carbon coating is to be done to facilitate interaction with electron beam. For topographical analysis, SE mode should be used and to obtain compositional contrast, BSE mode needs to be used. Additionally, to analyze the composition of the sample under study, Energy dispersive X-ray spectroscopy can be performed. One can also opt for various options using EDS, such as spot analysis, line scanning and even area scanning. For better understanding of the sample features, it is recommended to start from lower magnification and proceed gradually to higher magnifications, while recording images. However, images recorded at a certain magnification should be first focused at relatively higher magnification. Additionally, for recording a high quality SEM image, the image should be recorded under slow scanning mode.

5.4.3 | **Transmission electron microscopy (TEM)**

At this juncture, it is instructive to mention that while SEM is perhaps the most widely used electron microscopy technique in research on various materials, its application to probe into extremely finer microstructural features at submicron length scale (in particular at nanoscale) is limited due to resolution. From the preceding discussions, it should be clear that the limited resolution is mostly due to the use of 20–30 KV accelerating voltage. The operating voltage of TEM is usually 200–300 kV. The combinatorial approach of using electron transparent thin film together with higher acceleration voltage of 200–300 kV can be useful to obtain images with much higher resolution. This has been the rationale for the use of TEM, wherein the transmitted beam from an electron transparent thin sample is analysed to obtain microstructural details at nanoscale.

In particular, TEM is advantageous over SEM in that diffraction, spectroscopy and imaging can be carried out from the same region of an electron transparent thin sample, primarily using the following techniques:

(a) Imaging- grain size, defects, dislocations, phase/mass contrast.
(b) Diffraction- phase identification, crystal/grain orientation, defect characterization.
(c) Spectroscopy- EDS, EELS for compositional information.

As in SEM, a heated tungsten filament or a field emitter is the source of electrons. Generally, a parallel beam of electrons is incident onto the sample and the transmitted electron signals are detected.

For the specimen to be electron transparent, the sample must be thinned down to 80-100 nm by dimpling or precision ion milling using argon beam or focused ion beam.

5.4.3.1 | Working principle

It can be mentioned here that two important aspects distinguish TEM from SEM. These include: (a) use of an electron-transparent thin sample in TEM to transmit a higher energy electron beam to reveal finer scale microstructural details, as opposed to much lower energy electron beam interacting with a rather bulk sample with finite thickness in SEM (energy of e-beam in SEM ~ 20 kV and that used in TEM ~ 200 or 300 kV); and (b) the efficacy of TEM to identify crystalline or amorphous phase at nanoscale as well as identifying the structure of the crystalline phases at finer scale. Both of these aspects are accomplished with the use of specific imaging mode and careful analysis of diffraction patterns, as briefly described in this sub-section with the help of Figs. 5.23–5.26. In reference to Fig. 5.23, an image is formed in a TEM due to the spatial intensity variation after the electron beam passes through the specimen, while a diffraction pattern results from the angular intensity variation. In particular, bright field (BF) images are formed when the direct beam is used to form the image, while dark field image (DF) is obtained when any diffracted beam(s) is used to form the image. In a bright field image, heavier atoms appear dark and the lighter atoms appear bright. This is due to the greater interaction of the electron beam with the heavier atoms leading to lesser transmitted signal.

> For biological samples, the samples are negatively stained with salts of heavy atoms, like uranyl acetate, to improve the contrast during electron microscopy analysis.

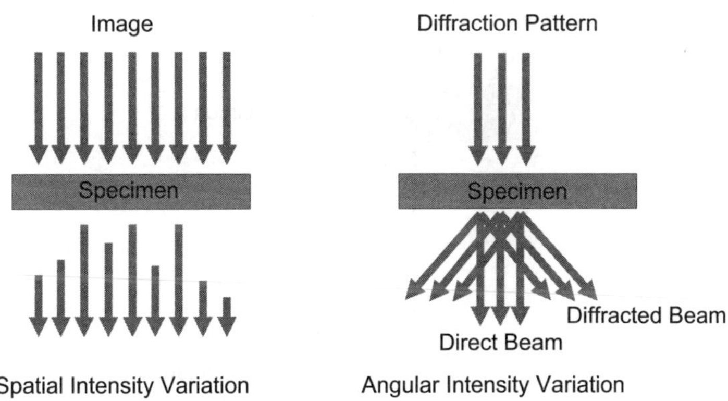

Fig. 5.23 Basis for obtaining image and diffraction pattern using a transmission electron microscope [Adapted from Ref. 554].

In order to view a diffraction pattern (in the back focal plane of the objective lens), one has to insert an selected area aperture into the image plane of the objective lens (Fig. 5.24). While analyzing the diffraction patterns, the reciprocal of the interplanar spacing (d) of a crystalline phase needs to be determined with the highest level of accuracy. The diffraction pattern formation can be better understood by the use of electron ray diagrams, as shown in Figs. 5.24 and 5.25. While the direct beam will form the brightest spot in a diffraction pattern, the diffracted beams cause spots of different brightness. In reference to Fig. 5.25, the relationship between d and L (the camera length) can be described by the following relationship:

$$Rd = \lambda L \tag{5.16}$$

where, R is the distance between the diffraction spots, λ depends inversely on the accelerating voltage, L is the camera length, and d is the interplanar spacing.

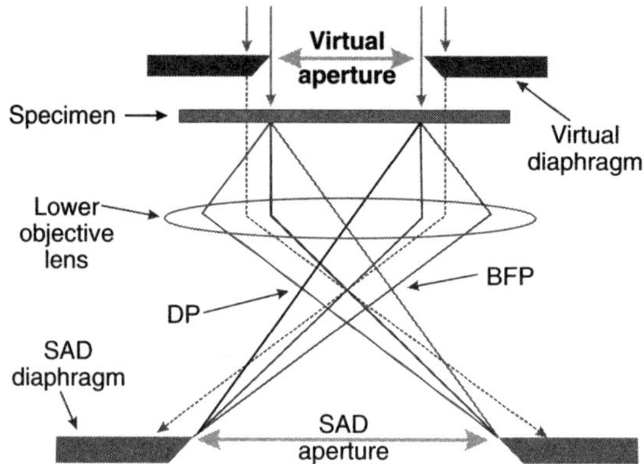

Fig. 5.24 Ray diagram illustrating selected area diffraction pattern (SADP) formed at the back focal plane of the objective lens [Adapted from Ref. 554].

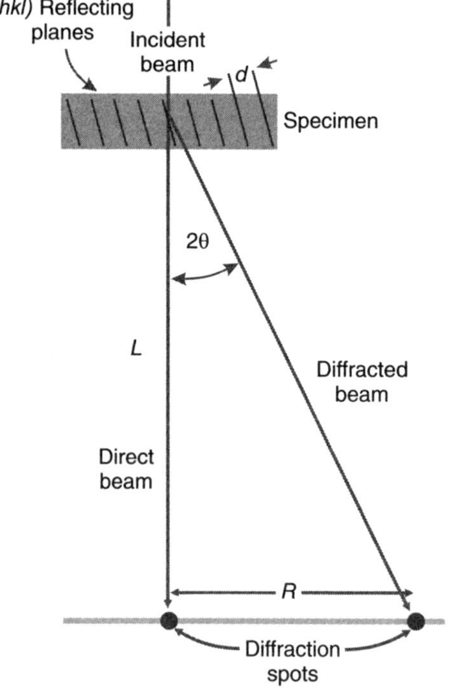

Fig. 5.25 Relation between diffraction spots and reflecting planes for constant camera length and energy of the electron beam [Adapted from Ref. 554].

The ratio of any two R-spacing gives the inverse ratio of d-spacing, which is characteristic of the material. Similar to XRD analysis, $F \neq 0$ holds true for allowed reflections and $F = 0$ is valid for forbidden reflections.

For polycrystalline samples, diffraction from crystallites in all orientations leads to the formation of ring patterns (Fig. 5.26). The breadth of the ring is inversely related to the crystallite/grain size. It is important to note that, broader the rings are, smaller is the grain size. A spot pattern is obtained from a single grain by appropriate tilting of the sample so as to achieve the correct orientation/direction that contains all the diffracting planes. Zone axis [U, V, W] is the direction perpendicular to all the planes (h, k, l) lying in the diffraction pattern. One of the important aspects of TEM analysis is to index the SAED patterns. For this one has to effectively use Weiss zone law. The (hkl) reflection must lie on the [UVW] zone axis such that the relation hU + kV + lW = 0 for the allowed reflections is satisfied. Obtaining specific zone-axis diffraction patterns usually helps in the analysis. In order to accurately identify the material, usually more than one diffraction pattern (more than one zone axis pattern) is required.

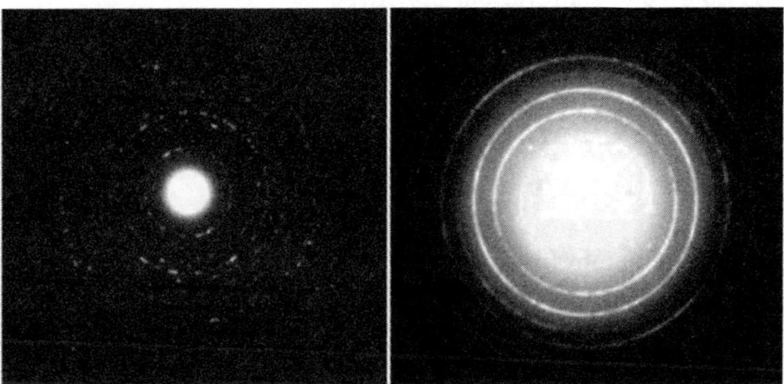

Fig. 5.26 Illustrative examples of TEM diffraction showing the formation of rings for polycrystalline samples (right) and spot pattern for a single grain/crystallite (left) [Adapted from Ref. 554].

5.4.3.2 | Practical guidelines for TEM

Formvar coated Cu (or carbon) grids should be used for powder samples and for other samples a very thin electron-transparent section should be cut from the bulk sample, using conventional precision machining techniques. For metallic sample, the electro-polishing of a thin sample of micron thickness can be used to obtain sample with <100 nm thickness. For non-conducting ceramic sample, precision ion polishing system (PIPS) is used for a machined sample of thickness close to 1–2 µm. Prior to electro-polishing, dimpling is often adopted. In view of the brittle nature of ceramic samples, the focused ion beam (FIB) based machining of localized region in a bulk sample can be used prior to PIPS. It is also important to note here that the preparation of thin samples with thickness of less than 100 nm is by far the most difficult step in TEM analysis. The quality of TEM results is significantly dependent on uniform thin nature of the sample. The difference in thickness across a TEM sample is reflected in additional contrast or artefact (thickness fringes) in the image. The electron beam should be aligned at the central region of the screen. Initially, a bright field mode should be used and one should look for the thinnest region (preferably next to hole). One should always start from lower magnification and proceed to higher magnifications. Also, to study inclusions or precipitates of distinct morphology, dark field mode should be used. In order to identify various phases present, Selected Area Diffraction Pattern (SADP) should be taken. For compositional analysis, energy dispersive X-ray analysis (EDX) should be performed. Similarly, for light element analysis electron energy loss spectroscopy (EELS) should be performed and high-resolution transmission electron microscopy (HRTEM) can be used to identify grain boundary phases.

It is worthwhile to mention that a different sample preparation technique is to be adopted for soft polymeric sample or biological samples (tissue/cell/bacteria). The ultramicrotome is widely used to obtain thin sections of thickness between 50 nm and 100 µm. For TEM analysis, the thickness of less than 100 nm is to be used and lower the thickness, the better is the resolution. In microtome, the steel blades are used to prepare animal/plant tissue for light microscope/histology analysis. In contrast, diamond glass knives are used to slice thin sections of bone/teeth/plant tissue/cell/bacteria for electron microscopy analysis. Another point that must be mentioned here that any tissue section needs to be stained with osmium tetraoxide (OsO_4), which acts as a contrasting agent by absorbing electrons. OsO_4 is also used to stain various polymers.

5.5 | 3D Structural Characterization using X-ray Micro Computed Tomography (micro-CT)

X-ray micro computed tomography (micro-CT) is a non-destructive experimental technique, which enables us to virtually reconstruct the internal microstructure of the sample with high spatial resolution in 3D with the use of X-ray shadow images of the samples at different orientation.

Fundamentally, micro-CT is based on the principle of X-ray absorption. Micro-CT is the most effective tool in quantifying microstructural architecture of 3D porous scaffold.

This technique is also equally effective for visualizing bone-tissue ingrowth, *in vivo*[544,545,546,547]. The main advantage of micro-CT is that it is a non-destructive technique[548]. Compared to conventional techniques, like scanning electron microscopy (SEM), flow porosimetry, gas pycnometry and gas porosimetry, micro-CT has an edge because it allows the user to measure a large pool of biologically relevant parameters in a quantitative manner[549].

Concerning other complementary or alternative characterization techniques (e.g., SEM, flow porosimetry, etc.), each of them has advantages and disadvantages. For example, the sample handling for SEM is easy, but the information we get is only limited to 2D surface from where one cannot quantify pore interconnectivity. Flow porosimetry cannot detect the closed or blind pore, because the pores must be interconnected to allow intrusion of fluid in pore. Therefore, the total pore volume fraction cannot be obtained using this method[550]. Also, pre-treatment in a vacuum oven for volatile substance removal is a pre-requisite in gas pycnometry for samples prior to measurement via gas pycnometry. In contrast, the sample handling in micro-CT is relatively easy and some commercial instruments (e.g., Skyscan and Bruker) can even scan organs of live animals[551]. As long as the sample shows considerable X-ray attenuation, any kind of material with any degree of complexity can be imaged using micro-CT, but the minimum feature size detected is dependent on the resolution of the instrument.

As far as the appearance of different tissues from live animal scan on a micro-CT image is concerned, it is worthwhile to remember that X-rays are electromagnetic waves. The main reason why X-rays are used in diagnostic purposes is that all tissues differ in their ability to absorb X-rays. Some tissues are more permeable to X-rays, while others are impermeable. Owing to such differences, different tissues appear differently, when the X-ray film is developed. For example, the dense tissues, such as the bones appear white on a CT film, while the soft tissues, such as the brain or kidney appear gray. The cavities filled with air, such as the lungs appear black.

As far as the functioning of micro-CT is concerned, the sample is fixed on a stable stage and rotated with fixed steps determined by the number of projections (see Fig. 5.27). As discussed earlier, X-ray diffraction typically uses X-ray source of 20 kV. In contrast, micro-CT uses high energy X-ray source (80 or 100 kV). During the interaction of X-rays with a 3D material, X-rays get deflected and absorbed to different extent and the sample is scanned at different angles and the intensity of transmitted X-rays is measured. In high resolution micro-CT, low-energy X-ray source is used together with high-resolution detectors.

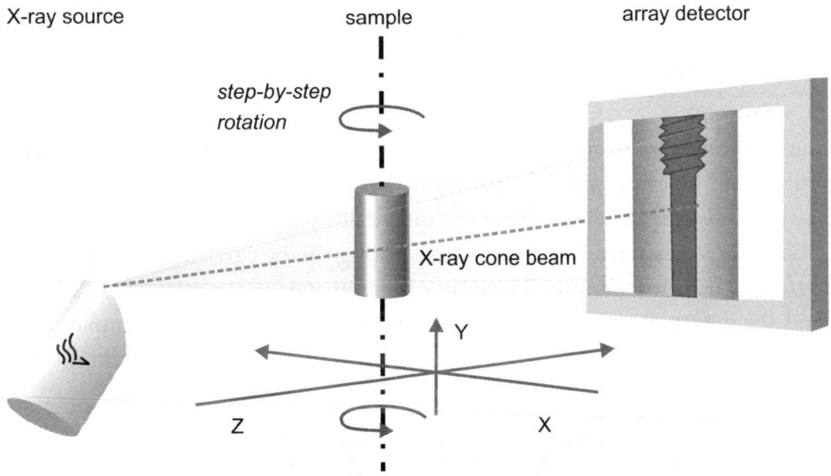

Fig. 5.27 Schematic illustration showing the working pinciple of micro-CT.

It is particularly important to mention here that greater the number of projections, better is the quality of the tomogram. Various image reconstruction protocols are used to intelligently stack the 2D projection around of the rotational (e.g. cone beam projection) axis to get the 3D tomogram.

Typically, 2D slices are reconstructed from 1D projections and for the same, number of angular views are needed. Also, the reconstruction is based on 'back projection' of all views. The image formation in micro-CT is facilitated using CCD, which converts the optical image into electronic signal.

It may be worthwhile to define the spatial resolution, which is the measure of how closely lines can be resolved in an image. In fact, spatial resolution refers to the number of independent pixel values per unit length. A pixel is defined as the smallest distinguishable element in a display device. These are arranged in 2D grid. In contrast, the image resolution describes the details that an image is characterized for. According to standards, the number of effective pixels that an image sensor or camera has is the count of elementary pixel sensor that contribute to the final image. To this end, a voxel is defined as the volume element, representing a value on a regular grid in 3D.

In general, there can be different modes of imaging, e.g., attenuation (absorption), phase contrast, etc., but most of the micro-CT instrument around the world is based on X-ray absorption. Typically, the acceleration voltage used in an X-ray source of a micro-CT machine is much higher than that used in a conventional X-ray diffraction machine. Concerning the image acquisition and analysis, both the 3D (volume rendering, iso-surface rendering) as well as the 2D (slices of one voxel thickness) images are used to study different phases, which can also include pore, fracture, etc. Depending on the absorption of different elements in a given material, they appear with different grey level. This is the basis to distinguish different phases in absorption based tomography.

Many useful parameters e.g., porosity, pore interconnectivity, pore wall/strut thickness, shortest distance of pores from periphery to pore through interconnection (SDI), depth of periphery (DP), detour index (DI), volume fraction of open and closed pores can be estimated from a single reconstructed 3D image. This is the major advantage of micro-CT.

The porous network can be visualized using inverse threshold and useful parameters, like pore interconnectivity can be evaluated, as reported by Wang et al.[552]. The scanning of same sample before and after *in vivo* implantation allows the quantification of tissue regeneration at the implant site. In addition, the *in situ* tensile or compressive testing can be carried out on the porous scaffold and the progressive deformation under loading can be imaged using micro-CT. This enables one to understand the deformation characteristics of porous scaffolds. Some examples of tomograms obtained with two different kinds of materials are shown in Figs. 5.28 and 5.29.

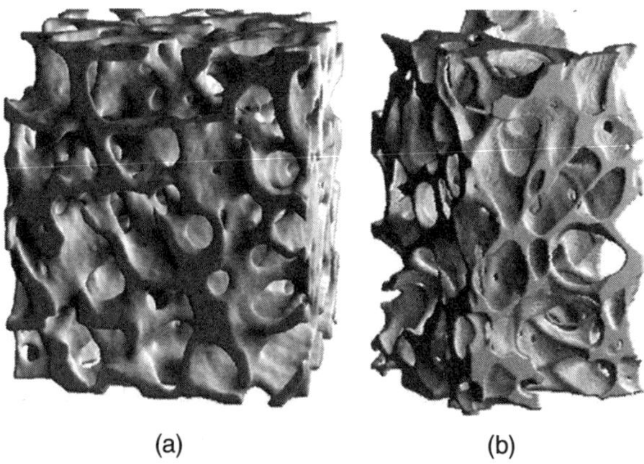

(a)   (b)

Fig. 5.28 (a) Typical micro-CT images revealing the natural trabecular bone and (b) the synthetic porous polymer mimicking the trabecular bone structure[555].

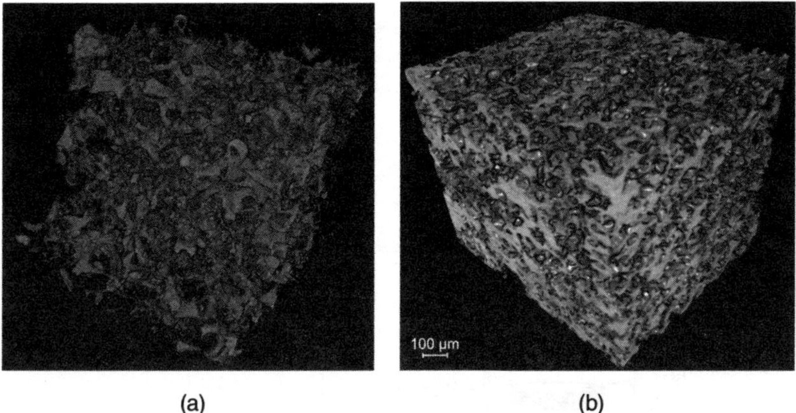

Fig. 5.29 Porosity and pore interconnectivity, shown after image-processing of micro-CT tomogram of (a) highly porous biodegradable polyester and (b) porous hydroxyapatite.

5.6 | Electrical Characterization

In the field of biomedical engineering, it has been widely recognised that biophysical cues in the form of electric or magnetic fields can stimulate multiple cellular responses and therefore, enable us to manipulate cell fate processes of certain cell types and in particular, mesenchymal stem cells[34]. In particular, the magnetic stimulation via static or pulse magnetic field is reported to accelerate bone repair in case of delayed fracture union or to treat various bone disorders (e.g. osteoporosis, healing of osteotomies, etc.). Recent research conducted in the author's group (see Ref. 34) has conclusively established the synergistic interaction of the electric field stimulation with electroconductive substrates and magnetic field stimulation with magnetoactive substrates towards cell functionality modulation, respectively. Therefore, it is important to develop electroconductive or magnetoactive biomaterials. This rationalises the need to have an understanding as how to characterise the electrical and magnetic properties. In the above backdrop, the following section discusses the principles and measurements of such properties.

5.6.1 | Electrical impedence spectroscopy

The dielectric properties of dense materials can be determined by measuring the frequency vs. capacitance and dielectric loss vs. frequency responses at a particular temperature using an electrical impedance analyser. In this method, the silver coating is laid over a sample surface and a copper wire is connected to the anode part of the sample surface, as shown in Fig. 5.30.

The dielectric constant (ε_r) can be calculated from the capacitance vs frequency curves, recorded by the precision impedance analyser at a particular temperature, as per the following relation.

$$\varepsilon_r = Cd/\varepsilon_o A \tag{5.17}$$

where, C is capacitance, d is thickness and A is area of the sample electrode without silver paste/resin, ε_o is the permittivity of free space (8.854×10^{-12} F m^{-1}).

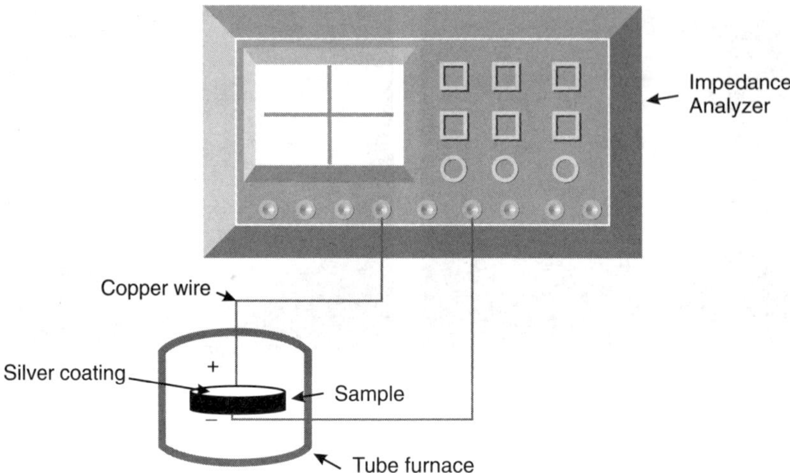

Fig. 5.30 Schematic showing the experimental set up of impedance analyser for performing dielectric measurements.

The dielectric loss factor (tanδ) is also measured as a function of frequency by the instrument. The AC conductivity (σ in ohm^{-1} cm^{-1}) can be calculated from the measured dielectric data by the following relationship,

$$\sigma_{AC} = \omega\varepsilon_o\,\varepsilon_r\,\tan\delta \qquad (5.18)$$

where, ω is $2\pi f$ and f is the frequency in Hz.

For highly insulating materials, the conductivities are determined by impedance analyser, while more accurate values are obtained by 4-probe method in case of conducting and semi-conducting materials, as described below.

5.6.1.1 | Electrical conductivity measurements

The classes of materials based on electrical conductivities are tabulated below along with their ranges of resistivity and conductivity at room temperature (Table 5.4). Next, the theory and the measurements of the electrical conductivity are explained.

Table 5.4 Classification of materials based on electrical conductivity.

Material classes	Resistivity (Ω m)	Conductivity (S/m)
Insulators	$10^6–10^{17}$	$10^{-18}–10^{-7}$
Semiconductors	$10^{-3}–10^5$	$10^{-6}–10^3$
Conductors	$10^{-9}–10^{-2}$	$10^2–10^8$

5.6.1.2 | Two probe measurements

As shown in Fig. 5.31, this method is used for electrical characterization of near insulator or high resistivity materials. The measured conductivities are of the order of $10^{-8}–10^{-18}$ S/m (beyond the range of 4-probe method). In this method, the sample is electroded between two copper plates and with a simple voltmeter–ammeter circuit, the resistivity is measured.

In this method, the current is sourced at two terminals and the potential is measured at the other two terminals to determine the resistance (Fig. 5.32). From the above resistances (R_A and R_B), the sheet resistance (R_s) is calculated by the Van der Pauw method:

$$\exp\left(-\pi R_A/R_s\right) + \exp\left(-\pi R_B/R_s\right) = 1 \tag{5.19}$$

Bulk resistivity $\rho = R_s/d$, where d is the sample thickness. Subsequently, the conductivity can be obtained using $\sigma = 1/\rho$ (S/m). It may be further noted that for the measurement of surface resistivity, four probe method can be adopted whereas, two probe method is ideally suited for the determination of through thickness resistivity.

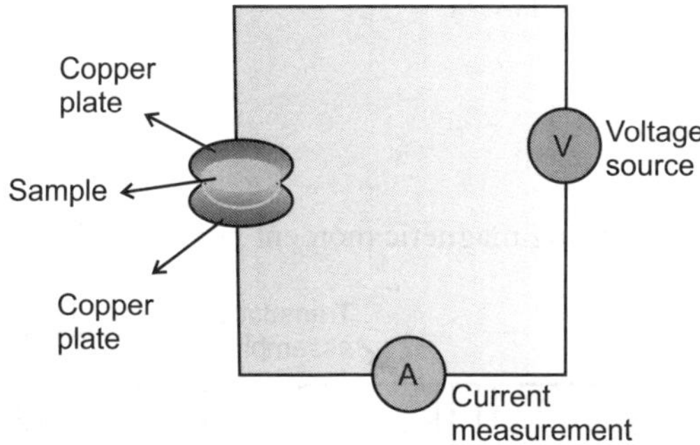

Fig. 5.31 Two-probe configuration for measuring the electrical conductivity of nearly insulating samples.

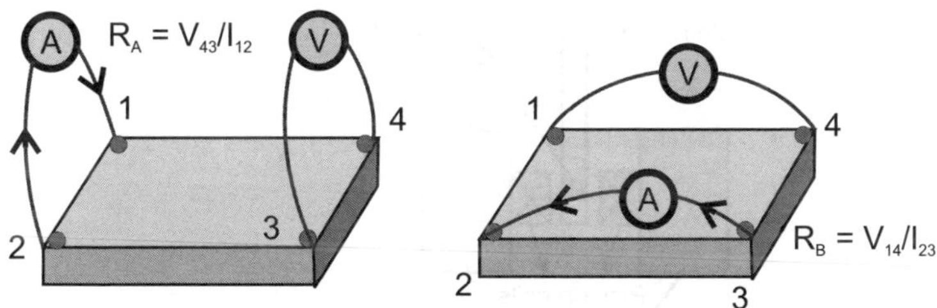

Fig. 5.32 Experimental configurations for four-probe conductivity measurements of conducting and semiconducting materials.

5.7 | Magnetic Characterization

5.7.1 | Vibrating sample magnetometry (VSM)

VSM is based on Faraday's law of induction, which states that a change in magnetic flux density generates an induced voltage in a coil,

$$U_{ind} = -\,\partial\Phi/\partial t = -n_w A\,\partial\mathbf{B}/\partial t \tag{5.20}$$

where,

U_{ind} is the induced voltage,

Φ is the magnetic flux,

n_w is the number of windings and A is area of the pick-up coils.

In the experimental set up (see Fig. 5.33), the oscillator generates a sinusoidal signal which is translated into a vertical vibratory motion by the transducer assembly. The sample is fixed to a sample rod, which is placed between the poles of an electromagnet and the rod vibrates with an amplitude and frequency of ~ 1 mm and 60 Hz, respectively. The sample is magnetized and demagnetized by the sweeping field produced by the electromagnet (varied by changing the current in the coils of the electromagnet) from -2 to +2 Tesla. For a ferromagnetic sample, the magnetization is related to the applied magnetic field by the following equation.

$$B = \mu_o(H_o + M) \tag{5.21}$$

where,

B is Magnetic flux density,

H_o is homogenous field,

M is Magnetization,

M is d$\mathbf{m}$/dV where $\mathbf{m}$ is the resulting magnetic moment per unit volume.

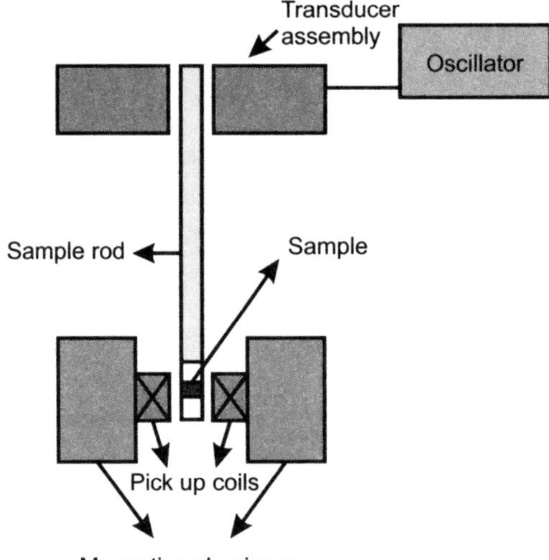

Fig. 5.33 Schematic of a vibrating sample magnetometer (VSM) consisting of an electromagnet and a transducer assembly.

The relationship between magnetization and the magnetic field can be described by the relationship, $M = (1+ \chi)$ H; where χ is the magnetic susceptibility. The ferromagnetic materials have large values of χ ($\gg$ 1). Depending on the temperature, a material can undergo magnetic phase transition and the temperature for ferromagnetic to paramagnetic transition is known as the Curie temperature (T_c). Similarly, the antiferromagnetic to paramagnetic transition takes place at Neel's temperature (T_N). In a constant magnetic field (H_O),

$$\partial\mathbf{B}/\partial t = \partial\mathbf{M}/\partial t \tag{5.22}$$

However, the vibration of the sample and the change in magnetic flux density around the pick-up coils lead to an induced voltage which is proportional to the below parameters

$$U_{ind}(t) \propto -\mathbf{m}\,\omega\,Z\,y_o\,n_w\,n_c\,G\cos(\omega t) \tag{5.23}$$

where,

ω and Z are frequency and amplitude of vibration,

y_o is distance to the pick-up coils,

n_c is the number of pick up coils,

G is the geometric factor,

m is the magnetic moment of the sample,

n_w is the number of windings.

Two categories of experiments can be performed using a VSM:

(i) Magnetization of a material as a function of applied field at constant temperature provides information on the nature of the magnetic material (diamagnetic, paramagnetic, and ferromagnetic). Fig. 5.34a is a typical hysteresis loop of a ferromagnetic material at room temperature.

(ii) Magnetization of a material as a function of temperature under constant field provides information on magnetic phase transformations of materials at characteristic temperatures. Figure 5.34b shows characteristic Curie temperature (T_c) for ferromagnetic to paramagnetic transition and Neel temperature (T_N) for antiferromagnetic to paramagnetic transition.

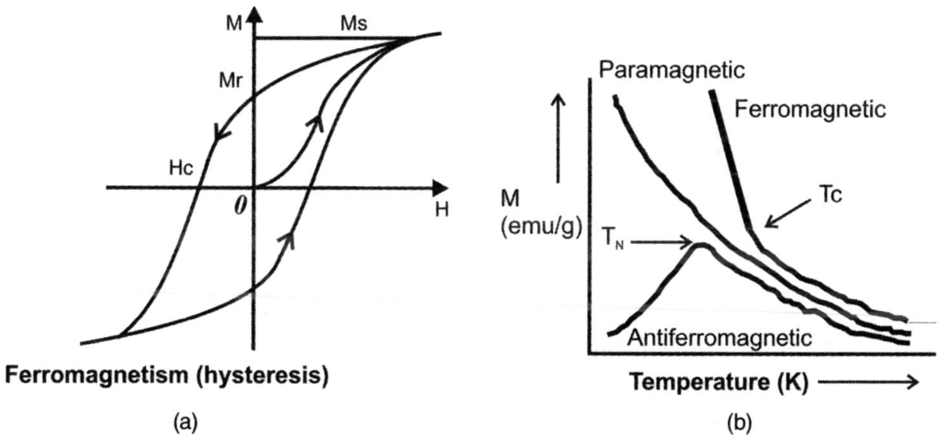

Fig. 5.34 (a) Typical magnetic hysteresis response of a ferromagnetic material, which can be recorded using VSM [M$_s$: saturation magnetization; M$_r$: remnant magnetization; H$_c$: coercive field]; (b) Magnetic phase transitions detected by measuring sample magnetization as a function of temperature under constant magnetic field.

5.7.1.1 | Practical guidelines for VSM

The sample dimensions depend on the geometry of the coils in the VSM. The sample must be securely fastened for noise reduction, which is generally challenging. Generally, a small vertical size compared to the baseline of the sample is necessary for noise minimization. Different sample holders are designed for different samples. For example, the fragile quartz paddle is used for samples with low moments, while brass trough may be used for other samples.

5.7.2 | **Mössbauer spectroscopy**

Mössbauer effect is the recoilless emission and resonant absorption of γ-rays from nuclear transitions.

As shown in Fig. 5.35, these transitions are affected by the electronic and magnetic environment due to splitting of the energy levels. The source, usually ^{57}Co/Rh matrix is oscillated towards and away from the absorber (sample) and the resonant emission and absorption of frequencies as in Doppler effect leads to the generation of spectral lines (singlet, doublet and multiplet).

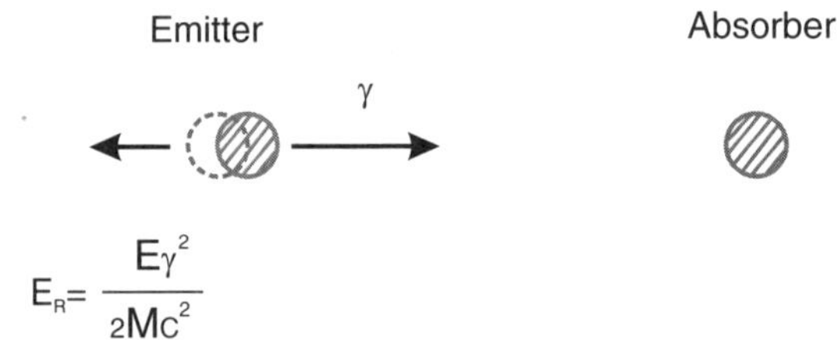

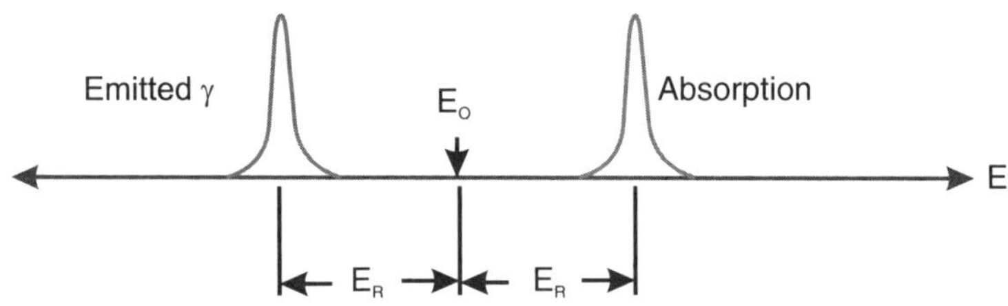

Fig. 5.35 Recoilless emission and absorption of γ-rays leading to Mössbauer effect [Adapted from Ref. [556]].

The recoil energy of the emitting nucleus is related to the energy of the γ-rays, as shown in Fig. 5.35. Mössbauer spectroscopy is generally used for studying the magnetic structure of Fe-containing compounds or their composites . The high sensitivity for ^{57}Fe isotope is due to the γ-ray photon of energy 14.4 keV with a line width of 5×10^{-9}eV, which enables detection limits of upto 1 in 10^{12}.

In reference to Fig. 5.36, Mössbauer spectral parameters and information provided by them is summarized below:

(a) Isomer shift (IS in mm/s) provides chemical information (electronic environment/oxidation state). IS arises due to differences in s-electron density around the nuclei. It gives information about valency states, ligand bonding states, electron shielding and electron-drawing power from electronegative groups. For example, IS (mm/s): $Fe^{2+}(3d^6) > Fe^{3+}(3d^5)$.

(b) Quadrupole splitting (QS in mm/s) provides structural information. This arises due to asymmetric electronic charge distribution around the nucleus. The extent of splitting is given by the following relation.

$$\Delta = eQV_{zz}/2 \tag{5.24}$$

where,
Q is Nuclear quadrupole moment,
V_{zz} is Electric field gradient (EFG),
QS is 0 for cubic symmetry and QS increases as symmetry decreases.

(c) Hyperfine field (H_f in Tesla) reveals information on magnetic ordering. The splitting of spectral lines due to dipolar interaction of the nuclear spin moment with magnetic field results in Zeeman splitting. The hyperfine fields due to Zeeman splitting follow the relationship.

$$B_{eff} = (B_{contact} + B_{orbital} + B_{dipolar}) + B_{applied} \tag{5.25}$$

The line intensities are split in the ratio $3:(4\sin^2\theta)/(1+\cos^2\theta):1$, where the middle term averages to 2 for polycrystalline solids in the absence of an applied field.

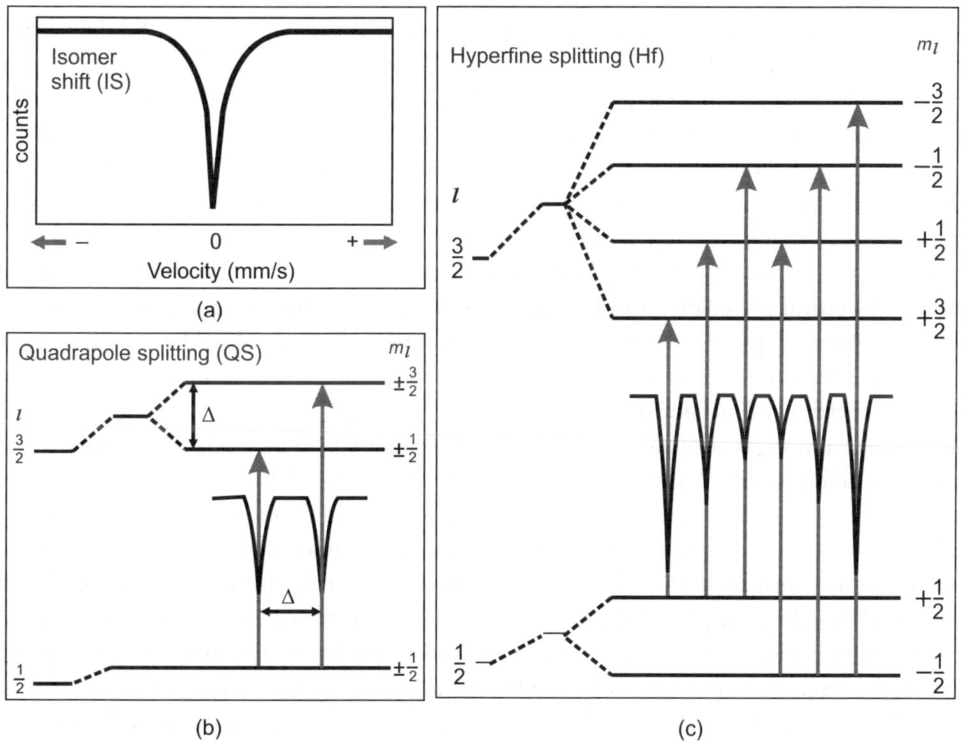

Fig. 5.36 (a) Simple Mössbauer spectrum showing zero isomer shift (mm/s) for identical source and absorber, (b) quadrupole splitting (mm/s) arising from asymmetry in the electronic environment leads to the formation of a doublet separated by Δ(mm/s), (c) hyperfine splitting of the nuclear energy levels into multiplets in case of magnetic samples[557].

5.7.2.1 | Practical guidelines for Mössbauer spectroscopy

Since Mössbauer spectroscopy is used to quantify the amount of iron in the sample, the spectrum obtained depends on the amount of sample. Too much sample can affect the area, width and intensity of the spectrum, while too less sample may not give reliable results. The incident γ-rays may not encounter an iron atom if the sample is of small weight. Hence, optimization of the amount of sample to be used is important in Mössbauer spectroscopy. This necessitates the use of large quantity of samples, though recent developments in the technique allow effective evaluation for samples as light as 1–5 mg. Samples may also be mixed with some inert material and spread evenly across the surface of the holder. A cellophane tape, which is non-absorbent to γ-rays may be used to keep the sample in place.

5.8 | Closure

A biomaterial is a substance or a construct which is naturally derived or chemically synthesized to interact with biological systems. It encompasses a broad range of materials including metals, ceramics, polymers, or composites and is increasingly used in various tissue replacements. So the fabricated material is not only expected to be biocompatible, but also recreate a congenial environment to obtain a desirable cell-biomaterial interaction. Thus, in addition to the composition, the physical, chemical and mechanical properties of such materials should be carefully assessed prior to implantation. Increasing research over the years has resulted in the development and usage of a variety of advanced techniques, aimed at analyzing material properties and biocompatibility. This chapter summarizes such state-of-the-art techniques used for compositional/phase analysis (XRD, XPS, FT-IR, Raman), surface topography (AFM), microstructural analysis (SEM, TEM, tomography), electrical (Electrical impedance spectroscopy, 2/4-probe) and magnetic characterization (VSM, Mössbauer spectroscopy). Some of these characterization techniques should be implemented by any researcher working with biomaterials in order to understand and establish correlations between substrate properties and biocompatibility.

Microscopic investigation of biomaterials has become important in the field of biomaterial research, owing to its ability to reveal a lot of information about cell-material interactions. Among several microscopic techniques, the scanning electron microscopy provides the most reliable and readable data on adhesion of cells to the surfaces of biomaterials. Unlike fluorescence/confocal microscopy, SEM allows more direct imaging of the microstructural features of metallic and ceramic biomaterials with cells adhered on it. In principle, SEM provides a resolution of between 1 and 5 nm (albeit somewhat restricted for biological samples), while TEM provides a resolution of down to 0.2 nm or better. However, the resolution cannot be improved beyond 2 nm for biological samples due to the fixing process. Even though it is the most effective technique to decipher ultrastructural details of cells, the preparation of a biological sample, cells or tissue, for TEM observations requires several stages, some of which are quite complicated and critical at the same time. On the other hand, AFM has several advantages over electron microscopy in biomaterials research, including the ability to image cells in liquid medium with minimal sample preparation without labelling, fixing, or coating. AFM, in particular allows the topographic characterization of surfaces at resolutions not achievable with optical microscopes. Optical and electron microscopes can generate two dimensional images of a sample surface, with a magnification as large as (1000X) for an optical microscope and a few hundred thousand (~100,000X) for an electron microscope. However, these microscopes cannot measure the vertical dimension (z-direction) of the sample, the height (e.g., particles) or depth (e.g., holes, pits) of the surface features. In contrast, AFM, which uses a sharp tip to probe the surface features by raster scanning, can image the surface topography with very high magnifications, up to ~1,000,000X, comparable or even better than some of the electron microscopes. The measurements by AFM are made in three dimensions, the horizontal X–Y plane and the vertical Z dimension. The resolution (magnification) at Z-direction is normally higher than that achievable in X–Y plane.

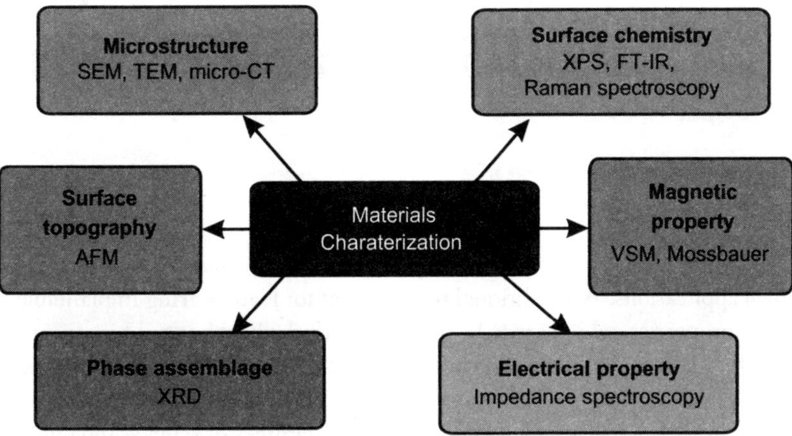

Fig. 5.37 Schematic illustrating the various methods to probe structure and properties of materials.

Mechanical Properties: Principles and Assessment

The requirement of desired biocompatibility property is of primary importance for materials in biomedical applications. An additional requirement for load-bearing implantable biomaterials is its mechanical properties, which are to be matched with the host bone. This aspect is particularly most important for implants in articulating joints. To this end, this chapter discusses the fundamental principles of the mechanical properties of metals, ceramics and polymers. Considering the participation of researchers from widely disconnected disciplines of science and engineering in the field of biomaterials or in a broader sense, biomedical engineering, this chapter starts with the basic definition of stress and strain as tensorial quantities. The difference in mechanical behaviour, in terms of the stress–strain response of materials under different modes of loading is also explained. This is then followed by a brief discussion on the deformation and strengthening mechanisms of metals. While metals are ductile (permanent deformation under stress), the ceramics are inherently brittle. The relevant theories to explain the brittle behaviour of ceramics are explained with sufficient details. The concept of the criticality in fracture conditions, leading to the definition of the fracture toughness has been emphasized. Thereafter, the viscoelastic deformation of thermoplastic polymers is described in reference to the necking and crystallization during tensile deformation. Various experimental methodologies together with guidelines to conduct mechanical property evaluation are also summarized. Overall, this chapter establishes a theoretical framework to understand and experimentally determine the mechanical properties of the materials used for biomedical applications. Some contemporary discussion can be found elsewhere[558,559,560].

6.1 | Conceptual Understanding of Stress and Strain

First, let us understand the concept of stress and strain. When an implant or scaffold experiences or is subjected to mechanical load, stress is developed. If the implant is assumed to have a complex shape, like the one shown in Fig. 6.1a, one may be interested to know the stress at a plane as well as stress at a point 'O'. The definition of stress on a plane and that at a point needs to be differently expressed.

The stress at a point depends on the orientation as well as cross-sectional area of the hypothetical plane containing the point of interest, with respect to the net force acting on the plane. This needs to be determined while constructing the free body diagram of the part of the solid containing that specific point. For example, one can consider two hypothetical planes, mm and nn around the point 'O' in Fig. 6.1a. In defining stress, one has to first determine net force from the equilibrium of the free body diagram of the solid under the action of the external forces. In reference to Fig. 6.1a, the area of 'mm' or 'nn' plane is different as well as the orientation with respect to the net force.

One such example of defining stress at point O with hypothetical plane (mm) is shown in Fig. 6.1b. First, one has to consider different forces acting on the free body diagram and then to determine the net force, under equilibrium of the external forces acting on the solid.

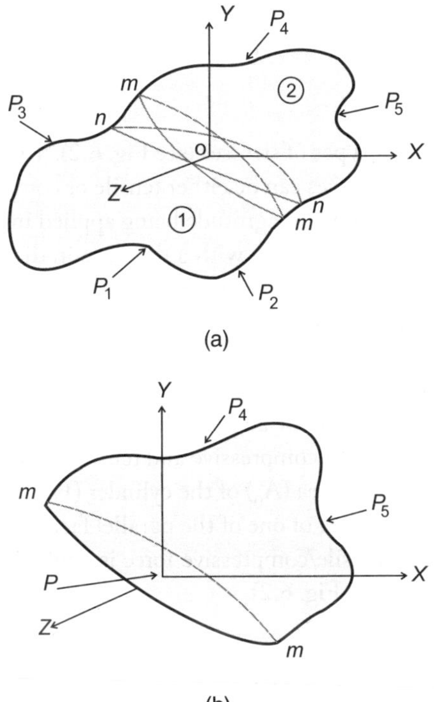

Fig. 6.1 Conceptual definition of stress at a plane in a solid under equilibrium of external forces: (a) solid experiencing forces from various directions, (b) free body diagram of a section with plane mm [Adapted from Ref. 579].

The stress on a plane can be described as the ratio of the net force acting on the cross-sectional area of that plane. In simplistic terms, stress at a point can be defined as the limiting ratio of the resultant force acting on the hypothetical plane containing the specific point to an infinitesimal area around a specific point in a solid.

Mathematically, it can be expressed as,

$$\sigma = \lim_{\Delta A \to 0} \frac{\Delta P}{\Delta A} \tag{6.1}$$

where, ΔP is the resultant of all load components (P_1, P_2, P_3) that the solid experiences under equilibrium (see Fig. 6.1). The stress at a point 'O' can therefore be defined as a measure of net force on a hypothetical plane around the point of interest, when the area of the plane is assumed to be negligible or zero.

> The stability of any material is dictated by the equilibrium of mechanical forces acting on it from different directions and NOT by the equilibrium of stresses.
>
> $$\sum_{x,y,z} P_i = 0$$
>
> where, P_i represents the components of the different forces acting on the solid along x, y and z directions.
>
> $$\sum_{i,j} \sigma_{ij} \neq 0$$

Next, we define three different types of stresses (see Fig. 6.2). Two generic variants of stress are normal and shear stress. Normal stresses can be either tensile or compressive in nature. The tensile force can be defined as the force of equal magnitude being applied in opposite direction and under such loading, a cylinder will extend in length with a decrease in diameter (Fig 6.2a). The reverse happens in case of compressive force, which is applied with equal magnitude in the same direction (see Fig. 6.2b). Under compression, a cylinder will experience an increase in diameter, but will shorten in length. The shear loading is however different with equal forces being applied at two opposite faces of a cube. While both tensile and compressive stresses will cause volume change, the shear stress leads to a shape change or angular distortion, measured by angle 'θ' (Fig. 6.2c). Following the definition of stress, both compressive and tensile stress can be defined as the ratio of force (F) to the initial cross-sectional area (A_0) of the cylinder (F/A_0). Similarly, shear stress can be quantified as F/A_0, where A_0 is the area of one of the parallel faces of a cube along which the shear force is being applied. Note that tensile/compressive force is applied perpendicular to the marked plane of a cylinder with area, A_0 (see Fig. 6.2).

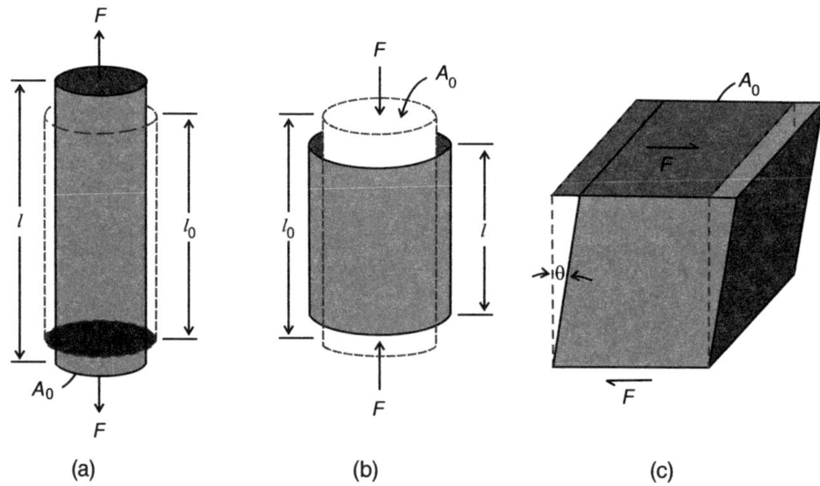

Fig. 6.2 Different types of stress: (a) tensile, (b) compressive and (c) shear [Adapted from Ref. 579].

It should be clear that the stress is to be represented not only by magnitude and direction, but also by its orientation with respect to a plane.

This brings us to define stress as a tensorial quantity. A scalar needs only magnitude, while a vector is to be represented by both magnitude and direction. However, a tensor requires magnitude, direction and a reference plane. While defining stress, one therefore needs to know the magnitude, direction as well as the plane across which it acts. Since stress is a tensor of rank 2, it needs a total of $3^2 = 9$ terms. With reference to Fig. 6.3, the stress state at a point (O) can further be described by constructing a cube around point 'O' and then one has to consider various stress components along each face of the cube.

Stress is not a vector, but a tensor of rank 2.

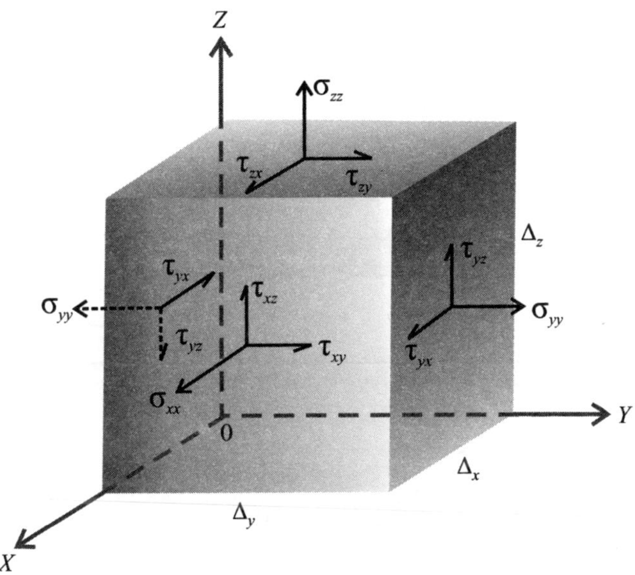

Fig. 6.3 Stress acting on an elemental unit cube [Adapted from Ref. 579].

In defining various stress components of stress tensor (σ_{ij}), one has to remember that the first subscript refers to the reference plane and the second subscript denotes the direction. The normal stress component has double identical subscript, indicating that force/stress is acting on the planes along the plane normal.

In contrast, the shear stress has different subscripts. In reference to the volume element shown in Fig. 6.3, the three stress components acting on the X-plane (the plane perpendicular to X-axis) are, σ_{xx}, τ_{xy}, τ_{xz}. Similarly, specific stress components for Y-plane or Z-plane are also shown in Fig. 6.3. If one considers the areas of the cube faces small enough (change in stress over faces therefore negligible), then the moments created by the shear forces would be equal. Thus, one would be able to realize that $\tau_{xy} = \tau_{yx}$; $\tau_{yz} = \tau_{zy}$; $\tau_{xz} = \tau_{zx}$. Therefore the stress tensor, σ_{ij}, will have six independent stress components. The stress tensor, σ_{ij} can therefore be defined as,

$$\sigma_{ij} = \begin{pmatrix} \sigma_{xx} & \tau_{xy} & \tau_{xz} \\ \tau_{yx} & \sigma_{yy} & \tau_{yz} \\ \tau_{zx} & \tau_{zy} & \sigma_{zz} \end{pmatrix} \tag{6.2}$$

Similar to stress, the strain at a point in a solid continuum will now be explained. In simplistic sense, strain denotes the ratio of change in the dimension to the original dimension, when a solid experiences a stress state. With reference to Fig. 6.4, the strain can be described in terms of displacement of a point, P in a solid continuum. While P has a coordinate (x, y, z), the displaced point P′ has a different coordinate (x + u, y + v, z + w). Therefore, the displacement component along X, Y and Z direction is u, v and w, respectively. In accordance with the convention, the normal strain components are defined as,

$$e_{xx} = \frac{\partial u}{\partial x}, e_{yy} = \frac{\partial v}{\partial y} \text{ and } e_{zz} = \frac{\partial w}{\partial z} \tag{6.3}$$

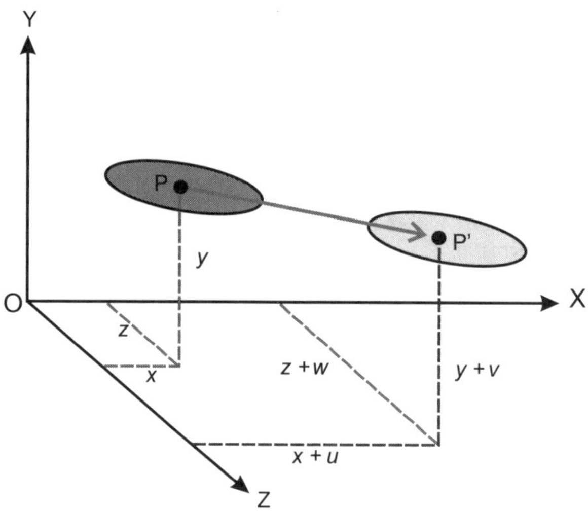

Fig. 6.4 Defining strain due to the displacement of a point P to P' in a solid continuum [Adapted from Ref. 579)].

Similarly, the shear strain components are defined as the rate of change of displacement component along X-axis to the y-component position and likewise as follows:

$$e_{xy} = \frac{\partial u}{\partial y}, e_{yx} = \frac{\partial v}{\partial x}, \tag{6.4}$$

$$e_{yz} = \frac{\partial v}{\partial z}, e_{zy} = \frac{\partial w}{\partial y} \tag{6.5}$$

$$e_{xz} = \frac{\partial u}{\partial z}, e_{zx} = \frac{\partial w}{\partial x} \tag{6.6}$$

Based on the above definition of strain components and similar to stress tensor, the entire displacement tensor can be defined as follows,

$$e_{ij} = \begin{pmatrix} e_{xx} & e_{xy} & e_{xz} \\ e_{yx} & e_{yy} & e_{yz} \\ e_{zx} & e_{zy} & e_{zz} \end{pmatrix} = \begin{pmatrix} \dfrac{\partial u}{\partial x} & \dfrac{\partial u}{\partial y} & \dfrac{\partial u}{\partial z} \\ \dfrac{\partial v}{\partial x} & \dfrac{\partial v}{\partial y} & \dfrac{\partial v}{\partial z} \\ \dfrac{\partial w}{\partial x} & \dfrac{\partial w}{\partial y} & \dfrac{\partial w}{\partial z} \end{pmatrix} \tag{6.7}$$

Like stress, strain is also a tensor of rank 2.

The displacement tensor can be further described as,

$$e_{ij} = \frac{1}{2}\left(e_{ij} + e_{ji}\right) + \frac{1}{2}\left(e_{ij} - e_{ji}\right) \tag{6.8}$$

or
$$e_{ij} = \varepsilon_{ij} + \omega_{ij}$$

In the above, strain tensor is,

$$\varepsilon_{ij} = \frac{1}{2}\left(\frac{\partial u_i}{dx_j} + \frac{\partial u_j}{\partial x_i}\right) \tag{6.9}$$

and rotation tensor is,

$$\omega_{ij} = \frac{1}{2}\left(\frac{\partial u_i}{\partial x_j} - \frac{\partial u_j}{\partial x_i}\right) \tag{6.10}$$

In terms of the complete description, the strain (ε_{ij}) and rotation (ω_{ij}) tensor can be further expressed as,

$$\varepsilon_{ij} = \begin{pmatrix} \varepsilon_{xx} & \varepsilon_{xy} & \varepsilon_{xz} \\ \varepsilon_{yx} & \varepsilon_{yy} & \varepsilon_{yz} \\ \varepsilon_{zx} & \varepsilon_{zy} & \varepsilon_{zz} \end{pmatrix} = \begin{pmatrix} \dfrac{\partial u}{\partial x} & \dfrac{1}{2}\left(\dfrac{\partial u}{\partial y} + \dfrac{\partial v}{\partial x}\right) & \dfrac{1}{2}\left(\dfrac{\partial u}{\partial z} + \dfrac{\partial w}{\partial x}\right) \\ \dfrac{1}{2}\left(\dfrac{\partial u}{\partial y} + \dfrac{\partial v}{\partial x}\right) & \dfrac{\partial v}{\partial y} & \dfrac{1}{2}\left(\dfrac{\partial v}{\partial z} + \dfrac{\partial w}{\partial y}\right) \\ \dfrac{1}{2}\left(\dfrac{\partial u}{\partial z} + \dfrac{\partial w}{\partial x}\right) & \dfrac{1}{2}\left(\dfrac{\partial v}{\partial z} + \dfrac{\partial w}{\partial y}\right) & \dfrac{\partial w}{\partial z} \end{pmatrix} \tag{6.11}$$

$$\omega_{ij} = \begin{pmatrix} \omega_{xx} & \omega_{xy} & \omega_{xz} \\ \omega_{yx} & \omega_{yy} & \omega_{yz} \\ \omega_{zx} & \omega_{zy} & \omega_{zz} \end{pmatrix} = \begin{pmatrix} 0 & \dfrac{1}{2}\left(\dfrac{\partial u}{\partial y} - \dfrac{\partial v}{\partial x}\right) & \dfrac{1}{2}\left(\dfrac{\partial u}{\partial z} - \dfrac{\partial w}{\partial x}\right) \\ \dfrac{1}{2}\left(\dfrac{\partial v}{\partial x} - \dfrac{\partial u}{\partial y}\right) & 0 & \dfrac{1}{2}\left(\dfrac{\partial v}{\partial z} - \dfrac{\partial w}{\partial y}\right) \\ \dfrac{1}{2}\left(\dfrac{\partial w}{\partial x} - \dfrac{\partial u}{\partial z}\right) & \dfrac{1}{2}\left(\dfrac{\partial w}{\partial y} - \dfrac{\partial v}{\partial z}\right) & 0 \end{pmatrix} \tag{6.12}$$

Following the above analysis, the shear strain, which determines the shape change involved in the plastic deformation of metals, is defined as,

$$\gamma_{xy} = \frac{\partial u}{\partial y} + \frac{\partial v}{\partial x}$$

$$\gamma_{xz} = \frac{\partial w}{\partial x} + \frac{\partial u}{\partial z} \tag{6.13}$$

$$\gamma_{yz} = \frac{\partial w}{\partial y} + \frac{\partial v}{\partial z}$$

Similar to the definition of stress components, any strain component with identical subscripts defines normal strain, whereas shear strain is represented by unequal subscripts. Like stress tensor, the shear components have the following relationship,

$$\gamma_{xy} = \gamma_{yx},\ \gamma_{yz} = \gamma_{zy},\ \gamma_{xz} = \gamma_{zx} \tag{6.14}$$

Therefore, the final description of strain tensor is,

$$\varepsilon_{ij} = \begin{pmatrix} \varepsilon_{xx} & \frac{1}{2}\gamma_{xy} & \frac{1}{2}\gamma_{xz} \\ \frac{1}{2}\gamma_{xy} & \varepsilon_{yy} & \frac{1}{2}\gamma_{yz} \\ \frac{1}{2}\gamma_{xz} & \frac{1}{2}\gamma_{yz} & \varepsilon_{zz} \end{pmatrix} \tag{6.15}$$

The above analysis indicates that strain tensor has six dependent components. Further, total volume change is zero during plastic or permanent deformation of a metal and this can be expressed as:

$$\Delta V = \varepsilon_{xx} + \varepsilon_{yy} + \varepsilon_{zz} = 0 \tag{6.16}$$

Summarizing, the strain tensor has five independent components with two normal and three shear components.

6.2 | Stress–Strain Response of Metals

During biomedical application, an implant may experience different loading modes, i.e., tension, compression and shear, and sometimes a combination of all three (see Fig. 6.3). Reiterating, in a typical tension mode, equal force is applied normal to the surface of the two ends of a specimen in opposite directions, whereas in compression mode, the force is applied normal to the two ends of a specimen in the same direction. In contrast, in shear loading mode, the forces are applied tangentially on opposite surfaces (for example of a solid cube) and the application of such forces causes angular distortion.

The stress–strain response of a ductile metal under tension, compression and shear is schematically shown in Figs. 6.5a, 6.5b and 6.5c. Clearly, a metal exhibits initial linear stress–strain response, irrespective of the mode of loading (tensile/compressive/shear). In the linear elastic region, stress is proportional to strain and this is known as Hooke's law of elasticity. In tensorial notation, this law can be expressed as,

$$\sigma_{ij} = C_{ijkl}\varepsilon_{kl} \qquad (6.17)$$

Alternatively,
$$\varepsilon_{ij} = S_{ijkl}\sigma_{kl} \qquad (6.18)$$

In the above expressions, C_{ijkl} and S_{ijkl} are the stiffness and compliance tensor, respectively.

Unlike stress/strain, the elastic stiffness/compliance is a tensor of rank 4.

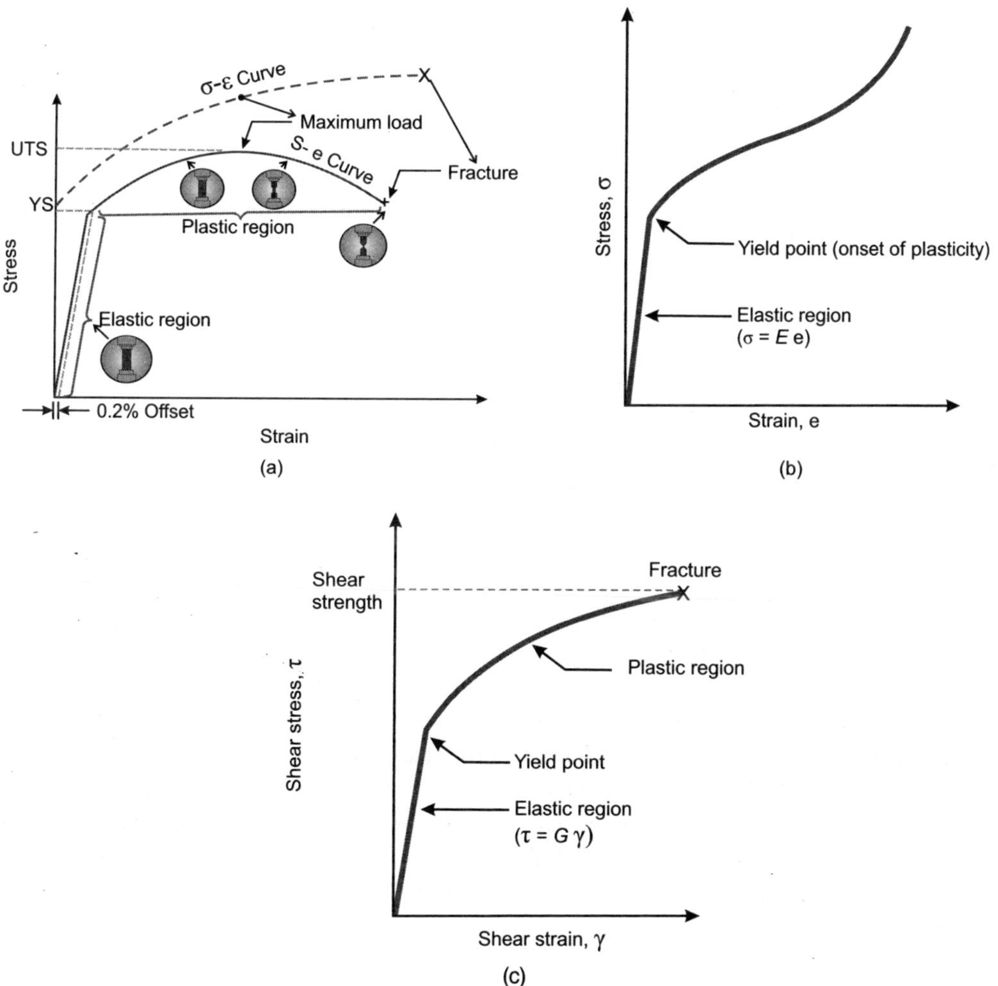

Fig. 6.5 (a) Schematic illustration of true stress-strain (σ-ε) and Engineering stress-strain (S-e) response together with the various transition points and their significance on deformation behaviour of metals; (b) typical stress–strain response of metals under compression; and (c) shear.

Therefore, C_{ijkl} or S_{ijkl} would require 81 terms to define them. More discussion on the description or number of independent terms of these two tensors is beyond the scope of this book. Irrespective of tensile/compressive/shear mode of loading, the linear deformation response of a metallic implant is always observed and the slope in the linear elastic region determines the corresponding elastic modulus. At the point of yielding (transition from elastic to plastic region), the non-linear deformation starts till the point of fracture. The transition from linear to non-linear deformation is defined as the yield strength. In simplistic terms, linear deformation is known as elastic deformation and non-linear response as plastic deformation.

> Also, the term 'yield' means that a metal surrenders at yield stress and would start to deform permanently.

Prior to yield stress, the strain is proportional to stress and in the post-yield region, the metal will have residual strain even after complete removal of load. Another point to note is that non-linear stress strain response qualitatively depends on the mode of loading. For example, the stress monotonically increases both in compression and shear, but the increase is more significant in the last stage of compression loading. For shear loading, the stress appears to increase at a lower rate to reach almost a steady value. Let us now discuss more on tensile deformation in metals as crack propagation leading to failure is more easy under tensile than under any other loading mode.

In materials science or solid mechanics, a distinction between engineering stress/strain and true stress/strain is made. In Fig. 6.5(a), tensile deformation is plotted in terms of both engineering stress(S)–strain (e) and true stress (σ)–strain (ε). These terms are defined as follows,

Engineering stress, $\qquad\qquad S = P/A_0$ $\qquad\qquad\qquad\qquad\qquad\qquad\qquad$ (6.19)

True stress, $\qquad\qquad\qquad \sigma = P/A_i$ $\qquad\qquad\qquad\qquad\qquad\qquad\qquad$ (6.20)

Engineering strain, $\qquad\qquad e = \Delta l/l_0$ $\qquad\qquad\qquad\qquad\qquad\qquad\qquad$ (6.21)

True strain, $\qquad\qquad\qquad \varepsilon = \ln l_i/l_0$ $\qquad\qquad\qquad\qquad\qquad\qquad\qquad$ (6.22)

In the above definitions, P is the given load, A_0 and A_i are respectively the original and instantaneous cross-sectional area, Δl is the change in length, l_i and l_0 are the instantaneous length and original length of the test sample, respectively. Since the length and area of the cross-section constantly changes with load during mechanical deformation (e.g., tensile loading), the 'true' stress/strain realistically captures the dynamic changes in the stress state of a material under load. Since the strain in the elastic regime is usually very low with respect to plastic strain in case of metals, the engineering stress-strain relationship is normally used. However, the stress – strain response, when plotted in terms of true stress/strain will be shifted upwards and will also show an increasing trend (particularly in necking region) when compared to the same being plotted in terms of engineering stress–strain (see Fig. 6.5a).

> Since a metal always strain hardens under stress, the true stress–strain response therefore describes 'true' deformation response of a metal experiencing dynamic changes in stress state.

Also engineering stress–strain response in the post-UTS region shows a decrease in stress, although cross-section decreases faster than increase in length (see Fig. 6.5a). Such behaviour is certainly not the true intrinsic response.

In particular reference to the engineering stress–strain response in Fig. 6.5a, the non-linear deformation of metals (plastic deformation, occurring above YS) consists of uniform deformation as well as non-uniform deformation. The transition from one to another marks the ultimate tensile strength (UTS), i.e., the maximum load the sample can withstand. The region of uniform deformation (spanning between the yield strength and UTS) can be described by a power law relationship between stress (σ) and strain (ε) as,

$$\sigma = K\varepsilon^n \tag{6.23}$$

where, K is the strength co-efficient and n is the strain hardening coefficient.

Considering the fact that volume is constant during plastic deformation and using simple mathematical calculation, one can arrive at the following relationship between the strain (ε_u) at the point of maximum load, i.e., at UTS and strain hardening coefficient as,

$$\varepsilon_u = n \tag{6.24}$$

In the description of the stress–strain response in terms of engineering stress/strain, the decrease in cross-sectional area above UTS is much faster than the increase in the length of the sample. This phenomenon is known as necking. This also leads to a decrease in the stress–strain plot, since engineering stress is defined on the basis of the original cross-sectional area. However, the metal, in reality, gets strain hardened due to extensive deformation. Therefore, one would need to apply more stress so that the metal can be continually deformed till the point of fracture. This physical concept is realistically captured in the stress–strain plots, when true stress/strain is used (see Fig. 6.5). Also, if one closely follows the definition of true stress and engineering stress, then one would perceive why the true stress, corresponding to the point of maximum load is at the higher level and also at the right of the UTS co-ordinate (see the definition of true strain). Similarly, the tensile failure takes place at a larger true strain value compared to that when defined in terms of engineering strain. Nevertheless, the extensive necking in the non-uniform deformation region finally leads to the fracture of metals.

Summarizing, the mechanical performance of a metallic biomaterial in a given biomedical application, would be determined by elastic modulus and yield strength. Clinically, a large difference in elastic modulus between natural bone and a biomaterial leads to stress shielding.

6.3 | Tensile Deformation Behaviour

In this section, the difference in the mechanical response of metals/ceramics/polymers under tensile loading will be discussed. Prior to this, it is instructive to recall the nature of bonding in these materials. The metals are characterized by metallic bonds with atoms floating in a cloud of electrons. While such bonding explains good electrical and thermal transport properties, this specific bonding is also isotropic in nature, implying the mechanical properties to be isotropic as well. The same is not true for ceramics, which predominantly either have ionic (oxides) or covalent (non-oxides) bonding. The ionic bonding requires electro-neutrality conditions to be maintained during deformation. The covalent bonding, being directional in nature, would impart anisotropic mechanical properties. Polymers, on the other hand, have mixed bonding. Along the backbone chains in a polymer, the 'mer' units are covalently bonded, while two neighbouring chains are weakly bonded by either H-bond or van der Waals bonds.

> In view of the differences in bonding characteristics, the three different classes of materials viz. metals, ceramics and polymers respond in a distinguishable manner, when mechanically loaded.

A schematic illustration of such difference in the tensile response, in terms of engineering stress vs. strain plot, is shown in Fig. 6.6. While a ceramic implant mostly exhibits linear response up to the point of fracture, a metallic implant experiences initial linear response followed by non-linear deformation behaviour upto the point of fracture. The slope of the linear part of such response determines the Young's modulus or elastic modulus of the materials. In contrast to metals and ceramics, a thermoplastic polymer exhibits a constant deformation at a lower stress level, which is preceded by linear stress strain response, like metals/ceramics. The slope of initial linear response, a measure of elastic modulus, is however markedly different in metals/ceramics/polymers.

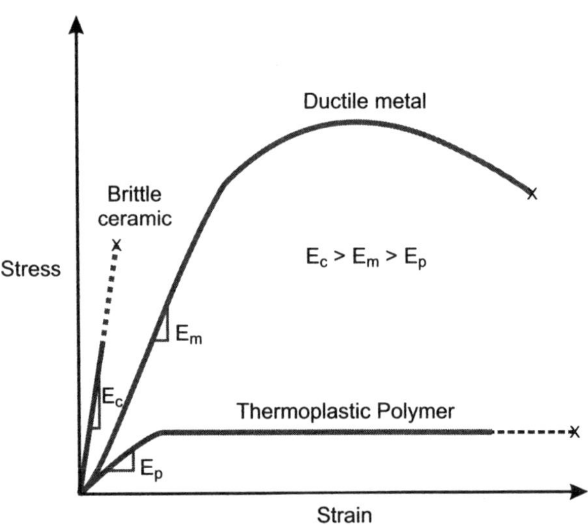

Fig. 6.6 Schematic illustration comparing the stress–strain response of three primary classes of material under uniaxial tension.

The common point in the deformation of the three distinct class of materials is the initial linear response. However, the difference lies both in terms of the elastic modulus as well as in the total strain to failure or the maximum load bearing capability. As shown in Fig. 6.6, the elastic modulus of ceramics is greater than that of metals and is much larger than that of polymers. Also, the tensile failure strain of thermoplastic polymers can be an order of magnitude larger than that of a ductile metal. In addition, the transition from linear response, i.e., yielding takes place at a much larger stress value in case of metals than that of polymers. With reference to Fig. 6.6 as well as in the light of the above discussion, it should therefore be clear that for large load bearing capability under tensile mode, a metallic implant will always be preferred. However, a polymeric implant can be selected for biomedical applications requiring low load bearing capability, with the ability to sustain large deformation without fracture.

> Although tensile strength is low for ceramics, the compressive strength is typically 8 times larger than the tensile strength.

The better compression strength is particularly advantageous for ceramics in load bearing implants. Also, the elastic modulus is the highest for ceramics and this is often much higher than natural bone. This can be compensated by the use of porous ceramics to match the elastic modulus of host bone.

6.4 | Strengthening of Metals

Next, we discuss the origin of the ability of metals to undergo plastic deformation. The non-linear deformation of metals is facilitated by the motion of linear defects, i.e., dislocations. These defects can glide on specific crystallographic planes (slip planes) along specific crystallographic directions (slip directions) to cause displacements among atomic planes. Slip planes are the most close-packed planes with the highest planar density of atoms, while slip directions are the most close-packed directions with the shortest interatomic distances between two adjacent atoms on a slip plane. The movement of dislocations is only possible in case of metals, as any such movement under shear stress requires localized bonds to be ruptured or remade. This is not possible in ionic ceramics so easily due to the restriction of electro-neutrality change conditions. The restriction of directional or non-deformable covalent bonds similarly restricts dislocation movement in covalent ceramics. In polymers, dislocations are not available and a different defect type 'crazes' evolve during deformation, as will be discussed later. The gliding of a dislocation can be described with reference to Fig. 6.7. In the ball and stick model, the dislocation is shown as the extra half-plane of atoms and a crystal defect is usually described as a departure from the long range ordering of atoms. The dislocation line lies perpendicular to the plane of the paper. In the presence of shear force acting to the parallel faces of the crystal, the extra half plane of atoms move from left to right and finally creates a step at the surface. The step formation is a measure of plastic deformation and in a real solid, a large number of dislocations glide simultaneously under shear stress, leading to plastic deformation. More discussion on the dislocation motion of various types of dislocation is beyond the scope of this book.

One of the important aspects in the mechanical behaviour of metals is the strengthening mechanisms, which are essentially physical mechanisms activated as dislocations glide on slip planes. These mechanisms can potentially increase the strength often at the expense of ductility.

> The fundamental basis for strengthening mechanisms lies in restricting dislocation motion through design of microstructures.

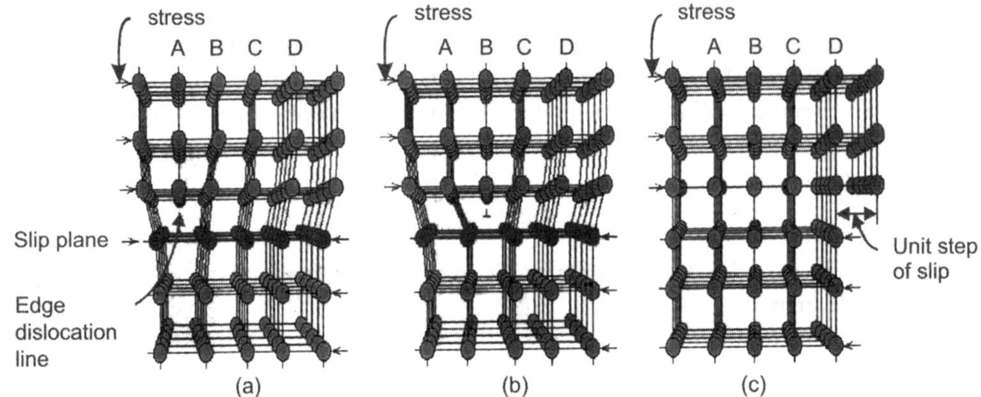

Fig. 6.7 Schematic illustration of the gliding of edge dislocation on a slip plane: (a,b) dislocation movement and formation of slip step, a measure of plastic deformation in metals[580].

Different strengthening mechanisms are briefly discussed below.

(a) **Solid solution strengthening:** In metallic alloys (e.g., Fe-C alloys), the small interstitial solutes (here, C-atoms) can diffuse to the dislocation core thereby restricting the dislocation motion in the matrix (here, Fe) under stress. As shown in Fig. 6.8, the solute segregation at the dislocation core makes them immobile, as dislocation has to glide with the solute atmosphere. In case of solutes of larger/smaller size, but within 15% of the atomic size of the parent or base metal, the parent atom can be substituted by the foreign atom. This results in lattice strain that will contribute in impeding the dislocation motion. As a result, larger stress would be needed to propagate dislocations.

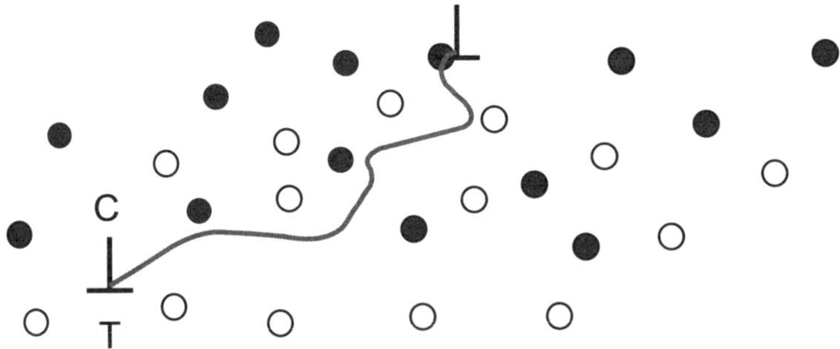

Fig. 6.8 Solute atmosphere around a dislocation causing solid solution strengthening.

The solid solution strengthening linearly scales with solute addition up to a critical alloying addition, beyond which intermetallic formation is likely to take place.

(b) **Precipitation hardening:** In case of metallic alloys with specific composition (e.g. Al-4%Cu alloy), the heat treatment of the alloys often results in the formation of precipitates in the matrix. Depending on size, shape, volume fraction of precipitates as well as nature of matrix, precipitate interface, strengthening can be realized. For better strengthening, coherent/semi-coherent interface with minimal lattice plane bending at the interface would be desired. Also, if the precipitates grow in size or lose coherency, then also the magnitude of strengthening decreases.

To substantiate the above statement, precipitation hardening is most effective, when precipitates are non-deformable and also would have finite inter-precipitate spacing. In a microstructure with such precipitates, a dislocation can only overcome the barrier by forming a dislocation loop around such precipitates. This is possible as dislocation has elastic properties, meaning it can change its configuration, when needed.

(c) **Dispersion strengthening:** In case of deformable second phase, added intentionally during processing stage, a dislocation can cut through the dispersoids, while gliding through the matrix (see Fig. 6.9). During such interaction, the surface area increases and more work is to be done to allow dislocation glide. Therefore, the dispersion strengthening increases with the second phase amount. The last two mechanisms are sensitive to the heat treatment conditions. Both the temperature and time are to be carefully tailored within a narrow window to optimize strengthening contribution.

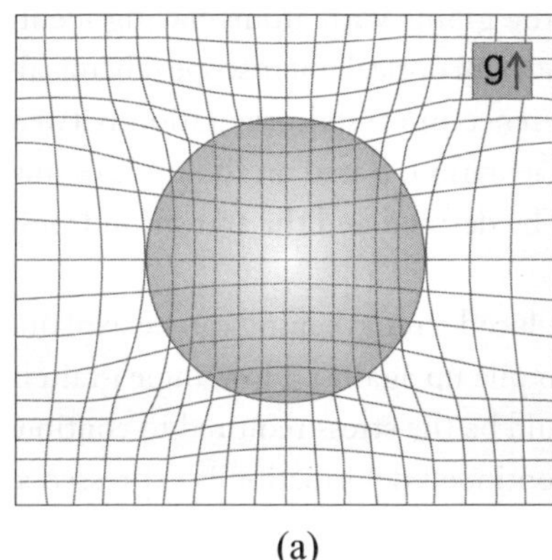

(a)

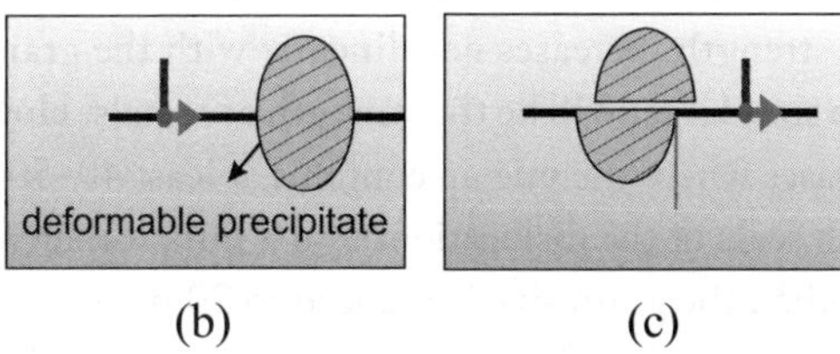

(b)   (c)

Fig. 6.9 (a) Schematic illustration of the coherent interface, and (b, c) the dislocation-precipitate interaction.

(d) **Work hardening/strain hardening:** When a metal undergoes repeated straining during fabrication, a large number of dislocations are generated, leading to the formation of 'dislocation forest'. The formation of deformation substructure leads to a decrease in the mobility of dislocations. For sustained deformation, a mobile dislocation has to cut through dislocation forest; this would arguably impede dislocation motion.

Any mechanism which can potentially make the dislocation motion difficult will make the metal stronger.

(e) **Grain size strengthening:** Let us first define the concept of a grain. During the solidification of a metal either by conventional casting or by other advanced manufacturing processes, the metallic atoms are arranged in crystal lattices to form a solid. In crystalline materials, atoms are arranged in a number of lattice planes in 3D space with both long and short range order. In amorphous materials, like glass, only short range order exists. In crystalline material, the atomic arrangement at a microscopic length scale varies from region to region.

A grain is defined with a characteristic microstructural length scale, wherein the atoms are arranged with identical lattice orientation. All the atomic planes within a grain have identical lattice orientation. On moving from one grain to a neighboring grain, the lattice orientation changes. The region of misorientation between two neighboring grains is defined as the grain boundary.

In polycrystalline materials, the grains with atoms having identical orientation on chemically etched surfaces, can be viewed under an optical microscope (chemical reagent etches grain boundary and grain differently). The difference in crystal orientation between two neighbouring grains is determined by the angle of misorientation. Based on the angle of misorientation, the low angle and high angle grain boundaries can be defined and the transition from one to another typically takes place at 5–10°.

The origin of the grain size induced strengthening can be explained by Hall–Petch relationship, which is based on the dislocation pile up against a low angle grain boundary (see Fig. 6.10). More the dislocation pile up, more would be the stress required to continue deformation. The associated increase in the stress to ensure continued deformation is expressed as,

$$\sigma = \sigma_0 + Kd^{-0.5} \tag{6.25}$$

where, d is the grain size; σ_0 is the single crystal strength, i.e., the strength of a crystal without any grain boundary. Although strength increases non-linearly with the grain size, the above equation explains that finer the grain size, larger will be the strength of metals. However, the dislocation pile-up model is valid only for cases where the pile up contains at least 40–50 dislocations. Considering the magnitude of the length scale of the dislocations (~ 0.5 nm), it therefore appears that the above equation will not be valid when the grain size is reduced to 20 nm or below!

Summarizing, the above section briefly introduces various strengthening mechanisms, which can be thoughtfully considered while designing and manufacturing metals for various biomedical applications. It should be borne in mind that larger the strength, more will be the ability of a metal to sustain higher load during biomedical applications. Also, the elastic modulus, another factor determining the stress shielding of a metallic implant, should not be compromised while introducing various strengthening elements in the microstructure.

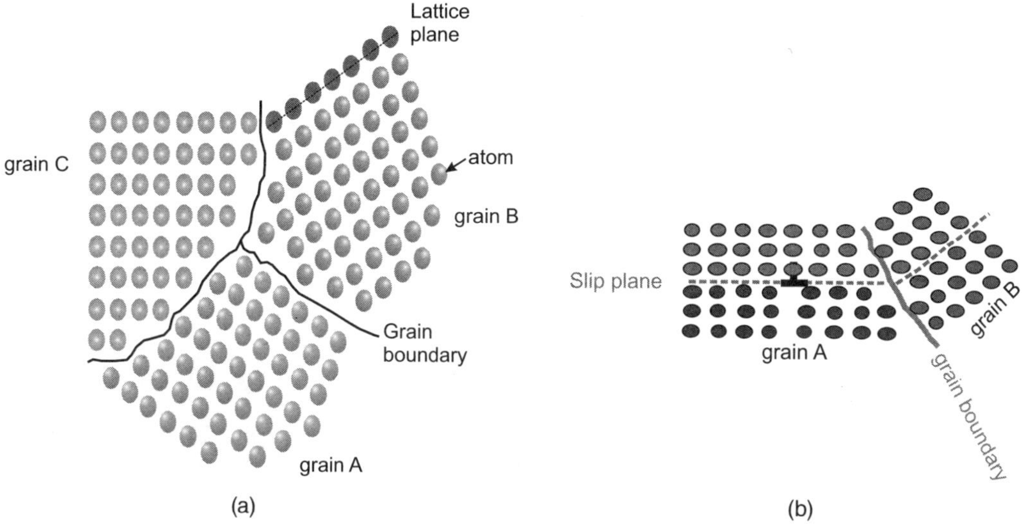

Fig. 6.10 (a) Atomic model of grain and grain boundary (b) Schematic of dislocation propagation on a slip plane in a grain and possibility of getting stuck at a grain boundary.

6.5 | Brittle Fracture of Ceramics

It has been mentioned earlier in this chapter that ceramics, under tension, show linear elastic response upto fracture. The absence of non-linear deformation is the signature of brittleness. In simplistic sense, the ceramic microstructure, unless otherwise designed with toughening elements, cannot resist the crack propagations, i.e., the pre-existing cracks or cracks, developed during loading, will not find any resistance in its path. This leads to catastrophic or ungraceful failure of ceramics.

It is well known that the use of ceramics as implantable materials is not a widely acceptable treatment option for many clinicians around the world. Despite higher hardness, compressive strength, wear resistance properties (favourable for load bearing orthopedic/dental applications), such widespread apprehension is primarily due to the brittleness of ceramic materials.

In this section, an attempt will be made to explain the origin of such brittleness. It is worthwhile to reiterate here that the non-linear deformation in metals, as shown in Fig. 6.5 is primarily due to the motion of dislocations. As explained in the following sections, the propagation of cracks as opposed to dislocation motion like in metals, play an important role in determining the mechanical response of ceramic materials. Ceramics are possibly best known for their brittleness rather than for some of their outstanding properties like excellent compressive strength properties or extremely good biocompatibility/bioactivity (e.g., hydroxyapatite).

Let us now revisit the atomistic description of the fracture process. As shown in Fig. 6.11, the interatomic bonds can be described as 'spring-like' structures. The elastic response of the stretching of such structures can be described as, $F = -Kx$, where F is external load, K is the spring constant and x is the spring displacement. As shown in Fig. 6.11b, the bonds stretch from the equilibrium spacing of a_0 to a (a > a_0) and as explained below, the stress concentration at the crack tip will subsequently rupture the spring (bond), leading to continued crack propagation. This entire scenario of physical events, as shown in Fig. 6.11, takes place ahead of any advancing crack in a ceramic microstructure under external tensile loading.

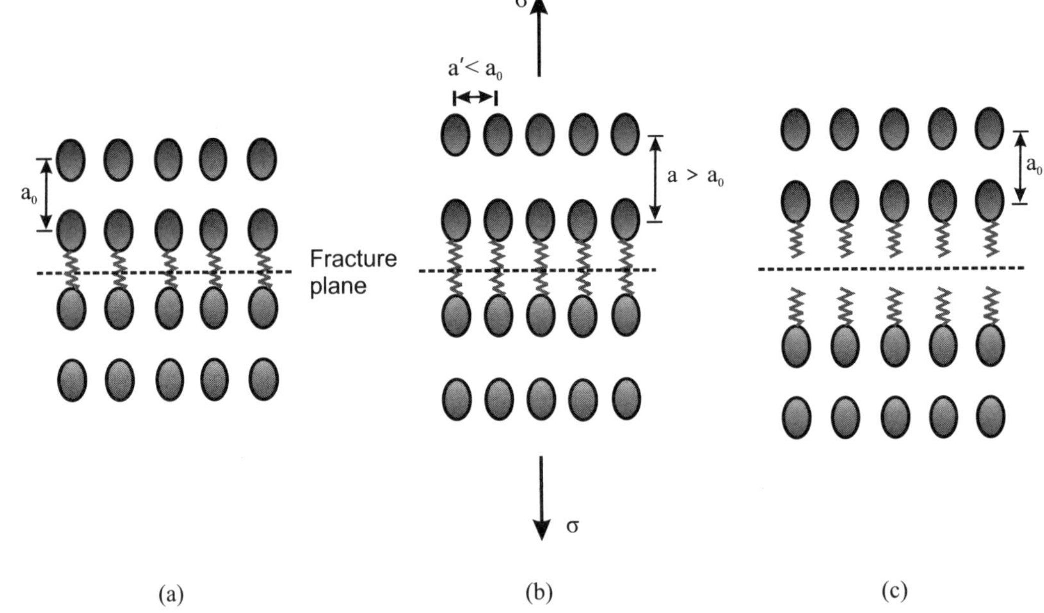

Fig. 6.11 Atomistic description of fracture of a solid:(a) undeformed state, (b) deformed state under tensile load (c) fracture due to the breakage of interatomic bonds. The interatomic bonding access the fracture plane is represented by spring-like behaviour [Adapted from Ref. 581].

The microscopic view of the crack size distribution, their orientation w.r.t loading axis and their influence on fracture progression/failure under tensile and compression loading is shown in Fig. 6.12. It is widely known that the defects or cracks are normally generated at the processing stage (i.e., sintering) and therefore, the sintering of ceramics needs to be carefully tailored to obtain implants with finer or smaller cracks. These cracks often vary in size and their orientation with respect to tensile axis. The largest crack with most favourable orientation (close to 90° with respect to tensile axis) would propagate in a direction perpendicular to tensile axis. Once such crack grows to criticality, it leads to failure. The scenario is strikingly different in case of compression loading. Under compression, the cracks, oriented closer to the axis of compression are more likely to grow (see Fig. 6.12b).

If the pre-existing flaws are oriented at an angle to compression axis, then it will first grow in a manner to orient itself to as close to that of compression axis before further growth. This is shown in the formation of secondary cracks (see Fig. 6.12d). The difficulty in crack propagation under compression also explains why ceramics have exceptional compression strength, i.e., 8 times that under tension. It may be worthwhile to mention that for metals, tensile and compression strength is nearly equal. For example, the tensile strength of a stainless steel is 500–550 MPa, meaning their compression strength also around 500–550 MPa. In contrast, the tensile strength of Al_2O_3 ceramic can be 150–200 MPa, but the compressive strength of Al_2O_3 can be as high 1200–1600 MPa. Therefore ceramics can outperform metals in terms of compression strength.

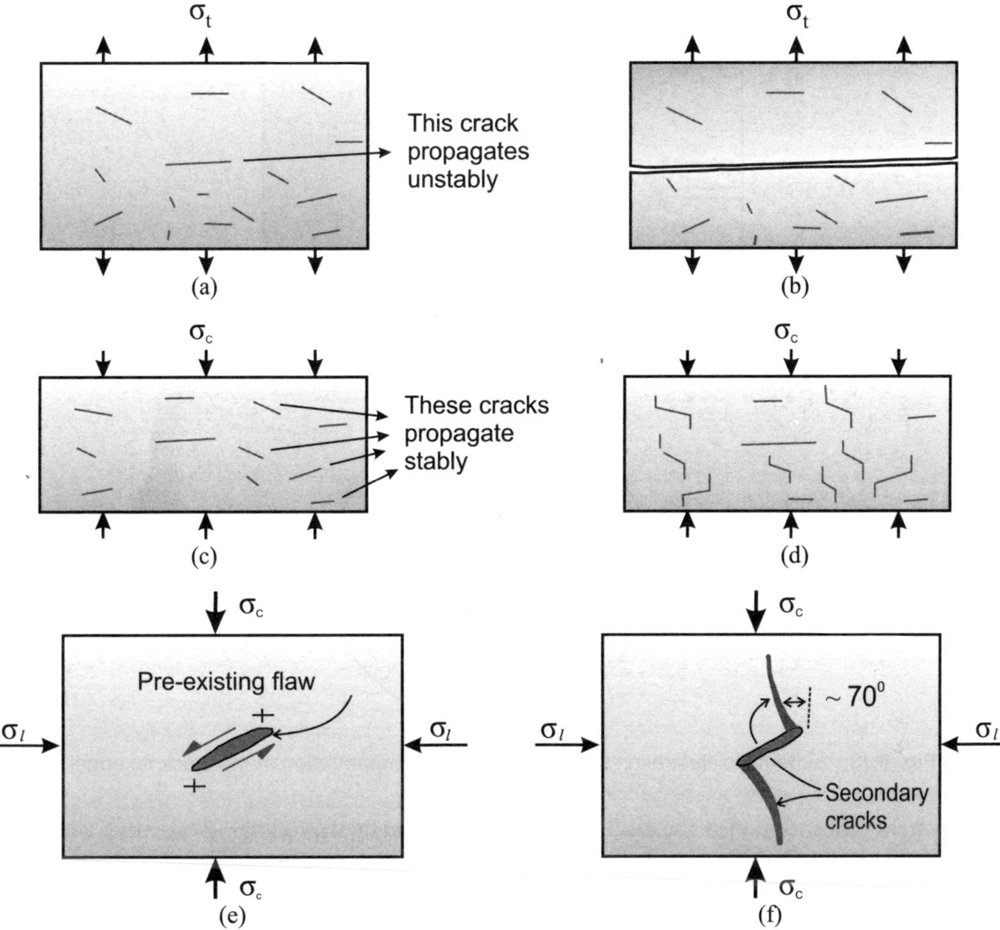

Fig. 6.12 Schematic illustration of the crack propagation from pre-existing cracks (a and b) under tension and (c and d) under compression. (e and f) The growth of secondary cracks from pre-existing primary cracks, inclined at an angle to the compression axis is also shown.

In the following, some of the important theories explaining the brittle fracture of ceramics/glass materials will be reviewed, in particular reference to quantitatively explain criticality of fracture with respect to critical crack size.

The initial theory of brittle fracture is based on interatomic bond breakage/rupture which gives rise to two additional surfaces.

In one of the early theories, this aspect was considered by Inglis, who proposed a theory based on stress concentration at crack tip[561] (see Fig. 6.13). In such a theory, a crack is physically considered as an elliptical hole or cavity in an otherwise solid continuum and a crack essentially represents localized area of bond rupture or discontinuity in interatomic bonding.

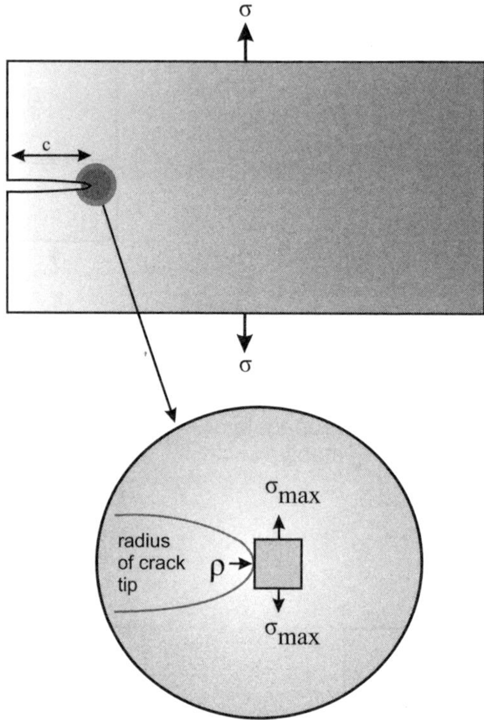

Fig. 6.13 Schematic illustration showing the stress concentration at the crack tip edge.

The stress concentration at the crack tip can be expressed by the following equation, wherein σ is the externally applied stress, and c, ρ are respectively the edge crack length and crack tip radius of curvature,

$$\sigma_{max} = \sigma\left(1 + 2\sqrt{\frac{c}{\rho}}\right) \tag{6.26}$$

From geometrical considerations, it is evident that the crack length is much greater than crack tip radius (i.e., $c \ggg \rho$) and therefore, the maximum stress at crack tip edge is,

$$\sigma_{max} = 2\sigma\sqrt{\frac{c}{\rho}} \tag{6.27}$$

From the above expression, it should be clear that the crack tip will experience much higher value of stress than externally applied tensile stress ($\sigma_{max} > \sigma$). According to Inglis theory, fracture will take place when the stress at crack tip is just sufficient to break interatomic bonds ahead of the crack tip.

$$\sigma_{max} = \sigma_{th} \tag{6.28}$$

$$2\sigma\sqrt{\frac{c}{\rho}} = \left(\frac{E\gamma}{a_o}\right)^{1/2} \tag{6.29}$$

$$\rho = a_o \tag{6.30}$$

Orowan approximated that the crack tip curvature radius is approximately of the same magnitude as interatomic distance (a_o)[562], i.e., $\rho = a_o$. This leads us to the expression for the critical fracture stress,

$$\sigma_f = \left(\frac{E\gamma}{4c}\right)^{1/2} \tag{6.31}$$

From the above expression, it is clear that larger the crack length (c) or sharper the crack with reduced radius of curvature (ρ), the lower will be the fracture strength.

A different expression for fracture strength of ceramics can be obtained following Griffith's theory, which is based on the total energy minimization of a ceramic plate with central through-thickness crack of length 2c,

$$\sigma_c = \sqrt{\frac{2\gamma E}{\pi c^*}}$$

or,

$$c^* = \frac{2\gamma E}{\pi\sigma_c^2} \tag{6.32}$$

Based on the above theory, it can be said that fracture will occur if one of the following conditions is satisfied:

(a) for a particular crack length, external stress, $\sigma \geq \sigma_c$
(b) at a particular stress level, intrinsic flaw size, $c \geq c^*$

It can be further stated that cracks of sizes less than c^* will not grow at a given external stress, as any infinitesimal growth of crack will lead to an overall increase in energy of the system. When $c \geq c^*$, any infinitesimally small increase in crack length can lead to an overall decrease in energy leading to crack growth. The critical cracks of finer sizes will be able to grow at higher stress, leading to delayed fracture of brittle solids.

An important point to be noted here is that in glasses and some ceramics, like Si_3N_4 the cracks with size lower than critical crack size $(c < c^*)$ can grow in an unstable manner at a given stress level in moist/humid environment, ultimately leading to fracture.[563] This is known as subcritical crack growth and attributed to the environmental interaction, which leads to chemical attack at the crack tip and finally results in easy bond breakage. Although not widely reported, it is plausible that any degradation in strength properties of ceramic/glass implants *in vivo* can be attributed to the corrosive attack at the crack tip in the dynamically changing physiological environment inside human body.

It can therefore be rightly stated that the fracture of brittle solids does not depend only on one parameter, but depends on a combination of both, critical stress (σ_c) and critical crack length (c^*).

It is therefore obvious that a combination of crack size and stress will determine the fracture of brittle materials. Based on this observation, the stress intensity factor can be defined as

$$K = Y\sigma\sqrt{\pi c} \tag{6.33}$$

where, Y is a factor dependent on location/orientation of crack and loading condition.

In the classical fracture mechanics theory, different modes of loading of the crack faces are recognized as tensile or crack opening mode (mode I), shear mode (mode II) and tearing mode (mode III). Corresponding to the three different modes, the stress intensity factor can be defined as K_I, K_{II}, and K_{III}. Tensile mode is considered to be the most dangerous among the three modes and most failures of brittle solids are largely due to mode I failure. In view of this, researchers are mostly concerned with K_I values. At the critical condition, the crack tip stress intensity factor under mode I loading is defined as,

$$K_{I_C} = Y\sigma_c\sqrt{\pi c^*} \tag{6.34}$$

where, c^* is the half of the through-thickness critical crack length.

The value of the geometric factor Y varies around 1–1.1. The K_{Ic} defined in Eq. 6.34 is widely known as a parameter, which describes fracture toughness. From the above discussion, it should be clear that K_{Ic} is a material property, i.e., each biomaterial has a unique K_{Ic} value. In various biomedical applications, the external force induced K_I value should not exceed K_{Ic} of the biomaterial and therefore, K_{Ic} is one of the important parameter to assess the structural reliability.

6.6 | Mechanical Properties of Polymeric Biomaterials

As discussed earlier in this chapter and in reference to Fig. 6.14, the stress–strain response or deformation characteristics of polymeric biomaterials are distinctly different from that of ductile metals and brittle ceramics. In view of the wider use of thermoplastic polymers in biomedical applications and the development of HDPE or UHMWPE (one such thermoplastic polymer) based hybrid composites as acetabular socket in total hip joint replacement application, this section discusses the deformation mechanisms of thermoplastic polymers. In Fig. 6.14a, the typical stress–strain response of a thermoplastic polymer is shown, wherein the deformation of polyethylene (PE) can be described as 'pseudo-ductile' behaviour. In Materials Science literature, the deformation dominated by dislocation motion is strictly referred to as 'ductile' behaviour. In case of polymers, the dislocation as a defect is not present and instead the deformation is primarily dominated by necking and crystallization, as will be discussed later. Therefore, the deformation of polymer is referred to as 'pseudo-ductile' behaviour.

The characteristic deformation of polymers is widely referred to as 'viscoelastic' or 'pseudo-ductile' deformation. Such deformation can be described as an integrated response of an elastic solid and shearing of parallel layers of polymeric chains.

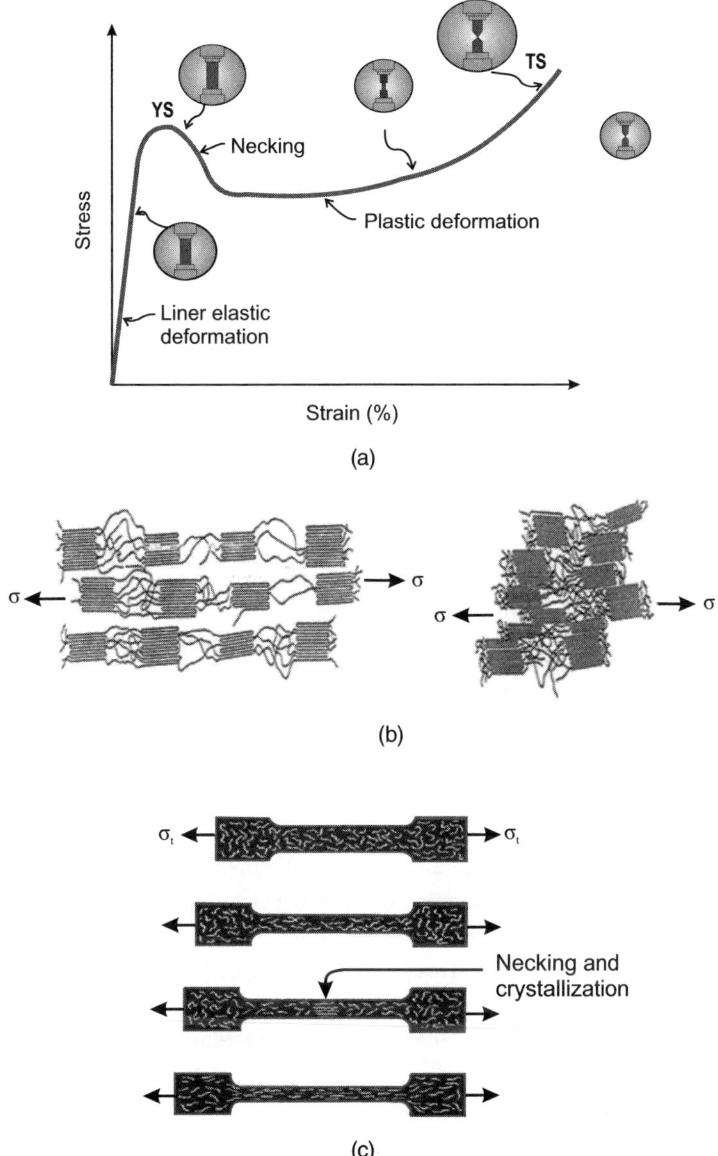

Fig. 6.14 (a) The uniaxial stress-strain response of a typical thermoplastic polymers, which are widely used in biomedical applications. (b) Schematic illustration of deformation of a semi-crystalline polymer with the alignment of the chains in the amorphous region along the stress direction. (c) The phenomenon of necking and crystallization in polymers during viscoelastic deformation.

In Fig. 6.14b, the deformation mechanisms are schematically illustrated. Considering a typical tensile sample of a polymer consisting of randomly oriented short chains loaded in tension, the short chains would align themselves along the tensile axis during the loading. Eventually, the parallel alignment of short chains is accomplished at specific location in the tensile sample, leading to a localized orderly arrangement of polymeric chains. This phenomenon, known as crystallization leads to necking in polymers. Once necking starts, the thermoplastic polymer maintains steady deformation almost at constant load, which leads to a large deformation region in the stress –strain response, as shown in Fig. 6.14c. The alignment of short chains across the entire gauge length (length of the tensile sample with a constant cross-section) is a time-dependent phenomenon, as it involves a steady rotation and alignment of the polymeric chains along tensile axis. This requires considerable time and this is the origin of the viscoelastic deformation. While the crystalline region with more ordered arrangement of chains readily aligns along the tensile axis, the alignment of chains in the disordered or amorphous region is a rather slow time-dependent process (see Fig. 6.14c). With continuous loading, a situation arises when most of the disordered region is aligned along the tensile direction. The above explains the viscoelastic response of large deformation with little increase in stress.

6.7 | Experimental Assessment of Mechanical Properties

In the following section, a brief discussion on the measurement of various mechanical properties is provided. Such a discussion would be extremely useful for the researchers in the interdisciplinary area of biomaterials.

6.7.1 | Metals

6.7.1.1 | Tensile properties

The tensile stress–strain response of a metal is conveniently measured using a dogbone geometry on universal testing machine. From such testing, one can get stress–strain plot, like the one shown in Fig. 6.5. From such plot, one can quantitatively determine (a) elastic modulus/modulus of elasticity/ Young's modulus, E (slope of linear) stress–strain response, (b) yield strength (stress at 0.2% deformation a measure of transition from elastic to plastic response), (c) ultimate tensile strength, UTS (a point of maximum load on Fig. 6.5a), (d) fracture strength (stress at point of fracture in Fig. 6.5a) , (e) ductility (% strain to failure), and (f) toughness (total area under entire stress–strain response). It is instructive to note that toughness of metals, as determined above, is conceptually different from fracture toughness of ceramics, as defined in the preceding section. Therefore, the toughness property should be measured differently for ceramics.

6.7.1.2 | Hardness

The resistance against permanent deformation, i.e., hardness of a metal can be experimentally measured using Brinell or Rockwell hardness tester, as shown in Fig. 6.15. The principle of hardness measurement in these two techniques is fundamentally different. In Brinell technique, steel or WC sphere of 10 mm diameter (D) is pressed against a smooth flat metallic surface with an indent load (P) and the diameter of the indentation impression (D_i) is measured using an optical microscope. Then, Brinell hardness value is determined using the following formula,

$$BHN = \frac{2P}{\pi D(D - \sqrt{D^2 - D_i^2})}$$

(6.35)

In contrast, Rockwell hardness number is measured in terms of the depth of the indentation, when a cone shaped indenter is pressed against a flat metallic surface. In Rockwell scale, larger the depth of indent, lesser is the hardness number. Also, the thickness of the test sample should be at least 10 times the depth of indentation to ensure reliable hardness measurement. During such measurement, an initial preload is given to make a good contact between indenter and metal. Depending on the hardness of the metal, three different scales, i.e., Rockwell A (R_A), Rockwell B (R_B) and Rockwell C (R_C) are widely used. The indent load is lowest for R_A scale measurement (60 kg) and the highest for R_C scale (150 kg). For example, R_C of full hardened martensitic steel is around 65.

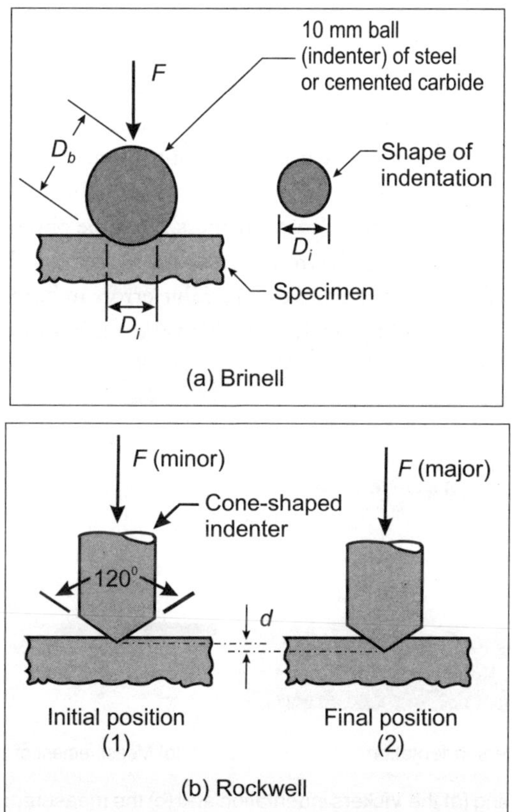

Fig. 6.15 Hardness testing of metals: (a) Brinell and (b) Rockwell.

6.7.2 | Ceramics

In view of inherent brittleness, the preparation of standard tensile samples from ceramics is extremely difficult. Therefore, different test methods are adopted to measure basic mechanical properties, as discussed below.

6.7.2.1 | Vickers hardness

Conventionally, the hardness is defined as the resistance to permanent (plastic) deformation. This property is measured by taking indents on flat polished surfaces (Fig. 6.16). Mostly for ceramic materials, hardness is measured using Vickers indentation technique. The following expression is used to assign the hardness value of a material after the indentation,

$$H_v = 1.584\left(\frac{P}{d^2}\right) \tag{6.36}$$

where, H_v is Vickers hardness (also called Vickers pyramidal number), P is the applied load; d is the average length of two diagonal of indentation (e.g. average of d_1 and d_2 in Fig. 6.16).

The following aspects need to be considered in obtaining reliable values of hardness of engineering ceramics:

(a) The indent load should be taken in such a way that it does not cause cracking from indent corners as well as stable indentation develops without any spalling (damage) around the indentation.

(b) It is suggested that hardness of a new ceramic composition, processed using a new synthesis (sintering) route, be measured using various indent loads. This could reveal 'indentation size effect' and a conservative estimate of 'true hardness' could be obtained.

(c) It is recommended to use scanning electron microscope to measure the indent diagonal (length scale of the order ' μm') in order to avoid considerable errors in hardness measurement, which is expressed in GPa. This would provide a reliable value for hardness of a ceramic material.

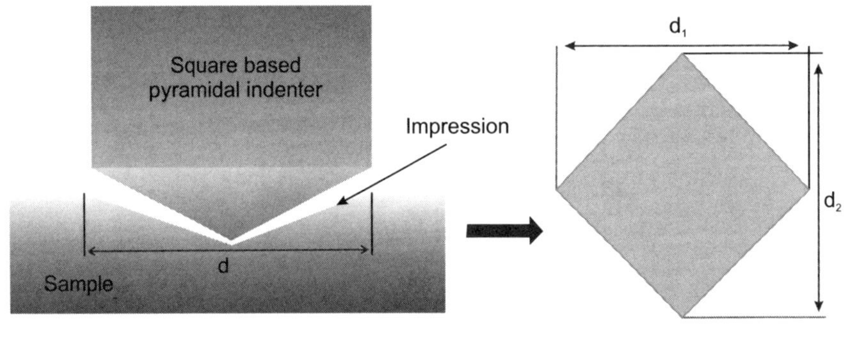

Fig. 6.16 Schematic showing (a) the Vickers indentation and (b) the measurement of impression diagonals for the hardness measurement.

6.7.2.2 | Compressive strength

Although ceramics are extremely weak in tension, they have superior compression property. The difference can be attributed to the difference in microstructural resistance to the crack growth and nature of crack propagation under two different loading conditions. In Fig. 6.17a, the test configuration followed to evaluate compressive deformation properties is schematically illustrated.

Typically, cylindrical samples with aspect ratio of 1.0 or above are used. Ceramics behave like a perfectly linear elastic material up to fracture and in a non-linear manner, after reaching a peak load (much higher than that in tension) in compression. In contrast to tensile crack growth, the cracks extend vertically along the compression loading direction. As shown in Fig. 6.17b, the serration in compression stress–strain response is essentially due to spalling of a small test volume, as the growing cracks either coalesce with each other or, meet free/unconstrained surface of the material. Quite clearly, delayed fracture behaviour is realized in compression and typically, compressive strength is around eight times higher than the tensile strength.

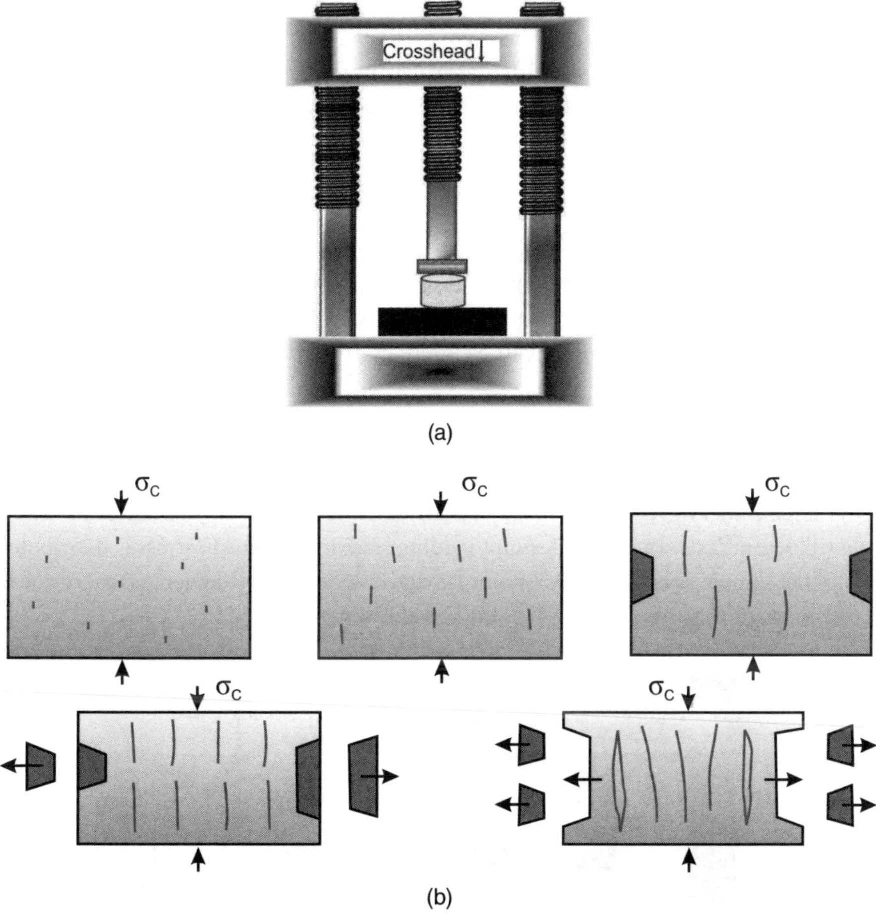

Fig. 6.17 (a) Typical test set up showing the compressive strength measurement. (b) The compression failure mechanisms are also shown schematically in terms of vertical crack propagation.

Compressive strength can be measured using a universal testing machine (UTM). The samples are placed in between two parallel plates of the machine and force is applied on the appropriately aligned samples with a constant crosshead speed (typically around 0.05 mm/s). During the entire compression test, the load-displacement response can be recorded using a computer attached to UTM. The compressive strength of the samples can be calculated from the fracture load and dimension of the samples using the simple formula,

$$\sigma_{cs} = \frac{P}{A} \tag{6.37}$$

where, P is the maximum load (fracture load) and A is the cross sectional area.

6.7.2.3 | Flexural strength

Considering the fact that tensile sample (dog-bone geometry) preparation for ceramics is a challenging task, the strength of the ceramics is measured under flexure mode, either by 3-point or by 4-point bend configuration. For this purpose, either rectangular or circular cross-section samples of beam geometry are placed in a bend fixture. Either a concentrated load is applied in 3-point configuration or a distributed load is applied at two different places in four point configuration (see Fig. 6.18). The flexural strength thereafter is calculated on the basis of measured fracture load and dimension of the test sample. For 3-point loading (Fig. 6.18b), the fracture strength can be calculated from the following expression:

$$\sigma_f = \frac{3PL}{2bd^2} \tag{6.38}$$

where, σ_f is the flexural strength of the material, P is the fracture load, L is the span length, b is the width of the sample and d is the thickness of the specimen. Similarly, the flexural strength for 4-point bend configuration (Fig. 6.18c) can be obtained from the expression.

$$\sigma_f = \frac{3PL}{4bd^2} \tag{6.39}$$

During flexural testing, the loading surface is placed in compression, while the opposite surface is placed under tension. Also, the stress value linearly decreases along the thickness ('z' direction) of the sample till neutral axis. In case of 4-point loading, maximum tensile stress is distributed over a larger area of the sample as opposed to 3-point flexural mode. Hence, lower and more conservative estimate of the strength is obtained in 4-point flexural tests.

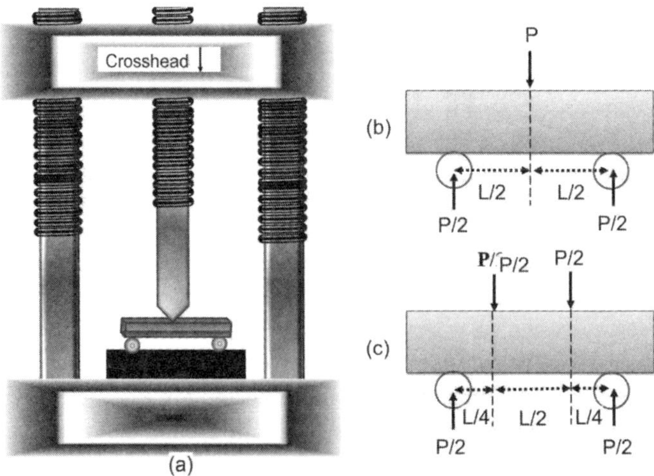

Fig. 6.18 Schematic illustration of the experimental measurement of flexural strength (a). Also, the loading configuration for the three point (b) and four point (c) flexural mode are shown.

6.7.2.4 | Tensile strength

For the metallic materials, the determination of tensile strength is readily obtained from uniaxial tensile test. Since the preparation of a standard tensile sample from brittle ceramics is extremely difficult, an alternative approach can be adopted to measure tensile strength of ceramics (see Fig. 6.19). In this method, a cylindrical sample with much larger diameter than thickness is loaded in pure compression along its diameter. At a peak load, the entire disc fractures into multiple fragments with failure predominantly along the compression axis. The beauty of this simple yet elegant method is that one actually applies compressive forces, but in reality measures the tensile strength of brittle ceramics. The tensile strength (σ) can be calculated according to the following expression,

$$\sigma = \frac{2P}{\pi Dt} \tag{6.40}$$

where, P is the break load, D is the diameter of the disc and t is the specimen thickness.

In order to reduce shear failure resulting from point loading in diametral compression tests, two techniques may be used. The first is to use packing strips, or another approach to distribute the load over a controlled volume. Thin sheets of metal, such as copper, have been found to be effective in giving reproducible tensile-mode results[564]. A second method involves grinding opposite edges of the sample. A recent work reported that for a flat length <0.2D, the tensile strength value should be more reliable[565].

Fig. 6.19 Schematic of tensile strength measurements of ceramics under diametral compression mode.

6.7.2.5 | Elastic modulus

The elastic modulus can be determined using dynamic elastic properties analysis instrument. This instrument works on the basis of impulse excitation technique and the measured frequency of vibration is utilized to estimate Young's modulus. A steel ball impulser is used to excite the sample at rest. The sample undergoes transient vibrations, similar to spring mass system, in response to the excitation. The frequency of vibration is strongly dependent on shape, dimension, mass and stiffness of the sample. The vibration response of the system is recorded using accelerometer (contact sensor) or microphone (non-contact sensor). Typically, the sample locations which experience maximum displacement (anti-nodal points) are used for excitation and sensing purpose, while locations with zero displacement (nodal points) are used to support the specimen in order to avoid rapid damping. The impact location, sensor location and support locations are selected such that desired mode of transient vibration is experienced by the sample. Impulse excitation technique has been extensively used in measuring elastic modulus and damping properties of various metals[566], glasses[567], ceramics[568,569,570], etc.

Fig. 6.20 shows the nodal and anti-nodal locations for bar/rod samples under flexural mode of vibration, wherein the displacement is in the plane normal to the length dimension. Out-of-plane flexure and in-plane flexure refer to flexural mode of vibration in which displacement is perpendicular and parallel to the major plane of the specimen, respectively. Figure 6.21 depicts the nodal and anti-nodal locations along with impact and sensing positions pertaining to disc shaped specimens. Mode 1 corresponds to anti-flexural mode of vibration, wherein displacement in cross-sectional plane is symmetrical about two orthogonal diameters in the plane of the disc. In contrast, mode 2 corresponds to axisymmetric flexural mode of vibration, wherein displacement in cross-sectional plane is uniform for specified radial distance originating from the centre. The above discussion is summarized in Fig. 6.22.

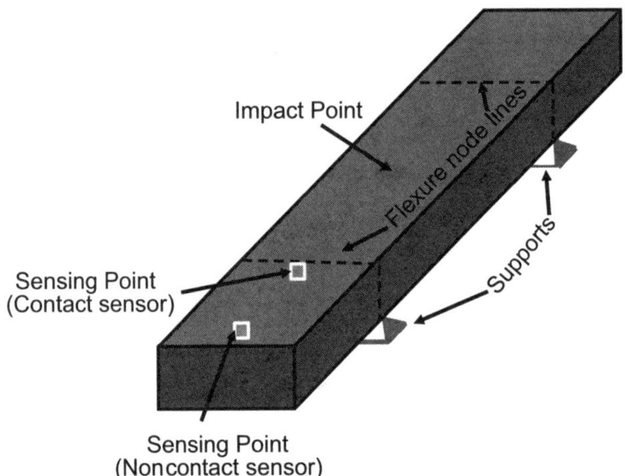

Fig. 6.20 Nodal and anti-nodal locations in bar/rod for out-of-plane flexural vibration.

Once the vibration frequency is measured, Young's modulus (E) values can be determined using any of the following formula (6.41 for bar, 6.42 for rod and 6.43 for disc)[571]:

$$E = 0.9465\left(\frac{mf_f^2}{b}\right)\left(\frac{L^3}{t^3}\right)T \tag{6.41}$$

$$E = 1.6067\left(\frac{L^3}{D^4}\right)\left(mf_f^2\right)T_1 \tag{6.42}$$

$$E = \frac{37.6991 f_i^2 D^2 m\left(1 - v^2\right)}{\left(K_i^2 t^3\right)} \tag{6.43}$$

where, m is mass, f_f is flexural frequency, b is width, L is length, t is thickness, D is the diameter, f_i is the natural resonant frequency (i=1, 2 which corresponds to first and second mode), v is the Poisson's ratio and K_i is the geometric factor, obtained from empirical relations. The correction factor, T and T_1 is given as,

$$T = 1 + 6.585\left(\frac{t}{L}\right)^2, \text{ if L/t} \geq 20 \tag{6.44}$$

$$T = 1 + 6.585\left(1 + 0.0752v + 0.8109v^2\right)\left(t/L\right)^2$$

$$-0.868\left(t/L\right)^4 - \left[\frac{8.340\left(1 + 0.2023v + 2.173v^2\right)\left(t/L\right)^4}{1 + 6.338\left(1 + 0.1408v + 1.536v^2\right)\left(t/L\right)^2}\right], \text{ if L/t} < 20, \tag{6.45}$$

$$T_1 = 1 + 4.939\left(\frac{D}{L}\right)^2, \text{ if D/t} \geq 20 \tag{6.46}$$

$$T_1 = 1 + 4.939\left(1 + 0.0752v + 0.8109v^2\right)\left(D/L\right)^2$$

$$-0.4883\left(D/L\right)^4 - \left[\frac{4.691\left(1 + 0.2023v + 2.173v^2\right)\left(D/L\right)^4}{1 + 4.754\left(1 + 0.1408v + 1.536v^2\right)\left(t/L\right)^2}\right], \text{ if D/t} < 20, \tag{6.47}$$

Alternatively, Elastic modulus (E) and Poisson's ratio (v) can be evaluated by ultrasonic method using lithium niobate crystals for reusing transmitting and receiving signals. Such signals are usually generated at 10 MHz resonant frequency. The thickness of the specimens and travel time of the waves across the thickness/height of specimens can be used to estimate the velocities of the longitudinal and shear waves. The following relationships can be further used to determine the elastic modulus (E, GPa) and Poisson's ratio (v),

$$E = \frac{(1 + v)(1 - 2v)}{(1 - v)}(\rho C_L{}^2) \tag{6.48}$$

$$v = \frac{1/2(C_L/C_S)^2 - 1}{(C_L/C_S)^2 - 1} \tag{6.49}$$

where, ρ is the density of specimen (g/cc), C_L is the velocity of longitudinal wave (m/s) and C_S is the velocity of shear (transverse) wave (m/s). It is important to mention here that the ultrasonic method can only be used for metallic samples to get statistically relevant data. In ceramic samples, this particular method may give erroneous values due to the presence of cracks, pores and other flaws. Such defects can reflect ultrasonic waves, leading to shorter travel time for the given sample thickness, ultimately resulting in overestimation of elastic modulus.

It is worthwhile to note that Young's modulus of ceramics can also be determined using static methods, like instrumented indentation, three point bending, diametral compression, etc. While the initial slope of the unloading response (P-h plot) in case of instrumented hardness measurement provides an estimate of elastic modulus, the slope of the initial linear loading response under uniaxial compression/flexure/diametral compression can be used to determine the elastic modulus under respective loading condition. Generally, all the static methods result in permanent deformation to the sample, whereas dynamic methods are non-destructive in nature. Though, both the methods can be used to experimentally estimate elastic modulus, the values obtained from both the methods differ, typically about 10%, due to inelastic effects.

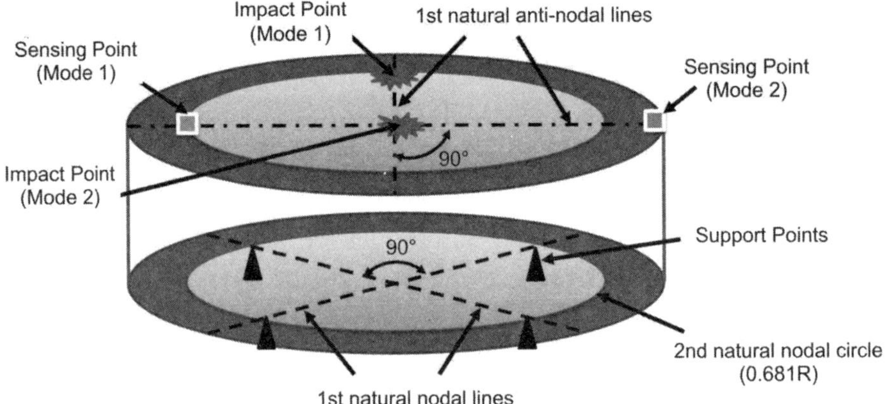

Fig. 6.21 Nodal and anti-nodal locations in discs, while measuring elastic modulus.

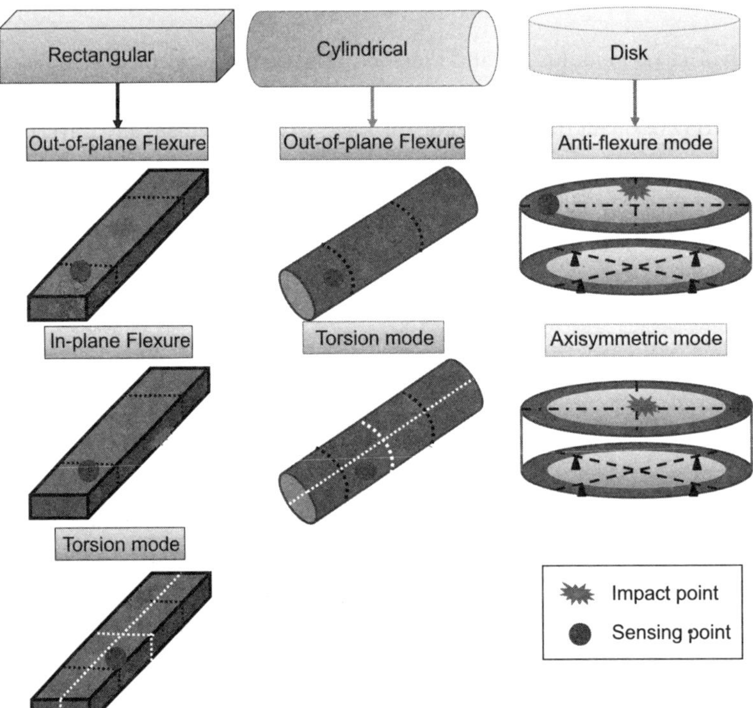

Fig. 6.22 Impact and sensing locations for various specimen configurations, used in elastic modulus measurements.

Analytically, the rule of mixtures is used to determine the elastic modulus. Since ceramic samples are inherently associated with pores, corrections are to be incorporated in the modulus values estimated through the rule of mixtures, which are typically based on exponential/linear variation of elastic modulus with respect to sample porosity. Since porosity is one of the important parameters in the context of biocompatibility, it is important to know how porosity influences the elastic modulus. It is also important to reiterate that large difference in modulus property between implant and host bone causes osteolysis. In literature, some empirical relationships are reported to analyse porosity dependent elastic modulus and those are as follows[571],

$$E = E_0(1 - aP) \tag{6.50}$$

$$E = E_0(1 - aP + bP^2) \tag{6.51}$$

$$E = E_0 e^{-bP} \tag{6.52}$$

where, E_0 is the elastic modulus of dense material, P is the porosity volume fraction, a and b are empirical constants and e is the natural logarithm base. The above relationships are sensitive towards pore volume fraction, interconnectivity and also material, etc.

6.7.2.6 | Fracture toughness

As far as the toughness measurement is concerned, it is important to note that the toughness of brittle materials is dependent on the test technique employed. These techniques are widely classified into long crack and short crack methods. Long crack methods consist of single edge notched beam (SENB) and single edge V-notched beam (SEVNB) techniques. Short crack techniques involve measurement of the crack lengths (radial/median) around hardness indentations, from which the toughness data can be approximated.

It is important to note here that the absolute toughness values of brittle materials cannot be measured by indentation technique, for which one has to adopt more reliable long crack fracture toughness measurements techniques, e.g., SENB, SEVNB, Chevron Notch Beam (CNB), etc.

However, in order to compare the toughness property of the newly developed composites, indentation technique can be used. Additionally, it is now widely recognized that a careful use can provide reproducible results for indentation toughness measurements. It should be mentioned here that the indentation method is now extensively used for convenience to determine the fracture toughness of small and relatively brittle specimens, which are otherwise hard to machine into standard tests samples (for example, SENB, SEVNB, etc.). Broadly, there are two established ways to calculate the fracture toughness of brittle materials. One is to use toughness evaluation with large cracks and the other one is to evaluate by short crack methods.

6.7.2.6.1 | *Long crack method*

It can be reiterated here that it is indeed a challenge to introduce sharp cracks of specified geometry at the notch tip in a ceramic sample. In general, two controlled methods are used for pre-cracking the ceramic material:

(a) Single Edge Notched Beam (SENB) method: In this case, compressive loads are applied in cycles to a SENB specimen with a cut notch. This results in damage accumulation leading to crack growth in the zone ahead of the notch, when loaded in flexure mode. The testing geometry as well as various parameters is shown in Fig. 6.23. The evaluation of mode I critical stress intensity factor (K_{Ic}) by SENB is based on the following classical expression:

$$K_{Ic} = Y\,(3PL/h^2 d)\,c^{1/2} \tag{6.53}$$

In the above expression, $Y = 1.99 - 2.47(c/h) + 12.97(c/h)^2 - 23.17(c/h)^3 + 24.8(c/h)^4$ where, P is fracture load, h is the specimen thickness, d is the specimen width, c is the crack length and L is the span length.

(b) Single Edge V-Notched Beam (SEVNB) method: This is a modified version of SENB technique where a sharp 'V' notch is produced by polishing the notch root with a razor blade coated with diamond paste. The notch radius can be made smaller than 10 µm (essential to ensure quick crack propagation) by carefully polishing. The samples are then fractured using a four point bending set up on a universal testing machine. Figure 6.23 depicts shape of a typical notch. The fracture load is recorded and fracture toughness can be calculated from the following equation[572,573]:

$$K_{Ic} = \frac{P_f\left(L_0 - L_i\right)}{BW^{3/2}} \frac{3\alpha^{1/2}}{2\left(1 - \alpha\right)^{3/2}} f\left(\alpha\right) \tag{6.54}$$

where P_f, Lo, L_i, B, W are fracture load, outer span, inner span, specimen width, and specimen depth, respectively, $\alpha = a/W$, where a is the pre-crack size, and $f(\alpha)$ is :

$$f(\alpha) = 1.9887 - 1.326\alpha - \frac{\alpha(1 - \alpha)\left(3.49 - 0.68\alpha + 1.35\alpha^2\right)}{\left(1 + \alpha\right)^2} \tag{6.55}$$

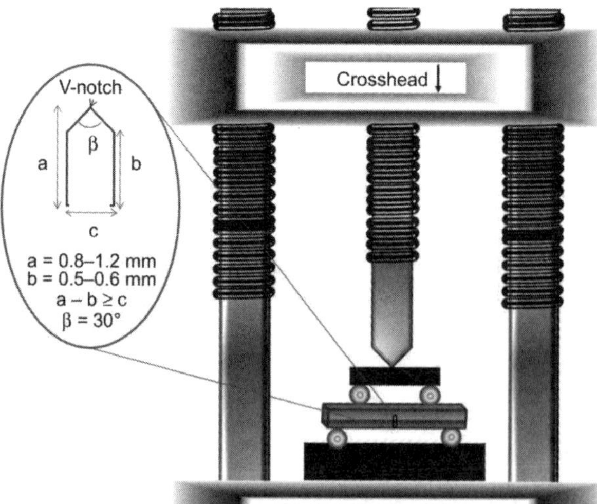

Fig. 6.23 Typical geometry and loading configuration involved in SEVNB testing to evaluate fracture toughness of ceramic samples. Typical dimension of the V-notch is also mentioned (left).

6.7.2.6.2 | *Indentation microfracture method*

Vickers indenters can be used to induce small surface cracks of controlled size and sharpness in brittle materials, like ceramics. In view of intrinsic brittleness, the cracks emanate from indent corners and the crack length measurement is the basis for indentation toughness estimation. As discussed below, indentation toughness can be measured using various formulae. A schematic of indentation cracking is shown in Fig. 6.24. It is also recommended that the toughness of ceramics with new composition be measured at different loads to check whether toughness increases with crack length, leading to 'R'-curve behaviour.

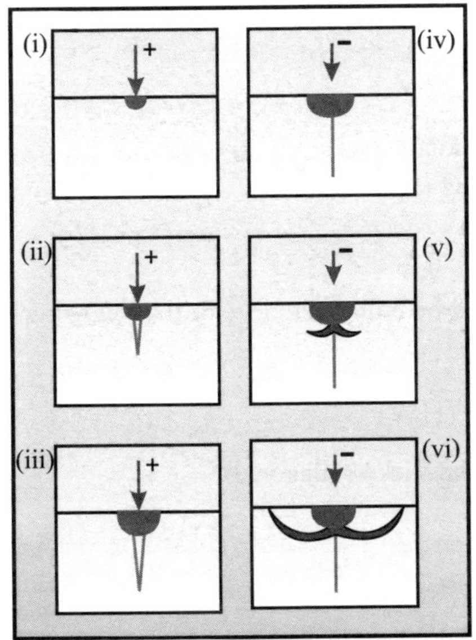
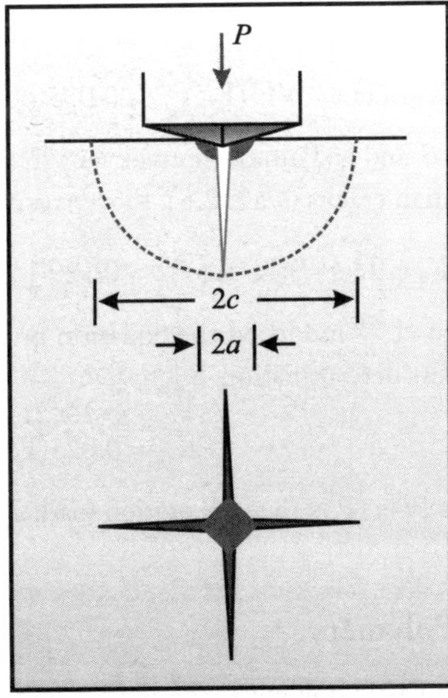

Fig. 6.24 (i–vi) Schematic illustration of the development of radial-median (R) and lateral cracks (L) around Vickers indentation in a brittle ceramic. While '+' sign indicates loading during indentation and '-' sign indicates the unloading during post-indentation. The right panel defines the indent diagonal and crack length, as used in various equations in this chapter [Adapted from Ref. 581].

The median cracks, which extend from the corners of a Vickers indentation are arrested when the residual stress driving force (K_{res}) at the crack tip is in equilibrium with the fracture toughness,

$$K_{Ic} = A(E/H)^n(P/c^{3/2}) = \chi(P/c^{3/2}) = K_{res} \tag{6.56}$$

where K_{Ic} the indentation fracture toughness (MPa.m$^{1/2}$), E the elastic modulus (GPa), H the Vickers hardness (GPa), P the indentation load (N) and χ is the residual stress factor and c the crack length measured from the centre of the indent impression (m). Furthermore, fracture toughness (K_{Ic}) values at different indent loads can be calculated from the measurement of radial cracks around the Vickers indentations, according to the formula proposed by Anstis et al.[574].

$$K_{Ic} = 0.016 \, (E/H)^{1/2}P/c^{3/2} \tag{6.57}$$

Anstis et al.[574] proposed an expression for toughness estimation which was recently modified by Kaliszewski and co-workers[575] to account for the effect of the compressive stresses due to the surrounding transformation zone:

$$K_{Ic} = 0.019 \, (E/H)^{1/2}P/c^{3/2} \tag{6.58}$$

As stated by Evans and Wilshaw[576], the Palmqvist cracks were formed in the low-load regime for a number of brittle materials. The dimensions of the Palmqvist and median cracks are related by:

$$l/a = c/a - 1 \tag{6.59}$$

For the Palmqvist cracks ($0.25 < l/a < 2.5$), the fracture toughness (K_{Ic}) can be obtained from the expression given by Niihara et al.[577]:

$$(K_{Ic}\varphi/H \, a^{1/2})(H/E\varphi)^{2/5} = 0.035(l/a)^{-1/2} \tag{6.60}$$

where, $\varphi = 3$ and l= Palmqvist crack length for median cracks

For median cracks ($c/a \geq 2.5$), the corresponding expression is,

$$(K_{Ic}\varphi/H \, a^{1/2})(H/E\varphi)^{2/5} = 0.129(l/a)^{-3/2} \tag{6.61}$$

Shetty at el.[578] modified the equation proposed by Nihara et al.[577] to the following expression for toughness determination:

$$K_{Ic} = 0.025(E/H)^{0.4}(HW)^{1/2} \tag{6.62}$$

where, W= P/4a (P is the indentation load, 2a is the Vickers diagonal).

6.7.3 | Polymers

For thermoplastic polymers, the tensile sample of dogbone geometry, like metals, can be used to measure elastic modulus, strength and viscoelastic properties. In view of the intrinsic deformation characteristics, polymers exhibit unique strain rate dependent tensile properties, whereas metals and ceramics do not. The strength and elastic modulus increases with strain rate, while strain to failure decreases with strain rate. Therefore, one can design tensile experiments with varying strain rates.

6.8 | Practical Guidelines for the Experimental Measurements

6.8.1 | Hardness

As described earlier, hardness of a material can be defined as its resistance to permanent (plastic) deformation. Due to their excellent ductility, hardness measurements in metals can be obtained with ease band accuracy. Usually Brinnell and Rockwell indentation tests are used to quantify the hardness in metals. Ceramics, due to their inherent brittleness show very little plastic deformation and hence the tests which are applicable for metals may not provide accurate results in case of ceramics. Usually, Vickers indentation method is used for hardness measurement in ceramics, where the diamond pyramidal indentor is used. In order to get appropriate measure of hardness in ceramics, the indentor load should be chosen such that there is no cracking around the indent. The hardness should be measured at various loads to obtain the load independent value for the property, concurrently ensuring that there is no cracking around the indent. As cracking is a result of release of the stored elastic energy, this leads to the underestimation of hardness values in ceramics.

6.8.2 | Strength

Strength is the load bearing capacity of a material. As in the previous case, the strength measurements for metals can be easily obtained through a standard tension test conforming to ASTM standards. In ceramics, due to the difficulty in machining a ceramic into a standard dogbone shaped sample used for tension tests, flexural tests or diametral compression tests are used to evaluate strength properties. In flexure tests, the 4-point flexure is preferred over the 3-point configuration, as larger volume of the sample will be under tension in the former case. In diametral compression tests, the diameter of the sample should be much larger than the thickness of the sample to obtain reliable results. A considerable number of samples are usually subjected to these tests to obtain the Weibull modulus for a particular material to account for the difference in the flaw distribution and density of the samples used in the tests.

6.8.3 | Fracture toughness

Toughness of a material is its resistance to crack propagation under the influence of external load. In ceramics, toughness in the form of critical stress intensity factor under mode I loading (K_{Ic}) can be measured either by indentation method or by the standard Single Edge V-Notch Beam (SEVNB) test, which is also used for metallic materials. Indentation toughness is usually used to obtain an estimate of the fracture toughness of a ceramic and is especially useful when comparing the toughness properties of two or more ceramics[578]. This method is similar to a Vickers hardness test except that to obtain toughness estimates, the indentor load is chosen so as to ensure uniform cracking at the indent corners. Again, this should be done at various loads to obtain the relationship between crack length and toughness (higher the crack length, lower is K_{Ic} at a given indent load). The SEVNB test has been known to provide a more reliable estimate of toughness in ceramics (K_{Ic}). Here, a V-notch is made in a rectangular cross section of the ceramic sample and tested in 4-point flexure configuration. The preparation of the V-notch has been found to be critical as parameters, like crack length and crack tip radius strongly affect the K_{Ic} values obtained from this test. For metals and polymers, the area under tensile stress–strain plot is a measure of toughness. It should be mentioned here that the fracture toughness can be reliably measured for metals and polymers using SEVNB technique.

6.8.4 | Elastic modulus

The following guidelines can be used to determine the elastic modulus in non-destructive manner:

(a) Prior to carrying out impulse excitation experiment on desired material, the validation experiments are to be performed on standard samples of similar sample configuration to ensure reliability.

(b) The desired sample dimensions for impulse excitation technique are, b/t > 5 for rectangular samples, L/D >10 for cylindrical samples and D/t >4 for disk type samples.

(c) All the faces of the specimens need to be flat and parallel for samples with rectangular cross section, while constant diameter and roundness need to be ensured for samples with circular cross section.

(d) Measurement of length, width, thickness, diameter and mass of the samples are to be done with an accuracy of 0.1%. Care must be taken while recording diameter and thickness, as these influence the elastic modulus estimation by factor D^4 and t^3, respectively.

(e) In case of accelerometer based contact sensor for acquisition of vibration signal, the placement of sensor is to be made such that it does not significantly mass load the sample. To avoid this, the sensor should be placed only as far from the nodal location as necessary to record the desired output.

(f) For disc shaped specimens, striking for first mode (anti-flexure mode) is to be made such that it will minimize the excitation of second mode (axisymmetric mode).

(g) The frequency must be determined to an accuracy of 10% for rectangular and cylindrical samples and 1% for disc shaped samples.

(h) If desired repeatability in frequency measurements is not achieved, iterative changes in sensor location, excitation location, sample support locations, etc. are to be made. Generally, repeatability is affected when two vibration modes exists with similar frequencies or due to inhomogeneity in the sample.

(i) In case of rectangular (or cylindrical) samples, if L/t (or D/t) $\geq$ 20, elastic modulus can be calculated directly based on the measurement of flexural frequency. Else, torsional frequency needs to be measured and shear modulus is to be calculated, based on which elastic modulus will be estimated through iteration of Poisson's ratio.

(j) For disc shaped specimens, the ratio of mode 2 and mode 1 vibration frequency is used to estimate Poisson's ratio, followed by calculation of elastic modulus for both the modes as represented in Fig. 6.21. The dynamic shear modulus can also be calculated by knowing elastic modulus and Poisson's ratio.

6.9 | Closure

This chapter provides a conceptual understanding of stress and strain as tensorial quantities. The principles of mechanical properties of three major material classes, metals, ceramics and polymers, in relation to the stress–strain response are also reviewed. Both the strengthening mechanisms and the theory of the brittle fracture are equally emphasized. As far as the metallic biomaterials are concerned, the elastic modulus of biocompatible metals (stainless steel, Ti-alloys, etc.) are far more than that of natural bone and this issue can be overcome with the use of designed porous metallic implants or the use of porous HA coating on the metallic implant. The tensile or compressive strength of metallic biomaterials are much better than polymers and is a clear advantage. The failure strain i.e., total deformation prior to failure for polymers is better than ceramics and do not allow any unpredictable or sudden failure during application. In case of ceramics, the larger compressive strength (much better than metals), as well as good hardness is advantageous for biomedical applications. However, poor fracture toughness remains a major concern. Also, with the use of porous ceramics, the elastic modulus can be matched with that of natural bone at the expense of lower strength properties. While the large deformation to failure for thermoplastic polymers can be an advantageous property, the enhancement of elastic modulus and strength clearly demands the development of polymeric biocomposites with tailored addition of ceramic fillers.

- **Hardness:** Resistance against permanent deformation
- **Strength:** Load bearing capability
- **Fracture toughness:** Resistance against crack growth
- **Elastic modulus:** Ratio of stress to strain in the elastic regime of the stress-strain curve.

Fundamentals – Biological Science

Cells, Proteins and Nucleic Acids: Structure and Properties

In the context of biomaterials science, it is important to introduce the relevant elements of cell biology. In view of the central importance of the cell–material interaction, this chapter will discuss the structure and characteristics of biological cells, proteins and nucleic acids. The various levels of protein structure are described. The differences between eukaryotic (truly nucleated) and prokaryotic (primitive) cells are also highlighted. This is followed by the description of eukaryotic cell structure. In view of the seminal importance of stem cells in human healthcare, some distinguishing features of stem cells are mentioned in this chapter. Various cellular adaptation processes are also highlighted. In a subsequent chapter, the cell signalling mechanisms and the cell fate processes are discussed in the context of describing the biophysical aspects of cell–material interactions. In another chapter in this book, the structural details of bacteria are described in relation to the bacteria–material interaction leading to biofilm formation involved in the prosthetic infection. A summary of the fundamental concepts and definitions can be found in the author's recent book[582].

7.1 | Introduction

The biological structure of an organism is the reflection of the highly ordered hierarchy of complex architecture. The inherently well-organized biological system follows a reductionistic approach, an essential characteristic of all living beings. Each level in the hierarchy represents an increase in organizational complexity with each 'system' being primarily composed of the basic structural units. The basic principle behind such biological organization is that the properties and functions are found at a hierarchical level (Fig. 7.1). The analysis of levels of hierarchies and their relationship imply that each hierarchical level has a determining influence on its lower hierarchical level that make up the upper level. In unicellular organisms, the single cell performs all life functions independently. However, every multi-cellular living organism has highly structured, hierarchical organization with each level having progressively increasing structural complexity. In the order of increasing complexity, these levels are cell, tissue, organ, and organ system, respectively[582]. Various organ systems work together to maintain life of a living organism. This can be further explained with an analogy to molecular structure of a material. The smallest and most fundamental unit of matter is an atom, which consists of a nucleus surrounded by mobile electrons in different orbitals. Two atoms, when held together by a chemical bond, form a molecule. A cell is the smallest unit of a biological system and it contains various biomolecules in the form of either macromolecules or their respective monomers. The most abundant and important macromolecules is protein and the monomer is amino acid. An important example of a protein is collagen, which is the most abundant protein in the human body.

> A cell is defined as a self-contained unit, capable of replicating itself given the proper nutrients and environment.

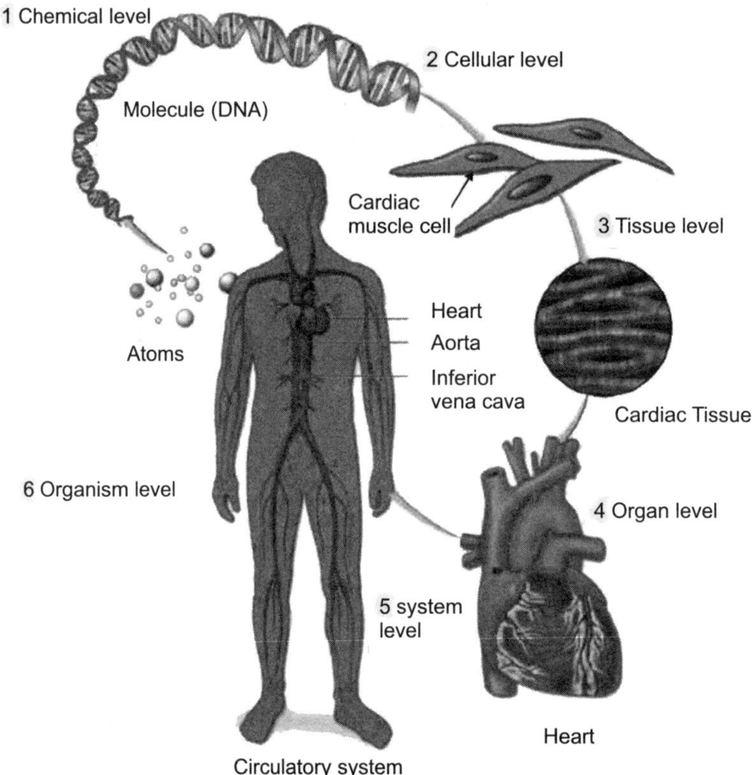

Fig. 7.1 Hierarchical organization of a human body in which each level is of increasing complexity[592].

Further, all living organisms are made of cells and based on the number of cells, organisms can be classified as unicellular or multicellular. Unicellular organisms are classified as the living entities, those comprise of a single cell which can be either prokaryotic type (that lacks membrane-bound cell organelles and nucleus) or eukaryotic type (well defined membrane-bound cell organelles along with a true nucleus having nucleoplasm, nuclear reticulum, nucleolus limited within nuclear membrane). Multicellular organisms are more complex living systems composed of eukaryotic cells with increasing complexity as a result of the evolution. A tissue is a self-organized array of similar types of cells and a tissue can perform specific physiological functions in a multicellular organism. Example of tissues in a human include muscle, nervous, connective, epithelial etc.

At the next level, organs are formed when different type of tissues having common integrated functions are grouped together. The functionally related organs constitute a higher level of organization, known as an organ system. Vertebrate animals have many organ systems. For example, digestive system includes organs such as the oesophagus, intestine, liver, gall bladder, etc.

7.2 | Protein: Structure and Characteristics

Proteins are one of the most abundant (50% of all major organic molecules) intracellular organic molecules, which have diverse range of functions in a particular cell type. Each cell in a living system may contain thousands of different proteins. It is important for one to know the total number and types of protein molecules that a cell typically contains. This is more important in the context of biomaterials as the protein molecules first adhere to a biomaterial substrate within less than a minute of its placement in a cell culture medium. Typical protein concentration in cytoplasm is 180 mg/ml and considering the volume of 15 µm sized fibroblast cell as 2×10^{-9} cm^3, the average number of protein is 4×10^9 molecules per eukaryotic cell. Since the size of a prokaryotic cell is less (considering 2 µm length and 0.8 µm diameter of *E.coli*, the volume 1×10^{-12} m^3), the number of proteins is 2×10^6 molecules per prokaryotic cell.

> Proteins are biological heteropolymers made up of 20 genetically coded amino acids, arranged in a linear manner, forming polypeptide chain.

In general, amino acids are structurally characterized by a carbon (the alpha carbon) bonded to the four groups- a hydrogen atom (H), a carboxyl group (-COOH), an amino group ($-NH_2$) and a 'variable' group or 'R' group. The reaction of two amino acids to form a peptide bond is shown in (Fig. 7.2). The 'R' group varies among amino acids and determines the differences between these protein monomers (Fig. 7.2).

The order of amino acids in a polypeptide chain is unique and specific to a particular protein. Alteration of a single amino acid is caused by a mutation, which most often results in a non-functional protein. The variety in their size, shape, and charge all add up to an extremely versatile set of building blocks to act as some of the most important working molecules of the cell.

One or more polypeptide chains get folded and form a 3D shape known as protein. Folding in proteins happens spontaneously. Chemical bonds are present between portions of the polypeptide chain and this can help in holding the protein together, while giving it a specific shape. Proteins can be divided into two general classes: (i) globular proteins, which are generally compact, soluble, and spherical in shape and (ii) fibrous proteins, which are typically elongated and insoluble.

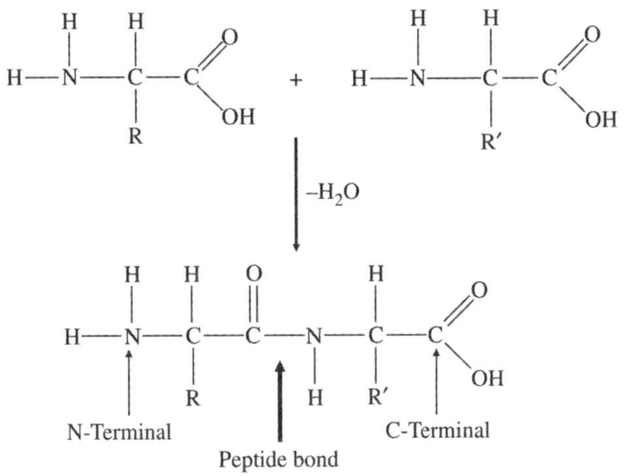

Fig. 7.2 Schematic showing the reaction between two amino acids leading to covalent peptide bond formation along the backbone chain of a protein molecule.

The amino acid sequence of a protein is determined by the information found in the cellular genetic code . Typically, the sequence is written always with N-terminal at the left and C-terminal at the right.

The structure of protein is specified at different levels, namely as primary, secondary, tertiary, and quaternary structure (see Fig. 7.3). The secondary, tertiary, and quaternary structure of proteins can be differentiated from one another on the basis of their degree of complexity in the polypeptide chain morphology.

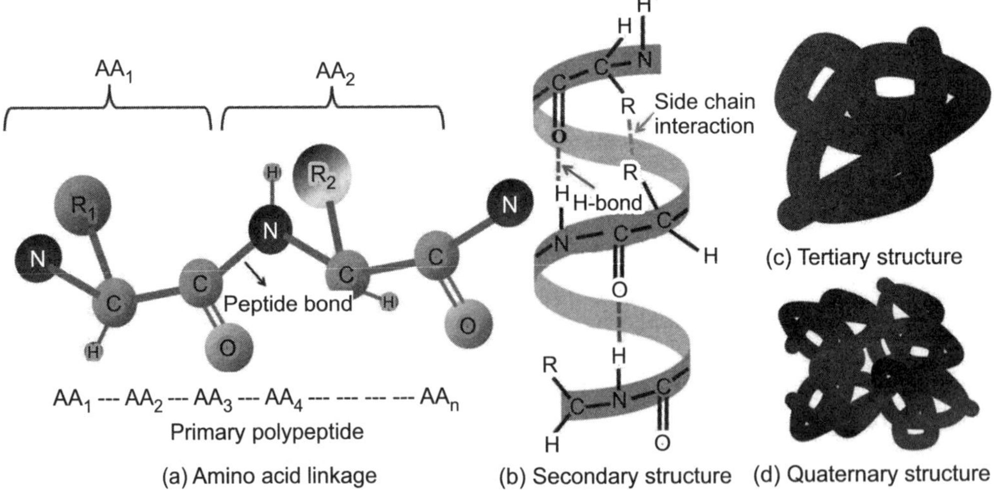

Fig. 7.3 Pictorial representation of different levels of protein structure [Adapted from Ref. 593] [see Colour Plate].

7.2.1 | Primary structure

This level of structure describes the linear polypeptide sequence in which amino acids are linked together to form a protein (see Fig. 7.2). Proteins are constructed from a set of 20 amino acids. Depending on the chemical properties of the side chains, the amino acids may be positively charged (basic), negatively charged (acidic), polar or non-polar. It is important to mention that the charges on the amino or carboxyl end of each amino acid do not play a major role in the overall character of any particular region of the protein. This is in view of the fact that because they are effectively neutral with the linkages between the amino group of one amino acid to the carboxyl group of another, by a peptide bond.

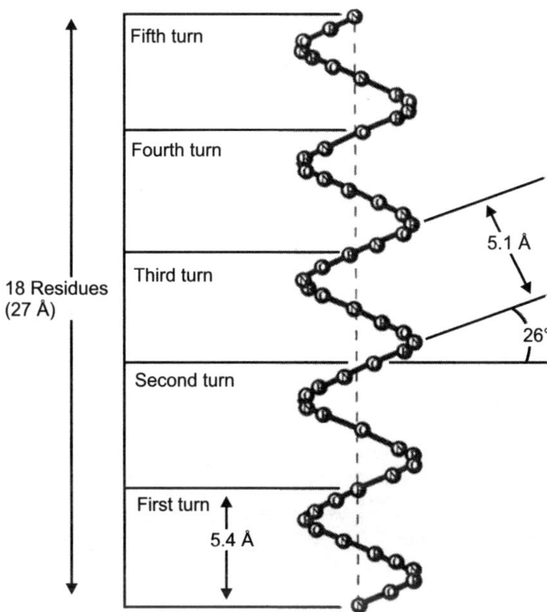

Fig. 7.4 The length scale and other morphological features of the helical structure of a typical protein molecule [Adapted from Ref. 594].

7.2.2 | Secondary structure

The secondary structure of protein refers to the coiling or folding of a polypeptide chain, which determines specific 3D shape of the protein. Two major types of secondary structures are the alpha (α) helix and beta (β) pleated sheet structure. The former resembles a coiled spring and the formation of such structure is facilitated by hydrogen bonding in the polypeptide chain (Fig. 7.3). Each helix contains five turns with each turns bends at an angle of 26° with the axis of helix. The linear elevation due to each turn is 5.4 Å and total length of a protein molecule is therefore 27 Å. The second type of secondary structure, i.e., beta (β) pleated sheet is formed by hydrogen bonding between polypeptide units of the neighbouring folded chains. The characteristic helical structure of a protein is shown Fig. 7.4.

7.2.3 | Tertiary structure

The tertiary structure of a protein molecule refers to the comprehensive 3D structure of the polypeptide chain of a protein, which is guided by several types of bonds and forces. For example, the hydrophobic interactions largely contribute to the folding and shaping of a protein. It can be remembered that 'R' group of the amino acid can be either hydrophobic or hydrophilic. The amino acids with hydrophilic 'R' groups can establish contacts only in aqueous environment, while amino acids containing hydrophobic 'R' groups would avoid water. Hydrogen bonding in the polypeptide chain and between the 'R' groups helps to stabilize protein structure. Due to protein folding, ionic bonding can be established between the positively and negatively charged 'R' groups. The protein folding can also result in covalent bonding among the 'R' groups of cysteine amino acids. This type of bonding forms a disulfide bridge. All such interactions, either attractive or repulsive, can together assist in the stabilization of protein structure. Secondary structure is established by more local interactions, whereas tertiary structures are determined by more global interactions within the same polypeptide chain.

7.2.4 | Quaternary structure

The quaternary structure of a protein macromolecule is essentially established through interactions among multiple polypeptide chains. Proteins with quaternary structure are characterized by more than one protein subunit. Haemoglobin is an example of a protein with quaternary structure, containing two alpha subunits and two beta subunits. The two characteristic representations of quaternary structure, i.e., wire-type and space filling type are shown in Fig. 7.5.

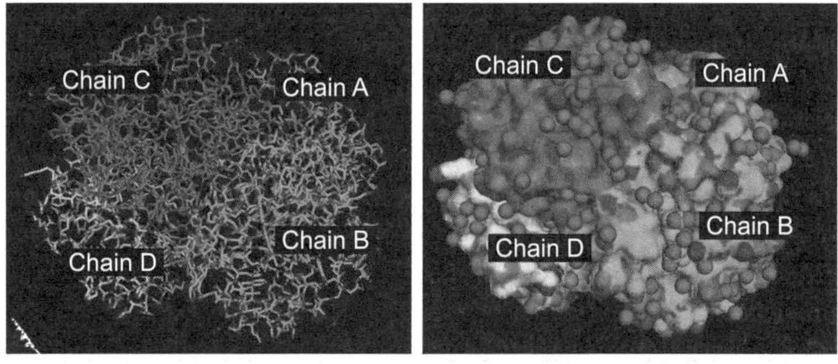

(a) Stick model of Hemoglobin   (b) Space filling model of Hemoglobin

Fig. 7.5   Illustration of the quaternary structure of protein, human deoxyhemoglobin at 1.74 Å using PyMOL (TM) 1.7.4.5 Edu - Educational Product [see Colour Plate].

7.3 | Protein–Protein Interaction

The unique characteristics and properties of various protein molecules can be attributed to the nature of bonding as well as flexibility rendered by light elements (C, H, N, O, etc.) constituting the backbone chain. In Fig. 7.6, the linkages established by two neighbouring protein molecules are shown. While the backbone chain can be identified as a hollow tube structure (see Fig. 7.6a) with R or R^1 being two interacting proteins, the bonding may be mediated either by ionic or hydrogen bonds. Also, the central schematic in Fig. 7.6a shows the possibility of van der Waals force of attraction to hold an atom with the backbone chain. The possibility of H-bonds in large protein structure with helical shape between two closely spaced backbone chains are also shown in Fig. 7.6b.

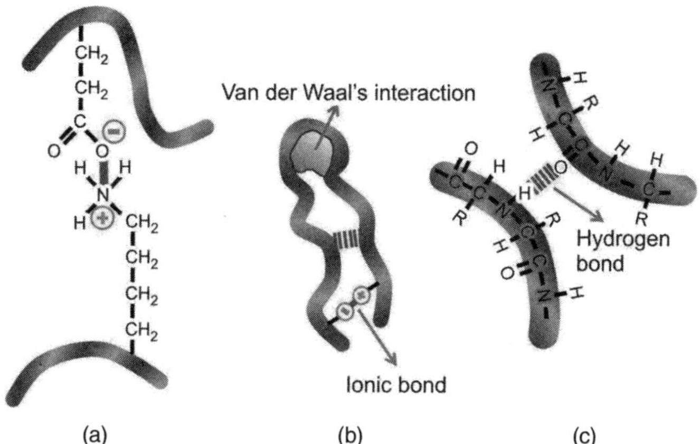

Fig. 7.6 Pictorial representation of protein-protein interaction mediated through different bond formations [Adapted from Ref. 593] [see Colour Plate].

Overall, H-bonds can not only form between side chains, but also between two peptide bonds. The weak nature of H-bond/van der Waals bonds makes the protein structure to stretch, squeeze or change its orientation, depending on its micro-environment. For example, Fig. 7.7 shows the uncoiled and coiled configuration of protein molecule. At this juncture, it can be mentioned that the most stable configuration corresponding to energy minimization is the folded state, as shown in Fig 7.7b. If the protein configuration goes through transition from a folded state to an unfolded one, then it is known as protein denaturation. Denaturation of a protein is often the result of the addition of detergent or urea to a protein solution.

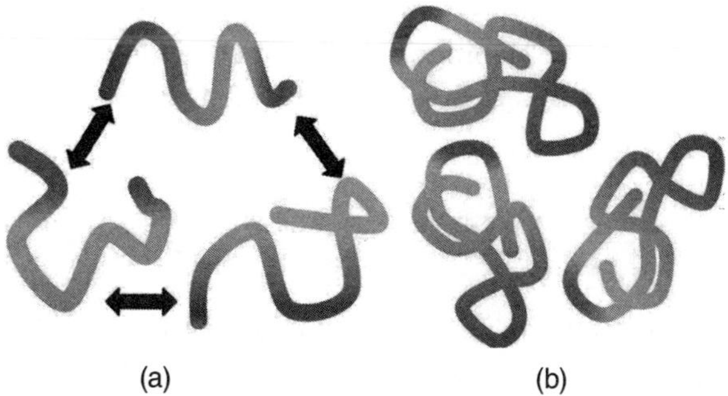

Fig. 7.7 Schematic illustration of protein conformations: (a) uncoiled protein due to denaturation and (b) coiled structure of the protein [Adapted from Ref. 594] [see Colour Plate].

It is instructive to mention that a protein performs its desired function only in folded, native state.

When protein gets adsorbed on a biomaterial, the protein molecules immediately adopt coiled configuration and the bonding or interaction of the different type of protein molecules can be or mediated via different types of bonding, as shown in Fig. 7.8. It will be discussed in a later chapter that cell–material interaction is established via adsorbed proteins on a material substrate. Also, the cell surface receptor interacts directly with adsorbed proteins and in this context; Fig. 7.8 describes how a ligand interacts with a protein molecule.

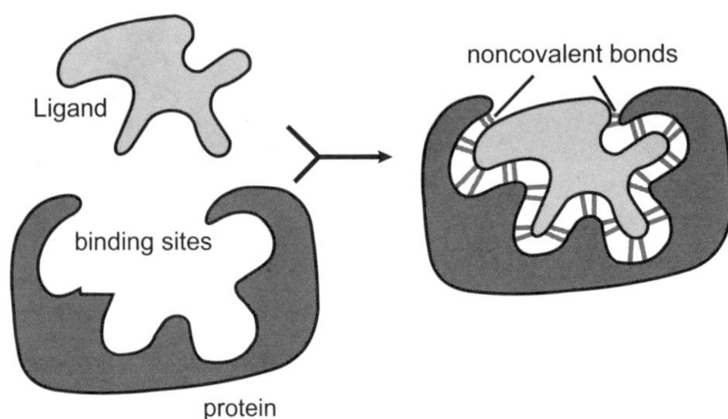

Fig. 7.8 Schematic demonstration of ligand–protein interaction through the formation of non-covalent bonds[593].

The specific configuration of the protein provides some characteristic binding sites, which allows the ligands to be specifically recognised through non-covalent interaction.

This acts as a basis for cell surface receptor-adsorbed protein interaction. In reality, cell–material interaction can be described as 'N' number of such ligand–protein interaction. Also, such interaction is involved in the attachment of signalling molecule, like growth factors to the cell surface receptor, before the signalling molecules are internalized in the target cell. Often such interaction is target-specific, i.e., a signalling molecule of a specific chemistry can interact with a specific cell surface receptor before signal transduction process in initiated. Summarizing, the above discussion is useful to understand the protein structure stability as well as protein–protein interaction in the overall context of cell–material interaction as well as cellular functionality.

7.4 | Cell

It can be reiterated here that cells are smallest structural and functional units of life that make up all the plants, animals and other self-sustaining organisms, such as bacteria and yeast. At this point, one needs to know the number of cells in a body. In natural tissues, the cell density is 1×10^9 to 3×10^9 cells/ml. An adult, weighing 50 kg, will therefore have approximately 10^{14} cells. All the living cells whether unicellular micro-organisms or just a tiny part of a multi-cellular organism look widely different, but share certain characteristics in common. They are small 'pockets' composed mostly of cytoplasm (consisting of cellular machinery and structural elements such as free amino acids, proteins, carbohydrates, fats, and numerous other molecules) enclosed within a phospholipid bilayer membrane. As shown in Fig. 7.9, a typical eukaryotic cell contains a number of organelles with each having characteristic structural features.

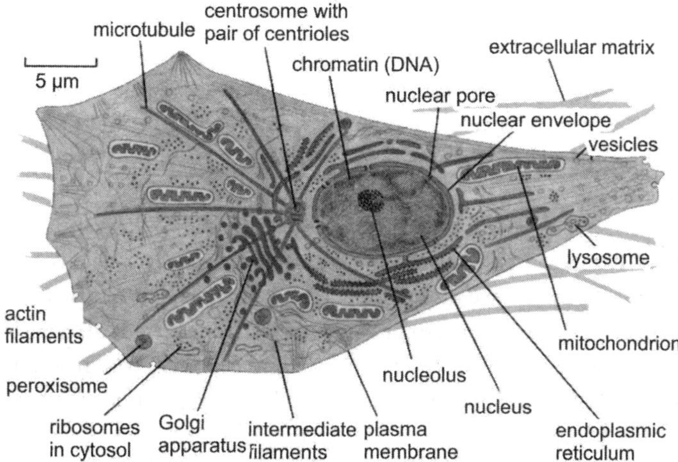

Fig. 7.9 Structure of a typical eukaryotic cell showing various cellular organelles[593] [see Colour Plate].

> In biological terms, cellular organelles are defined as subcellular structures, composed of aggregates of biomacromolecules, and are enclosed by lipid bilayers, with the capability to perform specialised cellular functions.

In general, eukaryotic ('truly' nucleated) cells are enclosed by a lipid bilayer and contain the necessary genetic material, needed to direct the continued propagation of the cell. Phospholipid bilayer membrane is semi-permeable membrane, which allows only some of the selective materials to pass in or out of the cell (Fig. 7.9). These membranes are also studded with proteins those serve various functions. All the cells contain necessary genetic information which is needed to regulate and produce its own parts for their continued propagations and for continuity of life. Nucleic acids (15% of all the molecules) are the key genetic material (macromolecules) that contain and help in expressing a cell's genetic code. The two major classes of nucleic acids are deoxyribonucleic acid (DNA) and ribonucleic acid (RNA). DNA is the genetic material found in all living organisms and DNA contains the necessary information required to build and maintain the cell. The other type of nucleic acid, RNA, has several roles, which are associated with the expression of the information stored in DNA. DNA is present in the nucleus of a cell and uses a RNA intermediate to communicate with the rest of the cells. Nucleic acids use proteins to help in preservation and expression of genetic material during cell division. All the living cells are made up of major elements: hydrogen (59%), oxygen (24%), carbon (11%), nitrogen (4%), and others (2 %) such as phosphorus (P), sulphur (S), etc.

7.4.1 | **Eukaryotic and prokaryotic cells**

All the cells can be categorized into two basic types: prokaryotes (*pro-* = before; *-karyon-* = nucleus) and eukaryotes (*eu-* = true; Karyon=nucleus). Predominantly, single-celled organisms such as bacteria and archaea are classified as prokaryotes, while the eukaryotes are either single cellular organism (e.g. amoeba etc.) or multi-cellular organisms (e.g. yeast, human etc.). The difference between prokaryotic and eukaryotic cells is very simple and can be easily distinguished under light microscope. A prokaryotic cell is a simple, single-celled organism, which lacks a complex level of organization, a true nucleus and any other membrane-bound organelle (Fig. 7.10). In sharp contrast, eukaryotic cells contain well-defined intracellular membrane-bound organelles and nucleus containing genetic material (DNA). The size of eukaryotes is much larger (25–30 μm) than prokaryotes (2–3 μm). Further, the doubling time of prokaryotes is around 20–60 minutes, while that of eukaryotes is around 12 hours or longer.

In a prokaryotic cell, the genetic material is a single circular DNA strand and is located within the central part of the cytoplasm (darkened region), known as a central nucleoid (Fig. 7.10). While in eukaryotic cells, there is a presence of membrane-bound nucleus. Prokaryotic cell is defined by a cell membrane separating a single compartment (cytoplasm) from surrounding environment. Outside the cell membrane, prokaryotic cells have a relatively rigid cell wall, which acts as an extra layer of protection. The cell membrane helps the cell to maintain its shape, and prevents dehydration. It is usually made up of peptidoglycan, a network comprised of sugars and amino acids. Also, the cell membranes have a polysaccharide capsule, which enables these unicellular microorganisms to adhere to their surface or to other individuals in a colony. The thickness of this wall varies depending on the type of prokaryotic cell (bacteria). Gram-positive bacteria have simpler but thick walls with a relatively large amount of peptidoglycan. Gram-negative bacteria have a thinner layer of peptidoglycan, which is located in a layer between the plasma membrane and an outer membrane.

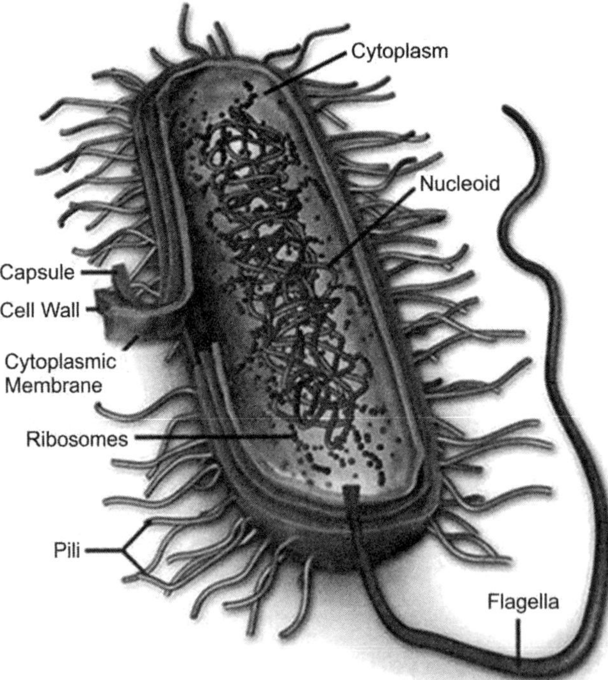

Fig. 7.10 Structure of a typical prokaryotic cell, *E.coli*[595].

Eukaryotic cells (10–100 μm in diameter) are typically 1–2 orders of length in size than prokaryotic cells (0.1–5.0 μm in diameter). The size of prokaryotes generally depends on their rate of efficient metabolism, whereas that of animal cells (eukaryotic) is limited by surface area to volume ratio. Sizes of different kind of cells and their components are shown in Fig. 7.11.

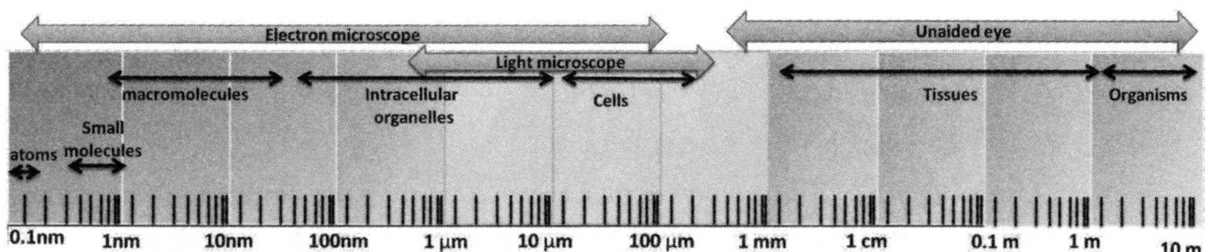

Fig. 7.11   Typical characteristic length scales of the spectrum of biological components of an organism together with imaging techniques to observe them. Fluorescence and confocal microscopes fall under the category of light microscope, while SEM/TEM belong to the class of electron microscopes.

7.4.2 | Structural details of a eukaryotic cell

An individual cell is too small to be seen by our naked eyes. A microscope is an instrument which is generally used to study the structural details of a cell. The images taken with a microscope are called as micrographs.

A biological cell is most widely imaged with optical or fluorescence or confocal microscope or TEM, while SEM is used to study cell morphology on a biomaterial substrate.

The organelles in a eukaryotic cell can be stained with specific fluorochromes. The fluorochromes are molecules of fluorescent dyes which absorb light and gets excited to a higher energy state. As they decay from this excited state, they emit fluorescent light. For example, the nucleus of a cell can be stained using DAPI (4′,6-diamidino-2-phenylindole), which binds strongly to A-T rich regions in DNA, actin filament by phalloidin and mitochondria using mitochondria specific marker (e.g anti-HSP60 antibody), etc. Various structural features can therefore be imaged using fluorescence/confocal microscope. A scanning electron microscope (SEM) can be used to observe the cell morphology and cell spreading. However, the finer details of the cellular substrate can only be imaged, when the cellular organelles can be tagged using specific fluorochromes. In SEM, a beam of electrons moves back and forth across a cell's surface, providing the details of cell surface characteristics by elastic scattering. A transmission electron microscope (TEM) can be used to provide details of a cell's internal structures (e.g., ribosomes, DNA). In TEM, the electron beam is transmitted through the cell rendering fine details of the ultrastructure organelles, but it requires tedious sample preparations.

A typical eukaryotic cell is a dynamic structure which consists of three basic components as observed under the light microscope:

7.4.2.1 | Cell membrane

Eukaryotic cells have a cell membrane or plasma membrane, which is a protective sheath having total thickness of 7–10 nm (Figs 7.12).

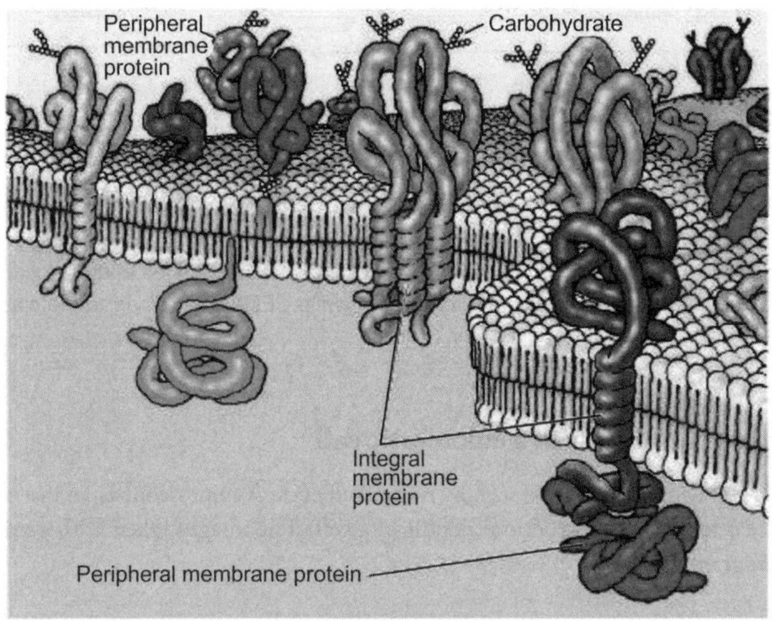

Fig. 7.12   The eukaryotic cell plasma membrane, characterised by a phospholipid bilayer embedded with proteins and cholesterol[596] [see Colour Plate].

The biological membranes can be best described as a phospholipid bilayer with several transmembrane proteins and voltage gated ion channels. Such membrane physically separates the intracellular organelles from the cytoplasm and the cell from the extracellular matrix.

In addition to phospholipids and proteins, other components, such as cholesterol and carbohydrates, can be found in the membrane. Biochemically, a cell membrane is composed of lipids (40%), proteins (55%) and carbohydrates (5%). A phospholipid is a lipid molecule, which primarily consists of two fatty acid chains, a glycerol backbone, and a phosphate group. The lipid bilayer of plasma membrane makes it semi-permeable regulating the passage of some substances, such as organic molecules, ions, and water (Fig. 7.12).

7.4.2.2 | Cytoplasm

Cytoplasm is an aqueous colloidal solution (cytosol) containing numerous functional and structural elements, which exist in the form of molecules/organelles, and can be broadly categorized into three groups: organelles, inclusion bodies and cytoskeleton (Fig. 7.9).

7.4.2.2.1 | *Cellular organelles*

These are membrane bound permanent cellular components, which regulate metabolic activities of the cells. These include:

- Mitochondria- They are major sites for adenosine-tri-phosphate (ATP) production (convert foods into usable energy) through aerobic respiration. They are limited by two membranes. The outer membrane is very smooth, while inner membrane forms incomplete septa/projections, called cristae studded with oxysome particles. The mitochondrion carries its own genetic material (DNA).
- Endoplasmic Reticulum (ER)- These are a series of interconnected membranous tubules and vesicles called cisternae that collectively play an important role for protein synthesis. There are two types:
  - Rough ER- characterized by having ribosomes on their surfaces and transports proteins made by cisternae.
  - Smooth ER- devoid of having ribosomes and helps in lipid and steroid synthesis.
- Golgi apparatus- The Golgi complex or bodies are collections of membrane bound vesicles, sacs and tubules, which are physically independent but continuous with ER. Here, proteins and other molecules are prepared for their transport to other parts of cells or outside the cell.
- Ribosomes- Cellular structures that contain 80–85% of RNA of the cell and are responsible for protein synthesis. They may be present on the endoplasmic reticulum or in the cytosol as free component, where the RNA goes for translation into proteins. Also, ribosomes help in synthesizing membrane bound proteins.
- Lysosomes- Rounded to oval shaped structures, found only in animal cells and are called as the cell's 'garbage disposal.' These organelles are the main point of digestion as they contain various digestive enzymes, which aid in the breakdown of proteins, polysaccharides, lipids, nucleic acids, and even worn-out cell components. Lysosomes also destroy disease-causing organisms with the help of their hydrolytic enzymes.
- Peroxisomes- Spherical structures enclosed by single membrane containing enzymes, such as oxidases (use oxygen to carry out catabolic reactions) and catalase (breakdown of hydrogen peroxide into water and oxygen gas).
- Centrosome- These are cellular organelles that consist of two short cylindrical shaped structures, known as centrioles. These are situated close to the nucleus near the centre of the cell. They are responsible for the movement during cell division.

7.4.2.3 | Cytoplasmic inclusions

These are the temporary structures of certain cells, which may or may not be bounded by cell membrane. The examples of cytoplasmic inclusions are lipid droplets (adipose tissue), glycogen (liver and skeletal muscles) and melanin pigment (epidermis and retina), etc.

7.4.2.4 | Cytoskeleton

Cytoskeletons are certain protein fibres present in the cytoplasm. Their presence helps to execute from different functions: a) maintaining the shape of the cells, b) securing certain organelles in specific positions, c) allowing cytoplasm and vesicles to move within the cell, and d) enabling unicellular organisms to move independently. There are three types of cytoskeletal fibres which include actin filaments, intermediate filaments, and microtubule (Fig. 7.13). All these cytoskeletal elements have unique characteristics and perform various structural roles and functional roles, as described below.

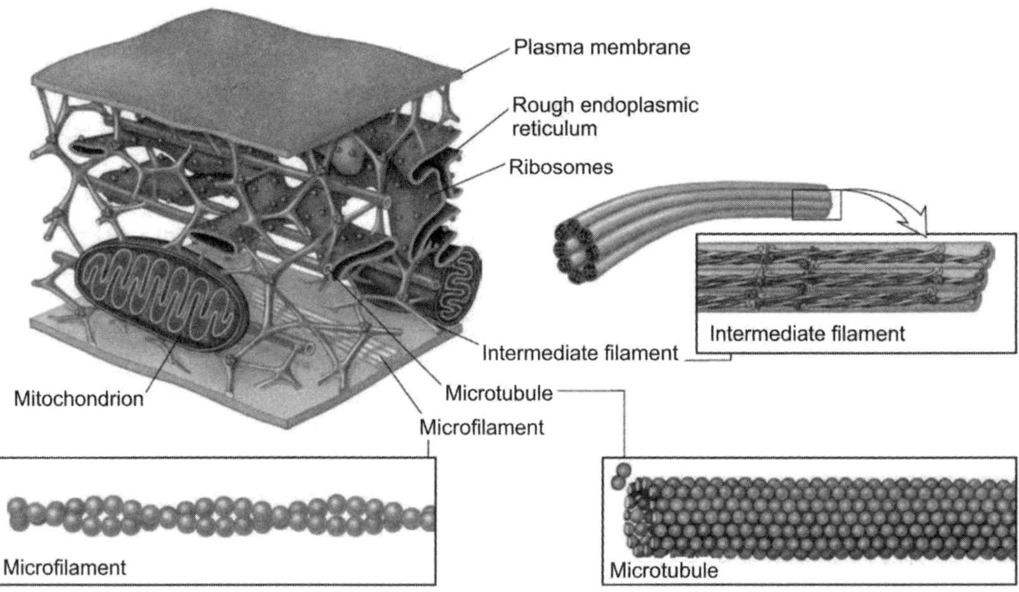

Fig. 7.13 Typical morphological features of different structural elements of cytoskeleton[597] [see Colour Plate].

7.4.2.4.1 | *Actin filament*

These are the thinnest of the cytoskeletal fibres and help in moving cellular components, for example, during cell division. In animal cells, actin filaments are abundant in muscle cells (responsible for muscle cell contraction). These filaments get attached to proteins present on membrane by forming a network structure. The actin filaments in a cell are also involved in the formation of the pseudopodia for cellular locomotion.

The monomers of microfilaments are globular actin (G–actin), which polymerize to form the filaments of 7 nm diameters. The structure of this monomer protein is similar in many organisms including amoeba, plants and humans. G–actin filaments have an ATP/ADP binding site. When ATP binds to this site [ATP-G actin], stable polymerization of G–protein occurs. However, after polymerization, ATP is hydrolysed to ADP, which results in instability and depolymerization of polymer. After depolymerization, G-actin monomer once again becomes stable to polymerize, when ADP is replaced by ATP. In this manner, G-protein monomers are recycled. Polymerization and depolymerization of actin filaments take place more accurately and quickly *in vivo*. This is because of the presence of some actin-binding proteins, which bind to actin filaments to regulate the polymerization, as compared to *in vitro* environments.

7.4.2.4.2 | *Microtubules*

These are thickest of all cytosketal fibres. These are polymers of two dimers known as α-tubulin and β-tubulin. Microtubules are hollow tubes of approximately 25 nm in diameter, with each turn of helix containing 13 dimers. Polymerization and depolymerization of microtubules in cells occur in same fashion as actin filaments. Heterodimers, whose β-tubulin is bound to GTP, are more stable to polymerize and after polymerization, hydrolysis of GTP into GDP occurs (Fig. 7.13, 7.14, and 7.15). This results in destabilization of dimer ending into depolymerization of polymer. Centrosomes, localized near nucleus, functions as a microtubule-organizing centre and serves as polymerization origin of microtubules. The protein complexes essential for polymerization of α-tubulin and β-tubulin dimers are found in centrosome. Centrosome is composed of a pair of microtubules. Each centrosome is a cylinder of nine triplets of microtubules and both lie perpendicular to each other. Microtubules play an important role in cell division by segregating the chromosome (Fig. 7.15). Centrosomes replicate itself before a cell divides and centrioles form spindle fibres, which play a role in pulling the sister chromatid to two opposite poles of dividing cells[583].

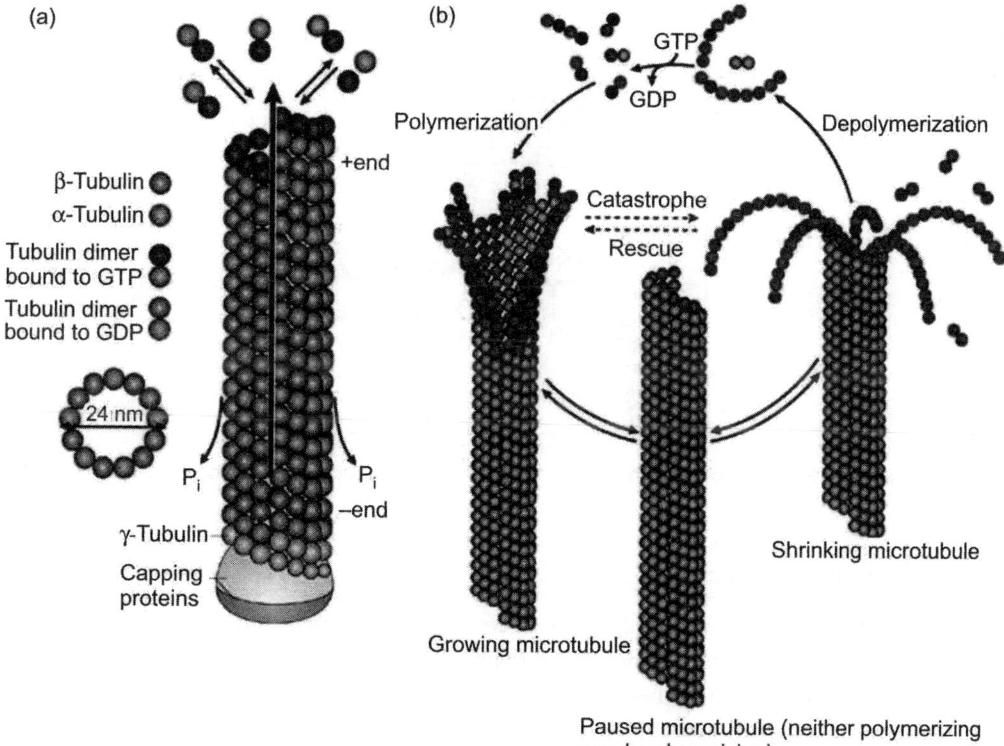

Fig. 7.14 Ultrastructural features of tubulins and the formation of microtubules[598].

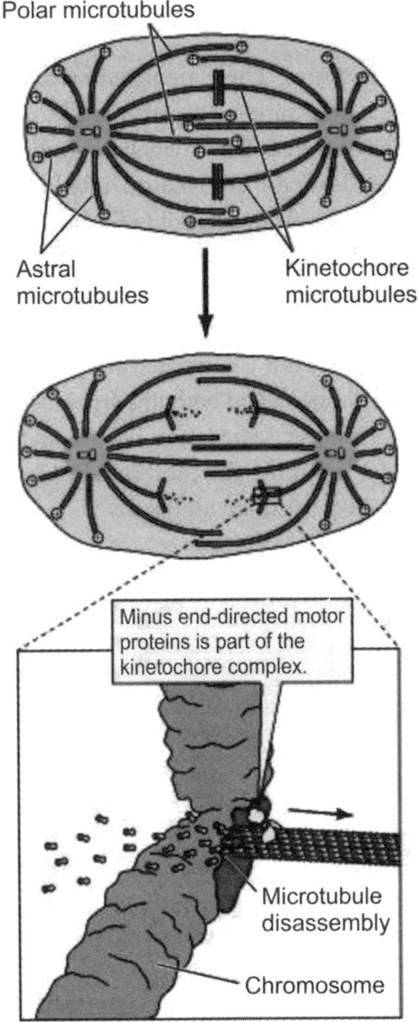

Fig. 7.15 Schematic depicting the chromosomal segregation and microtubules[596].

7.4.2.4.3 | *Intermediate filaments*

Intermediate filaments are of intermediate sizes between actin filament and microtubules and they are of diameter of approximately 10 nm. Also, their polymerization does not require nucleotides, such as ATP or GTP (Fig. 7.16). They form the complex networks with other cytoskeletons in cells and play various structural functions. They maintain the shape of the cell and help in anchorage of cell organelles as well cells to cells and matrices. They are mostly abundant in those cells on which physical tension is applied, for example muscle cells and neurons. These filaments are stable, but their degradation and reconstruction become very active, when cell division occurs. One example of intermediate filament is keratin, which strengthens hair and nails.

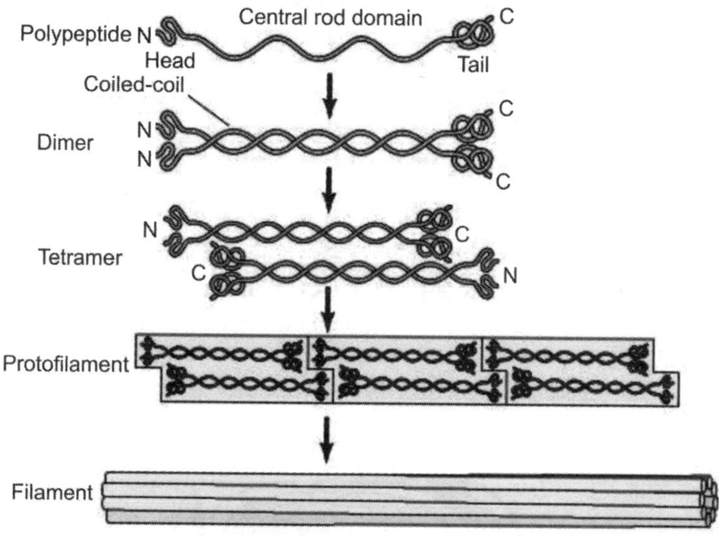

Fig. 7.16 The polymerization steps involved in the formation of intermediate filaments[596].

7.4.2.5 | Nucleus

The nucleus is the most prominent membrane bound organelle as it accounts for most of (10 percent) the cell's volume as compared to other cellular organelles. It is separated from the rest of the cell components by nuclear membrane (Fig. 7.17). The nucleus (plural = nuclei) houses the cell's DNA in the form of chromatin and directs the synthesis of ribosomes and proteins. Typically, the nucleus of a eukaryotic cell consists of a nuclear membrane (nuclear envelope), nucleoplasm or karyoplasms (matrix present inside the nucleus), nucleolus and chromosomes.

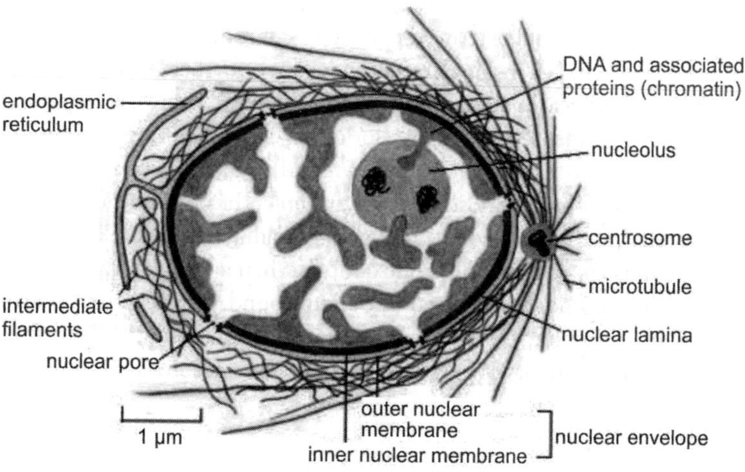

Fig. 7.17 Schematic diagram showing the nucleus and its components[593].

7.4.2.5.1 | *Nuclear membrane*

The nuclear membrane, also called as nuclear envelope is a double-membrane structure that encloses all the contents of nucleus (Fig. 7.18). The outer membrane of nucleus is connected to the endoplasmic reticulum. A perinuclear space filled with fluid is present between the two layers of a nuclear membrane. Nuclear membrane is punctuated with several openings, known as nuclear pores that control exchange of large molecules (ions, molecules, proteins and RNA) between the nucleoplasm and cytoplasm, permitting some to pass through the membrane, but not others. All these pores form nuclear pore complex. At the nuclear pores, the outer membrane is continuous with the inner membrane and thus renders a bi-membrane nuclear envelope.

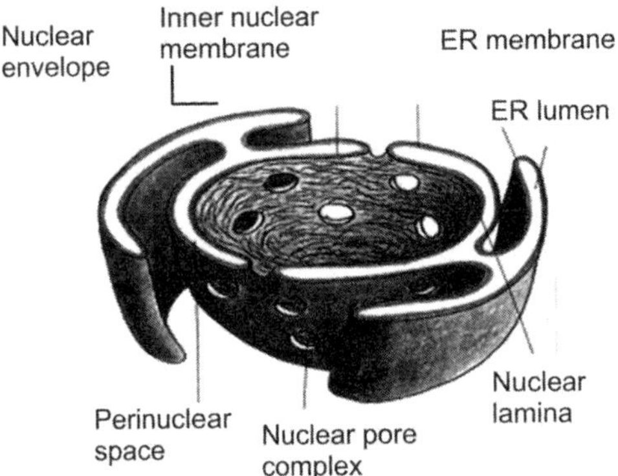

Fig. 7.18   Schematic illustration of bi-membranous nucleolemma, showing the presence of pores [Adapted from Ref. 593].

7.4.2.5.2 | *Chromosomes*

Chromosomes are packed inside the nucleus of every cell and DNA is the hereditary material with proteins ultimately forming a complex structure, called as chromosome (Fig. 7.19)

DNA and proteins are packaged inside chromatin, which is further classified as heterochromatin and euchromatin, based on their functions. The heterochromatin is present adjacent to the nuclear membrane and is a highly condensed and transcriptionally inactive form. In contrast, euchromatin is less condensed and is a delicate form of chromatin found abundantly in a transcribing cell. Every species own a specific number of chromosomes. For example, the chromosome number is 8 for fruit flies (*Drosophila*); whereas, the chromosome number of human is forty six. Chromosomes are clearly distinguished, when cell is about to divide. For example, when a cell is in the growth/maintenance phase of its cell cycle, the chromosomes appear like an unwound, jumbled bunch of threads. During cell division process, in particular, these chromosomes are duplicated as DNA is replicated. This process results in each cell having its own complete set of chromosomes.

7.4.2.5.3 | *Nucleolus*

Nucleolus (plural nucleoli) is a dense, spherical membrane-less organelle also known as cell's protein-producing structures within the nucleus (Fig. 7.17). Nucleolus also plays an indirect role in protein synthesis by producing ribosomes. Nucleolus disappears when a cell undergoes division and is reformed after the completion of cell division, when chromosomes are brought together into nucleolar organizing regions. Nucleolus looks like a large dark spot within the nucleus under a microscope. A nucleus may contain many (four) nucleoli, but their number is fixed within each species and increases in cancer cells.

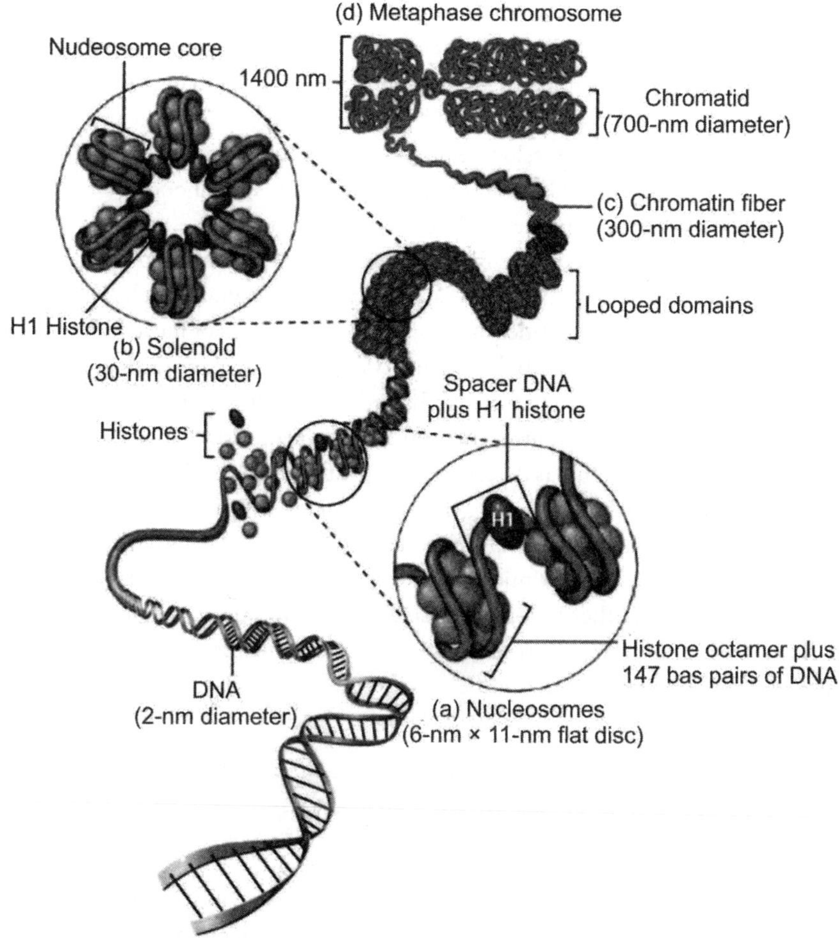

Fig. 7.19 Ultrastructure of chromosomes; extended DNA molecules can be very long (human DNA would be two meters long), but DNA is coiled and packaged in structures known as chromosomes; chromosomes are contained in the nucleus[599].

7.5 | Structure of Nucleic Acids

DNA and RNA are by far the two most important nucleic acids and these along with proteins and complex carbohydrates constitute the essential biological macromolecules for all known forms of life. Some basic understanding of the way the DNA and RNA structure are built up is an essential precursor to probe into the cell differentiation and gene expression. In this section, the structure of DNA and RNA will first be described and this will be followed by the translation–transcription process.

7.5.1 | Structure of DNA

Deoxyribonucleic acid (DNA) is a biomacromolecule, that contains genetic codes or instructuctions, which are utilized in the propagation or replication of all known living organisms, including eukaryotic, prokaryotic cells and some viruses. The functioning of DNA is also involved in cell growth, functioning and organism development. Eukaryotic cells store most of their DNA inside the cell nucleus and some of their DNA in organelles, such as mitochondria or chloroplasts. In contrast, prokaryotes (bacteria) store their DNA only in the cytoplasm.

As shown in Fig. 7.20, most DNA molecules are structurally characterized by two long biopolymer (polypeptide) strands uniquely winded around each other to form a double helix structure and each of these strands, known as polynucleotides, is composed of four types of nucleotides, i.e., nitrogen-containing nucleobase, either cytosine (C), guanine (G), adenine (A), or thymine (T) as well as a sugar called deoxyribose and a phosphate group linked covalently into polynucleotide chain with sugar–phosphate backbone. This unique structural arrangement results in a characteristic sugar–phosphate backbone. DNA structure is determined by the famous base pairing rules (A with T, and C with G), and hydrogen bonds bind the nitrogenous bases of the two separate polynucleotide strands to make double-stranded DNA. The complementary base pairing enables packing in energetically most favourable arrangement. To maximize efficiency of base-pair packing, two sugar–phosphate backbones wind around each other to form a double helix, with one complete turn every ten base pairs. Also, the shapes and chemical structure of bases allow hydrogen bonds to form efficiently only between A and T and between G and C.

> DNA double helix is stabilized essentially by two forces, hydrogen bonds between nucleotides and base stacking interactions among aromatic nucleobases.

As far as the structural length scale is concerned, each helical coil has a pitch of 3.4 nm and a radius of 1 nm. Despite being so small, DNA polymers can have large molecules with millions of nucleotides (220 million base pairs with nominal length of 85 mm). As far as the structural properties are concerned, the two strands in DNA align in opposite directions to each other (anti-parallel arrangement) having non-static characteristics and backbone is resistant to cleavage. It is the ATGC sequence of the backbone encodes the biological information, which is replicated as the two strands are separated. It is important to note that the transcription process allows the formation of RNA strands using DNA strands as a template. In particular, RNA strands are translated to specify the sequence of amino acids within proteins. It is worthwhile to note that a significant portion (> 98% for humans) does not have any ability for transcription and therefore do not serve as patterns for protein sequences.

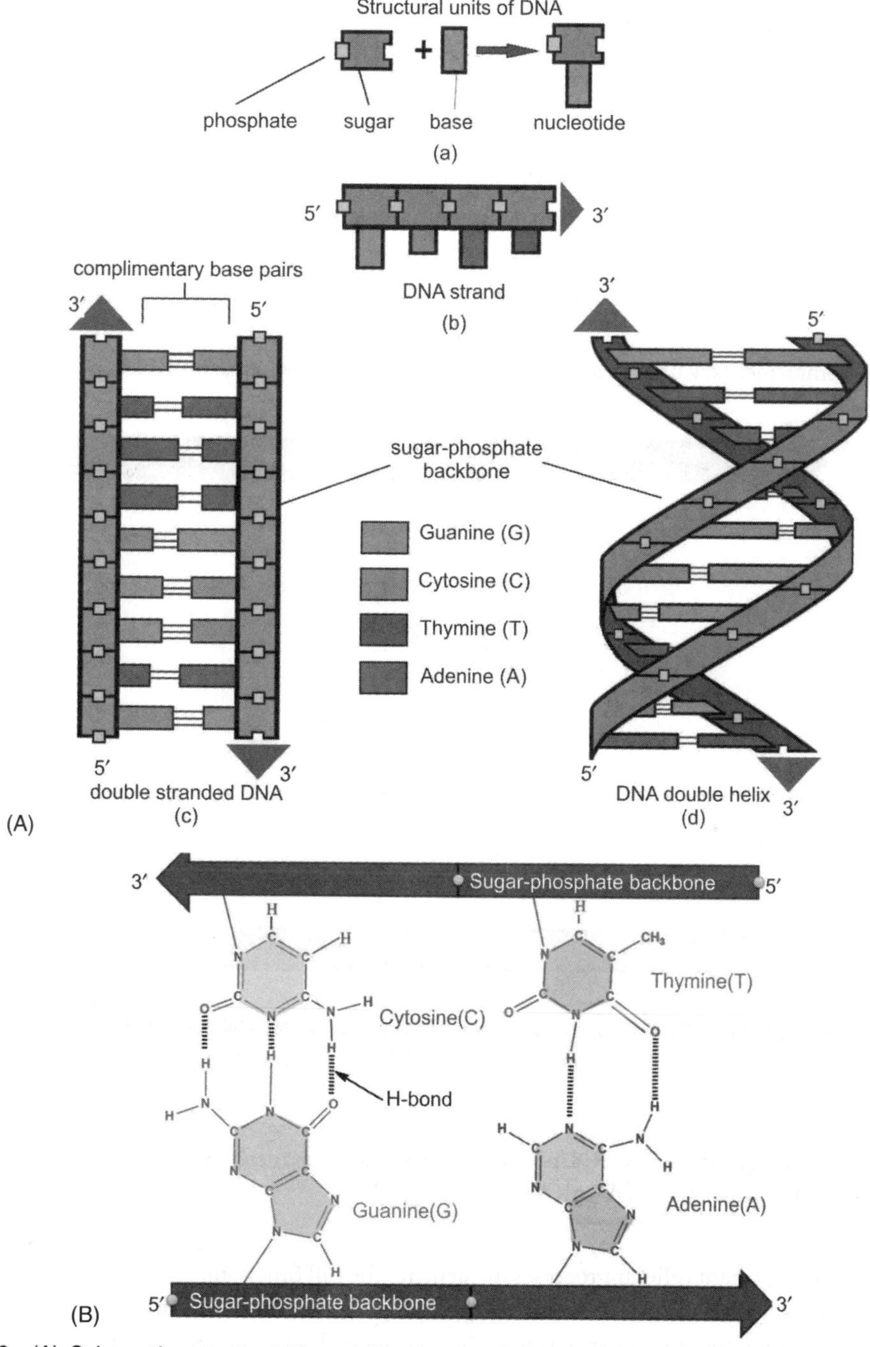

Fig. 7.20 (A) Schematic representation of the characteristic structure of DNA: (a) covalent bonding between the structural units to form nucleotide, (b) typical DNA strand, (c) complimentary base pairing in double stranded DNA, and (d) double helix structure of DNA; (B) bonding of nucleotide between two parallel backbone chains as per base pairing rule [Adapted from Ref.593] [see Colour Plate].

7.5.2 | **Structure of RNA**

Ribonucleic acid (RNA) is another important biomacromolecule involved in gene expression and regulation. Like DNA, RNA, structurally contains a chain of nucleotides. However important distinction is that RNA more often exists in nature as a single-strand folded onto itself, rather than a paired double-strand in case of DNA. While the chemical structure of RNA has a large similarity with DNA structure, the fundamental structural difference exists in reference to the following three aspects (see Fig. 7.20 and 7.21):

(a) While DNA contains deoxyribose, RNA contains ribose, which also makes RNA less stable than DNA.
(b) The complementary base to adenine in DNA is thymine, whereas in RNA, it is uracil.

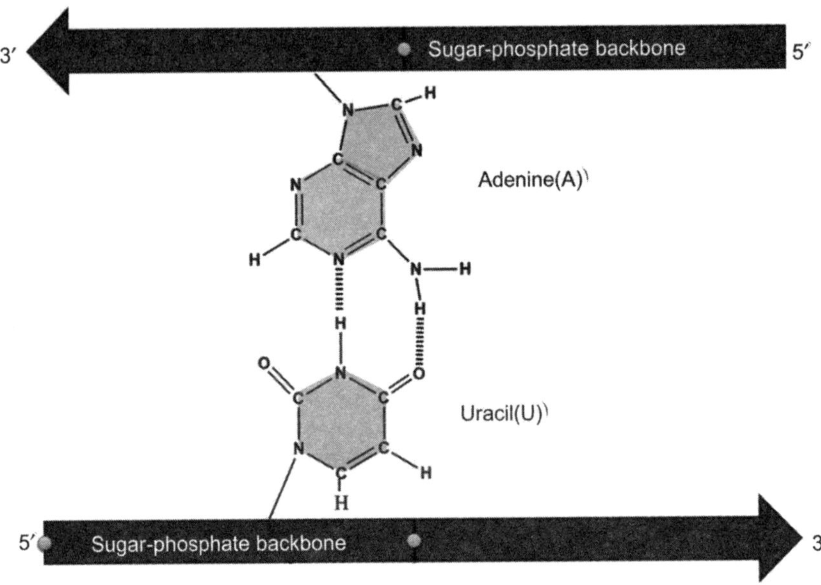

Fig. 7.21 Base pairing in RNA: A-U instead of A-T (as in case of DNA), while G-C pairing remains identical in both DNA and RNA [Adapted from Ref. 593] [see Colour Plate].

The most biologically active RNAs can be classified as transfer RNA (tRNA), ribosomal RNA (rRNA), messenger RNA (mRNA), etc.

One of the important cellular processes influencing the cell functionality is the protein synthesis, wherein mRNA molecules direct the assembly of proteins on ribosomes. In simple terms, mRNA carries the code for the protein structure to be synthesized, i.e., the amino acid sequence. Some RNA molecules also take part in stimulating some of the biological reactions, e.g., gene expression, or sensing/communicating cellular signals. As the name suggests, rRNA forms the basic structure of ribosome and also catalyzes the protein synthesis. Typical responsibility of tRNA is to deliver amino acids to the ribosome, while that of rRNA is to link amino acids together to form proteins. Like

DNA, different RNA types (mRNA, tRNA, and rRNA) contain self-complementary sequences. Such sequences enable the formation of double helices with the involvement of parts of the RNA, which can fold and pair with itself. Unlike DNA, the RNA structure does not consist of long double helices, but is characterized by the collections of short helices packed together into structures akin to proteins.

7.6 | Transcription and Translation Process

Transcription is a process, wherein the information contained in a part of a DNA sequence is replicated in the form of a newly assembled messenger RNA (mRNA) and this process is facilitated by specific enzymes, RNA polymerase and transcription factors. Another contemporary process, translation is defined as the transformation of mature mRNA to a protein.

> In simplest terms, DNA transcription produces a single stranded RNA molecule that is complementary to one strand of DNA.

In eukaryotic cells, such a process involves the splicing, which leads to the formation of a mature mRNA chain.

In prokaryotic cells without any well-defined compartmentalized organelles/nucleus, the processes of transcription and translation can be linked together in cytoplasmic space. In contrast, the transcription usually takes place inside the nucleus, while the translation occurs in the cytoplasmic space (outside the nucleus) in eukaryotic cells (see Fig. 7.22). Therefore, the transcription–translation process in eukaryotic cells requires mRNA to be transported out of the nucleus into the cytoplasm, where it can be bound by ribosomes (see Fig. 7.22 and 7.23).

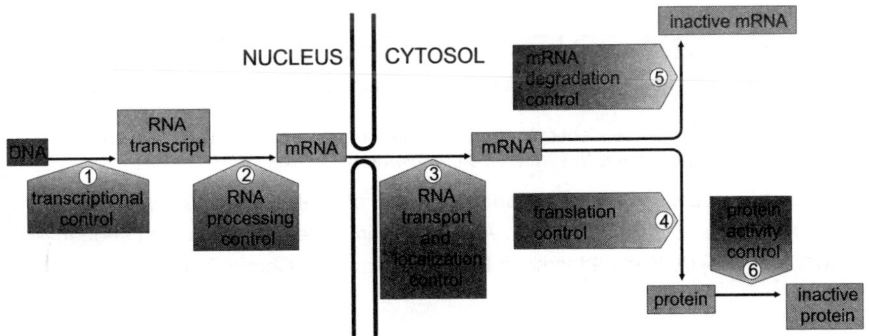

Fig. 7.22 The process of transcription–translation in a eukaryotic cell[593].

The above described transcription–translation process has a special place in the field of cell molecular biology and is commonly referred as 'central dogma'. In simplest terms, this theory was originally postulated as 'DNA makes RNA and RNA makes protein,' as originally proposed by Crick. However, as the overall understanding in molecular biology has evolved over last few decades, it has been established now that the reverse flow of genetic sequence information from RNA to DNA is also possible, which is known as reverse transcription process and this was originally not precluded by Crick. In other words, transcription and reverse transcription are indeed possible in living organisms. However, the transformation of protein to RNA does not commonly occur. Nevertheless, this important dogma is widely regarded as a framework to understand the transfer of sequence information between biopolymers containing genetic information in living organisms. The above theory will be utilized in subsequent chapters in the context of discussing differentiation and RT–PCR analysis of gene expression.

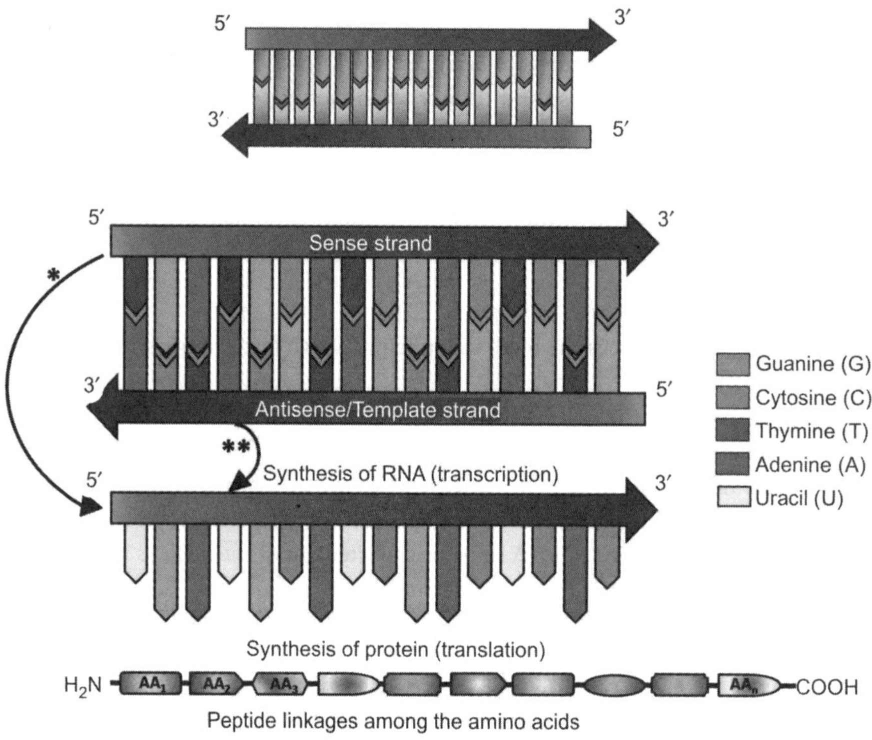

Fig. 7.23 Schematic representation of central dogma of molecular biology involving replication, transcription and translation. AA_1, AA_2,...AA_n are different amino acid sequences. [Adapted from Ref. 594] (Acknowledgement: Deepak Saini) [see Colour Plate].

7.7 | Stem Cell and Other Cell Types

Currently, stem cells are the focus of cutting edge research in biotechnology, bioengineering and biomedical applications all over the world. Stem cells and their promising applications in healing and regenerative therapies will be nothing short of revolutionary the field of medicine in the coming years. In view of its predominant importance in the modern biomedical engineering, the stem cells are briefly introduced in this section. As will be discussed, the stem cells have the unique ability to differentiate to various cell types and some of those cell types together with their associated functionality are summarized in Table 7.1.

Table 7.1 Various cell types and their functionality with all can be potentially differentiated from different types of stem cells under various biophysical or biochemical cues.

Cell Type	*Functionality*
Adipocyte	Cell that makes and store fat compounds
Astrocyte	A type of glia (glue) cell that provides structural and metabolic support to the neurons
Cardiomyocyte	Cell that forms the heart; also called myocytes
Chondrocyte	Cell that makes cartilage
Endothelial cell	Cell that forms the inner lining (endothelium) of blood vessels
Hematopoietic cell	Cell that differentiates into red and white blood cells
Keratinocyte	Cell that forms hair and nails
Mast cell	Associated with connective tissue and blood vessels
Neuron	Cell that forms the brain, spinal cords, peripheral nervous system
Osteoblast	Give rise to osteocytes or bone-forming cells
Pancreatic islet cells	Endocrine cells those synthesize insulin
Smooth muscle	Muscle that lines involuntary visceral organs like blood vessels and the digestive track

> Stem cells are unspecialized cells that have the remarkable ability to differentiate into cells of specific function and can also divide to produce more stem cells.

They are characterized by the following properties that are fundamentally different from other cells.

(a) Self-renewal- The ability of stem cells to divide, usually after long periods of time to produce more stem cells while maintaining their undifferentiated state.

(b) Potency- The ability to differentiate into specialized cell types. Stem cells can be totipotent (can differentiate into any kind of cell of a total organism), pluripotent (derived from totipotent cells, can differentiate into all cell types), multipotent (can differentiate into a number of close related cell types), oligopotent (can differentiate into only two cell types) and unipotent (can differentiate into a single cell type). Various cell types are mentioned in Table 7.1.

(c) Clonality- This implies the state of a cell being derived along a specific lineage or from one source. In cell biology, a clone represents a group of identical cells sharing a common ancestry.

There are two major types of stem cells: (i) embryonic stem cells (ESC), which are obtained from the human embryo by the process of *in vitro*-fertilization. ESCs can differentiate into all the cell types. (ii) adult/somatic stem cells, which are found in the differentiated cells of tissue or organ that can renew itself and differentiate into major cell types. These stem cells are mainly involved in repair and maintenance of tissues/organs.

Based on the source, the stem cells can be classified to three types: (a) hematopoietic stem cells (HSCs) that can differentiate into blood cells and (b) mesenchymal stem cells that are extracted from the bone marrow and can differentiate into multiple cell types (see Fig. 7.24); (c) adipose stem cells (ASCs)- adipose derived stem cells extracted from fat cells.

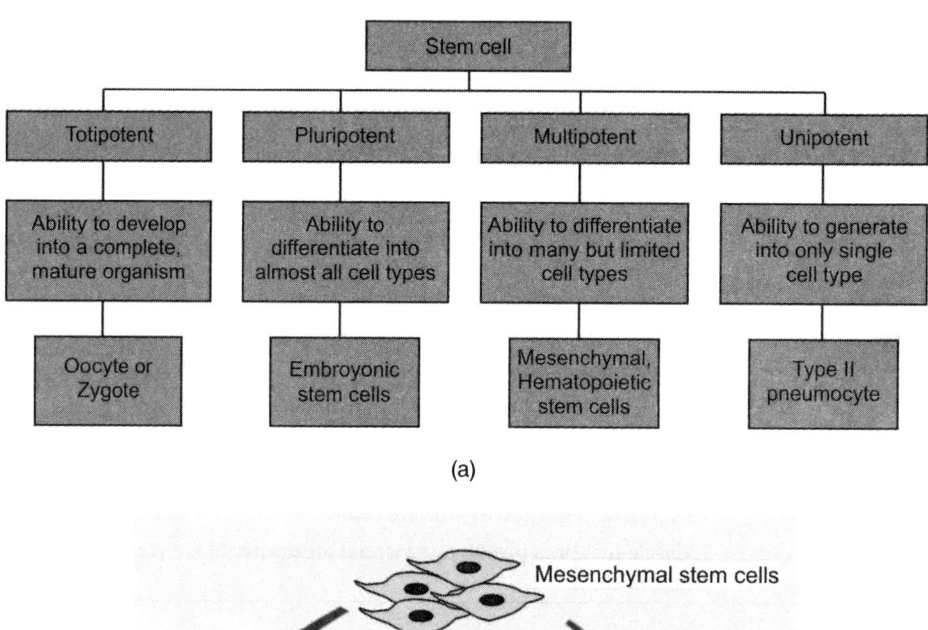

(a)

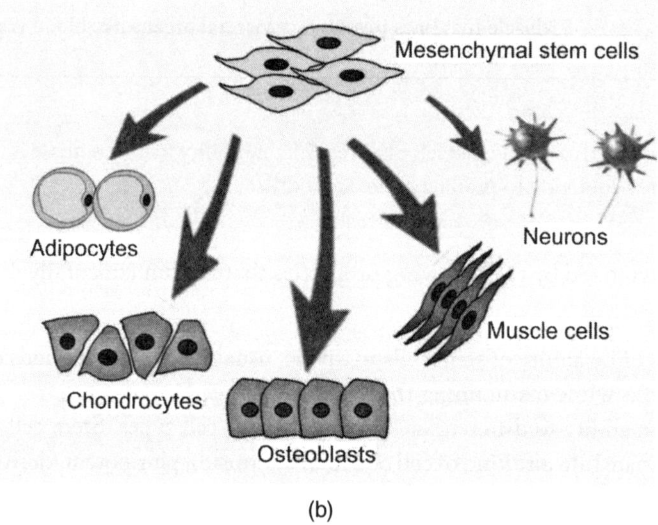

(b)

Fig. 7.24 (a) Classification of stem cells and (b) multi-potency of human Mesenchymal Stem Cells (hMSCs)[600].

There are two other stem cell types- (iii) amniotic stem cells found in the amniotic fluid and (iv) Induced Pluripotent Stem cells (iPSCs), which are mature cells genetically reprogrammed by certain transcription factors to a pluripotent stem cell state. The seminal work by Takahashi and Yamanaka[584] on the discovery of induced pluripotency was awarded 2012 Nobel prize in Physiology and Medicine jointly with Sir John B. Gurdon. As stem cells can potentially differentiate into multiple cell types, their relevance is mainly in healing and regenerative therapies. In the field of medical science, research currently is underway to use stem cells in the treatments (stem cell therapy) to host of diseases and conditions, such as diabetes, rheumatoid arthritis, Parkinson's disease, Alzheimer's disease, Osteoarthritis, anti-cancer drugs, repairing hearing and vision loss, cardiac infarction, Crohn's disease and in wound healing[585].

The fact that the stem cells have received global attention in treating many human diseases is that the above-mentioned stem cells can be differentiated to different cell types, as shown in Table 7.1. All these different cell types have different functionality under appropriate cellular micro-environment and the multi-lineage differentiation is possible only in the presence of biophysical/biochemical stimulus/cues. As far as different cell types are concerned, adipocytes help in storing/synthesizing fat compounds and constitute adipose tissues. While neurons are important cells of central and peripheral nervous system, astrocytes are a type of cell providing structural support to neurons. As far as the bone is concerned, three types of cells are present. Osteoblast is known as bone forming cell, while osteoclast as bone resorbing cells. Osteocytes are the matured osteoblasts and are the end member of the osteoblastic lineage during osteogenic differentiation. In a healthy bone, there should be an optimal balance between osteoblasts and osteoclasts. In osteoporotic bone, the activity of osteoclasts is more than osteoblasts. Also, all the three bone cell types are morphologically distinct. In skin tissue or endothelium, the endothelial cells with close spacing are present and they protect the underlying cellular structure. The cartilage tissue contains the characteristic chondrocytes and the differentiation of hMSc to chondrocytes is known as chondrogenesis. The myogenesis of hMSc leads to the formation of smooth muscle cells, which are present in the musculoskeletal system, e.g., lining of blood vessels or digestive tracks. Similarly, cardiomyogenesis of hMSc can result in the cardiomyocytes, which form the cardiac tissue and this cell type is one of the known cell types which does not grow easily in culture. Also, hematopoietic cells are those which can be differentiated into red and white blood cells. In patients suffering from blood cancer, one of the clinical treatment option can be to transfuse hematopoietic cells. Another important cell type is pancreatic islet cells, which has larger relevance in controlling blood sugar level and therefore relevant for diabetes disease. Popularly, stem cells are advertised to treat hair loss and the clinical background is that stem cells can be differentiated to keratinocytes, which form human hair.

One of the ways in which this can be achieved is through the concept of tissue engineering. Tissue engineering involves the use of cells in combination with suitable engineering materials (usually scaffolds) and implanted in a particular tissue site. This is usually done by providing appropriate growth factors for the stem cells to differentiate into the particular lineage/cell line. At present, the tissues that can be engineered using stem cells include various epithelial tissues such as skin, cornea and mucosal membrane and skeletal tissues such as bone and muscle. While the formation of skin tissue requires a 2D scaffold, bone tissue substitute is more complex as it needs to develop a 3D architecture with internal hierarchies and needs a bioactive intermediary such as HA or tricalcium phosphate to fully integrate into the tissue.[586] It is graphically represented in the Fig. 7.25. Scaffolds that simulate parts of bone architecture are produced efficiently by rapid prototyping technologies, such as 3D printing and 3D plotting with the ability to include growth factors in the scaffold to aid the differentiation of stem cells towards osteogenic lineage increasing the effectiveness of the intended therapy.[587] In addition, many of bone grafts and extracellular matrices which have potential applications in bone regeneration/healing are designed to promote differentiation of stem cells towards bone-like cells (osteogenesis), which helps in the healing of the damaged tissue.

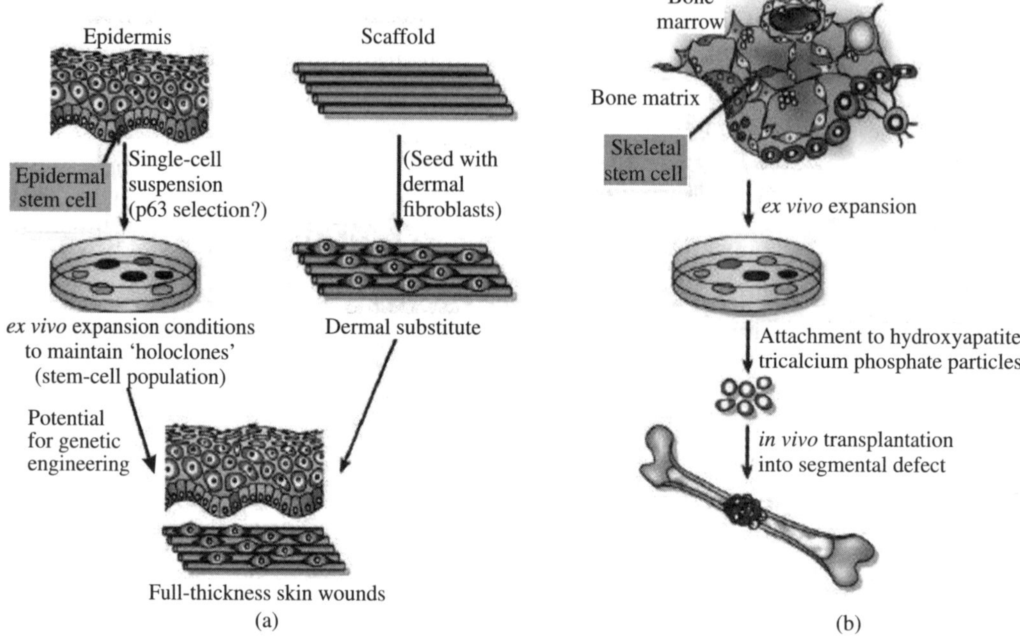

Fig. 7.25 Concept of stem cell-based (a) skin and (b) bone tissue engineering[586].

Adult skeletal stem cells (bone-marrow derived) can be differentiated into neurogenic lineage in the presence of appropriate chemical/physical cues in the cell microenvironment. This has potential applications in treating many nerve/spinal cord injuries and neuro-degenerative disorders. The differentiation of hMSC into neuronal cells is generally aided by the presence of a fibrous scaffolds or conducting substrates. For the purpose of tissue engineering, scaffolds of biodegradable materials are always preferred as they can be easily integrated into the host tissue. For nerve cell regeneration and related applications, electrospun fibre scaffolds have been found to be the most effective. In addition, electrospinning also allows for the addition of multiple components, so that a fibre composite can be obtained. In one such example, the effectiveness of poly(L-lactic acid)-co-poly-(3-caprolactone)/ collagen composite electrospun-fibre as a potential material for tissue engineering is shown by demonstrating the differentiation of hMSCs to neuronal cells (see Fig. 7.26) with the addition of relevant growth factors[588,589]. Recent studies have also shown that electrical stimulation (100 mV/ mm, 60 min) of neural stem cells on conducting fibre scaffolds composed of poly L-lactide and polyaniline resulted in enhanced neurite outgrowth, which reflects the synergy between substrate properties, electrical stimulation and stem cell fate process clearly[590]. Apart from the applications in epithelial, bone and nerve tissue engineering, stem cells (mainly embyonic and mesenchymal) can also be made to differentiate to cardiomyocytes (heart muscle cells)[591].

All or most of the afore-mentioned stem cell applications involve altering some part of the stem cell micro-environment by supplying biochemical cues, by changing the substrate properties (stiffness, conductivity, topography, geometry) or by external stimulation (mechanical, electrical, magnetic).

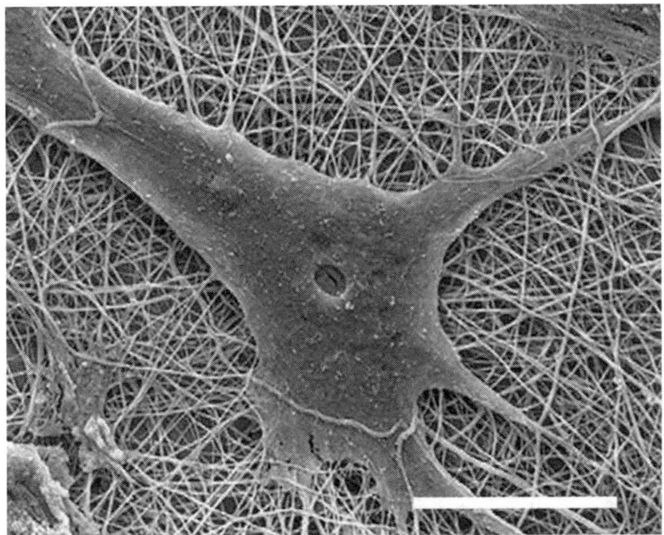

Fig. 7.26 A differentiated hMSC on electrospun fibres, scale bar 20 μm[589].

7.8 | Cellular Adaptation

Cellular adaptation refers to various changes, those manifest better survival at the level of individual cells, tissues, or whole organism in response to prolonged exposure of cells to adverse or exaggerated normal stimuli. Once the cause is removed, most cells that have adapted to chronic stimulation revert to normalcy, unless it is detrimental. The cellular adaptation can be classified further as physiological (normal) or pathological (abnormal). Cell adapts to changes by one of five major types of the following processes:

7.8.1 | Atrophy

Atrophy is characterized by a decrease in the size of a tissue, organ, or the entire body. Some examples of physiologic atrophy include a) thymus undergoing involution during early human development and b) the atrophic bones and muscles in the elderly become thin and prone to fracture, c) ischemic (lack of blood supply) organs are typically small (e.g., kidneys involved with atherosclerosis, hardening of the arteries) and d) Alzheimer dementia: brain shrinks 2–3x each year and neurons are replaced by plaques.

7.8.2 | Hypertrophy

It is characterized by an increase in the size of tissues or organs due to enlargement of individual cell. In hypertrophy, there is an increase in intracellular protein rather than cytosol of cell. Hypertrophy can be caused by trophic signals such as growth factors or mechanical signals (e.g., stretch). An example of physiologic hypertrophy is the enlargement of skeletal muscles in body builders due to sustained weight bearing exercises and cardiomegalysis/pathological hypertrophy of the heart that occurs as an adaptation to increased workload.

7.8.3 | Hyperplasia

Hyperplasia is an adaptive increase in the number of cells resulting from increased cell mitotic division, which can cause enlargement of tissues or organs. The physiologic hyperplasia can be of two types; compensatory and hormonal. The compensatory hyperplasia results in tissue and organ regeneration. It is commonly observed in epithelial cells of the epidermis and intestine, liver cells, cells of bone marrow and fibroblast. Hormonal hyperplasia, as the name indicates is seen mainly in oestrogen dependent organs. For example, the oestrogen-dependent hypertrophy is accompanied by hyperplasia in uterine lining during pregnancy. A common pathologic hyperplasia is endometriosis, which occurs as growth of endometrium outside the uterus.

7.8.4 | Dysplasia

Dysplasia refers generally to disordered growth or cellular changes in cellular shape, size, and/or organization, resulting from chronic irritation or infection. It is not considered as a true adaptation. It seems to be related to hyperplasia and is called as 'atypical hyperplasia' generally occurring in cervical and respiratory tissues. Dysplasia is also considered as a precancerous condition, which is reversible. But if stress persists, then it can progress into irreversible carcinoma occurring in the vicinity of cancerous cells. It may also result in the development of breast cancer.

7.8.5 | Metaplasia

Metaplasia is a reversible adaptive change of a differentiated cell of one type to another in order to adapt to the environment. One popular example is squamous metaplasia of the bronchial epithelium due to smoking. The pulmonary bronchial cells convert from ciliated, columnar mucus -secreting epithelium to non-ciliated, squamous epithelium, which does not secret mucus. These changed cells may develop dysplasia or become cancerous on prolonged exposure to the stimulus (e.g., cigarette smoking).

7.8.6 | Cell shape change

Apart from the above-mentioned cellular adaptation processes, another process which a biological cell continuously adapts while interacting with a synthetic biomaterial substrate or a tissue engineered scaffold is the cell shape. As shown in Fig. 7.27, in response to an external stimulus, the actin filament undergoes the depolymerization to produce numerous monomers, which can get distributed in various regions of the cytoplasmic space. The external stimulus can be substrate stiffness, which facilitates the mechanical signal transduction process. Depending on the substrate stiffness, the cytoplasmic reorganization can take place and this process is also involved in the cell migration process. In further response to the external stimulus, the monomers can be re-polymerized to form actin filamentous structure. The dynamic process of depolymerization–repolymerization is inherent in any cellular structure and the end result is the cell shape change mediated by the cytoplasmic reorganization. As discussed further in a subsequent chapter of this book, the above-described process is also involved in cell-material interaction.

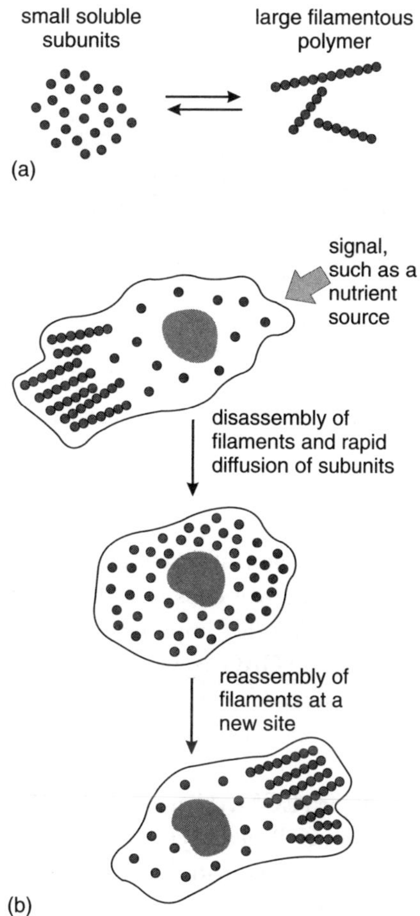

Fig. 7.27 Schematic of biological processes depicting the formation of cytoskeletal elements (a) extracellular signal mediated depolymerisation-repolymerisation involved in cytoskeletal dynamics resulting in cell shape changes (b)[593].

7.9 | Extracellular Matrix (ECM)

This chapter will not be completed without a brief discussion on the structure of extracellular matrix, i.e., the cellular microenvironment and the tissues. The interstitial space in between the cells within a given tissue remains filled with extracellular matrix and this contains types of collagen, elastic fibres, proteoglycans and hyaluronan, fibronectin and laminin (see Fig. 7.28). Therefore, ECM is biologically described as a collection of extracellular molecules secreted by cells via exocytosis. It is important to note that the ECM composition varies between different tissues.

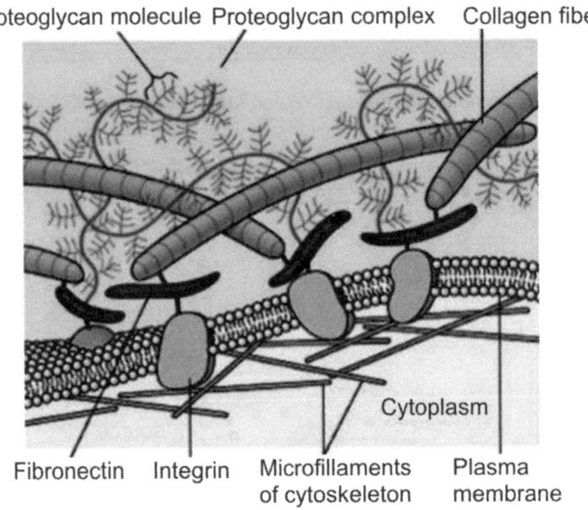

Fig. 7.28 Structural characteristics of extracellular matrix (ECM).

7.9.1 | ECM composition

The ECM is composed of an interlocking mesh of fibrous proteins and glycosaminoglycans (GAGs). Glycosaminoglycans (GAGs) are essentially a class of carbohydrate polymers and are linked to ECM proteins to form proteoglycans. Another important component of ECM, i.e., Hyaluronic acid or hyaluronan is the only unsulphated GAG, consisting of alternating residues of D-glucuronic acid and N-acetylglucosamine. The presence of hyaluronanin ECM essentially enables the tissues with the required ability to resist compression and therefore is found in abundance in the ECM of load-bearing joints.

Among the fibrous proteins contained ECM, collagen, elastin, fibronectin, laminin are important and therefore, briefly described below. Collagens, synthesised by the fibroblast, by far are the most abundant protein as it accounts for 90% of bone matrix. In particular, collagens are present as fibrillar proteins to provide structural support to the cells.

> The collagens exist in different structural forms, like fibrillar (type I, II, III, V, XI), short chain (type VIII, X), basement membrane (Type IV), etc.

In contrast to collagens, the presence of elastins, which are synthesized by fibroblasts and smooth muscle cells, provide elasticity to tissues so that the tissues can stretch, when needed and recoil back to original state in the absence of mechanical stimulus. As far as other fibrous proteins are concerned, fibronectins are glycoproteins, which bridge cells with collagen fibres and such characteristic arrangement allows cells to move through the ECM. In particular, fibronectins bind collagen and cell-surface integrins and such binding enables the communication with the cytoskeleton and facilitates cell migration. Another fibrous protein, laminins are found in the basal laminae in all the organisms and laminins form networks of web-like structures, which can resist tensile forces. Like fibronectin, laminins also assist in cell adhesion and establishes the binding with collagens. They also constitute basement membrane along with non-fibrillar collagen providing mechanical support and signalling cues to epithelial cells.

7.9.2 | ECM properties

The characteristic structure of many of the above-mentioned fibrous proteins and other biological macromolecules provide ECM with elastic stiffness property, varying from soft brain tissues to hard bone tissues. In fact the wide ranging elastic property of ECM can be attributed to the distribution of collagen and elastin amount.

Also, ECM provides structural and biochemical support to surrounding biological cells. In particular, the cell adhesion, cell-to-cell communication and differentiation are common functions of the ECM. The formation of the extracellular matrix is essential for processes like growth, wound healing, and fibrosis.

> The stiffness and elasticity of ECM have important implications in various cellular processes, including cell proliferation and differentiation.

In the contest stem cell behaviour, as briefly discussed in a preceding section of this chapter, ECM elasticity can direct cellular differentiation (differential gene expression). In particular, naive mesenchymal stem cells (MSCs) are reported to exhibit specific lineage commitment with extreme sensitivity to tissue-level elasticity. During cell adhesion, many transmembrane proteins, extending from intracellular to extracellular face of the cell, bind to ECM components. The cell adhesion is mediated either via focal adhesions or via hemidesmosomes (connecting the ECM to intermediate filaments). It can be reiterated that while focal adhesions primarily connect ECM of actin filaments to a cell, hemidesmosomes facilitate ECM to get connected to intermediate filaments. Overall, the cell-to-ECM adhesion is facilitated through specific cellular adhesion molecules (CAM), as integrins. During cell attachment, the integrins are bound to fibronectin and laminin.

7.10 | Tissue

Tissue can be described as an intermediate between cells and a complete organ in the hierarchy.

> In biological terms, a tissue can be defined as a self-assembly of similar cells of identical origin with the ability to perform a specific physiological function.

Various cell types and their functionality are summarized in Table 7.1 and these cells are contained in various tissue types.

The tissues can be classified to four basic types, connective, muscle, nervous, and epithelial. Among these four tissue types, the connective tissues are fibrous tissues to provide structural support to an organ and are made up of cells separated by ECM. The spindle-shaped fibroblasts are contained in connective tissues. Typical examples of connective tissues are blood, bone, tendon, ligament, adipose tissues. As shown in Fig. 7.29, the extensive proliferation od myoblasts followed by alignment, fusion and differentiation, together leads to the formation of matured myotubes, which are grown into adult muscle tissues. The characteristics of thus formed muscle tissue include active contractile nature and typical functions are to produce force and to cause motion, (locomotion or movement). Among different types of muscle tissues, muscle is found in the inner linings of organ, while the muscle is found to be attached to bone. Another type, i.e., cardiac muscle is found in the heart and it allows the heart to contract and pump blood throughout an organism. The neural tissue contains the characteristic neurons. In the central nervous system, neural tissue is contained in the brain and spinal cord, while in the peripheral nervous system, neural tissue forms the cranial nerves. The epithelial tissues contain the closely placed epithelial cells, which cover the organ surfaces as well as their interior and are involved in organ functions. The presence of semi-permeable tight junctions in epithelial tissue provides a barrier between the external environment and the underlying organ.

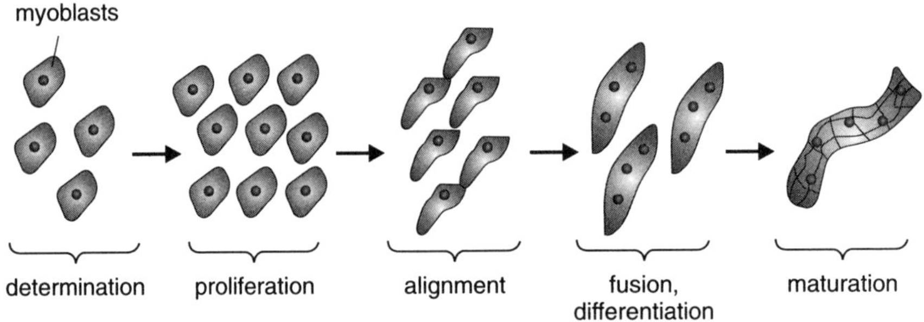

Fig. 7.29 Schematic illustration of the formation of mytotube in muscle tissue involving the fusion of the myoblast cells.

According to another classification, the tissues can be grouped to soft and hard tissues and such distinction is primarily based on the elastic stiffness or mechanical strength property. As the name suggests, hard tissues, are by far mechanically superior to soft tissues, which are essentially epithelial or neural tissues. The hard tissue, also known as the mineralized tissues, contains hard minerals and soft collagenous matrices. Some examples include bone, tendon, cartilage, tooth enamel and dentin. Some of these mineralized tissues have the characteristic adaptability and multi-functionality property.

> Due to the presence of minerals (inorganic component) and collagenous proteins (organic component), mineralized tissues combine an outstanding combination of stiffness, strength and toughness.

In view of the characteristic structural arrangement of collagen fibres and the calcium phosphate minerals, the mechanical stresses are transferred through several length-scales, from macro to micro to nano and these results in energy dissipation and damage tolerance. Despite several decades of research to develop synthetic materials for biomedical applications, often via biomimicking approaches, the researchers are yet to discover a synthetic material with matching mineralized tissue properties and therefore newer manufacturing techniques are constantly under development. Some of these approaches will be described in subsequent chapters.

7.11 | Closure

Starting from the basic characteristic of the structure and characteristic of a cell, this chapter provides a biological foundation for any non-biologist. Considering the role of protein adsorption in cell–material interaction, an emphasis has been placed to describe various levels of protein structure. A discussion on the structure of cellular organelles and the nucleic acids together with various cellular adaptation processes would be helpful to understand the cell–material interaction as well as cell fate processes on a synthetic material surface. Importantly, the structure and properties of ECM and tissues are briefly introduced.

Cell–Material Interaction and Biocompatibility

In this chapter, a holistic understanding of biocompatibility is scientifically developed starting with how a single cell interacts with a synthetic material substrate, *in vitro*. Since *in vivo* environment is much more complex with dynamic changes in various physiological parameters, it is important to develop such an understanding at a much simpler *in vitro* level. After a single cell establishes physical adhesion on a biomaterial through focal adhesion complexes, the next level is to understand how a single cell communicates with other cells in a cellular microenvironment. This requires knowledge of cell-signalling mechanisms. Once a group of cells co-ordinate their efforts to communicate among themselves and the 'cross-talk' among those cells is established, one has to consider how those cells alter their functionality, at the interface with a biomaterial substrate. This requires an understanding of cell fate processes (proliferation, migration, differentiation, etc.). After sufficiently describing the above aspect both qualitatively and quantitatively, this chapter will discuss as how the cells favourably change their fate in a substrate-property dependent manner. The quantitative analysis of cell morphological changes using fluroscence or confocal microscopy is highlighted. When implanted in an animal model, a material comes in contact with the host tissue. How the interface is dynamically established over implantation time is the paradigm of the host response. The entire cascade of biological events constituting the process of host response is discussed towards the end of the chapter.

8.1 | Introduction

It is important to recapitulate that the basic requirement of any biomaterial concerns the fact that the material and the host tissue should coexist without having any undesirable or inappropriate effect on each other. Such a requirement is broadly described by a concept, known as 'biocompatibility'[601]. It is widely perceived that a biomaterial disrupts normal body functions as little as possible and therefore it is important to assess how a biomaterial interacts, when it is implanted in a human subject/animal model. From biological perspective, the need for biocompatibility emerges from the differences in the characteristics/properties of living and non-living materials.

> *'Biocompatibility refers to the ability of a biomaterial to perform its desired function with respect to a medical therapy, without eliciting any undesirable or systematic effects on the recipient or beneficiary of that therapy, but generating the most appropriate beneficial cellular or tissue response in the specific situation, and optimizing the clinically relevant performance of that therapy.'*

Three important aspects are to be considered while assessing the concept of biocompatibility. First, biomaterials should be biochemically compatible and non-corrosive under physiological conditions. They must be inert or non-toxic, irritation as well as allergen resistant, and non-tumorigenic. Secondly, they should have acceptable biomechanical compatibility with host tissues. Thirdly, a bio-adhesive contact must be established between the biomaterial and host tissue. All the above aspects are sensitively dependent on the anatomical site of implantation in a living system.

Ideally, biomaterials should not induce any change or provoke any reaction in neighbouring or distant tissues. An alternative objective for an implant would be the formation of a structural bond between the material and host tissues. In practice, however, the formation of an ultra-thin capsule or fibrous tissue around the implant is regarded as a signature of desired biocompatibility.

> As the subject of biomaterials science grows, deeper understanding into how cellular signaling processes are involved in the functionality of the cell population on a material substrate is being more appreciated in the community.

The implantation of any biomaterial always elicits a local or systemic response, which is termed as the host response[602]. Such responses, measured by tissue reactions can be either local or systemic. Systemic responses include toxic or allergic reactions *in vivo*. The products of corrosion from metallic implants and biodegradation of polymers, once released can elicit systemic reactions in the body. The negative tissue/material interactions at the local level include tissue necrosis or proliferation, which may progress to granuloma.

8.2 | Biophysical Processes Involved in Biocompatibility

It is important to reiterate once more that biocompatibility requires an integrated understanding of various biophysical processes at the interface with a biomaterial, both *in vitro* and *in vivo*. In line with the abstract of this chapter, the mechanisms of the interaction of a single cell with a biomaterial is now discussed.

8.2.1 | Cell–material interaction

In an earlier chapter in this book (chapter 7), the characteristic properties and structural features of proteins and eukaryotic cells are discussed. At various points of discussion in this chapter, a reader has to recall those aspects. To start with, the biophysical interaction between a cell and a material will now be analysed, and to develop a simplistic understanding, we shall consider a material being placed in a culture medium containing a single cell. The growth medium, in an experimental scenario, contains antibiotics, proteins (e.g. fibronectin, vitronectin, etc.). In the above backdrop and given the fact that the living system of a human subject contains a large number (10^{14} - 10^{16}) of different proteins, it can be easily perceived that when a biomaterial is placed in a culture medium or in a host system, the proteins are adsorbed within seconds of implantation on the biomaterial surface and form a layer[603]. Cell-material interaction is thereafter mediated via adsorbed proteins on a biomaterial substrate. The protein adsorption kinetics generally depends on the biomaterial's surface properties. The characteristics of the adsorbed protein layer on a biomaterial surface are not static due to the competition among various proteins. Once a protein is adsorbed on the surface, it is not necessary that it will stick to the surface permanently. It can be desorbed off and other higher affinity proteins can be adsorbed at the same site.[604]

Phenomenologically, the first molecules to reach the implant surface are water molecules (within nanoseconds), forming a water shell on the biomaterial surface. The proteins and the biomolecules in plasma/cell culture media also have a hydration shell of water molecules. The water shell on the biomaterial interacts with the hydration shell of the biomolecules, and this interaction decides the orientation and coverage of the adsorbed proteins and whether they denature or not. This interaction is governed by the kinetic and thermodynamic processes at the interface. They may be expressed in terms of hydrophilicity–hydrophobicity at a macroscopic scale as in terms of water–water (cohesive) and water–surface (adhesive) forces at the microscopic level[605]. This process of protein adsorption, which is better known as 'biofouling'[606], has been illustrated schematically in Fig. 8.1.

The consequences of protein adsorption are manifold, as this step occurs even before the arrival of cells on a biomaterial substrate. When the cells reach the surface of a biomaterial, they 'see' the adsorbed layer of proteins, whose properties were determined by the water shell. The surface bound protein layer is influenced by the properties of the biomaterial. If the fragile quaternary structure of the proteins is not retained in the adsorbed layer, a loss in their activity is observed. Synthetic materials do not show the presence of immunologically recognizable biological motifs, as are found in their natural counterparts, such as organs or tissues.

Figure 8.2 shows the schematic of protein adsorption on the biomaterial surface in a medium[607]. The stability, concentration and conformations of the adsorbed protein layer depend on the surface properties of the materials, i.e., surface energy, roughness, wettability, the presence of surface functional groups, surface texture, etc. The type, concentration and conformation of the adsorbed proteins on a biomaterial surface decide the adhesion and survival of the cells, especially macrophages and monocytes[608]. Some of these adsorbed proteins have the potential to recruit additional macrophages to the biomaterial–tissue interface. At the implant site or biomaterial surface, the macrophages can adhere and engage in the subsequent events of the foreign body reaction. The monocytes/phagocytes fail to maintain adhesion over time; if not, they will degrade the biomaterial and this leads to the failure of the device. This emphasizes the core importance of a better understanding of the cell – material interaction.

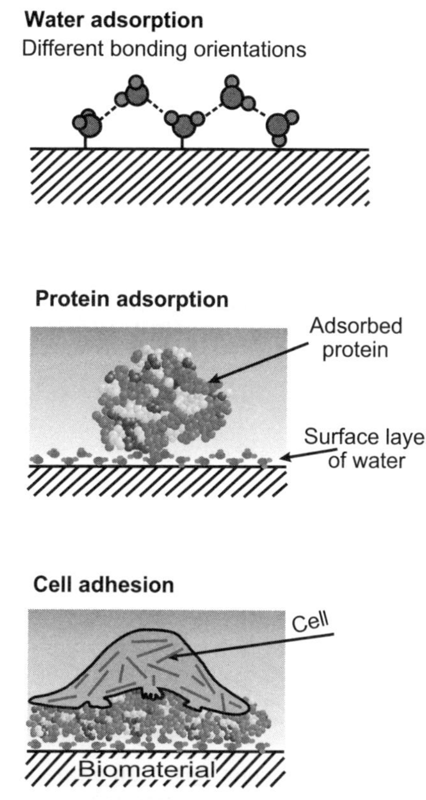

Fig. 8.1 A schematic depiction of the succession of events prior or during the interaction of a cell with a biomaterial; the primary moieties to cover the surface are water molecules (ns time scale); the water layer regulates protein adsorption on micro to millisecond time scale, and proceeds for even longer times; eventually cells reach the surface; denaturing of the proteins lowers their activity [Adapted from Ref. 605] [see Colour Plate].

8.2.1.1 | Protein adsorption isotherms

(a) *Langmuir isotherm*: As far as the protein adsorption is concerned, it is presumed that proteins are adsorbed on most of the material surfaces and the fractional surface coverage is strongly dependent on the bulk protein concentration. Based on the postulate that the adsorption is a signature of the reaction between adsorbate molecules and active sites of adsorbent, the most widely used Langmuir isotherm is developed for 1:1 matching of the adsorbate and active site. The key assumptions of such theory include: (a) monolayer adsorption of proteins, (b) the adsorbing surface has a number of discrete, identical, non-interacting sites, and (c) the ability of a protein molecule to get adsorbed on any such discrete site is independent of the occupancy of the neighbouring sites.

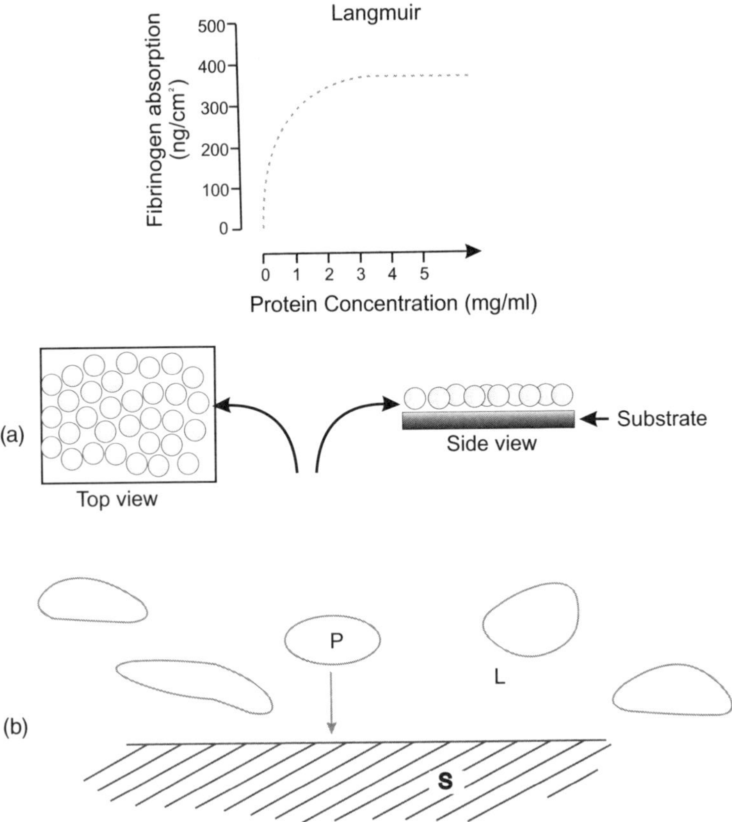

Fig. 8.2 (a) Protein adsorption process and isotherm, (b) Initial protein adsorption on a biomaterial surface, a precursor to cellular response for implanted biomaterial [Adapted from Ref. 610].

When time is kept constant, the protein adsorption is described by,

$$C_S = \frac{M}{AN_A}\left(\frac{KC_b}{1+KC_b}\right) \tag{8.1}$$

where, C_s stands for adsorbed protein density, C_b is concentration of seeded protein, M is molecular mass of protein, A is area of sample, K is adsorption coefficient.

When the seeding concentration is kept constant, the rate of protein adsorption follows the following relationship

$$\frac{dp}{dt} = K_a C\left(1 - \frac{\rho}{\rho_{monolayer}}\right) - \rho K_d \tag{8.2}$$

where ρ is the adsorbed protein density, $\rho_{monolayer}$ stands for monolayer adsorbed density, K_a is adsorption rate constant, K_d is desorption rate constant, C is bulk concentration of protein.

(b) *Spreading particle model*: In this model, the conformation/orientation changes of the surface adsorbed proteins are considered and this model is based on the assumption that protein molecules act as particles, which are adsorbed on a surface sequentially and randomly without any overlap. Once adsorbed, those particles may spread symmetrically and instantaneously to develop into a large diameter. The spreading can occur if space allows and this represents a post-adsorption transition in conformation or orientation. The key assumptions of this model are, (a) proteins interact laterally through a hard core potential, and (b) only a single altered state is possible.

The above discussion is significant in the backdrop of the fact that the adsorption of proteins on the surface of biomaterials plays a key role in determining the biocompatibility of an implant.

8.2.2 | Cell adhesion and cell morphological changes

Cellular recognition becomes possible due to the interaction of the adhesion receptors (integrins) with the adhesion proteins. Such interactions among cell receptors and ligands adsorbed on the substrate lead to the formation of focal adhesion points. Figure 8.3 depicts the various steps that occur during cell adhesion on a biomaterial substrate. Cells gradually spread and attach firmly to the substrate through ligand–receptor binding sites, as shown in Fig. 8.3.

> The focal adhesion points exert control on cell growth and proliferation through mechanical forces that cause changes in cell shape and cytoskeletal tension.

The mechanical properties (elastic stiffness) of the substrate are widely reported to influence the focal adhesion structures through mechanisms that sense matrix elasticity[609,610,611].

While interacting with the adsorbed protein layer, the cells reorganize the cytoskeleton over the surface (Fig. 8.3). The physical and chemical properties of the implanted material decide the type, concentration and conformation of the surface adsorbed proteins, which consequently control the adhesion and proliferation of cells[604]. If cells coming in contact with the surface find the surface compatible, they adhere to it and these adhered cells, in turn, send signals to other cells to adhere to the biomaterial surface. The cytoskeleton plays a major role during cell organization on the biomaterials surface. As discussed in an earlier chapter, the cytoskeleton consists of three types of polymers: actin filaments, microtubules, intermediate filaments. Together, the cytoskeleton controls the physical shape and mechanics of eukaryotic cells. It is well reported in literature that the filamentous proteins are organized into networks to resist deformation, but can reorganize in response to externally applied forces/stimuli[612].

It is important to mention that cell adhesion is one of the important cell fate processes. The interaction of cells with each other as well as with their substrates (extracellular matrix), is a primary feature of the architecture of many tissues. Cells interact with a specialized multi-protein adhesive structure, also known as cell surface protein (CAMs–cell-adhesion molecules). These proteins are commonly present on plasma-membranes that contact other cells and the cytosol- facing domains of these proteins are usually connected to the elements of the cytoskeleton. Despite the differences among these cell-surface proteins, their primary functions remain the same that is to enable the cellular communication and transduction of mechanical signals.

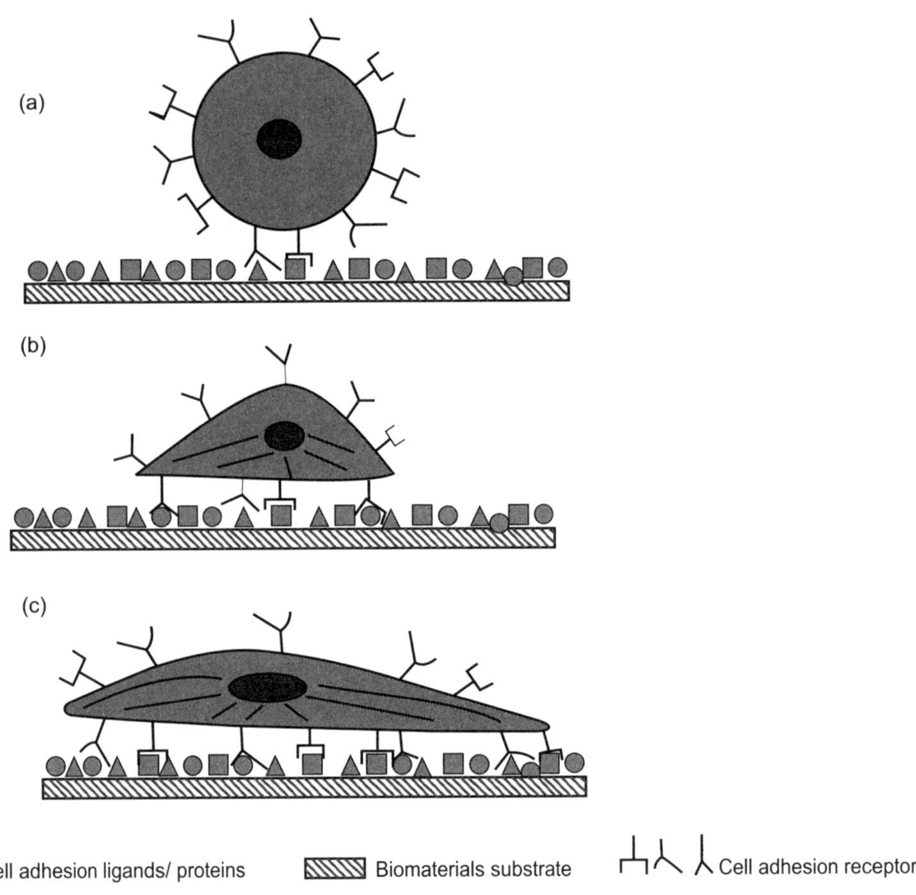

Fig. 8.3 A schematic depicting the steps in the progression of anchorage-dependent mammalian cell adhesion; (a) the initial contact of cells with the adsorbed protein layer, (b) the interactions between cell surface receptors and cell adhesion ligands that are adsorbed on the biomaterial, (c) extensive cell spreading on the biomaterial substrate with enhanced ligand–receptor interactions [Adapted from Ref. 610] [see Colour Plate].

Once a single cell adheres to a biomaterial substrate, it can communicate with other cells in the cellular microenvironment using various cell signaling processes.

Through the signalling pathways, an adhered cell also establishes 'cross-talk' with neighbouring cells and such biophysical mechanisms would have a significant influence on cell morphological changes as well as cell fate processes.

8.3 | Cell Signalling Mechanism

As mentioned in the preceding section, cell signalling involves the way a group of cells communicate with each other in an integrated manner at the interface with a biomaterial.

> The cell signalling can be defined as part of a complex communication network, which determines the cell fate processes in a coordinated manner.

In a cell culture medium, cells often receive multiple signals from the immediate microenvironment and they then integrate the information they receive into a unified action plan. Cell signalling involves highly complex and branched pathways interconnected with each other via different types of signalling proteins and small intracellular mediators. Ultimately, signalling molecules activate target effector proteins leading to a specific cell response. These pivotal cellular fate processes include proliferation, apoptosis, cytoskeletal re-organization, migration and so on, which are described in the next section. Usually, signalling in multicellular organisms is a sophisticated process, in which millions of highly specialized cells may need to act in a coordinated fashion. It is well reported that various signalling pathways not only transmit, but also encode and integrate internal and external signals[613].

Therefore, cell communication results in coordinated cellular activities and occurs in three principal ways, such as (a) secretion of soluble signals, (b) secretion of insoluble signals, and (c) direct cell–cell contact. The cellular communication by extracellular signals usually involves six steps: (1) synthesis of signal molecules; (2) release of the signalling molecule by the signalling cell; (3) transport of the signal to the target cell; (4) detection of the signal by a specific receptor protein; (5) a change in cellular metabolism, function, or development triggered by the receptor–signal complex; and (6) removal of the signal, which often terminates the cellular response. These will be discussed in more detail in one of the following sub-sections.

8.3.1 | Soluble signals

The cell-signaling processes are facilitated by millions of soluble signaling molecules, which are essentially, proteins. More specifically, these proteins exist as growth factors. The growth factors are essentially required for simulation of various cell fate processes e.g. proliferation, migration and differentiation. Many growth factors are quite versatile, stimulating cell growth in numerous different cell types, while others are highly specific to a particular cell type.

> The growth factors, in biological terms, are defined as small proteins of 15–20 kDa in size and those essentially control cell growth and differentiation. The growth factors stimulating cell proliferation/differentiation are known as cytokines, while those assisting in cell migration are known as chemokines.

The most widely used growth factors in biomaterial research include TGF-β, IGM, FGF, etc. (see table 8.1). Soluble signals secreted by cells can either be water soluble or lipid-soluble (Fig. 8.4). Hydrophilic molecules, such as the polypeptide hormone insulin, bind to receptors at cell surfaces. On the other hand, hydrophobic molecules, pass through the lipid bilayer of the cell membrane to reach receptors within the cytoplasm.

Table 8.1 Examples of most commonly known growth factors and their principal activities.

Cytokine	*Biological activity*
Hepatocyte growth factor (HGF)	Stimulates division in hepatocytes, epidermal keratinocytes, renal tubular epithelial cells and melanocytes
Fibroblast growth factor (FGF)	Mesodermal and neuroectodermal cell stimulator family of about 19 similar proteins that play a role in skeletal and nervous systems development
Interleukin-2 (IL-2)	Stimulates growth of T lymphocytes
Interleukin-3 (IL-3)	Stimulates proliferation, differentiation and survival of pluripotent hematopoietic stem cells
Epidermal growth factor (EGF)	Induces proliferation of various epithelial tissues
Platelet-derived growth factor (PDGF)	Induces growth of fibroblasts and smooth muscle cells
Insulin-like growth factors (IGF)	Stimulates proliferation, differentiation of various cell types
Transforming growth factor-beta (TGF-β)	Regulates cell growth and differentiation of many cell types, involved in regulating extracellular matrix proteins
Vascular endothelial growth factor (VEGF)	Specifically induces proliferation of endothelial cells

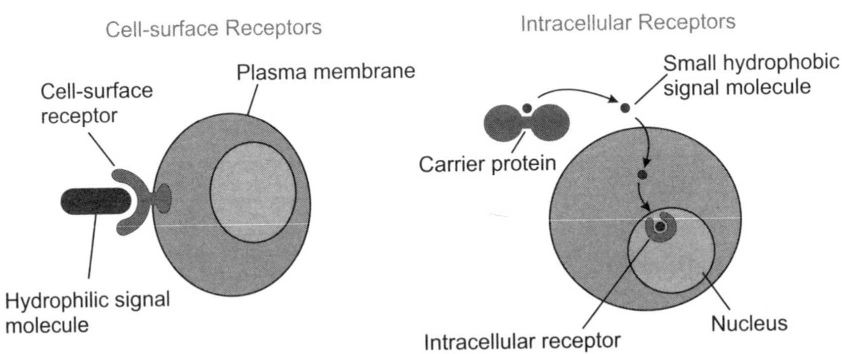

Fig. 8.4 Binding of extracellular signal molecules; most signal molecules or growth factors are hydrophilic and therefore unable to cross the plasma membrane directly; they bind to cell surface receptors, which in turn generate one or more signals inside target cell (left); some small molecules diffuse across plasma membrane and bind to receptors inside the target cell- either in cytosol or in nucleus (right) [Adapted from Ref. 684].

8.3.2 | Classification of signalling mechanisms

The cellular signalling process can be classified into three types based on the mode of mechanism and the distance over which the signal acts[614].

Autocrine signalling

In this mode, the soluble factors secreted by a cell signal to the same cell. When the signal from a group of identical cells binds back to a receptor on the same cell type, it encourages the coordinated cellular response (Fig. 8.5a).

Paracrine signalling

In this mode, a single cell transmits signal molecules to other cells in its neighbourhood. For instance, muscle cells receive chemical signals from adjacent nerve cells, by the paracrine mechanism. In both autocrine and paracrine signalling, the chemical signal works in the immediate vicinity of the cell that produces it and is present at high concentrations (Fig. 8.5b).

Synaptic signalling

This is similar to paracrine signalling, but there is a special structure called the synapse between the cell originating and the cell receiving the signal. It is a highly specific and localized type of paracrine signalling between two nerve cells or between a nerve cell and a muscle cell (see Fig. 8.5c).

Endocrine signalling

In this mode, the cell secretes growth factors into the blood stream, which are carried into the target cell at distant body sites. Endocrine signalling uses chemicals called hormones to send messages that can reach all parts of the body (Fig. 8.5d).

Direct cell–cell contact

Contact-dependent signalling demands the source and target cells to be in close membrane–membrane contact (see Fig. 8.5e). In such a scenario, some of the molecules on the plasma membranes may bind together in specific ways. Contact signalling can either occur via plasma membrane bound receptors or via gap junctions.

The last mode of cell signalling is more common in endothelial cells contained in epithelial tissue. A gap junction is a specialized cell junction that directly connects the cytoplasm of two cells. Gap junctions allow various molecules and ions to pass freely between cells. The transfer of signalling molecules communicates the current state of the cell that is directly near the target cell; this allows a group of cells to coordinate their response to a signal that only one of them may have received (Fig. 8.5e). Cells connected by gap junctions share small molecules, and can respond to the extracellular signal in a coordinated manner. Some of the molecules involved in direct cell–cell contact (known as cell-junction molecules) allow for direct cytoplasmic communication. Even though they are expressed in virtually all tissues and cells, they are most notably present in cell types that are involved in direct electrical communication, such as neurons and cardiac muscles.

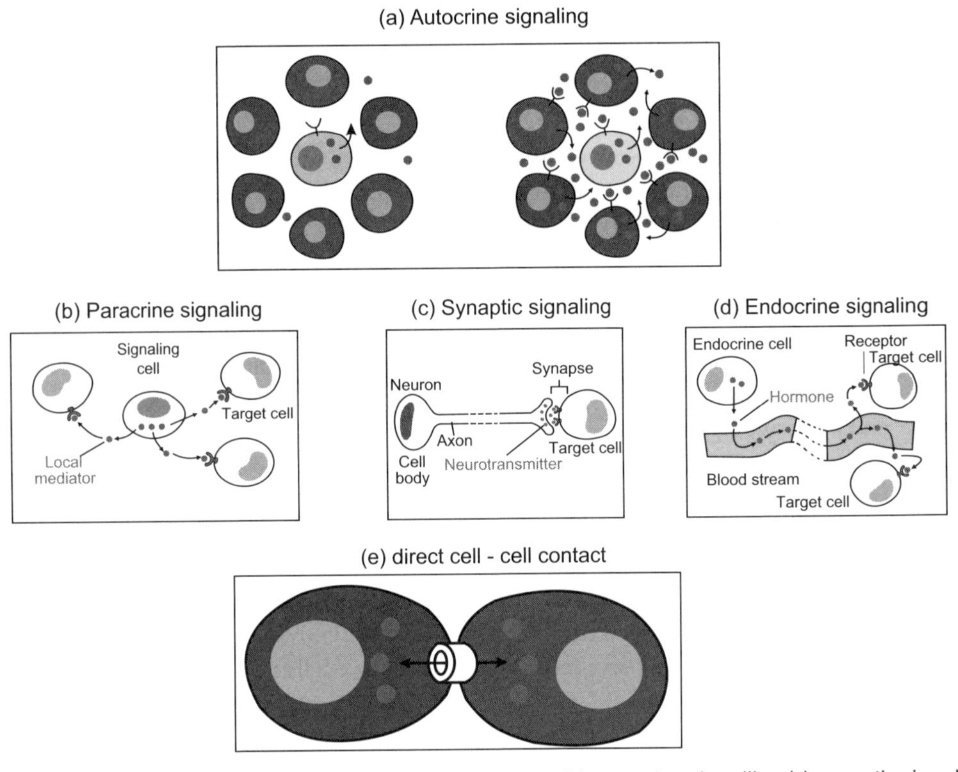

Fig. 8.5 Types of soluble signalling. (a) autocrine signalling, (b) paracrine signalling (c) synaptic signalling (d) endocrine signalling, and (e) direct cell–cell contact [Adapted from Ref. 684] [see Colour Plate].

8.3.3 | Quantitative analysis of cell signalling

From the preceding discussion, it is apparent that signal molecules are transported between closely-spaced cells or through a distance. It is, therefore, instructive to quantify the intercellular flux for direct cell–cell contact as well as the spatial extent through which such flux can diffuse away to another cell in its vicinity. For the latter, the threshold distance between the source and target cell for threshold signalling molecule flux needs to be estimated. The direct cell–cell junctions are typically on the order of 1.5 nm in diameter and allow molecules below ~1000 dalton to pass between cells. According to Fick's law, the flux through a hole of diameter 'd' of a solute with a diffusion coefficient 'D' that is present in the signalling cell at a concentration $[C]_1$ and the receiving cell at $[C]_2$, is

$$J = 4D/\pi d \,([C]_1 - [C]_2) \tag{8.3}$$

If d = 4 nm, D = 10^{-5} cm^2/sec, $[C]_1 - [C]_2$ = 100 μM, a flux of 2.4 × 10^5 molecules/pore/second can be estimated. Thus, with approximately 100 pores between cells, the flux will be 2.4 × 10^8 molecules/cell–cell boundary/second. Similarly, the flux can be estimated for other signalling modes, as shown in Fig. 8.5. This is addressed in the next sub-section.

Quantitatively, one has to know what would be the signaling molecule flux that a source cell would generate so that a target cell, located at some finite distance from a source cell, would be able to perceive and thereafter, internalize it onto itself in a manner to activate a chain of downstream signaling processes, thereby influencing cellular fate of the target cell.

Cell-to-cell signalling by extracellular chemicals occurs over distances from a few micrometres in autocrine/paracrine signalling to several metres in endocrine signalling. For example, a chemotactic migratory cell senses a gradient of a chemokine to properly respond to a signal. As shown in Fig. 8.6, the target cell senses the signal depending on its location relative to the signalling cell and its sensitivity to the signalling molecule (K_m). Once a sufficient number of signalling molecules bind to receptors on the surface of a target cell, they can induce a signal transduction process.

For a given cell of size (10–25 µm) and a low concentration of signalling molecules (sometimes >1000 molecules), it is difficult for a cell to sense a concentration difference across the cell body. However, a cell can extend filopodia/lamellipodia to travel the distance over which a concentration gradient is set up. For example, filopodia on neurites are highly dynamic structures, that are used to guide the migration and growth of a cell in a concentration field.

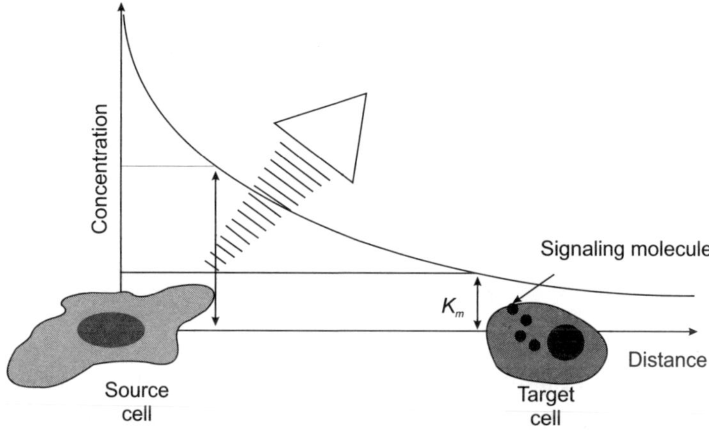

Fig. 8.6 Schematic showing the diffusive mass transfer from signalling cell to the target cell in response to concentration gradient of signal-molecule.

The highest concentration of a signalling molecule is found at the source cell surface and it decays in an inversely proportional manner with distance from the source (Fig. 8.6). A reasonable estimate of signal-propagation distance is the characteristic distance at which the signal strength is half of its maximum, leading to a time-constant estimate of about 20 minutes and a maximal distance of about 200 µm. This distance increases proportionally with the secretion rate (F) and inversely with the dissociation constant (K_d).

Under a steady state condition, the concentration ([C]) of a secreted signal as a function of the distance (r) from a cell can be described as,

$$[C]/K_d = a\,(R/r), \text{ where } a = (R^2/D)/(K_d R/F) \tag{8.4}$$

where, R- radius of the cell, F- secretion rate, K_d- dissociation constant of the signalling molecule to receptors on the receiving cell, D- diffusion coefficient of a signalling molecule.

From the above expression, it is clear that when $[C] = K_d$, the distance the signal can reach is, $a = r_{critical}/R$. The maximal secretion rate is given by the ratio of two time constants i.e. the time constant for diffusion away from producing cells, τ_{diff} (R^2/D) and the secretion time constant, $\tau_{secretion}$ ($K_d R/F$).

At the target cell site, the secreted signals bind to the receptors on target cell membranes with high affinities and the association constant (K_a) can be in the range of 10–100 pM. In reference to Fig. 8.4, the cell signalling to a target cell is accomplished only when a signalling molecule or a growth factor (G) is biochemically bound to a cell surface receptor (R) and such binding can be described by the following simple equation,

$$G + R <=> G : R \text{ and } [G][R]/[G:R] = K_d \tag{8.5}$$

where [G], [R], and [G:R] are the concentrations of the growth factor, receptor, and bound complex, respectively, and K_d is the dissociation constant.

The total number of receptors (R_{tot}) for a single cell is constant,

$$R_{tot} = [R] + [G:R] \tag{8.6}$$

Assuming that the binding is at equilibrium, one can write,

$$R_{tot} = [R] + [G:R] \tag{8.7}$$

$$R_{tot} = [G:R] K_d/[G] + [G:R]$$

$$= [G:R](1 + K_d/[G])$$

From the above, the receptor occupancy can be described as

$$[G:R]/R_{tot} \text{ or } [G]/K_d+[G]$$

To give an example, in the case of EGF receptor occupancy, ($[G:R]/R_{tot}$) needs to be on the order of 0.25–0.5 to reach a significant stimulation. In reference to the earlier estimated value of K_d of 10–100pM, growth factor concentrations as low as 10 pM at the receiving cell surface can be sufficient to generate acellular response.

8.3.4 | Intracellular signalling mechanism

Next, the signalling mechanism, involving the activation or switching ON of a series of intracellular proteins is to be analyzed, and Fig. 8.7 illustrates the same schematically. The receptor activation caused by ligand binding to a receptor is directly coupled to the cell's response to the ligand (Fig. 8.7).

The signal transduction process involves the activation of a large number of downstream intracellular proteins in the target cell.

The cell surface receptors are integral membrane proteins and have three domains, namely extracellular domains, transmembrane domains and cytoplasmic or intracellular domains. Based on the mechanism of action, there are different types of receptors such as gated ion channel receptors, receptors with enzymatic activity, steroid receptors, adhesion receptors, serpentine receptors and receptors without enzymatic activity. As shown in Fig. 8.7, once a receptor protein is activated, a series of intracellular signalling proteins are switched on. The end result is that a specific target protein is activated. These target proteins can be either metabolic enzymes or cytoskeletal proteins, which can alter either cell metabolism or cell shape/motility, respectively. A more complex description of intracellular signalling mechanisms is provided in Fig. 8.8. The bio-physicochemical process of signal transduction involves the transfer of an incoming signal to a target cell so that the intracellular signalling molecules are activated and the intracellular signalling pathway executes a given cellular function in a coordinated manner.

Cells respond to external stimuli by activating several signal transduction pathways which can diverge and be relayed to several different intracellular targets (Fig. 8.8). The ligand is the primary messenger. As the result of binding the receptor, other molecules or second messengers are produced within the target cell. Quantitatively, the speed of a response of a target cell to an incoming signal is not only dependent on signal delivery mechanism, but also on the nature of the target cell response. The timescale of such response can vary from a few seconds to a few hours with the latter involving changes in gene expression. This also partially explains why cell differentiation takes over a few weeks, and that is the reason stem cells are to be cultured for a long time to study the differentiation through a specific lineage.

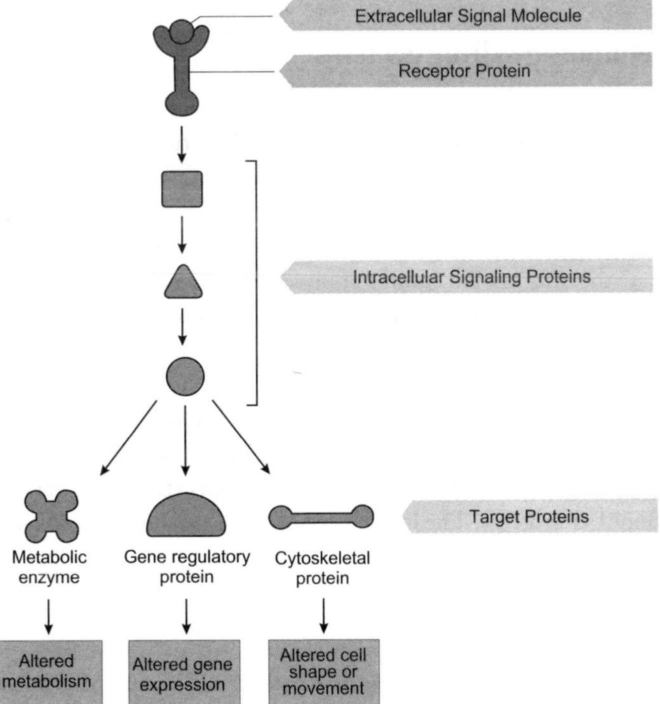

Fig. 8.7 Intracellular signalling pathway activated in a target cell by an extracellular signal molecule [Adapted from Ref. 684].

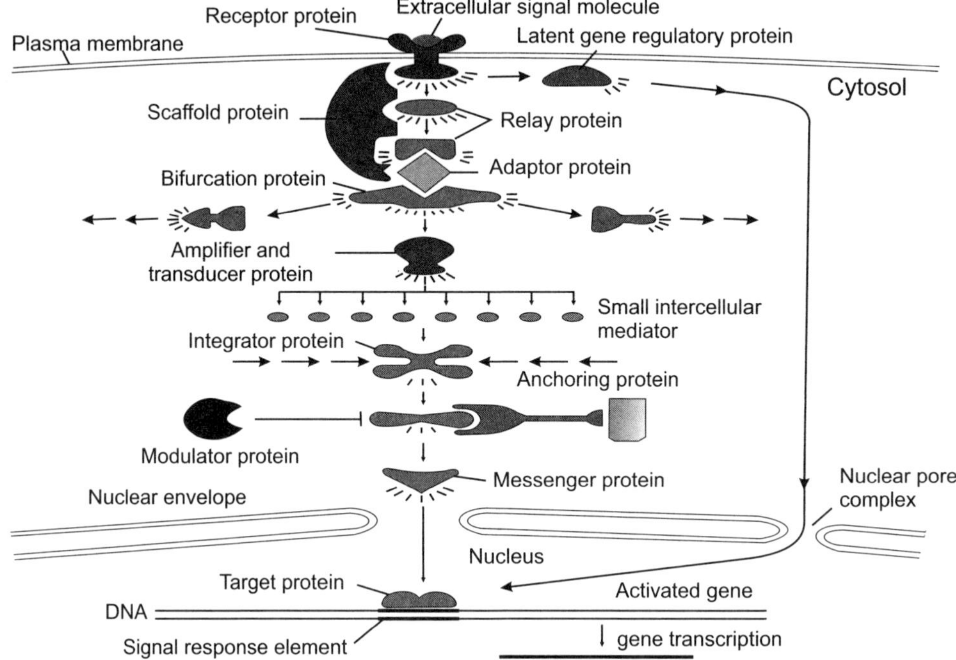

Fig. 8.8 Intracellular signalling proteins together with signalling pathway [Adapted from Ref. 684].

8.3.5 | Intracellular signalling proteins

A signal-transduction network, schematically shown in Fig 8.8, depicts several components of a complex, interconnected circuit in electrical sciences. The detection, transmission, and amplification of signals are interconnected, allowing precise control of cell fate processes.

Various cell fate processes, such as cell survival and cell proliferation, are generally stimulated by a specific combination of extracellular signals, and not by a single signal (see Fig. 8.9). This demands a given cell to integrate the information coming from separate signals before an appropriate response. This integration usually depends on integrator proteins and the integration of signalling response can be thought to be equivalent to microprocessors in a computer. The reason for such equivalence is that they both require multiple signal inputs to produce an output.

Signalling proteins generally, but not exclusively, function by activating the next signalling protein in the signal transduction cascade (see Fig. 8.8). Thus, signal transduction pathways involve the functionality of signalling proteins such as, (a) relay proteins (b) messenger proteins, (c) adaptor proteins, and (d) amplifier proteins. The last kind of proteins can be either enzymes or ion channels. These proteins greatly increase the signal intensity by activating numerous downstream intracellular signalling proteins.

A signaling cascade refers to multiple amplification pathways in a relay chain of signal transduction occurring in a target cell in response to the internalization of extracellular signals.

Other proteins, which take part in cell signal transduction process include, (a) integrator proteins, which collect signals from multiple biochemical pathways and consolidate the signal information before transmitting an onward signal; (b) anchoring proteins, which maintain specific signalling proteins at a precise location by tethering them to a membrane/cytoskeleton; and (c) scaffold proteins, which are proteins that simultaneously bind two or more other proteins, and organize binding partners into a functional unit to enhance signalling efficiency and fidelity. Scaffold proteins may therefore share the attributes of both anchoring and adaptor proteins.

In reference to the discussion based on Figs. 8.6–8.8, one can clearly perceive that cell viability itself depends on the integration of several signalling pathways. Any additional cellular functionality, e.g., cell division or differentiation, requires some additional signalling pathway to reinforce the combination of pathways required for cell survival.

When all the signaling pathways are terminated, a given cell activates its own suicidal mechanism and undergoes apoptosis.

In summary, signalling molecules control several cellular processes, growth and differentiation of tissues, synthesis and secretion of proteins, and the composition of intracellular and extracellular fluids (Fig. 8.9). The signalling systems discussed here are only some of the key players and even the description of this complex mechanism is greatly oversimplified in this section for beginners. From the above discussion on cell signalling mechanisms, it should be clear now that cell fate processes are essentially dictated by a combination of cell signalling mechanisms.

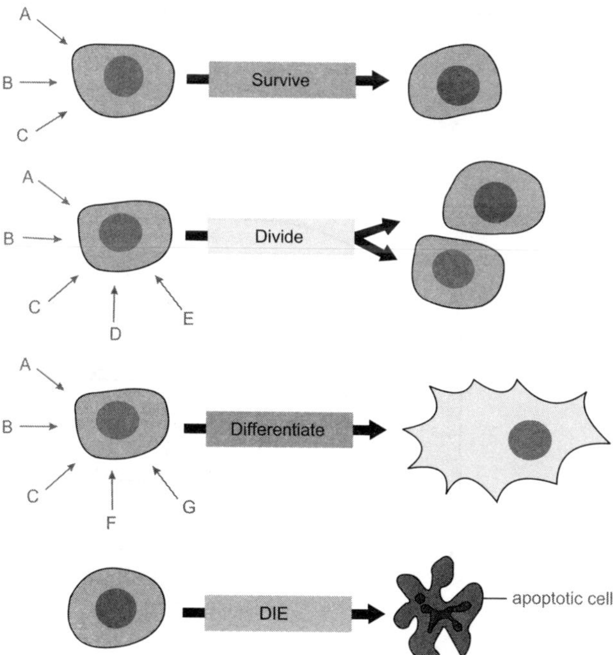

Fig. 8.9 Schematic illustration showing the role of multiple signalling molecules towards different cell fate processes [Adapted from Ref. [684]].

8.4 | Eukaryotic Cell Fate Processes

All the cells coordinate and communicate through principal cellular fate processes constituting the dynamic states of tissue function. These cellular fate processes are described below.

8.4.1 | Cell differentiation

Cell differentiation is a biophysical process derived from differential gene expression in which a cell undergoes phenotypic changes to get specialized into a matured cell type, which performs the destined physiological function.

The process of cell differentiation begins with irreversible changes in a set of genes that are expressed in the cell. More specifically, process involves the switching off and on of gene families in a co-ordinated manner. The changes in gene expression illustrate the changes in the ultrastructural features of the cell as well as its biochemical functions such as changes in the transcription factors and other regulatory proteins (Fig. 8.10). During the differentiation, those genes are expressed, which are more specific than that of the mature cell type.

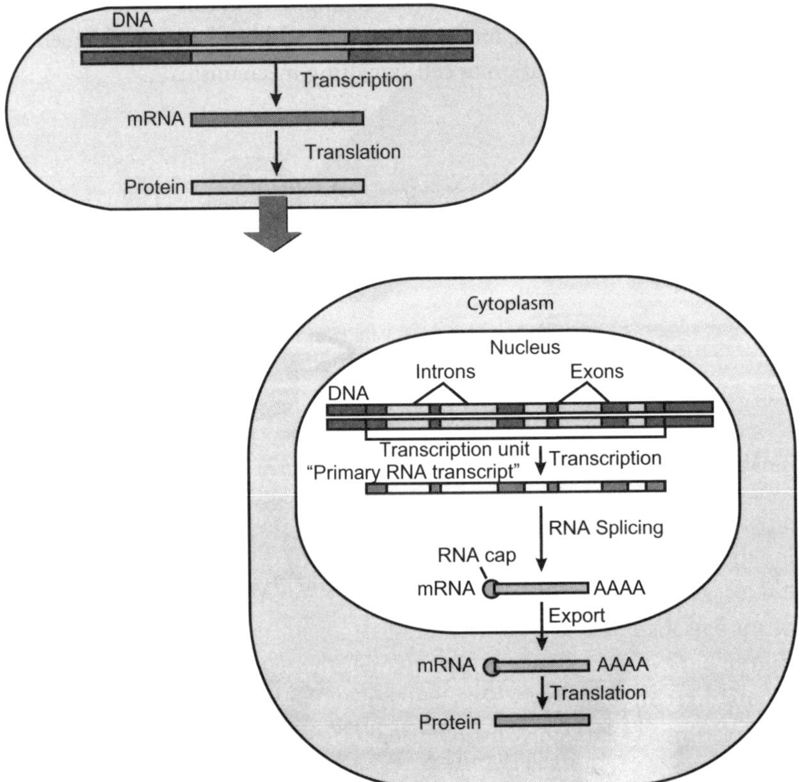

Fig. 8.10 Transcription and translation process involved in cell differentiation [Adapted from Ref. 684].

The process of cell differentiation can be directly observed using light microscopy (for morphological observation) or using fluorescent surface markers in a flow cytometry. The characteristic surface proteins are found on the cell surface at its various stage of differentiation as the cell matures and these surface markers can be used as binding sites for fluorescently conjugated monoclonal antibodies. The expression of such multiple surface markers can be traced using flow cytometry.

As pointed out in the previous section, the control of stem cell differentiation is essential for an intended regenerative/healing therapy. This control over stem cell differentiation is usually achieved by providing appropriate growth factors [(bone morphogenic protein-2,4, transforming growth factor (TGF) β1, hepatocyte growth factor (HGF), β nerve growth factor (NGF), fibroblast growth factor (FGF), retinoic acid)] and chemical supplements (Dexamethasone, ascorbic acid, β-glycerol) to aid in the differentiation process towards a particular lineage. For example, it was found that embryonic stem cells exposed to retinoic acid and TGF-β induced differentiation change into nerve and muscle-like cells, respectively. The three embryonic germ layers (ectoderm, mesoderm, and endoderm) were activated by exposure to HGF and NGF and it is reported that in the presence of BMP-2, hMSCs will differentiate towards osteoblast-like cells [615,616]. However, it is not always feasible to provide growth factors and chemical supplements as they might cause other systemic effects. But stem cells themselves are sensitive to their surrounding micro-environment, and it is observed that a change in any of these factors might act as a cue for stem cell differentiation.

8.4.2 | Cell migration

This process plays an important role in the physiological as well as the pathological process of tissues during organogenesis and embryonic development. In adult organisms, cell migration plays a key role in eliciting the proper immune response. Aberrant cell migration is found in various pathological situations, like cancer metastasis. Indeed, the migratory process is also important in wound healing and angiogenesis. It is known that cells migrate in response to a variety of stimuli (Figs. 8.11 and 8.12). Chemotaxis is the word used to describe cell movement in a concentrated gradient of soluble factors. Cell migration can be initiated by changes in cell–cell adhesion. The biochemical processes of cell migration are explained in Fig. 8.12[617].

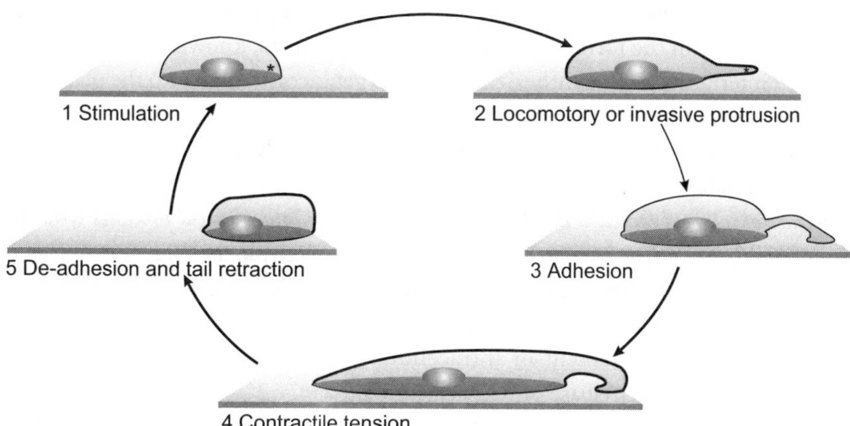

Fig. 8.11 Different steps involved in cell migration [Adapted from Ref. 617].

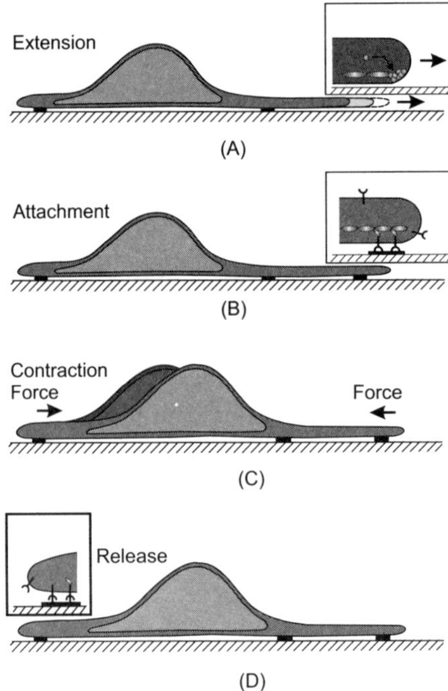

Fig. 8.12 Different stages of cell motility/migration on a biomaterial substrate (hatched plate): (a) cellular extension involving actin depolymerization, (b) attachment at the extended part involving actin repolymerization, (c) realizing contraction force at rear end, and (d) release of focal adhesion to enable cell advancement in a linear direction.

The molecular components involved in cell migration can be conceptualized in a five-step process:

(a) *Membrane lamellipodia protrusion* occurs by local actin polymerization, involving an increase in the number of sites for actin polymerization. This is followed by the addition of an actin monomer on the filament growth sites near the membrane.

(b) *Attachment of integrins to an extracellular matrix* provides a link to the actin cytoskeleton. The Integrin cytoskeleton link can be influenced by ligand binding, integrin cross-linking and matrix stiffness, resulting in the formation of specialized complexes, known as 'focal adhesion' at the cell/matrix interface. Newly synthesized integrins get incorporated into the plasma membrane and are transported to the leading edge.

(c) *Contraction of the actin cytoskeleton by myosin-based motors* leads to traction force generation against a substrate.

(d) *Rear release of cell and forward displacement by enzymatic and physical processes* would cause the preferential release of attachments from the rear end can enable the force transmission through fewer matrix contacts in the rear rather than in the front, leading to greater force per contact.

(e) Also, the dynamic formation of integrins in the rear can be involved in enzymatic processes, which include the requirement of intracellular Ca^{2+} for breaking some integrin/matrix complexes, favouring the release of rear attachments.

(f) *Recycling of remaining integrins* occurs by endocytosis into vesicles followed by intracellular diffusion and directed transport or integrins transport along the cell surface to the leading edge.

8.4.3 | Cell division

The process of eukaryotic cell division is perhaps the most distinguishable feature in biological systems. A eukaryotic cell, which grows and divides, is known to undergo a repeating series of events, called cell cycles comprising four phases (Fig. 8.13). During the first phase (G_1), the cell grows and prepares for DNA replication, which occurs in the subsequent S phase. The cell can stay in the G_1 phase for a variable length of time, while the duration of the S phase is about 8 hours. In the G_2 phase, a further growth of the cell takes place. A cell stays in this phase for about 2–3 hours before entering into the M phase, where, finally, mitosis occurs. The duration of S+G_2+M is constant (about 12 hours), while the time that a cell spends in the G_1 phase is highly variable.

The phases of the cell cycle are unidirectional and are controlled by checkpoints [Fig. 8.13]. At the checkpoint, the cell's size, certain environmental conditions and DNA integrity are checked before the cell enters the S phase. If the cell fails to pass these checkpoints, it may initiate apoptosis and cell will die. The G2 checkpoint senses for unreplicated DNA, which can potentially decide cell cycle arrest, unless DNA replication is complete. The cell cycle progression can be stopped at the G2 checkpoint in response to DNA damage. DNA damage arrests the cell cycle at G1 too. Cells typically divide at a rate proportional to the number of cells at a given point of time. For unconstrained growth, the growth rate is proportional to the number of cells;

$$dX/dt = \mu X = X(t) = Xo \exp(\mu t) \tag{8.8}$$

Based on the above equation, the growth rate can be defined as, $\mu = \ln (2)/t_d$; where t_d - doubling time. For human cells, $t_d \sim 12$ hours and therefore $\mu_{max} \sim 0.06$/day. The ability of cells to undergo division varies with the cell type. For example, the doubling time of hematopoietic progenitors, dermal foreskin fibroblasts and adult chondrocytes are 11–12 hours, 15 hours and 24–48 hours, respectively.

8.4.4 | Cell death

Cell death may occur for many reasons. For example, cells that die from tissue damage undergo a cell death process, known as necrosis (Fig. 8.14). The defects in transmembrane transport lead to cell swelling and lysis, whereby intracellular contents are released. Cell necrosis usually takes place, when a tissue is injured due to a sudden mechanical shock or an external electromagnetic field. But a major part of tissue development experiences programmed cell death (apoptosis), which occurs as a part of the co-ordinated function of tissue morphogenesis (Fig. 8.14). Cells undergoing apoptosis, first shrink, condense and then fragment into apoptotic bodies. These apoptotic bodies are membrane-bound structures and do not spill into the extracellular space to cause local peripheral damage. Thus, this form of cell death is highly regulated, and a failure often results in tumour formation. Cellular apoptosis often takes place, when a cell is stuck at one of the checkpoints, as described in Fig. 8.13.

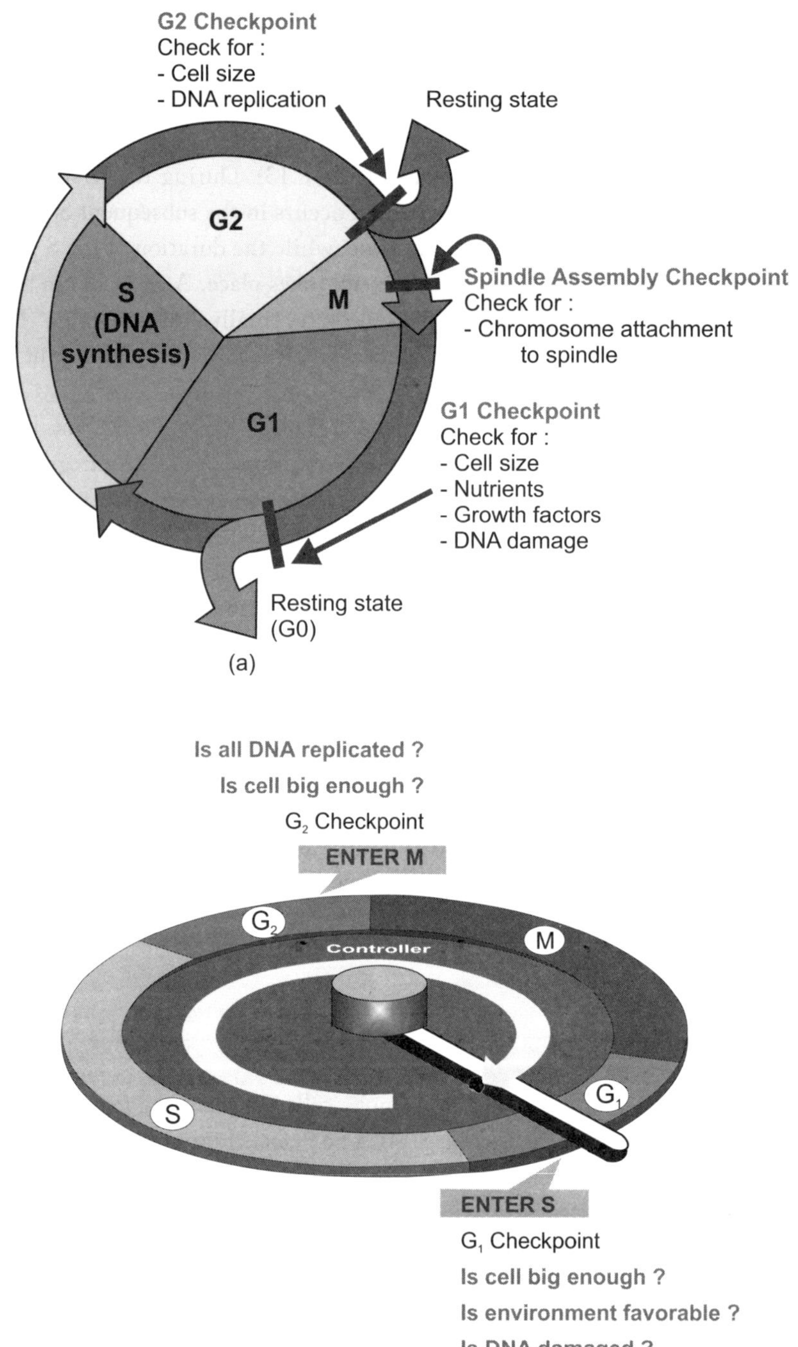

Fig. 8.13 (a) Description of eukaryotic cell cycle and (b) various checkpoints regulating the cell cycle [Adapted from Ref. 684].

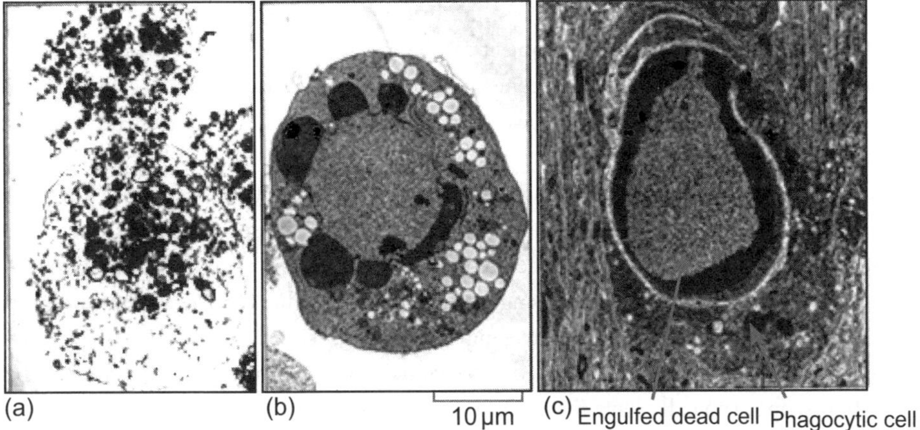

Fig. 8.14 Morphological changes during cell death: (a) necrosis and (b and c) apoptosis[684].

Due to unfavourable cellular microenvironments, if a cell is not able to cross the checkpoint in a cell cycle, then the cell activates its own suicidal mechanism.

Morphologically, cell apoptosis is characterized by a) shrinkage in cell size, (b) membrane blebbing and (c) nuclear condensation and fragmentation, as shown in Fig. 8.14.

The above discussion signifies that cell-morphological changes provide some insights into various cell fate processes and in the following section, a microscopic analysis of such changes together with their quantitative assessment is described.

8.5 | Qualitative and Quantitative Assessment of Cell Morphological Changes

It can be perceived from the above discussion that a qualitative and quantitative analysis of cellular morphological changes is of major relevance in the context of the cytocompatibility of a biomaterial. Specific morphological changes such as cell elongation, cell shrinkage and changes in membrane morphology are the consequences of cellular proliferation, differentiation, senescence, toxicity or pathology. Thus, the alterations in cell morphology are events of particular importance and play a key role in establishing a closer look at the cell-biomaterial interface. At first, we will discuss some relevant fundamentals of microscope-based analysis.

8.5.1 | Some fundamentals

In order to realize the full potential of microscopic techniques in biomaterial research, one must understand some of the important concepts in microscopy, such as resolution, magnification, numerical aperture, depth of field, etc. These are briefly mentioned below. The magnification is typically defined by a factor by which an object can be magnified, while observing it through a set of lenses on a microscope. All the microscopes, with the exception of AFM, use a set of objective lenses and an eyepiece. The overall magnification is determined by the product of the magnification of an objective and eyepiece lens.

Normally, the magnification is designated as times magnified with respect to that with the naked eye. The highest magnification that can be attained with an optical microscope is typically 1500x (1500 times larger as compared to the unaided naked eye), and their best resolution is 0.2 μm or 200 nm, which is of the order of wavelength of UV/visible light. On the contrary, high energy electrons possess a significantly smaller wavelength than UV/visible light (10^5 times lower, at 0.004 nm). This implies that electrons can be used to generate images with resolution as high as 0.1 nm. Electron Microscopes typically have magnifications of around 500000x.

The **numerical aperture** (NA) is defined as the ability of the microscope objective to collect light and to resolve image details at a fixed object distance. The NA determines resolving power, depth of field, and contrast of the image. It is defined by the following equation,

$$NA = n^* \, Sin \, \theta \tag{8.9}$$

'n' is the refractive index and θ is the half-angle of the maximum cone of light that can enter or exit the lens. The higher the NA, the greater is the resolving power and the smaller the depth of field. In light microscope, oil is used to tailor NA values.

Resolution is defined as the smallest distance below which two discrete objects are indistinct. It is the ability to distinguish between two points on an image, i.e., the amount of detail. The resolution of an image is limited by the wavelength of radiation used to view the sample. The resolving power (R) of a system is determined by N.A. and the wavelength of light (λ), as shown below,

$$R = 0.61^*\lambda/N.A \tag{8.10}$$

The **depth of field** is the range of distance along the optical axis in which the specimen can move without the image appearing with uncompromised sharpness. This depends on the resolution of the microscope and is dependent on the wavelength of radiation and acceptance angle of the lens. The latter varies with the focal length, which in turn regulates the numerical aperture. A microscope with a narrow depth of field will need to be constantly focused from top to bottom to image a thick specimen.

The **depth of focus** is the axial depth of the space on both the sides of the image plane within which the image appears acceptably sharp, while the positions of the object plane and of the objective are maintained. In short, it is the range of acceptable focus for the image. The depth of focus is as important as the depth of field, but for one important difference, which is related to the magnification. With higher magnification, the depth of field becomes shorter; however, higher magnification increases the depth of focus for the image. Given the above background, the microscopy techniques to characterize cell morphology are discussed below.

8.5.2 | Fluorescence microscopy

In contrast to optical microscopy, the technique of fluorescence microscopy has received wider attention in the biomaterials field, due to its ability to identify cells and sub-microscopic cellular components with a high degree of specificity.

Fundamentally, fluroscence is defined as the emission of light within nanoseconds after the surface absorption of excitation light, having a shorter wavelength.

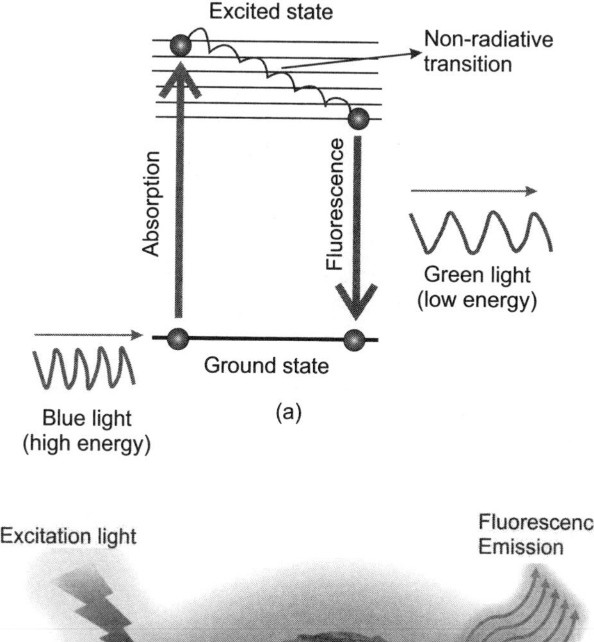

Fig. 8.15 (a) Simplified Jablonski diagram showing the energy state change of a fluorophore's electron as it undergoes fluorescence, with the corresponding change in the colour of light; (b) schematic showing the cell surface labelled with fluorophore, emitting red fluorescence upon exciting with green light.

The principles involved in fluroscence microscope can be understood using Fig. 8.15a, 8.16 and 8.17. The ability of a fluroscence microscope to reveal cellular organelles is demonstrated in fig.

8.15b. Molecules used for their fluorescent property are known as **fluorophores**[618]. When a molecule (fluorophore) with luminescent properties absorbs the light of one wavelength (excitation wavelength), it emits light of a different wavelength (excitation wavelength), rapidly after the absorption of the excitation light. The difference in wavelength is known as 'Stokes shift'. This has been shown in Fig. 8.16 for DAPI, a fluorescent dye used to stain the nucleus. DAPI intercalates with the DNA of a cell. Also, Table 8.2 summarizes a host of fluorochromes, which are widely used in cell biology research to stain various cellular organelles.

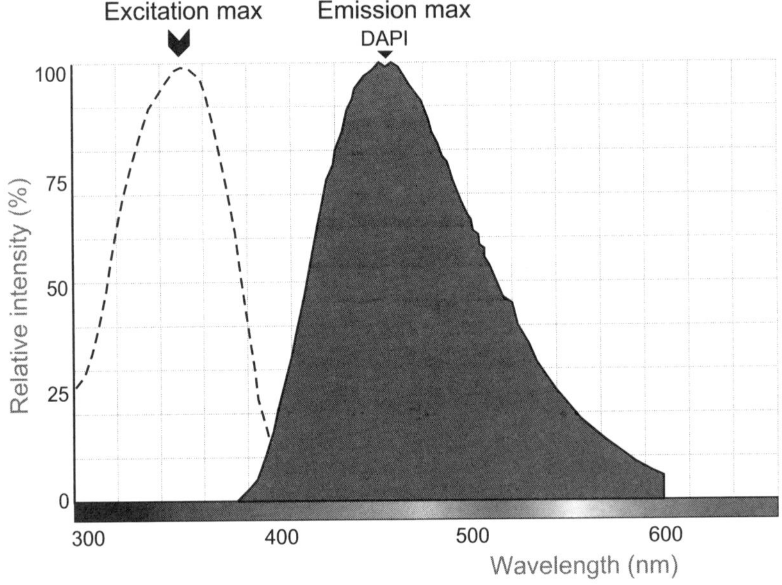

Fig. 8.16 Excitation and emission spectra of a nuclear dye (DAPI). Note the range of wavelengths emitted when DAPI fluoresces (from approximately 400 nm, violet, to 600 nm, red).

Physics of image formation:

The fluorescence process involves three important events,

(a) Excitation by an incoming photon (10^{-15} seconds)
(b) Vibrational relaxation of excited state electrons to the lowest energy level (10^{-12} seconds).
(c) The emission of a longer wavelength photon (10^{-9} seconds).

The key to fluorescence microscopy is the selection of appropriate filters to segregate the intense excitation light from the much weaker secondary emission generated by a fluorophores. As presented in Fig. 8.17, the fluorescence microscope has a filter that only lets through radiation with the specific wavelength with respect to the fluorescing material. The radiation collides with the atoms in the fluorescently tagged cellular organelles and the electrons are excited to a higher energy level. When they relax to a lower level, they emit light. The emitted fluorescence is separated from the much brighter excitation light in a second filter and then can be either seen with the naked eye or captured with a camera.

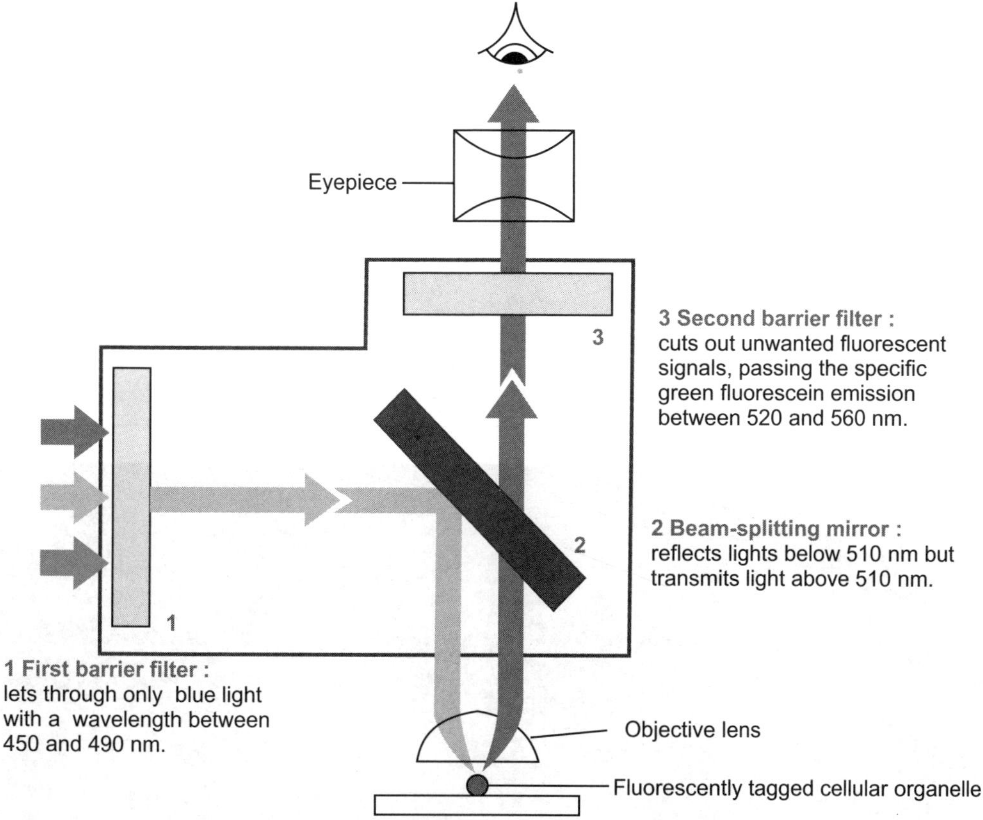

Fig. 8.17   Fluorescence microscopy set-up illustrating the suitable filter set for the fluorochrome of interest, and the detection mechanism [Adapted from Ref. 684].

Different fluorochromes are used to tag different cellular organelles (see Table 8.2). The fluorochrome can be conjugated directly to the primary or secondary antibody, or to streptavidin. Immunofluorescence is commonly used for the simultaneous visualization of multiple cellular targets. For example, DAPI is used as a general fluorescent dye to intercalate DNA in the nucleus that absorbs UV light and fluoresces bright blue. Alexa-flour FITC is used to stain actin filaments, while amicrotracer is used for the mitochondria of cells. Two representative fluorescent images of the cells are shown in Fig. 8.18.

Table 8.2   Examples of fluorescent molecules, which vary among themselves in excitation and emission maxima.

Dye	Absorbance wavelength	Emission wavelength	Visible colour
Alexafluor 488	494	517	green (light)
Fluorescein FITC	495	518	green (light)
Texas Red	595	613	Red

Contd.

Contd.

Dye	Absorbance wavelength	Emission wavelength	Visible colour
TRITC	547	572	Yellow
Cy3	550	570	Yellow
DAPI	345	455	Blue
Hoechst 33258	345	478	Blue
SYTOX green	504	533	Green
Propidium iodide	536	617	Red
Ethidium bromide	493	620	Red
Alexafluor 594	590	617	Red

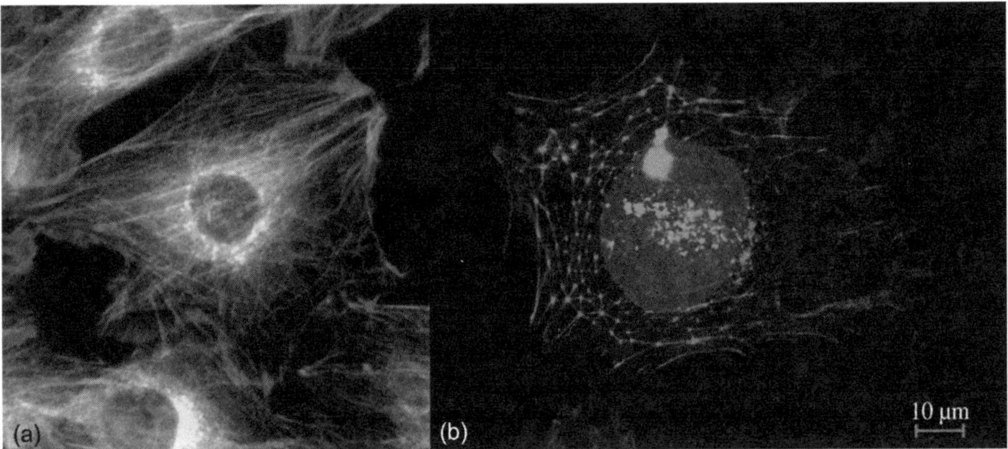

Fig. 8.18 (a) Fluorescence image of a biological cell with nucleus appearing in blue, cytoskeletal actin filaments in red and alpha-tubulin in green[685]; (b) Fluorescence image of a cell revealing the binding sites on actin filamentous structure of a cell [see Colour Plate].

8.5.2.1 | Time-lapse imaging

Time-lapse fluorescence imaging requires the repeated imaging of a labelled biological specimen at defined time intervals. Such careful study allows one to analyze the dynamic distribution of fluorescently labelled components in living systems. The time scale for sequential image collection can range from sub-seconds to days or even months[619]. Live time lapse imaging is performed to track the migration, proliferation, and apoptosis of cells, which have adhered to a biomaterial surface. Some widely used fluorescence molecules are listed in Table 8.2.

The following aspects of cell–biomaterial interaction can be investigated using fluorescence microscopy, which include, but not limited to;

(a) Cells: Cell number, morphology, viability, growth, cell–matrix interactions, differentiation.
(b) ECM proteins: fibre size and spatial orientation.

(c) Scaffold: Surface topography, inner structure, nanoparticle distribution in composites.
(d) Post-implantation processes: Tissue integration, biomaterial degradation and vascularization[620].

8.5.2.2 | Guidelines for fluorescence microscopy analysis

It is always recommended to choose a bright and photo-stable fluorophore with a high quantum yield and a high extinction coefficient. An undesired background owing to cellular auto-fluorescence (cells on a polymeric substrate) needs to be appropriately addressed. Hence, it is advisable to use band-pass filter sets that block auto-fluorescence. Appropriate fluorescence filter sets, that match the fluorophore spectra, should be selected. In order to minimize nonspecific labelling, the fluorophore labelling protocol has to be thoroughly optimized. Moreover, the specimens should be mounted in a minimally fluorescent medium (e.g., without phenol red). It is worthwhile to remember that fluorophores fade (irreversibly), when exposed to excitation light. Almost all fluoropores photobleach, when exposed to excitation light in the microscope. This can perhaps be avoided with the use of a high numerical aperture lens to collect more fluorescence light and thus use less excitation light.

8.5.3 | **Confocal microscopy**

The confocal laser scanning microscope (CLSM) is one of the most popular microscopy technique for producing 3D images of cells and thin tissues. Confocal microscopes are much more complex than wide-field fluorescence microscopes (see Fig. 8.19). Rather than exciting the entire field at once, the light is focused on a very small spot and this is scanned across the sample and an image is built up. Most often, an image reconstruction program sections the multi-level image data together into a 3D reconstruction of the targeted sample. Unlike the wide-field fluorescence microscope, the CLSM reduces the amount of out-of-focus fluorescence in the image by illuminating the specimen with a focused light source, while one or more corresponding pinholes at the image plane block out-of-focus fluorescence from reaching the detector. Confocal illumination and detection are used to reduce the background fluorescence from out-of-focus planes to obtain a better image quality than that achieved in conventional fluorescence microscopy. The ability of a confocal microscope to create sharp optical sections makes it possible to build 3D renditions of the specimen. It has several advantages over fluorescence microscopy, such as optical sectioning ability, 3D reconstruction, excellent resolution (0.1–0.2 μm) and high sensitivity. At the same time, it also has drawbacks. For example, it requires complex operating procedures, use of chemical labelling, high intensity laser light, etc. Overall, the distinguishing differences between fluorescence and confocal microscopy, with reference to Fig. 8.19, can be summarized as follows.

In wide-field microscopy, the sample is broadly illuminated with excitation light with a limited focus of its energy dictated through the numerical aperture and focal depth of the objective. Conversely, the CLSM focuses a single point of laser light through a small aperture (pinhole) and scans sequentially across the sample point by point. In a wide-field versus a confocal image of a mouse kidney section of intermediate thickness of about 15 μm, the elements of glomeruli and convoluted tubules are labelled in green, F-actin prevalent in glomeruli and the brush border were stained in red and the nuclei are stained with DAPI (blue)[621,622].

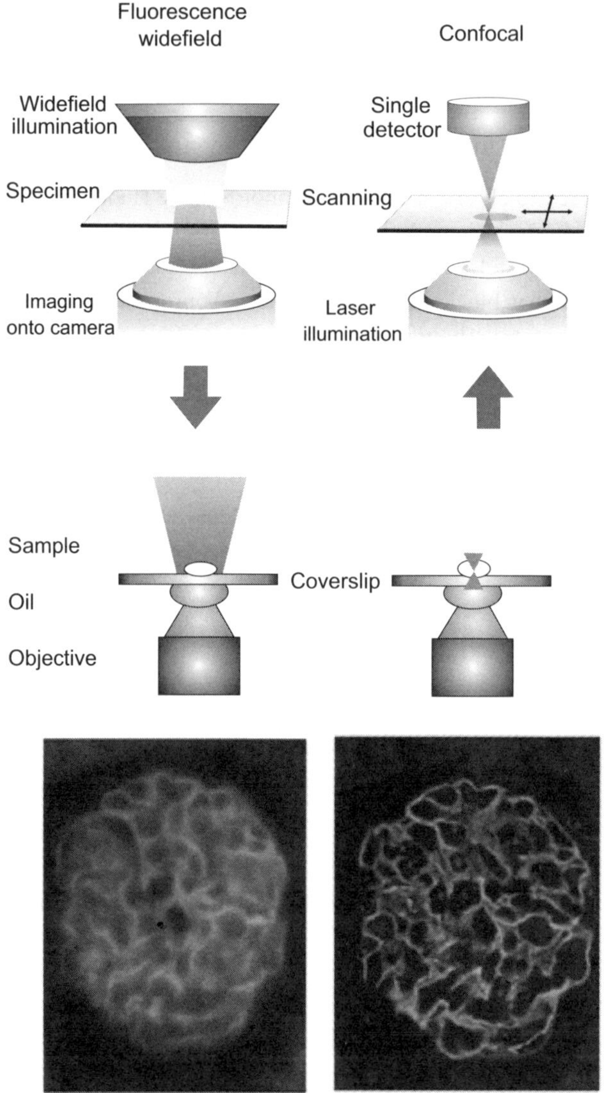

Fig. 8.19 Schematic illustration of the pattern of illumination in conventional wide field fluorescence microscopy vs CLSM[621,622].

8.5.3.1 | Physics of image formation

As shown in Fig. 8.20, the image formation essentially is characterized by the way the light travels through the optical system and in the case of confocal microscope, the light emerging from the pinhole passes through a beam splitter, before the laser rays are focused by an objective lens on to a spot in the focal plane. The objective lens forms an image of the detector pinhole and the illuminating pinhole at the same spot on the focal plane; hence, these are said to be confocal with each other[623].

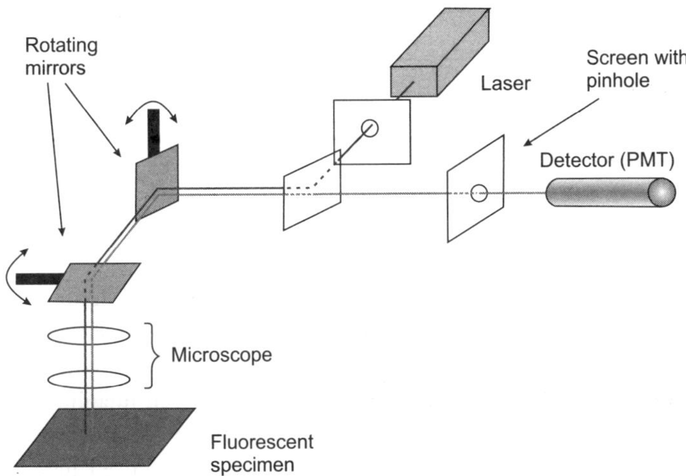

Fig. 8.20 Basic setup of a confocal microscope; light from the laser is scanned across the specimen by the scanning mirrors; optical sectioning occurs as the light passes through a pinhole on its way to the detector.

The CLSM is also capable of creating excellent images using backscattered or reflected light. Confocal backscatter imaging is particularly useful for creating 3D images of the surface of materials. In addition to the image data of cell morphology, microstructural data of scaffolds and cell–scaffold interaction can be readily obtained using this method[624]. Since the images are acquired by scanning and the image data are available in digital form, subsequent image enhancement by adopting suitable processing techniques without the loss of resolution can be followed.

As one example, the CLSM is widely used to investigate the internalization of nanoparticles within the cell. Figure 8.21 shows the actin cytoskeleton of the cell stained with a fluorescent green colour. It is indicated that GNPs (shown as red dots) have a good dispersion within the cytosol of the cell without any visible agglomeration within the resolution of the CLSM.

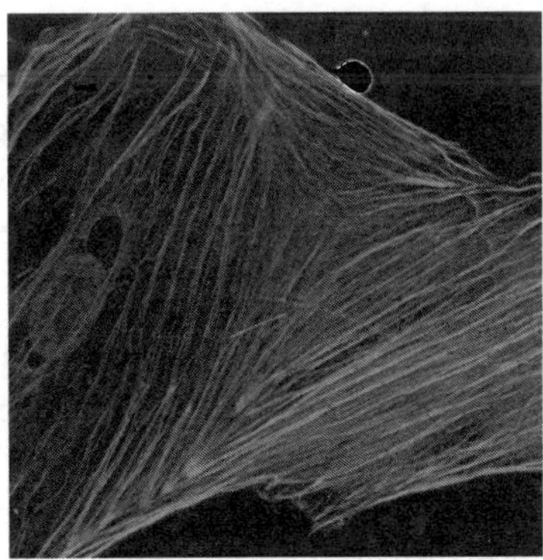

Fig. 8.21 Laser confocal fluorescence microscopic image of mesenchymal stem cells treated for 24 hours with citrate capped gold nanoparticles; cells were labelled with Alexa fluor-488 phalloidin for F-actin (green) and the nucleus was counterstained with DAPI (blue); co-localization of gold nanoparticles was observed in cells as red dots [see Colour Plate].

One of the major limitations of this method is the inability of visible light to pass properly through the thick sample. In this regard, the distances travelled by the probe should be small (micrometres in length) and the samples need to have a thickness from 1 to 20 microns, to be viewed with a CLSM. Therefore, cell morphological changes on thick metallic/ceramic/non-transparent samples can not be investigated using a CLSM.

8.5.3.2 | Live cell confocal imaging

Confocal microscopy has been used effectively for the 3D study of structural dynamics in living cells. However, the imaging of biological specimens adds the challenge of maintaining the life and normal function of the organism. Live-cell imaging is particularly useful to monitor dynamic events associated with morphological changes. For the purpose of live-cell imaging, a standard inverted tissue culture microscope with the capability of epi-fluorescence illumination is used and such a system is coupled to a CCD camera along with appropriate fluorescence filters. In order to ensure that the cells are still viable during extended timeframe for live cell imaging, one has to have a close control over a number of variables including temperature, atmosphere, humidity, osmolarity, and pH. It is important to note that live-cell imaging using a CLSM offers several advantages over conventional wide-field fluorescence microscopy. Such advantages include, controllable depth of field, the elimination of image degrading out-of-focus information, and the ability to collect serial optical sections from thick specimens, etc. Nevertheless, there is also a larger risk factor from intense laser irradiation, which leads to phototoxicity in live cells. The main phototoxic effects result from fluoro-photobleaching, which not only decreases the available fluorescence signal with each exposure, but also generates free radicals and other highly reactive breakdown products[625]. Thus, the major challenge for live-cell imaging using a CLSM is to generate a sufficient contrast, particularly when the cells are imaged for prolonged periods[626].

8.5.3.3 | Guidelines for confocal microscopy analysis

The major issue with the case of a CLSM is the photobleaching and phototoxic effects to cells (invasiveness). These adverse effects are due to high laser intensities in the focal plane and the presence of out-of-focus excitation. Especially with live-cell microscopy, there must be a compromise between acquiring scientifically useful images and collecting data that provide a high enough signal-to-noise ratio (S/N) to make meaningful quantitative measurements of a living specimen.

In order to limit light exposure and photobleaching, some of the design considerations that have to be followed are:

(a) Optimizing the wavelengths of excitation and emission filters to match the fluorophor used to limit unnecessary light exposure and also to optimize the detection of fluorophor emission. It is also important to reduce the background light by turning off the room lights, and to minimize scattered light in wide-field imaging by closing the field diaphragm as much as possible.

(b) Use of fast, motorized shutters to turn off the excitation light, when not needed to take an image.

(c) Limiting photodamage to the specimen, by using the lowest possible magnification as determined by the experimental question.

(d) Using cooled scientific grade cameras with the lowest readout noise available, to detect dim fluorescent signals

8.5.3.4 | Quantitative aspects of fluorescence/confocal microscopy analysis

The quantification of cell density and morphology provide scientifically meaningful information after long-term experiments on biomaterial scaffolds. The quantitative results are critically analyzed to gain an insight into the fate of cultured cells. If may be that any change in cell morphological traits or proliferative characteristics directly correlates with its dynamic interaction at the biomaterial interface. In fluorescence/confocal microscopy, the intensity value of a pixel is a measure of the number of fluorophores present at a cellular region. Therefore, fluorescence microscopy images essentially provide two types of information:

(a) spatial (to calculate such properties as a function of distances, areas, and velocities)
(b) intensity (to determine the local concentration of fluorophores), directly correlating with the expression level of specific proteins or molecules of interest.[627]

Cell density is simply given by the number of adhered cells per unit area. For this, a large number of images each with identical magnification are to be used and for statistical reasons, at least 200–300 cells are to be counted to report any reliable value of cell density. Also, in the case of images with auto-fluorescence effect, the DAPI stained images with a clearly visible nucleus can be used to count the number of viable cells.

Cell shape index: The shape index is a measurement of how circular or linear a cell is, and is given by,

$$\mathrm{CSI} = \frac{4\pi A}{P^2} \tag{8.11}$$

where, A and P are the area and perimeter, respectively, of a cell. Usually, the CSI assumes values between one (circular shape) and zero (elongated, linear morphology). As seen in Fig. 8.22, the larger the value of the shape index, the more circular a cell is. A similar quantification can be performed to calculate the **nuclear shape index** (NSI) from the nuclear projected area and the nuclear perimeter.

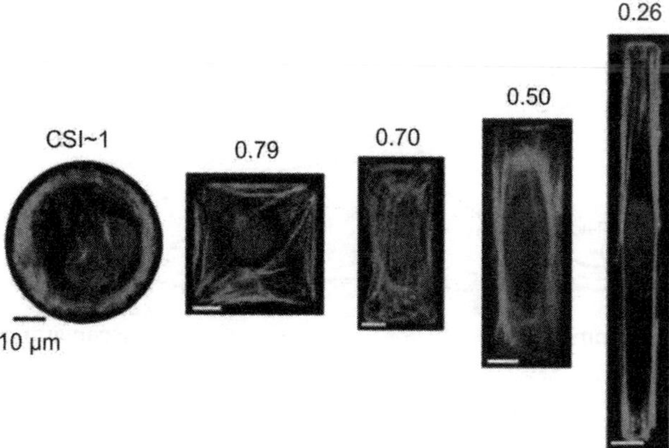

Fig. 8.22 CLSM image of cells, showing CSI ranging from 1 in circular-shaped cells to 0.26 in highly elongated cells; the nucleus is stained in blue with DAPI and actin is labelled with phalloidon Alexa 488 after 24 hours in culture[686] [see Colour Plate].

In similar line **Cell elongation factor** (CEF) is defined as the ratio of the major to minor axes, which are, respectively, the longest and the shortest length of an elongated cell. For cell elongation, borders are manually drawn around cells before segmentation to extract the cell's major and minor axes, using image software tools.

Cellular orientation: The quantification of cellular alignment and organization is significant in assessing the organization of cells on engineered tissues and biomaterials. Cellular alignment evaluation generally involves one or a combination of the following processes:

(1) identification of cell outlines and determination of the relative orientations of the cells,

(2) identification of elongation and cumulative orientation of the cellular nuclei, and

(3) identification and assessment of orientation of cytoskeletal components[628].

The orientation angle of the cell can be determined by measuring the acute angle extending the major axis of the cell with respect to the extracellular matrix direction, underlying topographical patterns, mechanical stimulation, electrical stimulation, bio-printing, etc. The longest aspect ratio of the cell can be traced by connecting two coordinates on the major axis and another point to find an acute angle. The measured polarization angle distributions ranging between –90° and 90°, while positive and negative signs are assumed to be in the counter clockwise and clockwise directions, respectively. An angle of 0° can be considered for the cells to be perfectly aligned parallel to the scaffold. The angle of orientation should also be measured with respect to an arbitrary direction (same baseline for consistent measurements).

From the preceding discussion, it is apparent that cell fate processes are largely dependent on substrate properties. In the following, the results from some published papers are reviewed to substantiate this point.

8.6 | **Illustrative Results of Cell Fate Processes**

Having described the cell fate and the cell morphological aspects, both qualitatively and quantitatively, this section discusses some studies to illustrate how various substrate properties can influence the fate processes with stem cell as a model. As discussed in an earlier chapter, stem cells have the unique ability to differentiate to various cell types.

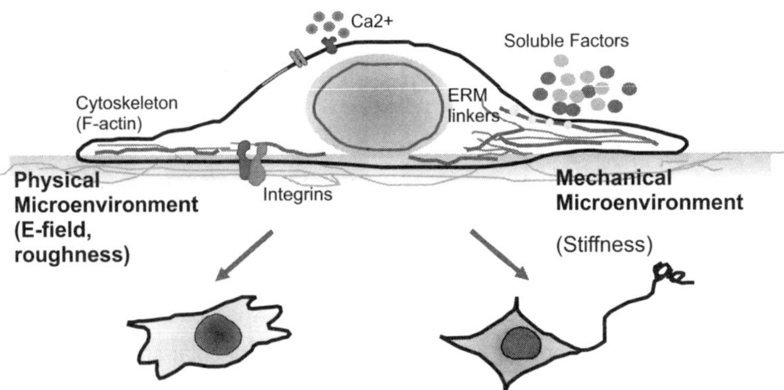

Fig. 8.23 Physicochemical regulation of matrix to direct stem cell differentiation for tissue engineering applications.

The fate of stem cells is strongly determined by the microenvironment in which they grow. The spatial and temporal cues produced by the microenvironment direct the cells to undergo self-renewal or differentiation (Fig. 8.23). With the rapid advancements in medical technology, it is possible to engineer biomaterials in such a way as to mimic the tissue microenvironment physically, chemically and biologically, thereby providing signals to direct stem cell growth and differentiation[629].

8.6.1 | Effect of matrix stiffness on stem cell behaviour

It has also been proved over the past few years that stem cells respond better to physical stimuli, *in vitro*. Physical stimuli are considered to be more beneficial than chemical or biological stimuli due to their non-toxic nature, thereby inducing less harm to neighbouring tissues. The intrinsic mechanical properties of a biomaterial at the nano/micro scale regime influence the forces exerted by the cells on the substrate (Fig. 8.24).

In the human body, the ECM of tissues exhibits a series of diverse stiffness, ranging from soft matrices in the brain, stiffer matrices in muscles, to comparatively rigid matrices that mimic bone[630]. The elastic modulus of softer brain tissue is in the tenths of a kilopascal (kPa), whereas for rigid bone, it is in the order of megapascals (MPa) to gigapascals (GPa) showing at least three to six orders of magnitude difference in the stiffness values. Therefore, the cells, which reside in such a matrix, are also tuned to the specific mechanical environment that dictates their morphological and functional characteristics.

In a pioneering study by Engler et al.[630], it was demonstrated that matrix elasticity directs MSCs to morphological changes and commitment towards a specific cell lineage. In this study, they illustrated that naive MSC grown on matrices with elastic moduli typically in the range of ~E_{brain} (0.1–1 kPa), ~E_{muscle} (8–17 kPa), ~$E_{bone\ matrix}$ (25–40 kPa), exhibited branched, spindle or polygonal shapes, respectively. Using a well-defined elastically tuneable gel system, they were able to provide evidences of MSC differentiation into osteoblast, skeletal muscle and neural lineages in a stiffness dependent manner. Furthermore, they also showed that cytoskeletal motors, NMM IIA–C (nonmuscle myosin II isoforms) are involved in sensing matrix stiffness, henceforth guiding specific lineage commitment. At the molecular scale, matrix sensing requires the tensioning of actinomyosin structures, that pull against the matrix, linked to focal adhesions that provide a pathway of force transmission from inside the cell to the elastic matrix with an interplay of various signalling molecules, that act as mechano-transducers[630].

Analogous to this work, Rowlands et al. designed gel substrates coated with ECM proteins, with varying stiffness from 0.7 to 80 kPa[631]. Osteogenic differentiation occurred significantly only on collagen I and fibronectin-coated gels with the highest modulus (80 kPa), but not on collagen IV and laminin coated substrates. In contrast, myogenic differentiation occurred on all gel-protein combinations that had stiffness greater than 9 kPa, but to varying extents. The maximum myogenic marker expression (MyoD1) was observed on gels with a modulus of 25 kPa coated in fibronectin, with similar levels of expression observed on 80-kPa collagen I-coated gels. Such results imply that substrate stiffness alone does not direct stem cell lineage specifications, suggesting the need of physiologically relevant elastic moduli in combination with relevant ECM molecules for directing stem cells to differentiate.

Similar studies about regulating MSC differentiation by ECM stiffness were carried out by Park et al. In this work, they investigated how ECM stiffness modulates MSC differentiation into other cell types, such as smooth muscle, adipocytes and chondrocytes in response to a growth factor, TGF-β, indicating the need for specific growth factors along with microenvironment cues to define a particular lineage commitment[632]. A stiff matrix promoted MSC differentiation into MSC lineage, while a soft matrix (~1 kPa) promoted MSC differentiation into chondrogenic and adipogenic lineage. Recently, Wang et al. fabricated polydimethylsiloxane (PDMS) with a stiffness gradient of 190 kPa–3.1 MPa to screen osteogenic differentiation in rat mesenchymal stem cells[633]. The substrate stiffness showed a noticeable effect on the cell morphology, and colony formation of rMSCs, while extracellular matrix proteins reabsorbed on PDMS influenced the attachment, proliferation and osteogenesis of rMSCs.

Likewise, Pek et al.[634] carried out a similar set of experiments on a 3D scaffold made of thixotropic polyethylene glycol-silica (PEG-silica) nanocomposite gel. The results obtained were in agreement with the earlier reported data on the effect of matrix stiffness on stem cell differentiation in 2D gels. The highest expressions of neural (ENO2), myogenic (MYOG) and osteogenic (Runx2, Osteocalcin) transcription factors were obtained for gels with stiffness of 7, 25 and 75 Pa, respectively. The immobilization of the cell-adhesion peptide, RGD, promoted both proliferation and differentiation of MSCs, especially in the case of the stiffer gels (>75 Pa)[634].

Developing substrates with gradient stiffness serves as an effective tool for studying the influence of substrate stiffness on MSCs culture and differentiation[635]. Wang et al. fabricated stiffness gradients on PDMS with an indentation modulus of 190 kPa–3.1 MPa, and screened for rMSC proliferation and osteogenesis (in the presence of differentiation media). The substrate stiffness showed a conspicuous effect on the cell morphology, colony formation and osteogenesis of rMSCs, whilst ECM proteins influence the attachment, proliferation and osteogenesis of rMSCs[635].

Cells are thought to migrate in response to soluble signals present under physiological conditions. In a recently published paper by Engler et al., it was described that undifferentiated MSCs tend to migrate towards greater stiffness regions in the absence of any external stimuli[636]. Also, from the lineage marker assessment, it was stated that most of the migrated MSCs expressed myogenic markers along with a few cells having neuronal expressions due to some degree of 'memory' of the previously soft environment from which they migrated. Thus, stiffness variation is also considered as an important parameter regulating stem cell fate.

Amongst the numerous natural biomaterials investigated, silk-based scaffolds with higher elastic moduli are known to support the growth and differentiation of MSCs efficiently[637]. MSCs seeded on porous three dimensional silk scaffolds generated zonal architecture, similar to native cartilage tissue, suggesting their application in cartilage repair[638]. Another group developed a unique combination of silk and elastin biomaterial to investigate the influence of elasticity and surface roughness on the myogenic and osteogenic differentiation of cells[639]. They elucidated that MSCs preferred rougher surfaces and tended to proliferate and differentiate on substrates having elastic modulus in the region of 1–30 MPa. But this report seems to be contrary to the existing literature results, which suggests that hMSCs typically sense stiffness in the range of 0.1–800 kPa.

Another important challenge in stem cell research is to maintain the stem cells in an undifferentiated state to get enough stock out of them, i.e., to maintain the stemness of a stem cell population. A study by Lee et al. demonstrated the use of porous PET membranes with optimized pore density (which dictates the hardness of the membrane) as an efficient culture substrate for the proliferation of hESCs[640]. Such PET membranes, even without any surface coatings or modifications allowed good cellular adhesion and the maintenance of uniform populations of hESCs that were free from spontaneous differentiation and apoptosis. This study clearly depicts that surface rigidity not only modulates stem cell differentiation, but also plays a key role in stem cell attachment, renewal and proliferation.

Cells actively probe the surrounding ECM, by pulling against the matrix. They sense the resistance of the matrix to deformation, that will sequentially activate the downstream mechano-transduction pathway (Fig. 8.24)[641]. Such types of biological mechanisms are elucidated as a feedback effect that works on the basis of inside-outside-in phenomena. Several studies have examined a number of different mechanisms involved in the mechanotransduction pathways that control stem cell differentiation. But as of now, it appears to be due to the synergistic role played by several processes, like actinomyosin contractility, force-sensing protein conformational changes and the recruitment of more integrins at the site of focal adhesions, changes in Rho activity and the regulation of mechanosensitive calcium channels[642].

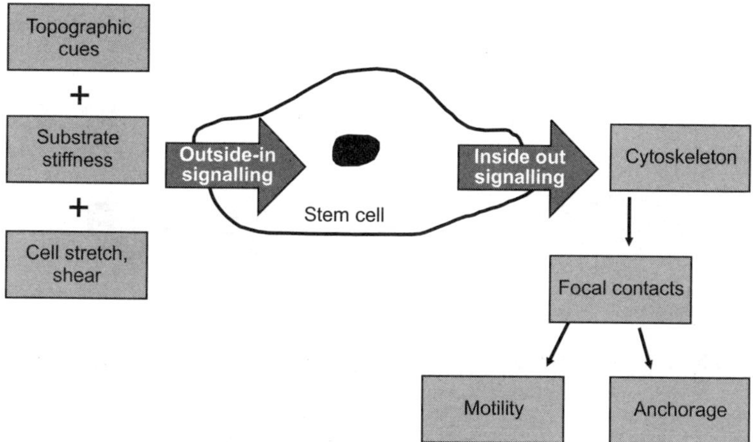

Fig. 8.24 Flow diagram representing the simplistic view of mechanisms involved in matrix stiffness associated cell signalling pathway [Adapted from Ref. 687].

8.6.2 | Effect of surface engineering on stem cell behaviour

Several results have come up in recent years showing the role of nanoscale features in eliciting biomechanical signals to cells, and leading to alteration in cellular behaviour (see Fig. 8.25). It has been found that nanoscale features help in the clustering of integrin molecules and the modulation of focal adhesion complexes (see Fig. 8.26). This kind of adaptation activates signal transduction pathways and directs cells to migrate, proliferate and differentiate distinctly on patterned matrices. A comparative study made using hMSCs cultured on nanopatterned and unpatterned PDMS and TCPS showed that hMSC matrix interaction is not only influenced by mechanical stiffness, but also by surface nanotopography[643]. Another advantage of nanotopography is its resemblance to the structure of ECM, which is in the nanoscale range. These nanoscale textures of the ECM provide cellular anchorage points and present instructive clues to guide cell behaviour[644]. Different types of nanotopographical modifications, like grooves, pores and pillars were investigated to examine the interaction of stem cells with the matrices[645]. For example, micro/nano fabricated silicon substrates coated with gold were evaluated to understand the effect of surface features on differentiation mechanisms. For this study, researchers used flat, micropillar and nanopillar features on silicon wafers. The measurements of cell viability along with immunofluorescence and lineage-specific-matrix staining depicted enhanced proliferation and differentiation of MSCs, with intense differentiation towards an osteogenic lineage. Further, they deduced that cell aggregates formed on hydrophobic nanopillars initiated osteogenic differentiation[645].

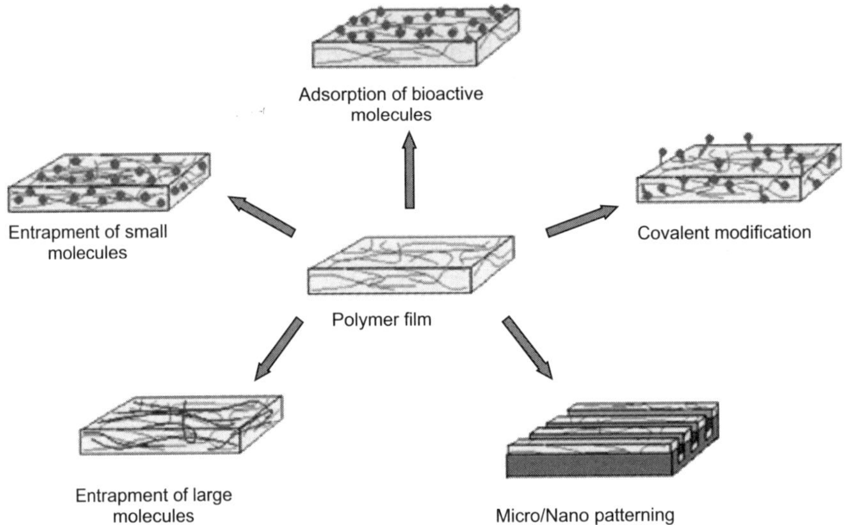

Fig. 8.25 Examples of modification strategies used with conducting polymers[651].

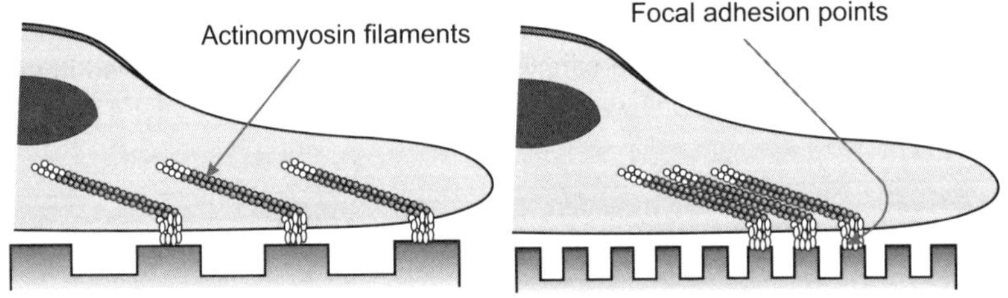

Fig. 8.26 Influence of nanotopography in stem cell response [Adapted from Ref. 645].

The surfaces of conducting polymers can be modified chemically and physically using a number of approaches as a means to change conductivity, bioactivity, and physical topography/geometry. A few surface modification approaches are reported, including non-covalent chemical methods (molecule entrapment and adsorption), and covalent chemical conjugation, and micropatterning using standard lithography techniques[651]. The surface energy of conducting polymers can be changed by changing the redox state of the polymer. The effect of the change in the redox state was studied on neural stem cells and on epithelial cells. It was shown that cell adhesion and cell seeding density were influenced by the change in the redox state[646].

It is also possible to maintain stem cells in their quiescent state and guide them to a specific lineage whenever it is required by altering the nanotopography. In another report, MSCs cultured on TiO_2 nanotubes of varying diameters of 30, 50, 70, and 100 nm on Ti substrates revealed a dramatic change in MSC adhesion and differentiation in a relatively narrow range of nanotube dimensions[647]. Furthermore, their results indicated that the hMSCs were extraordinarily more elongated and probably laid stress on the 100-nm nanotubes, which accelerated stem cell differentiation towards an osteogenic lineage. They correlated the effect of protein adsorption on cellular adhesion phenomena. On TiO_2 nanotubes having a diameter of <50 nm, the cells were able to attach more easily due to high protein adsorption, whereas in hMSCs cultured on 100 nm TiO_2 nanotubes, the cells had to extend to search for areas where more proteins are adsorbed. Such phenomenon results in it an elongated morphology and are forced to differentiate specifically into osteoblast cells to accommodate the stress. But, the behaviour of MSCs on large diameter (>100 nm) TiO_2 nanotubes due to the difficulty in fabrication of large diameter nanotubes remained unexplored. It has also been reported that cell shape regulates the switch in the lineage commitment of human MSCs (hMSC), by modulating endogenous RhoA activity. MSCs become osteogenic, if they stretch their filopodial and lamellopodial extensions to spread on the surface, whereas it becomes adipogenic, if they are unspread and rounded[648]. Moreover, manipulating surface topography not only supports protein adsorption, but also influences the shape of the cells that attach to the substrate. The mechanical stresses developed on the cytoskeleton affect gene transcription, with resulting changes in cell fate processes[649]. Therefore, nanopatterning on matrix surfaces acts as an efficient tool to understand the regulation of stem cell fate by the microenvironment.

Phillips et al. used self-assembled monolayers of ω-functionalized alkanethiols on gold to study the effect of surface chemistry on stem cell lineage commitment[650]. Their findings showed the importance of surface chemistry on the long-term functional differentiation of stem cells along osteogenic, adipogenic and chondrogenic lineages. They also highlighted that surface chemistry alters the conformation of adsorbed ECM proteins, which, in turn, direct specific integrin–ECM protein interactions that ultimately induces signal transduction pathways leading to specific lineage commitment.

8.6.3 | Substrate conductivity dependent stem cell fate

Specific tissues like the brain, heart and skeletal muscle in the body are electrically active in nature. These tissues make use of electrical conductivity for tissue homeostasis, development, regeneration, etc. Therefore, it has been envisaged that electrically conductive biomaterials can simulate the normal electrophysiology of the body due to their ability to interface with bioelectric fields in cells and tissues. In the following, the results of some relevant studies are briefly summarized to illustrate the modulation of stem cell response by tailoring substrate conductivity.

Conducting polymers (CP) like polypyrrole, polythiophene, poly-(3-methylthiophene), polyaniline, poly-(3,4 ethylene dioxythiophene) are reported to be cytocompatible[651]. The conductivities of such polymers have been found to vary with doping. The conductivities are found to range from 1 to 1000 S/cm, depending on the polymer and the type of dopant.

Electrical charges have been shown to enhance nerve regeneration. The electrical stimulation of the conducting polymer, polypyrrole, has also been shown to enhance the serum adsorption of proteins (especially fibronectin) on the surface of the polymer[652]. It has also been shown that electro-adsorbed poly lysine enables long term neuron survival[653].

In a review article, Ravichandran et al.[654] described the application of conducting polymers in tissue engineering applications. Dorsal root ganglion explants cultured on conducting polymeric nanotubes showed branched neuritic growth and extensions. hMSCs cultured on Polypyrrole-incorporated collagen-based fibres and stimulated with exogenous pulsed electrical signals for 5 and 10 days exhibited neuronal-like morphology on day 10, along with increased expression of neural markers, like MAP2, neurofilament, β tubulin III and nestin.[655] They also observed that electrical stimulation for extended duration adversely affects hMSC viability and this detrimental effect is significantly greater, when polypyrrole particles are dispersed in the fibre matrix to intensify the electrical signalling inside the cells. C2C12 skeletal myoblasts exhibited enhanced proliferation and differentiation, when grown on PANi-incorporated PLCL fibres[656]. The addition of PANi improved both the electrical properties as well as the differentiation capacity of the electrospun fibres. The discussion above therefore, establishes the effect of the substrate conductivity on mesenchymal stem cell growth, orientation and differentiation.

8.7 | Host Response

The host response is the response of the host tissue to the implanted device/biomaterial[657]. However, the appropriate host response can be defined as the beneficial cellular or tissue response towards an implant material.

All the vital organs of an animal model have their specific defence mechanism against foreign materials (e.g. synthetic biomaterials).

According to Ratner et al.[607], the placement procedure of any implant material initiates a response to the injury by the body, and mechanisms are activated to maintain homeostasis. The degree to which homeostatic mechanisms are perturbed and patho-physiologic conditions created and resolved are a measure of the host's reaction to the biomaterial and may ultimately determine its biocompatibility. Also, it must be noted that host reactions are dependent on tissues, organs and species. Later, Black et al. described the host response as the local and systemic response, other than the intended therapeutic response, of living systems to the implanted material. The appropriate host response can be considered as the resistance to blood clotting, normal healing and resistance to the growth of bacterial colonies, etc.[658].

When an implant material is bioinert, a fibrous capsule-like enclosure develops around the implant to keep it secluded from other components of the physiological system. For instance, inert bioceramics such as Al_2O_3 or ZrO_2, develop fibrous capsules at the material/tissue interface. It may be noted that the fibrous capsule layer thickness is governed by the degree of fit at the implantation site. Taken together, a bioinert material is not the best choice for the purpose of long term tissue replacement application. In many orthopedic surgeries, bioinert implants made of stainless steel or Ti-alloys are used and after a few years, the patients may need to undergo revision surgery. As mentioned in Chapter 2 of this book, the most significant class of biomaterial is bioactive material, which can potentially form biological bond with the neighbouring host tissues and thereby behave as the part of a living body. A few examples of bioactive materials are 45S5 bioglass, calcium phosphates, etc. For bioactive materials, the interfacial bond essentially prevents any motion at the implant–tissue interfaces, which are also similar to the kind of tissue interfaces during natural healing and tissue repair[659]. The triennial category of biomaterial is bioresorable or biodegradable that degrade with time under the physiological environment of the body.

As far as the terminology is concerned, 'bioresorbable' is more used to describe the characteristics of bioceramics, which undergo controlled dissolution in simulated body fluid, *in vitro* or in an osseous system, *in vivo*. Among the ceramics, Tricalcium Phosphate (TCP) is a popular example of a bioresorbable ceramic. In parallel, 'biodegradability' refers to the ability of certain biopolymers to exhibit continuous structural weight loss, *in vitro* or post-implantation in an animal model. The bone cement, PMMA is an important example of a biodegradable polymer. Importantly, the rate profile of dissolution or degradation should be synchronized with tissue regeneration kinetics.

8.7.1 | Consequences of the host response to biomaterials

Injury (which is inevitable during an implantation process) disrupts the vasculature at the implantation site and triggers a cellular cascade that ultimately invokes an inflammatory response. This leads to an influx of inflammatory cells such as neutrophils, macrophages and foreign body giant cells (FBGCs), which attempt to get rid of the foreign material. Due to the size of the implant, these cells are unable to engulf it. They send out cellular signals, indicating that a large foreign body needs to be walled off. Fibroblasts arrive and ward off the implant by synthesizing a collagen capsule around an implant. This is the fibrous capsule that constitutes the foreign body response by the host[660]. In brief, all foreign materials are surrounded by collagen deposition in the form of a fibrotic scar at the material–tissue interface as shown in Fig. 8.27. Thus, the goal of biological response evaluation is to predict whether a biomaterial, medical device, or prosthesis presents potential harm to the patient or user by evaluating conditions that simulate clinical use[661].

The foreign body response, which results in implant encapsulation, is highly detrimental to optimum implant function. This is because of the insulating nature of the fibrous capsule, which serves to block mass transport and electrical communication between the implant and the body[662]. This is the fate of all materials that are implanted in the body. Such fibrotic reactions are undesirable for next generation biomedical applications such as electrical sensors, drug delivery devices and tissue engineering scaffolds. In fact, fibrotic reactions are devastating as far as tissue integration with the host and nutrient supply are considered. The major challenge is to create a material that should not be encapsulated by the fibrous tissue, *in vivo*.

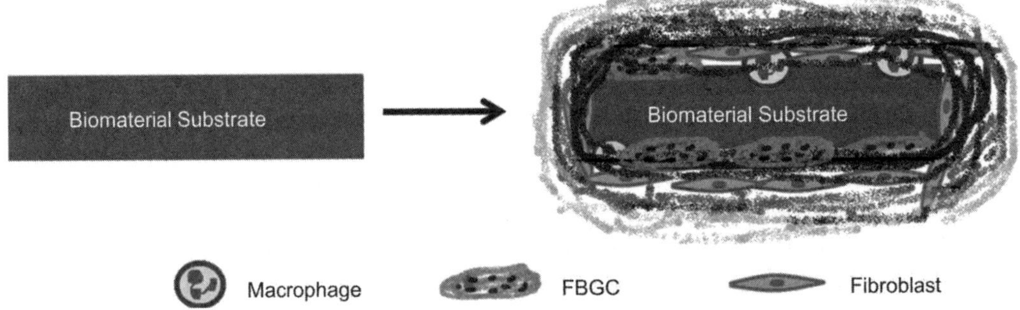

Fig. 8.27   A schematic depiction of the collagen deposition encapsulating an implant/biomaterial after implantation; also seen are the inflammatory cells at the material/tissue interface (not to scale).

To this end, it has been emphasized the need to create super-biocompatible materials should be emphasized. With the advent of tissue engineering, there is a demand for next-generation materials which can fulfil demanding requirements such as biocompatibility, degradability and tuneability for such many applications.

Although a large number of biomaterials and medical devices have been implanted successfully in humans, it has not been possible to create a stealth material that can surpass the highly efficient human surveillance system that is almost always flawless in distinguishing self from foreign. This is triggered by protein adsorption on the material (biofouling). The end result is that the implants are surrounded by a dense collagen capsule.

The steps that lead to the formation of the fibrous capsule are discussed here. The host response to an implant is evaluated by morphological and histological observations of the implant site and relates to the biocompatibility of the material.

> The different stages of host response include granular tissue development, blood-material interactions, provisional matrix formation, acute inflammation, chronic inflammation and fibrous capsule development.

Each of these events is very complicated and involves a complex interplay of inflammatory cells, mitogens, chemoattractants, cytokines and other bioactive agents.

(a) Injury to the vascularized connective tissue activates the initial inflammatory response. This is regardless of the tissue in which the biomaterial is implanted. In fact, inflammation is the reaction of vascularized, living tissue to local injury. The main purpose of inflammation is to contain or dilute or wall off the injurious agent. At the same time, inflammation also initiates a cascade of events that may heal the injured tissue[663]. The injury caused during implantation leads to changes in vascular flow and permeability, leading to a leakage of blood, fluid and proteins at the site of injury through a process known as exudation. A cascade of biopathological stages following injury to the development of fibrosis is shown schematically in Fig. 8.28.

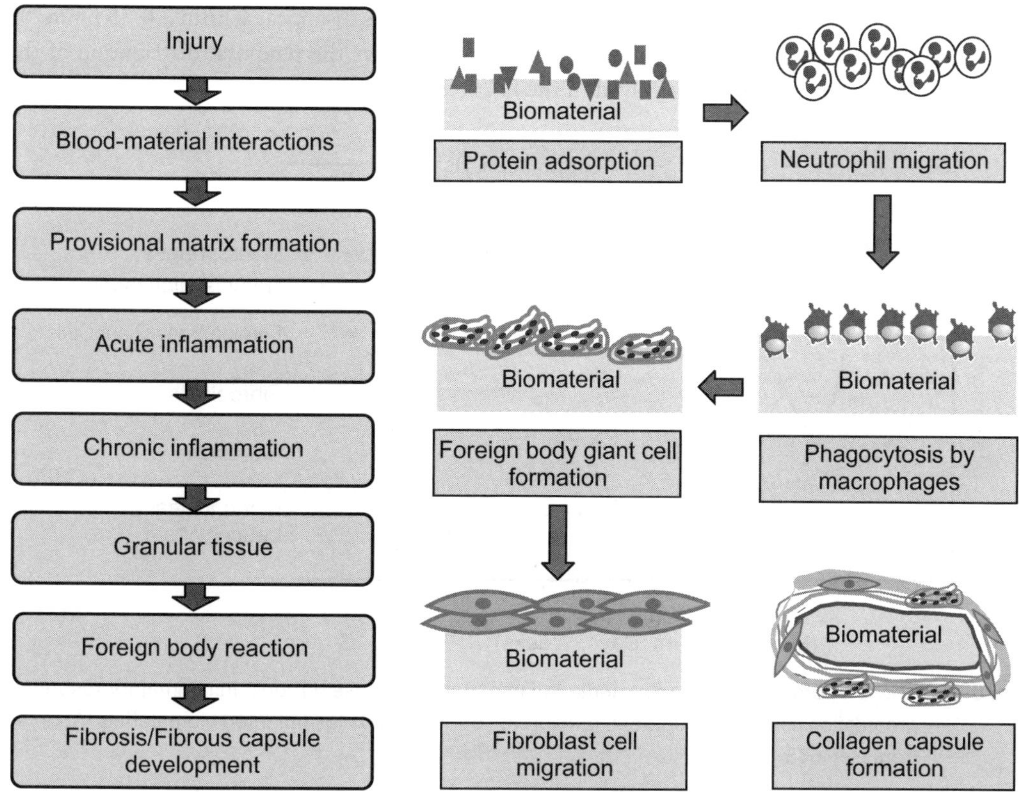

Fig. 8.28 A flowchart showing the sequence of host reactions following the implantation of a biomaterial/ medical device, along with a schematic representation of some of the stages involved in acute inflammation and chronic inflammation that ultimately lead to the formation of the fibrous capsule around the implant [see Colour Plate].

(b) The injury to the connective tissue leads to the development of a blood-based transient provisional matrix around the biomaterial[664]. Blood protein deposition, which occurs after injury at the biomaterial's surface, may be regarded as a provisional matrix from the wound healing perspective[665]. The protein cascades involve a dynamic displacement of initially adsorbed proteins by more surface active proteins, i.e., proteins with higher affinity. This is known as the Vroman effect[666].

Although the inflammatory response is initiated by injury, the response is mediated by chemicals released by plasma, cells and the injured tissue. These include cytokines, growth factors, vasoactive agents, lysosomal proteases, oxygen free radicals, etc. It is important to note that these mediators are controlled by feedback mechanisms, and that their action is predominantly local. The lysosomal proteases and oxygen-free radicals play a role in the degradation of a biomaterial[665]. An acute inflammatory response is characterized by the presence of neutrophils (polymorphonuclear leukocytes). Depending on the extent of injury at the implant site, the acute inflammation to biomaterials subsides quickly (within a week), as the neutrophils disappear within 24–48 hours[665]. The neutrophils are the first line of defence. Figure 8.29 shows the temporal distribution of the different cells and stages that are involved in the host response.

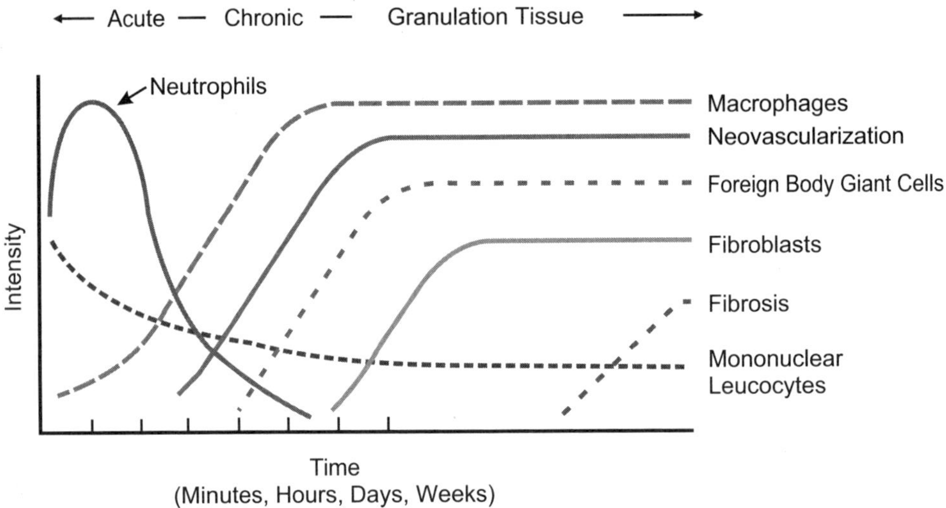

Fig. 8.29 The temporal variation in the acute inflammatory response, chronic inflammatory response, granulation tissue development, and foreign body reaction to implanted biomaterials [Adapted from Ref. 665].

As part of the host response, the acute inflammation is followed by chronic inflammation. The latter is marked by the abundance of mononuclear cells (monocytes and lymphocytes) in the area surrounding the implant. It differs from acute inflammation in a wider variety of cellular responses and non-uniformity in histology[667]. Monocytes migrate to the wound site and differentiate into macrophages. They constitute the second line of defence of the body[668]. The macrophages release mediators of inflammation such as oxygen-free radicals and reactive oxygen intermediates, degradative enzymes and acids on the surface of the biomaterials[665]. They also adhere to the material and attempt to phagocytose it. They are unable to digest or phagocytose a large mass such as an implant even after a final desperate attempt, which is known as 'frustrated phagocytosis'. Macrophages frequently undergo fusion to improve their efficiency and form multi-nucleated giant cells known as foreign body giant cells (FBGCs). It is believed that these FBGCs send out signals for fibroblasts to arrive[669].

The development of granulation tissue is also a part of chronic inflammation[665]. The arrival of fibroblasts and vascular endothelial cells at the implant site is a sign of healing inflammation. Histologically, granulation tissue is characterized by the presence of fibroblasts and new blood vessels. Small new blood vessels are formed from the branching of pre-existing vessels by a process known as neo-vascularization. Fibroblasts are active in synthesizing collagen. Granulation tissue is named due to its characteristic pink, granular appearance on the surface of wounds after 3–5 days, depending on the extent of injury[663,664,665].

The foreign body reaction comprises foreign body giant cells and granulation tissue along with the presence of inflammatory cells, such as macrophages and fibroblasts. Due to the collagen deposition by the fibroblasts, the implant is effectively isolated from the rest of the body. This condition, in which the fibrous capsule encapsulates the implant, along with the interfacial foreign body reaction is known as fibrosis[665]. Traditionally, an implant is considered to be biocompatible if it is encapsulated by an avascular layer of collagen[670]. However, as will be demonstrated later, this is not desirable as far as sensors, tissue engineering scaffolds and controlled release devices are considered[669]. A recent hypothesis also does not consider this a hallmark of biocompatibility[669].

The different stages of the host response to the implantation of a foreign body are schematically depicted in Fig. 8.30. The intensity and time scale of the response are dependent upon the extent of the injury created and on the size, shape, topography, and chemical and physical properties of the implant.

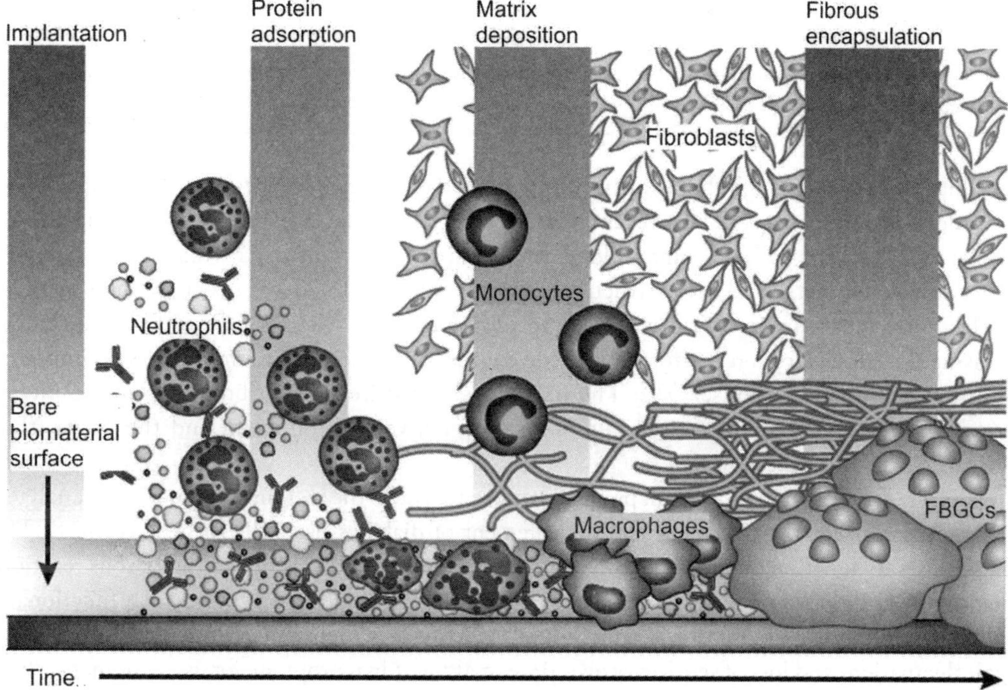

Fig. 8.30 A schematic representation of the host response upon implantation of a biomaterial (reproduced from Ref. 688] [see Colour Plate].

8.7.1.1 | Foreign body response

Apart from biocompatibility and biodegradability, an important requirement is that a scaffold should not generate a foreign body response. Hence, it is necessary to design strategies that can mitigate this response. However, it may be worthwhile to cast a quick glance at the purpose of the foreign body response that addresses the moot question- why has the foreign body response evolved?

8.7.2 | Consequences of the foreign body response

First and foremost, it is important to understand that immune and inflammatory reactions occur in order to protect the osseous system from foreign objects[671]. In a world full of pathogenic bacteria and microbes, this foreign body response has evolved over millions of years to protect the body against such disease-causing and unknown substances. The importance of this impeccable ability of the human/mammalian body to distinguish itself from foreign bodies has allowed humankind to survive and evolve over the years. In some cases, it is also advantageous to patients. For instance, lead poisoning was very rarely observed in patients, who had retained lead bullet fragments. The development of an impermeable collagen capsule around the bullet is thought to be the likely reason behind this[672].

Despite these benefits of the foreign body response, it has disastrous consequences as far as biomaterials and implants are concerned.

> The body's immune system recognizes an implant as foreign. Implants that do no elicit an adverse immune response are desirable.

This is achieved through the set of events mentioned earlier that leads to implant encapsulation by a dense collagen capsule. In recent years, there has been an explosion of bioengineered implantable devices such as joints, blood vessel substitutes, sensors, hernia mesh materials, heart valves, cosmetic and reconstructive implants, and artificial organs. For some of these applications, the formation of the collagen capsule is a liability. Also, in spite of the advancement in tissue-based sensors and scaffolds for tissue engineering, their optimum function is impaired by the foreign body response[673,674]. The impermeable nature of the fibrous capsule results in poor mass transport and electrical communication between the implant and the tissue, thus impairing implant function.

For instance, although significant progress has been made in the real time monitoring of analytes, a glucose sensor for the treatment and management of diabetes has still not been realized. The sensor loses its functionality in some time, and it is evident that this is due to the fibrous reaction, that occurs at the tissue–implant interface[675]. Currently, the longest *in vivo* time of a functionally active, FDA approved biosensor is only 7 days[671]. Also, it is highly inaccurate and suffers from high fluctuations and loss of sensitivity over time, making it inappropriate for use in humans[676,677]. The foreign body reaction is also the main reason why no tissue engineering product based on the encapsulation of cells by a polymer has seen the light of day. In the *in vivo* environment, these cell loaded constructs are encapsulated by the fibrous capsule[678]. Apart from this, the fibrous capsule can also cause tissue distortion and pain[673,674,679].

At this juncture, it would be of interest to know how a diffuse pattern of collagen deposition can aid in a better communication and transport between the implant and the body. In an elegant series of experiments performed by Sharkawy and his colleagues, the diffusion properties of the fibrous capsule have been investigated. According to their diffusion model predictions, a 200- µm thick capsule increases the time required for the diffusion of a small molecule such as glucose threefold. Hence, they conclude that the fibrous capsule poses a serious barrier to diffusion even for small molecules[680]. For larger molecules, this barrier will be even more pronounced. Researchers such as Woodward et al.[681] have hinted that a thick, but granular and well vascularized capsule can permit diffusion to a greater extent. Such a capsule is preferred over a thin, avascular and tightly packed capsule.

8.7.3 | Strategies to overcome the foreign body response

To develop effective strategies to reduce the foreign body response, Ward has advocated the study of the foreign body response in detail to gain an in-depth understanding[675]. On similar lines, Ratner has compared normal wound healing with wound healing in the presence of a biomaterial[669]. He has attempted to find an explanation to the question: Why do so-called 'biocompatible' materials shut off normal healing? He observes that while the human body has an excellent capacity for healing, wounds with implants do not heal like normal wounds. In this seminal shift in the paradigm of biocompatibility, he has critically assessed the differences in protein adsorption on the biomaterial, which is one of the initial stages in the foreign body response.

Wound healing was assessed by Ratner and his co-workers in an experiment where mammals were implanted with implants made of materials commonly used in medicine such as Teflon, polyurethane, silicone rubber, polyethylene, polymethyl methacrylate (PMMA), PHEMA, Dacron, gold, titanium and alumina[669]. All these implants healed in a similar manner after 1 month.

However, these materials showed differences in protein adsorption and hence, different cell growth behaviour *in vitro*. The common aspect in all the materials *in vivo* was the adsorption of a complex protein mixture (>200 proteins) in different orientations and conformations. Ratner has hypothesized that this is the key cause of the foreign body response. In nature, such non-specific protein adsorption is never observed. Hence, he proposed the reduction/elimination of this non-specific protein adsorption on the biomaterial as one of the main strategies to reduce the foreign body response. This approach has been schematically depicted in Fig. 8.31.

Different methods have been adopted to reduce non-specific protein adsorption on biomaterial surfaces. The most common is the use of surface coatings. Biocompatible coatings or surface modifications are used to mask the underlying implant surfaces and to improve the surface properties of implants. Generally, a coating that creates a hydrophilic surface is used. Zwitterions and hydrophilic polyethylene glycol are two major classes of non-fouling materials that are based on the total reduction of protein adsorption[682].

8.8 | Closure

The discussion in this chapter provides an important foundation to understand and to realize how the functionality of biomaterials is characteristically different from other material classes. The concepts of biocompatibility and host response need to be greatly emphasized, while designing any new material for biomedical applications. Overall, the discussion in this chapter provides a necessary theoretical framework to perceive biological aspects in the context of biomaterials science. Some contemporary discussion on the subject of this chapter can be found elsewhere[683].

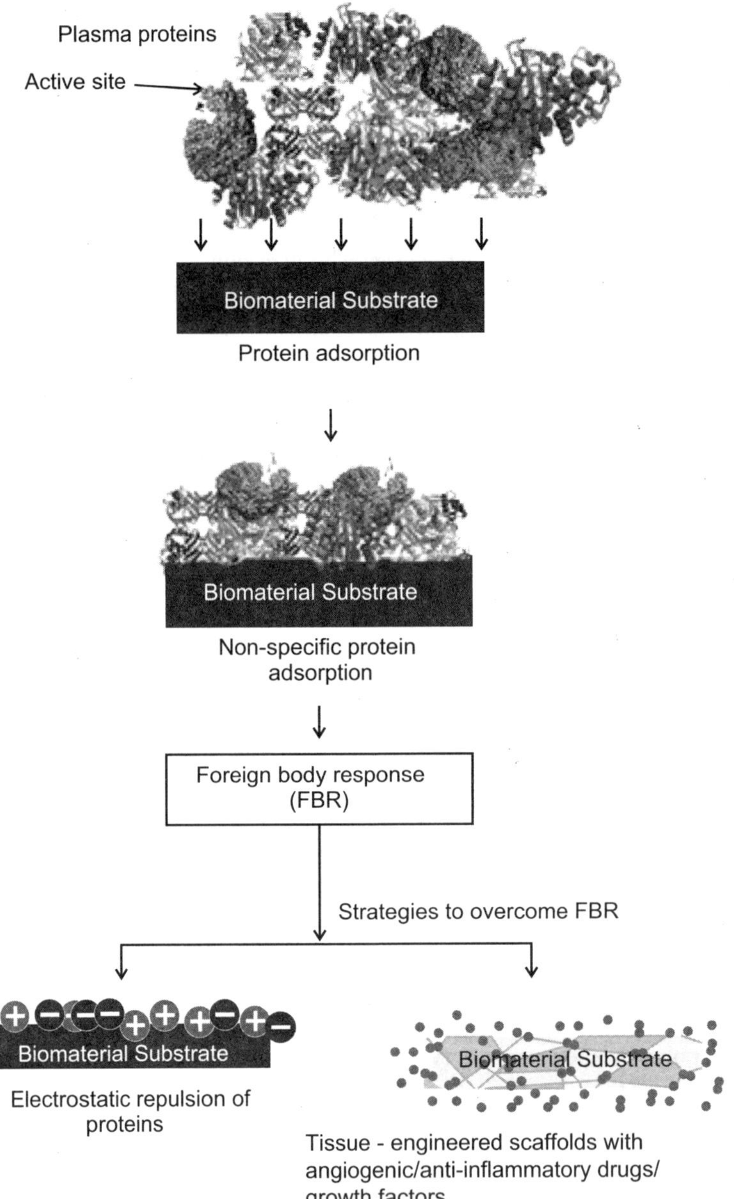

Fig. 8.31 A schematic showing the hypothesis that non-specific protein adsorption causes a foreign body response; the conformation and orientation of proteins are also distorted during non-specific protein adsorption; the foreign body response can be overcome by preventing protein adsorption or by modifying the biomaterial to release bioactive agents, such as anti-inflammatory drugs/ angiogenic drugs/Vascular Endothelial Growth Factor (VEGF) [see Colour Plate].

Probing Cell Response, *in vitro*

In an interdisciplinary research area like biomaterials science, researchers need to fuse together the techniques or expertise from multiple disciplines. In chapter 5 of this book, a multitude of characterization techniques were discussed in reference to their ability to unravel the structure and properties of materials. Some of those techniques can also be used to observe structure and characteristics of biological cells and tissues at multiple length scales. The central theme in biomaterials science is to probe cell response, *in vitro* and tissue response, *in vivo*. In this chapter, the fundamental principle and standard protocols for a number of *in vitro* assays for biocompatibility assessment are discussed in sufficient details for the benefit of non-biologists. In particular, the biological assays to quantify cell viability, toxicity and differentiation are described. In addition to colorimetric assays, the flow cytometry technique is also introduced in reference to its unique ability to quantify cell fate processes in a rapid manner. Additionally, the maintenance of cell culture facility or the requirements of specific physiological environment to culture a cell line are discussed. The ethical guidelines related to stem cell culture are also mentioned. Overall, this chapter will demonstrate the kind of expertise, knowledge, ethics and laboratory set up required to conduct *in vitro* experiments with biomaterials and how these are widely connected to the core theme of materials science. In a subsequent chapter of this book, various *in vivo* biocompatibility assessment protocols are described.

9.1 | Introduction

In a research program on biomaterials, once a material is synthesized for lab-scale experiments, they are characterized for various physical properties, like porosity distribution, elastic stiffness, strength or fracture toughness. The test samples exhibiting the property combination in a desired window, commensurate with the targeted biomedical application, are subsequently selected for *in vitro* cytocompatibility experiments. The scientific analysis of various *in vitro* assays, as described in this chapter, together provide a platform to probe the cell response, both qualitatively and quantitatively. The best material composition exhibiting the desired *in vitro* response is further taken for biocompatibility assessment, *in vivo*.

in vitro assays (Latin: in glass), also known as 'test tube experiments', are the type of scientific experiments performed with cells or biological molecules in physiologically simulated environment, but outside the normal biological context.

The above-described logical flow of experimental research on biomaterials is described in Fig. 9.1. It is worthwhile to explain the implication of cell response on the biomaterial in simple terms. In biological science discipline, the cells are normally cultured with growth factors or other biochemical cues in culture flasks. In the context of biomaterials science, cells are seeded on 2D material surface, which is placed in a culture well plate containing growth medium. Thus, the surface composition and properties of biomaterials are likely to influence the way a cell population would grow or function. In case of tissue engineering, the cells are seeded onto a scaffold and therefore scaffold structural properties (elastic stiffness) together with 3D microenvironment within a scaffold would influence cell response. In the last two cases, one can additionally employ some of the bioengineering approaches, like exposing cells to external biophysical stimulation (shear stress in dynamic culture, electric/magnetic/laser pulses, etc.). Irrespective of all these approaches, the cellular response is to be semi-quantitatively or quantitatively analysed using a biological assay (bioassay). Qualitatively, the cell response is to be studied using morphological analysis through fluorescence (confocal) microscopy, as briefly discussed earlier in this book. Although such morphological analysis often indicates as possible change in functionality or viable state of a cell, further quantification is always recommended for specific investigation in biomaterials.

As the field is growing, a more consensus thought currently is to probe the cell functionality change on a biomaterial surface rather than simply quantifying percentage of viable cells on a biomaterial, as was the case during the early years of growth of the field of biomaterials science. To substantiate further, the first generation researchers in the field of biomaterials were very enthused to show SEM images of cells on a material together with MTT data confirming cell viability. This approach is not appreciated in the global community anymore. First of all, one has to show fluorescence/confocal microscopy images to reveal structural details of morphological changes of cytoskeleton or cellular organelles. Secondly, one needs to use complementary assays, e.g., MTT assay to quantify metabolically active cells and LDH assay to quantify dead cells. Thirdly, the researchers are more interested to assess whether a given cell population changes its functionality over longer time in culture. For example, mesenchymal stem cells can differentiate along specific lineages, like osteogenic (bone), neurogenic (nerve), myogenic (muscle) or cardiomyogenic (cardiac). In order to confirm functionality changes, different differentiation assays involving up/down regulation of specific marker proteins at different timepoint in culture are to be conducted. The above discussion clearly emphasizes the critical role played by bioassays. It is worthwhile to mention how a biological assay is different from various experiments conducted to determine material properties in the field of Materials Science. For example, the strength of a metal is determined using a well calibrated universal testing machine in either tension or compression mode with a defined test parameters. The results of such experiments usually provide the property (strength) value for that specific material. In contrast, bioassay or a biological assay is a type of scientific assessment, which involves the use of live animal or plant (*in vivo*) or tissue/cell (*in vitro*) to determine the biological activity of a substance, such as a biomaterial or drug w.r.t. 'control' activity.

In contrast to testing of material properties, all biological assays require a 'control' of known biological activity.

Therefore, *in vivo* (Latin phrase meaning 'within the living') studies are those, wherein the effects of various biological entities are tested on whole living organisms (animals including humans, and plants). This is in contrast to *in vitro* ('within the glass') studies, which are conducted in a laboratory environment using test tubes, petri dishes, etc.

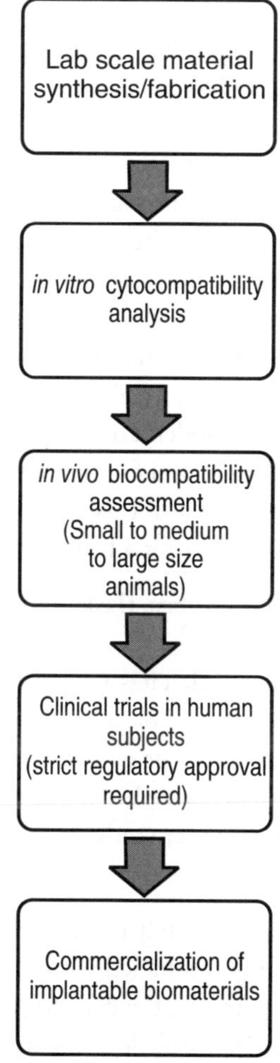

Fig. 9.1 Flowchart showing the rationale approach to be adopted to take labscale research towards commericalisation of biomaterials through biocompatibility assessment.

Therefore, the results of bioassays can be carefully analysed to measure the effects of a biomaterial on a living organism and to compare its effect on a test organism (control sample/ blank cell population or control animal). Such analysis is an important step in the development of new materials for biomedical applications. Quantitative bioassays are typically analysed using the methods of biostatistics and therefore, the involvement of biostatisticians is necessary in any pre-clinical study and clinical trial.

Depending on whether the assay is performed at a single time point or timed readings taken at endpoint or intermittently, the biological assays are either endpoint assays (fixed assay incubation period) or kinetic assays (readings at fixed time intervals).

> While developing a new biomaterial, the thoughtful and reliable assessment of biocompatibility properties should be given the highest priority.

It is well accepted that the *in vitro* assays must be performed as a first step prior to *in vivo* test to avoid the unnecessary use of animals. The *in vitro* cell culture assays typically make use of an organ specific cell line, [e.g., L929: mouse fibroblast (for general toxicity analysis); Saos-2: sarcoma osteogenic (for bone), hFOB: human foetal osteoblast (for toxicity analysis using human cells), N2a: mouse neuroblastoma (for neuronal cells)] or primary cells extracted from an animal tissue. The *in vitro* study is basically performed to collect the information about the cytotoxicity, bioactivity, genotoxicity, cell proliferation, cell apoptosis/necrosis, cell cycle and cell differentiation in response to exposure to the biomaterial under analysis. Although *in vitro* tests have been performed in laboratory under simulated body conditions, the absence of multiple proteins, enzyme, regular blood flow and other variables limit the realistic simulation of physiological condition, *in vivo*. In contrast, the *in vivo* tests are performed in appropriate animal model. For example, rabbits are used as an animal model for the study of osseointegration of a biomaterial for orthopedic application under ISO guidelines. This will be discussed in a subsequent chapter of this book.

Two criteria that must be satisfied in order for a test to qualify as an *in vitro* assay include:

(a) The test must be performed using test tubes, flasks and petri dishes or other sterile containers, rather than in a living organism.

(b) The test must be an assay, which means that it is a measure of the activity of a drug or implantable biomaterial on a sample of organic tissue or cells. The assay results are either quantified in measurable quantity with physical units or expressed in arbitrary units with respect to the results obtained with control/reference samples. In general, *in vitro* assays permit an enormous level of simplification of the system under study, as compared to extraordinary complexity of living organisms.

Similar to *in vitro*, a new term, *in cellulo* is lately introduced in the scientific literature. Currently, *in cellulo* is defined as any biological study conducted within cells, but outside the biological system (organism) or not in a glassware. For example, the cells together with proteins/growth factors can be printed along with a tissue engineering scaffold using a 3D bioprinter. One can further assess the role of physical cues towards the cell functionality modulation, when those printed cells are allowed to grow. Such study does not follow strict definition of '*in vitro*' or '*in vivo*' and hence called as '*in cellulo*' study.

The organization of this chapter is as follows. The next section will introduce various cell viability and toxicity assays. This will be followed by a discussion on differentiation assays together with PCR and flow cytometry technique for cell fate quantification. Subsequently, the experimental facilities to conduct cell culture experiments in reference to specific requirements to maintain desired physiological environment during cell culture are mentioned. The ethical guidelines related to stem cell culture are described before the closure section, which summarises the key points to be followed while probing cell response.

- The cellular functionality/activity in a culture medium, in isolation (without any substrate) is different from the cells grown on a biomaterial substrate/tissue-engineered scaffold.
- Cell morphological features after growing on a material substrate often provide the first signature on the cell-level compatibility or cellular functionality.
- Depending on cell type or substrate composition, a cell culture experiment is conducted for 24-72 hours to quantify cell growth/proliferation and 7–14 days for any change in cellular functionality (differentiation).
- The protocols for some of the biological assays are mentioned with specific details (type and amount of reagents to be added in stepwise manner) in this chapter. It is instructive to note that these details may alter depending on the experimental design or the biological questions to be addressed and even on the cell type and reagents in use.

9.2 | Assessment of Cytocompatibility

Analysing the effects on cell growth and/or cell death has been an important component of biological research.

The term 'cytotoxicity' refers to the ability of a biomaterial or a scaffold or a drug to cause toxicity to the animal cells.

The cytotoxicity can be quantitatively measured using a host of biochemical assays. Qualitatively, the cells may reveal the signatures to lose membrane integrity and die rapidly as a result of cell lysis (necrosis). Alternatively, a cell population, when growing on a biomaterial substrate, can stop actively dividing (a decrease in cell viability), or can activate a genetic program of programmed cell death (apoptosis). Overall, the assessment of cell membrane integrity is one of the most common ways to measure cell viability and cytotoxic effects. While performing all the assays described below and in general for any biological assay, the use of control is mandatory. The control or reference can be either a sample or a treatment which can produce predictable or reproducible results in terms of cell response. The control can be either positive control or 'negative' control. For any given assay, either positive or negative and only in few cases, both controls are used. In most of the assays, tissue culture polystyrene (TCPS) or plastic/glass cover slip is used and (most) cell types are known to grow on such control. In quantifying dead cells, H_2O_2 treatment is used as positive control, as this is expected to induce ~100% cell death. The results of any assay can be reported in either or combination of the following approaches:

(a) Optical density (O.D) data vs time in culture or material composition, in case of colorimetric assay.

(b) Cell viability or cell response indicator, normalized with respect to control (% control) vs. culture time or substrate/material composition.

In addition to the above approaches, another alternative could be used to normalize the test data with respect to data obtained with another control, e.g., hydoxyapatite (HA). To substantiate this point, one can use HA as a control material, while evaluating cytocompatibility of HA-based composites. In such cases, both HA and HA-based composites are to be processed under similar conditions. As will be shown in some of the chapters that follow in this book, the biological assay data are to be analysed using error bars and are to be analysed using students T-test and/or ANOVA analysis using SPSS software.

In view of the stochastic response of a biological system to any foreign material (e.g. implant/ scaffold), all the experimental results should be obtained in reproducible and reliable manner from multiple biological and experimental replicates. To infer any conclusive trend against any hypothesis, it is necessary to establish statistical significance using parametric and/or non-parametric analysis.

9.2.1 | MTT assay

This assay has widely been used to measure the viability/proliferation of cells under standard culture conditions to determine the cytotoxicity, when the cells are grown on a biomaterial substrate or grown in a culture flask containing a specific drug.

Principle: MTT (3-(4,5-Dimethylthiazol-2-yl)-2,5-diphenyltetrazolium bromide) is a water soluble tetrazolium salt. As shown in Fig. 9.2, the working principle of this assay depends on the fact that MTT is converted to an insoluble purple formazan by cleavage of the tetrazolium ring by succinate dehydrogenase enzyme of mitochondria[689].

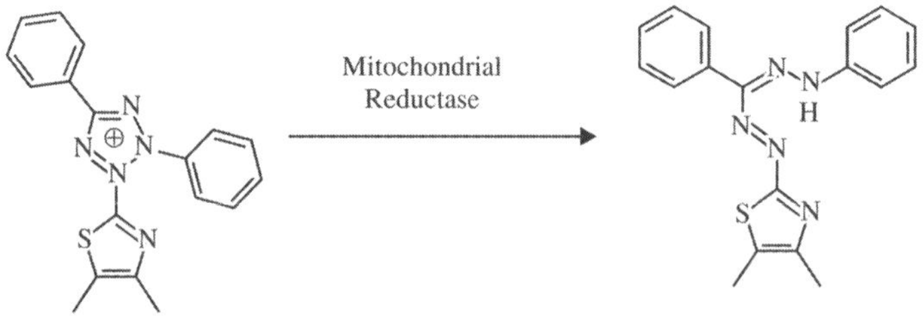

Fig. 9.2 MTT reduction in live cells by mitochondrial reductase results in the formation of insoluble formazan.

The formazan product is not permeable to the cell membrane and therefore it accumulates in healthy cells. The amount of formazan product is directly related to the percentage of viable cells. For a beginner in the field of biomaterials, the standard protocol for MTT assay is summarized below (Fig. 9.3).

Protocol for MTT assay, three replicates of each sample should be prepared and a fixed concentration of cells in each well should be seeded. The MTT assay protocol includes the incubation of the samples with cells (5000 cells/well) in the complete culture medium in 96 well plates (each well contains a 100 µl cell suspension) for 2, 4 and 8 days at 37°C under 5% CO_2 and 95% humidified air. After every second day, old media is replaced with the fresh media, followed by the addition of a 15 µl dye solution. The samples containing dye are incubated at 37°C for 4 hours in a CO_2 incubator, followed by gentle mixing with a 100 µl solubilization/stop solution after incubation. The samples are kept for incubation at 37°C in a CO_2 incubator for 12 hours. After the incubation, the absorbance (optical density) are recorded at 570 nm using the ELISA automated plate reader. MTT test should be repeated at least three times.

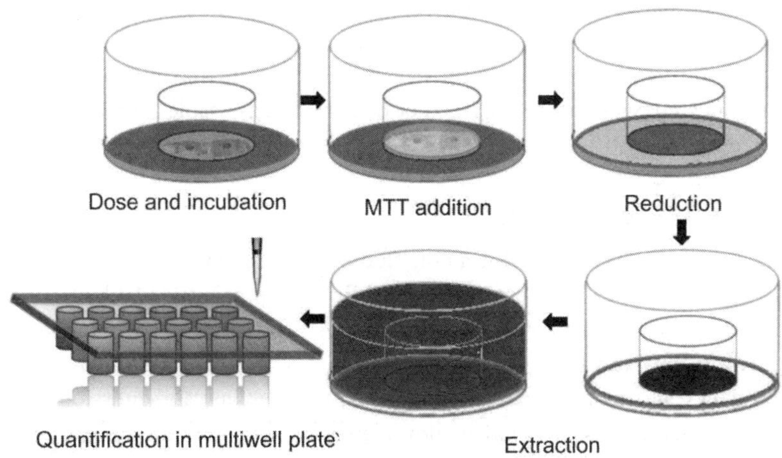

Fig. 9.3 Schematic diagram showing sequential steps of MTT assay [see Colour Plate].

9.2.2 | Alamar blue assay

In cell biology research, a number of assays of complementarity nature need to be used to quantify the cellular activity in a comprehensive manner. For example, the metabolic activity of cells in culture can also be determined using the Alamar Blue assay. After incubation of cells on a biomaterial substrate for desired period, the culture medium is removed from the well plates and washed with sterile phosphate buffered saline (PBS). Thereafter, 1 ml of fresh culture medium is usually added to 100 µl of AlamarBlue® solution (5 mg resazurin salt/40 ml PBS). This assay involves the incorporation of a fluorometric and colorimetric growth indicator[690]. The cell samples can be further incubated at 37°C with 95% humidity and 5% CO_2 for 2 hours and 30 minutes. The fluorescence is to be measured at an excitation of 530 nm and emission at 590 nm using the Fluorescence Microplate Reader. One of the disadvantages of this specific assay is that this is not a direct cell counting technique. High cell number and prolonged culture times often show reversal of the reduction process[691]. Also, if the molecule or compound in question modulates the cell metabolism without affecting the viability, the assay might provide misleading interpretations.

9.2.3 | WST-1 assay

It is important to note that MTT assay should NOT be used to quantify cell viability on materials, which interfere with the MTT reagent's functionality with the cells.

To substantiate the above statement, MTT assay can provide non-reliable results for carbonaceous materials, e.g., CNT or graphene or carbon substrates. In such scenario, WST-1 assay needs to be used, which involves the use of 4-[3-(4-iodophenyl)-2-(4-nitrophenyl)-2H-5-tetrazolio]-1,3-benzene disulphonate[692]. The advantages of this particular assay are that no volatile organic solvent is required for solubilization and the assay involves shorter reaction time. Also, there is no need for washing or harvesting the cells for this assay with the additional advantages being accelerated colour development and a greater sensitivity than MTT assay in terms to quantifying the metabolically active cells. The disadvantage however is that this assay is the cell-specific and media-specific and therefore, one must know how much enzyme can be regarded or disregarded as usable protein.

9.2.4 | Calcein AM cytotoxicity assay

Calcein-Acetoxymethyl or Calcein AM is a non-fluorescent cell permeant dye that can be used to obtain quick and accurate measurements of cell viability/ cytotoxicity. The hydrophobic Calcein AM easily permeates the cell membranes and is hydrolysed by intracellular esterases in the living cells to give a hydrophilic and fluorescent Calcein, which is retained within the cell membrane. The reaction product Calcein has an excitation wavelength at 490 nm and emission wavelength 520 nm. The fluorescent nature of Calcein also allows labeling, tracing and sorting of live cells using Calcein AM dye through flow cytometry.

9.2.5 | LDH assay

Lactate dehydrogenase (LDH) is a soluble cytosolic enzyme present in most eukaryotic cells. It is useful for monitoring cell death, as LDH is released into in the extracellular fluid (culture medium) upon cell death due to damage of plasma membrane.

Principle: Fundamentally, LDH catalyses the reduction of NAD+ to NADH in the presence of L-lactate, while the formation of NADH can be measured in a coupled reaction in which the tetrazolium salt INT [2-(4-iodophenyl)-3-(4-nitrophenyl)-5-phenyl-2H-tetrazolium chloride] is reduced to a red formazan product (Fig. 9.4). The amount of the highly coloured and soluble formazan can be measured at 490 nm wavelength spectrophotometrically. Similar to MTT assay, the protocol for LDH assay is briefly discussed below.

Protocol: Cells are seeded in a 96-well culture plate in culture medium with test compounds. The cells are cultured in a CO_2 humidified incubator at 37°C for the desired period of time. At least 3 replicates for each test sample are prepared. The 96-well culture plate is centrifuged at 400×g for 5 min and known amount of supernatant is transferred from each well to the corresponding well of the 96-well test plate. The working reaction mixture (2 µl of sodium lactate, 2 µl of INT and 20 μl of substrate mix to 36 μl of PBS) is added into each well of the 96-well test plate containing test samples followed by incubate for 20 minutes at room temperature in dark. The reaction is stopped with sodium oxymate solution (stop solution) per well of 96-well plate. The absorbance is read at 490 nm with an ELISA plate reader.

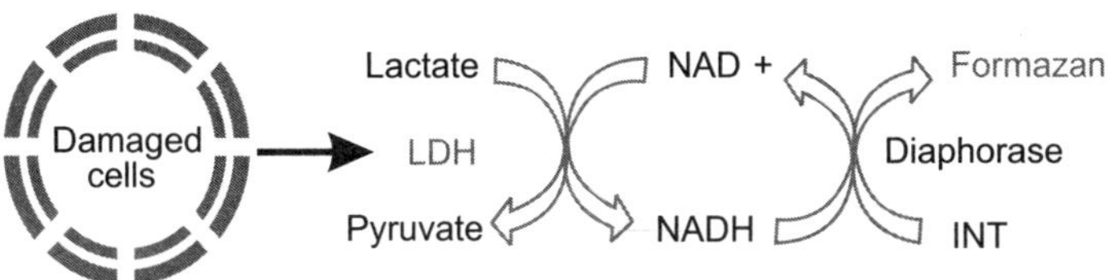

Fig. 9.4   Metabolic and chemical basis of the LDH cytotoxicity assay.

9.2.6 | Picogreen assay

The picogreen assay is used to quantify small amounts of double stranded DNA (dsDNA) and provides a more direct measure of cell numbers. The picogreen reagent is an ultra-sensitive fluorescent nucleic acid stain for quantitating dsDNA in molecular biology procedures such as cDNA synthesis for library production, DNA fragment purification for sub-cloning, and diagnostic applications, such as quantitating DNA amplification products and primer extension assays.

Principle: The assay contains a proprietary dye, which is highly specific to dsDNA and only fluoresces when dye is attached to dsDNA. As each cell contains a very specific amount of ds DNA, the picogreen expression can be correlated with cell numbers, when applied to a standard curve. For tissue engineering applications, this assay relies on lysing the cells and removing all dsDNA content from the cells on the scaffold.

Some basic advantage of picogreen assay over other conventional assays in the context of measuring nucleic acid concentration includes:

(a) Simpler, only a fraction of sample is needed (as little as 25 pg/ml)and single concentration of the picogreen reagent allows detection over the full dynamic range of the assay.

(b) No effect of contaminate such as salts, urea, ethanol, chloroform, detergents, proteins and agarose present in nucleic acid preparation.

(c) The picogreen assay can quantitate dsDNA in the presence of equimolar concentrations of ssDNA and RNA with minimal effect on the quantitation results.

Protocol: For the picogreen assay, the cells are to be seeded on a material substrate at a density of $5000/cm^2$ and cultured over different time points before measuring DNA content. After stipulated time period of incubation, samples are to be rewashed twice with PBS and subsequently incubated in 1 mL of 0.1% Triton-X 100 in 1 × TE buffer (10 mMTris-HCl, 1 mM EDTA, pH 7.5) for 48 h. The sample supernatant is to be spun down by centrifugation for 5 min. After centrifugation a 25 μL aliquot of supernatant needs to be taken from each sample and placed into 96-well plate with wells containing 75 μL of 1 × TE buffer. One hundred μL of a 1:200 dilution (200-fold dilution of the concentrated DMSO solution in TE) of PicoGreen reagent is to be then added to each well. The fluorescence of the sample is to be measured using a spectrofluorometer or fluorescence microplate reader and standard fluorescein wavelengths (excitation = 480 nm, emission = 520 nm).

Fluorescence value of the reagent blank is to be subtracted from that of each of the samples. The DNA concentration of the sample is then determined from the standard curve.

- Assessment of cell adhesion/growth/proliferation on biomaterial substrate should be done using targeted application-specific cell line at different time points in culture.
- Results from a particular biochemical assay should not be used to extrapolate the observations at a later time point.
- Complementary assays with adequate controls (both positive and negative) should always be done to avoid false positive results.

9.3 | Immunofluorescence Techniques

Immunofluorescence (IF) is one of the microscopy-based techniques, which enable the visualization of a specific protein or antigen in tissue sections or cell samples. The general working principle involves binding a specific antibody (chemically conjugated) with fluorescent dyes, such as fluorescein isothiocyanate (FITC) or tetramethylrhodamine isothiocyanate (TRITC). These fluorescent dyes (fluorochromes) make labelled antibodies glow in a specific colour, when observed microscopically under ultraviolet light. The technique is used widely in both scientific research and clinical laboratories, both on fresh and fixed cells.

Fluorescence is the property of certain molecules or fluorophores to absorb light of a shorter wavelength (excitation wavelength) and emit light at a longer wavelength (emission wavelength).

As far as the fundamental principle is concerned, the emission of light takes place rapidly once the excitation light is absorbed by fluorescent materials. It is well known that numerous electrons revolve around the nucleus in different orbitals, having discrete energy levels. Once a photon is absorbed by the electrons at the outermost orbitals, the electrons get 'excited' and jump to a higher, less stable energy level, generally in the timescale of nanoseconds. As a result of such interaction, the excited electrons would lose a fraction of energy as heat and rest is given off in the form of a photon[693,694,695]. A number of factors can affect the quality of a fluorescence image.

(a) Photobleaching: This phenomenon can be defined as the photochemical destruction of a fluorophore in the presence of reactive oxygen species, generated due to fluorescence excitation. It is thought that the primary causative mechanism of photobleaching appears to be photosensitization of singlet oxygen ($^1O_2$). Photobleaching can be minimized by, (i) reducing the intensity or time-span of light exposure, (ii) reducing the availability of singlet oxygen ($^1O_2$) by the addition of singlet oxygen scavengers (= antifade reagents), (iii) increasing fluorophore concentration, and (iv) employing fluorophores with less affinity to bleaching.

(b) Autofluorescence: It is often experienced that the fixation with aldehydes, particularly glutaraldehyde, results in high levels of autofluorescence. This may be avoided by washing the cells with 0.1% sodium borohydride in phosphate-buffered saline. Other factors that limit IF include the performance of the instrument (i.e., calibration), the specificity of the antibodies, and the specimen preparation.[7]

Two types of immunofluorescence staining are briefly described below.

9.3.1 | Direct immunofluorescence (DIF)

DIF uses fluorescent-tagged antibodies (primary antibodies) to bind directly to the target antigen. The following steps are involved and a schematic is shown in Fig. 9.5 to describe DIF. Target protein (Antigen or 'Ag') is fixed on the slide. Subsequently, the fluorescein labelled antibodies (Ab) is layered over it before incubation. Then, the slides are carefully washed to remove unattached Ab's and are examined under UV light in a fluorescent microscope. Those sites, wherein Ab is attached to its specific Ag, will reveal apple green fluorescence.

DIF assays are used to detect non-antibody targets such as infectious organisms (viral, parasitic, tumor antigens from patient specimens) or monolayer of cells. In this case, a fluorophore-labelled primary antibody, directed against the suspected antigen, is used to detect the presence or absence of the organism. DIF assays are used for identification of anatomic distribution of an antigen within a tissue or within compartments of a cell. Overall, DIF assays are rapid and quite specific. Shorter sample staining time is required. However, the disadvantages include lower signal, generally higher cost, less flexibility and difficulties with the labelling procedure, when commercially labelled direct conjugates are unavailable.

9.3.2 | Indirect immunofluorescence

This is essentially a two-step technique. The primary step involves the tagging of unlabelled antibody to the target. This is followed by the use of a fluorophore-labelled second antibody (directed against the Fc portion of the primary antibody) to detect the first antibody (Fig. 9.5). The presence of specific antibody bound to the antigen is detected by adding another antibody (fluorophores-labelled antibodies), which is detected by careful fluorescence microscopic observations.

This technique having better sensitivity is often used to detect autoantibodies in immunobullous diseases. A sample of the same (labelled or tagged) batch of secondary antibody can bind to many different (unlabelled) primary antibodies. However, some of the disadvantages include the potential for cross-reactivity and the need to find primary antibodies that are not raised in the same species or of different isotypes when performing multiple-labelling experiments. Samples with endogenous immunoglobulin may exhibit a high background[696].

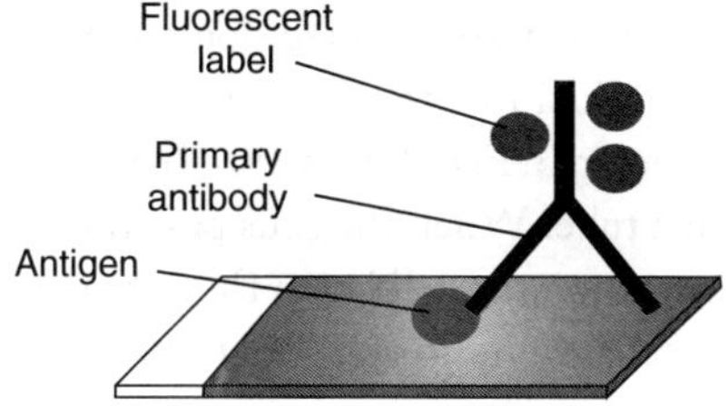

Fig. 9.5 Schematic of the principle involved in the direct immunofluorescence technique[717].

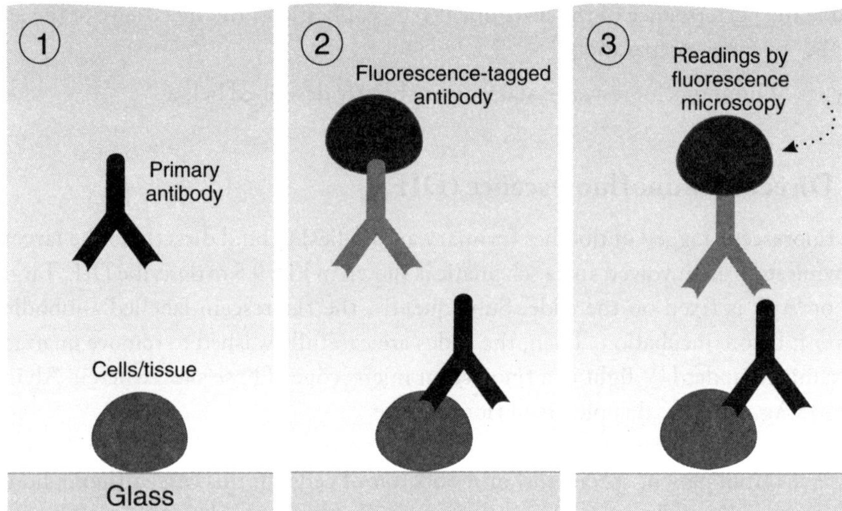

Fig. 9.6 Schematic illustration of the indirect immunofluorescence technique [Adapted from Ref. 718].

9.4 | Flow Cytometry

Flow cytometry, often known as Fluorescence Activated Cell Sorting (**FACS**) is a technology that simultaneously measures and then analyzes multiple cell fate processes, as the fluorescently tagged cell population flows in a fluid stream through a beam of laser light.

The functioning of any flow cytometry setup largely depends on the characteristics of an optical-to-electronic coupling system (fluidics, optics, and electronics), that records how the cell or particle scatters incident laser light and emits fluorescence. The outstanding advantage of flow cytometry is the simultaneous multiparametric analysis of the physical and chemical characteristics of up to thousands of cells/particles per second. In a commercial flow cytometer, the fluidics system transports cells through a guided system and a laser beam is made to focus on its path. The optics system consists of lasers to interact with the cell stream and optical filters to direct the resulting light signals to the appropriate detectors. The electronic system converts the detected light signals into electronic signals for further processing by a computer (Fig. 9.7).

Principle: Flow cytometry is perhaps one of the most powerful techniques to obtain cell subgroups of different characteristics with a greater precision and this can be pursued by tagging the cell population with an antibody linked to a fluorescent dye, as shown in Fig. 9.7. Typical results obtained during a FACS run are shown in Figs. 9.8 and 9.9. In flow cytometry, the cells, often tagged with fluoropores, travel through a tube. When the cells pass through the nozzle, a laser beam of a specific wavelength is made to focus on the cell laden fluid. Some of the laser light is scattered by the cells and the intensity of scattered light is used to determine cell size and cell count. If a cell is tagged with a specific antibody, it can be uniquely expressed in the cells that one wants to separate. The laser light excites the dye which emits a colour to be detected by the photomultiplier tube or light detector. This requires the careful optimization of cell to stain ratio in a given cell population. Depending on the cell type, a staining procedure needs to be established, often by trial-and-error approach with appropriate positive and negative controls. A typical protocol is shown in Fig. 9.10. The final step in flow cytometry-based cell is accomplished by electrical charge.

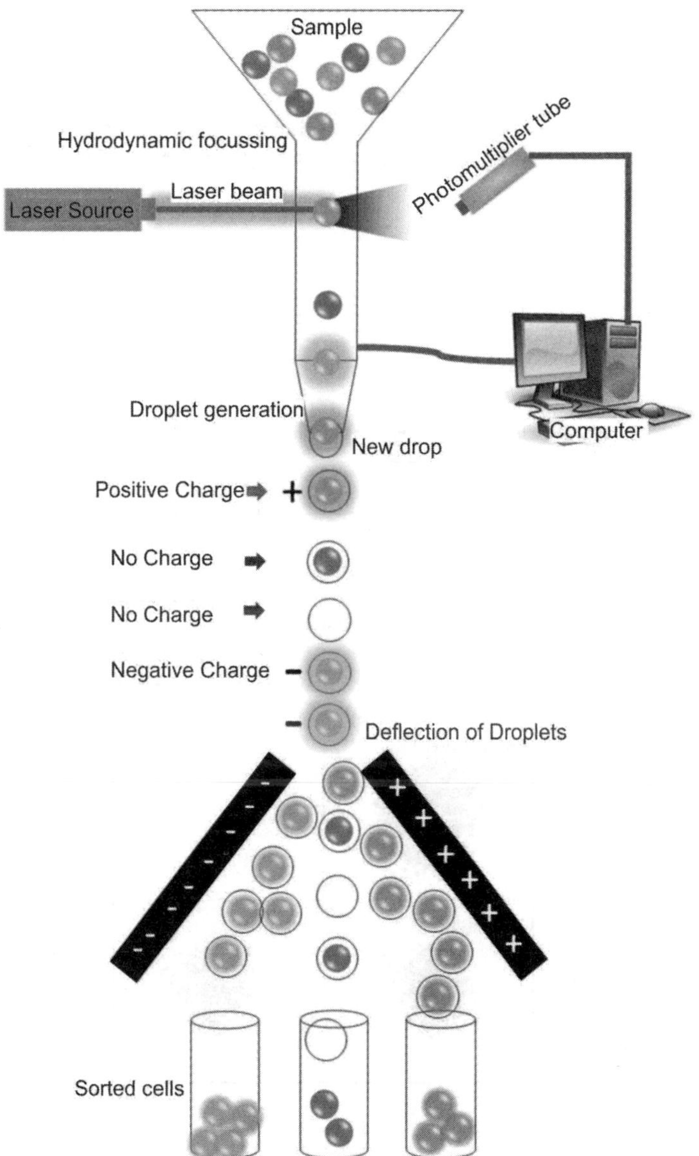

Fig. 9.7   Schematic showing the working principle of flow cytometry [see Colour Plate].

9.4.1 | Quantifying FACS data

FACS data can be displayed either by histogram or by dot plot. What we want to know is how many cells of each colour are sorted. In the first method, the intensity of the green or red fluorescence is plotted along X-axis with the number of cells with each level of fluorescence along the Y-axis. This method is the best if all cells are green, red or unlabelled and no cells are labelled by both colours (Fig. 9.8).

FACS analysis can be done by plotting the same data in different manner (Fig. 9.9). In one such matter, the X-axis plots the intensity of green fluorescence, while the intensity of red fluorescence can be plotted along Y axis. Such a plot enables the differentiation among those cells that express only one of the particular fluorescent markers, those that express neither, and those that express both. Frequently, such assessment helps to discriminate dead cells from the live ones.

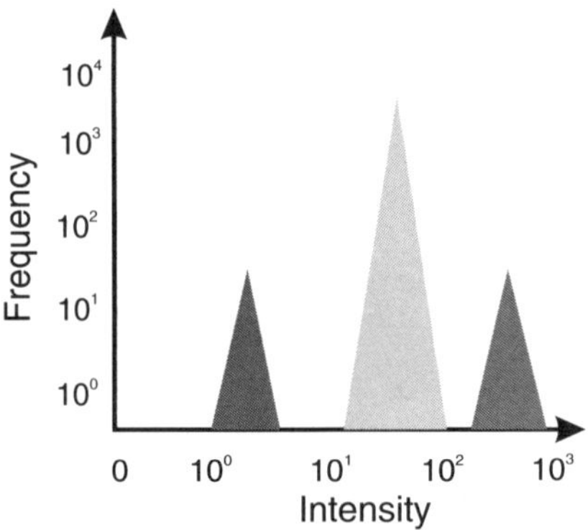

Fig. 9.8 Hypothetical plot to show the number of cells (Y-axis) and the level of fluorescence emitted (X-axis) by the labelled cells.

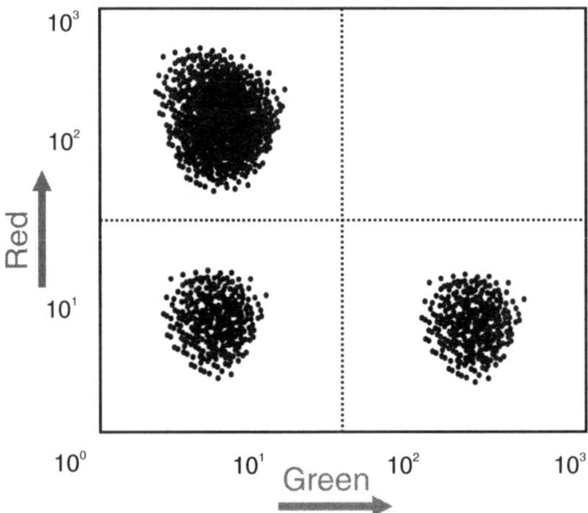

Fig. 9.9 Schematic representation of the dot–dot plots to analyse the flow cytometry results; the intensity of the emitted light increases as indicated by the arrows; the number of cells at each intensity is shown by the number of dots, where each dot refers to a single cell.

9.4.2 | **Flow cytometry based quantitative analysis**

A summary of protocol used in flow cytometry study is shown in Fig. 9.10 and the following describes how FACS analysis can be useful to study specific cell fate processes. Also, flow cytometry has been lately used in the cytocompatibility assessment of many new biomaterials[697].

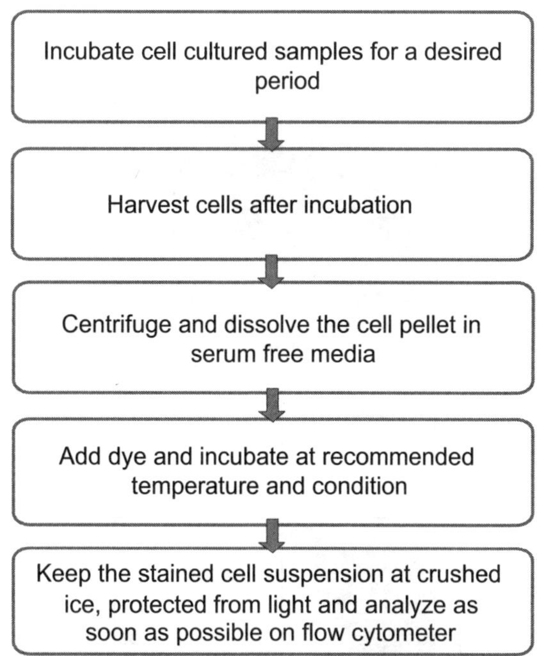

Fig. 9.10 Flowchart showing the steps for staining, prior to flow cytometry.

9.4.2.1 | Reactive oxidative stress

CellROX™ Deep red reagent is used to determine the reactive oxygen stress in the cells grown on biomaterial after for various time points (upto 1 week). It is non-fluorescent and is able to penetrate the cell membrane during cell apoptosis in the culture medium. Once it enters the cells, it is cleaved by endogenous esterase, which becomes fluorescent after being oxidized by ROS (Fig. 9.11). The quantification of ROS, based on fluorescent intensity, is carried out using flow cytometry. Specifically, 2 µl/ml of the dye solution is added to cell population and then incubated for 30 minutes at 37°C. After 30 minutes, the cells are washed three times with 1 X PBS and this step is followed by trypsinization using a trypsin-EDTA solution. 10^5 cells/ml are suspended in 1 X PBS and kept on ice until analysis. 50,000 cell events can be recorded for each sample using a 633 nm laser and 660/20 filter set. The collected data are plotted as fluorescent intensity versus number of cells. The level of ROS needs to be calculated by multiplying the median and number of events, corresponding to the peak in fluorescent intensity. The fold generation of ROS for each sample is the ratio of ROS level in treated cell population to negative controls.

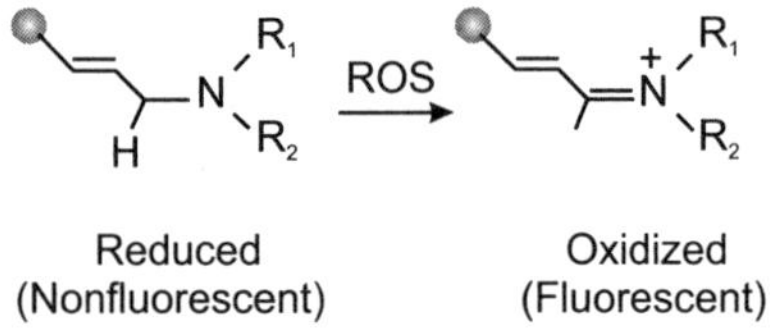

Fig. 9.11 Reduction of CellROX™ deep red reagent in the presence of reactive oxygen species (ROS).

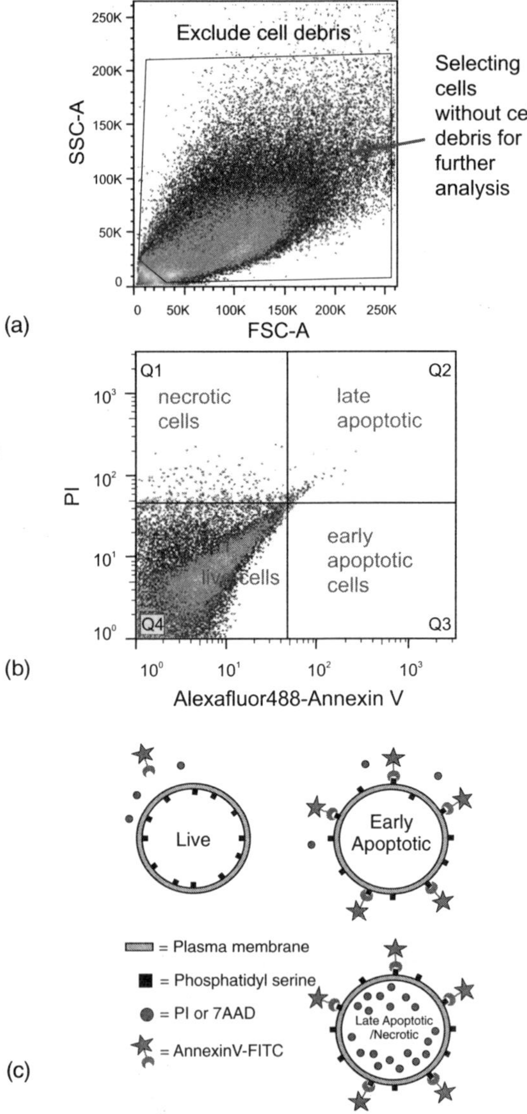

Fig. 9.12 Schematic illustration of the flow cytometry based quantification of cell apoptosis; the apoptotic cells expose phosphatidyl serine (PS), which can bind to annexin V; PI is impermeant to live cells and apoptotic cells, but stains dead cells with red fluorescence, binding tightly to the nucleic acids in the cell (c), the typical FACS plot to show how to exclude cell debris (a) and typical FACS plot to show the distinction of the apoptotic and necrotic cells (b) [see Colour Plate].

9.4.2.2 | Cell apoptosis/necrosis

A cell apoptosis dye is used for the quantitative analysis of the apoptosis using flow cytometry (Fig. 9.12). After the desired incubation period, the harvested cells are washed in cold 1XPBS (phosphate-buffered saline). The cells are diluted in 1×annexin-binding buffer to make the cell density equal to ~1 × 10^6 cells/mL. Then, 5 µL of Alexa Fluor 488 annexin V and 1 µL of a PI (propidium iodide) working solution are added to each 100 µL of cell suspension. After incubation at room temperature for 15 minutes, this solution is mixed with 400 µL 1Xannexin-binding buffer and kept on ice. This is followed by an analysis of the stained cells by the flow cytometer, involving the measurement of fluorescence emission at 530 nm and 617 nm using a 488 nm excitation laser. For this, flow cytometry tubes are prepared with each containing 1 ml of a stained cell suspension in the 1X annexin-binding buffer at a final cell concentration of 1 × 10^6 cells/ml and 50,000 events can be recorded for each sample.

9.4.2.3 | Cell proliferation

The CellTrace®violet/CFSEdye is used for flow cytometry analysis of cell proliferation. Prior to starting the culture with proliferation dye, each ml of the pre-cultured cells (10^6cells/ml) is incubated with 1 µl CellTrace violet stain for 20 minutes at 37°C, while keeping the entire solution protected from light. After 5 minutes of incubation, the reaction is quenched by replacing the dye containing PBS with ice cold complete culture medium, followed by incubation for 5 minutes at room temperature. After this, each sample surface is immersed in a fresh warm (37°C) complete culture medium. The washing is repeated two times to remove the excess stain from the extracellular matrix. Then, the cells are incubated for various desired timepoints at 37°. After each second day, old media is replaced with fresh media. After the incubation, cells are harvested and the flow cytometry tubes are prepared with each containing 1 mL of the stained cell suspensions in the 1×PBS at a final cell concentration of 1 × 10^6 cells/mL. The cell population is subsequently analysed by a flow cytometer at excitation/emission = 405 nm/450 nm (Fig. 9.13).

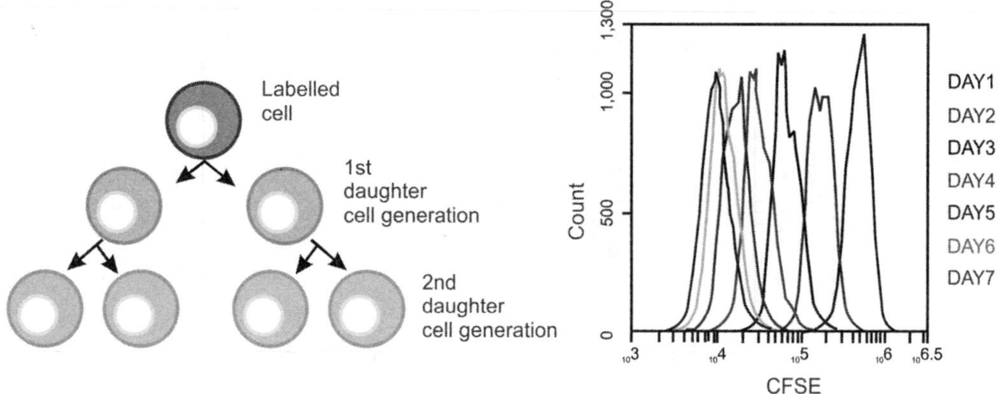

Fig. 9.13 Schematic illustration of cell division (left) and CFSE proliferation assay (right), which works on the principle that as dye molecules, once inherited equally by daughter cells during the cell division over different timpoint in culture, would result in the sequential halving of dye.

9.4.2.4 | Cell cycle

A cell cycle dye (Vybrant® dye cycle™ orange stain/PI/DAPI) is used for the flow cytometry study of cell cycles. The flow cytometry tubes are prepared with each containing 1 mL of cell suspension in 1×PBS at a final cell concentration of 1×10^6 cells/mL. This is followed by the addition of 2 µL of the Vybrant dye cycle orange stain and the entire solution is gently mixed. The stained cells are incubated at 37°C for 30 minutes. The cytometry tubes containing stained cells are kept on ice, protected from the light until acquisition. The cells are analysed by the flow cytometer at excitation/ emission= 488 nm/570 nm (Fig. 9.14).

Appropriate assays should always be chosen in such a way that the end-products do not interfere with the material/compound tested. All biological assays are prone to high variability and hence independent experiments are to be conducted multiple times and on multiple samples of the same biomaterial composition together with rigorous statistical analysis to establish any significant difference among the measured data.

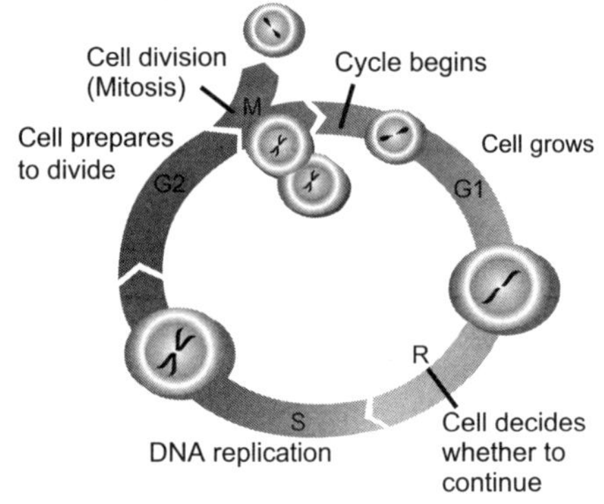

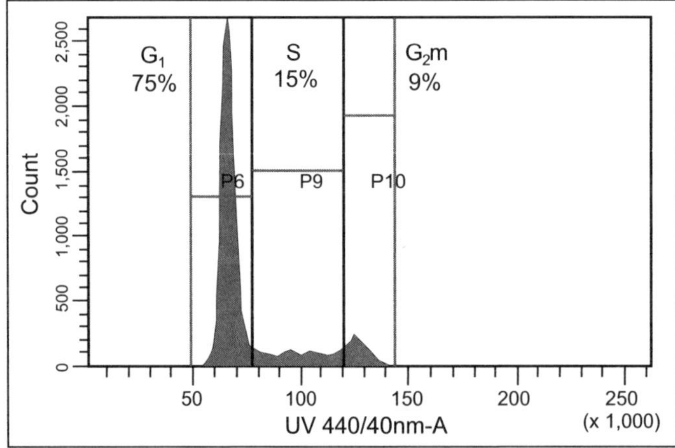

Fig. 9.14 Schematic diagram of the cell cycle (top) and a typical flow cytometry plot to quantify the fraction of cells in different phases of cell cycle (bottom).

9.5 | Reverse Transcription–Polymerase Chain Reaction (RT–PCR)

The understanding of the science of biomaterials typically requires an assessment of the changes in gene expression of the cells after they are grown on a biomaterial substrate and/or grown under stimulated culture microenvironment containing various biophysical cues. To this end, Reverse Transcription–Polymerase Chain Reaction (RT–PCR) assay is the most powerful technique as it captures the earliest reaction of cell to different treatment/materials. Thus, the change in cell fate, such as adhesion, migration, proliferation, differentiation, cell cycle control and apoptosis can be effectively detected by this tool. The scientific basis and working principles of PCR with related details are described in this section in reference to Figs. 9.15–9.21.

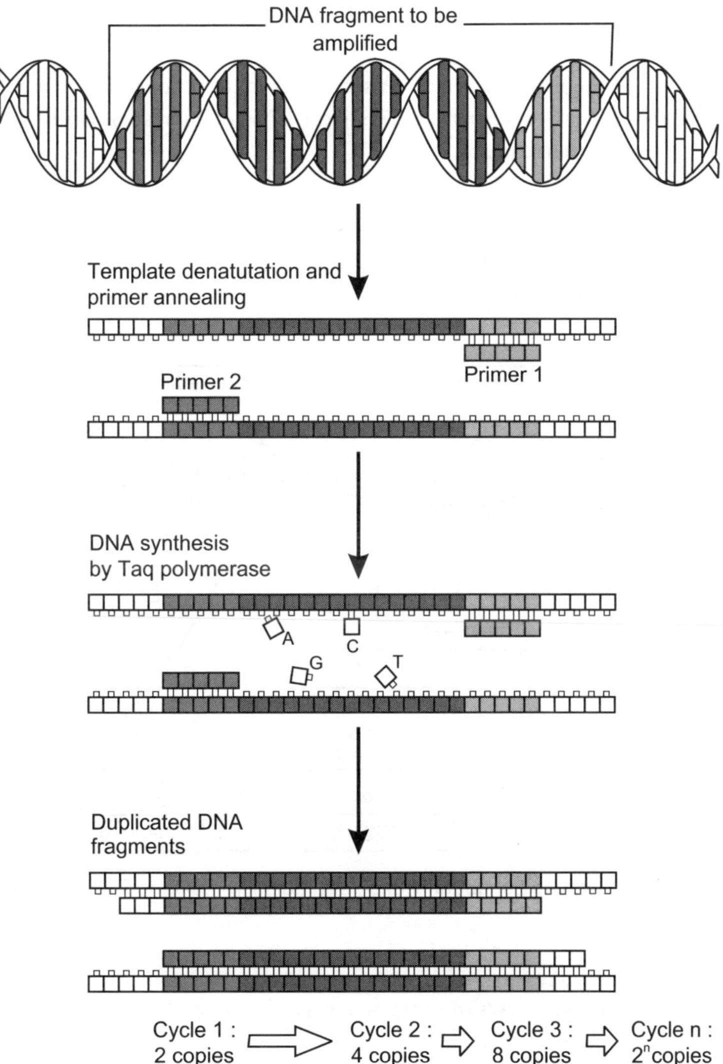

Fig. 9.15 Schematic diagram of the thermal cycles involved in Polymerase Chain reaction (PCR)[721].

For example, one of the important aspects in the study of cell differentiation is to establish the changes in expression levels of gene regulatory proteins. For differentiation along a specific lineage, a set of genes will either be activated or repressed. Considering the significantly low level of such expression changes in absolute scale, there is a need to use suitable assay, which can potentially magnify such changes to detectable extent. This is the fundamental aspect based on which polymerase chain reaction (PCR) based detection of gene expression is established (see Fig. 9.15). For example, when stem cells undergo neurogenic differentiation, the expression level changes of β III tubulin, nestin and neuro filament are detected using PCR technique. Similarly, to confirm the osteogenic differentiation of stem cells, the expression of ALP, osteocalcin, RunX2 are to be determined using PCR.

The underlying working principles for RT–PCR and the traditional PCR are fundamentally different.

While, traditional PCR amplifies target DNA sequences in an exponential manner, RT–PCR is used to clone expressed genes by reverse transcribing the RNA of interest into its DNA complement through the use of reverse transcriptase.

It is important to mention that RT–PCR can be used to compare levels of mRNA extracted from various cell populations and to characterize patterns of mRNA expression and to discriminate between closely related mRNAs (Fig. 9.16).

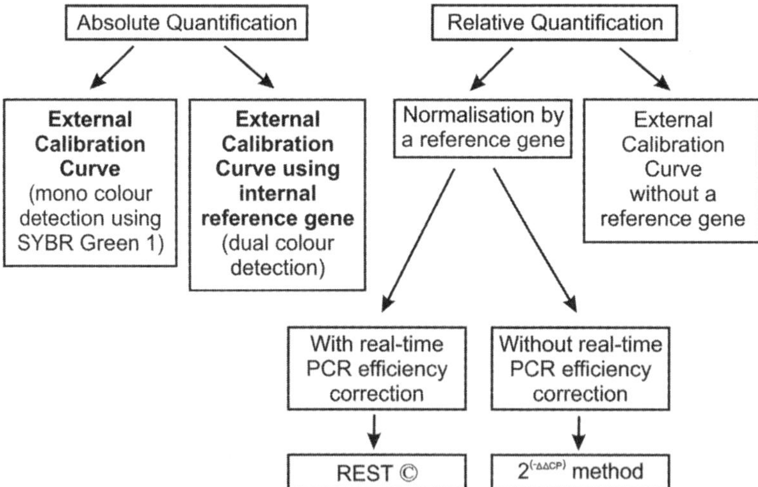

Fig. 9.16 Different types of quantification strategies in real time PCR technique[719].

During PCR analysis, the expressions of the gene regulatory proteins/markers are to be normalized with respect to a housekeeping gene (GAPDH, β-actin etc.), while comparing undifferentiated and differentiated states of a cell population.

One of the most important steps is to isolate good quality RNA from mammalian tissues and cells for RT–PCR analysis. Once the cells are grown on a substrate, the cell population is treated using a combination of phenol and quanidine isothiocyanate (TRIZOL). This approach denatures all of the cellular proteins, including RNAase. Subsequently, RNA will be separated from the remaining cellular macromolecules by differential extraction and precipitation with alcohol. The following steps are followed.

Step 1: Extraction of Total RNA using TRIzol Reagent

In order to attain a sufficient amount of RNA, at least 10^6 cells are to be pooled in a microcentrifuge tube and TRIzol reagent is to be immediately added to the cell population. The cells are then lysed and homogenized (approximately 20 strokes) with chloroform addition to ensure that the cells are sufficiently disrupted. Subsequently, the mixture is physically separated into a colourless upper aqueous phase (60% of the initial volume of TRIzol), an interphase and a lower red phenol–chloroform phase (Fig. 9.17).

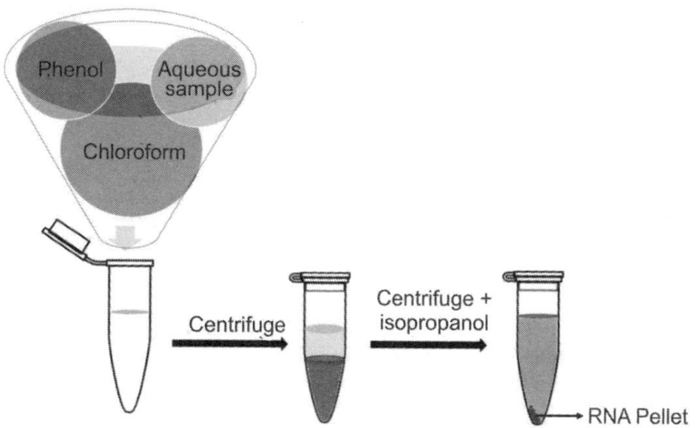

Fig. 9.17 Schematic depicting the phase separation by centrifugation of a mixture of the aqueous sample and a solution containing water-saturated phenol and chloroform.

The addition of isopropanol allows RNA to precipitate to form a gel-like pellet on the bottom of the tube after centrifugation. Without disturbing the pellet, the supernatant should be removed and washed with ethanol. After centrifugation, ethanol is removed and completely dried. The pellet needs to be re-suspended by adding 100 µl of RNAse-free water. The RNA sample quality and quantity are to be checked using a NanoDrop ND-1000 spectrophotometer.

Step 2: cDNA Synthesis

In view of the fact that RNA is relatively unstable and easily prone to cleavage by RNAses, most protocols enable the quantification of gene expression using a cDNA product that is directly synthesized from an RNA template. The resulting cDNA can be stored for subsequent analysis of gene expression by PCR (Fig. 9.18). This process is defined as reverse transcriptase-mediated synthesis of single-stranded DNA (complementary DNA) using single-stranded RNA as template. In this process, total RNA, either mRNA or specific RNA or *in vitro* transcribed RNA can be reverse transcribed as long as a single-stranded DNA primer is hybridized to the RNA. Alternatively, cDNA can be synthesized in laboratory using various enzymes and a typical synthesis protocol is described in Fig 9.18 together with structural components in Fig 9.20.

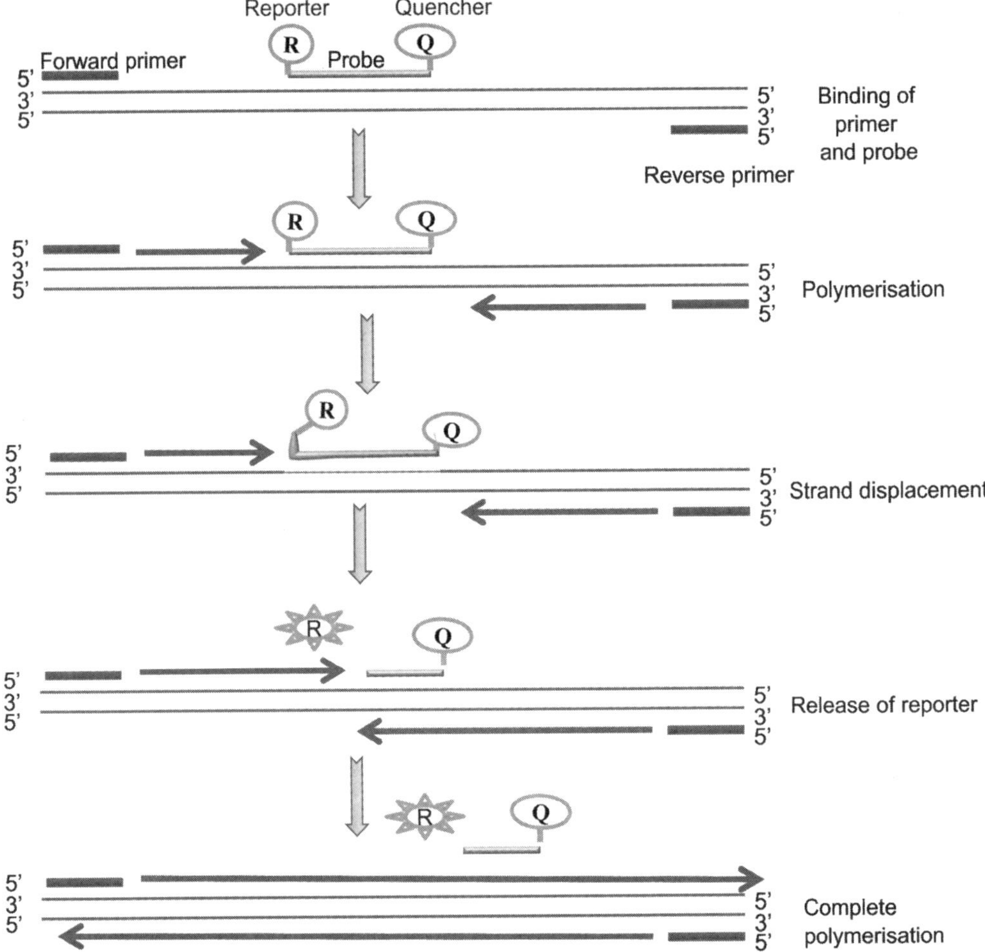

Fig. 9.18 Overview of steps involved in the conversion of RNA to cDNA for quantitative PCR assay[720].

Protocol: In order to extract 1–5 µg of total RNA per reaction of final 20 µl volume, oligo-dT primer or random-hexamers (dN_6), or gene-specific primer is to be added. Subsequently, DEPC-treated or RNAse-free water of appropriate volume is to be introduced prior to heating at 70°C for 10 minutes. This step is followed by ice chilling unfolds RNA secondary structures. Add dNTPs for final 0.5 mM concentration of each dNTP. After the addition of reaction buffer, the polymerase reaction is allowed to take place at 42°C and reverse transcriptase enzyme is added after 2–3 minutes. The entire stock is incubated at 42°C for 30–60 minutes for reverse transcription to proceed. After this, the reaction can be stopped and the enzyme can be denatured by heating at 70°C for 10–15 minutes. After cooling down from the reaction temperature, RNAase H (usually from *E.coli*) is added before incubating at 37°C for 20 minutes. Subsequently, the reaction mixture is frozen and 5–10% of the product of each reaction is used as a template for subsequent reaction.

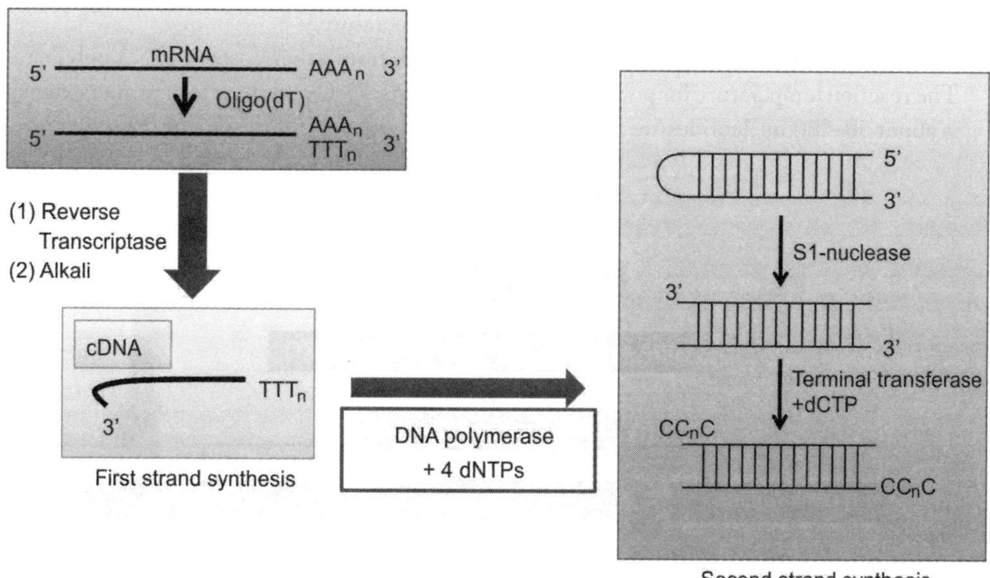

Fig. 9.19 Different approaches to synthesis first strand cDNA[720].

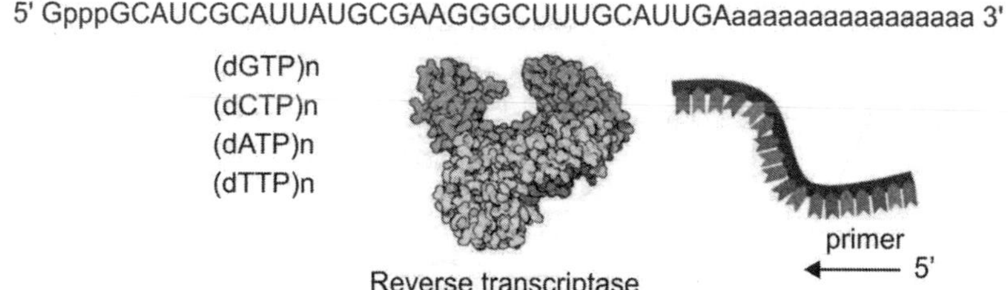

Fig. 9.20 Four basic components required to synthesize cDNA.

Step 3: Semi-quantitative PCR

The amount of cDNA can be detected using semi-quantitative PCR or more accurately by RT–PCR. For the former, the amounts of PCR products are measured during the exponential (i.e., log) phase of the PCR reaction, before the saturation stage. The DNA amplification involves 30–35 cycles of PCR, which are described in Fig. 9.21.

(a) Denaturation: This step involves high temperature (92–95°C) to denature the dsDNA. It is to be noted that the reaction temperature needs to be closely monitored within the strict temperature window to obtain reliable PCR data.

(b) Annealing: The second step involves annealing of the primers to the single stranded DNA. The optimal annealing temperature depends upon the melting temperature of the primer-template hybrid. The annealing temperature is typically 5°C below the melting point (T_m) of the primers. In order to ensure adequate specificity, the length of the primers should be 20–30 nucleotides and the close control over annealing temperature.

(c) Extension: In the third step, DNA synthesis takes place from a thermostable DNA polymerase. The reaction temperature for polymerization is about 70–75°C and the rate of primer extension is about 50–100 nucleotides/sec.

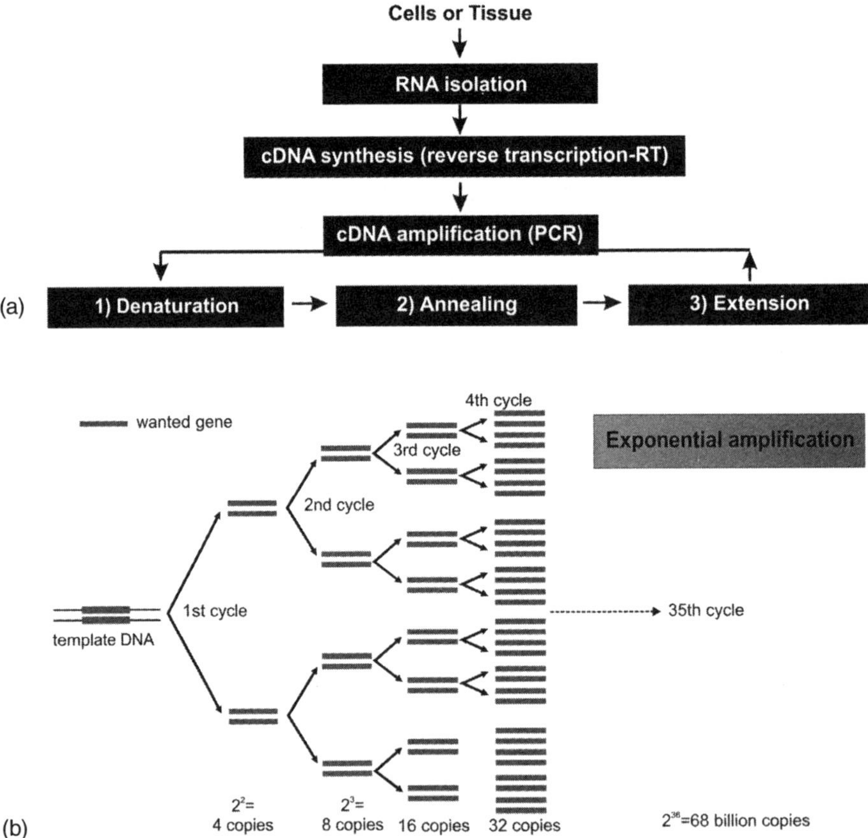

Fig. 9.21 (a) Fundamental principles involved in RT–PCR and (b) stage-by-stage amplification of target DNA[721].

These repeated thermal cycles are performed automatically in a thermocycler. The purity and yield of the PCR products depend on the following parameters: (a) temperature during each step of the cycle, (b) number of cycles, (c) purity and sequence of the DNA template, (d) $MgCl_2$ concentration, (e) primer concentration, (f) DNA polymerase used, and (g) primer selection and determination of T_m.

In summary, PCR analysis can be appropriately used to quantify the changes in expression of a specific gene, and therefore is used to establish molecular changes at gene level. Typically, the gene expression is normalized with respect to housekeeping gene, GAPDH or β-actin.

In the next few sections, we will discuss various differentiation assays to quantify the gene/protein expression related to bone, muscle, cardiac and neural cells.

> Cell morphological changes, a signature for cell functionality modulation, need to be confirmed using gene expression analysis in differentiation study. For specific cell type, the expressions of select group of gene regulatory proteins are to be determined with respect to housekeeping genes.

9.6 | Biological Assays for Osteogenic Differentiation

Cell differentiation is a multi-step cell fate associated with the expression of a specific set of cellular markers in each step; an undifferentiated cell acquires a feature of specialized cell. For example, a pluripotent stem cell progresses to a progenitor cell and a fully differentiated cell type. The morphology of the cell also changes during differentiation. Identifying, understanding and mastering cell differentiation processes are of key interest in the development of innovative therapeutic strategies. Different assays need to be followed to assess the progression of differentiation through a specific lineage. For example, the osteogenesis assessment requires one to follow early (ALP, RunX2) or late stage (Osteocalcin, Collagen type I, etc.) gene expression markers. Similarly, the up regulation of neuro-filament (NFL), β-III tubulin is to be assayed to evaluate neurogenic differentiation. In the following, some typical differentiation assays for osteogenesis are summarized:

9.6.1 | Alkaline phosphatase (ALP) assay

The alkaline phosphatase activity is considered as an early stage differentiation marker for bone cell. Alkaline phosphatase (ALP) catalyses the hydrolysis of phosphate esters in alkaline buffer and produces an organic radical and inorganic phosphate. The changes in alkaline phosphatase level and activity are associated with various disease states in the liver and bone. Bone alkaline phosphatase (BAP) is the bone-specific isoform of alkaline phosphatase. A glycoprotein that is found on the surface (plasma membrane) of osteoblasts and BAP reflects the biosynthetic activity of these bone-forming cells. BAP has been shown to be a sensitive and reliable indicator of bone metabolism.

> During *in vitro* differentiation studies, the ALP specific activity is used as marker for determining osteoblast phenotype and is considered to be an important factor in determining bone mineralization.

Principle: The alkaline phosphatase (ALP) can be described by the following reaction in accordance with a standardized method:

$$\text{p-Nitrophenyl phosphate} + H_2O \xrightarrow{\text{ALP}} \text{Phosphate} + \text{p-Nitrophenol}$$

As shown in Fig. 9.22, p-nitrophenyl phosphate is hydrolyzed by phosphatases to form phosphate and p-nitrophenol. AMP (Adenosine Monophosphate) serves as transient phosphate acceptor. The release of yellow coloured p- Nitrophenol is proportional to the ALP activity and can be measured photometrically at 405 nm absorbance.

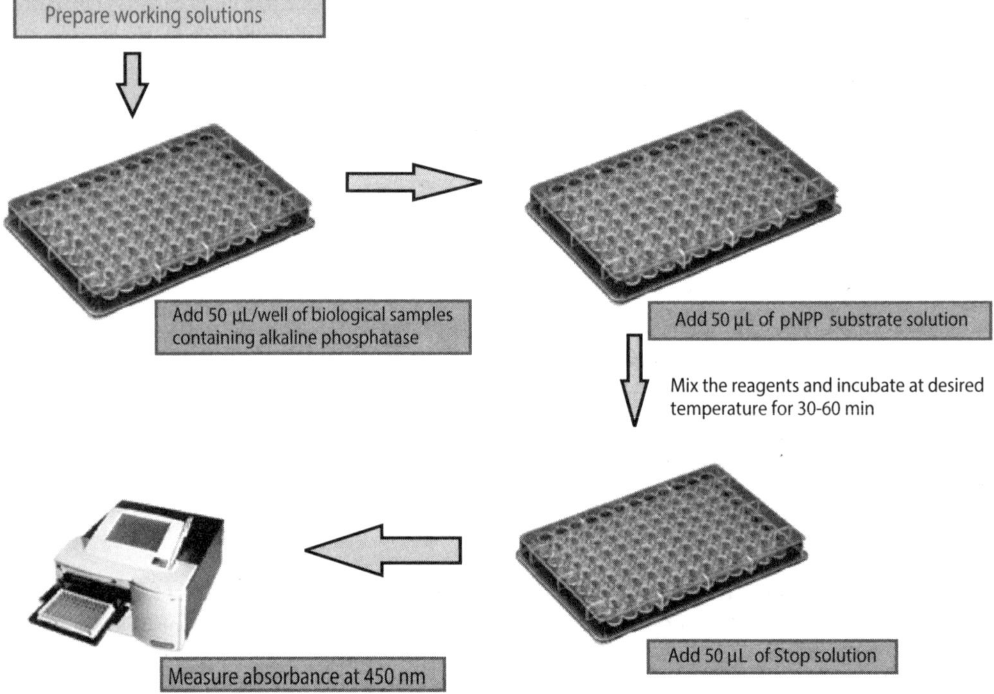

Fig. 9.22 Reaction scheme showing hydrolyzed p-nitrophenyl phosphate phosphatases to form phosphate and p-nitrophenol.

Fig. 9.23 Schematic diagram showing the protocol involved in alkaline phosphatase (ALP) assay.

Protocol: As described in Fig. 9.23, ALP assay protocol can be completed in three steps.

(a) The cell samples are added into 96-well plate. The total volume of samples is made up to 80 μl with assay buffer (1.0 M Diethanolamine Buffer with 0.50 mM Magnesium Chloride, pH 9.8 at 37°C).

(b) 50 μl of the 5 mM p-NPP solution is added to each well containing the test samples. Sample is mixed well. The reaction is incubated for 60 min at 25°C and protected from the light.

(c) All reactions are stopped by adding 20 μl stop solution into each sample reaction and plate is shaken gently. Optical density is measured at 405 nm in a microplate reader.

9.6.2 | Osteocalcin assay

Osteocalcin is a late marker of bone formation. It is a Vitamin K and Vitamin D-dependent protein produced by osteoblasts and is the most abundant non-collagenous proteins in bone. After production, it is partly incorporated in the bone matrix and the rest is found in the blood circulation. Circulating levels of osteocalcin reflect the rate of bone formation.

Principle: Like other immunoassays, Osteocalcin is performed on a monoclonal antibody (MAb 1) coated multiwell plate. This assay uses a mixture of monoclonal antibodies (MAbs) directed against distinct epitopes of osteocalcin. As far as the working principles are concerned, the cell population reacts with the monoclonal antibody (MAb 1) and with a monoclonal antibody (MAb 2) conjugated to horseradish peroxidase (HRP). After an incubation period allowing the formation of MAb 1-osteocalcin-MAb 2-HRP based sandwich (Fig. 9.24), the multiwell plate needs to be washed. This procedure enables unbound enzyme conjugated antibody to be removed. Subsequently, the bound enzyme conjugated antibody is measured through a chromogenic reaction. For this, cheomogenic solution (TMB ready for use) is added and incubated and subsequently, the colour change of the solution is read at 450 nm wavelength is assumed to be directly proportional to the concentration of osteocalcin.

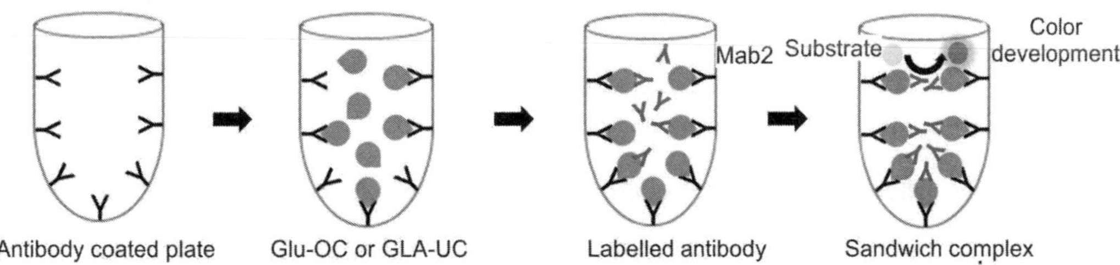

Fig. 9.24   Sandwich type model for a late differentiation marker, osteocalcin assay.

Protocol: The osteocalcin assay is generally performed by the following procedure (see also Fig. 9.25):

(a) Determine the number of wells for the assay run. 25 μL of each sample, to be tested, is added after determining the number of wells.

(b) 100 mL Monoclonal Anti-Osteocalcin-HRP is added and conjugated to all wells.

(c) Plate is taped gently to mix and covered followed by incubation for two hours at room temperature on a horizontal shaker set at 700 ± 100 rpm.

(d) Solution is thoroughly aspirated or decanted from wells and liquid is discarded. The wells are washed for a total of 3 times following washing instructions.

(e) 100 µL of stabilized chromogen is added into all wells within 15 minutes following final wash, which is followed by incubation for 30 min at room temperature on a horizontal shaker set at 700 ± 100 rpm, avoiding direct sunlight.

(f) 200 µL of stop solution is added to each well. This stops the reaction.

(g) Absorbance is read within 3 hours.

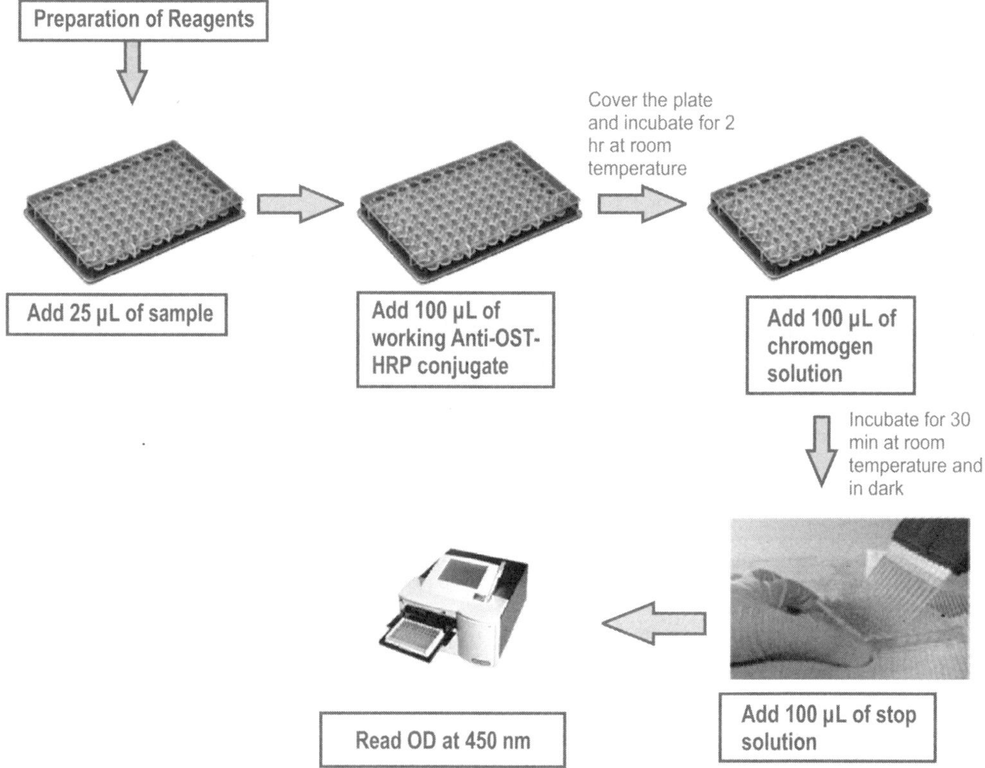

Fig. 9.25 Schematic diagram showing the protocol for Osteocalcin or bone Gla protein (B.G.P.) assay.

9.6.3 | Runt Related Transcription Factor 2 (RUNX2) assay

Runt Related Transcription Factor 2 is a protein that is encoded by the RUNX2 gene in humans. RUNX2 expression is largely restricted to osteoblasts and mesenchymal condensations forming bones, cartilages and teeth. This protein is essential for osteoblastic differentiation, skeletal morphogenesis, chondrocyte maturation, bone formation and remodelling. *RUNX2* protein also acts as a scaffold for nucleic acids and regulatory factors involved in skeletal gene expression. Concomitantly, the mutations in this gene have been associated with the bone development.

Principle: The test principle is based on sandwich enzyme immunoassay (Fig. 9.26). As far as the functional principle is concerned, the cell population, at an intermittent time-point of a differentiation study, is added to the appropriate microtiter plate wells. As per the recommended practice, the cells are pre-coated with an antibody specific to Runt Related Transcription Factor 2 (RUNX2). This step is followed by the addition of a biotin-conjugated antibody specific to Runt Related Transcription Factor 2 (RUNX2). Subsequently, Avidin conjugated to Horseradish Peroxidase (HRP) is introduced to each microplate well and the entire culture system is incubated. Once TMB substrate solution is added, those specific cells, tagged by Runt Related Transcription Factor 2 (RUNX2), biotin-conjugated antibody and enzyme-conjugated Avidin, will exhibit a colour change. For quantitative purpose, the colour change is determined spectrophotometrically by comparing the O.D. of the samples to the standard curve at a wavelength of 450 nm ± 10 nm.

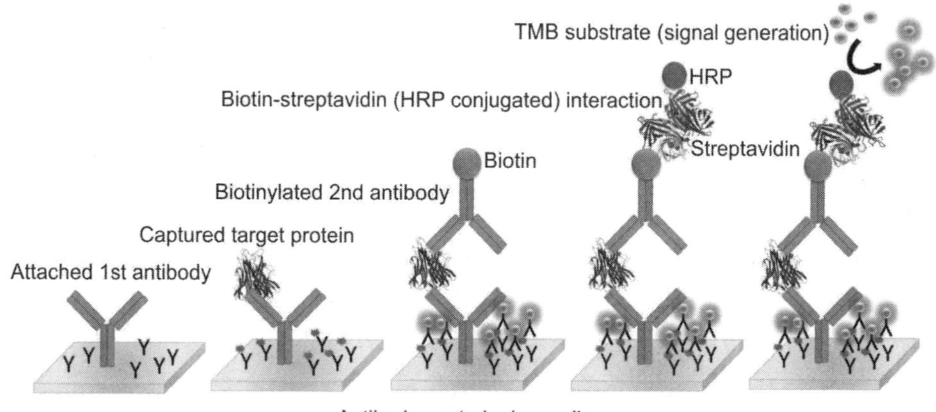

Fig. 9.26 Schematic of the underlying biological protocols for an early differentiation marker, RUNX2 (total sandwich ELISA assay).

Protocol: The protocol to be followed for (RUNX2) assay includes:

(a) 100 µL standard or sample is added to each well following incubation for 2 hours at 37°C.
(b) After aspiration, 100 µL prepared Detection Reagent A is added following incubation for 1 hour at 37°C.
(c) All wells are aspirated and washed 3 times.
(d) 100 µL prepared Detection Reagent B is added and followed by incubation for 30 minutes at 37°C.
(e) All wells are aspirated and washed 3 times.
(f) 90 µL TMB Substrate Solution is added followed by its incubation for 15–25 minutes at 37°C.
(g) 50 µL Stop Solution is added and OD is measured at 450 nm immediately.

9.6.4 | Osteopontin assay

Osteopontin (OPN) is essentially a highly acidic calcium-binding glycosylated phosphoprotein. This specific protein is secreted by osteoblasts, kidney tubule cells, macrophages, activated T cells, and vascular smooth muscle cells. OPN contains an Arg-Gly-Asp (RGD) amino acid sequence.

Principle: Osteopontin assay is an *in vitro* enzyme-linked immunosorbent assay for the quantitative measurement of Human Osteopontin in serum, plasma, and cell culture supernatants. As per the standard guielines, cell samples are harvested on a 96-well plate, coated with an antibody specific for Human Osteopontin with the hypothesis that Osteopontin, if present in a sample, is bound to the wells by the immobilized antibody. While performing this assay, biotinylated anti-Human Osteopontin antibody is added. After washing away unbound biotinylated antibody, HRP-conjugated streptavidin is pipetted to the wells. After subsequent washing, a TMB substrate solution is added to the culture plates. The stop solution changes the colour from blue to yellow, and the intensity of the colour is measured at 450 nm. The change in the colour of the culture solution is proportional to the amount of bound Osteopontin.

Protocol:

(a) Number of wells depending on number of test sample first determined. Then, 100 μL of each of the sample is put into the appropriate wells.

(b) Pre-coated plates are incubated at 37° for 1 hour after covering it with plate lid.

(c) After proper washing and aspiration, 100 μL of labelled antibody is pipette out into the each well followed by incubation of the pre-coated plate for 30 minutes at 4° after covering it with plate lid.

(d) After proper washing of plate. 100 μL of TMB solution should be pipette out into each of the well.

(e) The pre-coated plate is incubated for 30 minutes at room temperature in the dark. The liquid will turn blue by the addition of TMB solution.

(f) 100 μL of Stop solution is added into the wells. The liquid will turn yellow and the colour change is recorded by plate reader at 450 nm.

- Depending on cell doubling time, various timepoints of *in vitro* testing should be defined.
- Cell grown on biocompatible material will always change its shape due material-protein interactions. Failing of cells in doing so may indicate non-functionality of cells.
- In addition to generate qualitative understanding, more emphasis should be placed to obtain quantitative data on cell morphological changes or status of the cell fate.

9.7 | Biological Assays for Myogenic Differentiation

In *in vitro* culture, muscle cells have the ability to attach to the culture dish after seeding, followed by cell spreading. It will slowly acquire spindle-shaped morphology, which is an indicative of myoblast cells.

The fusion of myoblasts into multi-nucleated myotubes is the common path in differentiation of skeletal muscle, indicating successful progression of myotube differentiation.

The final morpho-functional features of the muscle cells showcase contractile structures and functions. During differentiation, myoblasts express various myogenic regulatory factors, which alternately appear and disappear in a space and time-correlated manner. The different stages of myogenic differentiation process can be monitored by regular microscopic observations. The genotypic and phenotypic expression patterns of myogenic factors can be determined using real time-PCR and western blotting analysis. Similarly, cytoskeletal rearrangements of actin and myosin filaments can be visualized under fluorescence microscopy. The morphology and cell surface features can be easily imaged using scanning electron microscopy, whereas the intracellular structural details can be efficiently analysed by transmission electron microscopy (TEM).

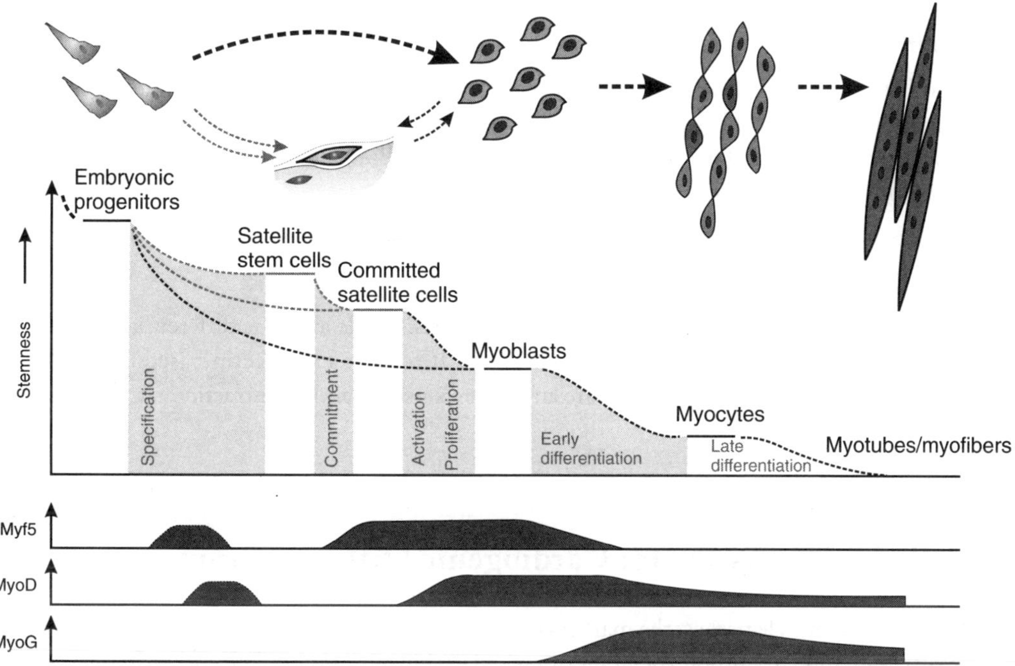

Fig. 9.27 Schematic showing myogenesis, a muti-step process involving the transcriptional activation and repression of several genes including MyoD, MyoG, Myf5, etc. [Adapted from Ref. 722] [see Colour Plate].

Some of the markers specific for myoblast cell differentiation are listed below. Typically in cultures, cells become positive for Myo-D. This is followed by cells expressing positive for both MyoD and myogenin (MyoG). The appearance of differentiated myocytes and myotubes expressing myosin heavy chain (MHC) can thereafter be distinguished[698].

Myo-D

It is an early differentiation marker. It is a basic helix-loop-helix transcription factor that plays key roles in the determination and differentiation of all skeletal muscle lineages. The upregulation of this gene in myoblasts activate transcription of MyoD target genes, resulting in irreversible cell cycle arrest and myogenic differentiation, whereas its down-regulation causes return of cells to quiescence state.

Myf-5

It is considered to be an early or commitment myogenic regulatory factor. Even though it is one of the earliest to express in muscle cells, it is also found in non-muscle tissues, such as preadipocytes and neurons. Of the various myogenic regulatory factors, Myo-D and Myf-5 are expressed in proliferating myoblasts.

Myogenin (MyoG)

It is a late differentiation marker. It is a muscle specific transcription factor important for the terminal differentiation and fusion of myoblasts into mature muscle fibres. It serves as a key developmental regulator for skeletal muscle formation. Myogenin deficient cells possess the ability to form myoblasts, but fail to fuse into myotubes.

MHC (Myosin heavy chain)

It is a terminal differentiation marker for muscle precursor cells and it is differentially expressed during development and maturation of muscle fibres. It belongs to the actin-based motor protein category of muscle thick filaments. They are largely responsible for the contraction rate of individual muscle fibres[699].

9.8 | Biological Assays for Cardiogenic Differentiation

Similar to bone or muscle tissues, the markers to confirm cardiomyogenesis or the pathway followed during differentiation to cardiomyocytes is discussed in this sub-section. During cardiomyogenesis, further maturation is indicated by increased calcium transients compared to primordial cardiomyocytes, as well as the formation of numerous ion channels[700]. Several unique transcription factors and proteins have been identified to characterize the stages of cardiac development (Fig. 9.28). Usually, cardiomyogenic events are characterized by the emergence of cardiac-specific markers, such as the expression of early and late stage markers namely Troponin T, NKX2.5, GATA-4, actinin 2, etc. The rhythmic beating of cells is considered as a functional marker of successful cardiac differentiation and this can be observed initially under an optical microscope. Through a series of molecular biology tools such as flow cytometry, immunofluorescence, real time PCR, microarray, ELISA and western blotting, the natural progression of cardiac maturation can be effectively studied. The changes in ultrastructure and well-developed sarcomeric structures with cell–cell coupling mainly via desmosomes and fascia adherens, can be precisely visualized under TEM. Moreover, intercalated discs containing gap junctions, desmosomes and fascia adherens can also be visualized in the differentiated cells using TEM or CLSM.

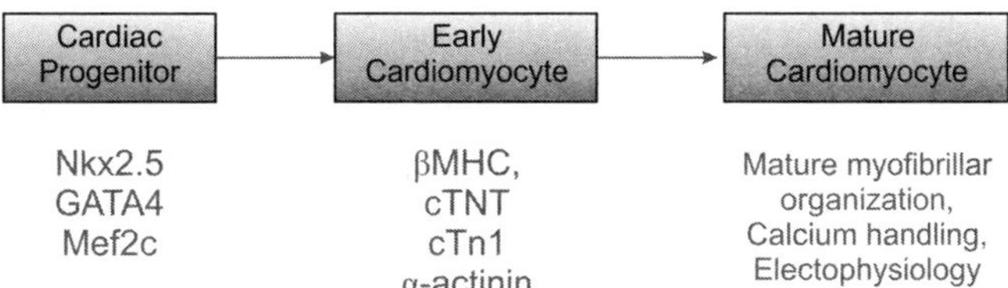

Fig. 9.28 Schematic displaying the three stages of cardiac tissue development along with associated expression of stage-specific differentiation markers.

Cardiomyogenesis is a multistep process that involves withdrawal of cells from the cell cycle, acquisition of a cardiac type-specific transcriptional program and morphological changes that include organized myofibrillar and sarcomeric patterning throughout the cell.

MYH6: myosin heavy chain 6, cardiac muscle, alpha

It is one of the late cardiac markers. It is one of the major components of the sarcomere, the building block of the contractile system of cardiac muscle. The MYH6 gene encodes the alpha heavy chain subunit of cardiac myosin (alpha-MHC), a fast ATPase primarily expressed in atrial tissue.

Nkx2.5: NK2 transcription factor related, locus 5

It is one of the earliest cardiac transcription factors seen in cardiac precursor cells or in immature cardiomyoblasts. It is a member of the NK2 class of haemoproteins and is detected in a tissue-specific manner, suggesting a role in tissue specification, differentiation, and patterning. It is even reported that Nkx2.5 is expressed in both undifferentiated cardiac precursors as well as in the differentiated myocytes later in the development[701]. Deletion of this gene in mice resulted in an arrest of cardiac development[702].

cTnI: cardiac troponin I

It is one of the late cardiac markers. It is one of the prime regulatory factors involved in cardiac muscle contraction and relaxation, by promoting actomyosin crossbridge formation. cTnI is the inhibitory subunit of the troponin complex that plays a central function in the intracellular calcium regulation during striated muscle contraction[703].

Cardiac α-actinin

It is an F-actin crosslinking protein that plays a pivotal role in the formation and maintenance of Z-line. The appearance of α-actinin as striated patterns, indicate the formation of sarcomeres. This binding protein is distributed at focal adhesion sites and along the cell–cell junctions.

Mef2c: Myocyte enhancer factor 2

It belongs to the family of transcription factors, which is expressed with the onset of myocyte differentiation in the cardiac mesoderm. It plays a key role in both cardiac development and hypertrophic remodelling. Even though Mef2c-null mice can undergo cardiomyogenesis, it will eventually die from a failure to undergo heart looping morphogenesis and proper development of the right ventricle[704].

GATA-4: GATA binding protein 4 transcription factor

The GATA binding proteins are zinc finger transcription factors that have been implicated in regulating the onset of cardiac myocyte differentiation. It is highly expressed in cardiac precursor cells, endocardial and epicardial cells of the heart, and is crucial for normal cardiac development. GATA4 can bind to the promotor sequences and can directly regulate the expression of several cardiac-specific genes, such as Myh6, cTnT, cTn1, etc. It is known to be a nuclear mediator of RhoA signalling and is involved in the control of sarcomere assembly in cardiomyocytes.

9.9 | Biological Assays for Neurogenic Differentiation

Neural progenitor cells can differentiate into various neural lineages including neurons, astrocytes, and oligodendrocytes. Generally, neural progenitor cells, terminally-differentiated neurons, and their differentiation intermediates express stage-appropriate markers, throughout the course of the differentiation protocol, which is orchestrated by many signalling pathways and transcription factors (Fig. 9.29). The differentiated cells exhibit specific function and morphology that are characteristic of mature neurons. The differentiated neurons can be analysed through various methodologies, including immunocytochemistry, flow cytometry, cellular imaging, gene expression and so on. Further functional evaluation, electrophysiology in particular, of these committed cells is necessary to characterize neural differentiation, as terminally differentiated cells exhibit typical characteristics of excitable neurons, including evoked action potentials.

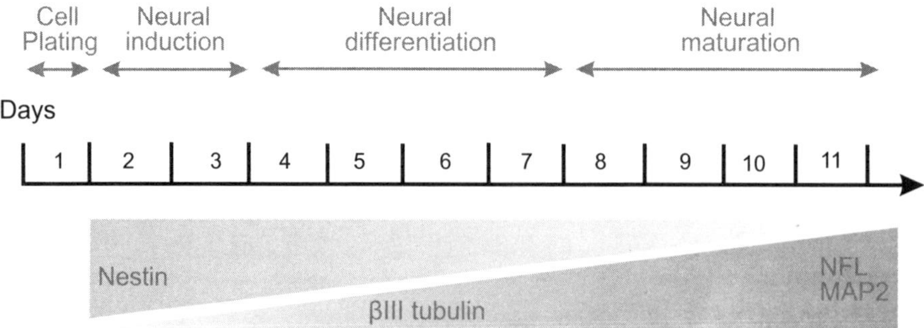

Fig. 9.29 Schematic representation showing the timeline of neuronal induction protocol *in vitro* and the phenotypic expression patterns of a few important neural markers during the differentiation and maturation phases.

Nestin

It is an early differentiation marker, which is expressed transiently in immature neural progenitor cells, and diminishes upon terminal maturation of neurons. It is an intermediate filament protein and it is a widely employed marker of multipotent neural stem cells. Mostly nestin-positive cells are identified to be engaged in active proliferation. Upon nestin downregulation, concomitant upregulation of other tissue-specific intermediate filament proteins occur, such as GFAP in astrocytes, and NFL in neurons.

Glial Fibrillary Acidic Protein (GFAP)

It is a type III intermediate filament protein, which is predominantly expressed in astrocytes. It has also been detected in the glial cells of the enteric nervous system and some Schwann cells in the peripheral nervous system.

Microtubule associated protein2

MAP 2 protein is a mature differentiation marker and it is thought to be involved in microtubule assembly, acting to stabilize microtubules growth by crosslinking with intermediate filaments and other microtubules. It is generally detected exclusively in mature post-mitotic neurons and reactive astrocytes, specifically in the perikarya and dendrites of these cells. It plays a key role in determining and stabilizing dendritic shape during neuronal development.

βIII tubulin

It is one of the major cytoskeletal proteins and it participates directly in axonal growth. It is one of the earliest indicators of microtubule reorganization that is required for neuronal differentiation. This protein is also regulated during peripheral and central nervous system development.

Neurofilament light

NFL is an intermediate filament protein which provides structural support for axons and regulates axon diameter. These cytoskeletal components also allow nerve cells to establish and generate highly asymmetrical cellular extensions. Of the 3 major components, [NF-L (light polypeptide), NF-M (medium polypeptide), and NF-H (heavy polypeptide chain)], NF-H appears later in neurons than do NF-L and NF-M and in particular, NF-M is reported to be synthesized first[705].

Summing it all, numerous regulatory factors play pivotal role in cellular maintenance, proliferation, and differentiation. However, it is difficult to determine the cell response/fate using a single marker. Therefore, it is necessary to apply multiple markers to the analysis, as molecular biology provides sophisticated and high-throughput screening tools to precisely find out the fate of cells.

9.10 | Cell Culture Laboratory-Testing, Safety and Ethical Issues

In this section, the general guidelines for cell culture experiments, related safety, and ethical issues are discussed. In general, a good biomaterials research group should ideally have all or most of the following laboratories:

1. Cytocompatibility evaluation laboratory
2. Tissue compatibility evaluation laboratory
3. Hemocompatibility evaluation laboratory
4. Sterility evaluation laboratory
5. Histopathology evaluation laboratory
6. Physiochemical property evaluation laboratory

The above set of laboratories would ensure the successful development of next generation biomaterials. As shown in Fig. 9.30, the representative cell culture lab facilities should have the most, if not all, of the array of equipment: (a) 3D culture bioreactor for growing cells into 3D scaffolds, (b) CO_2 incubators for cell culture on flat 2D samples, (c) fluorescence microscope, (d) micro-computed tomography, (e) class II, type A2 biosafety cabinet or higher level biosafety cabinets, (f) incubator and refrigerated centrifuge, (g) multimode plate reader/biospectrometer, (h) gel doc and gradient PCR, (i) deep freezers (-20 and -80°C), (j) Milli Q ultra purification system, and (k) state-of-art fume hood. In addition, the biomaterials processing facilities may include (a) rapid prototyping machines (e.g., 3D Printing, 3D plotting, selective laser sintering/laser engineered net shaping for designed scaffold fabrication), (b) high temperature furnaces for glass melting/ceramic sintering, (c) spin coating or dip coating to coat substrate with biocompatible coatings, etc. As far as the cell lines are concerned, the cell culture labs again can have, if not all, the following cell types: (i) muscle cells (C2C12 mouse myoblasts), (ii) bone cells (osteoblast-like cells :M3T3-E1, hFOB, SaOS2), (iii) L929 fibroblasts, (iv) Neural cells (Schwann cell line, RT4-D6P2T, mouse neuroblastoma, N2a cells) and stem cells (human bone marrow derived mesenchymal stem cells). In addition, the researchers need to use multiple biochemical assays/molecular biology techniques to solve various biologically/clinically relevant problems and these assays include, MTT assay, LDH assay, live/dead Staining, Annexin V/ PI apoptosis assay, cell cycle analysis by FACS (VybrantDyeCycle[TM] Violet Stain), cell proliferation by FACS (CellTrace[TM] Violet), ROS assay- DCFDA-CM, pro-inflammatory cytokine detection kit, PNPP substrate based detection of ALP activity, total protein content detection kit, ELISA kit for osteocalcin quantification. One needs to use various techniques, including immunofluorescence analysis for Actin, Vinculin, Vimentin, RUNX2, Nestin, βIII tubulin, desmin, etc., Alizarin red staining, BoCIP/NBT staining for ALP, SEM and biological TEM. In summary, a typical cell culture lab in a biomaterials group may appear to be very similar to a biology lab, but it is quite remotely connected to conventional materials science discipline.

9.10.1 | Good laboratory practice

It is worthwhile to reiterate here that, cell/tissue culture methods and related analysis, commonly referred to as cytocompatibility testing, are used as the first step to establish the biocompatibility of any new material. This is often accompanied by the toxicity assessment of a biomaterial, either in bulk or particle eluate form, using a battery of biological assays. It is possible to use some of the genotoxicity assays to confirm the capability of a biomaterial on the DNA fragmentation (gene-level toxicity), *in vitro*. Two widely used genotoxicity assays are single cell gel electrophoresis (SCGE) or Comet assay and the micronucleus (MN) assay[706]. The latter is conducted using flow cytometry. The use of such assays for HA-based biomaterials is reported elsewhere[707]. More discussion on the protocol of these two assays are discussed in author's recent book[708].

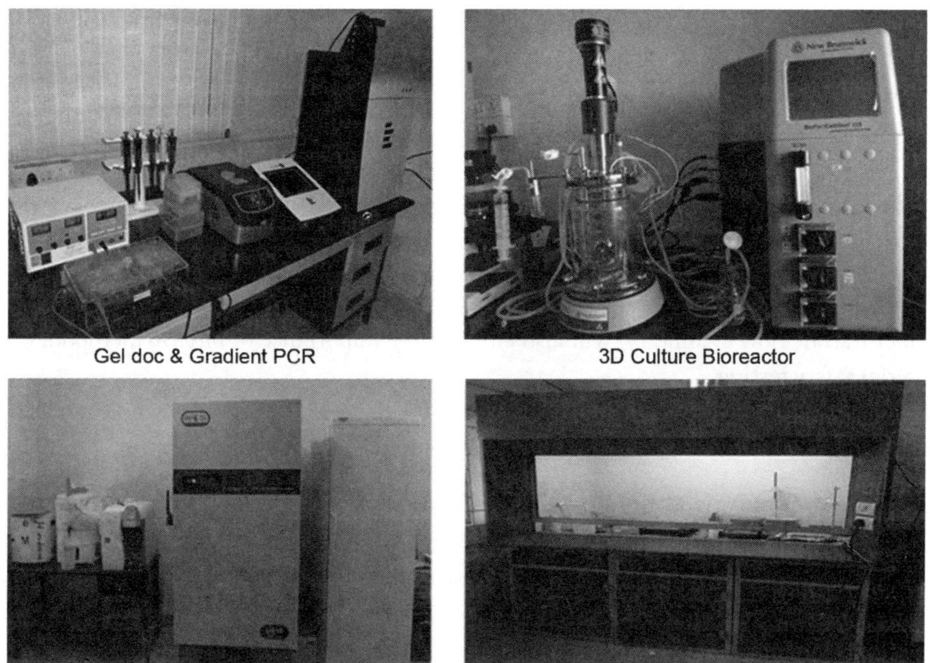

Gel doc & Gradient PCR

3D Culture Bioreactor

Deep Freezers (−20 & −80°C) &
water Ultrapurification system

State-of-art Fume hood

Fig. 9.30 Various basic research facilities related to *in vitro* biocompatibility assessment of biomaterials.

Good laboratory practice (GLP) is recommended in all leading biomedical research laboratories and GLP provides a framework of principles for performing cell culture studies. In particular, GLP guidelines prescribe as how to monitor cell culture test systems (i.e., standard operating procedures, SOPs) as well as how to record assay results and subsequently analyse such data from biostatistics perspective. GLP therefore assures regulatory authorities that the data in a report comply with the standard guidelines. GLP comes under FDA (Food and Drug Administration) regulation. GLP is sometimes confused with the standards of laboratory safety like wearing safety goggles, etc. Some of the important examples of poor lab practices include (a) use of improperly calibrated equipment, (b) incorrect/inaccurate accounts of the actual lab study, and (c) inadequate test systems.

GLP essentially refers to a set of principles to establish ethically acceptable framework as any *in vitro/ in vivo* study to produce reliable and high quality results/outcomes are to be planned, performed, monitored, reported and archived according to GLP guidelines.

In India, National GLP Compliance Monitoring Authority was established by the Department of Science and Technology, Government of India, with the approval of the Union Cabinet on April 24, 2002. Industries/test/facilities/laboratories dealing with manufacture of industrial chemicals, pharmaceuticals, veterinary drugs, pesticides, cosmetic products, food products, feed additives, etc. require approval from this regulatory authority.

Currently, attempts are being made to replace animal studies with *in vitro* studies for safety assessment. In order to ensure high quality data from these validated *in vitro* assays, the studies must comply with existing GLP norms.

In view of the growing concern over ethical issues to decrease the use of animals in safety testing, some attempts are made to correlate the *in vitro* and *in vivo* results. In a recent study, Hulsart–Billstrom et al.[709] reported the outcome of a multicentric study on this subject. The extensive analysis indicated a lack of any significant correlation between two sets of such studies, which establish biocompatibility of synthetic biomaterials for bone tissue engineering applications. Lately, the developments in the area of genomics, proteomics and various high throughput screening techniques (e.g., micro-array, flow cytometry) will also enhance the importance of *in vitro* methodologies for biomaterial safety testing.

Three common concepts as per the principles of GLP should be followed while conducting *in vitro* study in any laboratory and these are briefly mentioned below.

(a) The use of positive and negative controls and benchmarks: The *in vitro* experiments for biocompatibility assessment should always include both positive and negative control, which has to be similarly tested alongside to validate the consistency of the culture test results. Generally, the use of controls is vital for the critical assessment of results. A series of standard materials ranging from highly cytotoxic to nontoxic can be included to improve objective evaluation of biocompatibility testing. For example, while comparing the cytocompatibility of Hydroxyapatite-based biocomposites, cells grown on tissue culture polystyrene (TCPS) serves as positive control, whereas cells grown on HA will become the benchmark control. Cell culture medium without cells can be employed as a negative control.

In case of a given biochemical assay, the term positive control is associated with a maximum assay signal/expression and the term negative control is associated with a minimum signal/expression. A negative control (blank) helps to ensure that the false positives are not obtained. Benchmark control includes substances that respond in a similar way as the test articles or have similar chemical structure or physico-chemical properties as the test substances.

(b) The care and maintenance of test systems: During all the *in vitro* studies, the cell population must be inspected at various time points in culture, prior to and throughout the study. In order to avoid potential contamination, appropriate separation should be maintained between different test systems. In particular, the cell and bacteria culture rooms must be physically separated with a wider distance. Any slight variation between media composition (with respect to reagents) has the potential to alter the results of cellular activity.

(c) Quality assurance monitoring: Specific areas to be inspected may include, but not limited to, the procedures and measures in terms of the following aspects:
  • monitoring of batches of components of cell and tissue culture media (e.g., FBS, etc.)
  • assessing and ensuring functional and/or morphological status (and integrity) of cells and tissues

- monitoring for potential contamination by mycoplasma and other pathogens, as appropriate
- recommended use of sterile equipment and reagents, and wash hands, reagent bottles, and work surfaces with a biocide or 70% ethanol before entering cell culture room
- ensuring sterility for materials, equipments, etc., as well as proper cryopreservation and reconstitution of cells and tissues

9.10.2 | Cell culture maintenance

Extending the discussion on GLP norms, the researchers must be aware of the necessity to closely monitor the cell growth medium and culture conditions to ensure desired growth pattern. These aspects are discussed below.

Cell culture media

The growth of biological cells demands the use of specific cell culture medium and the composition of medium depends on cell type. Most often, DMEM medium is used which contains 13 essential amino acids, Vitamins, serum, antibiotics, sugar (glucose), trace elements (Fe, Zn, etc.), bulk ions (Na^+, Ca^{2+}, Mg^{+2}) or bicarbonates. For longer culture duration, the culture medium needs to be replaced 2–3 times per week. In cell culture medium, the use of fetal bovine serum, which contains a large number of growth promoting factors, helps in neutralizing trypsin/other proteases to ensure healthy growth.

Although, not required for cell growth, the antibiotics are added to cell culture medium in order to inhibit bacterial growth and fungal contamination. During the cell culture experiments, an inverted phase contrast microscope can be used to observe cell morphology. While live cells appear as phase bright, the suspension cells are typically rounded. In contrast, adherent cells form projections, when they attach to any biomaterials substrate. Alternatively, the cell viability can also be assessed using trypan blue. Such stains exclude live cells, but accumulate in dead cells. Quantitatively, the cell numbers are determined using a hemocytometer.

During an *in vitro* experiment, the cultures should be examined daily for signs of contamination and to decide if the culture needs feeding or passaging. A tissue culture log should be maintained, independent of the regular laboratory notebook. The log should contain the name of the cell line, the medium components and any alternation to the standard medium, the passage number and a calculation of the doubling time of the culture, and any observation relative to the morphology, etc.

The culture conditions during cytocompatibility must ensure an optimum growth of cells, and to this end, the three important parameters include, (a) temperature, (b) humidity, and (c) CO_2 level. In the following, each parameter has been briefly analysed.

Cell incubation conditions

The optimum growth of cells in order to achieve best quality cells (highest growth potential) are only attained when all the above parameters are maintained and controlled, independent of each other, inside a CO_2 incubator. The cells have a narrow temperature range for optimum growth. A hypothetical diagram of cell growth rate with temperature of incubation is shown in Fig. 9.31.

Among three representative temperatures, the cell growth rate is the highest at T_2 and is reduced at both lower/higher temperatures than T_2. Also, a given cell type has the highest growth arte potential only within a narrow temperature window of $\pm$ 2°C of T_2. Such a brief discussion clearly emphasizes the need for close temperature control in a CO_2 incubator, while allowing cells to grow on a material substrate. Any deviation from the optimum range results in a reduced cell harvest, in worst case to the death of the cell. For optimal growth condition, most mammalian cell lines are maintained at 37°C in 5% CO_2 medium.

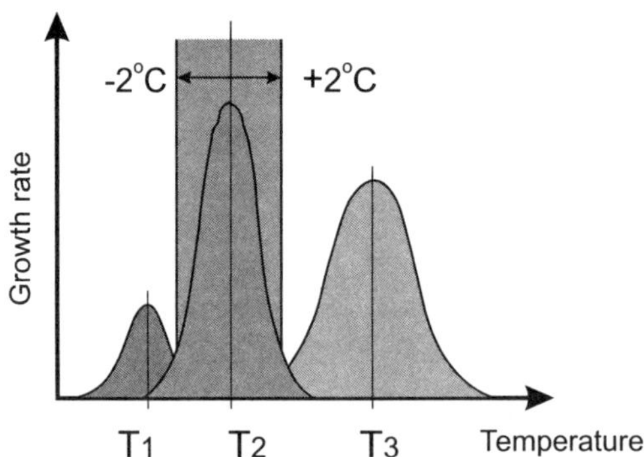

Fig. 9.31 Schematic illustration of cell growth rate at three hypothetical incubation temperature, showing maximum growth rate at T_2 with a sharp reduction within $\pm$ 2°C around T_2.

Moreover, the cells should be cultivated in the presence of a saturated humid atmosphere (95% relative humidity) in order to prevent the dehydration of media. The important factor is maintaining humidity in a controlled manner, while at the same time avoiding condensation in the interior. The optimal pH working range for mammalian cells is between 7.2–7.4. The growth medium and CO_2 buffers the cells against pH changes. Usually, this is ensured by adding bicarbonate based buffer to the culture medium. Because the pH of the medium is dependent on the delicate balance of dissolved CO_2 and bicarbonate, the changes in the atmospheric CO_2 can alter the pH of the medium. It is possible to control this by artificially supplying CO_2 to the incubator and preventing the gas from leaving the liquid, thereby reducing the hydroxyl ion concentration in solution. The researchers generally use 5–7% CO_2 in air for most cell culture experiments.

Growth pattern: During the cell passaging, a given cell type initially goes through a quiescent or lag phase depending on the seeding density, the media components, and previous handling. Finally, the cell population will undergo a transition through exponential growth while exhibiting the highest metabolic activity. Many adherent cell cultures will cease proliferating, and some will die, if they are left in a confluent state for too long.

Harvesting: Cells are harvested when they reach a population density, which suppresses growth. Ideally, it is the semi-confluent state, yet in the log phase, at which the cells are detached and re-plated. One point of caution is that a cell population, if not passaged and instead allowed to grow to a confluent state, can lag for a long period of time without recovery. It is important to ensure that cells remain healthy as much as possible. In general, most cells are passaged three times a week.

Depending on the ability of cells to attach onto a substrate, two types of culture are commonly used in cell biology experiments. The suspension cultures are those where cells are suspended in a culture flask and are constantly being fed by dilution into fresh medium. In contrast, adherent culture can be fed by replacing the fresh medium by the old medium. In order to harvest confluent monolayer of adherent cells, the cell culture media should be completely removed from the culture vessel. The culture vessel should be gently washed with phosphate buffered saline (PBS) to remove any traces of serum, calcium, and magnesium that would inhibit the action of the dissociation reagent. After discarding the wash solution, the culture vessel is incubated with pre-warmed dissociation reagent at 37°C for approximately 2 minutes. The cells can be observed under the microscope for detachment. Once more than 90% of the cells are detached, pre-warmed complete growth medium should be added to neutralize the enzymatic reaction. Subsequently, the cells are spun down and the cell pellet is re-suspended in fresh culture medium. The total number of cells and percent viability can be determined using a hemacytometer, cell counter and Trypan blue exclusion. Once the cells attain the state of semi-confluency, some of the following approaches can be utilized to remove the cells from the biomaterial substrate for the purpose of dilution and reculturing.

- Mechanical- This is relatively a faster method, but can be disruptive to the cells and may result in significant cell death, as this method involves a rubber spatula.
- Proteolytic enzymes- This technique uses the proteolytic enzyme trypsin to detach adherent cells from the surface of a cell culture vessel. Trypsin, collagenase, or pronase, usually in combination with EDTA, causes cells to detach from the growth surface. Among all these, trypsin is the most frequently used enzyme for passaging cells.
- EDTA- The use of EDTA is a gentler way to detach cells than trypsin. EDTA is a divalent chelator that will remove the Ca^{2+} ions that integrins require to maintain cell adhesion.

9.10.3 | Safety Considerations

The following safety precautions should also be observed by all biomedical researchers, while performing cell culture experiments:

- Creation of aerosols should be avoided
- Hands should be washed before and after working with cell cultures
- Sanitize the work surfaces with disinfectant (before and after) or whenever necessary
- Sterile flasks, bottle, petri dish, etc., should not be opened outside the hood
- Autoclave all waste
- All liquid waste should be disposed after each experiment and should be treated with bleach

If the appropriate safety considerations are not followed at various points while conducting the cell culture experiments, the researchers may need to spend longer than expected timeframe to obtain any biologically meaningful results from a given set of experiments. Therefore, the training on the biological safety issues should be imparted to the researchers, before they enter the cell culture lab. In view of the above safety considerations, wet bench experiments, before a culture is put into a CO_2 incubator, are conducted in biological safety cabinet. In this sub-section, various levels of biological safety are briefly discussed.

It is the responsibility of every researcher to follow existing ethical and safety guidelines, while performing cell culture experiments with biomaterials as well as while using materials of natural origin, in laboratory.

9.10.3.1 | Biosafety levels

Biosafety levels are a means to determine the level of danger that the work done in a facility poses to its users, other users present, as well as the environment. Biosafety levels 1 through 4 were established by the Centres for Disease Control (CDC) and the National Institutes of Health (NIH). These levels are a combination of laboratory practices and techniques, safety equipment and facilities. All of these levels are appropriate for the biohazard posed by the agents used and for the laboratory activity.

Biosafety Level (BSL)-1

BSL-1 can be defined as those practices, safety equipments and facilities, appropriate for working with well-characterized agents, which are not known to consistently cause disease in healthy adult humans and are also of minimal potential hazard to laboratory personnel and the environment. Neither special practices nor primary safety barriers are required, while working with these agents.

Biosafety Level (BSL)-2

It is suitable for work involving agents of moderate potential hazard to personnel and the environment. Generally, these agents are known to be associated with human disease in varying severity. Especially, it involves working with those agents that spread via percutaneous injury, ingestion, mucous membrane exposure. Class I or II Bio safety cabinets are used for the manipulation of agents that cause splashes or aerosols of infectious materials. Further, personnel protection equipments, such as laboratory coats, gloves, face protection are needed. It also requires lockable doors for facilities, that house restricted agents. The laboratories should be away from public areas. Also, furniture used in laboratory has to be covered with a non-fabric material that can be easily decontaminated. In addition, an eye wash station should be readily available.

Biosafety Level (BSL)-3

This safety level is applicable when culture experiments involve indigenous or exotic agents that may cause serious or potentially lethal disease as a result of exposure by the inhalation route. Safety equipment/primary barriers, such as biosafety BSL III cabinets are used along with personnel protection equipment. The laboratory should also possess special engineering and design feature.

Biosafety Level (BSL)-4

BSL-4 are those practices, safety equipment and facilities, which are appropriate to work with extremely dangerous/exotic agents having high risk of life-threatening disease, aerosol-transmitted lab infections or related agents with unknown risk of transmission. The members of the laboratory should have specific and thorough training in handling extremely hazardous infectious agents and they understand the primary and secondary containment functions of the standard and special practices, the containment equipment and the laboratory design characteristics. All procedures are conducted in Class III Bio safety cabinets in combination with full-body, air-supplied, positive pressure personnel suit. The facility will also have dedicated supply and exhaust, vacuum, and decontamination systems.

9.10.3.2 | Biological safety cabinet

Biological Safety Cabinet (BSC) and biohazard cabinet refer to a variety of containment devices equipped with high-efficiency particulate air (HEPA) filters, which are designed to provide personnel or both personnel and product protection from biohazardous materials. The term BSC can follow Class I, II or III specifications, based on their construction, air flow velocities and patterns, and their exhaust systems. There are two key elements to consider while choosing the correct class and type of cabinet for biomaterial research, such as the type of hazard containment and control required by the application and the exhaust capacity of the building, where the cabinet will be installed. The following tests are to be carried out to check the efficiency of BSC. (i) Impulsion filters integrity test, (ii) Exhaust filters integrity test, (iii) Inflow filters integrity test (Cytostar), (iv) Downflow air velocity test, (v) Inflow air test, (vi) Smoke test, (vii) Light test, (viii) Noise level test, and (ix) Alarm test. International Standards are mainly issued by two organizations: (a) International Standard Organization (ISO), based in Geneva and (b) Comitee Europeen de Normalization (CEN), based in Brussels. The aim of the standards is to construct a mechanism to provide the manufacturer and the user of a common site to specify technical specifications and features. In USA, NSF-49 was one of the first standards on this subject and was the reference point for a lot of manufacturers. NSF (National Sanitation Foundation) is an independent organization established in 1976, that works as neutral agency to help on the problems that could affect public health and environment. Later standards are upgraded version, but using some common concepts, such as, (a) American Standard NSF 49, (b) European Standard EN 12469, (c) Australian Standard AS 2252, and (d) Japanese Standard JIS K 3800. While or prior to the usage of BSC, the sterility inside the cabinet should be ensured and in particular, the biosafety cabinet air flow gauge should be checked for relative pressure drop across the HEPA filter to assure proper operation of the cabinet before placing any material.

9.10.3.3 | Risk assessment

The degree of hazard during cell culture experiments depends on the cells being used and the experimental protocol. While following safe practices in a biological laboratory, appropriate concern must be exercised for some of the most poorly appreciated risks, such as those arising from genetic manipulation. It is the duty of the researcher to exercise stringent practices based on laboratory specific conditions. The risk assessment process should consider a number of factors, including (a) virulence (the degree of toxicity potential), (b) infectious dose (that required to cause infection in humans), (c) antibiotic resistance, (d) transmission route (parenteral, ingestion, mucous membrane exposure, or inhalation), and (e) pathogenicity (the ability of an organism to cause disease). It is important to remember that use of recombinant DNA may modify any of the above risk factors and necessary modification should be considered when working with recombinant microorganisms[710].

9.10.4 | Ethical Considerations

The majority of *in vitro* tests carried out on evaluating the biocompatibility of materials or devices encompass the use of animal and human cell cultures, as suggested in the International Organization of Standardization (ISO) protocols. Two culture protocols can be followed, (a) direct culture of cells on the material and (b) extracting leachable materials from the material, before exposing them to cells. Such protocols can be followed by either quantitative or qualitative analysis of cell viability. In order to reduce some of the above mentioned risks, a thorough biosafety risk assessment is recommended, and those, in general, include implementation of appropriate containment measures and work practices. Such practices are expected to provide maximal protection of human health.

> A cell line is derived from a primary culture at the time of the first successful subculture. The term, cell line, implies a number of cell lineages originating from the primary culture. The terms, finite or continuous cell lines are used as a prefix depending upon whether it has limited culture life span or it is immortal in culture.

In the context of cell culture, a few terms need to be defined. A primary cell culture can be defined as a culture of cells, tissues or organs from organisms. Since these cells underwent very few population doublings compared to continuous cell lines, they are more representative of *in vivo* studies.

Even though mammalian cells are reasonably considered free of adventitious contaminating pathogens, precautionary measures are always advised for animal cell cultures. These measures are expected to reduce the risk of contamination with adventitious agents by ensuring protection of both researcher and cell culture.

Two main types of stem cells, that are used in biomaterials research include, (a) embryonic stem cells (ESC) (extracted from human embryos) and (b) adult stem cells (ASC). As expected, the source of isolation of embryonic stem cells causes major public concern in many countries worldwide, more so for clinical research involving ESCs. On contrary, the adult stem cells utilize the patient's own cells and thus, do not pose any serious ethical conflict. Also, the different ways to reprogram adult stem cells should pave the way to produce customized stem cells with the genetic material of the individual patient. Currently, the uses of human adult stem cells for biomaterials research are frequently being investigated[711].

For sustained availability of stem cells for regenerative medicine and therapy, the stem cell banks of ethically sourced quality cell lines are currently available worldwide to store stem cell lines under biologically appropriate conditions. It may be worthwhile to mention that a stem cell bank is composed of a stem cell repository, (biological specimens stored), and a stem cell registry (data bank containing information about stem cell lines). The ethical clearance requires one to comply with the following documentation, which includes

(a) Proof of institutional ethical and legal approvals;
(b) Donor informed consent documents;
(c) Confidentiality on the information of donor;
(d) All scientific details, like culture conditions, passage number, etc.

The stem cell guidelines can vary from country to country. In India, the 'Ethical Guidelines for Biomedical Research on Human subjects', released by Indian Council of Medical Research (ICMR) has provided certain requirements for carrying out 'stem cell research and therapy' (SCRT). The investigators should ensure that the cell lines are used in accordance with the existing guidelines of the country, as laid down in an appropriate Material Transfer agreement (MTA). While using stem cells, an investigator has to declare the level of manipulation, that a given stem cell line is expected to experience during culture on biomaterials.

9.10.4.1 | Cell/Tissue banks

A tissue bank (or biobank) is defined as a repository for a broad and heterogeneous range of tissues or cells, that share common biological features.

Broadly, the tissue banks store biological materials with varying amounts of medical, demographic and epidemiological data. A number of cell banks, as listed below, are available to collect, store and distribute human tissue derived cells with requisite ethical approval for subsequent downstream research. For various cell types, the growth medium, growth characteristics, plating efficiency, sterility and virus susceptibility are to be made available to the researchers, when they obtain a cell line from any of the following sources.

(a) Cell Bank Australia
(b) American Type Culture Collection (ATCC)
(c) Coriell Cell Repository
(d) Deutsche Sammlung von Mikroorganismen und Zellkulturen (DSMZ)
(e) European Collection of Animal Cell Cultures (ECACC)
(f) Health Science Research Resources Bank (HSRRB), Japan
(g) Japanese Collection of Research Bioresources (JCRB)
(h) NIH Stem Cell Unit
(i) RIKEN Gene Bank
(j) UK Stem Cell Bank (UKSCB)
(k) WiCell

One of the cell banks, ATCC maintains nearly 4,000 cell lines, such as HeLa cells, L929, hFOB, etc., as well as over 70,000 microbial cultures.

9.10.4.2 | Quality control tests

Once a cell line is procured from one of the above cell banks, the researchers must perform most of the following quality control tests (but may not be limited to)[712]:

(a) Viability and functional assessment testing,
(b) Cell culture morphology and dynamics,
(c) Karyotypic and genomic stability, and
(d) Phenotypic profiling of stem cell lines using a panel of antibodies, to a range of markers known to be expressed in undifferentiated and differentiated stem cell lines

9.10.4.3 | Material-transfer agreements (MTA)

Prior to the start of any stem cell based research on biomaterials, one has to identify the type of stem cells. Irrespective of whether the cell line to be obtained from a hospital/clinical research centre/a stem cell bank, the researcher has to get into the Material Transfer Agreement (MTA). This is a legally binding contract to allow the transfer of biological/non-biological materials between two organizations with the endpoint objectives of research purposes[713]. Biological materials, such as biochemical reagents, cell lines, plasmids and vectors are some of the materials, which come under MTA regulation. In all cases, MTA would define the intellectual property/patent rights of the provider and recipient w.r.t the materials and any derivatives[714].

If primary human tissue or cells are involved, the MTA should include a statement of ethical approval and informed consent. Also, the recipient should confirm that, he/she will be responsible for using, storing and tracking the material.

Three types of MTAs are most common at academic institutions: (a) transfer between academic or research institutions, (b) transfer from academia to industry, and (c) transfer from industry to academia. It is important to mention that once a cell line procured through MTA, should not be passed on to a third party without consent and neither be used for commercial exploitation.

The procurement or sharing of cell lines across institutions facilitates scientific progress. This is in view of the fact that cell lines may not remain the same or possess the same property over repeated passaging. Cross-contamination of cell lines is a longstanding problem and a repeated and frequent cause of scientific misrepresentation[715].

Authentication and standardization of cell lines are now possible by using the short tandem repeat (STR) profiling techniques developed for forensic applications. It is worthwhile to mention that while reporting the cell studies in a research paper, it is recommended to mention the source of a cell line as well as the state of a cell line prior to seeding on a biomaterial substrate. Also, detailed culture protocol must be mentioned in any research article.

9.11 | Closure

In reference to the discussion on the multitude of *in vitro* bioassays as well as consideration of ethical issues therein, the following aspects can be useful for a beginner in biomaterials science to design *in vitro* experiments and related analysis. The reader can follow additional references while using specific cell lines[716]:

(a) The necessary approval from institutional biosafety committee needs to be obtained while establishing any cell culture facility and conducting cell culture experiments for *in vitro* study.

(b) For the use of cells to be obtained from human tissues or to use human blood for haemocompatibility study, one has to take human ethics committee's approval.

(c) Depending on targeted endpoint biomaterial application, a cell line needs to be selected to study its compatibility or interaction with a given biomaterial substrate.

(d) For the use of stem cells, the necessary approval from institutional stem cell research committee is to be obtained.

(e) One has to carefully assess the health status of biological cells before culturing them on a biomaterial. It is recommended to use n<10 (n = passage number) so that one would know that the cells are on growing stage.

(f) For a regular routine study, a researcher can grow the chosen cell line for 3–4 time intervals, depending on the doubling time of the selected cell type and thereafter use MTT/LDH assay to quantify cell viability as well as to use fluorescence microscopy/SEM to record the cell morphological changes.

(g) One has to be careful in selecting a biochemical assay and the reagent used in an assay should not interfere/react with the materials substrate to give unreliable reading. For example, WST-1 assay is to be used to quantify metabolically active cells on carbon based materials, instead of widely used MTT assay.

(h) The cell numbers can be counted for relatively large number of cells at a given time point in culture and a systematic increase with time in culture is a signature of the fact cells are able to grow on the given substrate.

(i) In order to probe any cellular functionality change, one has to grow cells (e.g., stem cells) for 3–4 week time and observe the signatures of any visible change in morphological characteristics.

(j) On the basis of the careful fluorescence microscopy observations, one can further use various early or late differentiation markers to assess up or down-regulation of specific genes w.r.t housekeeping genes in terms of fold changes. This requires careful PCR analysis.

(k) For the study of blood compatibility, ISO guidelines can be followed to allow a material to be immersed at 37°C for 30 minutes and thereafter to assay blood parameters.

(l) For many studies, it is often designed that, two or three cell types of relevance towards final application can be used. For example, fibroblast and osteoblast can both be used for bone replacement applications and similarly for neural applications, neuroblastoma and schwann cells.

(m) For all *in vitro* experiments, appropriate control (e.g., tissue culture polysterene, TCPS) as well as benchmark controls (e.g., pure HA as benchmark control for HA based composites) are to be used and all the microscopic observations or quantified biochemical assay results are to be compared among test and control.

(n) An important aspect in the quantitative analysis involved in probing cell responses is the statistical analysis. It is therefore strongly recommended that the quantitative data related to cell responses are to be analyzed using statistical tools such as SPSS software or Anova or, students' t-test at different levels of significance, before reporting them in any research publication or scientific document. The involvement of a biostatistician may be necessary.

Bacterial Growth and Biofilm Formation

Microbial infection has been widely recognized as a major cause for numerous biomedical devices, particularly for orthopedic implants. In the pursuit to develop antimicrobial biomaterials with resistance towards prosthetic infection, one has to first understand the phenomenon of bacterial adhesion and growth on biomaterial surfaces. Starting with a recapitulation of the bacterial cell structure, this chapter discusses the thermodynamic aspects of bacterial adhesion on biomaterial surfaces, followed by various parameters influencing bacteria–material interaction. As a logical continuation, the subsequent discussion in this chapter focuses on bacterial growth kinetics and biofilm formation. The chapter concludes with the relevant assays to qualitatively and quantitatively assess antibacterial properties, *in vitro*. Some guidelines to conduct relevant *in vitro* experiments are also mentioned. It is expected that this chapter will provide the readers a platform to understand the fundamentals of bacteria–material interaction and such understanding is necessary to follow some of the case studies reported in this book and elsewhere[723,724].

10.1 | Introduction

One of the major issues limiting the long-term use of many implanted or intravascular devices (e.g. joint prostheses, heart valves, vascular catheters, contact lenses and dentures) is the prosthetic infection[725]. From a clinical perspective, this involves implant contamination via bacterial migration through surgical incision channels to biomaterial surfaces in the host. The endpoint outcome is the premature implant failure, necessitating the revision surgery. The microbiological profile of biomaterial associated infections is determined by the nature of surgery, implant surface properties/chemistry and the severity may vary depending on implantation site.

At a longer time scale, the bacterial colonization leads to the formation of biofilm. The biological entity to generate an entire biofilm is the micro-colony, comprising of microorganisms. In simplistic terms, biofilm can be considered as a community of micro-organisms held together by a matrix, wherein the microorganisms cooperate and interact with one another synergistically towards the pathogenesis of the infected site. The polysaccharide matrix, similar to extracellular matrix, is formed due to secretion of proteins from micro-colony. This matrix also contains channels to facilitate the necessary diffusion of nutrients and oxygen during biofilm growth, while the metabolic cellular waste products are moved out of the biofilm.

> It is widely recognized now that the bacterial adhesion and colonization on biomaterial surfaces is the first step in the pathogenesis of the prosthetic infections.

Like different cell types, as mentioned in previous chapters, the microbes are available in biological systems in a variety of strains/species. While the biofilm formation can be contributed by a single microorganism/species, the formation of biofilm due to the co-operative interaction among multiple species of microbes (polymicrobial communities) is also reported. In certain conditions, the cells of identical species embedded within the microscale layers of a biofilm can express diverse phenotypes. These microscale layers create gradients of oxygen partial pressure, pH and other environmental changes. The most common pathogens implicated in hospital-acquired and implant-associated infections are *S. aureus* and *S. epidermidis* (both, Gram-positive). *S. aureus* is typically implicated in infections of metallic biomaterial, bone-joint, and soft-tissue, while *S. epidermidis* is more commonly detected in polymer based implants.

Among all the pathogenic bacteria, *S. aureus* possesses a specific category of cell surface receptors, called microbial surface components, recognizing adhesive matrix molecules (fibronectin binding proteins, clumping factor binding proteins, collagen binding protein) [726,727,728]. These receptors facilitate bacterial attachment to biomaterials and to ECM proteins adsorbed on the biomaterial surface after implantation.

10.2 | **Generic Description of Bacterial Cell Structure**

Bacteria, a variant of microbes, can be described as unicellular prokaryotic microorganisms, typically with dimensions of a few micrometres (0.5–2 µm) in length. Morphologically, bacteria occur in a range of shapes (spherical to rod to comma-shape and spiral). Bacteria can grow faster than eukaryotic cells and bacterial cell division is normally accomplished by binary fission.

The morphological features of a bacterial cell are shown in Fig. 10.1. Typically, a bacterial cell contains a characteristic lipid membrane, or cell membrane and its structure varies, depending on bacterial strains. Being prokaryotes, bacteria do not have membrane-bound organelles and also contain a few intracellular structures (less complex structure than eukaryotes). Characteristically, bacteria lack a well-defined nucleus, mitochondria as well as other organelles such as the Golgi apparatus and endoplasmic reticulum. Unlike eukaryotes, DNA in bacteria is a supercoiled single circular chromosome and such a structure is suspended with associated histone proteins and RNA as nucleoid. Apart from genomic DNA, bacteria possess extrachromosomal DNA called plasmids, which can be transferred between the cells by horizontal gene transfer. For motility/locomotion, bacteria have flagella which are rigid protein structures, about 20 nm in diameter and up to 20 µm in length. Flagella operate by the energy generated from the movement of ions guided by an electrochemical gradient traversing the cell membrane[729]. Additional cell surface appendages, called fimbriae are believed to be involved in the attachment to solid surfaces or to other cells and are essential for the virulence of some bacterial pathogens. Fimbriae are ultrafine filamentous proteins, about 2–10 nm in cross sectional diameter and many micrometres in length. They are present on the outer surface of the cell, and appear as fine hairs under an electron microscope[730]. Pili (*sing.* pilus) are extracellular structures, marginally bigger than fimbriae. The functional role of Pili is genetic material transfer between bacteria during bacterial conjugation.

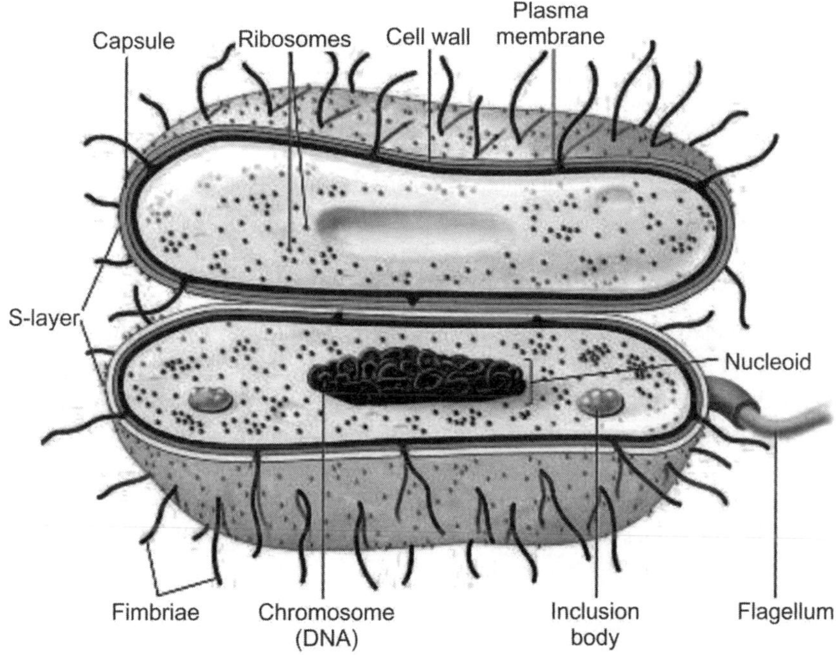

Fig. 10.1 Schematic illustration showing the anatomy of a bacteria cell[787] [see Colour Plate].

10.3 | Classification of Bacteria

There are numerous types of bacteria in the environment. Before the discovery of the DNA sequencing technique, bacteria were mainly classified on the basis of their shape, metabolism biochemistry and response to Gram staining (Gram-positive and Gram-negative).

10.3.1 | Classification based on shape

Some bacteria have different shapes, which are more complex than simple rod shapes, as shown in Fig. 10.2.

Most bacteria have four main shapes: rod (bacilli), spherical (cocci), spiral (spirilla) and vibrio (comma shaped).

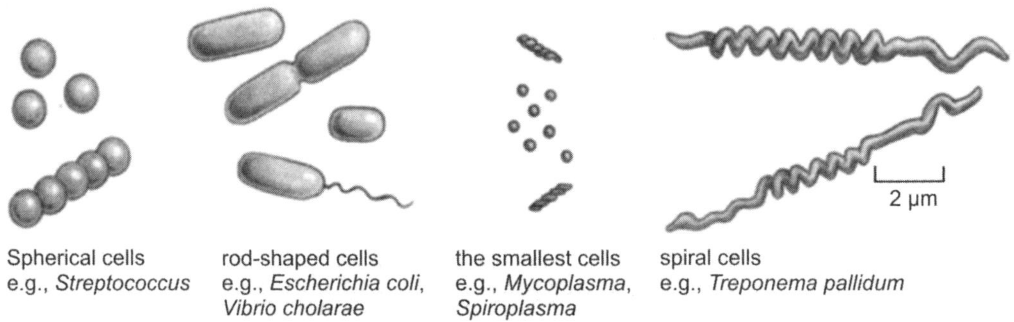

Fig. 10.2 Different kinds of bacteria based on their shapes[788].

10.3.2 | Classification based on energy of metabolism

Bacteria can also be classified on the basis of the requirement of oxygen for their metabolism and energy production. The type of bacteria, which need oxygen for metabolizing nutrients to release energy are called aerobic bacteria. Anaerobic bacteria are those which do not require oxygen for respiration. The latter bears oxygen and may die, if kept in an oxygenated environment. Under both categories, obligate and facultative aerobes and anaerobes exist. Obligate aerobes need oxygen for respiration as they cannot ferment to release energy, while obligate anaerobes are poisoned by oxygen. Facultative anaerobes can respire with or without oxygen, while microaerophiles need a specific oxygen, gradient for their respiration.

10.3.3 | Classification based on Gram staining

Bacteria can also be categorized as 'Gram-positive' and 'Gram-negative', based on their appearance after Gram-staining procedures. There are four steps in the Gram-staining protocol. First is the application of a primary stain (crystal violet) to a bacterial smear heat-fixed on a glass slide. This step is succeeded by treatment with a trapping agent (Gram's iodine), followed by decolourization of the primary stain with alcohol or acetone. Finally, the sample slide is counterstained with another dye, such as safranin. After the decolourization step, the Gram-positive cells remain purple due to the retention of crystal violet while the Gram-negative cells are decolourized and exhibit the colour of the counter-staining dye. The Counterstain is usually positively charged safranin or basic fuchsin, and imparts a pink or red colour to the decolourized Gram-negative bacteria. The differential response of the two types of bacteria arises from the differences in the cell wall structures.

> Gram-positive bacteria have a cell wall composed of a thick peptidoglycan (50–90% of cell wall) layer, which stains purple with Gram's stain. In contrast, Gram-negative bacteria have an outer lipopolysaccharide membrane and an inner thin peptidoglycan layer (10% of cell wall), which is delineated from the cell wall by the periplasmic space.

A comprehensive view of the 3D structure of peptidoglycan (responsible mainly for structural rigidity and osmotic balance) in the bacterial cell wall has been described by Samy et al.[731]. The differences in cell wall structure between the two bacterial types are shown in Fig. 10.3. A more comprehensive distinction between the two types of bacteria in terms of the cell wall structure, composition and function is summarized in Table 10.1.

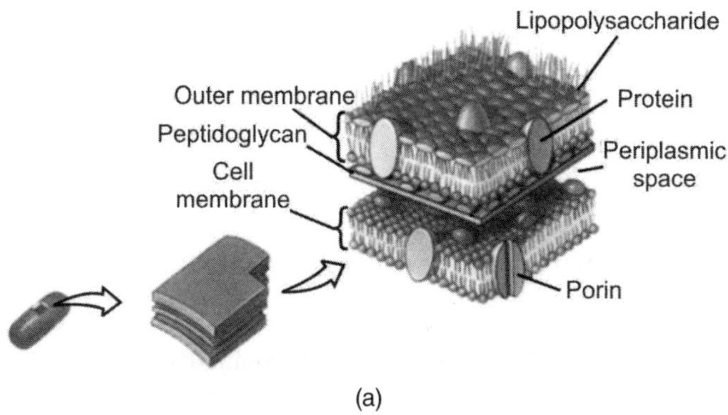

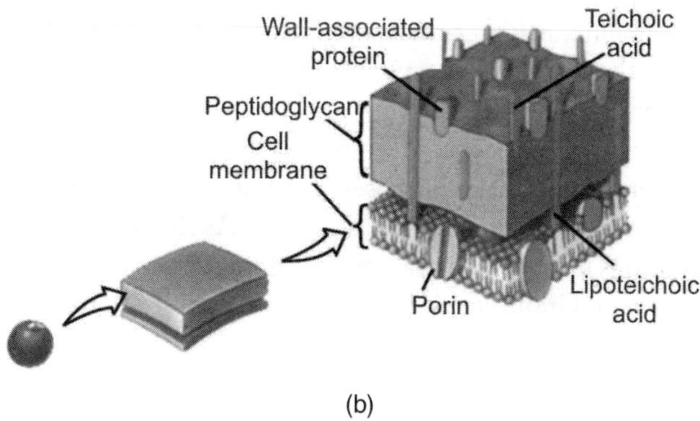

Fig. 10.3 Cell wall structure of (a) Gram-negative and (b) Gram-positive bacteria; adapted from[789] [see Colour Plate].

Table 10.1 Difference in the cell wall structure, composition and function in Gram-positive and Gram-negative bacteria.

Characteristics	Gram-positive	Gram-negative
Gram Staining	Retention of crystal violet dye leads to violet/purple coloured bacteria	Decolourization followed by counter staining with safranin attributes red colour to bacteria
Peptidoglycan Layer	Thick (Multi-layered)	Thin (mono-layered)
Teichoic Acid	Present in most	Absent
Periplasmic Space	Absent	Present
Outer membrane	Absent	Present
Lipopolysaccharide (LPS)	Negligible or absent	High
Amount of Lipid and lipoprotein	Generally Low (few acid-fast bacteria have lipids linked to Peptidoglycan)	High (due to presence of outer membrane)
Flagellar structure	2 rings in basal body	4 rings in basal body
Toxins	Principally exotoxins	Chiefly endotoxins
Resistance to physical rupture	High	Low
Inhibition by basic dyes	High	Low
Sensitivity to anionic detergents	High	Low
Tolerance to drying	High	Low

10.3.4 | Classification based on food/nutrient source

Based on the ability of bacteria to synthesize their own food or on external sources for their nutrient supply, they are classified into autotrophic and heterotrophic bacteria. Autotrophic bacteria or autotrophs obtain carbon from carbon dioxide (CO_2). With the help of light energy from sunlight or chemical energy, they convert CO_2 into carbohydrates/sugars. Hence, they are called photoautotrophs and chemoautotrophs, respectively. In chemoautotrophs, the chemical energy is obtained from the oxidation of nitrates, ammonia, methane and sulphur compounds. Heterotrophic bacteria obtain carbon and/or sugar from the surrounding environment. They may be further classified as parasitic and saprophytic organisms.

Among all the above classification of bacterial types, the categorization based on Gram straining is widely considered for general classification purposes. The primary difference between Gram-positive and Gram-negative bacteria lies in their cell wall structure. As the cell wall is the major barrier for the entry of antibiotics/drugs into the bacterial cell, it is important to devise antimicrobial methods that are effective against both Gram-positive and Gram-negative bacteria.

10.4 | Bacterial–material Interaction

As mentioned in the introductory section of this chapter, the first step in the pathogenesis of prosthetic infection is the adhesion and colonization of bacteria on an implant surface. Bacterial adhesion can be described in terms of physicochemical interactions which enable bacteria to adhere firmly to a material substrate or biological cells/tissues. The adhesion is a result of the balance among the attractive and repulsive forces at the bacteria/material interface. Bacterial adhesion and subsequent cell growth on a surface have important roles in a variety of biomedical fields, including implant/biomaterial/biomedical device development and bacterial transmittal for bioremediation[732,733]. The precursor to bacterial adhesion is slime formation and the consequence of a large number of bacteria adhering to a surface leads to the production of biofilms. A slime is predominantly composed of exo-polymers of polysaccharides that can be solubilized in aqueous media or buffers to separate the bacterial cells from the biofilm matrix.

- Slime is defined as an extracellular material secreted by bacteria.
- A biofilm is a dense biomass of bacteria along with the secreted exopolysaccharides (slime) assembled as a 3-dimensional structure on a solid substratum.

A substratum is any material surface to which a microorganism may attach and colonize. The receptors on the bacterial surface responsible for adhesive activity are called adhesins. Bacteria possess a wide variety of adhesins that can attach to different adsorbed proteins (fibrinogen, fibronectin, vitronectin, collagen) on biomaterial surfaces[734]. A receptor is a component (both known and putative) on the bacterial cell surface that can bind to ligands sported by adsorbed proteins on a biomaterial or host tissue. The specific ligand–receptor interaction by the adhesin at the active site leads to irreversible adhesion of bacteria on implant and tissue surfaces.

10.4.1 | Thermodynamics of bacterial adhesion

Several interactions must be taken into account in studying bacterial adhesion. Interfacial phenomena are guided by Lifshitz–van der Waals (LW) and Lewis acid–base (AB) interactions, which are important for the initial attachment of bacteria. Lifshitz–van der Waals interactions are non-polar, whereas Lewis acid–base interactions are polar, and encompass all electron-acceptor and electron-donor interactions. In developing an understanding of the thermodynamics basis of bacterial adhesion, the reader is reminded of the fact that any biophysical process is the thermodynamically feasible, if and only if that process lowers the total energy of the system.

The total surface energy involved in bacterial adhesion can be described using the following equation:

$$\gamma_i = \gamma_i^{LW} + \gamma_i^{AB} \tag{10.1}$$

The non-polar and polar constituents of the interfacial free energy are additive in nature:

$$\gamma_i^{AB} = 2\sqrt{\gamma_i^+ \gamma_i^-} \tag{10.2}$$

where, γ_i is the total surface free energy of component i, γ_i^{LW} is the Lifshitz–van der Waals energy component, γ_i^{AB} is the Lewis acid–base energy component and γ_i^+ and γ_i^- are the electron donor and acceptor components, respectively. The Dupré equation for the work of adhesion between a solid and a liquid is expressed as follows:

$$\delta_{SL} = \delta_S + \delta_L - W_{SL} \tag{10.3}$$

The electrostatic and van der Waals forces have been interlinked in the Derjaguin-Landau-Verwey-Overbeek (DLVO) theory of colloidal stability, wherein the energy of interaction varies as a function of distance of separation (Fig. 10.4). This theory is applicable here since bacteria possess a size of 0.5–2 μm, which falls in the range of colloidal particle dimensions. Some researchers explained the adhesion of bacteria to surfaces using the classical DLVO theory[735].

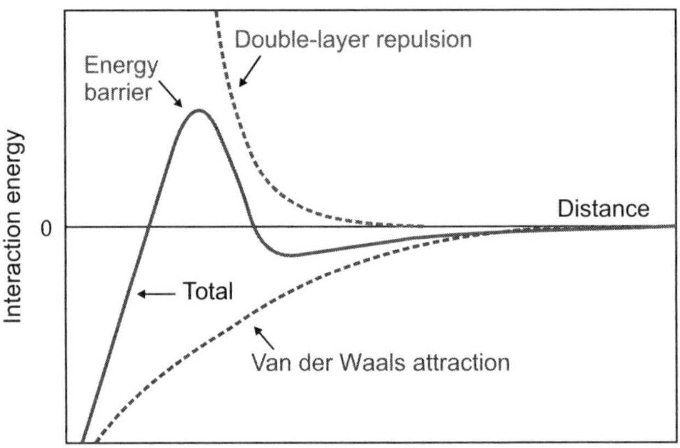

Fig. 10.4 Schematic energy profiles of DLVO interactions[790].

The DLVO theory has been extended by the inclusion of acid–base interactions (XDLVO), which can account for the hydrophobicity of the surfaces involved[736]. In addition to DLVO and XDLVO theories, bacterial properties such as water contact angles[737] and zeta potentials[738] are incorporated into these models, in addition to cell surface hydrophobicity, surface free energy and adhesive behaviour. Van Loosedrecht et al.[739] showed that the percentage of bacteria attaching to a solid surface such as glass was related to both the hydrophobicity of the bacterium and its charge. Nevertheless, they were not able to use the model to explain bacterial adhesion based on bacterial surface polymers, which interfere with other interactions. In view of this, microbial cell surface characteristics and bacterial adhesion are studied using different techniques, such as transmission electron microscopy (TEM) and atomic force microscopy (AFM) to better understand micro-scale interactions.

10.4.2 | Different factors influencing bacterial adhesion

Bacterial adhesion is considered as the initial step for bacterial infection and it depends on the various factors as discussed below:

10.4.2.1 | Material factors

(a) Material surface composition: Material surface composition greatly governs bacterial adhesion[740]. In the group of absorbable sutures, polydioxanone sutures exhibited the smallest affinity towards the adherence of both *Escherichia coli* and *S. aureus*. A study by Sugarman and Musher[741] reports that the adherence of bacteria to the gut was up to 100 times greater than to nylon and adherence to polyglycolic acid or silk was intermediate. According to Gristina et al.[742], *S. E pidermidis* preferentially adheres to polymers and *S. aureus* to metals. If the surface chemistry is changed or modified, such as with an antimicrobial peptide coating, bacterial adhesion to these surfaces is discouraged[743].

(b) Material surface characteristics: The physical contour of a material surface is another parameter, which governs bacterial adhesion on an implant surface. It is basically an external feature description of the designated arrangement of motifs on a material surface, such as a monofilament surface, a braided surface, a porous surface, or a grid like surface. Merritt et al.[744] reported higher implant site infection rates for porous compared to dense dental materials. This implies that bacteria preferentially adhere and colonize porous surfaces.

(c) Surface roughness: Surface roughness characterizes the topography of a material surface that influences bacterial adhesion. Typically, surface roughness is quantitatively expressed as the deviation of the peaks and valleys from the mean line on a material surface. McAllister et al.[745] found that the surface irregularities/unevenness on polymeric implants promote bacterial adhesion and development of biofilms. Baker and Greenham[746] found that the surface roughening of either glass or polystyrene with a grindstone greatly accelerated bacterial colonization in a river environment. The causes for this phenomenon may include a rough surface with a greater surface area and the depressions in the roughened surfaces provide more favourable sites for colonization. As clinically different prostheses or implant devices have different surface roughness, this factor plays a vital role in microbial adhesion and biomaterial associated infection.

(d) Surface hydrophobicity/hydrophilicity: Typically, metallic surfaces are hydrophilic and in contrast, polymers such as ultrahigh molecular weight polyethylene or Teflon are hydrophobic. Depending on the wettability characteristics of both bacteria and implant surfaces, bacterial adhesion is variable on materials with different hydrophobicities.[747]

Fletcher and Loeb[748] investigated the adherence of a marine species of *Pseudomonas* to different surfaces. A large number of bacteria were observed to attach to hydrophobic polymers with negligible or zero surface charge [Teflon, polyethylene (PE), polystyrene, and polyethylene terephthalate]; moderate numbers to hydrophilic metals possessing a positive or neutral surface charge and a small number attached to hydrophilic, negatively charged substrate (glass, mica, oxidized plastics). In a different study, Satou et al.[749] investigated the adhesion of two *S. sanguis* strains and two *Streptococcus mutans* strains to four surface-modified glass slides with different hydrophobicity. The *S. sanguis* strains (with more hydrophobic surfaces) adhered more on hydrophobic glass slides than on others.

10.4.2.2 | Bacteria related factors

(a) Characteristics of bacteria: For a given material surface, different bacterial species and strains adhere differently. This can be explained physicochemically, because the physical and chemical characteristics of bacterial cell walls/membranes differ between species and strains.

(b) Bacterial hydrophobicity: The surface hydrophobicity of bacteria is an important physical factor for adhesion, especially when the substrate surfaces are either hydrophilic or hydrophobic. The hydrophobicity of bacteria can be obtained by contact angle measurements, such as the sessile drop method[750]. The hydrophobicity of bacteria varies according to bacterial species and is influenced by growth media, age of the bacterial culture, and bacterial surface structure. Krekeler et al.[751] reviewed all these factors. Hogt et al.[747] found that one strain of *S. epidermidis* with hydrophobic characteristics showed a significantly greater adhesion to hydrophobic FEP than *S. saprophyticus*. Satou et al.[749] also found that *Streptococcus sanguis* strains with hydrophobic surfaces adhered more to hydrophobic glass slides than to others with a less hydrophobic character.

(c) Bacterial surface charge: The surface charge of bacteria may be another important physical factor for bacterial adhesion[752]. Most particles acquire an electric charge in an aqueous suspension due to the ionization of their surface groups. The surface charge attracts ions of opposite charge in the medium and results in the formation of an electric double layer. The surface charge is usually characterized by zeta potential isoelectric point pH at which the zeta potential is zero[753]. Bacteria in an aqueous suspension are generally negatively charged. A high surface charge is accompanied by the hydrophilic character of the bacteria, but a hydrophobic bacterium may still have the rather high surface charge. The surface charge of bacteria varies according to the bacterial species and is influenced by the growth medium, age of the bacterial culture, and bacterial surface structures. Long-range electrostatic forces may influence the initial phase of bacterial adhesion on solid surfaces.

10.4.2.3 | External factors

(a) Surface proteins: The role of protein adsorption on the material surface or outer membrane of bacteria is crucial to bacterial adhesion. In a microenvironment, the synthesis of proteins and relative enzymes may influence the electrochemical balance of the growth environment, which can affect its adhesion behaviour. Many proteins (serum or tissue proteins) have been studied for their effects on bacterial adhesion to material surfaces including albumin, fibronectin, fibrinogen, laminin, denatured collagen, and more. They promote or inhibit bacterial adhesion in either binding to substrata surfaces, binding to the bacterial surface, or being present in the liquid medium during the adhesion period. In the following, the role of some typical proteins is discussed.

(b) Fibronectin: Fibronectin (Fn), which is recognized for its ability to mediate the surface adhesion of eukaryotic cells, has also been shown to ligate to Fn-binding receptors on the surface of *S. aureus*[754]. Kuusela et al.[755] demonstrated a time-dependent and Fn concentration-dependent adhesion of *S. aureus* to Fn-coated cover slips. Staphylococci may be saturated with Fn at a level that suggests the presence of specific receptors (staphylococcal Fn-binding molecules) on bacterial cells and this Fn-binding molecule has been cloned in *E. coli* and purified[756]. The *S. aureus* binding domain of Fn was also found in the Fn molecule[757].

(c) Albumin: Albumin adsorbed on material surfaces has shown obvious inhibitory effects on bacterial adhesion to ceramic surfaces[758]. The mechanism of the inhibiting effect of albumin is not clear. Albumin may reduce bacterial adhesion by changing the substratum's surface hydrophobicity, because in the presence of dissolved and adsorbed BSA, substrata surfaces became much less hydrophobic[759].

(d) Fibrinogen: Fibrinogen is another important serum protein that mediates bacterial adhesion to biomaterials and host tissues. Most studies showed that adsorbed fibrinogen promotes the adherence of bacteria, especially staphylococci to biomaterials. In a study by Hermann et al.[760], fibrinogen markedly promoted the adherence of all *S. aureus* strains, but only a few coagulase-negative strains. The latter finding was supported by the study of Muller et al.[761]. Fibrinogen, bound to cover slips also increases streptococcal adhesion. In another *in vitro* study, the pre-treatment of bacteria or both bacteria and PE catheter surfaces with fibrinogen-enhanced bacterial adherence, suggest the presence of ligands for fibrinogen on the staphylococcal cell surface[762].

10.5 | Bacteria Growth

Once bacteria adhere to a biomaterial substrate, they will multiply to form a colony. Bacterial growth kinetics, when cultured in isolation, i.e., in a growth medium without any substrate, is now described below.

> Bacterial reproduction occurs by the division of a single bacterium into two daughter bacteria. This process is called binary fission.

In a favourable *in vitro* environment, a growing bacterial population doubles at intervals of 20–30 min. Bacterial multiplication follows a geometric progression in the exponential growth phase: 1, 2, 4, 8, etc. or $2^0, 2^1, 2^2, 2^3 \ldots 2^n$ (where n is the number of generations/divisions). In reality, exponential growth phase is only one of the four phases of the bacterial growth cycle, and not characteristic of the complete bacterial growth pattern[763].

When a freshly prepared nutrient broth medium is inoculated with a particular number of cells, and the bacterial growth is monitored over a time period, plotting the bacterial numbers versus time will yield a typical bacterial growth curve, as shown in Fig. 10.5.

When bacteria are cultured in a closed system (also called a batch culture), like a broth culture in a test tube or a sterile flask, the bacterial population invariably obey the following growth dynamics. In general, bacterial cells acclimatize to the new environment presented by the growth medium (lag phase). Next, they divide at regular intervals characteristic of the doubling times, by the process of binary fission (exponential phase). When the nutrients in the growth medium get depleted, the cells stop dividing (stationary phase), and finally, they show a loss of viability (death phase). The growth is expressed in terms of the number of viable cells as a function of time. The growth rate is determined from the slope of the exponential phase in the growth curve. The culture duration timescale is in hours for bacteria while the generation/doubling times are much shorter, of the order of 20–30 min. Four characteristic phases of the growth cycle are recognized and each individual phase is discussed below (see Fig. 10.5).

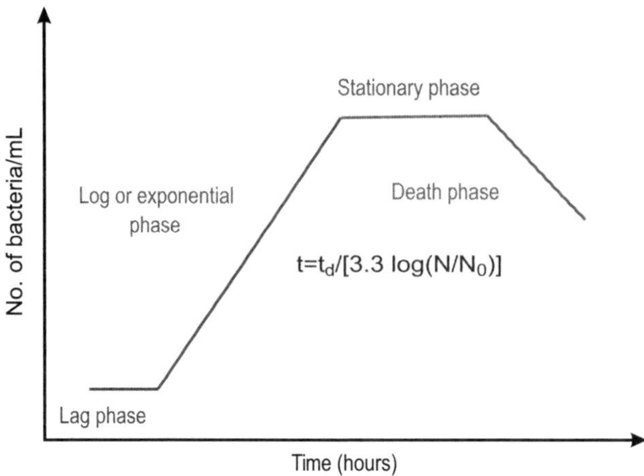

Fig. 10.5 A typical bacterial growth curve.

10.5.1 | Lag phase

For a considerable time period after bacterial inoculation into a fresh broth medium, the bacterial number remains temporarily unchanged. This is due to the minimal time taken by the microbes to adjust to the physicochemical condition of the microenvironment around the prokaryotic cells. Although cell division is supposedly not occurring, the cells grow in volume and mass, synthesizing necessary enzymes, proteins, RNA, etc., and preparing the cell for division. The lag phase time interval apparently varies with extrinsic and intrinsic factors, like the amount of inoculum, the recovery time from physical damage or thermal shock due to cryopreservation, and the time necessary for the synthesis of essential proteins, coenzymes or cell division factors.

10.5.2 | Exponential (log) phase

The exponential growth phase is a period of sustained growth, wherein all the bacteria actively undergo cell division periodically by binary fission, i.e., bacterial numbers increase in geometric progression. The bacterial multiplication generally follows first order growth kinetics, which is regulated by the composition of the growth media (normal versus enriched versus differential) and the incubation temperature as well as atmosphere. The exponential growth rate of a bacterial culture is characteristic of the generation time or doubling time of the cell population. Generation time (G) is described as the time (t) per generation (n = number of generations). More precisely, it can be mathematically formulated as:

$$G = \frac{t}{3.3 \log \dfrac{b}{B}} \tag{10.4}$$

where, 't' is time interval (hours or minutes), B and b are the bacterial numbers at the beginning and end of time interval 't', respectively.

10.5.3 | Stationary phase

Exponential bacterial growth cannot progress indefinitely in a batch culture (e.g., a closed system like test tube or flask culture). The cell growth is limited by one or more of following factors: (a) depletion of nutrients; (b) accretion of metabolic wastes and inhibitory factors, and (c) diminishment of growth space, in this case called a lack of 'biological space'.

In the stationary phase, the bacterial numbers are nearly constant as a function of time. This is due to equilibrium of actively dividing cells and rapidly dying cells in the growth media. On other occasions, the cells simply cease cell division, but are yet viable and multiply, when transferred to fresh media. Similar to the lag phase, the stationary phase is not necessarily a quiescent period. Bacteria produce many secondary metabolites, including toxins and antibiotics during the stationary phase of the bacterial life cycle. Also, in case of spore-forming bacteria, multiple genes governing the sporulation process are activated during the stationary phase.

10.5.4 | Death phase

If the bacterial culture is continued even beyond the stationary phase, the death phase follows, wherein the number of viable cells drops with time. During the death phase, the viable bacterial number declines geometrically (exponentially), essentially the inverse of exponential growth phase.

The growth kinetics described above is ideally followed in a growth medium and the presence of a synthetic biomaterial substrate can considerably alter growth kinetics.

10.6 | Biofilm Formation

Biofilm formation begins with the attachment of microorganisms to a surface or substratum. Once the cells have attached, they produce an extracellular polysaccharide matrix (EPS), which provides stability to the biofilm by enabling cell–surface and cell–cell interactions. After stable attachment of bacterial micro-colonies, the biofilm develops into a more complex environment. Additional planktonic cells adhere, micro-colonies develop, and complex biofilm architecture forms (see Fig. 10.6a).

> Biofilms are a dynamic environment, wherein cells from the growth medium attach to the biofilm.

Microbial biofilms are particularly resistant to the action of antibiotics and antimicrobial agents. For instance, the antibiotic dosage to eradicate planktonic bacteria generally has a lesser or negligible detrimental effect on a bacterium enveloped within biofilms. Often, the dosage of antimicrobial agents necessary for biofilm eradication is a thousand times greater than those necessary to kill planktonic microorganisms. Further, even antimicrobial transition metal ions such as copper and silver are potent against planktonic bacteria or suspension cultures, but are much less efficacious against bacterial biofilms. The biofilm matrix that envelopes the bacteria in the inner layers acts as a diffusion barrier to the entry of antibiotics and antimicrobial agents.

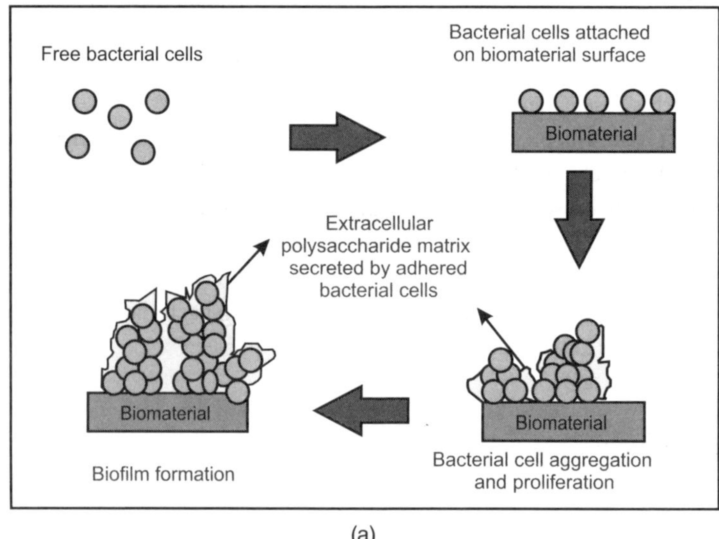

(a)

(b)

Fig. 10.6 (a) Schematic representation of biofilm formation on material substrate; (b) SEM image of a staphylococcal biofilm (author's own result); it is noted that the multiple layer of bacteria covered with a polysaccharide matrix.

It is well known that bacteria can transfer resistant genes between themselves via horizontal gene transfer, thereby conferring resistance properties to all the bacteria in the biofilm community. Microorganisms generally occur in two forms based on their motility. In the sessile form, the microbes are firmly embedded within the biofilm matrix and are less motile except for when the biofilm opens up and disperses onto newer surfaces. In the planktonic form, the bacteria are free-floating, motile and disseminate onto new biomaterial/implant surfaces. Among the pathogenic bacteria, staphylococci exhibit a high degree of natural variation in their phenotypes. This is referred to as **phase variation**.

This is a generic term for a set of genetic mechanisms, wherein the particular gene undergoes rapid modification of expression in a reversible fashion between generations[764]. Such variations in phenotype can lead to the secretion of multiple enzymes that catalyze exopolysaccharide slime accumulation, thus generating biofilms. Planktonic staphylococci also experience changes in their phenotypes that promote attachment, biofilm induction and enhanced resistance to antimicrobials/phagocytes/antibodies. In a biofilm environment, biomaterial adhesion mediated phase variation can lead to two sub-populations of adhesive and non-adhesive bacteria[732]. Planktonic bacteria can colonize an implant surface and cause systemic infection of the host, while sessile bacteria scrupulously endure antibiotic treatment, causing chronic infections.

> A single bacterium is sufficient to generate a bacterial colony and a single colony can develop into a biofilm on an implanted biomaterial.

The molecular basis of resistance of bacteria in biofilms may be correlated with the retardation of cellular growth within mature biofilms, to the physicochemical interaction of the glycocalyx and exopolysaccharides ('slime') with macromolecular antibiotics (via dipole–dipole, hydrogen, ionic bonds, complexes) and to modifications of the cellular envelope subsequent to colonization of host tissues[765]. Then, the retarded growth rate and alterations in cell wall structure and consequent cell density transcription activation[733] may majorly regulate the phenotypic resistance of sessile bacteria to many antibiotics[766].

Costerton et al.[767] reviewed various models of biofilm formation and reduced biofilm susceptibility. They also tried to explain the genetic and molecular mechanisms governing community behaviour in bacteria in reference to therapeutic targets that may provide a means for the control of biofilm infections.

10.7 | Experimental Assessment of Antibacterial Properties, *in vitro*

A host of biochemical assays and molecular biology-based techniques are being employed to ascertain and characterize bacterial viability after antibiotic/antimicrobial treatment[768]. Most of the newly developed drugs to treat microbial infections are first tested against planktonic cultures. Similarly, any new biomaterial with potential antimicrobial properties is to be tested against both Gram-positive and Gram-negative bacteria. The most common and traditionally performed assays are described in the following section, although newer methods for characterizing different viable states in bacteria, such as viable and culturable as well as viable and non-culturable are being developed. The following assays are based on the notion that viable implies the capability to grow and multiply[769].

10.7.1 | Minimum inhibitory concentration (MIC)

The minimum concentration of the antibiotic/antimicrobial at which there is no visible turbidity, indicated by the absence of the exponential phase in the bacterial growth curve is designated as MIC[770].

A bacteriostatic effect or growth inhibition arising from the cessation of bacterial multiplication is implicated at the minimum inhibitory concentration (MIC).

Figure 10.7 is a pictorial illustration of MIC determination, which is indicated by the lowest concentration at which there is no visible turbidity due to bacterial growth.

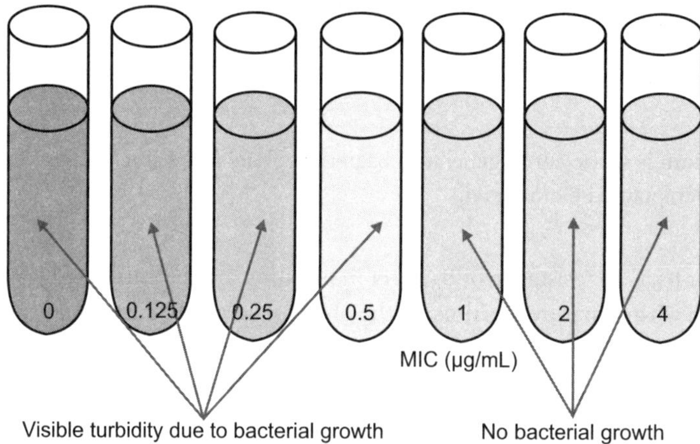

Fig. 10.7 Pictorial illustration of MIC indicated by the lowest dosage at which no bacterial growth occurs [see Colour Plate].

10.7.2 | Minimum bactericidal concentration (MBC)

The minimum concentration at which antibiotic treated cultures at $\geq$ MIC dosages lead to the absence of bacterial colony formation on nutrient agar is designated as MBC.

The bactericidal effect or killing of microbial cells is implicated at the MBC. It is instructive to note that the terms, bactericidal and bacteriostatic are not synonymous.

It may be noted that the MBC procedure is an extension of MIC. In terms of dosage, MBC is always greater than MIC. Figure 10.8 is a representation of MBC determination, which is indicated by the lowest concentration of the antibiotic/antimicrobial at which bacteria fail to form colonies on nutrient agar. Tolerance is defined as the dosage window in which the bacteria cease to divide in suspension cultures but can however remain viable when plated on nutrient agar. It is calculated as the ratio of MBC/MIC[771].

10.7.3 | Disc agar diffusion (DAD)/Zone of inhibition (ZOI) assay

The susceptibility or resistance of microorganisms towards antibiotics or antimicrobial surfaces is evaluated by measuring the zone of inhibition formed on a bacterial lawn spread on nutrient agar.

The diameters of the inhibition zones formed around the antibiotic disc or a 6 mm filter disc loaded with pre-determined amounts of the drug are measured (see Fig. 10.9). The absence of a ZOI is indicative of the resistance of the microorganism towards that particular antibiotic. This assay is also popularly known as the Kirby–Bauer (KB) test for determining the antibiotic sensitivity of a bacterium[772]. The diameter of the inhibition zone in the KB test is inversely related to the minimum inhibitory concentration (MIC) at which bacterial growth inhibition occurs.

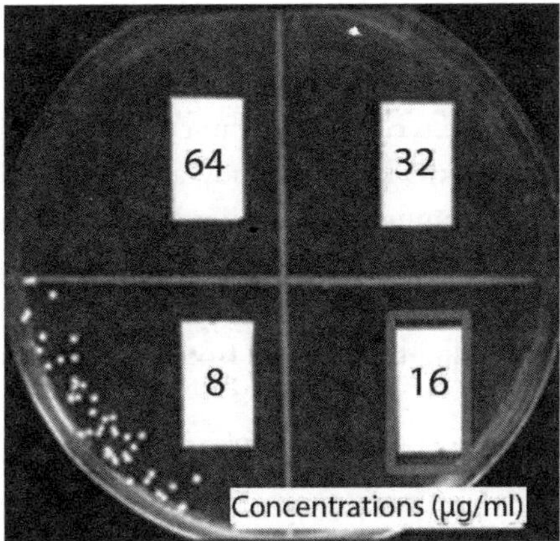

Fig. 10.8 Illustration for minimum bactericidal concentration (MBC) at which the treated cultures with dosage ≥ MIC do not form colonies on nutrient agar.

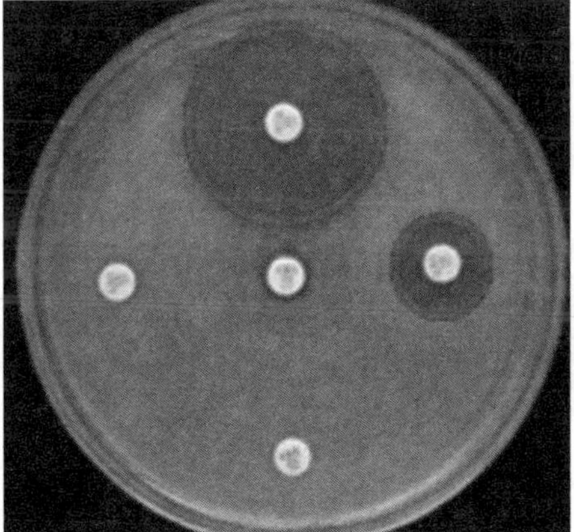

Fig. 10.9 Zone of inhibition (ZOI) formed by the diffusion of antibiotic/antimicrobial from the filter disc to the bacterial lawn spread on nutrient agar.

10.7.4 | Colony forming units (CFU) assay

The number of viable bacterial cells is enumerated directly by counting the number of bacterial colonies formed on nutrient agar plates from serial dilutions of bacterial cultures[773]. Each viable bacterium can actively divide and can give rise to a single colony.

Figure 10.10 is a schematic depiction of a ten-fold serial dilution of bacterial cultures, followed by plating on agar and counting the number of colonies formed to enumerate the number of viable bacteria/mL or CFU/mL

$$\text{No. of CFU/mL} = \text{No. of colonies/(Volume of dilution plated} \times \text{Dilution factor)}$$

The reduction in the number of viable bacterial numbers upon antibiotic/antimicrobial treatment is generally reported as Log_{10} CFU reduction.

$$\text{Log}_{10} \text{ CFU reduction} = \text{No. of CFU/mL in control/No. of CFU/mL in a treated culture}$$

The Log_{10} CFU reduction can be recorded as a function of antibiotic concentration at a fixed time interval or a time-dependent kill kinetics study at the MIC dosage of the antimicrobial. Another advantage of the colony forming units assay is that by enumerating the number of viable cells at different time points in the exponential phase, it is possible to calculate the doubling times of different bacterial species. The only disadvantage of this assay is that it is tedious, time-consuming and prone to large errors, while plating the bacteria from serial dilutions.

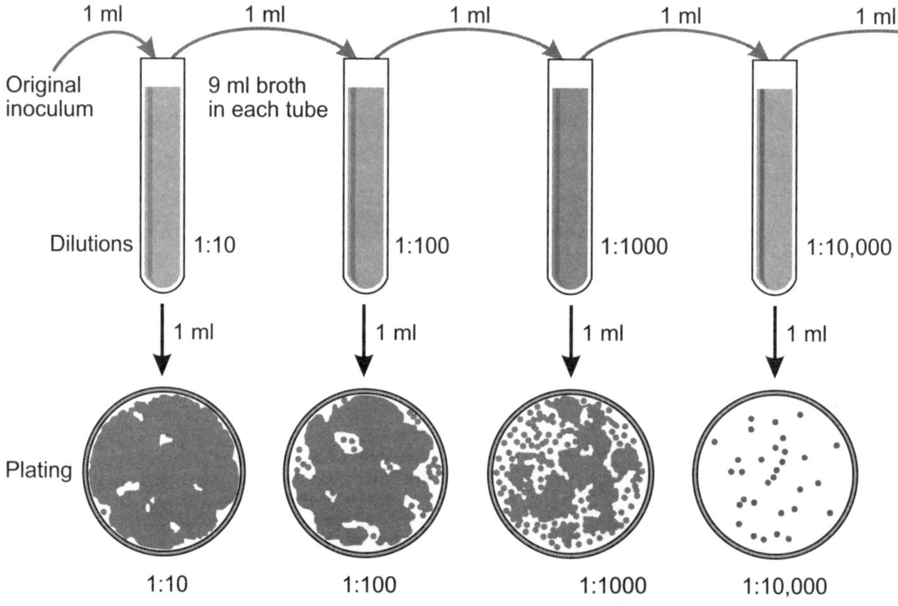

Fig. 10.10 A schematic illustration of the Colony forming units (CFU) assay by serial dilution method followed by plating of the dilutions; (adapted from Pearson Education, Inc. publishing as Benjamin Cummings).

10.7.5 | **Inner membrane permeabilization/ONPG assay**

The extent of bacterial membrane permeabilization by antimicrobial treatment (such as antibacterial peptides or pore forming aminoglycosides) can be measured by the catalytic conversion of o-nitrophenyl-D-galactopyranoside (ONPG) into o-nitrophenol by β-galactosidase enzyme released from leaky bacterial membranes. The absorbance of the culture supernatant measured at 420 nm (λ_{max} for o-nitrophenol) is a measure of inner membrane permeabilization[774]. Figure 10.11 is a reaction schematic for the enzymatic conversion of ONPG into ONP.

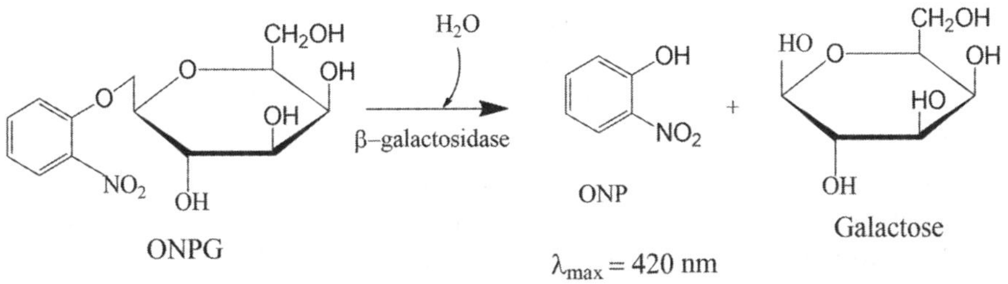

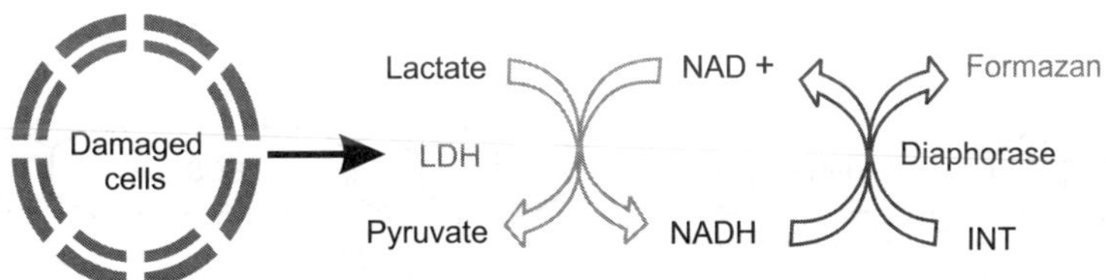

Fig. 10.11 Reaction schematic of the inner membrane permeabilization/ONPG assay showing the β-galactosidase catalyzed conversion of o-nitrophenyl-D-galactopyranoside (ONPG) into o-nitrophenol (ONP).

10.7.6 | Membrane integrity assays

The loss of bacterial cell envelope integrity can be determined through a host of biochemical/colorimetric assays. The release of intracellular DNA from bacterial cells can be determined by recording the absorbance at 260 nm (λ_{max} for DNA)[775]. The lactate dehydrogenase (LDH) assay is based on the reduction of NAD^+ to NADH by the released LDH. The NADH, in turn, reduces a tetrazolium salt (INT) to formazan (Abs. max = 450 nm), while itself getting oxidized back to NAD^+. Figure 10.12 is a reaction schematic for determining cytotoxicity based on the enzymatic activity of LDH released from lysed cells. The LDH assay is also used to characterize cytotoxicity in eukaryotic cells[776].

Fig. 10.12 Reaction schematic of the LDH assay showing the cascade of enzymatic reactions leading to the conversion of the chromogen (INT) to a coloured product (formazan).

10.7.7 | Microbial flow cytometry

Live, dead and injured bacterial populations in suspension cultures can be quantified by flow cytometry[777]. It can be reiterated that flow cytometry is a high throughput technique developed to count and characterize the behaviour of cell populations and the principle of flow cytometry is discussed in chapter 9. Around 10,000 to 20,000 scattering events and fluorescent signals are recorded from cells as they flow in a streamlined manner along a microfluidic channel. The flow cytometry data are represented in the form of dot plots, contour plots or histograms. In the dot plot, each dot represents a single cell/bacterium. For the live/dead assessment, combinations of dyes that specifically stain live and dead cells are used. The co-localization of the two dyes in the bacterium indicates injured cell populations.

In Fig. 10.13(a), the four quadrant dot plot shows bacteria stained by a combination of DNA intercalating dyes. Thiazole orange is a membrane permeable dye that intercalates with the DNA of all cells while propidium iodide (PI) replaces Thiazole orange in membrane-compromised cells[778]. The histogram (see Fig. 10.13(b)) represents an increase in the green fluorescent signal from bacterial cells as a result of reactive oxygen species (ROS) accumulation within the cells. Figure 10.13(c) is a schematic reaction of a non-fluorescent probe (DCFH-DA), which upon de-esterification reacts with intracellular ROS to give a green fluorescent signal due to DCF.

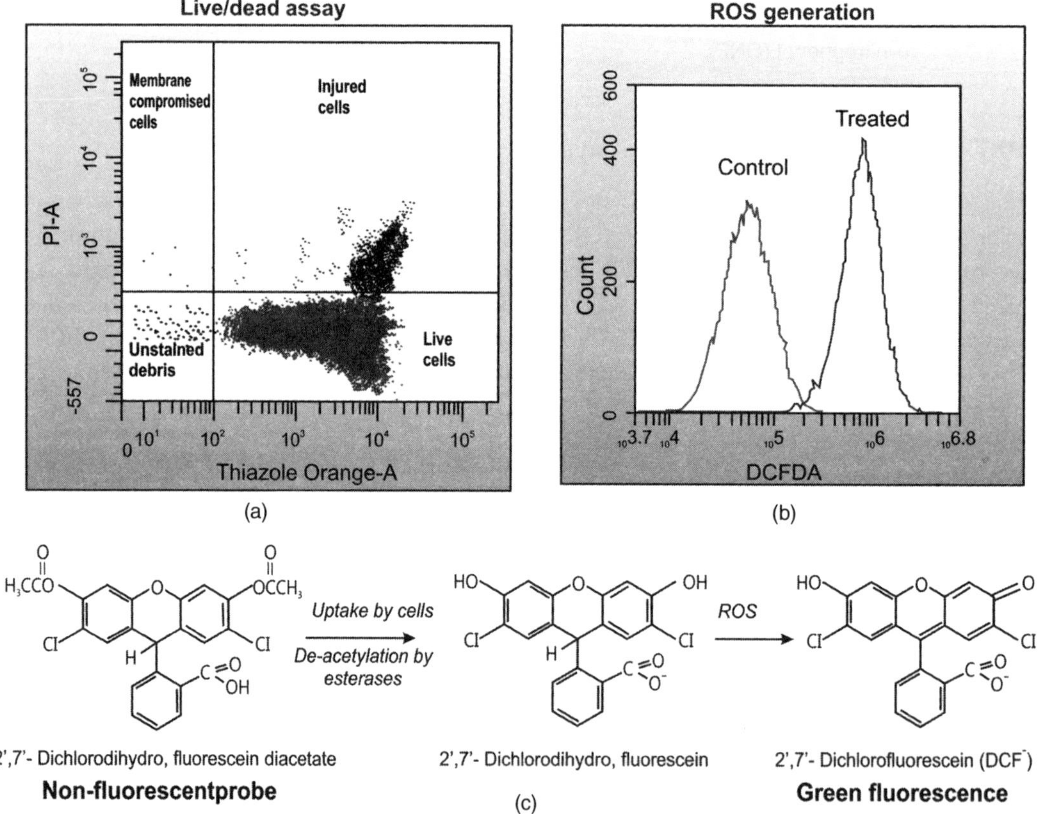

Fig. 10.13 Schematic illustration: (a) Flow cytometry data pertaining to the live/dead assay is represented as a four quadrant dot plot with each quadrant indicating the live, dead and injured cell populations based on differential staining by the vital dyes (thiazole orange- live and Propidium iodide- dead); (b) flow cytometry data pertaining to the detection of reactive oxygen species (ROS) by staining with DCFDA is represented as histograms with the x-axis indicating the fluorescent intensity and y-axis, the cell count; (c) reaction schematic for the conversion of the non-fluorescent ROS probe into green fluorescence signal upon reaction with ROS.

The quantification of ROS is important as various reactive oxygen species (superoxide $.O_2^-$, hydrogen peroxide H_2O_2, hydroxyl $.OH^-$ and singlet oxygen $^1O_2$) are implicated in lipid, protein and genomic DNA peroxidation and these together can affect bacterial viability.

Further, the detection of bacterial viability based on membrane potential sensitive probes such as bis-(1,3dibutlylbarbituric acid) trimethineoxonol $(DIBAC_4)$[779] and dihexyloxacarbocyanine iodide $(DiOC_6)$ provide information of about bacterial membrane depolarization by antibacterial peptides, such as gramicidin[780,781].

10.8 | Experimental Assessment to Characterize Biofilm

Once bacteria are grown on a biomaterial substrate, the following assays are used to characterize the biofilm.

10.8.1 | Resazurin dye reduction test

Resazurin (7-Hydroxy-3H-phenoxazin-3-one 10-oxide) is a blue-coloured dye, which is enzymatically reduced to a pink fluorescent molecule, resorufin by live bacteria (see Fig. 10.14). This is a redox based indicator assay for estimating the viability of planktonic as well as biofilm enveloped bacteria[782]. Bacterial viability is indicated by the fluorescence intensity measured with absorption and emission filters of 546 nm and 595 nm, respectively. The advantage of this assay is that it enables fast and efficient screening of antibiotics/antimicrobials aimed at reducing biofilm formation and growth.

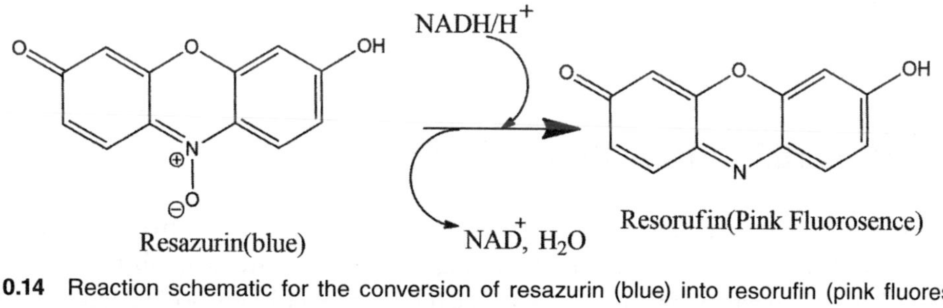

Fig. 10.14 Reaction schematic for the conversion of resazurin (blue) into resorufin (pink fluorescent molecule) by enzymatic activity of live bacteria.

10.8.2 | Total biomass quantification by crystal violet staining

The total bioburden/mass (viable and nonviable bacteria in a polysaccharide matrix) present in the biofilms can be determined by crystal violet staining in a 96-well microtiter-based assay. This assay is based on crystal violet dye binding to all the organic content (both viable and non-viable) in the well. The bound dye is eluted with 30% (v/v) acetic acid and the absorbance recorded at 600 nm is a measure of the total biomass[783].

10.8.3 | Live/dead biofilm imaging

For determining the viability of bacterial biofilms by fluorescence microscopy, a combination of live and dead stains is being used. These include thiazole orange–propidium iodide (TO–PI), syto 9–PI and Fluorescein diacetate FDA–PI, to name a few[784]. The TO–PI and syto 9–PI combination are DNA binding dyes, where TO and Syto 9 are membrane permeable dyes that intercalate with the DNA of all the bacteria, while PI binds to the DNA of membrane compromised cells. On the other hand, FDA–PI is a more potent combination for live/dead imaging in that FDA is de-esterified by cellular esterases into a green fluorescein molecule in viable bacteria, while PI binds to the DNA of membrane-permeabilized cells[785].

It may be emphasized that biofilm viability can be determined by the resazurin dye reduction test and live/dead biofilm imaging. On the other hand, crystal violet staining and biofilm thickness provide information on the biomass and phenotype of the biofilms, respectively.

10.8.4 | Biofilm thickness by optical/fluorescence microscopy

The thickness of bacterial biofilms can be estimated by both optical and fluorescence microscopy. By determining the vertical displacement of the sample in order to focus the biofilm–substratum interface to the biofilm–liquid interface, it is possible to estimate the thickness of the biofilm[786]. This can be experimentally determined from the number of divisions the coarse adjustment screw must be rotated in order to shift the focus from the substrate to the biofilm growing over it. It is important to characterize the biofilm phenotype in order to ascertain its response towards combinatorial antibiotics. Figure 10.15 summarizes the commonly employed methods for the evaluation of bacterial biofilms.

10.9 | Bacterial Culture Protocol

A sterility check must be performed as a prelude to bacterial culture. The freshly prepared agar plates and nutrient broth media must be left overnight at 37°C in a sterile incubator. The absence of bacterial/fungal growth on the agar plates and no turbidity in the broth medium indicate sterility or absence of contamination. For reviving the bacteria, clinical isolates from patients are streaked on freshly prepared agar (1.5%) plates with a sterile inoculation loop. Bacteria can also be revived from frozen glycerol stocks, where the frozen crystal is incubated within nutrient broth before streaking on agar plates. The streaked plates are incubated overnight at 37°C, until bacterial colonies are visible (~24 hours). For broth cultures, a single colony is inoculated in a sterile tryptone soya broth (TSB- 30 g/L) or luria broth (LB- 20 g/L) medium and incubated at 37°C, ≥100 rpm in an incubator shaker overnight. The overnight culture is inoculated in a fresh medium for preparing a sub-culture. The bacterial sub-culture in the logarithmic phase (~4 hours) is used for performing experiments after adjusting the optical density to a fixed value (OD~0.1) with the help of a UV-visible spectrophotometer.

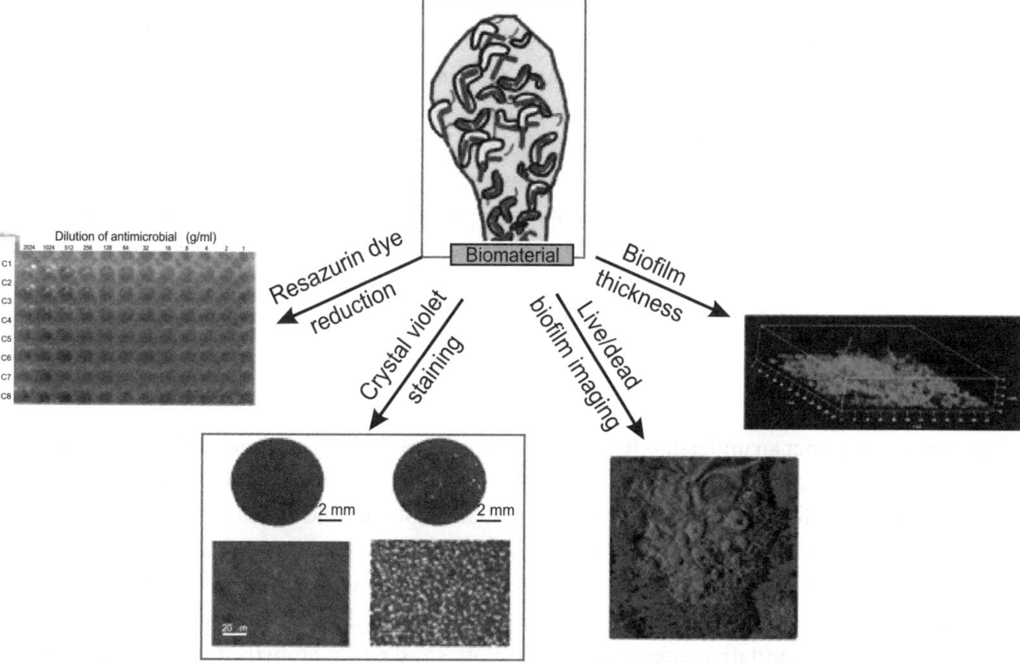

Fig. 10.15 Summary of the commonly employed methods for the evaluation of bacterial biofilms grown on a biomaterial; resazurin and live/dead imaging are indicators of biofilm viability, while crystal violet staining and biofilm thickness reflect on the biomass and biofilm phenotype, respectively [see Colour Plate].

The bacteria culture protocol as well as the testing protocol for antibacterial property are mentioned in this chapter with specific details. It is instructive to note that these details may alter depending on the experimental design or the biological questions to be addressed and even on the bacterial strains or reagents in use.

10.10 | Guidelines for Antibacterial Testing of Biomaterials

For evaluating antibacterial properties of biomaterials, both Gram-positive and Gram-negative bacteria must be used. All the pathogenic and drug-resistant bacterial strains must be handled carefully in a biosafety cabinet level II. As a prelude, the biomaterial (metal/ceramic or polymer) must be heat sterilized by autoclaving at 121°C, 15 psi for 20 min or UV-sterilized overnight. In the case of nanoparticles, the particle suspension can be filter sterilized with a 0.2 μm size filter. A fresh culture of bacterial suspension in the logarithmic phase with an optical density of ~0.1 (~10^6–10^7

bacteria/mL) is seeded onto the biomaterial substrate and cultured for 4 hours to check for the initial response of the bacteria to the antimicrobial surface. The adhesion of the bacteria on biomaterial substrates can be visualized under a fluorescence or a scanning electron microscope. Further, the bacteria can be detached by mild sonication for 30 s and the number of viable bacteria can be enumerated by a colony forming units (CFU) assay, flow cytometry and/or other biochemical assays. In the case of nanoparticles, the MIC, MBC and ZOI assays can be performed to characterize their efficacy in eliciting antimicrobial action. For evaluating the anti-biofilm action of the nanoparticles, mature biofilms of age $\geq$ 2 days are treated with 1x MIC, 2x MIC and so on. The viability of the biofilms can be characterized by the resazurin assay and live/dead imaging. All the protocols used for performing various antimicrobial assays must comply with the standard guidelines recommended by the National Committee for Clinical Laboratory Standards (NCCLS).

10.11 | Closure

This chapter brings out an understanding of bacterial cell structure, the thermodynamics of bacterial adhesion, the physical and chemical factors governing the bacterial colonization on biomaterial surfaces and assays for antimicrobial evaluation against planktonic strains. A knowledge of the physicochemical interactions occurring between bacteria and the material with a particular emphasis on biofilm formation can help in designing better biomaterials with antimicrobial properties. The chemical/enzymatic reactions involved in the biochemical based detection of bacterial viability aid in the qualitative and quantitative assessment of antibiotics and antimicrobials. Finally, a basic outline of the protocol for bacteria culture and a few guidelines for the antibacterial testing of biomaterials have been provided towards the end of the chapter.

From the biomaterials compositional prospective, it is important to develop material, which do not encourage the bacterial growth without compromising cellular functionality of eukaryotes. For example, ZnO/Ag is often added to various biomaterials in a tailored amount to induce bactericidal or bacteriostatic properties.

11

Probing Tissue Response, *in vivo*

The biocompatibility assessment of any implantable biomaterial involves the *in vitro* testing and pre-clinical study, which is to be conclusively validated using clinical trials. The *in vitro* testing to probe cell response is discussed in the previous chapter. The discussion in this chapter focuses on pre-clinical studies, which are conducted using experimental animals with ethically approved protocols. The readers will be introduced to some of the experimental methodologies to probe into the tissue response to synthetic biomaterials, *in vivo*. While discussing this aspect, the selection of appropriate animal model, design of implantation experiments as well as histology based analysis of tissue/implant interface are emphasized. In view of the sensitivities involved in the use of animals during *in vivo* experiments, it is of utmost interest to critically consider the ethical concerns while assessing safety of biomaterials. Moreover, the objective of introducing bioethics in the field of biomaterials is to make the scientists feel more responsible for the research they perform and to uphold those virtues to warrant good ethical conduct. These aspects are largely discussed in this chapter. The chapter concludes with some illustrative examples for *in vivo* biocompatibility testing and analysis of biomaterials for orthopedic, neural, cardiovascular, soft tissue regeneration applications. The *in vivo* toxicity of biomaterial nanoparticulates is also mentioned.

11.1 | Introduction

Biomaterials are materials that are used in contact with biological systems or often integrated into medical devices. Unlike other nonviable materials used for structural or non-biomedical applications, biomaterials are expected to have a favourable long-term interface with surrounding tissues. This issue is to be investigated first in animal models (*in vivo*) and thereafter in human subjects (clinical trials, often referred to as "first-in-man" study). The *in vivo* study is aimed to predict the safety and efficacy of any new biomaterial[838]. The local and systemic response of the tissues toward implants comprises an important aspect of biocompatibility. An ideal biomaterial does not cause inflammation, undesired immunological responses, cancer, cytotoxicity and mutagenicity.

Nevertheless, many synthetic biomaterials, particularly the bioinert materials are isolated with collagenous encapsulation after implantation.

> The biological reactions of the host tissues around a test implant are generally compared with the responses of controls (materials having proven biocompatibility).

The response of a control implant, e.g., HA in case of bioceramics research is critical and some response parameter, i.e., bone regeneration is to be quantified with respect to that for the test implant. The performance of a test implant in an experimental animal is therefore considered as the most direct means of evaluating potential effects on the surrounding living tissue.

The primary purpose of performing safety evaluation of a biomaterial is for protecting humans from the potential biological risk arising from its use. Hence, long before the clinical application, toxicological risk assessment, involving both *in vitro* and *in vivo* biocompatibility tests are performed to establish pre-clinical biological safety of the material or device. Throughout the biomaterials developmental process, several important criteria should be considered. This includes research design and experimentation, declaration of results and use of the results thereof, the means by which they are sought to be achieved, the anticipated benefits and hazards and finally, the potential uses and abuses of the experiment and its results. Unfortunately, most of the scientists pursuing biomaterials science seem to lack an overall understanding of the basic moral and ethical concerns associated with biological applications. As a result, there are many cases of conflicts of interest, research misconduct and random animal experimentation for new biomaterials, which are not needed. In view of this, new quality-control standards and institutional review boards are being introduced to ensure ethical issues in biomaterials research. It is imperative that biomaterial scientists/biomedical engineers acquire sufficient background knowledge about the principles of bioethics, whenever they develop a new biomaterial for human healthcare applications.

It may be reiterated from the discussion in the previous chapter that the ethical responsibilities of biomedical researchers particularly start when they begin cell culture experiments. First and foremost is that the researchers must perform such experiments using culture protocols as per the biosafety norms and such approval is more stringent when stem cells are used.

> The adherence to bioethics is more stringent as the research moves from *in vitro* to pre-clinical (animal models) and to finally clinical trials (human subjects). In the clinical trials, the ethical consideration is at the highest level, because such norms must combine those of biomedical engineers as well as medical professionals (combination of medical ethics and bioethics).

A number of ethical issues therefore arise in the field of implant design and regenerative medicine, such as tissue engineering. All biomedical engineers must adhere to the biomedical engineering society code of ethics. This code places additional emphasis on dealing with animal/human patients and healthcare, because of the close ties that biomaterial scientists have with those areas. In addition to ethical issues related to the use of biomaterials for human and animal experimentation, other general issues, e.g., truthfulness and the conflicts of interests are also important. While considering the consequences of the biomedical device designs for medical practice, the researchers must ensure that technologies and experimental protocols appropriately support the ethical principles for healthcare application (Fig. 11.1).

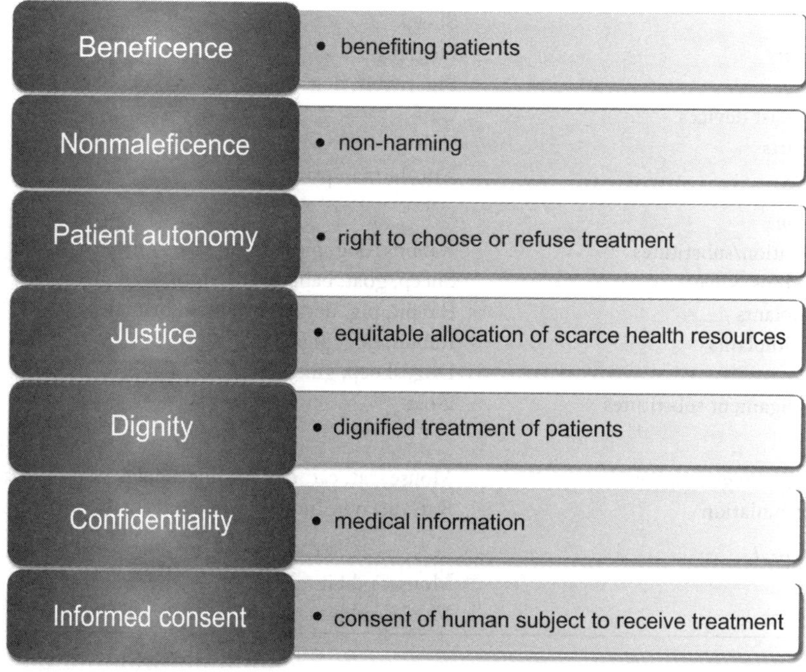

Fig. 11.1 Basic concepts of biomedical ethical principle [Adapted from Ref. 837].

Based on *in vitro* screening, which often involves preliminarily toxicity analysis with specific cell lines, only a few selected material compositions should be considered for *in vivo* biocompatibility testing.

11.2 | Tissue Compatibility Assessment

It is widely perceived that suitable animal models can be used to predict the safety and biocompatibility of medical devices in humans (Table 11.1). Therefore, the first step in the design of pre-clinical experiment is the selection of animal models for the *in vivo* assessment of tissue compatibility and at the initial stage, one must consider the advantages and disadvantages associated with animal model for human clinical applications (Table 11.1).

Pre-clinical testing in suitable animal model is an important step to determine the safety and efficacy of the devices prior to clinical trials in human subjects. The choice of the animal model and *in vivo* study design are governed by the intended use of the respective medical device.

Table 11.1 Animal models used for *in vivo* assessment of various medical devices.

Device Classification	Animal
Cardiovascular	
Heart Valves	Sheep
Vascular Grafts	Dog, pig
Stents	Pig, rabbit, dog
Ventricular assist devices	Calf
Artificial hearts	Calf
ex vivo shunts	Non-human primate, dog
Orthopedic/bone	
Bone regeneration/substitutes	Rabbits, dog, pig, mouse, rat, guinea pig
Total joints- hips, knees	Sheep, goat, baboon
Vertebral Implants	Rabbit, pig, dog, non-human primate
Craniofacial implants	Rabbit, dog, pigs
Cartilage	Dog, sheep, guinea pig
Tendon and ligament substitutes	Goat
Neurological	
Peripheral nerve regeneration	Mouse, rat, cat, rabbits, dog, sheep, non-human primate
Electrical Stimulation	Rat, cat, non-human primate
Ophthalmological	
Contact lens	Mouse, rabbit
Intraocular lens	Mouse, rabbit and non-human primate

Overall, the design of animal experiments requires careful considerations of multiple aspects and these experiments demand fairly different approaches, which are new to non-biologists, like engineers or material scientist. Keeping this in mind, an attempt has been made in this section to describe the design of animal experiments. Figure 11.2 describes the various steps involved in conducting the pre-clinical study with experimental animals.

As the first step towards assessing tissue compatibility of an implant in an animal model, the test samples are cut to desired size to fit into the defect model and prior to implantation, the test implants are to be sterilized either via steam/ETO/γ-ray. At different time points, post-implantation, the implant sites are examined. Prior to the retrieval of explants, local pathological effects on the structure and function of a living tissue around a test implant needs to be investigated. The most basic pathological evaluation is conducted at both the gross level and microscopic level.

The choice of timescale of the implantation as well as intermediate termination of the test is largely determined by the need to confirm a steady state host response to the implant. It is important to perceive that the injured tissue surrounding the implant will be in a state of enhanced recuperative activity, post-implantation. In general, tissue activity/functionality would increase during the early period, post-implantation (less than 1 week). The short term assessment of tissue responses, therefore, may provide a signature towards the biological response to the implanted material. Typical periods for sub-chronic implantation are 1, 3, 4, 9, and 12 weeks. A common study design uses 2, 4, and 12 weeks of implantation, prior to scientific assessment. ISO 10993-6 categorizes implant tests as acute (within 24 hours), sub-acute (14-28 days), sub-chronic (up to 90 days or less than 10% of animal's life span) and chronic (more than 90 days or more than 10% of animal's life span).

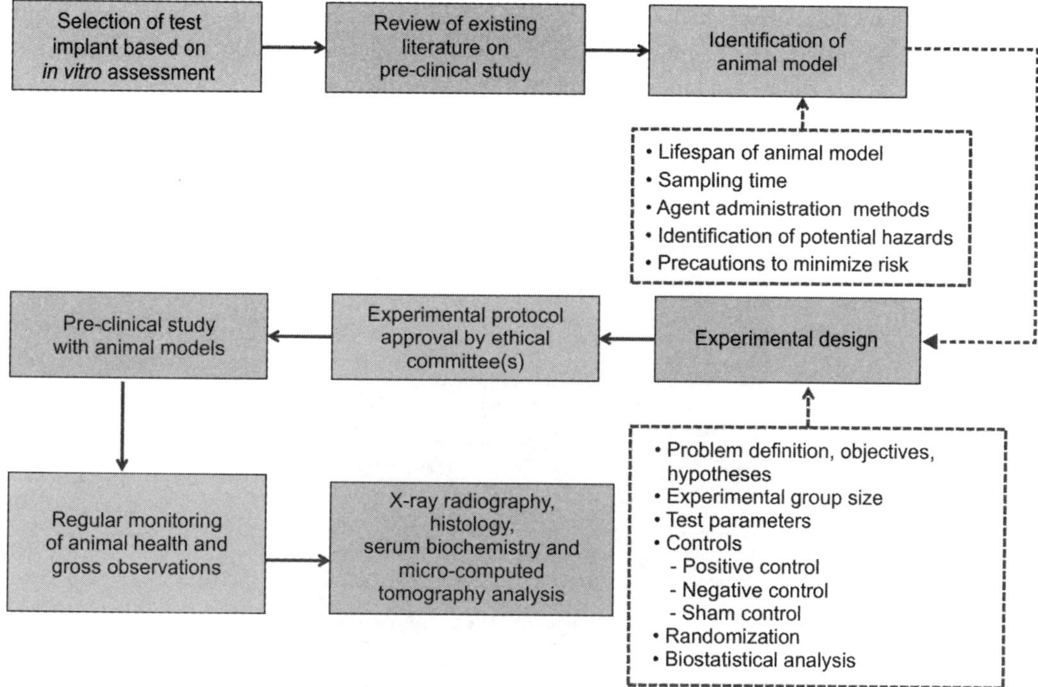

Fig. 11.2 Scientific approach in designing an *in vivo* experiment.

Mice, rats, guinea pigs or rabbits are the most common animals utilized for short- term implantation study. Long-term implantation (26, 52, 78 and 104 weeks) is carried out in subcutaneous tissue, muscle or bone of larger animals with longer life span (e.g., dogs, sheep, goats, pigs, etc.).

11.2.1 | Animal testing and tissue compatibility laboratory

Any animal testing lab must have a clean animal house, where experimental animals of different breeds/strains are kept under aseptic conditions with regular maintenance. Figure 11.3 shows a typical facility where experimental animals are kept in individual cages. Each cage has necessary food *ad labitum* and is maintained appropriately to ensure animal health.

In biological terms, a strain is defined as a group of animals that are genetically uniform.

Different strains of the same animal are used for specific experiments on account of their suitability for studying different aspects of human diseases, which may not be possible using a single strain. In order to substantiate this point further, Table 11.2 mentions a representative list of various strains used in biomaterials research involving small to medium size animals and Fig. 11.4 shows few of such strains. For cancer study, immunodeficient nude mouse model is used and those include the single-gene mutation models such as nude-mice (*nu*) strains and the severe combined immunodeficient (*scid*) strains. In contrast, a different mouse strain is used to study the inflammation. New Zealand white rabbit strain is widely used in bone implantation experiments.

Fig. 11.3 Animal room with individually ventilated cages.

> Most toxicity studies with biomaterials are initially carried out using mouse or rat model.

The animal testing laboratory should also have the facilities to monitor the health of the experimental animals, prior to or during the experiments with drugs or materials. The typical facilities include, but are not limited to, the serum biochemistry, haematology, microbiology, parasitology, histopathology and some physiological analysis, if possible.

Table 11.2 Various animal species used for tissue compatibility testing of biomaterials.

Wild type mice strains	
BALB/cJ	General purpose mouse strains, used for routine biocompatibility assessment.
Swiss albino	
C57BL/6	
Rat strains	
Wistar	General purpose rat strains used for toxicity of drugs/biomaterials.
Sprague Dawley	
Rabbit strain	
New Zealand White	Used for bone implantation study to assess bone regeneration

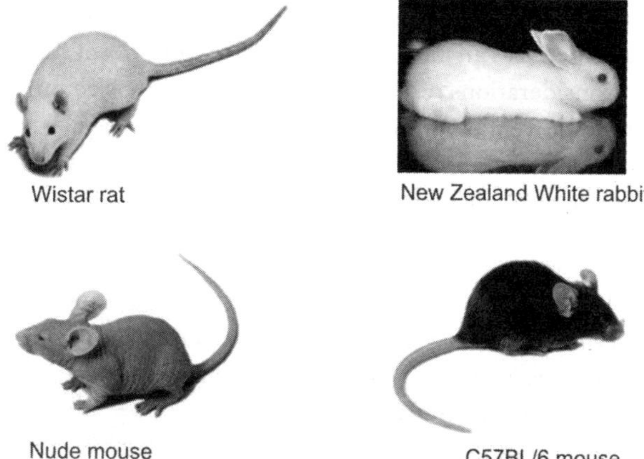

Fig. 11.4 Different types of animal models used in biomaterials and tissue engineering research.

11.2.2 | Selection of animal model

An ideal animal model should have similar anatomy/physiology of the implanted site, like that in humans. The pathogenesis and disease progression should be similar to that of humans and this should be based on the histopathology response. The costs associated with acquiring/maintaining the animal model during the experiment should also be considered.

It is worthwhile to mention that commonly used animals for experiments are rats, mice, rabbits, goats, sheep, pigs, dogs, cats and primates (see Table 11.1). Among all the animal models, rabbits are by far the most commonly used animal model in tissue engineering and in orthopedic research. This is because of the fact that rabbits are vertebrate animals with an appropriate size for surgical operation. Hence, it is possible to investigate healing of gap non-union in bones, repair of damaged articular cartilage, reconstruction of ligaments and tendons, and spinal fusion.

The selection of animal model for orthopedic applications can be qualitatively explained on the basis of four attributes, as summarized in Table 11.3. As far as the bone macrostructure is concerned, sheep or goat has the most similar bone macrostructure like human bone with the least similar for rabbit bone. Concerning the bone microstructure and composition, both pig and canine are close to that of humans. As far as the bone remodelling is concerned, the pig model can simulate anatomy/physiology close to that of humans. It has been lately perceived that while skipping the small models and using only one large preclinical model, it may be possible to restrict the number of experimental animal models.

Table 11.3 Four key attributes in terms of similarity between animal and human bone[836].

	Dog	*Sheep/Goat*	*Pig*	*Rabbit*
Bone macrostructure	++	+++	++	+
Bone microstructure	++	+	++	+
Bone Composition	+++	++	+++	++
Bone Remodelling	++	++	+++	+

Note: + least similar, ++ moderately similar, +++ most similar

From both an anatomical and physiological standpoint, the primates closely resemble humans. The primates are also considered as an ideal animal model for long-term studies, usually over several years. However, ethical considerations restrict their use in numerous countries. This is particularly justified when other species lower down the phylogenetic scale can be used for pre-clinical studies. Although large animal models, such as non-human primates closely resemble the human compared to smaller animal models such as rabbits, it is ethically incorrect to conduct initial experiments in larger animals.

It is generally well accepted to choose a small animal model for initial lines of investigations. However, final pre-clinical evaluation of a biomaterial implant or device requires the validation of biocompatibility results in a large animal model.

11.2.3 | Bone implantation experiments

The *in vivo* biocompatibility assessment of the implantable biomaterials for orthopedic and dental application demands experiments to be conducted in an animal model, whose bone structure and properties are similar to humans. The most common dental implant designs are either screw type (threaded) or cylindrical (rod shaped), while less commonly designs include disc or plate shaped.

Independent of the implant design, the implants should have an appropriate size for the defect to be created in the animal model chosen for the bone implantation site. While the screw type implants have the specific advantage of producing good initial stability, cylindrical implants can have exact fit with good stability within the bone. The latter design can provide results on better osseointegration. However, an analysis of rod or cylindrical shaped implants may be less complicated due to their more simple geometry.

As far as the implant sizes are concerned, the cylindrical implants, which are usually placed into femoral diaphyseal bone defects in rabbits, should be no larger than 2 mm in diameter and 6 mm in length. For cylindrical implants to be placed into femur or tibia of larger animals (such as sheep, goats and dogs), the ISO recommended dimensions are 4 mm in diameter and 12 mm in length. The implant dimensions may vary with the breed of animal. For example, large breeds of sheep can be used to investigate the implant stability/osseointegration of samples of 5 mm in diameter in certain locations (such as tibia and metatarsus). Orthopedic bone screw-type implants may range from 2–4.5 mm depending on the species chosen, with the 4.5 mm screws generally allowed only for the larger species (dog, sheep, and pig).

The use of control implants must be considered while designing any pre-clinical study. The control implant can be made of either materials already in clinical use (International Standard ISO 10993-6, 1994) or those with accepted *in vivo* biocompatibility (already published in literature). This will allow the outcome of any pre-clinical study to be validated or compared with reference/baseline biocompatibility property. In addition, the implant design would determine the experimental techniques to evaluate the material, (e.g., the mechanical testing). Some of the common mechanical testing includes torque removal tests (screw-type implants), pull-out tests and push-out tests (screw, cylindrical implants). The results of such tests enable one to evaluate the strength of the bone-implant interaction. Higher strength values reflect a good integration between the bones and implant surface or, a high degree of bone in-growth into the pores of the porous implant.

In this chapter, the protocols for the histological analysis, quantitative assessment of tissue compatibility and pre-clinical experiments with different materials are described with specific details. It is instructive to note that these details may vary depending on the experimental design or the biological questions to be addressed and even on the animal model and/or defect type.

11.2.4 | Preparation of tissue samples for histological analysis

Post-implantation, the explants along with host bone-tissue are collected for further histological analysis to establish any interfacial tissue reaction. It is important that any tissue slide preparation technique prevent damage to the bone or the tissue and at the same time maintain the integrity of the implant. A section of tissue around/adjacent to an implant must provide clarity, and good contrast to give maximum information. The sample preparation involves fixing in formalin, before being dehydrated in a graded series of ethanol. This is followed by treating in a graded series of acetone in ethanol. Subsequently, the tissue samples are embedded, cut into sections and stained prior to examination under the microscope[791]. The tissue sample preparation for histology analysis involves a number of sequential stages, which are described below (also see Fig. 11.5).

Fixation: A chemical solution (10% solution of neutral buffered formalin) containing a fixative at pH 7.0 is added to the tissue. The most commonly used fixative is formaldehyde at a concentration of 4%. Formaldehyde binds to and cross-links some proteins, and denatures others, but does not interact well with lipids. The overall effect is to harden the tissue and inactivate enzymes, preventing the tissue from degrading as tissues undergo cell death (autolysis) when removed from the body.

Dehydration: In order for sections to be cut, the tissue has to be embedded in wax. However, wax is not soluble in water. Therefore, the water in the tissue has to be removed and eventually replaced with a medium (alcohol) in which wax is soluble. This is achieved first by sequentially replacing the water with alcohol, placing the tissue in a series of solutions that contain increasing concentrations of alcohol, ending at 100%. This process is carried out gradually in order to minimize tissue damage. The tissue must then be 'cleared' before it can be embedded in wax. Subsequently, the section is placed in an organic solvent such as xylene, which replaces the alcohol. Wax is not soluble in alcohol. Finally, the tissue is impregnated with hot wax, which is soluble in this type of organic solvent.

Embedding: The tissue is placed in warm paraffin wax in a mould. On subsequent cooling, the wax hardens, and the tissue slices can now be cut.

Section cutting: Sections (slices) about 4–8 μm) thick are cut using a steel or tungsten blade fixed in a microtome. The wax sections are laid onto a glass microscope slide.

Staining: In order to observe microstructural details of cells/tissues, the components of the tissue have to be stained. Prior to staining, the wax has to be dissolved and replaced with water (rehydration), for the stains to be able to penetrate the tissue section. The sections are therefore placed in decreasing concentrations of alcohol, ending up at 0% alcohol (water). A number of different stains can be used, but the most common is hematoxylin and eosin.

Dehydration and mounting: The stained specimen is once again dehydrated, before placing it into mounting medium dissolved in xylene. Finally, a coverslip is placed on top of the sample to protect it, and the slide can be viewed under the microscope. In the following, some widely used stains and staining procedure for histology sections are mentioned.

Hematoxylin and eosin (H&E): Hematoxylin is derived from the logwood tree (*Haematoxylum campechianum*), and can only be used as a dye in its oxidized form (hematein). It is a basic dye that binds to acidic structures in cells (DNA in the nucleus, RNA in the cytoplasm and some extracellular materials) and stains them a purplish blue. Eosin is a negatively charged acidic dye. It binds to basic structures in cells (most proteins in the cytoplasm and some extracellular fibres) and stains them red or pink. The cells in tissue stained with H&E are therefore pink, with a purple nucleus.

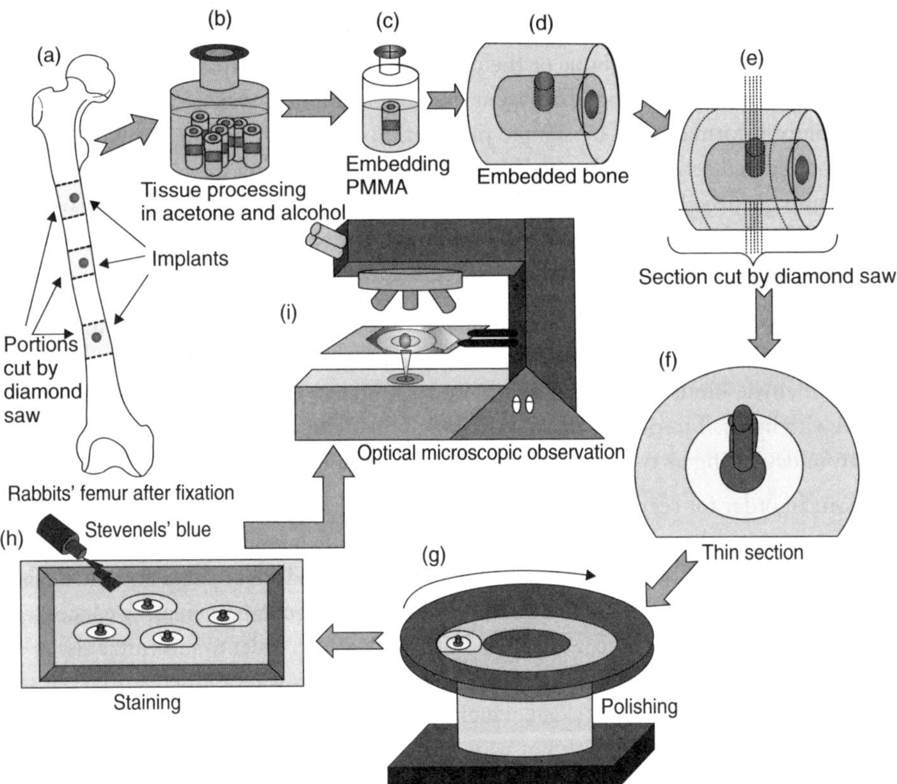

Fig. 11.5 Schematic showing step-by-step protocol of the sample preparation for histopathology analysis of the explants for the bone replacement applications. It is worthwhile to note that the basic tissue harvesting and sectioning protocol varies from tissue to tissue;. (a) showing the femur of the rabbit after fixation, the red circles indicate implants, (b) dehydration of bone pieces along with implants, (c) embedding in PMMA polymer, (d) embedded bone piece in PMMA, after removing from the bottle, (e) cutting of thin sections by diamond saw, thin dotted lines indicate the cutting path, (f) thin section showing the bone (yellow), bone marrow(pink) embedded in PMMA matrix, (g) polishing of thin section using diamond paste, (h) staining in Stevenels' blue, (i) optical microscopy observation; Note: Depending on the histological features to be examined in the thin tissue sections, one can use alternative stains instead of Stevenels' blue stain[839] [see Colour Plate].

Masson's trichrome method uses three different dyes (hematoxylin, acid fuchsin, and methyl blue), resulting in three colours in the stained section. During staining, the nuclei are stained blue; cytoplasm, red blood cells (erythrocytes) keratin are stained bright red; and collagen in the basement membrane, connective tissue, cartilage are stained bluish green. A related stain also used to stain connective tissue is Van Gieson[840]. In order to illustrate how different cells or tissues or other biological structures appear in a typical histology section, Fig. 11.6 shows the distribution of blood vessel, collagen, and muscle in subcutaneous section.

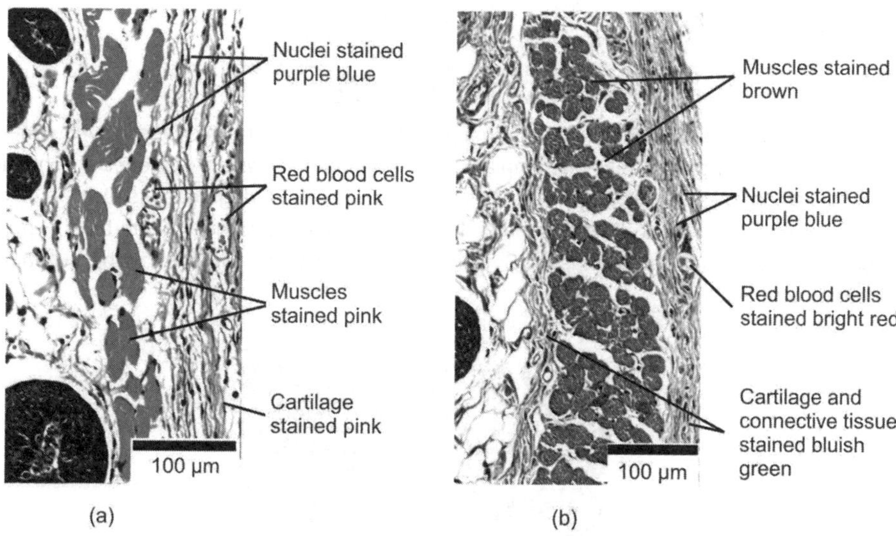

Fig. 11.6 Subcutaneous tissue sections stained with (a) Hematoxylin and Eosin stain, and (b) Masson's trichrome stain [see Colour Plate].

11.2.4.1 | Illustrative example: response of soft tissue upon implantation

During the implantation of any prosthetic device, a surgeon has to first injure the tissue, which leads to its destruction. The cells in the tissue die due to necrosis, leading to release of various cell constituents that attract different types of defence cells (present inside the small blood vessels in the tissues, close to the implants site) towards the site of injury. The WBCs (white blood cells) move out of bloodstream towards the site of injury. This directional movement is known as chemokinesis.

Neutrophils form the first line of defence through their cell constituents (cytokines, granzymes and other hydrolytic enzymes). Monocytes on activation by tissue injuries convert to macrophages, the scavenger cells of the body. These cells and resident macrophages in the extracellular matrix functionally take up or engulf dead tissue as well as any foreign body. The fact that the implant itself is a foreign body, macrophages make every attempt to engulf the implant and when they fail to do so, they coalesce together and form foreign body giant cells leading to an acute response, collectively referred to as 'inflammation' (Figs. 11.7 and 11.8).

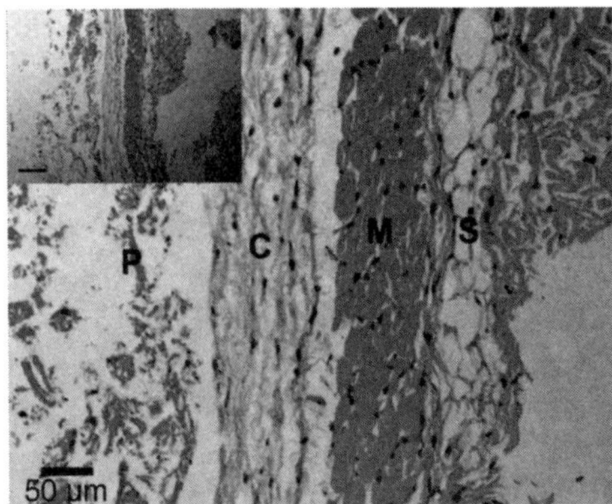

Fig. 11.7 Representative H&E stained histology image of the polymer-tissue interface in subcutaneously implanted PLGA at 16 weeks after implantation. The figure shows the presence of some tissue in the region where the polymer was present earlier. As PLGA degrades, its place is taken up by connective tissue. Collagen is stained pink while cell nuclei are stained blue. P-polymer site; C- fibrous capsule; M- muscle; S- skin. Inset magnification - 10 X, scale bar = 150 µm.[791] [see Colour Plate].

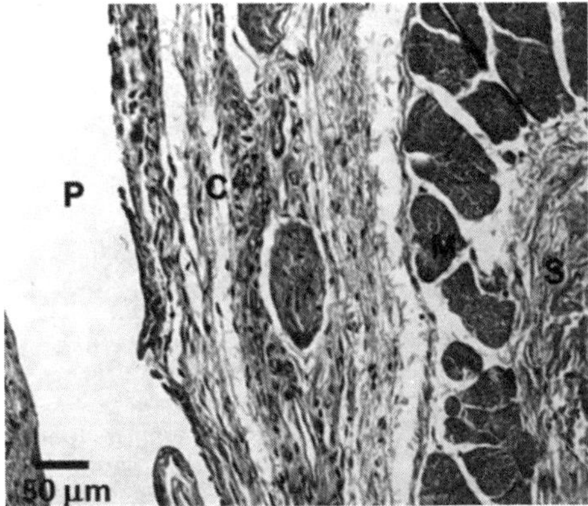

Fig. 11.8 Mason's Trichrome stained images [20 X magnification] of the tissue/implant interface of PLGA at 2 weeks post implantation. Numerous inflammatory cells are seen at the tissue-PLGA interface, while fewer cells were visible at the tissue-SAP interface.[791] [see Colour Plate].

When cell debris are removed, the fibroblasts, i.e., the interstitial cells in the extracellular matrix take part in the healing process or repair. The fibroblasts lay down collagen or more extracellular matrix, which participates in the healing process or repair. New small blood vessels are formed as offshoots of the existing blood vessels and the process is known as neo-vascularization. The tissue around the implant consisting of the migrated cells and new vessels is known as granulation tissue, which is gradually replaced by collagen deposition or fibrosis around the implant in the form of a layer enveloping or isolating the implant from the surrounding tissue (see Fig. 11.9). After this, a slow transformation of a thick fibrous layer occurs, which gets resolved into a very thin layer of collagen over a period of few months.

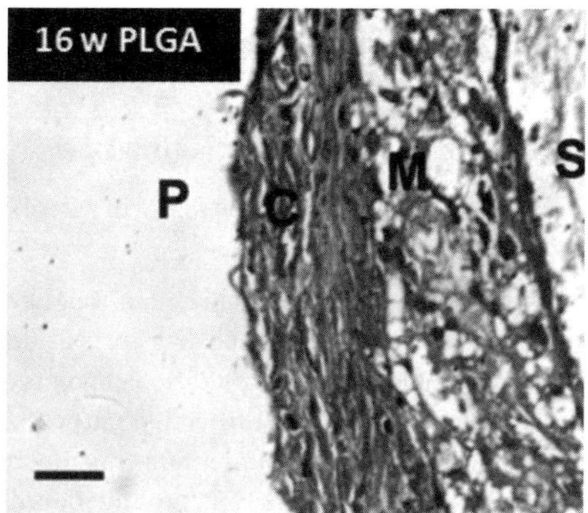

Fig. 11.9 Representative Mason's Trichrome stained images [50 X magnification] showing the implant/ tissue interface for PLGA at 16 weeks post-implantation. Muscles, collagen and cell nuclei are stained red, blue and black, respectively. P-polymer, C-collagen, M-muscle, S-skin. Scale bar = 25 µm.[791] [see Colour Plate].

Figure 11.10 depicts the tissue response towards injury which may vary widely according to site, species, presence or absence of infectious agents, etc. The intensity and time variables are dependent upon the extent of injury created in the implantation and the size, shape, topography, and chemical and physical properties of the biomaterial. In Fig. 11.10, the temporal variation in functional activity of the different types of cell proteins, involved in the inflammation and repair stages of post-implantation is qualitatively described.

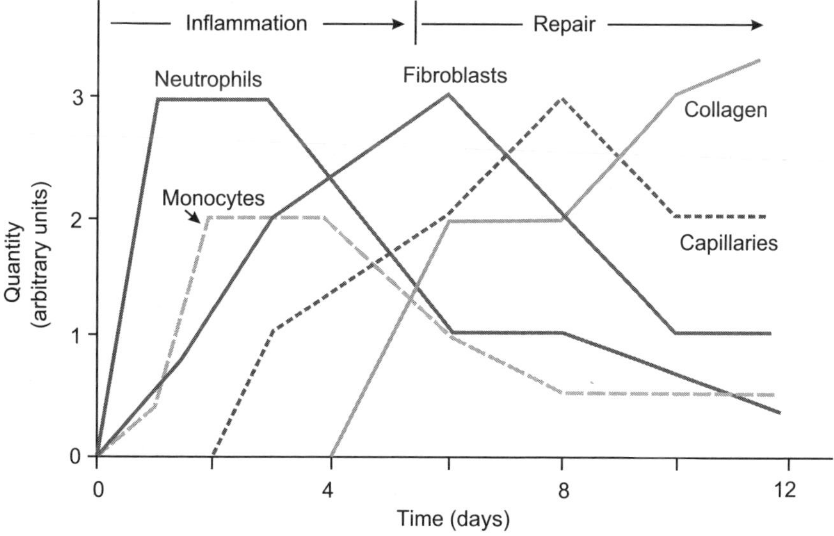

Fig. 11.10 Qualitative temporal variation in angiogenesis, connective tissue cells, collagen and immunological cells during foreign body response towards an implanted biomaterial.[841]

11.2.5 | Qualitative and quantitative assessment of tissue compatibility

11.2.5.1 | Histological analysis of tissue/material interface[792]

The histology analysis provides outcomes of some of the following attributes, which can be used to establish a scientific basis to assess the tissue response.

(a) *in vivo* **degradation:** This is determined by the change in weight of the implanted scaffold/polymer as a function of implantation time and such data are considered as a measure of *in vivo* degradation properties. The weight loss measured at different points of time, or the dimensional changes of the implant can be used for the quantification purpose.

(b) **Porosity of scaffold:** Two kinds of porosity are important- microporosity with the pore sizes of less than 10 mm being useful for initial cell adhesion, proliferation/differentiation, *in vitro* and macroporosity with pore sizes of 40–100 mm or larger for tissue in-growth and angiogenesis, *in vivo*.

(c) **Nature and distribution of vasculature:** Such quantitative assessment of the extent of angiogenesis provides the signature of tissue/wound healing response.

(d) **Number and distribution of inflammatory cells (including neutrophils, monocytes, macrophages, fibroblasts):** Such quantification is based on detailed investigation of histology slices taken at various timeframes and a comprehensive analysis indicates the nature of inflammatory reaction. It should be borne in mind that the initial acute inflammatory tissue response is predominated by the recruitment of neutrophils and monocytes. However, during chronic inflammation, macrophages take over the function of phagocytosis and antigen presentation.

(e) **Thickness and quality of fibrous tissue surrounding material:** Such evaluation provides an indication of the degree of acceptance of the implanted material in a quantitative manner.

(f) **Histomorphometric analysis:** In addition, micro-CT technique can be effectively used to qualitatively analyze the bone regeneration around implant in 3D space. In addition, the quantitative estimation of trabecule thickness or bone volume/total volume ratio is also widely adopted in biomaterials research.

Apart from the above parameters, the cytokine assays, as described below, are largely used to quantitatively assess *in vivo* biocompatibility.

Any *in vivo* study requires animal ethical committee approval at the institutional level and if necessary, at the national level. Also, the appropriate animal model should always be selected on the basis of the final targeted application of a biomaterial .

11.2.5.2 | Cytokine assays as tools for determining tissue health

The term 'cytokine' is derived from a combination of two Greek words- 'cyto' meaning cell and 'kinos' meaning movement. Cytokines are cell signalling molecules released predominantly by a broad range of immune cells that aid cell-to-cell communication during immune responses and stimulate the movement of cells (chemokinesis) towards sites of inflammation, infection or trauma. Cytokines are often released in a cascade manner, as one cytokine stimulates target cells to make additional cytokines. Also, similar immunological responses can be stimulated by different cytokines. Cytokines, as immunomodulatory agents, exist as peptide, protein and glycoprotein (proteins with sugar attached) forms and are the soluble factors that mediate acute and chronic inflammatory responses. Also, cytokines are involved in many physiological events from wound healing to autoimmune disorders. It is worthwhile to mention that cytokine is a generic name, which includes lymphokine (cytokine made by lymphocytes), monokine (cytokine made by monocytes), interleukin (cytokine made by one leukocyte and acting on other leukocyte), chemokine (cytokine with chemotactic activities) and tumor necrosis factor. However, cytokines do not include growth factors or hormones.

11.2.5.2.1 | *Tumor necrosis factor (TNF-α) analysis*

TNF-α is a multifunctional cytokine involved in homeostasis and pathophysiology of mammals. Multiple biological functions of TNF-α include apoptotic or necrotic cell death (including certain tumour cell lines), cellular proliferation and differentiation, immune regulatory activities.

> TNF-α is a signature marker for local or systemic inflammation.

It is known that high expressions of TNF-α after infection can result in septic shock, whereas sustained release of low levels can lead to cachexia and inflammation. Dysregulation of TNF-α is associated with viral infections and endotoxins, tissue injury, DNA-damage, etc. It is primarily expressed in macrophages, but also in monocytes, neutrophils, NK-cells, mast-cells, endothelial cells and activated lymphocytes.

Principle: The quantitative detection of TNF-α analysis is based an enzyme-linked immunosorbent assay (ELISA) (Fig. 11.11). As per recommended protocol, anti-mouse TNF-α coated antibody is adsorbed onto micro-wells. A biotin-conjugated anti-mouse TNF-α antibody (detection antibody) usually binds to mouse TNF-α, captured by the first antibody. Following the standard protocol, a coloured product is formed in proportion to the amount of mouse TNF-α present in the sample. For quantitative analysis, the absorbance is measured at 450 nm.

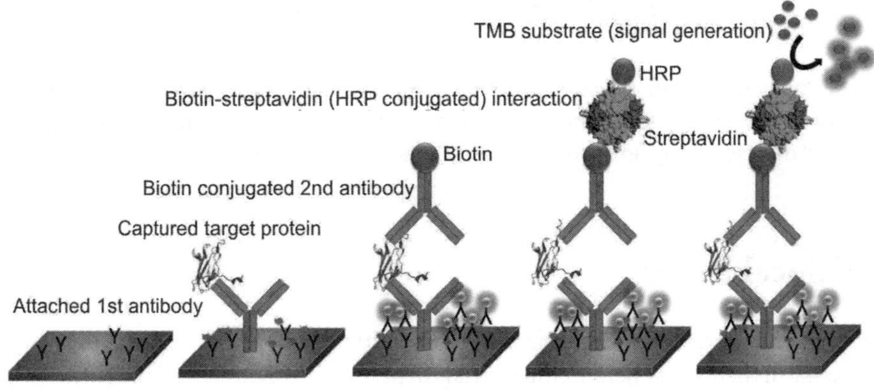

Fig. 11.11 Schematic illustration depicting the components of an ELISA used for tumour necrosis factor (TNF-α) analysis; the target protein in this assay is TNF-α [see Colour Plate].

11.2.5.2.2 | *Interleukin-1 (IL-1 β) analysis*

Interleukin-1 (IL-1), also known as catabolin was originally described in 1972 as lymphocyte activating factor (LAF). This is because of the fact that it was a lymphocyte mitogen and also is a polypeptide cytokine with two molecular forms (IL-1α and IL-1β).

> Both IL-1α and IL-1β mediate identical ranges of biological activity, which include synthesis of the acute phase proteins by hepatocytes, chemotaxis and release of polymorphonucleocytes from blood and bone marrow.

IL-1β is expressed by many cells including macrophage, NK cells, monocytes, neutrophils as well as injured and damaged cells. This cytokine is an important mediator of the inflammatory response, and is involved in a variety of cellular activities, including cell proliferation, differentiation, and apoptosis. An increased production of IL-1β causes a number of different auto inflammatory syndromes. The principle of IL-1β analysis is identically similar to that of tumour necrosis factor (TNF-α) analysis (Fig. 11.12). The experimental methodology to analyse IL-1β expression is shown in Fig. 11.13.

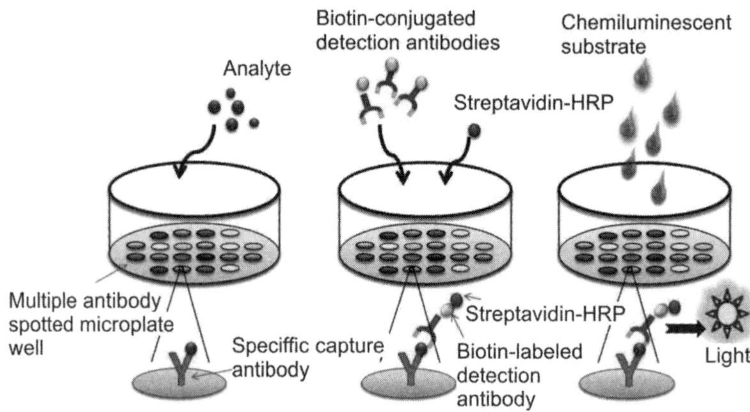

Fig. 11.12 Illustration depicting the components of an ELISA used for IL-1β analysis; the setup shown is for a multiplex system where capture and detection antibodies for multiple antigens are used and detected using an imager (ELISPOT assay) [see Colour Plate].

11.3 | Ethical Issues

Animal experiments or pre-clinical study is essential in the development of new implants/medical devices to establish their safety before clinical trials in humans. Although, such importance is widely recognized among most biomedical scientists, it is important to ensure the animal-welfare. In public/ Government funded biomedical research, we have the responsibility to disseminate such study as how animal research has contributed to the improvements in human healthcare.

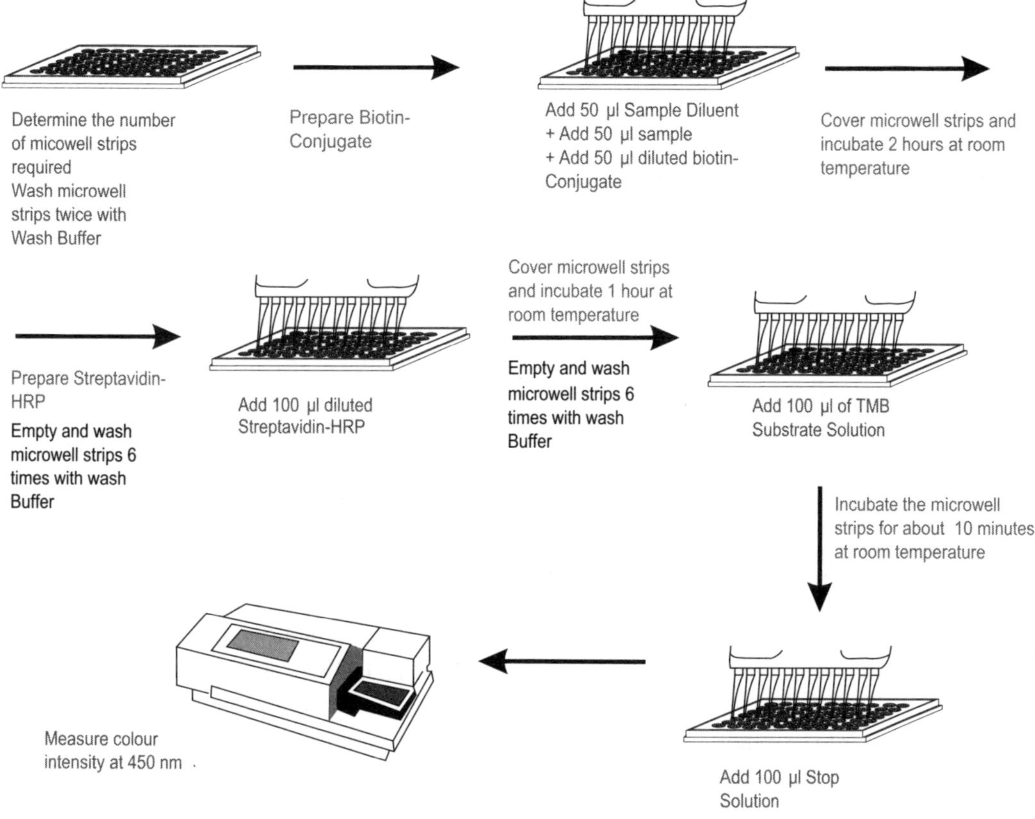

Fig. 11.13   Schematic diagram showing various steps of IL-1β analysis.

Any animal experiment must significantly advance the present knowledge and understanding of treatment modality under investigation, without compromising animal's healthcare.

There has been a greater social awareness of the peaceful use of animals in biomedical research. The ethical committee approves pre-clinical study only when such a study does not cause any pain to animals, while providing experimental models appropriate care during the study period with the endpoint objective to develop new treatment modalities of relevance to human healthcare. While designing any pre-clinical study, the following aspects are to be critically considered,

(a) Minimal number of animals while ensuring statistically relevant results/outcome
(b) Adherence to the established standards of animal care (minimal inflammation/post-operative pain)
(c) Justification for the use of a specific animal model such that the risks of implants/devices are sufficiently addressed. The institutional/national ethical committee, which approving such a planned pre-clinical study, must ensure that a minimum number of animals suffer during experiments, while animals survive in an acceptable living conditions during the study period.

(d) Similarities or differences between the implant prototype system and the human and the size or shape of the biomedical device systems, as compared to same characteristics of the version intended for use in humans.

> The pre-clinical experiments on most biomaterials and tissue engineering scaffolds involve small laboratory-bred animals and these experiments need to be approved by the Institutional Animal Ethics Committee (IAEC).

As defined in 'Breeding of and Experiments on Animals (Control and Supervision) Rules, 1998', IAEC essentially represents a committee of persons recognized and registered by the Committee for the Purpose of Control and Supervision of Experiments on Animals (CPCSEA). The latter is constituted in accordance with established procedures[793]. Current rules clearly allow IAECs to approve experiments only on small laboratory bred animals (rats, mice, guinea pigs, rabbits, hamsters, and invertebrates). For all other animals, the permission must be sought from a subcommittee of the CPCSEA[794]. Most organizations in India have formed their IAEC only since the start of 2000, and many still have to do so. The constitution of IAEC includes, (a) a biological scientist, (b) a veterinarian involved in the care of animal, (c) scientist in charge of animal facility of the establishment concerned, (d) a scientist from outside the parent institute (where *in vivo* tests are to be conducted), (e) a non-scientific socially aware member and (f) a nominee of CPCSEA. IAEC is expected to review, approve and monitor all research proposals involving small animals. For experimentation on large animals, the approval needs to be sought from CPCSEA after IAEC recommends a proposal. It is mandatory that the committee must ensure the compliance with the existing regulatory requirements (which may change from time to time), rules, guidelines and laws.

11.3.1 | Conditions for using animals for biomedical research

It needs to be categorically mentioned that there is always a reasonable societal expectation that the research outcome of animal study would enhance scientific knowledge, and improve the quality of health or welfare of humans or other animals. The scientific questions to address in animal experiments should therefore have substantial biomedical significance to justify the use of animals and that the number of animal to be used in such study should be minimized. However, a sufficient number of animals should be used to obtain necessary statistical analysis and results, preventing the consequent need to use more animals[795]. The following points deserve further attention,

(a) The pre-clinical study should be designed to address a significant basic science or potential therapeutic question for humans or animals.
(b) Prior *in vitro* research should justify the pre-clinical experimental approach.
(c) The potential scientific benefits must be assessed against the potential risks to the animal subjects.
(d) The animal model chosen should be the ideal one to answer the questions posed.
(e) The number of animals must be minimized.
(f) All animal welfare regulations should be strictly adhered to with a special emphasis on the species' typical behaviour and environment unless exceptions are scientifically justified.

11.3.2 | Elements of the animal study

The proposal for any animal study should include a discussion of each of the following key animal study features. Therefore, the following aspects need to be clearly mentioned/discussed in any animal study related proposal for evaluation by IAEC:

(a) a rationale for the selection of the particular animal model;
(b) the scientific objectives of each test protocol;
(c) the study schedule and in particular, treatment timeline;
(d) the methods and procedures, including pre-operative and post-operative care;
(e) prior *in vitro* study results;
(f) any *ex vivo* implant/device characterization, and
(g) any *ex vivo* tissue characterization.

The study animal should be provided with adequate veterinary care as per national regulations, standards, policies, and procedures. FDA recommends that the experimental animals must be monitored at a frequency for known risks posed by the biomaterial or implant. These routine monitoring parameters should be developed after consulting a well-trained veterinary staff at the study facility.

> FDA recommends that the study protocol include details of the terminal study and include all methodology for the collection and processing of tissue.

This section of the protocol should include the following information: (a) methods for end-period examination, (b) whether or not in-life radiographic analysis of blood flow or device positioning is needed, (c) methods for establishing end weight; and (d) anticoagulation, euthanasia, and tissue harvest methods. The comprehensive collection of all such information would be useful to determine if all observations made during necropsy and tissue preparation are related or unrelated to the method of euthanasia or harvest.

The final report for regulatory submissions should discuss all the important matters pertaining to each study, such as (a) the number of studies conducted, (b) the rationale for the model selected, (c) the similarity of the selected model compared to humans, (d) the general animal study methodology used, (e) whether there were quality systems in place during the study, (f) how the quality systems maintained independence and impartiality in the inspection of the data and the reporting of the results.

The report of any pre-clinical study using animal models must include the following aspects,

(a) the study groups with the number of animals in each group;
(b) study duration;
(c) the device design used, and
(d) a summary of study outcomes.

> The well-known 3R (replacement, reduction, refinement) approach is used as the guiding principle for the welfare and humane use of animals in scientific research.

However, animal welfare concerns not only what happens to the animals during the experiment in the narrow sense, but also their living conditions and housing. Any researcher planning to use animals in their research must first show, why there is no alternative and what will be done to refine their techniques to minimize animal suffering, by following the 3R principle, as shown in Fig. 11.14[796].

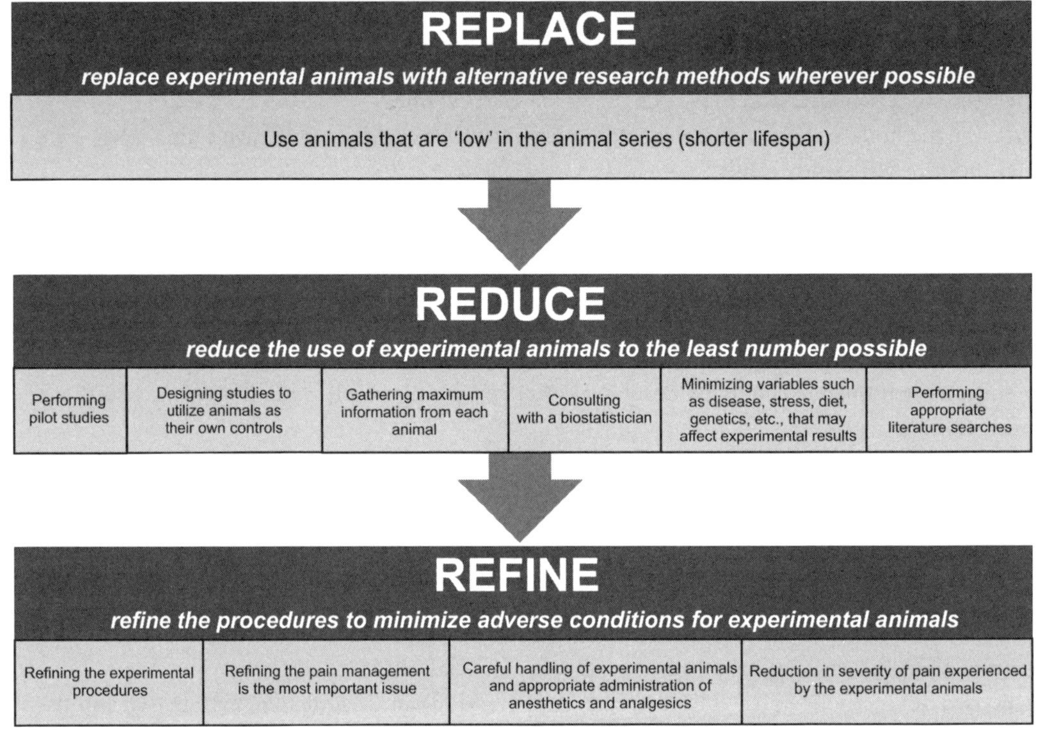

Fig. 11.14 Key principles of 3 'R's to be followed in any pre-clinical study using animal models in biomaterial research; adapted from[842].

11.4 | Illustrative Examples of Animal Experiments on Biomaterials

In this section, a few illustrative examples will be provided so that the readers will understand how to plan the animal experiments for implantable biomaterials, which are proven to be compatible with biological cells, *in vitro*.

It may be worthwhile to mention that only test material with the best cytocompatibility properties in terms of cell adhesion/proliferation or differentiation should be chosen for *in vivo* study to minimize the number of experimental animals to be sacrificed. In each experiment, one must use the control implant with proven *in vivo* biocompatibility property. Also, the institutional level clearance from the animal ethical committee for the use of small and medium sized animals is mandatory prior to the start of any *in vivo* experiment. For the larger animals, like monkey, sheep, pigs, one needs to obtain prior approval both from the institutional as well as at the national level ethical committee.

The following examples are chosen to demonstrate how to design the study for specific biomedical applications and should be considered for illustrative purpose only. In fact, the number of animals, timeframe of the entire study and the time interval for the termination of animals depend on the overall objective of the *in vivo* study as well as various endpoint scientific analysis, to be supported by quantitative results having statistical significance.

While designing and conducting pre-clinical experiments involving biomaterials, the following questions are typically to be addressed. (a) Does the animal remain healthy without any gross infection post-implantation, up to the time point of sacrifice? (b) How does the animal weight change with time, post-implantation? (c) In case of bone replacement material, how does the bone formation progress with time during post-implantation? (d) Is there any histological evidence of inflammation around bone tissue/material interface? (e) How does the foreign body response, qualitatively and quantitatively, change with post-implantation time? (f) How to describe the morphology of neobone formed at host bone/material interface in terms of various cell types (fibroblasts, osteoblasts, osteocytes)? (g) Quantitatively, how do you see the change in pro-inflammatory cytokine expression with implantation time, if some clear signature of inflammation observed histologically? (h) How does the implant integrate into the osseous system, 2D (histology) and 3D (micro-CT image or tomogram)? (i) How does the bone regeneration change with implantation time quantitatively in terms of BV/TV ratio? (j) If an implant exhibits biologically satisfactory response in a lower animal model, how such results are validated in larger animal model?

11.4.1 | Bone implantation in rabbit animal model

Two types of bone defect models are widely used in bone implantation experiments: (a) cylindrical defect and (b) segmental defect model. The implant size and shape therefore depends on the defect model to be used in a given study. In the following, a representative protocol for bone implantation in cylindrical bone defect model is described.

Prior to the implantation experiment, the implants are generally subjected to ultrasound cleaning in distilled water for 10 min, dried overnight at 70°C in a hot air oven and sterilized by autoclaving at a pressure of 15 psi at 121°C for 15 min. Alternatively, the sterilization can be done using Ethylene oxide (ETO). Both test and control materials are to be implanted in the femur of 10 New Zealand white rabbits of body weight more than 2 kg and of either sexes as per ISO 10993-6 guidelines. Typically, 10 test animals should be used, depending on the number of timelines at which the animals are to be sacrificed. The experimental animals are anaesthetized with xylazine (5 mg/kg body weight), ketamine (35 mg/kg body weight), and midazolam (0.3 mg/kg body weight) and maintained by continuous propofol infusion (0.4 mg/kg body weight/min). The animals are controlled on lateral recumbency and a linear incision is made on the cranio-lateral aspect of the thigh under standard aseptic precautions (see Fig. 11.15). The shaft of the femur is exposed after a blunt dissection through the junctions of the tensor Fascia Lata and Vastus lateralis muscle. Typically, the cortical defects of approximately 2.5–3 mm diameter can be created approximately 10 mm apart in the mid shaft of each femur with a 1.5 mm drill using surgical micromotor with continuous saline irrigation. This requires the test and control implants to be of similar size and shape. While such strict dimensional control is possible for metallic and polymeric biomaterials, this is undoubtedly difficult for ceramic samples. Each implant must be pre-soaked with saline and press fitted into the defect. The surgical wound needs to be closed in layers. Both ceftriaxone (20 mg/kg body weight) and meloxicam (0.25 mg/kg body weight) are administered intramuscularly for 5 days post-implantation.

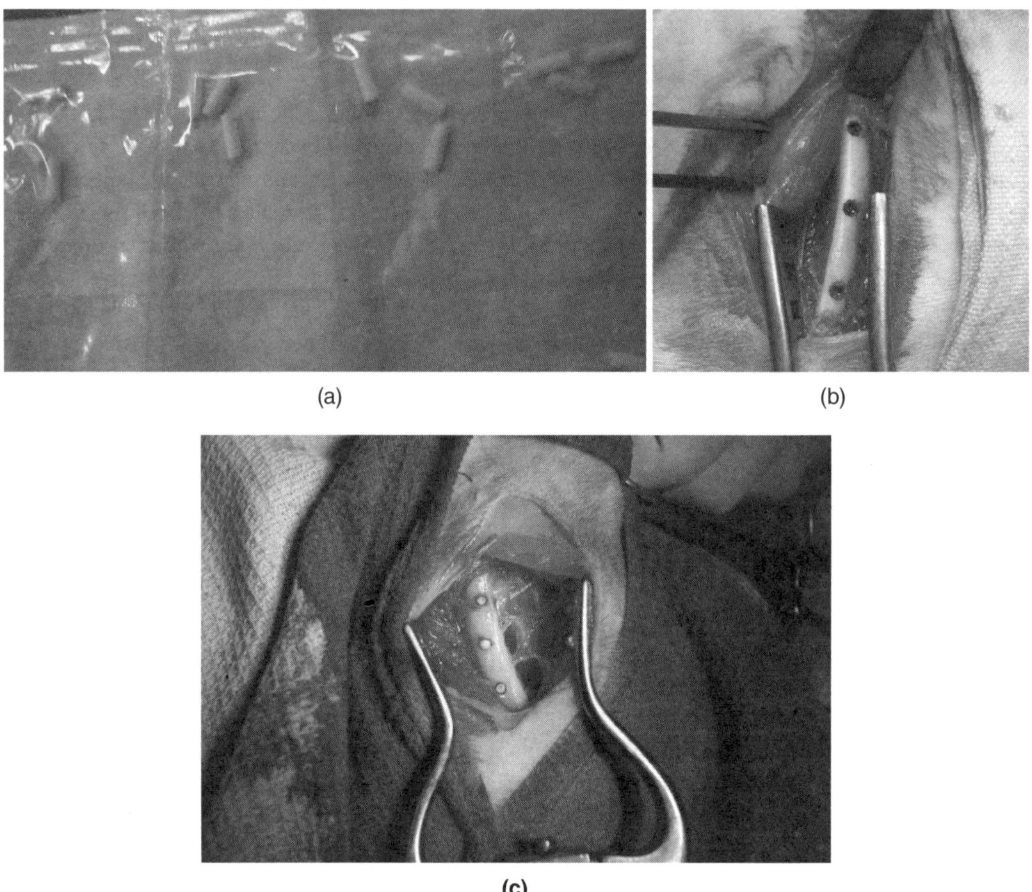

Fig. 11.15 Photograph showing (a) the sterilized cylindrical shaped implants prior to implantation, (b) creating cylindrical bone defects of a limb in an experimental rabbit during surgery and (c) three implants placed in the cylindrical bone defects, post-surgery[839] [see Colour Plate].

After the desired timeframe of implantation, the experimental animals are euthanized by an overdose of thiopentone sodium at each time period. Also, the implantation sites are grossly examined for healing of bone defects. The femurs with the implant materials need to be removed and fixed in 10% neutral buffered formalin. An identical procedure is to be conducted on all experimental rabbits. After the surgical procedure, all the animals should be given post-operative care. X-ray radiographies should be taken from after each time interval to monitor the implants at their sites (see Fig. 11.16). Also, the gross view of the implant with host tissue can be analyzed (see Fig. 11.17). Bone labelling can be adopted to identify site of new bone deposition post-implantation in one rabbit of each time period by sequential intramuscular administration of two different flurochromes, namely xylenol orange and alizarin red at specified time intervals.

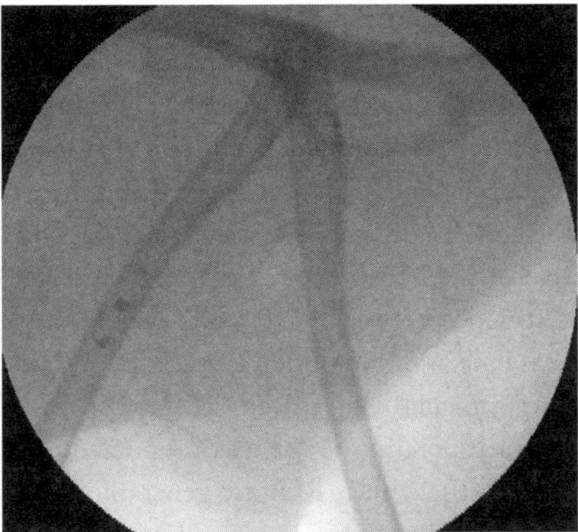

Fig. 11.16 X-ray radiographs of femurs with three implants (radio-opaque) on the left and three implants (non-radio-opaque) on the right limb, post-implantation (magnification- 1x)[839].

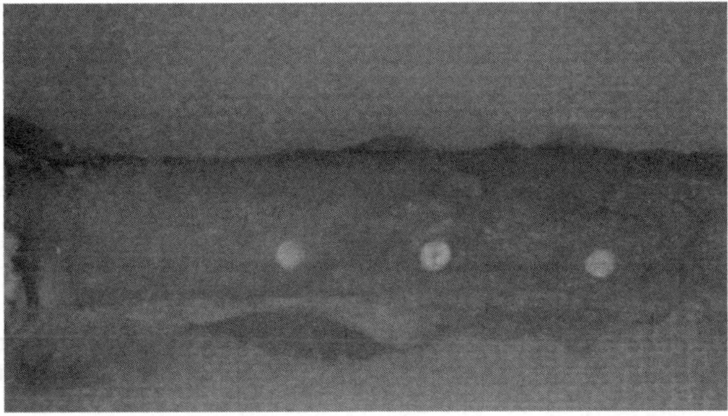

Fig. 11.17 Representative digital image revealing the implant site covered by neobone around the test implants (non-radio-opaque) after implantation for 26 weeks[839].

Histology: Significant biologically relevant information can be obtained by histopathological analysis of implant/host tissue interface. The sample preparation for such analysis follows a different protocol in contrast to conventional materials science samples. The entire process of tissue sample preparation for histology analysis is summarized in Fig. 11.5 by schematic drawings. Few thin samples should be utilized for SEM and AFM observations as well.

Tissue material interaction at the cellular and molecular level should be analysed qualitatively and quantitatively to understand the underlying mechanisms of biocompatibility.

The examination of tissue material interface should not be limited to microscopic evaluation alone. Cytokine secretion, cell signalling pathways and expression (pro and anti-inflammatory) at gene level around implants are some of the parameters, which can be analysed.

Micro-CT analysis: Mineralization of newly formed bone, overall morphology of the neobone around the implant and quantification of neobone fraction with respect to host bone can be quantified using high-resolution X-ray micro-computed tomography. The working principle of micro-CT is briefly described in chapter 5. Briefly, some salient details are mentioned. In particular, the specific area of the scanned region containing one implant is selected and the reconstruction of the selected region is to be made using the cone-beam convergence/back projection algorithm-based software. Based on the density variation of the host bone and the implant, the 3D image of the implant is extracted on the basis of the distinction between ROI (region of interest) and VOI (volume of interest). The overall scan time can be as long as 3,000,000 microseconds and the entire scan distance needs to be 12,288 μm. From the experimentally measured CT database, a cylindrical region of interest (ROI) is selected for analysis corresponding to the cortical bone during implantation. In order to evaluate bone regeneration within the defect, ROI are generally further sectioned transversely as top, medium and bottom and radially as outer, inner and core shells. Also, thresholds should be suitably applied to images of each sample in order to segment the newly formed bone from the residual implant. After thresholding, the bone volume (BV) should be determined by counting the total number of bone voxels and multiplying by their known volume, while the total volume (TV) is to be determined by counting the bone and non-bone voxels. Summarizing, an overview of the pre-clinical study design, including qualitative and quantitative analysis to assess tissue-material interaction is provided in Fig. 11.18.

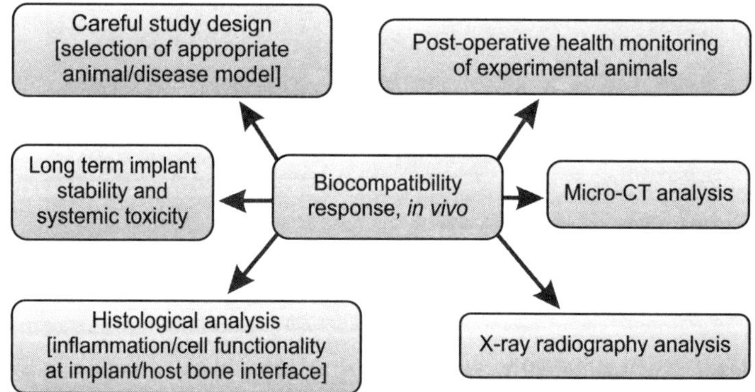

Fig. 11.18 Overview of different aspects of the *in vivo* biocompatibility assessment of biomaterials.

11.4.2 | Toxicity assessment of biomaterial nanoparticulates, *in vivo*

It is worthwhile to mention the origin of the wear debris particles in the context of the pre-clinical study and its relation to the toxicity, *in vivo*.

> The generation of wear debris and its accumulation in and around the implant are inevitable for most articulating joints and therefore, their toxicity in biological system is of significant concern!

Several factors influence the progressive wear of particles from the prostheses, such as the implant design, the physiological microenvironment and material parameters[797]. Severe wear takes place under higher contact pressures. When the local stresses exceed the tensile strength of the material, multiple grains will spall from the ceramic matrix, leading to increased wear rate[798]. Improper fixation of the prostheses and increased angle of inclination of the acetabular cups has been proved to be a possible reason for accelerated wear[799]. Moreover, the friction and wear of the implants strongly depend on the surrounding environment. Besides, ceramic wear particles may be trapped between two moving surfaces, causing three body wear in the physiological environment[800]. Inside the human body, the implant is continuously exposed to a corrosive mixture of extracellular fluid comprising of water, complex organic compounds, various ions, proteins, enzymes, etc.[801]. The exposed surfaces will undergo chemical dissolution/degradation either by the body fluids or by the foreign body reaction, elicited by invading immune cells. The degree and nature of the inflammatory cellular response depends on the material composition, surface structure, surface chemistry, and so on[802].

The important prerequisites required for achieving lower wear rate in load bearing and articulating (hip joint) implants are extreme hardness, finely grained microstructure and smooth surface finish[803]. The most clinically accepted wear couple is the combination of alumina head and alumina cup (hard/hard), articulating against itself. This ceramic-on-ceramic bearing surfaces offer considerable advantages over more traditional articulations, which utilize UHMWPE as a bearing material, both in terms of wear volume and osteolytic potential[803]. In contrast, the wear rate of zirconia articulating against itself was found to be high, almost 5000 times greater than that of alumina[804]. Similar catastrophic wear rates were found for the combination of medical grade alumina/zirconia[803]. As a first step towards the toxicity assessment of ultrafine particles, the synthesis of such finer particulates from bulk materials without compositional change is an important aspect. Once such particles are prepared, one can prepare eluate by dissolving such particles in phosphate buffered saline (PBS) or other medium.

There is no unambiguous evidence in the literature, which suggests the critical concentration of particles, irrespective of the chemical composition. Similarly, we do not have any clear literature proof indicating the critical size of the particle that will be toxic, if generated *in vivo*.

> The biological response to wear particles involves a complex interaction and multiple variables, like size, concentration, shape and composition, influence such interactions.

With regard to the composition, HDPE (high-density polyethylene) particles showed significantly higher inflammatory response compared with Alumina particles of similar size and concentration, implying that in addition to the ability of macrophages to respond to the size and concentration, these cells are sensitive to composition effect[805]. In another study, the volumetric concentration of alumina particles required to stimulate inflammatory effect was found to be extremely high (500 μm^3 per cell). However, it is unlikely to achieve such threshold concentration *in vivo*, due to the extremely low wear rates (< 4 mm^3 per million cycles) of ceramic-on-ceramic prostheses even under severe microseparation conditions[806]. Another report suggests that the osteolysis due to the wear particles is more likely to occur when the wear rate is more than 0.1 mm/year and its occurrence is least expected for the wear rate of less than 0.05 mm/year[807].

The wear products generated *in vivo* can disseminate via the lymphatic system and can either remain inert or can induce carcinogenic or mutagenic effect[808]. The material toxicity arises mainly from the ions leaching out from the wear debris. *in vivo*, the wear rate of ceramic-ceramic is 4000 times less than that of metal-polymer[809]. Due to the particle exposure, macrophages and fibroblast-like cells from the synovial fluid releases prostaglandins, metalloproteinases, and cytokines stimulating inflammation causing aseptic loosening. IL-1, IL-6 and the metabolite PEG_2 are potent inducer of osteoclast activation and resorption in the joint area. The long-term exposure of zirconia and alumina powders to synoviocytes slowed its proliferation rate without any significant change in the release of IL-1 and IL-6, but to a small extent inhibits the functioning of some of the metabolites associated with lipoxygenase pathways[810]. Zirconia and alumina are the two widely used ceramic material as load bearing implants due to its bio-inertness and stability that can meet the criteria at the loading parts of the human body. Wear particles that are of few micron sizes can cause cellular reactions in the biological environment. To this end, the larger release of yttrium and zirconium ions from the yttrium-doped zirconia in the dissolution test using lactic acid indicate the non-suitability of these materials in an lower pH environment, especially as dental implants[811].

11.4.3 | Subcutaneous implantation of biodegradable polymer in mice model

Biodegradable polymers, like PLA/PGA/PLGA can be used to deliver anti-inflammatory drug molecules *in vivo*. In order to establish such efficacy, the first step is to carry out *in vitro* study with specific cell line as well as *in vitro* degradation. Depending on the degradation profile, the protocol for *in vivo* biocompatibility assessment should be carefully planned. Similar experiments have been conducted on salicylic acid releasing poly(anhydride-esters). These polymers release salicylic acid very fast. For example, in a study by Ulrich and co-workers, 56 Sprague Dawley rats were used to study the impact of salicylic acid release on diabetic bone regeneration. An osteotomy defect was made in the mandible[812]. In another study, 10 Balb/C female mice were used to assess the effect of salicylic acid delivery on bone[813]. A polymer membrane was placed adjacent to the maxillary first molar. It was shown that the density of inflammatory cells and swelling were decreased in the tissue adjacent to the polymer. In a different study, polymer discs of 10 mm diameter and 1 mm thickness were implanted subcutaneously by small midline dorsal incisions and the polymers were placed in subcutaneous pockets created by blunt lateral dissection[814]. The skin was closed with staples. After 10 days of implantation, the inflammatory response for all the implanted polymers was evaluated. The rats were sacrificed and the tissues surrounding the implants were evaluated after 12 weeks of implantation. The commonly used polymer PLGA was used as the control.

In order to assess the host response of a given biodegradable polymer, the polymer discs (8 mm diameter and ~1.4 mm thickness), can be implanted in mouse model by three small midline dorsal incisions at subcutaneous pockets created by blunt lateral dissection. The BALB/c mice can be chosen as they are a commonly used multi-purpose animal model. Also, these mice are fast growing, and have also been used previously in similar experiments. The number of animals should be kept at a minimum to conduct experiments, while ensuring statistically significant result. Also, the use of the cured and the non-cured polyesters in *in vivo* studies can provide a comparative result of the effect of release of different concentrations. All the animals in the experiments should be identified by group and individual number with the help of potassium permanganate marking on the body. The polymer is to be implanted subcutaneously in the animals by small mid line dorsal incisions and the polymer pellets needs to be placed in subcutaneous pockets created by lateral incisions. Blood samples are collected at every time point. An implant of 8 mm diameter and 15 mm thickness can thereafter be implanted subcutaneously by small mid-line dorsal incisions. The polymer pellets are placed in subcutaneous pockets created by lateral incisions. After induction of general anaesthesia, (pentobarbital or ketamine/xylazine cocktail, I.P. Injection), the implant needs to be subcutaneously implanted in the body. The peritoneal walls are to be lifted with forceps. After placing the implant, the peritoneum and the muscular layer are sutured with 5/0 absorbable suture and the skin is closed with suturing clips. The incision is cleaned and dusted with Nebasulph antibiotic powder.

After the animal sacrifice at various points of time, the neutral buffered formalin fixed explant tissue samples are dehydrated in a series of gradient ethanol and embedded in paraffin blocks. Thereafter, the thin sections are cut at 5 μm and are stained with Hematoxylin and Eosin (H and E) for the analysis of the inflammatory response and with Masson's Trichrome Stain (MTS) for the analysis of fibrous capsule. For each time point, three slides are prepared per animal for each stain. The prepared histology slides should be observed by two observers independently without prior knowledge of the mice from which the section was obtained, under a light microscope to avoid any bias. The entire pre-clinical animal experimental procedure and post-surgery analysis are shown schematically in Fig. 11.19.

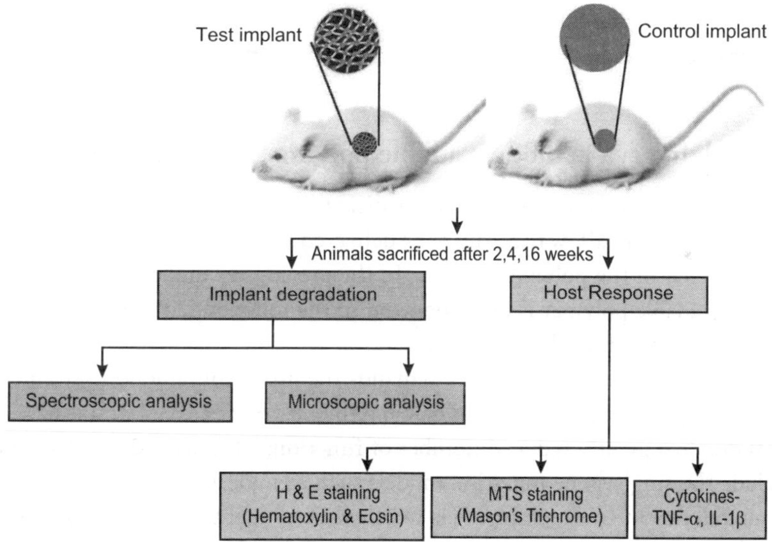

Fig. 11.19 Schematic illustration of the protocol to be followed with a biodegradable polymer in mouse model[843].

The thin sections used for histomorphometrical analysis can be analysed using ImageJ software. Digital analysis can be used to measure the area of the region of interest. Different parameters, like density of inflammatory cells (neutrophils, monocytes/macrophages, fibroblasts), blood vasculature and distribution of collagen surrounding the implant can be quantified to measure the tissue response. The cell density is averaged after measuring the number of cells in a region of interest for at least 5 fields of view per slide. The identification of different cell types is based on their respective morphologies in the H&E sections. Neovascularization is calculated by measuring the total number of blood capillaries in the entire fibrous capsule section and normalized to the area from both, H&E and Masson's Trichrome stained images. The collagen distribution around the implant is to be measured by calculating the blue pixel coverage using image J software in the Masson's Trichrome stained sections[815]. The blue pixel coverage are calculated first for 5 μm at the implant-tissue interface and subsequently at 10 μm intervals, throughout the capsule thickness for at least 10 points per field of view for at least 3 images per slide. The concentration of pro-inflammatory cytokines, viz. tumour necrosis factor-α (TNF-α) and Interleukin-1β (IL-1β) in blood serum can be quantified spectroscopically using commercially available ELISA kits (Platinum ELISA kit, eBioscience, Austria). These assays are to be carried out as per the manufacturers' instructions. After addition of the stop solution, the absorbance can be measured at 450 nm using a microplate reader or UV-Vis spectroscopy and the concentration is to be determined using a standard curve plotted for each cytokine. Summarizing, the scientific analysis for such a study involving sacrificing the test animals at different points of time, (e.g., 1 week, 4 weeks, 2 months, 4 months), would involve the following aspects: (a) histological analysis of the host tissue/implant interface, (b) serum biochemistry of blood samples from test animals, (c) kidney and liver function tests of animals, (d) haematology- WBC/RBC count, glucose of blood samples of animals and, (e) detection of pro-inflammatory cytokines- TNF α, IL-1β of tissue/implant interface.

As a follow-up plan, one can also use an inflammatory model to test the anti-inflammatory effect of sustained release of salicylic acid. For this purpose, the researchers can plan to use a chronic model as the release of salicylic acid is slow. Rheumatoid arthritis would be the model of choice here.

11.4.4 | Drug delivery via biodegradable polymer for cancer xenografts

One of the widely investigated approaches in biomedical technology is to load disease-specific drugs in a biodegradable polymer to allow the drug to be delivered *in vivo* at a rate in synchronization with the degradation kinetics of the polymer carriers. In the previous example, the *in vivo* test protocol for one such biodegradable polymer is described and in the following, a typical protocol to deliver anti-cancer drug in a mouse model is described.

To begin with, the specific disease model demands the use of immunodeficient mouse models, characterized phenotypically by an absence of hair and thymus. Such animal strains have an inhibited immune system due to a greatly reduced number of functional T cells and B cells. The strains of these mice include the single-gene mutation models such as nude-mice (*nu*) strains and the severe combined immunodeficient (*scid*) strains. It is worthwhile to mention that the primary treatment for the identified cancers is surgery, which subsequently needs to be followed up with administering systemically chemotherapic drugs to prevent recurrence. However, such treatment options have limited clinical effectiveness due to several reasons, such as the complex biology of tumours, cell types, doubling time, as well as tumour volume, heterogeneity and vascular barrier. Taken together, all these factors influence the desirable results of chemotherapy.

High systematic administration of chemotherapeutic agents often leads to several unwanted side effects towards healthy tissues. Consequently, a local delivery of therapeutic concentration of a drug in a sustained manner by the use of drug-releasing polymeric implants is required at the surgical site, which can minimize the rate of tumour repopulation.

Such methods of sustained drug delivery also enhance chemo-responsiveness by exposing tumours and adjacent metastases to high local concentrations of anti-neoplastic agents, while reducing drug induced potential toxicity to normal healthy cells. Similar approach has been reported in the literature as well. For example, Shikanov et al.[816] tested the effectiveness of paclitaxel–polymer formulation injected intratumourally in mouse bladder tumour heterotopic model. Here, 10 animals were used for each of the 5 groups (3 tests and 2 controls). To determine paclitaxel biodistribution in tumour at predetermined time periods, the tumour was excised, frozen and sectioned, and the paclitaxel concentrations were determined in the tumour tissue and in plasma by HPLC. Histopathological evaluation of the necrosis and inflammation was performed on tumour sections. Results showed that the treatment with local injection of polymer–paclitaxel formulation inhibited the growth of solid tumours, and a high drug concentration was found in tumour[816].

The submicron/nanoscale, paclitaxel-loaded polylactide-co-glycolide (PLGA) nanofiber discs were fabricated by Ranganath et al.[817] They had evaluated the pharmacokinetics and therapeutic efficacy of this submicron/nanoscale, intracranial implants for treating malignant human glioblastoma (U87 MG-luc2) in BALB/c mice. 15 animals were used for each of the four groups (2 tests and 2 controls). The polymer disc was intracranially implanted in mice and the coronal brain sections were analyzed for bio-distribution of paclitaxel on 14, 28 and 42 days post-implantation. Results revealed that the submicron/nanoscale implants were able to demonstrate optimal paclitaxel pharmacokinetics in the brain/tumour with significant tumour inhibition in a glioblastoma xenograft model in mice[817].

Jaiswal et al.[818] evaluated post-surgical antitumor efficacy and biodistribution of doxorubicin loaded semi-interpenetrating polymeric hydrogel in Ehrlich's ascites tumour model using albino mice. In this study, 12 animals were used for each of 5 groups (2 tests and 3 controls). The drug concentration in the tumour, tumour periphery and residual hydrogel was reported to decrease with increasing time. Importantly, no sign of tumour recurrence until the 25th day with 100% necrosis was recorded. Naraharisetti et al.[819] examined the responses of BALB/c nude mice, bearing C6 glioma tumours subcutaneously, to treatments by PLGA microspheres, microparticles and discs-delivering Paclitaxel and Etanidazole. Their results showed inhibited tumour growth over the blank placebo groups after 21 days of tumour growth for spray-dried microspheres, electro-hydrodynamic atomization microparticles and spray-dried discs.[819]

Gou et al.[820] had encapsulated curcumin into monomethoxy polyethylene glycol–poly-3-caprolactone (MPEG–PCL) micelles through a single-step nano-precipitation method, creating curcumin-loaded MPEG–PCL (Cur/MPEG–PCL) micelles. In their study with female BALB/c mice, they reported that, intravenous application of Cur/MPEG–PCL micelle in female mice inhibited the growth of subcutaneous C-26 colon carcinoma *in vivo*, and induced a stronger anticancer effect than that of free curcumin[820].

In the perspective of the above studies, the following protocol can be useful to deliver an anticancer agent (capecitabine/doxorubicin/5-flurouracil/irinotecan/oxaliplatin), released from subcutaneously implanted biodegradable polymer against cancer xenograft in immunodeficient (*nu/scid*) mouse animal model.

Initially, a solid tumour needs to be established upon subcutaneous injection of 0.1 ml suspension of any cancer cell line ($7{\times}10^6$ cells per mouse) to the dorsal side of nude (nu/nu) BALB/C mice (6–8 weeks, 22–34 g). The tumour is allowed to grow until it reaches the size volume of 48–100 mm^3. The tumour bearing mice are separated typically into 5–6 groups for different treatments (n = 10): Anticancer agent control, sham control, placebo newly synthesized polymer, placebo control polymer, anticancer agent loaded newly synthesized polymer and anticancer agent loaded control polymer. Animals in anticancer agent control group should receive anticancer agent only. No treatment is given to animals in the sham control group. Mice in the placebo groups must have implants with no anticancer agent. The newly synthesized polymer with anticancer agent needs to be implanted surgically into the tumour mass of nude mice, under anaesthesia, by making three midline dorsal incisions near the tumour. After implantation, wound will be sutured and all the animals will be sprayed with a disinfectant powder. They will be kept under observation until the last set of studies. Tumour growth must be documented by measuring the length of major axis and minor axis (perpendicular to minor axis) with dial callipers twice weekly. The animals should be sacrificed at different points of time to observe the efficacy of anticancer agent loaded newly synthesized biodegradable polymeric implant against cancer xenograft. Following surgery, suitable topical antibiotic such as Nebasulph should be applied to protect against microbial infection. After the animal gains consciousness, the animal needs to be allowed to rest in a cage provided with *ad libitum* food and water.

Following the above-described *in vivo* experiments, the scientific analysis must involve to assess the following aspects: (a) survivability of mice, (b) tumour size and growth delay, (c) histopathological preparation of implantation site and other tissue sample such as colon, liver, kidney, etc., (d) *in vivo* distribution studies of released drug polymer implant by UV-Vis spectroscopy or HPLC, (e) haematology and blood biochemistry analysis, (e) immonohistochemical analysis and (f) detection of pro-inflammatory cytokines-TNF-α and IL-1β.

11.4.5 | Peripheral nerve regeneration in rat model

From the clinical perspective, the regeneration of peripheral nerve (PN) involves the use of an appropriate guiding channel or suitable biomaterial substrate, which should support the nerve cell functionality.

> A biomaterial conduit for neural tissue engineering should encourage the axonal growth and should have sufficient mechanical flexibility and high porosity for neurotropic communication and degradability.

A number of materials tested for neural tissue regeneration include poly-L-lactic acid (PLLA)[821], polycaprolactone (PCL)[822], polydiemethylsiloxane (PDMS)[823], polyglycolic acid (PGA), polylactic-ε-carprolactone[824] and polypyrrole[825,826]. Many of these materials are reported to support axonal growth, however with a slow growth rate (0.5–1 mm/day) primarily due to poor functional recovery[827]. It is also well perceived that the axonal growth can be moderately enhanced using external stimulation (such as external electric field). It has been generally reported that the rate of axon regeneration on the conducting materials is high under the influence of external electric field as compared to non-conducting materials[828,829]. Such observations also drive the development of electroconductive biomaterials to realize faster axonal growth and complete functional recovery. For example, Kazuya et al.[830] reported the *in vivo* study involving the implantation of the polyglycolic acid (PGA)–collagen tube filled with laminin-coated collagen fibres in the peroneal nerves of dogs for the time period of up to 12 months. The results could not reveal any kind of inflammation at the implant-tissue interface. In a different study, Evans et al.[831] reported the outcome of *in vivo* study on PLLA based porous biodegradable nerve conduct for peripheral nerve regeneration using the sciatic nerve model in rats. Importantly, the significant increase in axon number/mm^2 in the distal sciatic nerve was recorded at 16 weeks, post-implantation.

The animal experiments to assess PN regeneration involve different type of implant preparation as well as protocol to be followed. For this, the typical samples having ~2 mm of diameter and 8–10 mm of length are used to carry out the implantation experiment in the chosen animals (rats), according to ISO recommended guidelines. The silicone conduits are to be filled with scaffolds and they are sterilized by Ethylene Oxide (ETO). 3 rats with nerve isografts to their right sciatic nerve from donor animals served as controls. 6 adult rats (either sex) are chosen for the experiments. The entire implantation experiments can be carried out for three different time periods; i.e., 1 week, 6 weeks and 16 weeks with 2 rats per time duration. The implantation procedure is carried out under clean and aseptic conditions, as per GLP protocol. A pre medication is to be always given to the experimental animal with atropine (0.15 mg/kg) and diazepam (3.0 mg/kg). The animals are anesthetized with Xylaxin (5 mg/Kg) and ketamin (35 mg/kg) by intramuscular injection. The skins of the anaesthetized rats will be lightly swabbed using 70% alcohol followed by beta dine solution. During such procedure, the sciatic nerves are usually exposed by a muscle splitting incision and as a result, the sciatic nerve is divided near its origin to create an adequate distal segment. As far as the implantation is concerned, typically 10 mm conduits are interposed between the sciatic nerve ends using nylon sutures during microsurgery. The nerves of the experimental animal are then inserted into the artificial conduits to ensure that 1 mm of the nerve end remained within the nerve conduit. As far as the assessment of the control behaviour is concerned, 10 mm nerve isografts harvest from donor animals, is utilized to repair defects in the rat sciatic nerve, following a similar microsurgical technique. Subsequently, the wound is closed using stitches. An identical procedure needs to be conducted on 6 rats to follow the biostatistics criteria. After the surgical procedure, all the animals are to be given post-operative care. Walking track analysis can be performed on all animals after conduit placement monthly through 16 weeks. At the end of the implantation period, the animals are euthanized by an overdose of anaesthetic agent. The sites of implantations are macroscopically examined for any evidence of tissue reaction and sections of the micro-conduit/isograft and distal sciatic nerve from the same rats used for functional evaluation is harvested and fixed with 3% glutaraldehyde solution.

After fixation, cross section of each implant needs to be embedded in resin and stained with different staining agents, like toluidine blue, which can be helpful to distinguish nerve sections. For such tissue sections, careful analysis should be conducted using standard image processing, which would involve analysis using threshold and segment analysis of individual axons. For quantitative analysis, the number of axons per at least three randomly selected fields should be counted to provide the number of axons/mm^2. In order to obtain nerve density, the area of each axon is determined and the cumulative area is to be normalized with respect to the image area.

11.4.6 | Cardiac tissue regeneration with cardiac patch

The pre-clinical animal study to simulate cardiovascular diseases and to probe into clinically relevant treatments is often reported to be carried out in small animal model (e.g., mouse).

> The animal study related to cardiovascular tissue regeneration on synthetic cardiac patches ideally should be carried out in large animal models (e.g., pig or sheep).

In the following, an example is described for the readers to understand how different methodology is to be adapted to assess the efficacy of cardiac patches, *in vivo*, when compared to bone implantation or tissue compatibility of degradable biopolymer. For the cardiac patch implantation, 10–14 domestic piglets either male or female can be used. Myocardial Infarction (MI) can be induced with the injection of collagen suspension into the middle of the selected DA for embolic occlusion. This protocol would allow an experimental infarction to be created, while restricting a discrete location in the anterior anteroapical position in the left ventricle free wall. This procedure also enables direct access of the heart from the median sternotomy. The sternal thoracotomy needs to be performed under general anaesthesia, while the size/location of MI needs to be assessed using intra-cardiac echocardiography. Subsequently, a partial cardiopulmonary bypass needs to be initiated with the beating heart and a transmural piece of the infarcted region of the left ventricular free wall needs to be excised. The synthetic cardiac patch, already tested to be cytocompatible in terms of allowing cardiomyocytes to grow and respond to electrical stimulation, can be attached to the endocardial aspect of the incised myocardial edge. During the next stage involved in the operation, the patch needs to be placed to close the pericardium. This step should be carefully followed to prevent the formation of fibrotic adhesion between the region of myocardium with the patch and the pericardium. After this step, the entire operated region (sternum, muscle and skin) needs to be closed, once the chest tubes are placed appropriately to drain fluid from substernal and pericardinal spaces.

The above-described surgery can be conducted on the chosen group of test animals with one control group, which do not receive any cardiac patch. Prior to the procedure, echocardiography and all electrophysiological studies are to be carried out. The study can be continued for different time scales of 1,2 or 3 months. After each month of surgery of patch implantation, all electrophysiological studies are to be reported. Once the animal is terminated, the heart of the deceased animal is removed and opened along the left anterio–septual and atrioventricular grooves to expose the endocardial aspect. The exposed part will be imaged photographically for pre-clinical evidences and the selected tissue sections will be stained with H&E prior to histological examination.

11.5 | Design of Pre-clinical Study with Biomaterials

It is important to reiterate at this juncture that once a material is proven to be cytocompatible after a series of *in vitro* tests, the next level of tests for biomedical implant/devices is the haemocompatability test. Any biomaterial, when implanted in any living system, comes in contact with blood and therefore, one has to consider haemocompatability of any synthetic biomaterial. Therefore, prior to the clinical study and after successful completion of the desired cytocompatibility study, the haemocompatibility, i.e., blood level compatibility of a biomaterial needs to be established. As mentioned in Fig. 11.20, when blood, a tissue composed of plasma, platelets, RBC, WBC, etc., interacts with a non-living material surface, three scenarios are the possibility. The ideal scenario is the absence of any thrombus formation and the next level is the non-adherent thrombus formation.

Here again, ISO recommended protocols of the blood compatibility tests are to be used. Once a material is screened to be non-thrombogenic, it enters pre-clinical study.

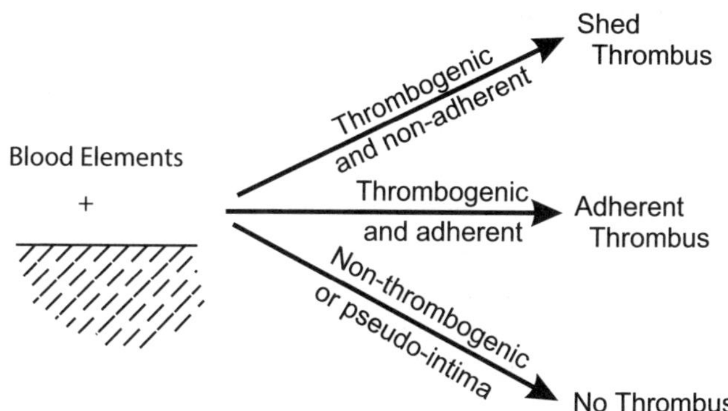

Fig. 11.20 Schematic of three possible scenarios to illustrate the interaction of material with blood issue, while assessing haemocompatability.

It is the moral responsibility of a researcher to comply with the existing ethical guidelines, while performing pre-clinical experiments with biomaterials.

Based on the discussion in the preceding sections, the following aspects are to be specifically considered while designing the pre-clinical study to obtain approval of the ethical committee and subsequently to conduct research to probe into histocompatibility of implantable biomaterials (see Fig. 11.2):

(a) selection of appropriate animal models considering end-point application of a given biomaterial;

(b) identifying appropriate defect model and to consider the possibility of creating such defect in the selected animal model, (e.g., bone grafting may use different defect model than nerve grafting/cardiac patch);

(c) selection of total timespan of implantation and intermediate timeline of animal sacrifice;

(d) preparation of test sample of appropriate size to be fitted into the defect site in animals;

(e) identification of control implant;

(f) total number of test and control implants;

(g) surgery procedure, depending on animal and defect model;

(h) monitoring of animal health during surgery (minimum pain) and post-surgery (weight change, inflammation, movement, behavioural changes);

(i) adopting appropriate ethically approved surgical procedures while sacrificing an animal;

(j) histology sample preparation and analysis using light microscope;

(k) tissue/material interface analysis using SEM, AFM, TEM, etc.

(l) biomechanical testing to analyze tissue/implant interface strength;

(m) micro-computed tomography analysis of bone regeneration in 3D space;

(n) validation of small animal (mouse, rat, etc.) test results in higher animals (pig, sheep, etc.).

11.6 | Closure

This chapter establishes a biological foundation necessary to conduct pre-clinical experiments with implantable biomaterials using clinically relevant animal models. In addition, the experimental assessment to qualitatively and quantitatively understand the tissue response to implanted biomaterials is also discussed. A number of case studies described in this chapter would also help a reader to plan animal experiments. More discussion on the many of the theme topics, mentioned in this chapter, can be found elsewhere[832]. Also, considering the importance of pre-clinical testing, a number of researchers reported newer approaches to conduct *in vivo* testing of drug delivery system or biomaterials together with new ways to characterize the host response[833,834,835].

It should be emphasized that the concept of "bioethics" has been at the centre of a long debate involving experts from the field of ethics, legal issues, science and medicine as well as governmental institutions and policy making officials. Especially, in case of *in vivo* studies, the ethical issues relate directly to the respect for animals, namely that they should not be injured and that nothing should be done to them without the consent from the regulatory bodies. Taken together, the biomaterials researchers should be aware of their moral responsibilities to the scientific community and to the society. The human healthcare aspect should be the main motivation in conducting pre-clinical research on biomaterials in an ethically appropriate manner. Once, the researcher completely understands his/her ethical duties, then the regulation of research and conducting it through humane and moral means become possible.

Section IV
Illustrative Examples of Biomaterials Development

Case Study: Corrosion and Wear of Selected Ti-alloys

The corrosion and wear of load bearing implantable metallic biomaterials in physiological environment limits their biomedical applications. This chapter discusses the published results from author's group on the electrochemical degradation and tribological properties of some Ti-based alloys (Ti6Al4V, Ti5Al2.5Fe, Ti13Nb13Zr, Ti6Al4Fe, and TiSiC). Some of the significant results that will be highlighted in this chapter would include the low coefficient of friction (COF) of 0.3 for Ti5Al2.5Fe/steel as well as the dominance of tribomechanical wear in simulated body fluid. An interesting observation regarding corrosion behaviour has been that the electrochemical corrosion behaviour of Ti6Al4V is not influenced significantly upon refining the alloy composition in terms of substitution of vanadium with niobium and iron. In addition, some results will be discussed in relation to the better *in vitro* corrosion of a relatively new titanium based α-alloy (Ti0.5 wt% Si0.65 wt% C) than Ti6Al4V in phosphate buffer saline medium. Furthermore, the *in vitro* bio-mineralization experiments in concentrated simulated body fluid (10×SBF) confirmed good bioactivity property, which is demonstrated by the deposition of fine, porous micron-sized calcium–phosphate (CaP) rich phase. The results summarized in this chapter are based on the studies conducted in the authors' research group (see the work by Choubey et al. and Raghunandan et al. in the reference list).

12.1 | Introduction

It can be reiterated here that one million patients worldwide are treated annually for total hip joint replacement (THR) of arthritic hips and knee joints. From materials perspective, THR usually consists of a metallic femoral stem and head, which are fitted into an ultrahigh molecular weight low friction polyethylene socket[844]. The metallic femoral stem and bone plates should possess high specific strength, high corrosion resistance and importantly biocompatibility. Very few materials, like Ti meet these requirements. Titanium and its alloys are considered the most suitable materials in medical applications as they possess the excellent biocompatibility, good chemical and mechanical stability, better specific strength and corrosion resistance. Such property combination for Ti-alloys is better than other competing materials, like stainless steels, Co–Cr alloys, commercially pure (cp) Nb and cp Ta[845,846,847,848,849,850]. In clinical settings, a large number of titanium-based devices of every description and function are implanted in patients annually.

Currently, Ti-based alloys are finding increasing applications in a wide range of medical devices including heart valves (as housing or outer ring), cardiac pacemakers (as housing), bone plates, artificial joints and dental implants[851,852,853,854,855]. The most widely used commercial scale titanium implant materials for orthopedic applications include unalloyed titanium and Ti6Al4V. Other candidate materials include Ti5Al2.5Fe, Ti13Nb13Zr and Ti12Mo6Zr2Fe, whose mechanical properties, biocompatibility and *in vivo* behaviour are not yet published [856,857,858].

Despite the fact that cp titanium exhibits very high corrosion resistance and tissue acceptance, its mechanical properties limit its use as biomaterial in some cases[859]. This is particularly true when high mechanical strength is critical, like in hard tissue replacement or in applications involving wear[860]. To overcome such restrictions, cp Ti was substituted by titanium alloys, particularly, the classic grade 5, i.e., Ti6Al4V $\alpha+\beta$ type alloy, that was initially developed for aerospace applications[861,862]. Among various Ti-alloys, the Ti6Al4V alloy has been successfully used in a number of biomedical applications[861,862,863]. There is a clear understanding of the effect of heat treatment on the microstructure of this alloy[864]. It possesses sufficient strength and ductility for use as implants in human body[865]. Moreover, the good corrosion resistance of this alloy in human body environment is established, both *in vitro* and *in vivo*[866]. Although this alloy is considered to be a better material for surgically implanted parts, more recent studies have shown that vanadium may have adverse reactions with the tissues of the human body[867]. While designing new Ti-alloys, Vanadium, a β-stabilizer element, was substituted by niobium and iron, leading to Ti6Al7Nb and Ti5Al2.5Fe $\alpha+\beta$ type alloys[850,856,868].

> Following a few studies indicating that aluminium may be a cause of neurological disorders and Alzheimer's disease, more varieties of Ti-based alloys are being investigated.

It is therefore important to select non-toxic elements for biomedical grade Ti alloys. Although it is difficult to evaluate different kinds of toxicity of alloying elements[869], earlier studies did not report toxicity of Ti-alloys with Nb, Ta, Mo, Zr, and Sn additions[870]. Recently, vanadium free Ti alloys such as Ti6Al7Nb and Ti13Nb13Zr have been developed to replace vanadium in Ti6Al4V[871], because of the reported toxicity and negative tissue response of vanadium ion, *in vivo*[872]. Release of vanadium ions by passive dissolution or other process involving wear can cause discoloration of the surrounding tissue or an inflammatory reaction causing pain and even leading to loosening owing to osteolysis[872]. It has been suggested by Khan et al.[873] that Ti6Al7Nb can be a better alternative to Ti6Al4V because of its resistance to corrosion and loss of mechanical properties with changes in pH in simulated body fluid environment. To this end, this chapter briefly discusses the corrosion of some Ti-based alloys.

It is widely accepted that the wear of biometallic materials is also one of the most important aspects of implant surgery[874,875,876,877]. The presence of particulate wear products in the tissue surrounding the implant may lead to a series of events leading to periprosthetic bone loss[878], excretion of excess metal ions (especially titanium, chromium, cobalt and nickel) and their suspected role in induction of tumours, e.g., malignant fibrous histiocytoma[879]. Wear particles are formed in a piece of bone if it rubs against the implant, or if two parts of an implant rub against one another. Therefore, implants of self-mated titanium are generally not used as joint surfaces. If titanium ion is released by the process of wear, the tissue reaction may vary. This reaction could be a mild response (e.g., a discoloration of the surrounding tissue) or a more severe one (e.g., inflammatory reaction causing pain and even leading to loosening owing to osteolysis) [880] and therefore wear of Ti-alloys has attracted wider attention[881,882].

> The natural selection of titanium for implantation is determined by a combination of most favorable characteristics that include immunity to corrosion, biocompatibility, strength, modest modulus, density and the capacity for being integrated with bone and other tissues (osseointegration).

Concerning fracture resistance, the fracture toughness of all implantable high strength alloys is above 50 MPa m$^{1/2}$ (like e.g., Ti-alloys) with critical crack lengths well above the minimum for detection by non-destructive testing[883]. Therefore, it is necessary to tailor the design of Ti-alloys to improve the fracture resistance, which would influence tribological behaviour. In the above perspective, the present chapter will partly discuss the results obtained from the author's group in the context of friction and wear of Ti-alloys and more details can be found elsewhere [884].

One of the relatively new alloy composition being highlighted in this chapter, i.e., Ti–Si–C alloy is chemically analogous to a dispersion strengthened titanium material produced by ball milling and spark plasma sintering[885,886]. Silicon and carbon were chosen as alloying elements in this study as no harmful effects of these two elements on the human body tissue are known.

12.2 | **Corrosion Behaviour of a Few Ti-alloys,** *in vitro*

The corrosion behaviour of cpTi, Co28Cr6Mo, Ti6Al4V, Ti6Al4Nb, Ti6Al4Fe and Ti5Al2.5Fe (all compositions are in mass %) in Hank's solution will be discussed in this section. Here, it may be recalled that one of the variants of commercial THR assembly consists of a femoral stem made of Ti-based alloy and a femoral head made of Co28Cr6Mo alloy. The composition, microstructure and mechanical properties of Co28Cr6Mo alloy meet the requirements of ASTM standard F 75 and ISO standard 5832/4. A comparison of some Ti-alloy with other metallic biomaterial is provided in Table 12.1. The electrochemical corrosion experiments on implantable biomaterials can be carried out in different physiological environments at 37°C. One such testing media is Hank's balanced salt solution, containing (in gm/l) 8 NaCl, 0.4 KCl, 0.14 $CaCl_2$, 0.06 $MgSO_4.7H_2O$, 0.06 $NaH_2PO_4.2H_2O$, 0.35 $NaHCO_3$, 1.00 Glucose, 0.60 KH_2PO_4 and 0.10 $MgCl_2.6H_2O$, with all amount in gm/l. Another test medium can be Phosphate buffered saline, PBS (8 g NaCl, 0.2 g KCl, 1.15 g Na_2HPO_4, 0.2 g KH_2PO_4 in 1 l of water). In preparing such medium, each chemical is usually added successively after the former was dissolved completely and the solution pH was adjusted to 7.45 by adding suitable amount of HCl/NaOH.

Table 12.1 A comparison of the mechanical properties of various metallic implant materials[900].

Material	E (GPa)	Hardness (HV)	Reference
Co-Cr alloys	210–250	420	[846,847]
316L stainless steel	200	140	[846,848]
Titanium (grade 4)	116	160	[847]
Ti6Al4V	124	300	[846,847]
Ti/1.3HMDS	130	320–420	[865,869]
TiSiC	132	360–440	[900]
Human bone	3–80	25	[906]

Figure 12.1 presents the chemically etched microstructures of Ti-based (α+β) alloys. Herein, beta phase appears dark and the alpha phase light. Vanadium, niobium and iron are beta stabilizers, while aluminium is an alpha stabilizer. Alpha was the dominant phase in all these alloys, as evident from the microstructures. The structure of Ti6Al4V was fairly fine grained with a grain size of 22.4 µm for primary alpha (Fig. 12.1). It has also been reported that, owing to a two-phase equiaxed microstructure, Ti6Al4V is more susceptible to corrosion as the compositional difference across the grain boundaries leads to the galvanic cell formation[863].

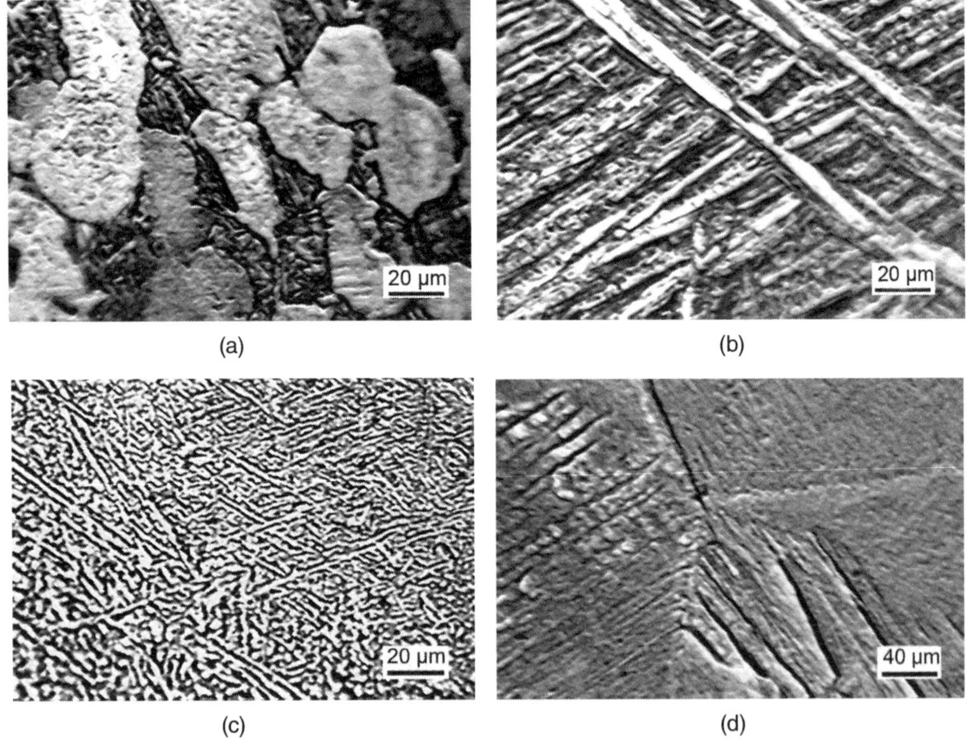

Fig. 12.1 Representative optical microscopy image of chemically etched surfaces of different Ti-alloys, (a) Ti6Al4V, (b) Ti6Al4Nb, (c) Ti6Al4Fe and (d) Ti5Al2.5Fe[884].

The addition of niobium increased the grain size considerably as evident from Fig. 12.1b. A typical Widmanstätten type of structure is observed. Grain boundary alpha can also be seen in the microstructure in a transformed beta matrix. The plate like alpha precipitates that nucleate and grow below the beta transus, produce the Widmanstätten structure.[864] The plates often precipitate in colonies of the same crystallographic orientation as can be noticed in Fig. 12.1b, because of autocatalytic nucleation[864]. The alpha phase appears to be fine needle shaped (acicular alpha) in Ti5Al2.5Fe alloy, which also possesses a coarse grained structure. It has been reported that the Ti5Al2.5Fe is a near alpha–beta alloy and the alpha phase dominates the properties of this alloy[864].

The passivation parameters like zero current potential (ZCP) breakdown potential (E_b), passive current density (i_{pass}) and the passivation range (E_b-ZCP) were estimated from the polarization curves in both PBS and Hank's solution and they are tabulated for selected Ti-alloys in Table 12.2 and 12.3. For reference, Fig. 12.2 plots a typical Tafel plot, that is reproducibly recorded with all Ti-alloys. Therefore, Fig. 12.2 is a typical representation of corrosion behaviour. All the important parameters are also marked on Fig. 12.2. For example, the zero current potential, the cathodic (β_c) and anodic (β_a) Tafel slopes and the corrosion current densities (i_{corr}) can be estimated from the Tafel plots and this procedure is outlined schematically in Fig. 12.2. The corrosion rate in penetration units, for example µm/y, can be obtained from the corrosion current densities using Faraday's law[887], as follows:

$$\text{Corrosion rate (in µm/y)} = 3.27\,(a\,i/n\,D) \tag{12.1}$$

where a is the atomic weight (47.9), i is the corrosion current density in µA/cm^2, n is the valence change on corrosion (4) and D is the density (4.51 g/cm^3). The corroding species is assumed to be Ti in the above calculations.

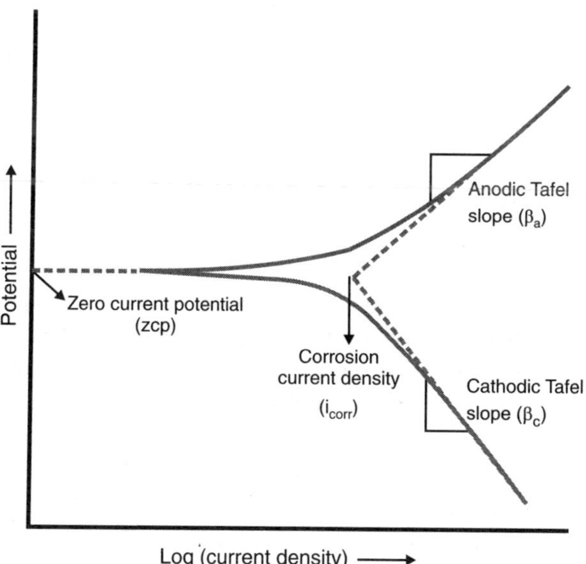

Fig. 12.2 Schematic outlining the estimation of corrosion current density (i_{corr}) from a Tafel plot; the parameters zero current potential (ZCP), the cathodic (β_c) and anodic (β_a) Tafel slopes have also been indicated[884].

In reference to table 12.2, the zero current potential (ZCP), obtained from the potentiodynamic polarization curves, of all the Ti-alloys is in the range of −400 to −650 mV except for Ti6Al4Nb, which is relatively more noble. The passive current densities (i_{pass}) are also listed in Table 12.2. The passive current densities of the alloys investigated were of the same order of magnitude (1.2–7.9 $\mu A/cm^2$). Ti6Al4Nb exhibits varying passive current densities in the range of 1.4 to 7.9 $\mu A/cm^2$. Addition of alloying elements like niobium, iron and vanadium does not significantly affect i_{pass}. The passivity range in the case of materials exhibiting stable passive behaviour is provided by the difference between breakdown and zero current potential. The passive range for Ti6Al4V, Ti6Al4Fe and Ti6Al4Nb are comparable although the range is lower for Ti6Al4Nb. It must be noted that ZCP is at a relatively noble potential for Ti6Al4Nb compared to others and this can be the reason for the apparent decrease in passive range for this alloy.

Table 12.2 Passivation parameters of different Ti-based materials in phosphate buffered saline of pH 7.4 at 37°C[900].

Material	*ZCP (mV)*	E_b *(mV)*	i_{pass} *($\mu A\ cm^{-2}$)*	*Passive Range (mV)*	i_{corr} *($\mu A\ cm^{-2}$)*	*Reference*
cp Ti	-190	1180	1.5	1370	0.036	[900]
Ti6Al4V	-460	2000	5.4	2460	0.104	[900]
NiTi	-180	400	-	580	0.20	[907]
NiTi (electropolished)	-330	1500	-	1830	0.06	[907]
Ti6Al7Nb	-200	-	1.6	-	0.30	[908]
Ti13Nb13Zr	-400	-	1.0	-	0.40	[908]
TiSiC	-160	1180	1.5	1340	0.033	[900]

12.3 | Corrosion Behaviour of Novel TiSiC Alloy

This section discusses the results obtained with a relatively less investigated Ti-alloy, Ti0.5Si0.65C (TiSiC) alloy, which was produced by induction melting followed by rapid solidification of the melt contained in a water cooled copper channel[900]. The starting materials, 98.85 wt% titanium (99.8% pure) pieces and foil, 0.5 wt% silicon (99.6% pure) and 0.65 wt% graphite flakes, were pre-alloyed (Ti+C at 1400°C for 5 min, Ti+Si at 1500°C for 5 min) by melting so as to obtain Ti_xSi_y and TiC_x secondary phase precipitates. In the final step, both these castings were melted together at approximately 1650°C for 10 min.

All the TiSiC samples showed identical density, and the average density was obtained as 4.525 g cm^{-3}, which is equal to the theoretical density of cp Ti (4.52 g cm^{-3}). This value also closely matches with the theoretical density of the alloy (negligible porosity). Figures 12.3(a–d) compare the microstructures of cp Ti, Ti6Al4V, Ti/1.3HMDS and TiSiC. Figure 12.3(a) shows an annealed single-phase alpha grain microstructure of commercially pure titanium (cp Ti). Figure 12.3(b) shows a two-phase alpha–beta structure of Ti6Al4V. Ti/1.3HMDS contained fine dispersoids of Ti_5Si_3

and TiC_x in an ultra-fine grained titanium matrix, as shown in Fig. 12.3(c)[888]. TiSiC is composed of a two phase microstructure, as shown in Fig. 12.3(d). The average volume percentage of the TiC_x in the matrix was calculated to be nearly 19%. High volume percent of the hard TiC_x precipitates results in an enhancement of hardness and thus, wear resistance of the alloy. The grain size of the as-cast TiSiC was approximately 2–20 μm (Fig. 12.3d). In contrast, spark plasma sintering (SPS) processed Ti/1.3HMDS has a grain size ranging from 40 nm to 1 μm. This reveals the effect of rapid cooling vs. short sintering time and low temperature treatment on grain size. TiSiC and Ti/1.3HMDS differ in microstructure and hardening mechanism. Whereas the first is rather a solid solution with a TiC_x phase in micrometre range, Ti/1.3HMDS represents an ultra-fine grained, dispersion strengthened material with dispersoid size of 200–300 nm[900]. Vickers micro-hardness values for the four TiSiC alloy samples varied between 360 and 440. It was noticed that the hardness of TiSiC alloy improved significantly compared to that of unalloyed titanium and matches with other conventional implant alloys, probably due to the presence of hard TiC_x precipitates. Also, the hard secondary phase precipitates pin down the movement of dislocations across the grains, results in an enhanced strength compared to cp titanium. The elastic modulus of TiSiC alloy was measured to be 132 GPa, which is higher than that of the human bone (3–80 GPa), but close to that of cp Ti (116 GPa).

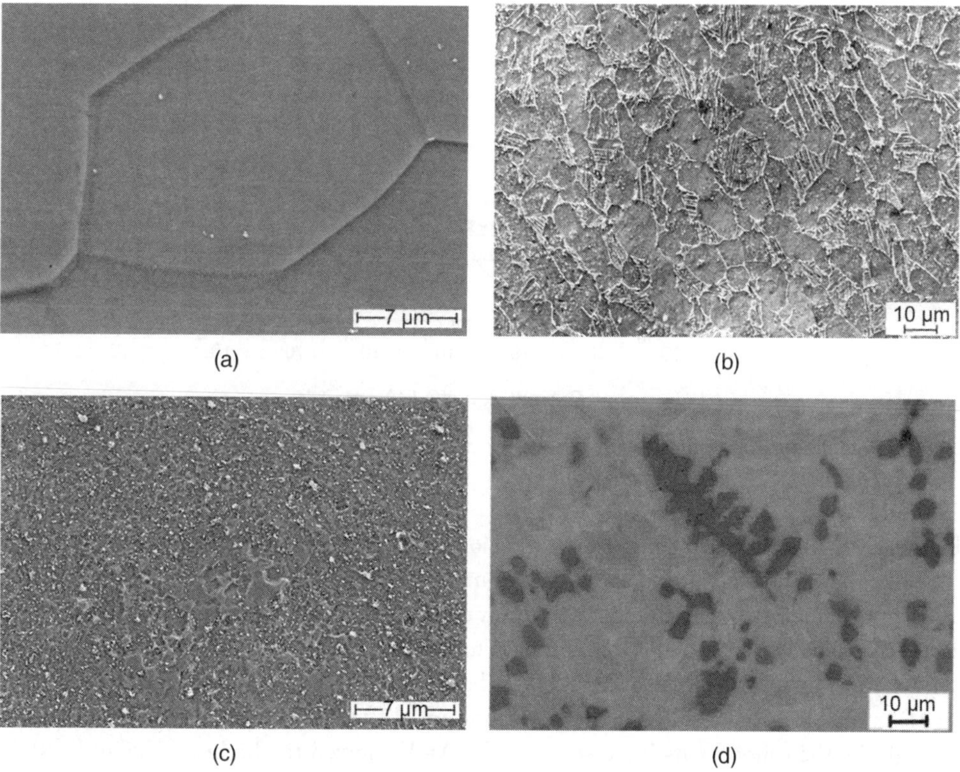

Fig. 12.3 Scanning electron microscopic images of the four Ti-based materials: (a) alpha grains and grain boundaries of cp Ti; (b) two-phase equiaxed structure of Ti6Al4V; (c) fine dispersion of silicides and carbides in titanium matrix of Ti/1.3HMDS; (d) precipitates of hard TiC_x phase in titanium solid solution matrix of as-cast TiSiC alloy[900].

> It is a well-known fact that any significant mismatch of elastic moduli of the implant and the host tissue would result in aseptic loosening, due to differential strains.

The passivation behaviour of the alloys was investigated by performing potentiodynamic polarization experiments. Figure 12.4 shows an overlay of the potentiodynamic polarization curves of cp Ti, Ti6Al4V and TiSiC in PBS at 37°C. All the tested materials showed a potential range with stable passivity under the experimental conditions, before the occurrence of breakdown. Various parameters, like zero current potential (ZCP), passive current density (i_{pass}), passive range (E_b-ZCP), and breakdown potential (E_b) were obtained from the potentiodynamic polarization data of each material immersed in PBS. From Fig. 12.4, the corrosion current density of TiSiC in PBS is the least (0.033 µA cm^{-2}) among other alloy systems tested. Further, the passive current density (i_{pass}) of TiSiC (1.5 µA cm^{-2}) is very low, signifying that the surface oxide layer has a highly protective nature, probably due to its very low ionic/electronic conductivity.

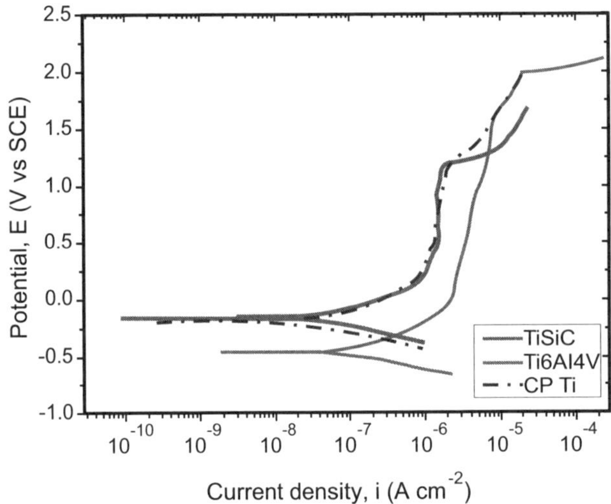

Fig. 12.4 Potentiodynamic polarization curves for cp Ti, Ti6Al4V and TiSiC in de-aerated PBS of pH 7.45 at 37°C[900].

Figure 12.4 demonstrates the highest i_{pass} for Ti6Al4V at a value of about 5.4 µA cm^{-2} (Table 12.3). On the other hand, both cp Ti and TiSiC show a much lower i_{pass} of 1.5 µA cm^{-2}. The addition of Si and C in minor amounts evidently did not have any effect on the passive current density of cp Ti. This indicates that a passive layer of TiO_2 might have formed on TiSiC alloy, which is similar in nature to that forming on cp Ti. In case of Ti6Al4V, it has been reported that the vanadium oxide in the passive film dissolved and resulted in the generation and diffusion of vacancies in the oxide layer[889]. This might be the cause for its increased i_{pass}. Ti6Al4V showed the highest value of breakdown potential (E_b) of 2000 mV in de-aerated PBS at 37°C. Similarly, many authors[851,871,890,891,892,893] reported that E_b of Ti6Al4V varied between 1500 mV and 2000 mV in de-aerated simulated body fluids. However, the E_b values of both cp Ti and TiSiC were measured to be 1180 mV, which shows that alloying Si and C in titanium did not influence the breakdown potential.

Table 12.3 Test results of corrosion rates of the Ti alloy determined by Tafel extrapolation method in Hank's solution at pH 7.4 and 37°C[884].

Sample	ZCP (mV vs SCE)		β_c (mV/dec)		β_a (mV/dec)		i_{corr} ($\mu A/cm^2$)		Corrosion rate ($\mu m/y$)	
	(1), (3)	(2), (4)	(1), (3)	(2), (4)	(1), (3)	(2), (4)	(1), (3)	(2), (4)	(1), (3)	(2), (4)
Ti6Al4V	-231	-271	-176	-181	168	106	0.16	0.16	1.39	1.39
Ti6Al4Nb	-596	-590	-158	-155	185	182	0.10	0.20	0.86	1.74
	-360		-149		448		0.48		4.17	
Ti6Al4Fe	-390	-478	-122	-120	181	120	0.04	0.20	0.35	1.74
Ti5Al2.5Fe	-588	-604	-102	-121	175	170	0.13	0.12	1.13	1.04

Overall, the corrosion characteristics exhibited by cp Ti and TiSiC were almost similar and in most aspects, better than that of Ti6Al4V. The alloying elements, Si (0.5 wt%) and C (0.65 wt%) did not seem to affect the passive film formation on titanium. The addition of carbon, in very low amounts, resulted in the formation of carbon-deficient, non-stoichiometric TiC_x precipitates that may have remained unaffected during the polarization experiments. On similar lines, an improvement in corrosion resistance of Ti-based composites containing TiC as secondary phase has been reported elsewhere[894,895,896]. The effect of silicon on corrosion behaviour could not be studied due to its very low concentration.

Corrosion current density (i_{corr}) is used as a direct measure of the rate of corrosion of a material. Hence, to calculate corrosion current density, both by Tafel extrapolation and by using Stern–Geary equation, Tafel plots are shown in Fig. 12.5 for a potential range of -250 mV to +250 mV with respect to the free corrosion potential of the alloys. The scan rate used was 0.166 mV s^{-1}. An overlay of the Tafel plots of cp Ti, Ti6Al4V and TiSiC is shown in Fig. 12.5. It can be observed that the corrosion current density of Ti6Al4V was the highest among all the tested materials, at a value of 0.104 μA cm^{-2}. Owing to the fact that corrosion rate is inversely related to R_p, it can be recalled that the linear polarization results also showed that Ti6Al4V possessed the lowest polarization resistance among the tested materials.

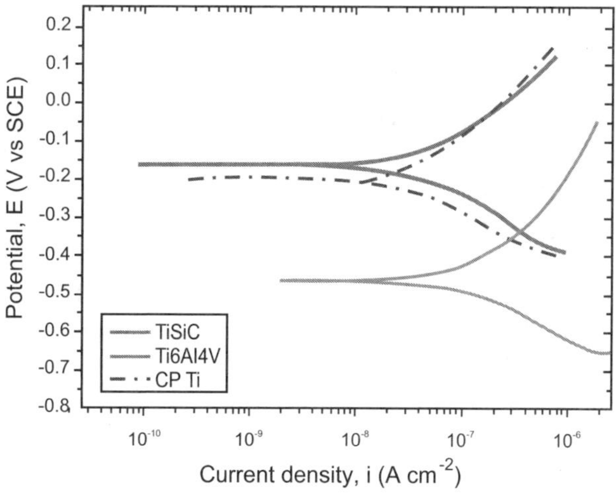

Fig. 12.5 Tafel polarization curves for cp Ti, Ti6Al4V and TiSiC in de-aerated PBS of pH 7.45 at 37°C[900].

The electrochemical impedance spectroscopy (EIS) experiments were also analysed at their respective free corrosion potentials in an effort to analyse the electrochemical interface of the materials in terms of an equivalent electrical circuit. The Nyquist plots of cp Ti, Ti6Al4V and TiSiC are presented in Fig. 12.6. The nature of the Nyquist plots for all the materials is similar, as can be seen from Fig. 12.6. Such incompletely resolved arcs, as shown by all the materials, indicate that RC is large. This response is a characteristic of passive metals[897]. The diameter of the arc, when extended to make a semicircle, represents the polarization resistance (R_p) of the respective material in the given electrolyte. It can be observed from Fig. 12.6 that the diameter of the extrapolated semicircle (when drawn) for TiSiC would be much larger, indicating that the resistance to polarization (R_p) for this alloy is more than that of the others.

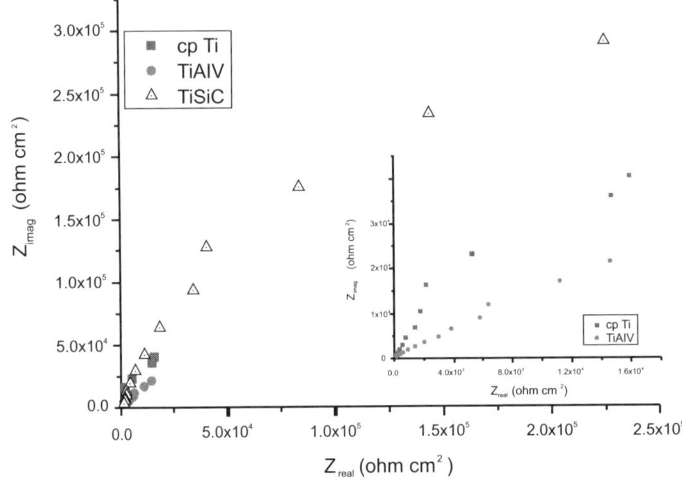

Fig. 12.6 Nyquist plots for cp Ti, Ti6Al4V and TiSiC in de-aerated PBS of pH 7.45 at 37°C recorded at respective open circuit potentials of the alloys; inset shows the Nyquist plots for cp Ti and Ti6Al4V more clearly[900].

Figure 12.7 shows the plausible equivalent electrical circuit, based on the nature of Bode plots (not shown here) and also on the basis of the least χ^2 value obtained by appropriate circuit fitting of the corresponding Nyquist plots (not shown here) of all the materials. The circuit model R(CR(QR)), with a capacitance and a non-ideal capacitance, is well fitting with the experimental data. The circuit diagram (Fig. 12.7a) represents the corresponding physical model (Fig. 12.7b), where R_s represents the solution resistance, R_{pore} and C_{pore} are resistance and capacitance of the outer porous layer respectively, R_b and Q_b are the resistance and constant phase element (CPE) of the inner barrier layer respectively. CPE represents a deviation from the ideal capacitor behaviour. Similar nature of the surface films forming on Ti-alloys in Hank's solution and also in PBS[898,899] have been reported earlier. Table 12.4 provides the values of various electrical components obtained by circuit fitting. It was observed that for TiSiC, the R_s value was slightly higher than that for cp Ti or Ti6Al4V although the electrolyte used was similar. This discrepancy may have resulted because of intangible reasons or experimental artefacts, but interestingly such behaviour was found to be reproducible in three independent experiments. The impedance of CPE with the value of n is often related to a non-uniform current distribution due to the porous oxide layer. CPE defines an ideal capacitor for $n = 1$, an ideal resistor for $n = 0$, and a pure inductor for $n = -1$. Resistance to polarization (R_p), obtained after circuit fitting, had the highest value for cp Ti (1032 kΩ cm^2). R_p of TiSiC was close to cp Ti (981 kΩ cm^2) and that of Ti6Al4V was the minimum at 622.5 kΩ cm^2.

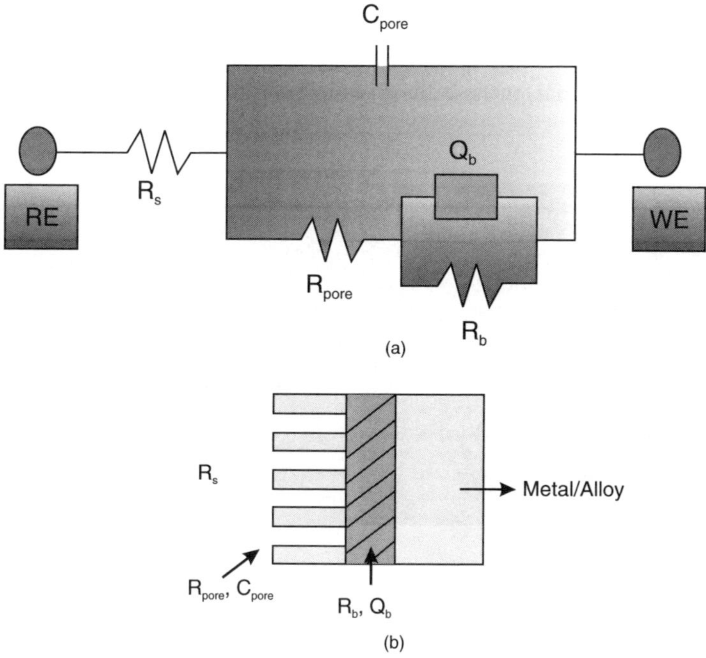

Fig. 12.7 (a) Equivalent electrical circuit model to describe electrochemical corrosion, (b) physical model of the interface between de-aerated PBS and metal at their respective open circuit[900].

Table 12.4 Various electrical components obtained after equivalent circuit fitting to the EIS data obtained at the open circuit potential of cp Ti, Ti6Al4V and TiSiC immersed in PBS of pH 7.45 at 37°C for 1 hour (% error is mentioned in the parentheses)[900].

Material	R_s (Ω cm^2)	C_{pore} ($\times 10^{-5}$ F)	R_{pore} (kΩ cm^2)	Q_b ($\times 10^{-5}$ ohm^{-1} s^n cm^{-2})	n	R_b (Ω cm^2)	χ^2 (Chi-squared)
cp Ti	0.57 (12.03)	6.50 (10.02)	1032 (261)	7.58 (18.19)	0.82 (10.98)	369.9 (102.5)	1.064×10^{-1}
Ti6Al4V	0.57 (9.99)	7.75 (8.89)	622.5 (486)	18.06 (13.85)	0.60 (17.3)	316.4 (145)	7.731×10^{-2}
TiSiC	1.75 (6.362)	1.04 (5.14)	981 (37.58)	1.24 (8.87)	0.75 (5.47)	875.5 (86.65)	2.23×10^{-2}

Scanning electron micrographs of TiSiC alloy, viewed after film breakdown, showed the signs of localized corrosion that might have been a result of chloride attack. The topography of the corroded areas indicates that selective dissolution may have taken place on the matrix, preferably on the grain boundaries. This explains the similar nature of potentiodynamic polarization behaviour observed for TiSiC and cp Ti (as observed in Fig. 12.4). The attacked areas of the TiSiC alloy are shown in Fig. 12.8. Few or no TiC_x dendrites seem to have been attacked above the breakdown potential, indicating that they might be inert. This observation is further corroborated by a number of energy dispersive spectra (EDS) collected individually from the matrix and the dendrites. On the other hand, both cp Ti and Ti6Al4V might have undergone uniform corrosion, as no local degradation was observed on the surface. It is also possible that oxygen evolution reaction might have occurred above their breakdown potential for these alloys, which would have resulted in an increase in current density above E_b.

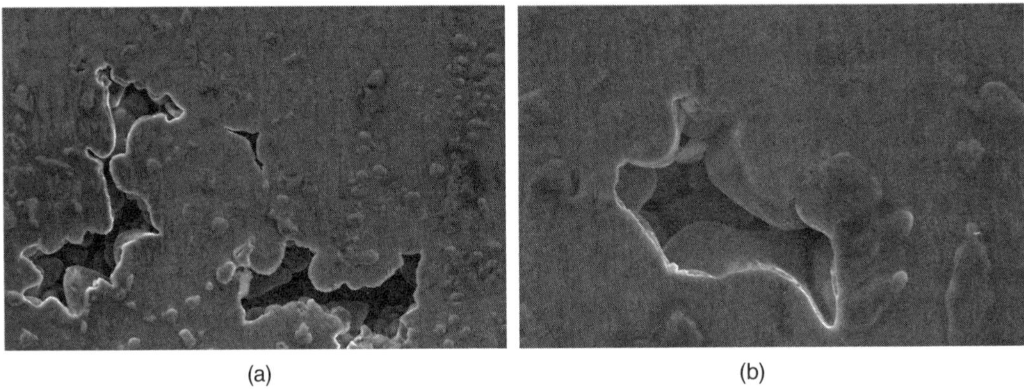

(a)   (b)

Fig. 12.8 Scanning electron micrographs (a and b) showing localized pitting attack on TiSiC alloy in PBS above breakdown potential[900].

Further, the pH of phosphate buffered saline was measured before and after each sample of cp Ti, Ti6Al4V and TiSiC was subjected to potentiodynamic polarization experiments. The pH of the initial solution was 7.45 at 37°C, while the pH of PBS after the corrosion tests, for all the samples, varied in the range of 7.39 and 7.42 at 37C. The absence of any considerable change in the pH of the solution indicates that there is little possibility for the degradation of corrosion resistance of the tested alloys due to pH changes. However, the extent of leaching of toxic/non-toxic corrosion species cannot be clearly indicated by the change of pH of the electrolyte.

12.4 | Bio-mineralization of Novel TiSiC Alloy in SBF

One of the approaches that will be discussed in this section is as how heat treatment of Ti-alloy can enhance the biomaterialisation. Cp Ti, Ti6Al4V and TiSiC samples were subjected to alkali+heat treatment before performing mineralization experiments. Pre-calcification was performed as an alternative to routine NaOH alkali treatment, since Na$^+$ ions introduced by the latter increases the external alkalinity, leading to inflammation and cell death[900]. Pre-calcification helps in the formation of Ti-OH and Ca-OH groups, thus speeding up the nucleation of Ca–P during the subsequent biomimetic deposition process[901]. Such a treatment, as performed in the present case might form Ca-OH, Ti-OH functional groups and TiO_2 groups on the surface. The biomimetic deposition or *in vitro* mineralization was carried out in 10×SBF, which corresponding to an artificial physiological solution that is 10 times more concentrated than normal simulated body fluid (without any proteins). The pre-treated samples were immersed in 10×SBF for 6 hours. The solution was replenished every hour to avoid crystallite formation within the highly concentrated solution. The samples were then taken out, gently washed with double distilled water and dried at room temperature.

For the biomaterialisation study, the pre-treated samples were immersed in 10×SBF for 6 hours, the solution being refilled every hour.

> When a pre-calcified Ti-alloy sample is immersed in 10×SBF, the surface is covered with negatively charged units of titania formed by condensation of Ti-OH. These interact with positively charged Ca2+ ions to form amorphous calcium titanate. Subsequently, calcium titanate attains positive charge and then interacts with the negatively charged phosphate ions in SBF to form amorphous Ca-P layer.

Such mineralized layer eventually stabilizes into bone-like crystalline apatite of Ca/P ratio 1.65[902]. Besides the hydroxide group, the presence of oxide on Ti surface (as a result of heat treatment following pre-calcification) also helps in easier apatite formation. After a period of 6 hours of immersion, a thick Ca–P mineral layer was found to be deposited on all the samples. The rates of calcium phosphate nucleation is faster in 10×SBF containing Ca^{2+} and HPO_4^{2-} ions in a concentration 10 times higher than that in a conventional SBF (c-SBF).

Within the first hour of immersion in concentrated SBF, the alkali+heat treated surface of TiSiC was rapidly covered with a micro-textured Ca–P globules layer, consisting of the globules formed by nano-sized crystals. The morphology of the mineral layer formed on the alkali+heat treated TiSiC alloy is shown in Fig. 12.9a. A magnified image of an individual Ca–P sphere is shown in Fig. 12.9b. The EDS pattern of the Ca–P layer (inset of Fig. 12.9a showing prominent peaks of Ca and P). A conventional SBF (i.e., 1.5×SBF) can only form a 20 µm thick layer of apatitic calcium phosphate after two weeks of soaking at 37°C, while the 10×SBF, used in the current study, achieves this in only 2 hours at room temperature. Thus, this technique can be used for a rapid formation Ca–P coating on biocompatible metallic implant substrates[900].

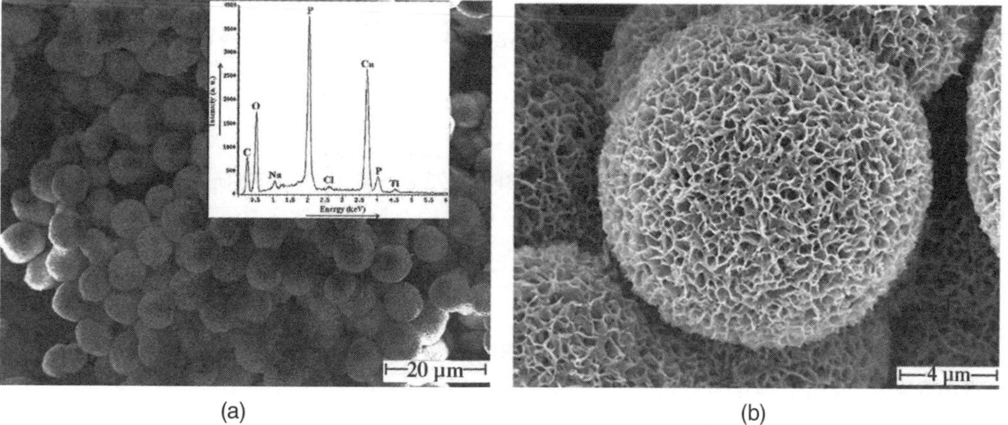

Fig. 12.9 Scanning electron micrograph showing (a) a cluster of Ca–P globules in the mineralized layer; (b) magnified image of a micron-sized Ca–P sphere showing an interconnected porosity, formed after biomimetic deposition on alkali + heat treated TiSiC alloy[900].

Table 12.5 summarizes the nature of the Ca–P layer formed on pre-treated TiSiC alloy. The table also provides the results of a few similar studies made on Ti/Ti-alloy substrates under different conditions[903,904,905]. One common feature observed from the morphology of the Ca–P layer is that, in most cases, irrespective of the solution concentration, Ca–P layers were deposited in the form of micron-sized spheres. Pre-treating the materials, prior to biomimetic deposition, induces a faster rate of deposition and might also be helpful in homogeneous nucleation of Ca–P phases over the surface. The biomimetic deposition in 10×SBF resulted in a dense Ca–P layer, while a few days to weeks are required if simulated body fluids of lower concentrations are used. It indicates that the kinetics of Ca–P layer formation on TiSiC alloy is faster. This may be attributed to the use of much stronger SBF (10×) as well as the alloy chemistry and the pre-treatment procedure adopted in the present work. On the other hand, no Ca–P mineral layer formation was observed on pre-treated cp Ti, Ti6Al4V, TiSiC after immersing in revised SBF (r-SBF) for duration of 1 week. Therefore, the influence of alloy chemistry and microstructure on kinetics of Ca–P nucleation should be investigated in future research.

Table 12.5 Summary table comparing the characteristics of Ca–P layers produced under different conditions[900].

Material	*Pre-treatment*	*Test medium*	*Test duration and temperature*	*Ca–P layer characteristics*
Untreated TiSiC	As-cast	10×SBF	6 h at room temperature	Porous spherical globules (size: 5–20 μm); not homogeneous
Heat treated TiSiC	Heat treated at 800°C for 1 h			Porous spherical globules (size: 5–20 μm); nearly homogeneous
Pre-calcified and heat-treated TiSiC	Immersed in boiling saturated Ca(OH)$_2$ solution for 24 h + heat treatment at 800°C for 1 h			Porous spherical globules (size: 5–20 μm); homogenous
Grit blasted Ti6Al4V	Grit blasted with corundum particles (~700 μm) at 4 bar pressure	5×c-SBF	1st stage: 24 h at 37°C 2nd stage: 24 h at 50°C	Spiky crystals with narrow cracks; homogeneous
Alkali and heat treated Ti6Al4V	Alkali treatment (in 10M NaOH) + heat treatment at 600°C for 1 h	c-SBF	72 h at 37°C	Near-spherical nodules (size: ~ 2 μm)
Alkali and heat treated titanium	Alkali treatment (in 10M NaOH) + heat treatment at 600°C for 1 h	m-SBF	3 weeks at 37°C	Apatite globules (size: 1–4 μm)

12.5 | Friction and Wear of Ti-alloys in Hank's Balanced Salt Solution

In addition to electrochemical corrosion and biomineralisation assessment, another performance-limiting property for structural biomaterials is the friction and wear resistance properties, *in vitro*. The lab-scale tests should ideally stimulate closely the physiological contact configurations. Extending our discussion further on Ti-based alloys, this section will summarise the published results from the authors group on fretting wear properties of selected Ti-alloys in SBF. In order to study the severe wear behaviour, the wear tests were performed at 10 N normal load for 10,000 cycles with relative displacement stroke between the flat and ball adjusted to 80 μm and the frequency 10 Hz. Such a combination of testing parameters resulted in a gross slip fretting contact with a linear sliding speed of 0.0032 m/s. Hank's balanced salt solution, the electrolyte to simulate human body fluid conditions, was prepared using laboratory grade chemicals and double distilled water. The solution pH was precisely maintained at 7.4. Freshly prepared solution was prepared fresh for each experiment. CP titanium, Ti13Nb13Zr and Ti6Al4V show similar frictional behaviour under the chosen experimental conditions. Figure 12.10 shows the steady state COF as a function of different material combinations. A comparison of the steady state coefficient of friction values (COF) shows a higher COF value of 0.5 for CP titanium; whereas COF value for Ti13Nb13Zr and Ti6Al4V was comparable at 0.48 and 0.46, respectively. Co28Cr6Mo/steel shows a steady state COF of around 0.4. Significantly lower COF is noticed for Ti5Al2.5Fe (COF-0.30) when compared to other investigated alloys.

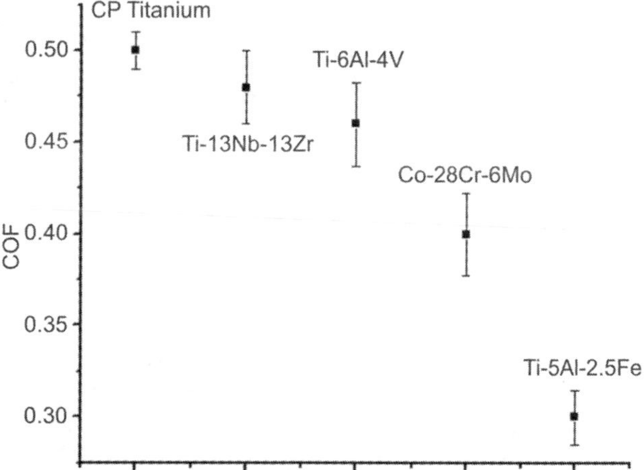

Fig. 12.10 Plot of steady state coefficient of friction for Ti-based materials measured during fretting against 8 mm diameter steel ball, at 10N load for 10,000 cycles with a frequency of 10Hz frequency for 80 μm displacement stroke[884].

Apart from recording COF values, the worn surfaces are investigated using electron microscopy techniques to unravel the wear mechanisms. Figure 12.11a shows the detailed morphological analysis of the fretting scar on commercially pure titanium. Extensive grain boundary cracking is clearly visible. It is observed that the cracks are formed preferentially along the grain boundaries and the material at the worn surface is observed to be plastically deformed.

The topographical features of the worn surface on Ti13Nb13Zr, reveal abrasive scratches, the transfer layer and extensive cracking (Fig. 12.11b). It is highly likely that high plastic deformation results in the cracking of the material during the wear process.

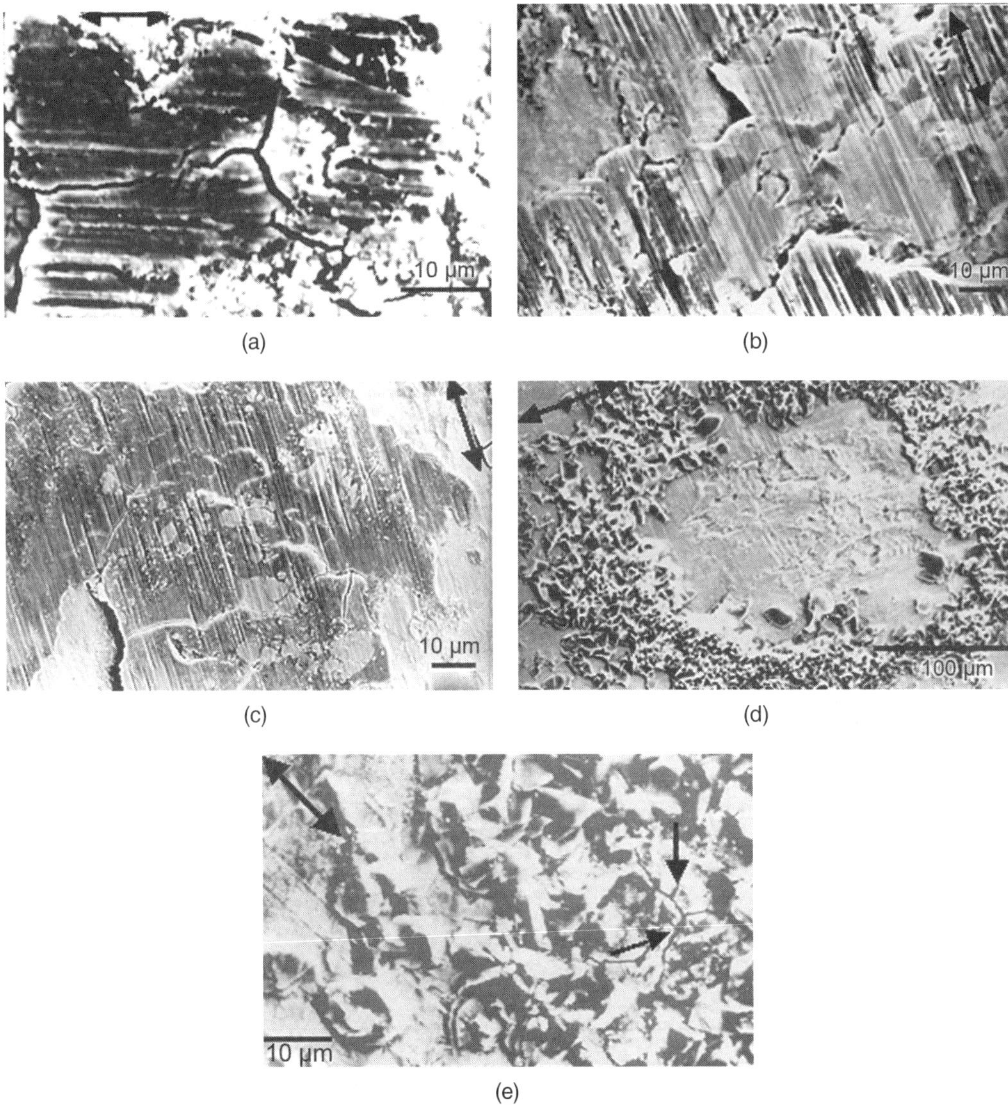

Fig. 12.11 SEM images revealing the details of the topographical features of the worn surfaces: CP (a) Titanium, (b) Ti13Nb13Zr, (c) Ti6Al4V, (d) Ti5Al2.5Fe and (e) Co28Cr6Mo after the fretting experiments against steel ball at 10N load, 10Hz frequency, 10000 cycles and 80 μm displacement stroke; double pointed arrows indicate the fretting direction and single pointed arrows indicate the presence of cracks[884].

Figure 12.11e shows the detailed morphological features of the worn surface of Ti6Al4V, characterized by non-uniform transfer layer together with the grain boundary cracking. Closer observation of the wear pits on Ti5Al2.5 Fe reveals smooth appearance of the worn surface together with signs of plastic deformation. Closely spaced deformation bands are also seen. Figure 12.11e shows the SEM micrograph of the fretted Co28Cr6Mo surface, revealing rough nature of worn surface, abrasive scratches with extensive plastic deformation and grain pull out.

Summarizing the wear results, it is clear that all the investigated materials with the exception of Ti5Al2.5Fe have undergone severe fretting wear, in simulated body fluid. The wear occurs mainly by abrasion, plastic deformation and cracking. Under the tribomechanical stress conditions, Ti-based materials undergo severe plastic deformation, which involves the formation of persistent slip bands. The extensive formation of slip bands possibly results in the observed cracking in the investigated materials. In the case of Ti5Al2.5Fe, the severity of the abrasion is much less compared to the other Ti-based alloys investigated. Moreover, the appearance of the smooth worn surface correlates well with the low COF obtained with this material. At tribological contacts, Ti-alloys undergo tribochemical oxidation to form TiO_2-rich layer during the process of wear. However, the cracking and spalling of the TiO_2-rich tribolayer during repeated fretting strokes is considered to be the major cause for severe wear of the investigated materials.

12.6 | Closure

The results of *in vitro* corrosion behaviour of several implantable metallic biomaterials are discussed in this chapter and all these results are based on past research of author's own group. Such a discussion reflects that a range of techniques as well as a depth of analysis is required to carefully characterize corrosion of implantable metallic alloys in physiological medium. Some results related to friction and wear of selected Ti-alloy are also discussed in this chapter. This chapter also demonstrates that Ti-Si-C alloy has a good potential for orthopedic applications, based on mechanical properties, *in vitro* corrosion and mineralization behaviour. It is noteworthy to mention here that the corrosion behavior of metallic alloys has significant impact on the biomechanical behavior of implants[906]. Besides orthopedic applications, the corrosion behavior of metallic alloys determine the long term functionality of metallic stents (e.g. NiTi) in cardiovascular applications[907]. Although significant effort has been invested to summarize the corrosion behavior of the major biomedical grade Ti-alloys, a reader can refer other scientific papers[850,908] and book[849].

13

Case Study: Calcium Phosphate–Mullite Composites

As mentioned in an earlier chapter, the development of new hydroxyapatite (HAp, $Ca_{10}(PO_4)_6(OH)_2$) based bioceramic composites demands improvement in physical properties (strength, toughness) without compromising biocompatibility. Using extensively published research results from author's own research group, this chapter will illustrate how mullite ($3Al_2O_3.2SiO_2$) addition (up to 30 wt%) can influence the phase assemblage, microstructure development and properties in CaP–mullite system. While employing a number of complimentary techniques to obtain reliable measure of mechanical properties, a combination of elastic modulus (~80 GPa), compressive strength of more than 350 MPa, 3-point flexural strength of 70–80 MPa, hardness of 4–5 GPa and modest SEVNB toughness of 1.5 MPa.m$^{1/2}$ were recorded with the investigated composites. In reference to *in vitro* cytocompatibility property, much emphasis has been provided to discuss the cell viability and proliferation as well as osteoblastic differentiation marker expression. Considering the potential toxicity of mullite, the submicron and nano-sized particle (NP) eluates of HA-20 wt % mullite (H20M) were evaluated for their effect on cellular viability and genotoxicity properties, *in vitro*. From the critical analysis of cytotoxicity (MTT) and genotoxicity (Comet and Micronucleus assay) data, it was evident that mullite as well as H20M nanoparticle eluate exhibit cytotoxicity and genotoxicity above 50% of eluate concentrations in time dependent manner, *in vitro*. Finally, the *in vivo* biocompatibility property will be discussed in reference to the implantation experiments with selected CaP-mullite composites using femoral bone defects in rabbit model. The critical assessment of histological analysis also establish that CaP-20% mullite composite is biocompatible, *in vivo* and also induces excellent healing of femoral bone defects. The extensive results on the development of CaP–mullite composites, as discussed in this chapter, are primarily a summary of the studies performed by the author's research group over last one decade (see the papers of Nath et al. in the reference list).

13.1 | Introduction

The last few decades have seen extensive efforts towards developing hydroxyapatite (HA)-based composites. The motivation for developing HAp-based composites stems from the requirement to fabricate materials with improved strength and toughness properties, while retaining bioactivity. To this end, hydroxyapatite has been used in combination with another metal/ceramic phase, which can improve the physical properties of HA without compromising biocompatibility. Some examples include HA–alumina[909,910] and HA–zirconia[911,912]. Among other HA-based composites, attempts have been made to introduce HA in a bioglass matrix[913] or HA whiskers to HA matrix[914]. In the above mentioned HA-based systems, the optimization of sintering parameters as well as phase stability has been a major issue. It is instructive to note that hydroxyapatite is abbreviated as HA or HAp in this book and similar terminology is used in other books or published research papers.

Mullite, an important structural ceramic material, is a solid solution of alumina and silica and has been shown to be a potential reinforcement to HA[915]. Mullite ($3Al_2O_3.2SiO_2$) has lower density (~3.05 g/cc) than Al_2O_3 (~3.95 g/cc) or ZrO_2 (~6.1 g/cc), higher hardness (~15 GPa) than HA (~7 GPa) or ZrO_2 (~12 GPa) and moderate fracture toughness (~3 $MPam^{0.5}$). But while developing HAp–mullite composites, the common hypothesis was that mullite will not react directly with HA or TCP. The results shown in this chapter however reveal that mullite reacts only with CaO, one of the dissociation products of HA[916]. Therefore, in HA–mullite composites, maximum retention of TCP is possible, which is unlikely for other HA based composites. For example, in case of HA–alumina and HA–zirconia composites, the second phases (alumina[909,910] or zirconia[911,912]) can directly react with CaP phases (HA or TCP). Another benefit using mullite is that good densification can be achieved due to liquid phase sintering[916], which is not possible in many HA based composites.

> The evaluation of cellular functionality (and cytotoxicity) and genotoxicity of a biomaterial (bulk/coatings/particle elutes) are important to ensure safety for long term biomedical application.

Genotoxicity and cytotoxicity are influenced not only by the chemical composition of implanted material, but also by its physical properties such as size, dose matrix, and method of testing and reference material[917,918,919,920]. It has been widely recognized that the implants, such as those made of metallic, polymeric and ceramics generate a large number of *de novo* particles in the range of nano/submicron[921,922], which cause short term effect such as cytotoxicity, apoptotic cell death followed by macrophages activation as well as long term effect such as genotoxicity, aseptic loosening at the implant site *in vivo*. It is therefore essential to understand the effects of *de novo* generated wear particles and get an insight into the physical as well as chemical basis of cytotoxicity and genotoxicity[923,924].

A significant part of this chapter presents the summary of the published results to illustrate *in vitro* cytocompatibility of bulk or particulated CaP-20% mullite composites as well as *in vivo* biocompatibility property. While *in vitro* property is largely analysed using cell viability/differentiation, genotoxicity assays, the *in vivo* biocompatibility is investigated by short term implantation in rabbit animal model, followed by extensive histological analysis.

13.2 | Sintering Reactions and HA Stability

One of the important aspects in HA-based ceramic composites, densified via high temperature sintering, is the stability of HA. It is however worthwhile to mention that biphasic calcium phosphates (a mixture of HA and TCP) have better biocompatibity property than HA or TCP alone.

Another challenging aspect from material science point of view is to propose thermodynamically feasible sintering reactions to explain the phase assemblage in sintered microstructure. This issue will be addressed in this section in reference to HA-mullite system.

13.2.1 | HA stability

Following conventional ceramic processing route, HA–mullite composites with mullite addition of up to 30 wt% are sintered in air and in this section, various   compositions are referred as HAp xM (where x = 10, 20, 30 wt %, M stands for mullite; e.g. HAp10M denotes HAp – 10 wt % mullite, and likewise). Similarly, biphasic calcium phosphate (BCP) composites are referred as BCPxM. The stability of HAp as a function of sintering temperature and mullite content will be discussed on the basis of XRD and microstructural investigation. Semi-quantitative analysis of the XRD results is presented in Fig. 13.1. Figure 13.1a shows the ratio of the most intense X-ray peaks of HAp to β- TCP as a function of sintering temperatures. In all the cases, the amount of HAp decreases with increase in mullite content as well as sintering temperature. But, the difference between HAp20M and HAp30M composites is insignificant. Figure 13.1b plots the ratio of the most intense X-ray peaks of mullite to total CaP phases with respect to sintering temperature and initial mullite content.

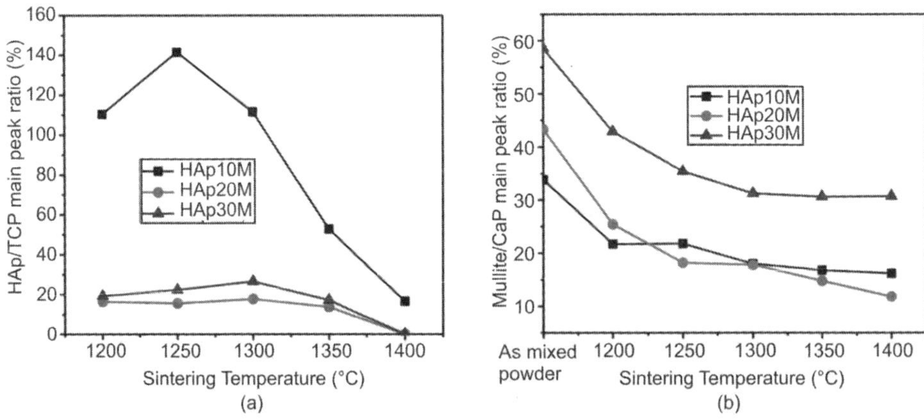

Fig. 13.1   (a) HAp and TCP ($\alpha+\beta$) main peak intensity ratio with variation of sintering temperature for HAp (10–30) mullite composite samples; (b) mullite to calcium phosphate main peak intensity ratio as a function on sintering temperature for HAp (10–30) mullite composite samples[916].

Some illustrative TEM results are shown in Figs. 13.2 and 13.3. In Fig. 13.2, the combination of bright field images reveals the presence of elongated mullite phase and the reaction product at grain boundary triple pocket. The SADP analysis further confirms the presence of HA, α-TCP and mullite. When higher mullite content is added, the presence of Al_2O_3 together with mullite needles along with HA/TCP are observed (see Fig. 13.3). Also, dislocation network can be seen and the analysis of SADP patterns also indicates the sinter reaction product to be gehlenite.

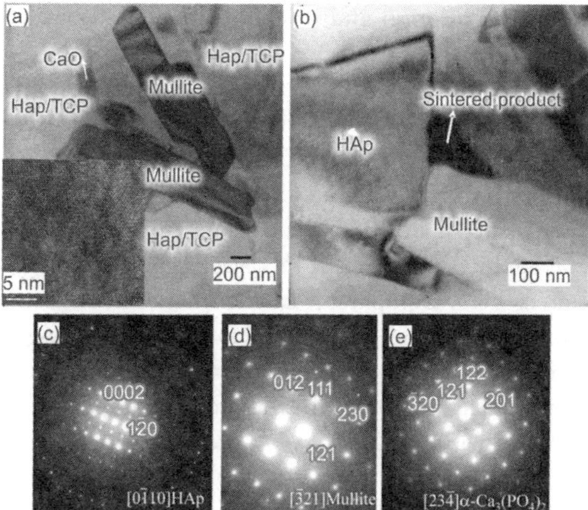

Fig. 13.2 (a) Low magnification TEM image of HAp-10wt% mullite sintered at 1350°C, revealing typical mullite needle along with HAp phase; (b) higher magnification image showing the presence of sintered product at triple pocket; (c), (d) and (e) are electron diffractions patterns from HAp, mullite needles and α-TCP phase respectively[916].

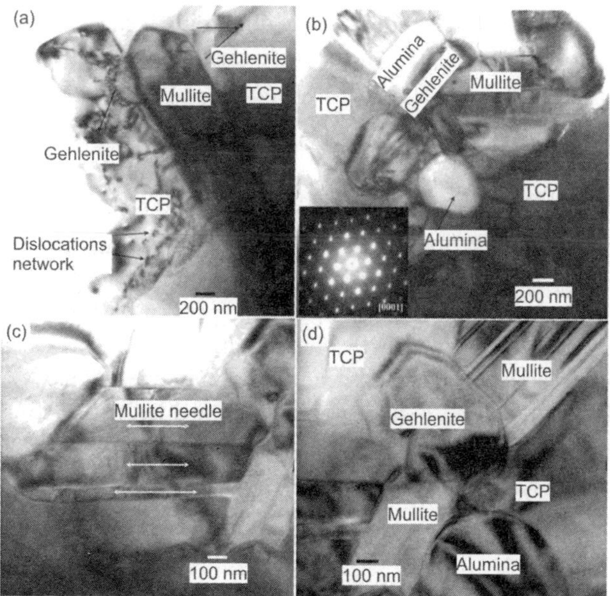

Fig. 13.3 TEM image of HAp-30 wt% mullite sintered at 1350°C for 2hrs in air; (a) gehlenite around the mullite needle, dislocation network in TCP grains, (b) alumina along with gehlenite ahead of mullite needles; the SAD pattern taken from rhombohedral alumina grain shown in the inset. (c) Parallel mullite needles are bridged in between TCP/HAp grains, (d) bright field image showing the presence of mullite, TCP, alumina and gehlenite[916].

One possible hypothesis for these observations is that in addition to the temperature, mullite promotes decomposition of HAp to CaO and TCP. Thereafter, mullite reacts with CaO to form calcium-alumino-silicate phase (gehlenite) and the favourable nature of this reaction accelerates the decomposition of HAp. The addition of mullite leads to more number of mullite/mullite contacts, which do not participate in the reaction and in contrast, acts as non-densifying inclusions in a sinterable matrix[925].

Our experimental results are in line with the earlier results reported in the literature. Yin and Stott[926] have shown theoretically as well as experimentally that β-TCP is much more stable than α-TCP. Pierre[927] discussed the β-TCP → α-TCP transformation under different experimental conditions. It was found that[931] normally β → α-TCP transformation took place at 1350°C and above. However, when β-TCP was heated up to 1348°C for 1 hour and then quenched, it fully transformed to α-TCP. Interestingly, at a temperature of 1294°C, similar heating and cooling schedule resulted in a mixture of β-TCP and α-TCP. Comparable transformation took place with the starting phase as α-TCP. But, no transformation occurred at and below 1250°C starting with either α or β phase. However, Mackay[928] reported that β → α-TCP transformation upon heating occurred at 1180°C, which was lower than that reported by Pierre. In the present case, β-→ α-TCP transformation in the composites took place at and above 1200°C.

13.2.2 | Sintering reactions

In this sub-section, the sintering reactions related to HAp dissociation have been discussed. Pure HAp starts to dissociate at about 900°C. However, HAp, synthesized in aqueous solution by chemical route, is found to be more stable[929]. From the XRD data (Fig. 13.1), it is revealed that pure HAp is quite stable up to 1200°C. In presence of mullite, it starts dissociating at much lower temperature, as indicated by XRD data. The possible reactions associated with the dissociation of pure HAp can be expressed as,

$$Ca_{10}(PO_4)_6(OH)_2 = 2Ca_3(PO_4)_2 + Ca_4P_2O_9 + \uparrow H_2O \tag{13.1}$$

$$Ca_{10}(PO_4)_6(OH)_2 = 2Ca_3(PO_4)_2 + CaO + \uparrow H_2O \tag{13.2}$$

Both the above reactions describe the formation of β-TCP, while the first reaction produces tetracalcium phosphate ($Ca_4P_2O_9$) as sinter product, the second one leads to formation of CaO. The presence of CaO is confirmed from the TEM diffraction pattern[916]. Therefore, in the present case, the latter reaction is more likely. Additionally, the formation of the other calcium-alumino-silicate grain boundary and the reaction products is the result of limited reaction of CaO with mullite. The calcium-alumino-silicate liquid phase can crystallize into gehlenite (C_2AS) and alumina (Al_2O_3) at grain boundary and the reaction can be written as:

$$4CaO + 3Al_2O_3.2SiO_2 = 2(2CaO.Al_2O_3.SiO_2) + Al_2O_3 \tag{13.3}$$

Another important aspect is that the sintering temperature has at least three effects on the sinter density: (a) it increases the density due to enhanced mass transport kinetics, (b) it reduces the density due to the transformation of HAp to TCP and (c) it enhances the reaction between mullite and CaO (from the decomposition of HAp). Thus, the densification kinetics and the final sinter densities are the interplay among these three factors.

13.3 | Mullite Dependent Enhancement of Flexural and Compressive Strength

Figure 13.4 plots compressive and flexural strengths for various CaP–mullite composites as well as for pure HAp. In both the experiments, few optimally sintered samples were selected. Pure HAp sample, sintered at 1200°C, showed the lowest compressive strength of 54 MPa. But, with an increase in mullite content, the compressive strength increased considerably. Much higher compressive strength of ~233 and 381 MPa were measured for CaP20M and CaP30M samples respectively, both sintered at 1350°C for 2 hours, respectively. Flexural strength data of CaP–mullite composite, measured using 3-point bending test, revealed that CaP10M possessed flexural strength of ~44 and ~51 MPa, when sintered at 1350 and 1400°C for 2 hours, respectively. The flexural strength values peaked at ~ 80 MPa for CaP20M samples, sintered at 1350°C for 2 hours. A little lower strength of ~75 MPa was recorded, when the mullite content was increased to 30 wt% (Fig. 13.4).

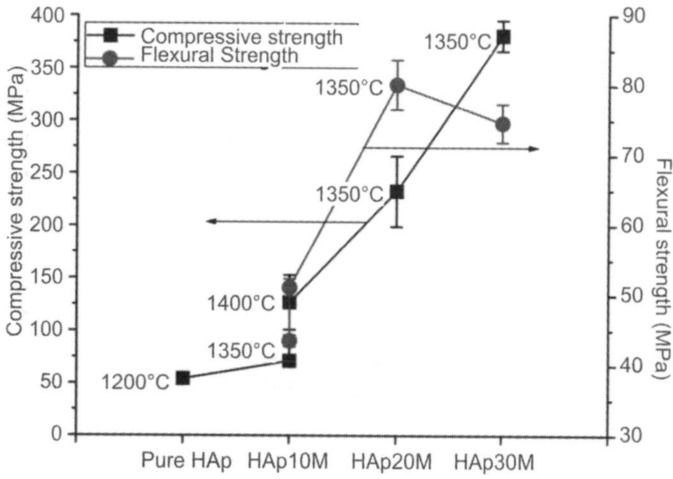

Fig. 13.4 Compressive strength and flexural (3-point bending) strength of various CaP–mullite composites are plotted against their compositions[916].

Like hardness, the elastic moduli of the composites were lower than pure HAp and mullite (120 and 230 GPa, respectively). Among the composites, CaP20M sample sintered at 1400°C, possessed highest elastic modulus of 94 GPa. For CaP30M composite, maximum elastic modulus (84 GPa) was obtained, when the sample was sintered at 1400°C. The elastic modulus of the few selected samples was measured by impulse excitation method and this showed almost similar results that of indent depth penetration method. Elastic modulus values of 66 and 73 GPa were obtained for CaP20M composites (sintered at 1350°C), when measured by indent depth-penetration and impulse excitation method, respectively.

Fracture toughness of the developed composites was evaluated using indentation method as well as by SEVNB method[930]. The fracture toughness results are plotted in Fig. 13.5. Among the composites, optimally sintered CaP20M showed highest indentation toughness of ~2.7 MPa.

$m^{0.5}$, calculated using Nihara formula. For pure mullite sample, the indentation toughness value, obtained using Anstis formula, was ~2.23 MPa $m^{0.5}$. It is well known that the pre-factors given in the equations of Anstis and Niihara are experimentally derived mean values. More details of various formulas to estimate indentation toughness can be found in chapter 6 of this book.

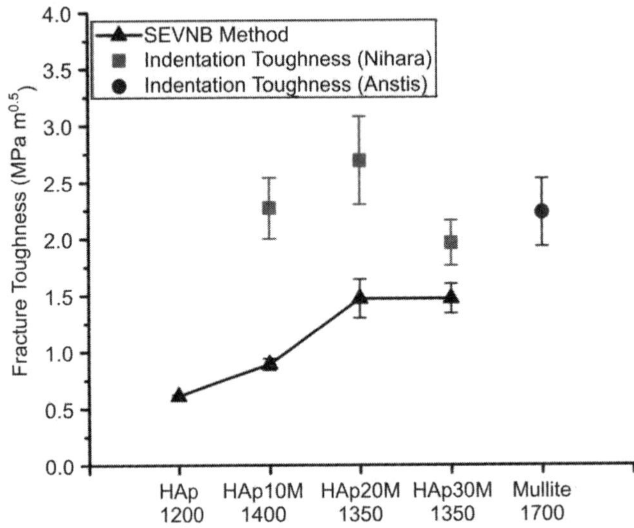

Fig. 13.5 Fracture toughness of pure HAp and CaP–mullite composites are plotted against the initial composite compositions; the results obtained from SEVNB method and indentation cracking methods are compared[916].

It is not possible to determine true fracture toughness of a new brittle material by indentation cracking method and long crack toughness measurement (e.g. SEVNB) needs to be adopted.

In contrast to indentation toughness values, a lower toughness value with smaller error bar was measured, when SEVNB method was adopted. In SEVNB method, the V notch tip was stress free as it was created by polishing and not by cutting. Here, the fracture toughness of pure HAp (sintered at 1200°C for 2 hours) was obtained as ~0.6 MPa.m$^{0.5}$ and the toughness value increased with increasing the mullite content from 0 to 20 wt%. For optimally sintered CaP10M and CaP20M samples, the fracture toughness values were calculated as ~0.9, and 1.5 MPa.m$^{0.5}$, respectively. Also, the toughness of CaP20M and CaP30M were similar.

The presence of mullite needles with a typical aspect ratio of 7–8, as revealed in Fig. 13.3, contributes to increased toughness in the composite. Such contribution should be similar to whisker induced toughening, i.e., crack deflection effect, as reported for ZrB_2–SiC_w whisker reinforced composites[931] and HAp–HAp_w composites[932]. In all such cases, the main toughening mechanism was the crack deflection.

Another important aspect of particulate reinforced composites is the residual stresses due to mismatch of Coefficient of Thermal Expansion (CTE). In the present case, major phases in the sintered composite were TCP and mullite. Mullite has a much lower CTE (5.3×10^{-6}/°C)[933,934] than

TCP ($14.9 \times 10^{-6}/°C$). Taya et al.[935] described how the apparent toughness of particulate reinforced composites can be increased due to residual thermal stress. Considering the dominant presence of TCP and homogeneous distribution of mullite in the investigated composites, the compressive stress field induced by the mullite needles at the vicinity of the TCP matrix was calculated to be 709 MPa and tensile stresses in matrix was 210 MPa for 20 wt% mullite content[935]. As the number of mullite particles/needles increase with increase in mullite content in the composites, the microstructure will contain more areas of compressive stress. Therefore, an advancing crack would be deflected by the randomly oriented compressive stress region, created by the mullite needle in the TCP matrix and also increasing tortuosity in the crack path. Despite the presence of elongated mullite needle, perhaps the grain boundary region was not weak enough for intergranular fracture and crack bridging to occur. From the above discussion, it is quite clear that the addition of mullite to CaP matrix would improve the apparent toughness and strength.

> The human cortical bone, depending on anatomical locations, can exhibit an array of mechanical properties including flexural strength of 50–160 MPa, compressive strength of 60–120 MPa, elastic modulus of 3–30 GPa and fracture toughness of 2–12 MPa.m$^{0.5}$.

While comparing the properties of cortical bone with the developed materials, it is clear that among many of the (though not all) bone replacement applications, the newly developed composites could be a suitable candidate material[936,937]. Based on the preceding discussion, it is apparent that CaP-mullite composites have better properties than matrix (i.e., TCP) and HAp. Also, the complete evaluation of various mechanical properties, like above, should be conducted while developing any new material for load-bearing biomedical applications.

13.4 | Cytocompatibility Properties, *in vitro*

Once the physical/mechanical properties are comprehensively evaluated, a selected group of materials exhibiting desired range of properties would enter into the *in vitro* cytocompatibility testing. The *in vitro* biocompatibility results, in reference to CaP-millite system, are interpreted in terms of the substrate compositional differences or phase assemblage. Such interpretation will be helpful to realize the following aspects: (a) *in vitro* cytocompatibility property on the basis of microstructural phase assemblages; (b) prediction of osteoconduction property based on biocompatibility with MG 63 cell line, and (c) bone-cell functionality and phenotypic marker expression based on ALP and OC expression assay.

Figure 13.6 shows the results of L929 fibroblast cell adhesion experiments over 3 days period and the cell seeding density was approximately 1×10^{5}/ml. SEM images reveal adhesion of L929 cells on BCP10M (Figs. 13.6c and d), BCP20M (Figs. 13.6e and f), BCP30M (Figs. 13.6g and h) substrates. Some interesting observations can be made. For example, Fig. 13.6e reveals how the cells were connected to each other by filapodia extension on BCP20M sample. In this case, some observable differences could be found, when the cell seeding density was lower. The cells were larger in size with an approximate dimension of 40–50 μm. The major difference could be found in cell morphology. In case of cell density being 1×10^{5}/ml, most of the cells were spread on the surface and intended to enhance the cell-material contacts. The filopodia of the cells were extended over longer distance than that was observed in the previous case (when cell seeding density was 5×10^{5}/ml) . Figure 13.6i shows an AFM image of a filapodia attached on pure HA sample. The branching of filapodia is clearly visible. This type of branching provided good anchorage with the material surface.

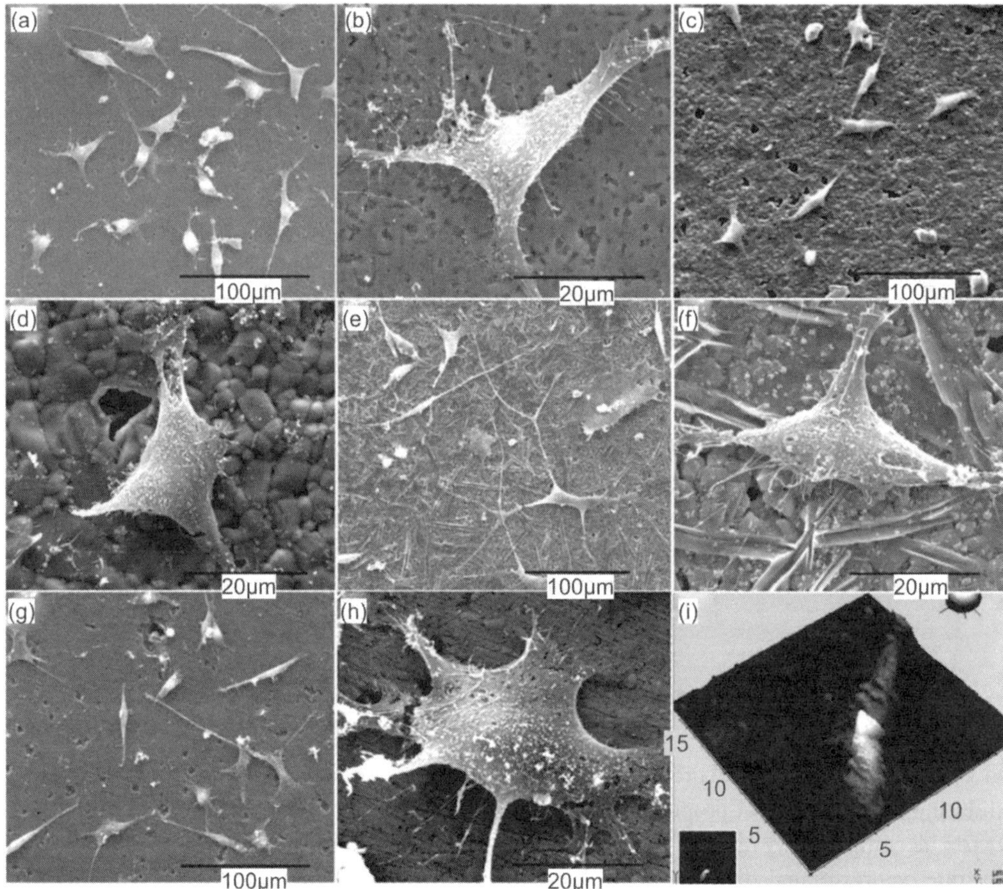

Fig. 13.6 SEM images of L929 cells adhered on (a and b) pure HA, (c and d) BCP10M, (e and f) BCP20M, and (g and h) BCP30M samples.; AFM image revealing extensive filopodia extension on pure HA sample is also shown in (i); results were obtained after 3 days of culture; the seeded cell density was 1×10^5 cells/ml[916].

ALP activity is considered as a phenotypic marker of differentiated cell. The ALP produced by metabolically active MG63 cells, after 3 and 7 days of experiments, is quantitatively plotted in Fig. 13.7. After 3 days of culture, the maximum ALP activity was measured with pure HA and for composites (10–30 wt% mullite) the ALP activity was very near to pure HAp. After 7 days of culture, the ALP expression of the cultured cells was significantly higher in all mullite containing composite compared to pure HA. However, almost no difference in terms of ALP expression was found among various mullite containing composites.

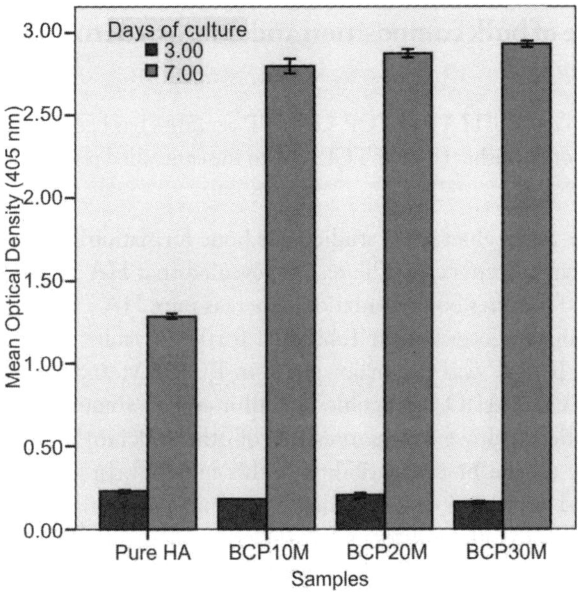

Fig. 13.7 Alkaline phosphatase (ALP) activity of MG63 cells on pure HA, BCP10M, BCP20M and BCP30M ceramics, after 3 and 7 days of culture in osteogenic medium[916].

OC is a later stage marker of bone cell differentiation. Declercq et al. [938] described calcification as a predictor of bone mineralization capacity of biomaterials in osteoblastic cell cultures. It is normally accepted that the OC measurement is important to substantiate the use of investigated ceramics as bone replacement materials. For BCP x M composites, the results of the OC assay are plotted in Fig. 13.8. The OC production after 7 days culture on control, pure HA, BCP10M showed significantly lower values, in comparison to BCP20M and BCP30M samples. However, no difference in OC production could be noticed between BCP20M and BCP30M samples.

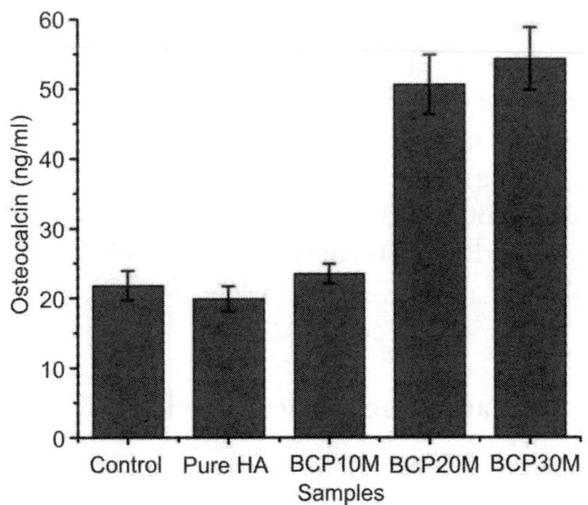

Fig. 13.8 Osteocalcin (OC) expression of cultured MG63 cells on various ceramics samples after 7 days of culture; the results were compared with control[916].

13.4.1 | Influence of bulk composition and microstructure on cytocompatibility

In recent years, the mixture of HA and TCP, i.e., BCP ceramics has been reported to be better than the monolithic forms of the either (HA or TCP) due to the controlled resorbability of BCP ceramics.

In an interesting work, Arinzeh et al.[939] studied the bone formation capabilities of BCP ceramics using human mesenchymal stem cells. The results revealed that HA : TCP ratio of 20:80 was the optimum combination for better bone formation, whereas pure HA and TCP phase had much less effect on bone formation. A closer look at Table 13.1 further reveals that while α-TCP dominated in BCP10M material, β-TCP was the major phase in BCP20M and BCP30M composites. The additional presence of CaO, Al_2O_3 and gehlenite, all in minor amount (much weaker X-ray peak intensity), was also noticed. However, the presence of other calcium-alumino-silicate phases may also have an influence on the biocompatibility of this material. In some earlier reported results [940,941], it was mentioned that the biocompatibility of the materials mainly depended on the leaching of ions. The adhered cells mainly proliferate and differentiate due to the activation of surface ions. In the present case, the possible leaching ions would be Ca, P, Si, and Al. The effect of Ca and P ions need not be discussed here, as these are already known marker for the bio-mineralization.[942] The effects of other ions therefore, need to be discussed with cited literature. For example, silicon containing materials were already investigated as silicon substituted HA, Si_3N_4, etc. Thain et al. [943] described the *in vitro* biocompatibility of silicon substituted HA thin film and reported that the presence of Si did not induce any toxic effect. In another study, Kue et al.[944] showed enhanced cell proliferation and OC production by human osteoblast-like MG63 cells on silicon nitride ceramic discs. Their results revealed that Si_3N_4 was a non-toxic biocompatible ceramic, which could be used as a potential biomaterial. This proves that the presence of Si in the composite, should not ideally degrade the biocompatibility. As far as the Al-leaching is concerned, a study by Ku et al.[940] on Ti6Al4V alloy suggested that the release kinetics of Al-ions could play a major role in influencing the osteoblast behavior.

The attachment of substrate-dependent cells, such as fibroblasts and osteoblast, to a substratum (e.g. biomaterial) is a synchronized process involving cytoskeleton reorganization, cell spreading, and formation of focal contact.[945,946] The cell adhesion behaviour of pure HA and composite samples showed presence of cytoplasmic extension, filopodia (Figs. 13.6a–h). From Figs. 13.6a–h, it can be said qualitatively that the filopodia was more visible in case of pure HA (Figs. 13.6a, b), BCP20M (Figs. 13.6e, f) and BCP30M (Figs. 13.6g, h) samples. However, the filopodia was less visible and the cells surface had minimum adhesion on BCP10M sample (Figs. 13.6c, d). BCP20M and BCP30M samples mainly contained β-TCP, whereas BCP10M contained mainly α-TCP (Table 13.1). Therefore, the difference in cytocompatibility behaviour could be attributed to the difference in phase assemblage.

13.4.2 | Osteoconduction and biochemical markers of bone turnover

As mentioned earlier, the aspects of bone cell differentiation and functionality could be explained on the basis of ALP and OC assay results. The synthesis of these biochemical markers of bone cell increases with the increasing expression of osteoblasts and decreases with the maturation of osteoblasts[947].

> OC and ALP are the phenotypic markers of the late and early stages of differentiation of osteoblast-like cells, respectively.

An increased specific activity of ALP in a bone cell essentially reflects a shift towards a more differentiated state.

Recalling ALP results, it is clear that all the mullite containing composites had comparable expression of early stage osteoblast differentiation marker, which was much higher than baseline HA. This means that osteoblast like cells were in a better functionally differentiated state in contact with the BCP–mullite composites. Furthermore, the extent of OC expression in case of BCP20M and BCP30M was much higher than both baseline HA and BCP10M composites. The above observations therefore confirmed that BCP20M composite could exhibit the best combination of OC and ALP expression.

The above described results need to be explained in the perspective of earlier literature reports. In several earlier research reports, it was mentioned that among CaP based materials, β-TCP containing composites showed better differentiation and *in vivo* bone formation/mineralization. For example, Shiratori et al.[948] studied the bone forming ability of β-TCP, when implanted in bone defects of rat femur. Their results revealed that β-TCP was an appropriate material for the treatment of bone defects. In another study, Matsuno et al.[949] compared the *in vitro* (ALP) as well as *in vivo* behaviour of β-TCP/collagen sponge composite with only collagen sponge (CS). In the *in vitro* experiments, human mesenchymal stem cell lines were used and for *in vivo* experiment, the samples were placed under the back skin of nude mice for up to 12 weeks. Their results revealed that the composite containing β-TCP show better osteogenic properties and had ability to promote bone formation. In an interesting study, Arinzeh et al.[939] tried to optimize the optimum HA/β-TCP ratio for better stem cell induced bone formation. For this purpose, they selected various ratio of HA: β-TCP, i.e., 100:1, 76:24, 63:37, 56:44, 20:80, and 1:100. In both *in vitro* (OC) and *in vivo* experiments (mouse), it was revealed that HA: β-TCP ratio of 20:80 was the best combination for new bone formation in bone remodelling process. It can be further noted that this combination showed better results compared to monolithic form of HA and β-TCP. In fact, all other combination of HA and β-TCP showed improved results compared to pure HA and β-TCP. At the molecular level, HA phase provided the required binding sites, whereas TCP, due to its higher dissolution properties, increased the concentration of Ca and P locally. All these led to the cascading of differential gene expression and positively increased the cell differentiation[950,951].

In the present case, BCP20M and BCP30M predominantly contained β-TCP phase, while BCP10M contained more α-TCP phase. On the basis of the above information, it should therefore apparent that the dominant presence of β-TCP, both BCP20M and BCP30M exhibited better expression of osteblastic phenotypic marker than BCP10M. The present investigation also reconfirmed that single phase HA had inferior expression of differentiated bone cell than BCP microstructure containing varying ratio of β-TCP. Additionally, it can be stated that the additional presence of gehlenite, alumina or CaO did not have any inhibitory effect as far as the osteoconduction property of BCP20M and BCP30M was concerned.

> In addition to the differences in phase assemblage, the difference in surface roughness at nanoscale can contribute towards modulation in the osteogenic property.

13.5 | **Cyto/Genotoxicity of Particle Eluates, *in vitro***

It is worthwhile to reiterate that the toxicity of biomaterial in bulk or particulate form towards cells is of prime concern and such study should be conducted with application–specific cell type. In discussing this issue in case of HxM composites (e.g. HA10M denoted here as H10M and likewise), MTT assay results are interpreted in terms of the ability to cause change in mitochondrial activity of nano/submicron eluate treated cells, depending on treatment time, concentration and particle size (Fig. 13.9). The relevance of such study with particulate eluates can be rationalized by two factors a) the toxicity of any biomaterial in particulate format is more severe than that in bulk form, and b) particles are released, *in vivo* during long term use of any load-bearing implanted biomaterial. Experimentally, one can prepare particulates by mechanical grinding / crushing a bulk biomaterial under sterile conditions and subsequently such particulates are filtered to get uniform size range (e.g. 0.2 µm or below) before dispersing them in different amount to physiological medium of relevance (e.g. PBS). Extending such discussion, 6 hours of eluate treatment did not show cytotoxicity and significant difference (p < 0.05) in any of the eluates. H20M (87±12 nm) eluate treatment causes cytotoxicity above 50% or more concentration and after 24 or 48h of treatment time (Fig. 13.9a). The reduction in percentage cell viability (H20M) with respect to control at 50% or more was 65–70% and further reduction was observed up to 29%, after 48 hours of treatment. Interestingly, H20M eluates (165±40 nm) show minimum of 78% viability even after 48 hours of treatment or after treatment with highest eluate concentration (Fig. 13.9b). The overall cytotoxic property of H20M therefore reveals size dependent toxicity. More discussion on particle toxicity towards L929 cells is made later in this chapter.

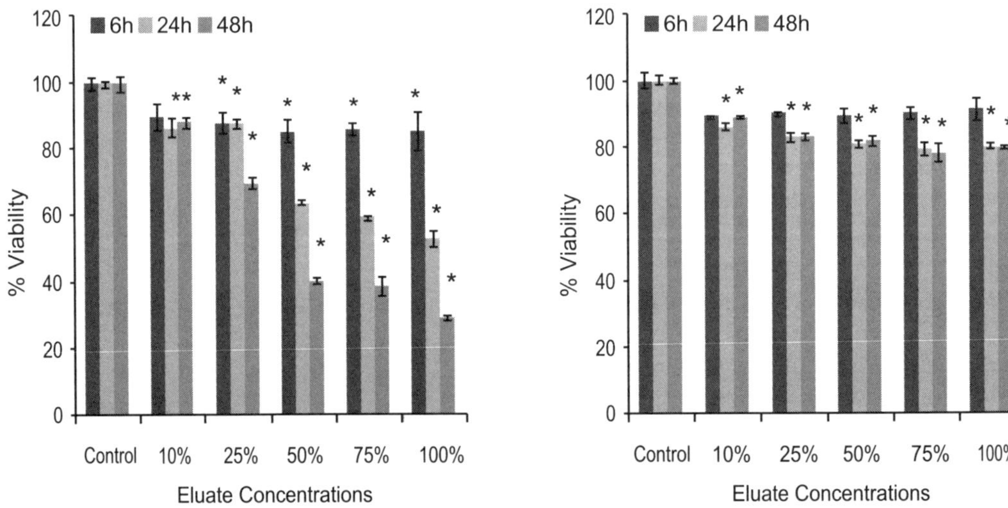

Fig. 13.9 MTT assay results to show the elute concentration and exposure time dependent L929 cell viability: (a) H20M, (87±12 nm), (b) H20M (165±40 nm) along Y-axis, percentage viability (OD at 540 nm) and along X-axis, the eluate concentration is plotted[916].

* signifies statistical significant difference at p < 0.05 with respect to control

13.5.1 | Genotoxicity assay methodology

Since the genotoxicity assays are less widely used or known in the biomaterials community than the cytotoxicity assays, two relevant assays to quantify the genotoxic effect of biomaterials are described below. It is important to note that these assays are carried out with particle eluates of biomaterials, rather than bulk biomaterials. For this, a given biomaterial is first subjected to extensive mechanical grinding or crushing to obtain finer particles and then those particles are dissolved in physiological medium, like PBS, like the way described above.

The first one is the single cell gel electrophoresis (SCGE) or comet assay, which is widely conducted according to the protocol proposed by Tice et al.[952]. It is recommended that as per the comet assay protocol, approximately, 70,000 cells/well are seeded in 12 well plate and after 24 hours of seeding, cells are treated with different concentrations of particle (10, 25, 50, 75, 100%) for 6 hours. Subsequently, the treated cells were washed with serum free medium and harvested with 0.065% trypsin. The cells are to be re-suspended in complete medium and slides can be prepared by the method of Singh et al.[953], modified by Bajpayee et al.[954]. Two slides are prepared from each well (one well/concentration). Briefly, cell suspension (100 µl) is to be mixed with 100 µl of 1% LMPA at 37°C. 80 µl of cell suspension is then loaded onto two microscope slides (75 mm × 25 mm with 19 mm frosted end), pre-coated with 1% normal melting agarose with cover slips being placed gently on the slides to allow uniform spreading of gel. The slides are kept on ice for 5 min for the gel to solidify. Third layer of 0.5% LMPA (90 µl) is to be added onto the slide, and kept over ice for 10 min. Finally, the cover slips are to be removed with the slides to be immersed in freshly prepared chilled lysing solution (2.5 M NaCl, 100 mM EDTA, 10 mM Tris; pH 10) with 1% Triton X-100 added just before use and kept overnight. After lysis, the slides are placed side by side in a horizontal electrophoresis tank containing freshly prepared chilled electrophoresis buffer (1 mM EDTA sodium salt and 300 mM NaOH; pH>13). Then the slides are left for 20 min for DNA unwinding and electrophoresis is performed at 0.7 V/cm and 300 mA at 4°C for 30 min. Subsequently, the slides are removed from electrophoresis buffer and Tris buffer (pH 7.4) is added drop wise to neutralize excess alkali. The slides are then dipped in chilled distilled water and stained with ethidium bromide (20 µg/ml) for 5 minutes. Finally, the slides are observed using fluorescence microscope, attached with image analysis system and appropriate filters.

The second relevant assay for genotoxicity is micronucleus (MN) assay, which is conducted using flow cytometry. For MN assay, the biological cells of relevance (e.g., L929 fibroblast cells or osteoblast cells in case of bone tissue engineering applications) are seeded with an approximate density of 7×10^4 cells/well in 12 well plate and treated with biomaterial eluates for 6 hours. The cells without eluate treatment can be used as negative control and EMS (6 mM) as positive control. After 6 hours of elute treatment, the eluate is removed and cells are washed with incomplete medium and grown further for 48 hours in complete DMEM medium. Subsequently, the cells are to be harvested with 500 µl of trypsin (0.125%), centrifuged at 250×g for 10 min. Following this, cells are re-suspended in 1 ml of solution I (containing 10 mM NaCl, 3.4 mM sodium citrate, 10 mg/l RNAse, 0.3 mg/l igepal, 25 mg/l EtBr), vortexed and incubated at room temperature for 1 hour. One ml of Solution II (1.5% citric acid, 0.25 M sucrose, 40 mg/l EtBr) is added and mixed. After 15 min, the suspension is filtered through 53 µm nylon mesh into polystyrene round bottom tubes and kept at 4°C, until analysis. The suspensions of nuclei and micronuclei can thereafter be analyzed using a flow cytometer with an argon ion laser (20 mW, excitation wavelength: 488 nm). During such analysis, the fluorescence is detected with 590 nm long pass band filter and data are acquired with commercial software. A log scale is usually used to register DNA and side scatter

(SSC) signals. G1-phase nuclei are stored around channel 20000 and micronuclei are counted in the region between 5% and 40% of the DNA content of G1-phase nuclei. A minimum of 30,000 events per replicate are to be analyzed for statistical significance. Percent MN per nucleus (%MN/N) can then be analyzed by dividing the number of micronuclei with number of nuclei. In case of both the assays, the negative control is the identical cell sample without any eluate treatment and the positive control is the EMS (ethyl methanesulfonate) treated identical sample. In case of former, null or negligible genotoxicity and the most significant genotoxicity can be recorded for positive control.

The comet assay data used for the DNA damage in cells include three specific parameters, (a) tail DNA (percent of DNA in the comet tail; %), (b) tail length (migration of the DNA from the nucleus; μm) and (c) Olive tail moment (OTM, arbitrary units). The last parameter is determined as the product of the distance of DNA migration from the body of nuclear core and total fraction of DNA in tail length.

13.5.2 | Genotoxicity results

Based on the analysis of comet assay results, DNA damage of L929 fibroblast cells after the exposure to H20M, HA and mullite eluates for 6 hours are shown in Figs. 13.10–13.12. H20M (87±12 nm) nanoparticle eluate shows statistically significant (p<0.05) DNA damage as apparent by higher OTM values at lower concentration of 10% of nanoeluate (Table 13.1, Figs. 13.10 and 13.11). In contrast, H20M (165±40 nm) eluate shows no statistical significant difference with respect to negative control at p < 0.05 (Table 13.2, Figs. 13.10 and 13.11). Comet data of H20M (87±12 nm) nanoeluate concentration from 10–100%, shows that DNA damage also significantly increased OTM ≤ 1.1 (control) to maximum OTM ≥ 2.4 (100% H20M). On the other hand, H20M (165±40 nm) eluate causes DNA damage in the range of 0.89 ≤ OTM ≥ 1.5. From the above data of H20M with two different nano/submicron sizes, it is apparent that genotoxicity of material is concentration as well as size dependent property. Summarizing the comet data, it is clear that H20M (87±12 nm) nanoeluate of below 10% do not exhibit genotoxic effect on L929 fibroblast cells, *in vitro*. Also, mullite eluate of particle size of 154±34 nm shows significant DNA damage (genotoxicity) with respect to control.

Table 13.1 Summary of the comet parameters, recorded while assessing the DNA damage of L929 mouse fibroblast cells after treatment with varying concentration of nanoparticle/submicron particle eluates of H20M (87±12 nm) composition [values are mean±standard error for three experiments, EMS: ethyl methanesulfonate was used as a positive control, * p<0.05, when compared to respective controls][916].

Submicron H20M (165±40 nm) concentrations	Olive tail Moment (OTM)	Tail DNA (TD)
Control	0.89±0.07	8.86±1.01
10%	1.10±0.04	8.79±0.79
25%	1.37±0.22	8.42±1.39
50%	1.64±0.20	10.36±1.54
75%	1.33±0.30	8.60±2.34
100%	1.50±0.34	10.45±1.75
EMS	4.0±0.46	18.10±2.33

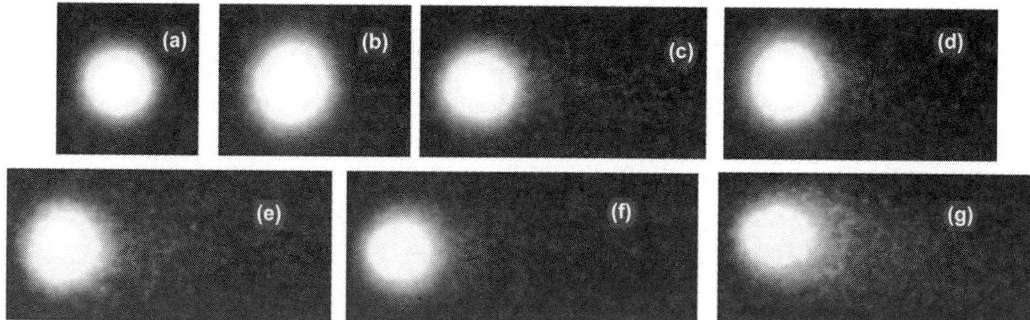

Fig. 13.10 Representative fluorescence microscopy images revealing DNA damage of treated L929 cells after treatment with H20M (87±12 nm) nanoparticle eluate of varying eluate concentrations: (a) negative control, (b)10%, (c) 25%, (d) 50%, (e) 75%, (f) 100% and (g) positive control[916].

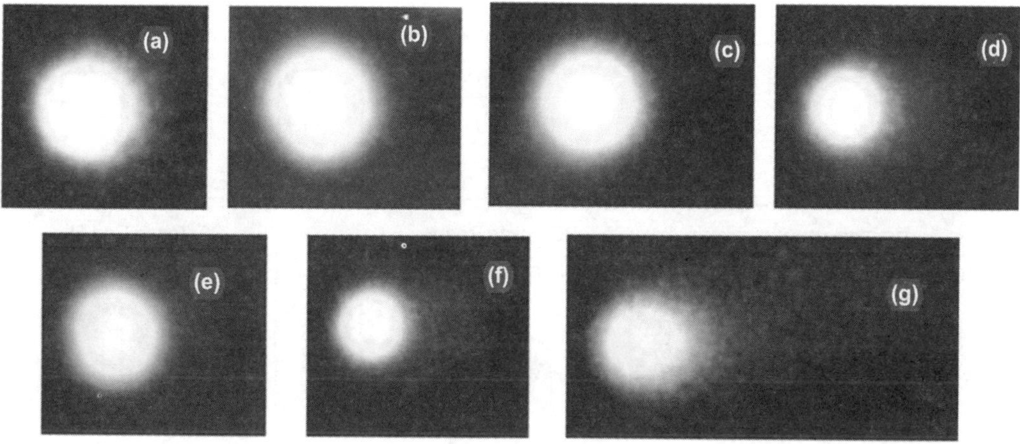

Fig. 13.11 Representative fluorescence microscopy images revealing DNA damage of treated L929 cells after treatment with H20M (165±40 nm) submicron particle eluate of varying concentrations: (a) negative control, (b) 10%, (c) 25%, (d) 50%, (e) 75%, (f) 100% and (g) positive control[916].

Micronucleus induction by the H20M, HA, and mullite in L929 fibroblast cell was detected by flow cytometry (Fig. 13.12). In Fig. 13.12, each dot represents information recorded from a single mononucleated cell. Total events of interest gated using flow cytometer are shown in larger box, marked as P2; while, the number dots within P4 marked box represents the number of cells forming micronucleus. The densely populated dots within P3 marked box reflect the total number of nucleus assessed. A closer observation of the results presented in Fig. 13.12 reveal that the degree/severity of MN formation increases with increasing concentration of eluate, independent of chemical composition of eluate. However, MN induced by 100% eluate of mullite or H20M is quantitatively lower than that induced by EMS, as revealed by intensity of dots within P4 gated area. Overall, 32–103% increase in MN formation compared to control is recorded for HA eluate treatment of 10–100%, respectively. At any given eluate concentration, percentage increase over control is highest in case of mullite eluate treatment and the lowest is with HA, while H20M eluate induces intermediate increase in MN formation with respect to control. Another important observation is the percent increase over control for 50% H20M treatment is comparable with that of 10% mullite eluate. This can be partly rationalized by the presence of 20% mullite in H20M eluate. Summarizing, only mullite shows the statistically significant (p<0.05) difference in the MN induction as compared to negative control. However, HA and H20M did not show statistically significant MN formation as compared to control.

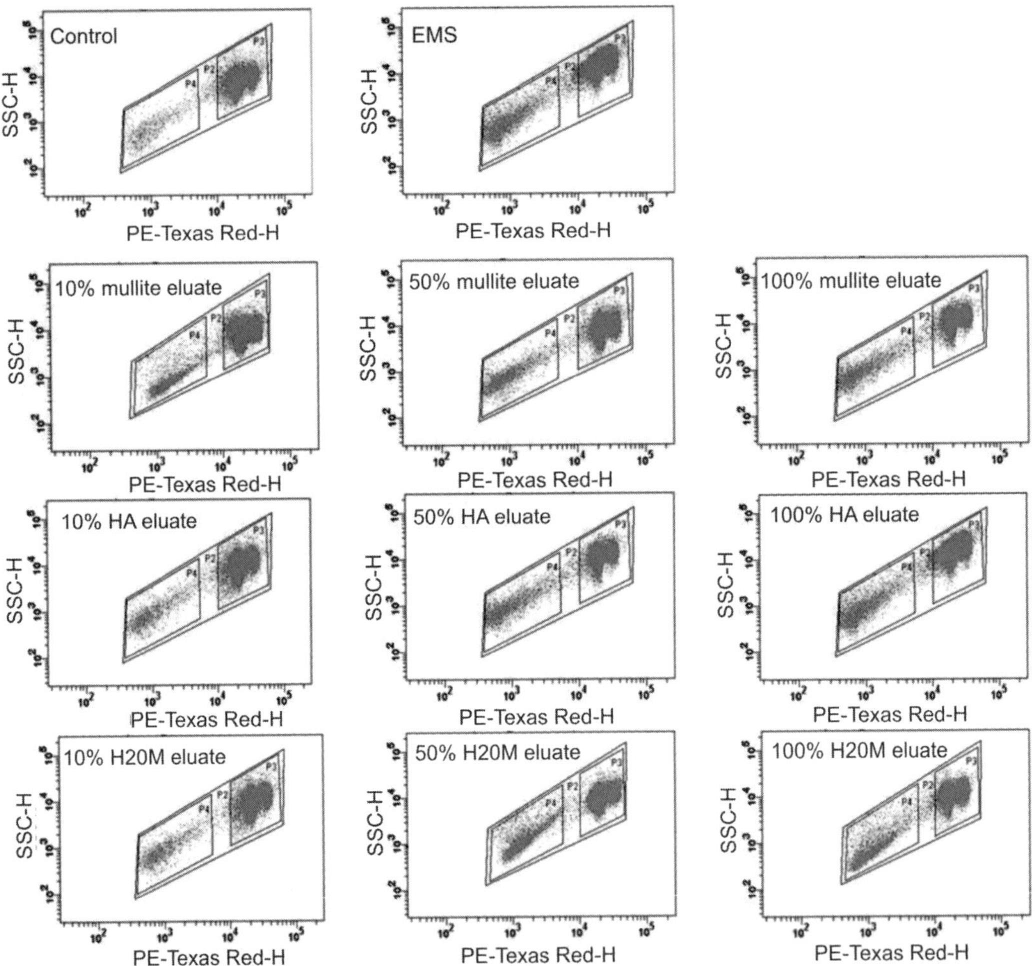

Fig. 13.12 Representative plots showing micronucleus formation (MN), as assessed by flow cytometer analysis of cultured fibroblast cells (L929) treated with H20M (165±40 nm), HA (171±28 nm) and Mullite (87±12 nm) particle eluate of different concentration: 10%, 50%, 100%; for comparison, the results of flow cytometry on control (negative) and positive control (EMS) are also shown; P2 gated for total events of interest (P3+ P4 + non-gated population), P3 gated for total nuclei and P4 gated for total micronuclei[916].

13.5.3 | Analysis of compositional dependent DNA damage behaviour

Mouse fibroblast cell line (L929), connective tissue cell of mesenchymal origin was used as a model cell type to understand the cellular and genotoxic effect of hydroxyapaptite-20 wt% mullite (H20M). It is reported that fibroblast cells play an important role in the osteolytic process at the bone/implant interface as well as in wound healing process[955].

Histocytological studies confirmed the presence of significant number of fibroblasts along with osteoblasts, macrophages and monocytes in the fibrous capsule, that form at the bone/implant interface. This rationalises extensive use of fibroblast cells as a first step to study cytocompatibility.

The fibroblasts participate in immune response by producing and responding to cytokines and chemokines[956,957] that enable the recruitment and retention of bone marrow-derived immune effecter cells. Also, particulate wear debris of phagocytosable size (<10 µm) activate macrophages, fibroblasts and osteoblasts more effectively rather than those of larger sizes[958]. In view of the above mentioned biological role, it is important that the cellular vitality of fibroblast cells be investigated, when exposed to new orthopedic biocomposite, i.e. HA-20% mullite[916]. In the following, we discuss the critical analysis of our experimental results in the context of dose and size dependent genotoxicity property.

The comet assay parameters, in the present case, are determined by measuring the migration of DNA fragments through an agarose gel electrophoresis. On the other hand, MN formation is determined by flow cytometry, a powerful technique that has the capability of analyzing thousands of events rapidly in 3D to provide robust statistics of the deflection of acentric fragments and lagging chromosomes[960,961].

Concerning the cytotoxicity or cell viability of L929 cells upon exposure to H20M, HA and mullite eluates, most interesting information can be summarized now. As expected, HA does not show any sign of cytotoxicity. However, H20M (85±12 nm) nanoeluate causes significant cytotoxicity at 50% or higher concentrations, while much less significant cytotoxicity measured with H20M (165±40 nm) eluate points towards the fact that H20M (85±12 nm) are highly cytotoxic. Further analysis reveals that an exposure of upto 48h causes extremely low value of MTT reduction of 40% or less in case of H20M eluate (87±12 nm) of 50% or more concentration. It is important to note that exposure of 48h does not cause the significant toxicity for H20M (165±40 nm) as 80% or more MTT reduction value is recorded for all doses. Since mullite particles have sizes of 154±34 nm does not show cytotoxicity even after 48 hours of treatment time, the above discussion signifies that the cytotoxicity of H20M composition is more of size dependent rather than entirely composition (mullite) driven.

It has been widely recognized that DNA damage is related to the size of particles and chemical nature of the biomaterial eluate. Two different particle sizes of H20M (87±12 nm and 165±40 nm) were used and the eluate concentration of those is also varied over a wide range up to 100%. The size dependent genotoxicity of H20M clearly reveals that DNA damage is more in the case of particles sizes of below 100 nm. Genotoxicity effect, evaluated in terms of percentage OTM and percentage tail DNA (Table 13.1) reveals that 10% or higher H20M treatment (87±12 nm) shows significant difference with respect to negative control. In contrast, H20M (165±40 nm) eluate does not induct any statistically significant DNA damage (Table 13.2). Importantly, the OTM values for H20M (165±40 nm) eluate are in the range of 1.1–1.5. It is therefore evident that genotoxicity is dependent more on size rather than on composition.

Table 13.2 Summary of the comet parameters, recorded while assessing the DNA damage of L929 mouse fibroblast cells after treatment with varying concentration of nanoparticle/submicron particle eluates of H20M (165±40 nm) composition [values are mean±standard error for three experiments, EMS: ethyl methanesulfonate was used as a positive control, * p<0.05, when compared to respective controls][916].

Nanoparticle H20M (87±12 nm) concentrations	Olive tail Moment (OTM)	Tail DNA (TD)
Control	1.15±0.12	10.70±1.26
10%	2.60±0.17 *	14.86±0.98 *
25%	2.47±0.16*	13.87±0.98
50%	2.59±0.16 *	15.54±1.20 *
75%	2.64±0.21 *	14.70±1.18 *
100%	2.40±0.17*	12.60±0.99
EMS	8.83±0.39*	35.73±1.40 *

Some representative images showing the fragmentation of DNA after treatment with varying concentration of biomaterial eluate are shown in Figs. 13.10–13.11. The difference in DNA damage can also be assessed from the comet shape. The comet patterns from L929 cells obtained at varying eluate concentration reveal regular shaped intact comet head and complete absence of DNA fragments in the form of tail. In case of H20M eluate (87±12 nm), the comet shapes were very similar to that of negative control at lower concentration of 10%. As expected, positive control (EMS) induces extensive DNA fragments, as consistently seen in Figs. 13.10 and 13.11. Although noticeable DNA damage is recorded at H20M eluate (87 ±12 nm) of higher concentration of more than 10%, the extent of DNA fragmentation is lower even at 100% dose, when compared to positive control (see Fig. 13.10). This corroborates well with the quantified data, as provided in Table 13.2. However, H20M eluate (165±40 nm) does not cause any observable change in comet shape at concentration of up to 50%, (see Fig. 13.11).

It is clear from Tables 13.1 and 13.2 that genotoxicity of H20M is dependent both on size as well as inherent chemistry of individual phases of H20M. A comparison of OTM trends, plotted in Fig. 13.10 reveals that significant genotoxicity of H20M (87±12 nm) of 50% dose or higher is necessarily due to mullite. However, the observed genotoxicity at 10% or 25% treatment of H20M (87±12 nm) nanoeluate must be due to size effect, since mullite does not show significant effect at such concentration. Nevertheless, it can be considered that mullite eluate has larger size of 154±34 nm than H20M (87±12 nm). Even at relatively larger particle size, mullite causes DNA damage, while H20M of similar size range does not show any genotoxic effect (DNA damage). As evident from genotoxicity data that H20M shows both size as well as chemical basis of genotoxicity (Table 13.1). Moreover, the chemical basis of genotoxicity is due to individual phases of H20M (HA and mullite). At this point, the limitation of comet assay in evaluating genotoxicity can be mentioned. The total DNA damage induced by particles/chemicals reagents cannot be entirely assessed by comet assay, as the cells undergoing early apoptosis can also show the comets. This is also observed in our experimental results. In Figs. 13.10 and 13.11, it has been shown that the length of tail DNA after treatment with 100% eluate concentration is shorter than that in case of 75% eluate treatment. This gives an indication that cells are possibly in early apoptosis stage in case of 75% eluate treatment.

The chemical basis of genotoxicity is further supported by MN analysis of eluate treatment L929 fibroblast cells, carried out using flow cytometry.

From biological perspective, MN is a small fragment of a genome, separated in the form of chromatinic body. Higher frequency of MN formation in cells treated by particles or chemical reagents can be correlated with greater severity of chromosome loss. Also, higher cellular chromosome breaks, accompanied by a larger MN ratio, is a good indicator of toxicity of biomaterials at genome level. Therefore, MN assay results can be used as signature of chromosomal damage in cells treated with biomaterials.

Mullite inducts statistically higher MN formation at all concentration, except at 10% dose. A critical analysis of Fig. 13.12 reveals that MN formation in contact with H20M eluate is much higher than mullite at an identical eluate dose, except at 50% concentration. Compared to negative control, percentage increase in MN formation always shows higher relative values in case of mullite, in comparison to H20M eluate, except at 75% dose. At 100% eluate, H20M composition inducts 129% more MN formation, while mullite induces 141% more MN in comparison to respective negative control.

The observed genotoxicity can be explained by the presence of mullite ($3Al_2O_3.2SiO_2$), as HA-20% mullite composition shows significant concentration dependent genotoxicity effect on L929 cells *in vitro* after treatment for 6 hours. From the application perspective, it is however necessary to assess the genotoxic effect of H20M in animal model (*in vivo*). The same biomaterial composition in bulk form is found to have good *in vivo* biocompatibility in rabbit model,[916] as discussed in the next section.

13.6 | **Tissue Compatibility,** *in vivo*

For the purpose of *in vivo* experiments in rabbit model, all the sintered pin shaped samples were cleaned ultrasonically and sterilized by Ethylene Oxide (ETO). Similar shaped commercially procured ultrahigh molecular weight polyethylene (UHMWPE) pins were used as control specimens. 12 adult (more than 2 kg weight) Albino rabbits (either sex) were chosen for the experiments. The entire implantation experiments were subjected for three different time periods; i.e., 1 week, 4 weeks and 12 weeks, and 4 rabbits per time duration, according to ISO-10993 guidelines. It needs to be pointed out here that the average time required for integration of orthopedic implants in human is three months, while it is 4–6 weeks in animal models, like rabbits[962]. Also, the faster the bone regeneration takes place at the implanted region; the better is the efficacy of the implant material.

Based on the histopathological analysis of the implant/bone interface, the *in vivo* biocompatibility can now be assessed. Phenomenologically, the bone implant interface gets filled with blood clots during the process of implantation. At the initial stage, the proteins from blood and tissue fluid are adsorbed onto the implant surface and process of healing includes inflammation followed by repair and bone deposition. In view of the above, the discussion largely is focused around the histological changes, in particular neobone formation, inflammation, granulation tissue formation and implant resorption.

In Fig. 13.13, the signature of early stage of osseointegration is provided with a set of histology images. Some salient observation are as follows, mild necrosis, cell debris and cellular infiltrate (black arrow) between host cortical bone (white block arrow) and implant (Fig. 13.13 a), fibrous tissue (white arrow) and fibrocytes (black arrows) at the interface (Fig. 13.13b), new bone trabeculae (black arrow) along host cortical bone (white arrow) and at endosteal opening (Figs. 13. 13c and 13.13d) cellular infiltrate, fibrocytes (black arrow) and macrophages (white arrows) between host cortical bone (white block arrow) and implant (Fig 13.13e) and occasional new bone trabeculae (black arrow) along host cortical bone (white arrow) (Fig. 13.13f). Although more histology images are not shown, the implant/host bone interfaces exhibit similar morphological features with more density over larger spatial extent after 4 and 12 weeks of post-implantation.

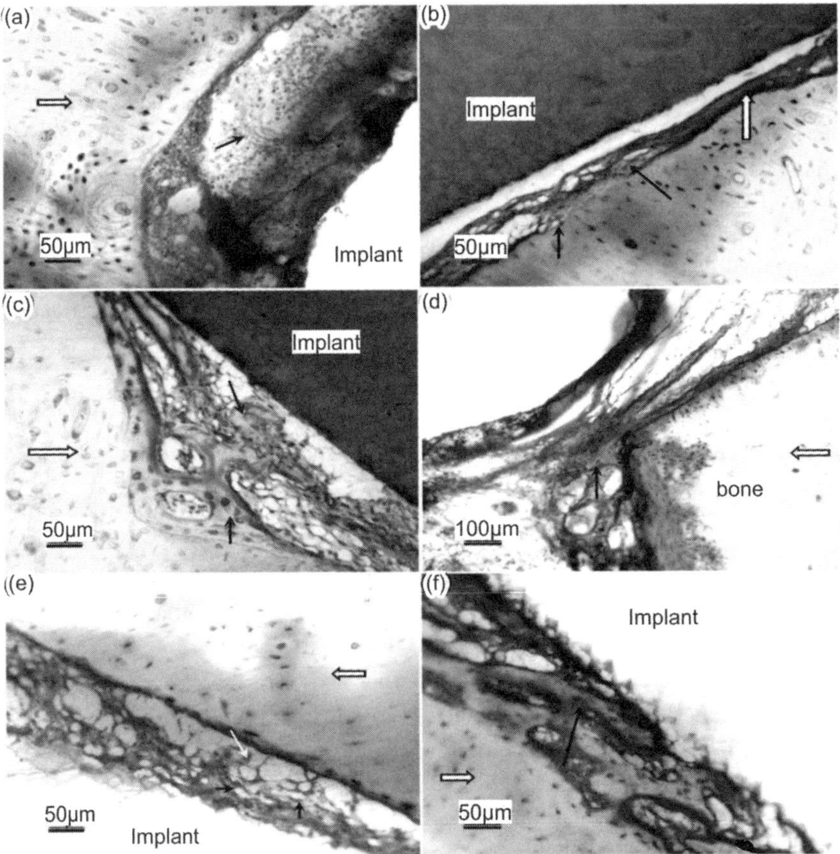

Fig. 13.13 Representative histology images of host bone-implant interface; (a–d) HAp 20 wt% mullite (test implant) and (e and f) UHMWPE (control) after one week of post-implantation[916] [see Colour Plate].

Biocompatibility of a material in bone at twelve weeks post implantation is usually determined by the absence of necrosis and inflammation, particularly macrophages and foreign body type of giant cells (FBG)[963]. The ability of the material to integrate with surrounding host bone, otherwise known as osseointegration, is determined by the presence of newly formed bone at the host bone-material interface.

> The absence of macrophages and FBG cells and deposition of new bone with minimal intervening fibrous tissue, are considered as signatures of good osseointegration.

Since the chemical and physical properties determine the macrophage activation, the formulated composition of the investigated biomaterials i.e. CaP–mullite appear to induce acceptable *in vivo* response. Among many factors influencing *in vivo* response, the surface composition of biomaterial is also important. For example, while 45S5 bioglass (a bioactive material), has been widely considered as a bone analogue material, the canasite glass ceramic formation (SiO_2-CaF_2-CaO-Na_2O-K_2O) are reported to be non-osteoconductive in rabbit as they degrade in the biological environment[964]. With the above discussion, the use of CaP–mullite composite with uncompromised *in vivo* biocompatibility is established.

13.7 | Closure

This chapter demonstrates how to design CaP–mullite composites in line with the existing motivation to enhance physical properties (strength, toughness, etc.) without compromising biocompatibility properties. The extensive results presented in this chapter also illustrate the breadth and depth of the experimental investigation that one has to follow while developing any new bio-ceramic composite. An important observation has been that the dissociation of HAp to TCP depends on both sintering temperature and mullite content. Higher mullite containing composites are more prone towards dissociation to β and α-TCP. The sintering experiments clearly reveal that dissociation of HAp to β/α-TCP is more favoured with increase in mullite content at 1350°C. The combination of large compressive strength approximately 400 MPa and modest SEVNB fracture toughness of ~ 1.5 MPa m$^{1/2}$ indicate that the optimally sintered light-weight CaP–mullite composite material could be a useful bone replacement material. Such toughness value is 2.5 times higher than monolithic HAp. The principal finding of the biocompatibility study is that the cell viability of fibroblast and osteoblast-like cells did not reveal any statistically significant difference in BCP–mullite composites in comparison with pure HA. This confirmed that mullite and additional presence of various phases (alumina, CaO, gehlenite) together with the presence of mullite did not degrade cell viability with respect to pure HA. Also, BCP-based composite substrates with 20% or 30% mullite supported superior osteoconduction than baseline monolithic HA ceramic. The presence of predominant β-TCP phase was found to be suitable for osteoblastic function and phenotypic expression.

Other than cytotoxicity, genotoxicity is also dependent on the size of particles. It has been noticed that HA-20% mullite eluate of size less than 100 nm induces genotoxicity at concentration of greater than 10%, whereas eluate of similar composition of size approximately 150 nm does not cause any statistically significant genotoxicity even at ≥50% concentration. The chromosomal damage, as evaluated using micronucleus assay, shows that the micronuclei formation after treatment with mullite eluate is significantly higher than that of HA and HA-20% mullite composition.

More importantly, CaP–mullite biocomposite exhibits a consistent pattern in deposition of neobone, which appears to remodel over survival period. On the basis of the absence of inflammation, the presence of new bone formation and minimal collagen at the interface, the suitability of the HAp–mullite to be used as bone replacement material for orthopedic applications can be established. Overall, the results of biocompatibility collectively support that hydroxyapatite-20% mullite should be considered as a new class of orthopedic implant material that can be safe at genome level without causing any significant DNA damage and mutagenic effect in fibroblast cells. More details of the genotoxicity related discussion can be found in author's recent book[965].

Case Study: Compression Moulded HDPE-based Hybrid Biocomposites

One of the important aspects related to the polymeric materials for hard tissue replacement applications is to enhance the physical properties (elastic modulus, strength, etc.) without compromising biocompatibility properties. This aspect is discussed in this chapter, while analyzing the published results of another's own research on HDPE-based composites. Through such analysis, it will be demonstrated as how physical properties of HDPE can be enhanced with simultaneous addition of HA and Al_2O_3. The processing approach adopted is compression moulding with 2% Titanium IV, 2- propanolato, tris iso − octadecanoato- O as coupling agent. As part of biocompatibility assessment, the cell adhesion and proliferation of human osteoblast or L929 mouse fibroblast cells are also qualitatively and quantitatively analyzed. The bone healing of HDPE-HA-Al_2O_3 as segmental bone defect in rabbit model is discussed in reference to bone remodeling and neobone formation, *in vivo*.

14.1 | Introduction

Natural bones have the unique capacity to remodel throughout the lifetime of an individual. The functions of bone tissue are to support and to protect the soft parts of various human organs. Bone tissues are of two types: cortical or compact bone and cancellous or trabecular bone.

The bone fracture initiates a complex healing process that includes the cell fate processes of fibroblasts, osteoblasts, etc. The gross instability in large bone defects necessitates stabilization and the healing of segmental bone defects remains a challenge in orthopedic surgery.

The biomaterials based clinical approaches aim to use structural implants and/or materials to replace the defective bone[967], however, major challenge remains the limited availability of bone replacement material with desired property combination to fill the defect and to promote bone growth.[966]

These approaches depend on the recruitment of endogenous osteoinductive factors and migration of osteogenic cells to regenerate the load-bearing segmental bone defect[968,969,970,971]. Even though quite a large number of biomaterial-based approaches have been developed for segmental bone defect repair and used with inconsistent success, none of them has proved ideal till date[972,973,974,975].

> Since natural bone is a collagen (polymer)-hydroxyapatite (ceramic) composite, polymers are one of the natural choices as bone replacement.

In view of the fact that polymer alone has inferior mechanical properties (E- modulus, strength) compared with other materials, the composites are being developed with enhanced mechanical properties[976]. The improvement in the polymer properties can be achieved by either the modification of the structure of the polymer, or the strengthening of the polymer with fibre and/or filler. Polymer-based bioactive ceramic composites are particularly attractive, because their mechanical behaviour can be tailored to match the properties of natural bone[977,978]. Additionally, these materials conform to the bioactive criterion, which is regarded as an essential pre-requisite for the bone bonding mechanism between implant and host tissue.

High density polyethylene (HDPE) is an important biocompatible semi-crystalline thermoplastic material with significant ductility. In view of the fact that natural bone is an anisotropic complex composite material, the semi-crystallinity in polyethylene has been suggested to provide the direction dependent characteristic properties, when used in a composite with bioactive Hydroxyapatite (HA)[979].

HA has been well recognized as the foremost biomaterial with orthopedic and dental applications due to its excellent biocompatibility, bioactivity and osteoconductivity[980]. The applicability of HA/HDPE composites in orthopedics has been clinically realized as low load bearing implants[981,982]. The reliability in terms of long term clinical survival is however questionable, because both the phases in this composite are only physically interlocked by the shrinkage of polyethylene during cooling from their processing temperature[983]. To this end, the results of a new class of HDPE composite are discussed and in processing such composites, the use of coupling agent will be mentioned.

Apart from HA, alumina is selected as a second reinforcing phase to HDPE, because of its excellent properties such as high hardness, elastic modulus, wear resistance, good biocompatibility and stability in physiological environments. A number of studies have established *in vivo* biocompatibility property of Al_2O_3. For example, different types of tissue reactions are reported around Al_2O_3 implants[984], while other researchers found mineralized bone at the interface[985]. Davies et al.[986] also reported that the formation of a fibrillary extracellular matrix on the surface of alumina implants.

The results discussed in this chapter will establish the cell-level and tissue-level compatibility of $HDPE-HA-Al_2O_3$ composites, while the efficacy of ceramics filler together with titanate coupling agent towards physical property enhancement will be demonstrated. More discussion on the results can be found elsewhere[987].

14.2 | Physical Properties

The compression moulding conditions, i.e., the combination of temperature of 130/140°C with curing time of 30/45 minutes and a pressure of 10 MPa were found to be optimum to consolidate HDPE-20% HA- 20% Al$_2$O$_3$[987]. In order to improve the mechanical properties of these composites, especially with the improved interfacial bond strength, the use of coupling agent was found to be a better option. Since the polyethylene is non-polar and hydrophobic, the introduction of polar group can effectively increase the affinity between polyethylene and HA. The interaction between a polymer matrix and filler can be mechanical or chemical or both; however the chemical bonding has important implication in improving the properties[983].

In the field of polymer-based composites, coupling agents (CA) have been shown to have an important contribution in promoting the molecular bonding and consequently, the improved interfacial adhesion[988]. Organo titanates and zirconates are two surface modifiers with similar structures that can be used as modifying filler[997]. Among all these, titanates are the fillers which can be used frequently, while zirconates are advantageous for certain applications[989] like grafting, etc. Several studies[990,991,992,993,994] demonstrate the possible application of titanate or zirconate surface modifiers[995,996] or zirconate alone or in combination with other coupling agents, to a wide range of composite systems[997]. Based on the chemistry and the molecular structure of HDPE as well as for the purpose of establishing chemical coupling between polymeric chains and ceramic fillers (HA or alumina), the titanates can be selected as coupling agents.

> The coupling agents play an important role in establishing the interfacial interaction during processing of polymer-ceramic hybrid composites.

In terms of processability to cylindrical shaped compact without any visible porosity, 2 wt.% of CA along with isopropyl alcohol (IPA) as a solvent (10 wt.%) was found comparatively better among all the trial batches. At a higher amount of coupling agent, the thermoplastic behaviour of HDPE was enhanced and therefore, the consolidation becomes difficult.

The analysis of FT-IR results (Fig. 14.1) provides more information on chemical bonding between ceramic fillers and polymer molecules. A comparison of FT-IR spectra obtained with polymer blend and compression moulded compact reveals that the intensity of FT-IR peak at 3417 cm^{-1} for OH$^-$ band changes during the compression moulding of composite mixture. The peak at 3417 cm^{-1} (–OH) from polymer backbone chain, those at 1093, 1027, 604 cm^{-1} for PO$_4$$^{3-}$ and peaks at 632, 3571 cm^{-1} for O–H bending of HA, and peak at 1726 cm^{-1} for C=O are clearly visible. This might be due to oxidation, which takes place during the compression moulding at ambient atmospheric conditions, producing carbonyl groups in the polymer-ceramic system. It is possible that (HA) Ca^{+2}– $^-$$_2O_2$P (O)–(CH$_2$)$_2$–OC–O–(Polymer) bridges as well as H-bonding are formed at the interface of polymer[998]. It can be said that the formation of monomolecular layer is formed as a result of alcoholysis reaction with the coupling agent. The surface active protons play a good role in this mechanism. Ti nucleus might act as electrophobic or electrophilic moiety, causing a catalytic rearrangement and redistribution of the molecular structure and molecular weight of polymeric phase. The prime advantage of titanate derived coupling agents is their reaction with free protons at inorganic interface with the formation of molecular mono-layer.

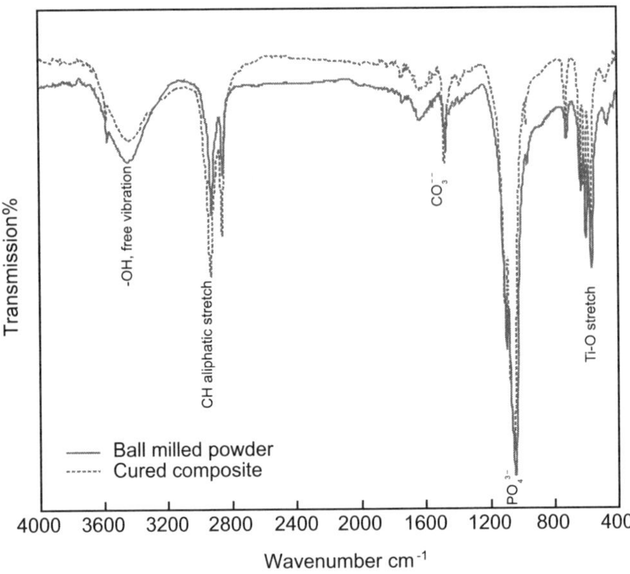

Fig. 14.1 Comparative FT-IR spectra of the ball-milled powder and compression moulded composite[987].

The E-modulus was measured using impulse excitation method and the dynamic elastic properties analysis. For the dynamic modulus measurements, the samples of known dimensions were impacted upon by a standard steel rod close to the high frequency sensor of the instrument to induce the vibrations in the sample[999]. For the disc shaped samples, two measures of dynamic Young's modulus, E (E_1 and E_2) are made independently from the two resonant frequency measurements, and then a final value E is determined by averaging the two calculated values E_1 and E_2, as per the following formulation,

$$E_1 = \left[37.6991 f^2_1 D^2 m(1 - \mu^2) \right] / (K_1^2 t^3) \tag{14.1}$$

$$E_2 = \left[37.6991 f^2_2 D^2 m(1 - \mu^2) \right] / (K_2^2 t^3) \tag{14.2}$$

$$E = (E_1 + E_2)/2 \tag{14.3}$$

where, E is Young's modulus; E_1 is the first natural calculation of Young's Modulus; E_2 the second natural calculation of Young's Modulus; f_1 is the first natural resonant frequency of the disc; f_2 is the second natural resonant frequency of the disc; D is diameter (mm) of the disc; m stands for mass (g) of the disc; μ is Poisson's ratio for the specimen as determined ; K_1 is the first natural geometric factor ; K_2 is the second natural geometric factor; t is the thickness (mm) of the disc.

From the dynamic elastic modulus values (Fig. 14.2), better properties in the presence of coupling agent can be again confirmed. The E- modulus values presented in Fig. 14.2 should be used for comparison purpose among the investigated composites only and should not be used for comparing with other composites or natural tissue/bone, where different test methods are used. For sample CS1, the dynamic elastic modulus was found as 0.96 GPa and it rises up to 8.5 GPa–11.3 GPa for the compositions filled with ceramic fillers without and with Titanium IV, 2-propanolato, tris isooctadecanoato-O, i.e., in samples CS6 and CS7, respectively. Likewise mentioned above, the statistical analysis of the data was performed using SPSS-13 software.

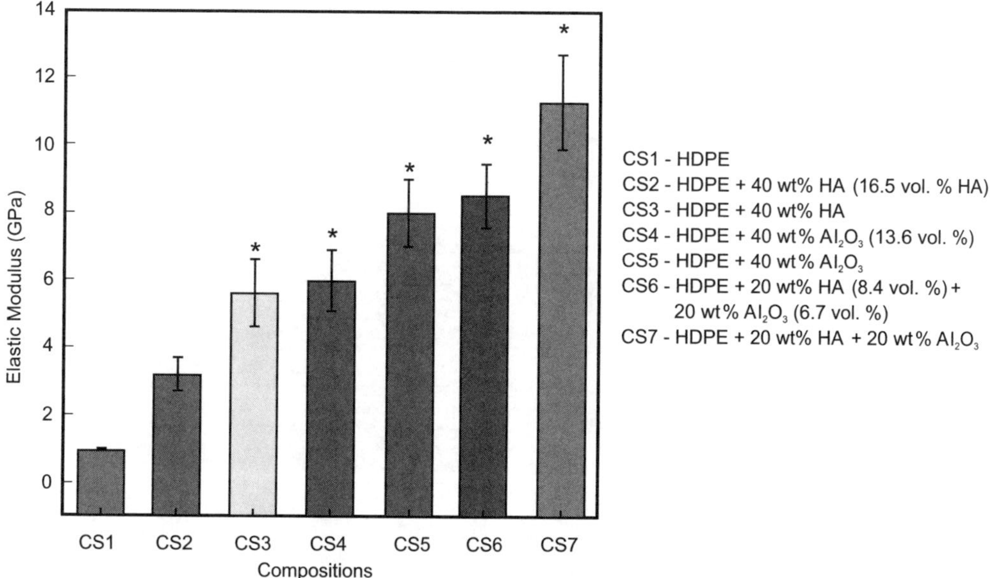

Fig. 14.2 Variation of elastic modulus of various HDPE composites; the asterisk mark (*) represent significant statistical difference at $p < 0.05$[987].

14.3 | Cytocompatibility Property

In order to assess the cell-level compatibility of the developed polymer-ceramic composites, cell culture study using mouse fibroblast cells (L929) and human osteosarcoma (SaOS2) cell lines was carried out under scientifically controlled condition as per ISO-10993-1 guidelines (Biological testing of Medical Devices-Part 1: Guidance on selection of Test (ISO 10933-1).

Figure 14.3 clearly shows that the SaOS2 cells were significantly spread with their multi-directional and elongated actin filament (filopodia) extension. Overall morphological features of SaOS2 cells suggest evidences of good cell-to-cell attachment and cell adhesion on composite surfaces. A closer observation of SEM images in Fig. 14.3 also reveals the formation of extracellular matrix (ECM). It is known that ECM provides support to anchorage-dependent cells as well as regulates the intercellular communication, and the dynamic behaviour of the cell. A closer observation of SEM images (Figs. 14.3) clearly reveals that apart from lamillipodia extension, the cell adhesion and spreading were facilitated through their filopodial extension on the substrates. In all the cases, cell-to-cell contact is clearly evident, which forms cellular network. The filopodia morphology is the evidence of healthy and migrating cell. The composite surface encourages similar cell behaviour due to the fact that composite substrate supports cell adhesion and proliferation. The flattened cells increase the cell-material interface area and hence, the biocompatibility of the developed composites is realized. Overall, Fig. 14.3 illustrates the expanded morphology of human osteoblast cell line on the prepared polymer-ceramic composites. Human osteoblast, SaOS2 cells, symbolizes fully mature osteoblastic cell morphology with wide spreading and good proliferation on the HDPE + 20% HA + 20% Al_2O_3 composite.

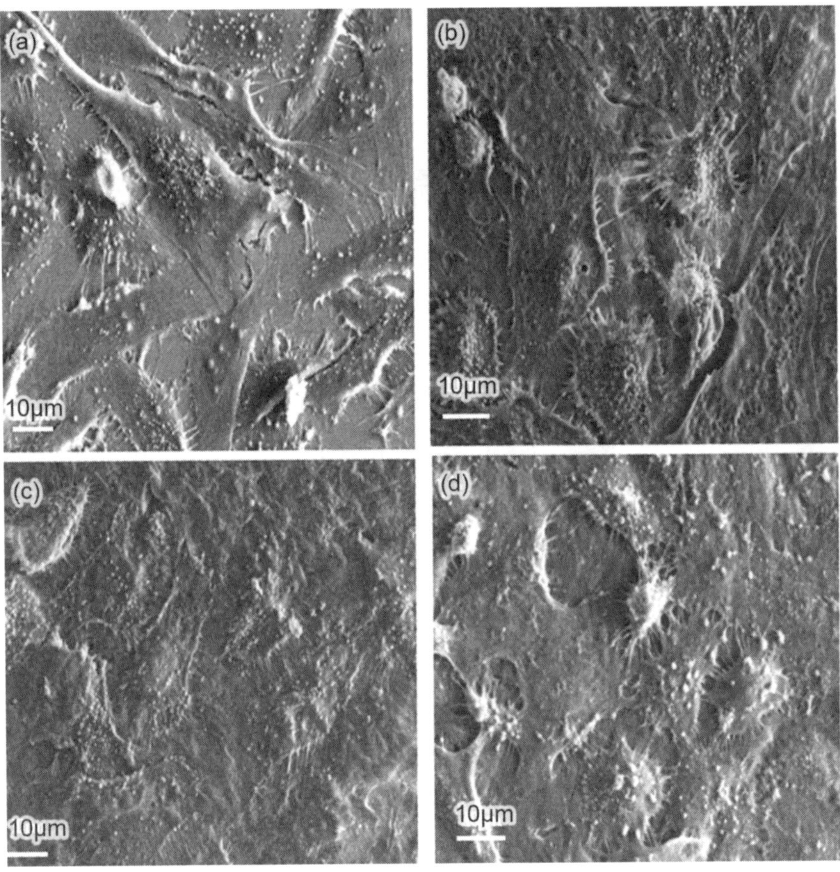

Fig. 14.3 Scanning electron micrographs revealing cellular functionality of SaOS2 human osteoblast cells (after incubation of 72 hours) on various sample surfaces: (a) HA, (b) HDPE + 40 wt% HA, (c) HDPE + 40 wt% Al_2O_3, (d) HDPE + 20 wt% HA + 20 wt% Al_2O_3[987].

In order to confirm cell viability, MTT assay results with SaOS2 cells are analyzed. MTT results in Fig. 14.4 confirm that the chemically coupled composites can support the metabolically active osteoblast cells and cell viability increases with incubation time. The mean optical density increases with the incubation period for all the analyzed samples. Also, the composite compositions CS3, CS5 and CS7 showed the significant difference in the cell density among the 3, 5 and 7 days of incubation periods. Therefore, the investigated materials can be used as a good substrate to support adhesion and proliferation of mitochondrically active osteoblast-like cells. MTT values obtained with L929 and SaOS2 cells, in combination, establish good cytocompatibility property of these composites in the context of bone replacement applications.

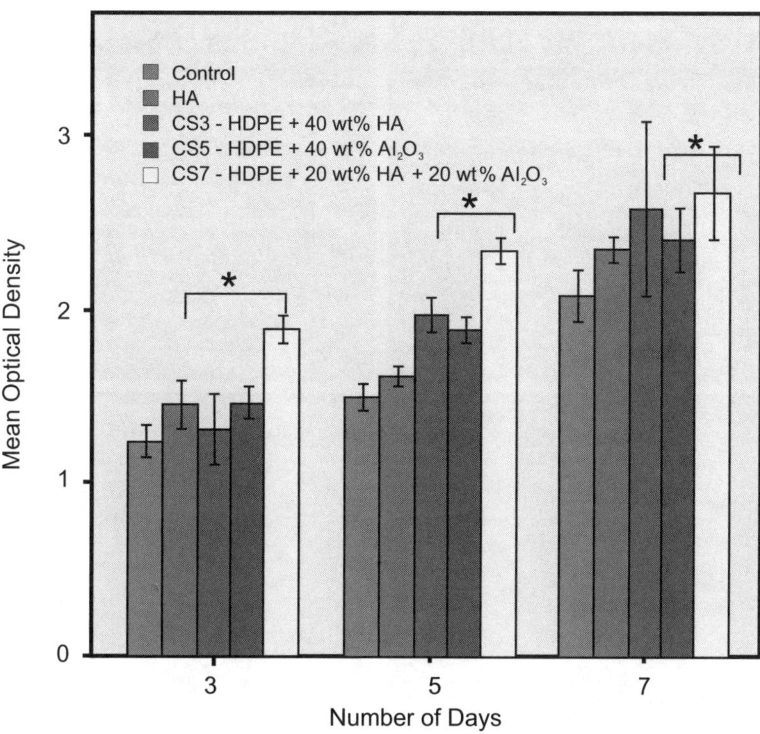

Fig. 14.4 MTT assay results showing the SaOS2 cells on control, HA and HDPE based hybrid composites after 3, 5 and 7 days of culture; *represent significant difference at $p<0.05$ with respect to compositions and Error bars correspond to +/- 1.00 SE for number of days of culture[987].

14.4 | Live/Dead Staining of Cells Treated with Finer Eluates

Next, the assessment of osteoblast proliferation in contact with the eluates of high density polyethylene, hydroxyapatite and alumina particles at 0.01, 0.1 and 1.0 wt% in DMEM culture medium is discussed. Some representative results after 72 hours of culture are presented in Fig.14. 5. A consistent increase in cell density, irrespective of eluate dose, was observed. The cell proliferation was found to be increased with higher concentration of eluates. Compared to the control culture (without particles), the cells cultured in the conditioned media were healthy. In order to evaluate the influence of the extremely low eluate concentrations on cellular functions, we reproduce all this set of experiment with 10 and 100 folds of initial eluate concentration, accurately.

In conclusion, it can be said that the live/dead assay showed high levels of cell viability as shown by green staining after treatment with different concentrations and this is comparable to the tissue culture plastic control (Fig. 14.5). Also, the cells were observed to be spread and polygonal in morphology, similar to those on the control surface. Cellular shape is closely related to the production or distribution of extracellular matrix proteins, such as type I collagen[1000]. These results are further correlated using cell viability assay and mineralization assay.

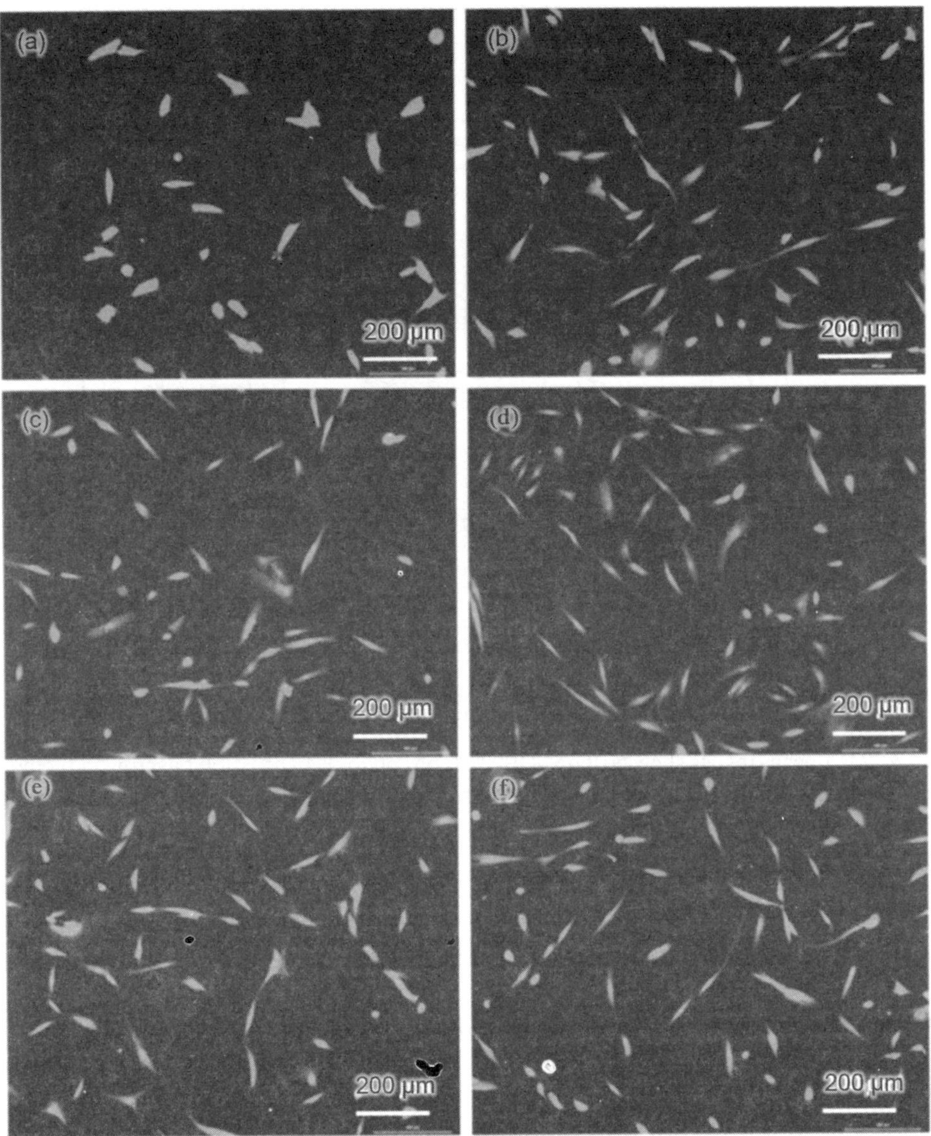

Fig. 14.5 Live/dead staining of HOB cells on (a) control (24 hours of culture), and (b) control (c) HDPE + 40 wt% HA (0.01 % dose) (d) HDPE + 40 wt% HA (1 % dose) (e) HDPE + 20 wt% HA + 20 wt% Al_2O_3 (0.01% dose) and (f) HDPE + 20 wt% HA + 20 wt% Al_2O_3 (1% dose). (Figs. 2b-f, all after 72 hours of culture)[987].

14.5 | *in vivo* Biocompatibility Property

The osseointegration of the HDPE-HA-Al_2O_3 based implant material with the host bone is assessed over 14 weeks of implantation using long segmental defect model.

An average time required for integration of orthopedic implants in human is 3 months and 4-6 weeks in animal models, like rabbit.

The cylindrical shaped (2 mm in diameter and 8 mm in height) test samples (HDPE-based biocomposites) were implanted on the left leg tibia bone without any synthetic bone cement (see Fig. 14.6a) and autologous bone segment (control) on the right leg tibia bone in rabbit model (see Fig. 14.6b). The rabbit was chosen as an experimental model on the basis that it has a very well defined cortical bone as well as Haversian system and the stimulation of bone remodelling after injury is more similar to that of other larger animals. From the implant perspective, a non-toxic surface is essential for the migration, attachment and proliferation of the osteoblasts *in vivo* [1001,1002] and this has been assessed below primarily using histology images.

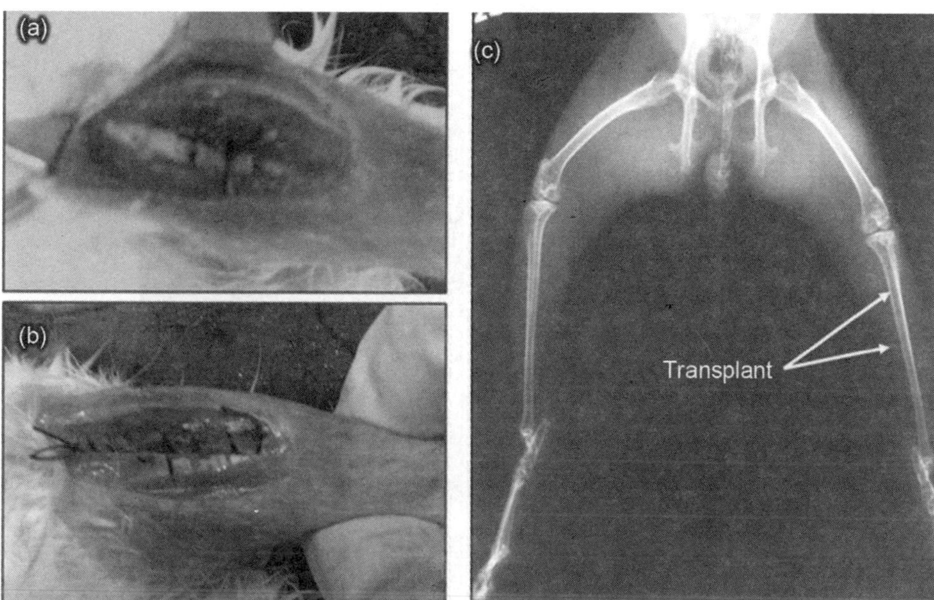

Fig. 14.6 Digital camera images showing (a) rabbit's tibia zone with cylindrical long segmental defect together with the biocomposite material before implantation, (b) rabbit's tibia zone after implantation, (c) X-ray radiographs of post-implant rabbit's tibia after 14 weeks of implantation of HDPE + 20 wt% HA + 20 wt% Al_2O_3 implant[987].

The physical activity of all the animals as well as external appearance of the implantation sites were monitored from time to time during short term implantation. The physical conditions of all the animals were normal throughout the experiments with normal feeding. There was no behavioural change or any abnormality of any of the experimental animals. X-ray radiography image reveals that the implant was radiopaque and could be distinctly visible after 14 weeks of implantation. Overall, the healing of segmental defect using synthetic material is comparable to that of autologous bone. Figure 14.7 represents the histological images of the bone/implant interface, which were captured using thin sections. The post-implantation common observation at both endosteal and periosteal spaces around the implant includes the bone ingrowth.

Figures 14.7a–d present the photomicrographs of HDPE-20 wt% HA-20 wt% Al_2O_3 at 14 weeks of post-implantation in the cortex region of host cortical bone. Bone forming osteoblast cells and new woven bone deposition are clearly evident along both sides of host cortical bone in Figs. 14.7b and c. Bone matrix is produced by osteoblasts. A number of osteoblasts are embedded into the matrix during bone formation and these cells become relatively inactive, which appear as osteocytes (Fig. 14.7c). Besides osteoblasts and osteocytes, two other cell types are present in bones: the bone lining cells, which can be considered as inactive osteoblasts on the surface of the bone and the osteoblasts, the bone resorbing cells.

> Morphologically, osseointegration is realized with direct contact between host bone and implant surface without an interposition of soft tissue.

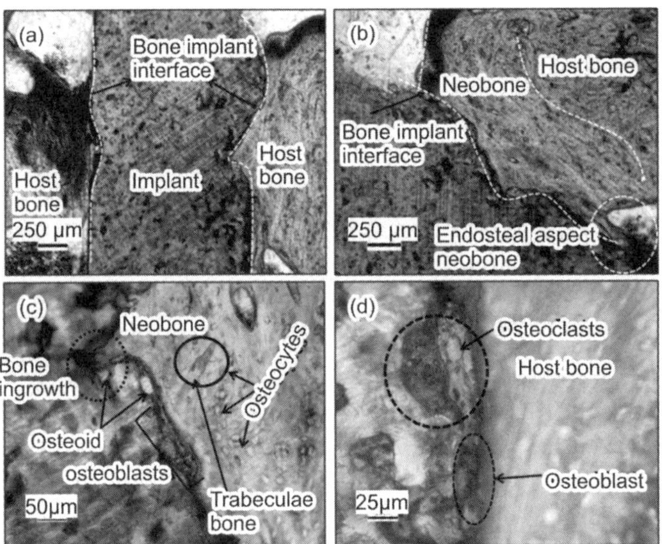

Fig. 14.7 Histology image of (a) host bone/implant interface (HDPE-20 wt% HA-20 wt% Al_2O_3), 14 weeks post-implantation, (b) neobone/host bone interface at endosteal aspect, (c) neobone and trabeculae bone, along with osteoids and osteoblasts, and (d) osteoblasts and osteoclasts together at bone/implant interface[987] [see Colour Plate].

Shors and Holmes[1003] stated that an ideal bone graft substitute would mimic the natural bone and help in the stimulation of osteoblast cells to differentiate and release the mineralizing matrix factors. The histological observations in the present case suggested that there was a strong tendency to rebuild the bone structure at the interface after implantation, independently of the implant composition. However, the direct bone bonding with the implant depends on the bioactive nature of the interface. The absence of macrophages and foreign body type of giant cells (FBC)[1004] at 14 weeks post-implantation around HDPE-HA-Al_2O_3 and deposition of new bone at the interface with minimal intervening fibrous tissue indicates good biocompatibility property of the material with excellent osseointegration property. Like other composite compositions, degenerative and necrotic changes were not observed at implant site (Fig. 14.7). There was no fibrous tissue and inflammatory cells at the host bone/implant interface. After 14 weeks (Fig. 14.7d), the empty spaces between the HDPE-20 wt.% HA-20 wt.% Al_2O_3 implants were almost occupied by de novo bone and small bone islands were observed at the site. After 14 weeks of post-implantation, new woven bone deposition was found around the implant at both sides of host cortical bone with the presence of osteoid and osteoblasts at bone/implant interface. As osteoclasts are also present at the implanted site, such observations indicate the process of bone remodelling. The observation of bone integration as well as the absence of any significant inflammation around the implantation site reveals that the investigated material can be considered as safe for biomedical applications[1005].

14.6 | Closure

As a concluding note, the results discussed in this chapter unambiguously confirms two important aspects, which are significant as far as the biomedical applications of the investigated HDPE-HA-Al_2O_3 biocomposites are concerned. First, the particles released due to long term use of such materials at articulating surfaces *in vivo*, when present in a lower concentration would not cause any *in vitro* toxicity of human osteoblast-like cells. It has been reported in a recent study[1006] that the HDPE-UHMWPE-TCP based hybrid composites can support the cell attachment, proliferation and mineralization of the human osteoblast cells. Second, the *in vivo* implantation of compression moulded HDPE-20 wt% HA-20 wt% Al_2O_3 as large segmental defect in rabbit model reveal bone cell activity as well as osseointegration with neobone formation adjacent to the host bone/implant interface, as evident from histopathological analysis. This is possibly a first proof of the *in vivo* biocompatibility of this composite system. It can be further noted that the *in vivo* studies on PE-based materials are scarce and in one such studies, Belanger and Marrois[1007] reported the histological results after implantation of low-density polyethylene (LDPE) in the paravertebral muscles of rabbits for 7, 30 and 90 days. Although LDPE implant exhibited mild inflammatory response during initial period of implantation of 30 days, the inflammation level is decreased after 90 days of implantation. The fibrous tissue encapsulation has been observed around implant/bone interface. In view of the above, the present results are important in the sense that despite addition of alumina to HDPE-HA based composites, the implant material exhibits favourable *in vivo* biocompatibility response in rabbit model. More details of the results summarized in this chapter can be found in a number of research papers published by author's research group in last one decade[987].

Case Study: Phase Stability, Bactericidal and Cytocompatibility of HA–Ag

In the context of prosthetic infection, the addition of anti-microbacterial agents (e.g., Ag, ZnO) to a biomaterial (e.g., HA) is considered as a clinically relevant approach. To this end, this chapter discusses a comprehensive part of author's own research results, published widely in last one decade, to unravel the influence of silver addition on the phase stability, bactericidal and the cell compatibility properties of HA composites. Two processing approaches are utilized with one involving wet chemical synthesis of Ag-doped HA and the other involving mechanical mixing (ball milling) followed by conventional sintering of HA–Ag composites. The extensive phase stability results, obtained using Raman spectroscopy and optical absorption spectroscopy are discussed together with conductivity measurements using impedance analyser. Apart from Ag doping, the influence of other heterovalent cations, like K^+ or Mg^{2+} on phase stability or conductivity property is also discussed. In the context of biomedical applications, the compositional dependence of the bactericidal and cytocompatibility property of Ag-doped HA (chemical synthesis route) are qualitatively and quantitatively established. The effectiveness of Ag addition to HA matrix (ball milling followed by sintering) is also established in the context of bactericidal property. Importantly, the specific compositions of chemically synthesized $Ca_{10-x}Ag_x(PO_4)_6(OH)_2$ $[x = 0.3, 0.4, 0.5]$ exhibit a reduction in mouse fibroblast cell viability, while HA–10% Ag composite (mechanically mixed and sintered) exhibits good cell adhesion and proliferation without any sign of cellular toxicity.

15.1 | Introduction

Impact associated bacterial infections are considered as a significant factor, which can lead to complication after implantation. Such problem, if results in extreme clinical complication usually requires revision surgery or removal of infected implants. From the clinical perspective, the bacterial adhesion to biomaterial surfaces is an essential step in the pathogenesis of these infections[1008].

> The clinical success of an implantable biomaterial depends not only on the bone-implant integration, but also on the resistance against bacterial infections.

Although clinicians perform surgery under the most sterile condition with sterilised implants, the bacterial infection can also occur during wound repair after surgery. Therefore, antibacterial properties in combination with good osseointegration property are highly desired for implantable orthopedic biomaterials.

Calcium phosphate-based biomaterials offer a great potential in many biomedical applications, due to their acceptable biocompatibility[1009,1010]. Among them, hydroxyapatite (HAp) [$Ca_{10}(PO_4)_6(OH)_2$] has been extensively researched as implantable orthopedic biomaterial, because of its structural and chemical similarities with the bone mineral composition[1011,1012,1013,1014]. It is stable in physiological pH solution and is known for its excellent bioactivity, good osseointegration and osseoconductivity[1015,1016]. As implantable biomaterial, it is mainly used in bulk porous form for filling in or reconstructing bone defects, and as a thin coating on metals, titanium and CoCrMo alloys, for hip, knee, and dental prostheses. Despite all its advantages, hydroxyapatite (HAp) does not show antibacterial properties. Therefore, the researchers attempted to improve the antibacterial properties of hydroxyapatite (HAp) by substitution of Ca^{2+} ions with several other ions like Ag^+, Cu^{2+}, Zn^{2+} and Ce^{3+}.

The antibacterial properties of silver have been used in different biomedical fields for a long time[1017]. It is well known that silver ions (Ag^+) and silver (Ag)-based compounds are capable of showing bactericidal effects on as many as 12 species of bacteria, that cause various human diseases[1018,1019,1020].

The combination of silver and hydroxyapatite (HAp) in the form of implantable orthopedic biomaterial seems to be an interesting approach to prevent bacterial adherence on biomaterial surface with good hard tissue regenerative properties. In addition to potential improvement of antibacterial properties, silver incorporation within the hydroxyapatite (HAp) matrix provides good mechanical properties[1021]. Lee et al.[1022] reported the comparison in terms of microstructure and fracture resistance property of HAp–5% Ag composites, fabricated using spark plasma sintering of electroless deposited powders. In this composite, nano-sized Ag particles of less than 200 nm in diameter were homogeneously dispersed in the HAp matrix without microcracks.

This chapter will demonstrate how to determine the critical/threshold level of silver doping in HA matrix to obtain a good combination of bactericidal and cellular functionality. It is important to note that the antibacterial property depends on the distribution and amount of dopant phase, apart from the type/chemical composition of dopant. To this end, silver doped hydroxyapatite (HAp) is synthesized using wet chemical method and the structural studies and properties of all the chemically doped samples are compared with that of mechanically mixed powders. The present chapter also discusses the influence of sintered HA–10% Ag (mechanically mixed powders) on the cytocompatibility and bactericidal property, *in vitro*. The results discussed in this chapter are the summary of the author's published papers[1023] on this topic.

15.2 | Structural Stability of Wet Chemically Synthesized Ag-doped HA

Raman spectroscopy provides the information about the structural transformations of the calcium phosphates due to the effect of ionic environment on the vibrational properties of the phosphate group.

In some of the earlier chapters, XRD results are discussed while analyzing the phase stability of HA composites. Since CaP – compounds are Raman active, here the structural stability in Ag – doped HA is discussed in reference to Raman spectroscopy results. The careful analysis of Raman results would provide critical insights into the role of these cations (K^+, Mg^{2+}, Ca^{2+}) on phase stability, while a comparison is being made with baseline HA. The Raman spectroscopic studies of the parent and doped compositions of hydroxyapatites together with Lorentzian fittings were used to find out the exact position of the peaks. While analyzing such results, it will be demonstrated as how the Raman shift of the mode frequency can be used to assess the local environment of the phosphate groups and to characterize the local structure of the calcium phosphate phases[1024]. Figure 15.1 presents Raman spectrum of the $Ca_{10-x}A_x(PO_4)_6(OH)_2$ (x=0.0, 0.5) (A = Ag, K, Mg), sintered at 1200°C together with the Lorentzian fitting of four distinct bands in the frequency range 425–450 cm^{-1}, 575–625 cm^{-1}, 955–965 cm^{-1} and 1020–1080 cm^{-1}. These bands are associated with the internal modes of phosphate ions surrounded by the Ca^{++} and OH^- ions in the hydroxyapatites. The isolated phosphate group has T_d symmetry and its 15 vibrational modes are $A_1+E+T_1+3T_2$, where A corresponds to a single degenerate band, E corresponds to a doubly degenerate band and T belongs to a triply degenerate band.

The major nine modes, which have non-zero frequencies, are A_1+E+2T_2[1025]. The symmetric stretching mode (v_1), due to stretching of P-O corresponds to A_1 representation and isolated phosphate group, frequency is 938 cm^{-1}[1026]. The doubly degenerate mode (v_2), due to the bending deformation of O-P-O bonds corresponds to E and for isolated phosphate group, the frequency is 420 cm^{-1}[1026]. The triply degenerate asymmetric stretching mode (v_3) involves P motion, which corresponds to T_2 representation and results in Raman band at the frequency of 1017 cm^{-1}. Other triply degenerate mode (v_4) corresponds to the bending deformation of O-P-O bonds at a frequency of 567 cm^{-1}. The local environment of the phosphate groups in the calcium phosphates affects the internal modes of phosphate groups and symmetric, asymmetric and bending modes shift towards higher frequency range or lower frequency range. For example, the most intense symmetric stretching mode frequency was found to be 917 cm^{-1} for octacalcium phosphate (OCP), 948 cm^{-1} and 970 cm^{-1} for tricalcium phosphate (TCP) and 961 cm^{-1} for hydroxyapatite[1027,1028,1029]. Summarizing, the critical analysis of Fig. 15.1 reveal marked signatures of distinct phase dissociation of HA to TCP/OCP in doped HA.

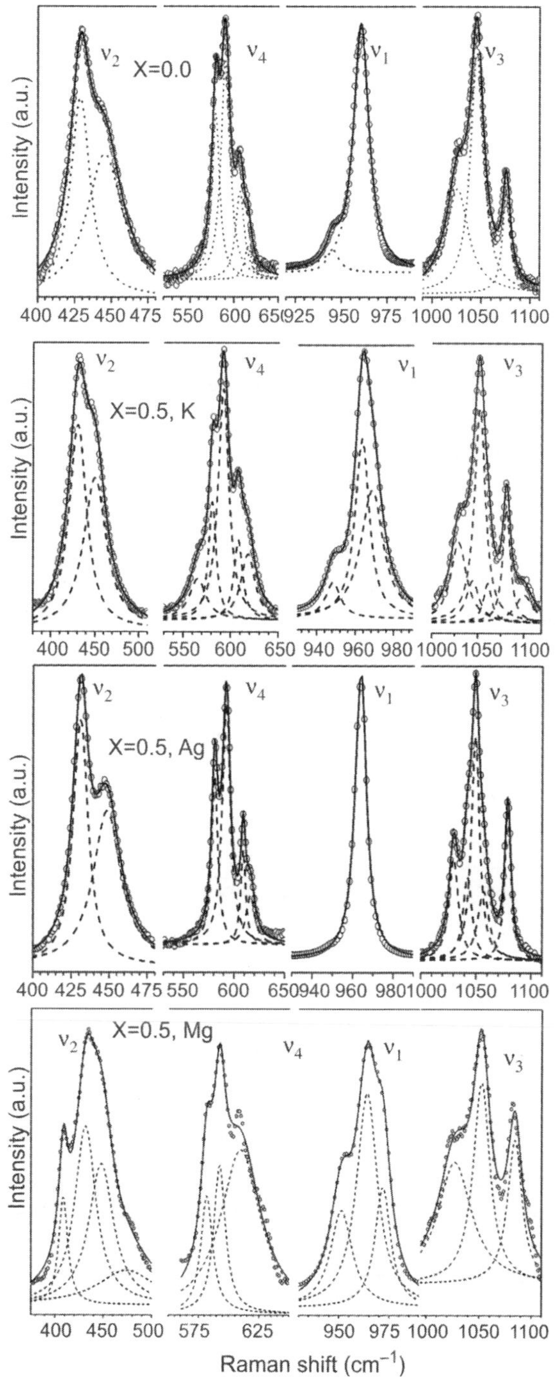

Fig. 15.1 Lorentzian fitting of Raman spectra of $Ca_{10-x}A_x(PO_4)_6(OH)_2$ (x = 0.0, 0.5) (A = Ag, Mg, K), sintered at 1200°C[1023].

15.3 | Electrical Conductivity of Wet Chemically Synthesized Ag-doped HA

One of the advantage of metal – ion doped HA is that one can tailor the electrical conductivity property by tailoring dopant content. The influence of $Ag^+/K^+/Mg^{2+}$ doping to HA on conductivity will be discussed now. In case of Ag-doped HA, one has to confirm the ionic state of silver in the chemically synthesized/doped HA as it will have an impact on the transport properties. For this, optical absorption data for $Ca_{9.5}Ag_{0.5}(PO_4)_6(OH)_2$ compositions in the wavelength range of 200–900 nm are shown in Fig. 15.2. The silver doped compositions show a broad absorption band starting from ~400 nm to 700 nm and a sharp absorption peak at ~220 nm. It has been reported that the metal nanoparticles of the (silver, copper and gold) have a broad absorption band in the visible region of the electromagnetic spectrum[1030,1031,1032]. In our observation, the 1000°C sintered x = 0.5 composition shows the sharp absorption peaks at ~221 nm and ~367 nm. These peaks are similar to those observed in aqueous solution of Ag^+ ions[1033,1034]. The absorption peaks at lower wavelength correspond to isolated Ag^+ ions. Such absorption was ascribed to parity-forbidden transitions involving electron promotion from the $4d^{10}$ ground state to some levels in the $4d^9\,5s^1$ configuration[1035]. The absorption features of x = 0.5 doped samples suggest the presence of Ag^+ ions in 1000°C sintered samples. The presence of Ag^+ ions in 1000°C sintered $Ca_{9.5}Ag_{0.5}(PO_4)_6(OH)_2$ motivated further to explore the ionic conductivity of these samples.

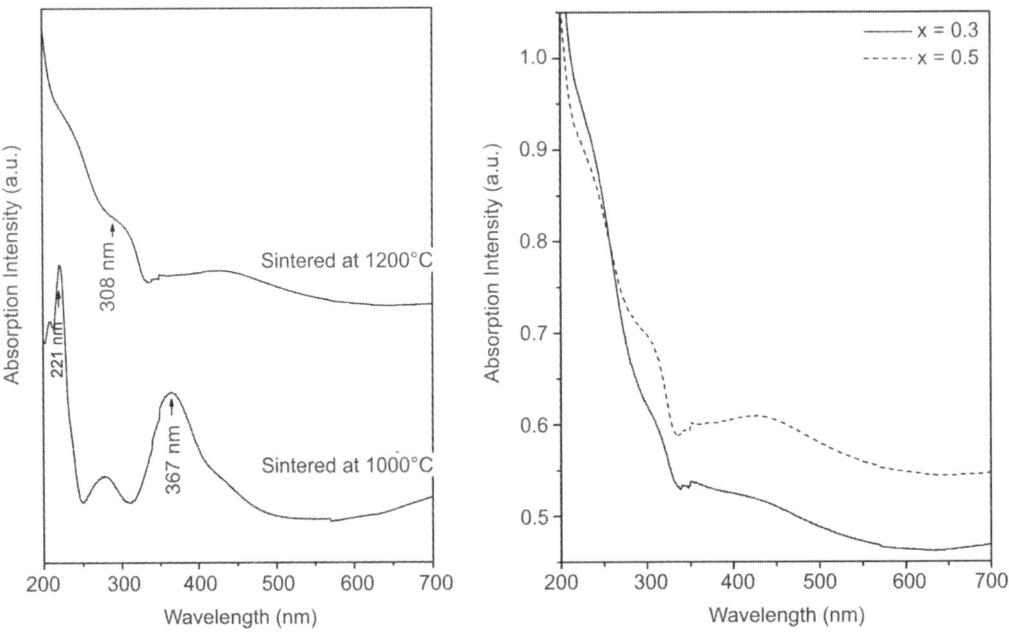

Fig. 15.2 Optical absorption spectra of $Ca_{9.5}Ag_{0.5}(PO_4)_6(OH)_2$, sintered at 1000°C and 1200°C[1023].

Several groups have explored the electrical conduction in HA and predicted the conduction mechanisms based on the activation energies calculated for higher temperature regions[1023]. Figure

15.3a shows the temperature dependent AC conductivity (100 kHz frequency) plots for the parent and $(Ca_{9.5}A_{0.5}(PO_4)_6(OH)_2$, A = Ag, K and Mg) hydroxyapatites, while Table 15.1 lists the experimentally measured electrical conductivity values. Figure 15.3b shows the temperature dependent conductivity plots for cooling and heating states for comparison. Silver and potassium doped samples exhibit the characteristic conductivity behavior of HA above 450°C, while magnesium doped sample does not show any such characteristic feature. The parent HA sample (sintered at 1200°C) shows ionic conductivity, σ ~2.8 x 10^{-5} S cm^{-1} at 650°C. The conductivity increases by two orders of magnitude in silver doped HA sample (sintered at 1000°C), when compared to the parent HA (σ ~ 3.5 x 10^{-3} S cm^{-1} at 650°C). Similarly, the sintered K doped HA $[Ca_{9.5}K_{0.5}(PO_4)_6(OH)_2]$ shows the conductivity σ ~ 1.1 x 10^{-4} S cm^{-1} at 650°C, which is one order magnitude higher than the conductivity value of the undoped HA. The silver doped composition (x = 0.5), sintered at 1200°C shows reduced conductivity compared to the parent HA. Similarly, Mg-doped HA composition also had less conductivity compared to HA.

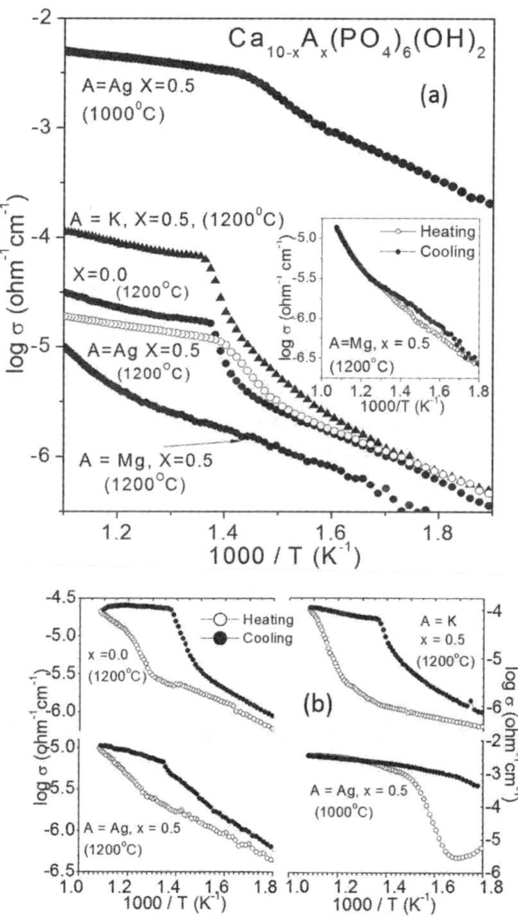

Fig. 15.3 (a) Temperature dependent ac conductivity data of Ca$_{10-x}$A$_x$(PO$_4$)$_6$(OH)$_2$ (x = 0.0, 0.5) (A = Ag, K and Mg) with 100kHz frequency; sintering temperature is given in the parenthesis; (b) conductivity data of Ca$_{10-x}$A$_x$(PO$_4$)$_6$(OH)$_2$ (x = 0.0, 0.5) (A = Ag, K and Mg) samples in cooling and heating states with 100kHz frequency[1023].

Table 15.1 Summary of the measured DC conductivity ($Ohm^{-1}cm^{-1}$) of hydroxyapatites; the number within parenthesis indicate the sintering temperature[1023].

$Ca_{10-x}A_x(PO)_4(OH)_2$	DC conductivity	
	at RT	*at 650°C*
x = 0 (1200°C)	3.7×10^{-11}	1×10^{-5}
A = Ag, x = 0.5 (1200°C)	-	1×10^{-5}
A = Ag, x = 0.5 (1000°C)	7.7×10^{-8}	8×10^{-5}
A = K, x = 0.5 (1200°C)	1.13×10^{-8}	9×10^{-5}
A = Mg, x = 0.5 (1200°C)	1.15×10^{-11}	2.2×10^{-6}

> Various factors (nature of charge carriers, the availability of vacant-accessible sites and the ease with which an ion can move to another lattice site) determine the conductivity of doped HA.

The ionic movement feasibility can be assessed from the activation energy barrier (E_a). The activation energy has been estimated to be 0.39 eV for parent HA in the temperature range 450°C to 600°C (Fig. 15.3). Similarly, activation energy in the same temperature range is 0.19 eV for K-doped HA and 0.11 eV for silver doped HA sample. The activation energy of the undoped HA has been observed to be comparable to the reported value for the H^+ ion conduction in HA. However, a large decrease in activation energy has been observed due to the doping of K and Ag ions at Ca-site in doped HA. These activation energy values are similar to the values obtained earlier for K^+, Ag^+ and Na^+ doped β-alumina samples, where the monovalent ions mediated conductivity depend on the moving dimensionality of the ions and the available space[1036].

Based on impedance measurements of undoped and doped hydroxyapatites ($Ca_{9.5}A_{0.5}(PO_4)_6(OH)_2$, (x = 0.0, 0.5) (A = Ag, K and Mg), the frequency dependence of the conductivity for the parent and ($Ca_{9.5}A_{0.5}(PO_4)_6(OH)_2$, A = Ag, K and Mg) hydroxyapatites are analysed in Fig. 15.4. The impedance data can be further assessed in three different ways: (i) frequency dependent ac conductivity, (ii) Cole–Cole plots of real (Z Cos θ) and imaginary (-Z Sin θ) parts of impedance, and (iii) imaginary part (M″) of the complex electric modulus versus log frequency plots. The parent and doped compositions show a typical ionic conductivity behaviour of apatites due to the hopping and vibrational mode of the ions[1023]. The increase in conductivity with the frequency follows the power law relationship, $\sigma(\omega) \propto \omega^n$ and such behaviour is attributed to the relaxation of the mobile ion hopping. The frequency dependent AC conductivity, $\sigma'(\omega)$ at a given frequency (ω), is well described by the Jonscher empirical expression:[1037]

$$\sigma'(\omega) = \sigma_{dc}[1 + (\omega/\omega_p)^n] \qquad (0 \leq n < 1) - (1)$$

where, σ_{dc} is the dc conductivity and n is a fractional exponent, characterizing the power law behaviour. The ω_p is the characteristic cross-over frequency marking the onset of the power-law regime, $\sigma'(\omega) \propto \omega^n$ and is found to be thermally activated with the same activation energy of the dc conductivity. Therefore, it can be assumed that the frequency dependent electrical conductivity, $\sigma'(\omega)$ originates from the migration of ions and at low frequencies ($\omega < \omega_p$). Also, the power-law dependence observed at high frequencies ($\omega > \omega_p$) is actually due to the same ionic transport mechanism in the wet chemically synthesized Ag-doped HA.

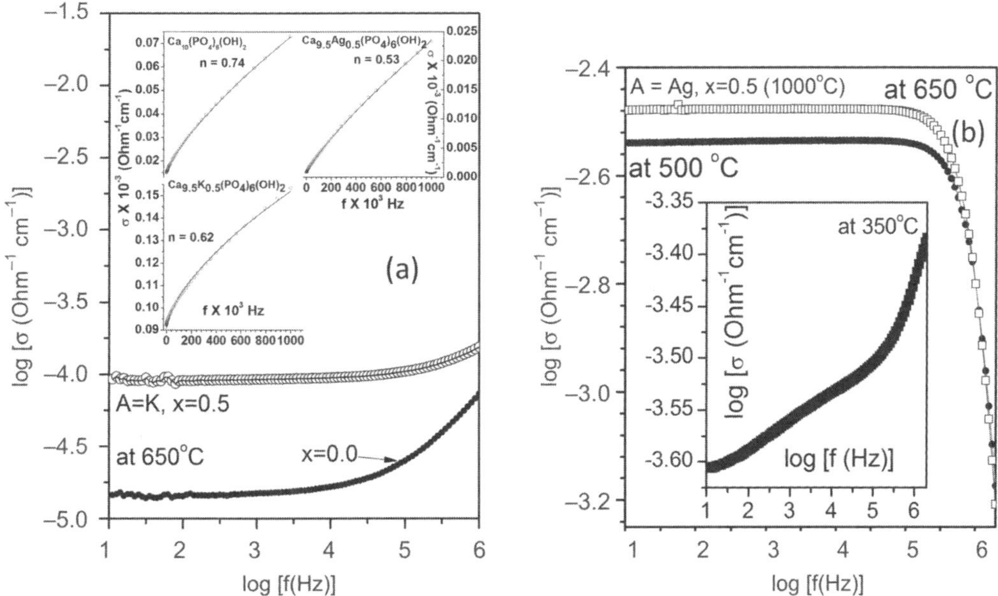

Fig. 15.4 (a) Frequency dependent ionic conductivity data of $Ca_{10-x}A_x(PO_4)_6(OH)_2$ (x=0.0, 0.5) (A = K) hydroxyapatite samples sintered at 1200°C; insets show the fitting of the frequency dependent data using $\sigma'(\omega) = \sigma_{dc}[1 + (\omega/\omega_p)^n]$ equation; (b) frequency dependent ionic conductivity data of 1000°C sintered $Ca_{9.5}Ag_{0.5}(PO_4)_6(OH)_2$ hydroxyapatite[1023].

15.4 | *in vitro* Biocompatibility Property of Chemically Doped HA

15.4.1 | Bactericidal property

The turbidimetric analysis have been used to analyze the antimicrobial activity of the powder samples shown in Fig. 15.5, while SEM is used to find out the bacterial adhesion on the hydroxyapatite pellet samples (Fig. 15.6). Figure 15.5 shows the statistical analysis of the turbidimetric results for *E. coli* bacterial strain on parent and silver doped hydroxyapatite $Ca_{10-x}Ag_x(PO_4)_6(OH)_2$ (0.0 ≤ x ≤ 0.5) samples after 4 hours of incubation. Silver doped samples show statistically significant decrease in the growth of gram negative bacteria. The scanning electron microscopic images show the bacterial adhesion on the surface of the parent and silver doped hydroxyapatite $Ca_{10-x}Ag_x(PO_4)_6(OH)_2$ (0.0 ≤ x ≤ 0.5) pellet samples (Fig. 15.6). On an increase in the silver content in the samples, colony forming ability of *E. coli* bacteria is significantly retarded and isolated bacteria are found on the surface.

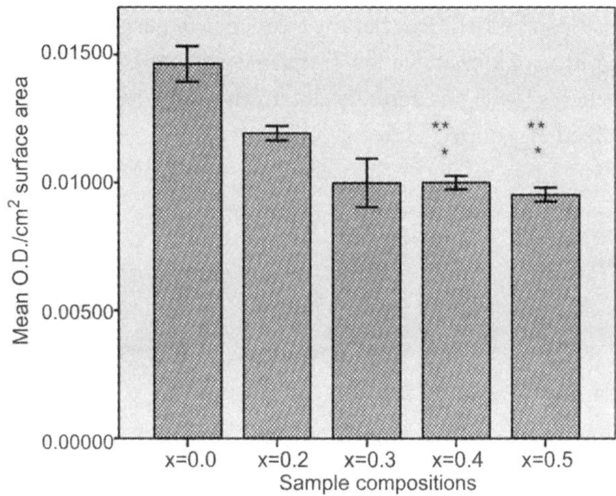

Fig. 15.5 Mean optical density (O.D.)/cm² surface area after 4 hours incubation of Escherichia coli (ATCC# 25922) on silver doped HAp samples (Initial inoculam was 1 x 10⁷ CFU/ml), *represents significant difference at p < 0.05 with respect to HAp[1023].

Note: **represents significant difference at p < 0.05 with respect to HAp–2%Ag (error bars correspond to +/- 1.00 SE)

15.4.2 | Cell proliferation

As mentioned earlier in this chapter, the cellular functionality should not be compromised while designing biomaterial composition for bactericidal property. In particular, one of the concerns, widely expressed in Ag/ZnO-doped HA is that the addition of antimicrobial agent can impact the cytocompatibility. This aspect is analysed for doped HA here using L929 fibroblast cells. As confirmatory evidence, some representative SEM images revealing cell adhesion behaviour of L929 fibroblast cells on silver doped samples are shown in Fig. 15.7. The general observation is that irrespective of compositions, the cells were found to adhere on each sample. In case of pure HAp, the cells were adhered on the surface with the multiple filopodia, extended from all the directions of the cell surface. These results show that ionic silver incorporation ($x \leq 0.2$) in doped HAp does not affect the L929 mouse fibroblast cell growth on the silver doped samples, but the cell proliferation is significantly affected with an increase in silver dopant in HAp.

Figure 15.8 shows the MTT assay results, obtained with undoped and silver doped hydroxyapatite $Ca_{10-x}Ag_x(PO_4)_6(OH)_2$ ($0.0 \leq x \leq 0.5$) samples after 3, 5 and 7 days of culture. The measured optical density, as recorded with ELISA plate reader is directly proportional to the number of viable cells adhered on the samples surface. The detailed univariate analysis using Post Hoc test among different compositions reveals that samples x = 0.3, 0.4 and 0.5 showed the statistically significant difference in comparison to the control sample (undoped HAp). The asterisk (*) represents significant statistical difference at $p < 0.05$ and error bars correspond to +/- 1.00 SE (standard error), compared to undoped HAp. The Post Hoc multiple comparisons among the samples showed that the samples cultured for three days could exhibit statistically significant difference in the viability with respect to the samples incubated for 5 and 7 days, respectively. Further, it can be seen that the mean optical density after 7 days decreases with the increase in the Ag content for (x = 0.3) as compared to undoped HAp. It shows the cell viability is significantly affected with time in the silver doped HAp, irrespective of their initial cell adhesion.

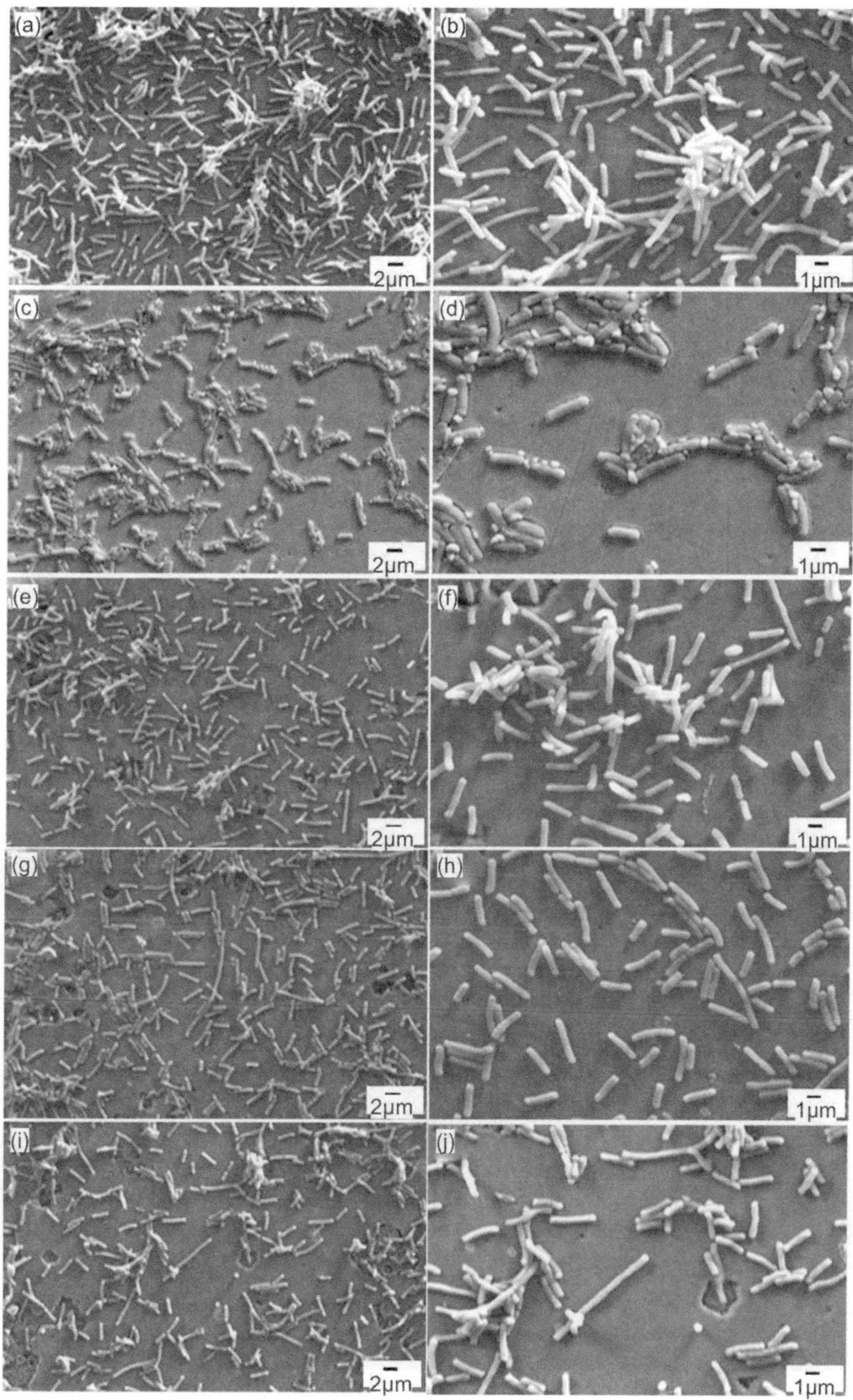

Fig. 15.6 SEM images of the $Ca_{10-x}Ag_x(PO_4)_6(OH)_2$ ($0.0 \leq x \leq 0.5$) pellets (sintered at 1200°C) [a and b- ($x = 0.0$), c and d- ($x = 0.2$), e and f- ($x = 0.3$), g and h- ($x = 0.4$), i and j- ($x = 0.5$)], after incubation in Escherichia coli bacterial suspension for 4 hours[1023].

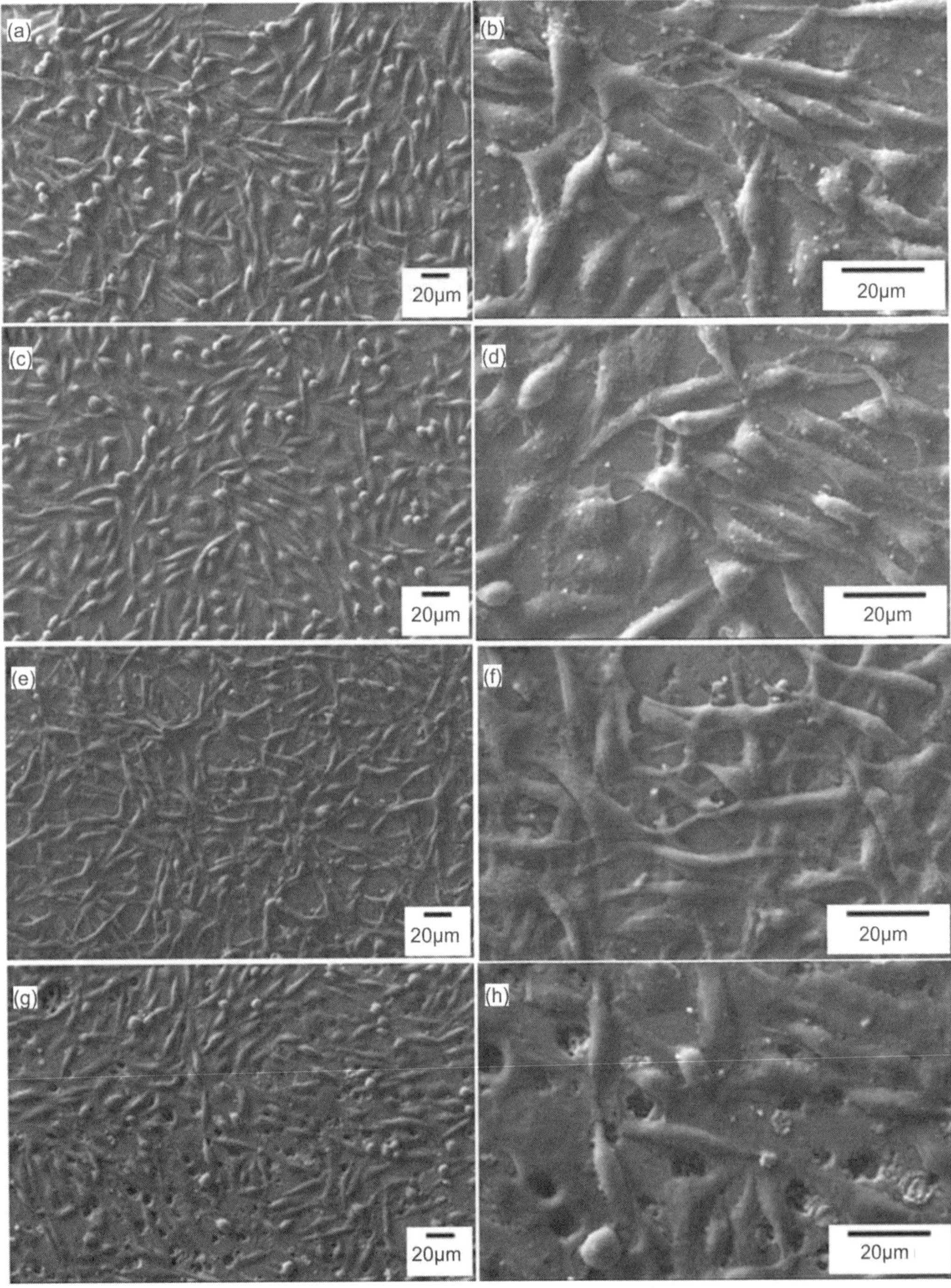

Fig. 15.7 SEM images of L929 cell morphology on Ca_{10}-xAgx$(PO_4)_6(OH)_2$ ($0.0 \leq x \leq 0.5$) pellets (sintered at 1200°C) [a and b- (x = 0.0), c and d- (x = 0.2), e and f- (x = 0.3), g and h- (x = 0.5)] after 48 hours in culture[1023].

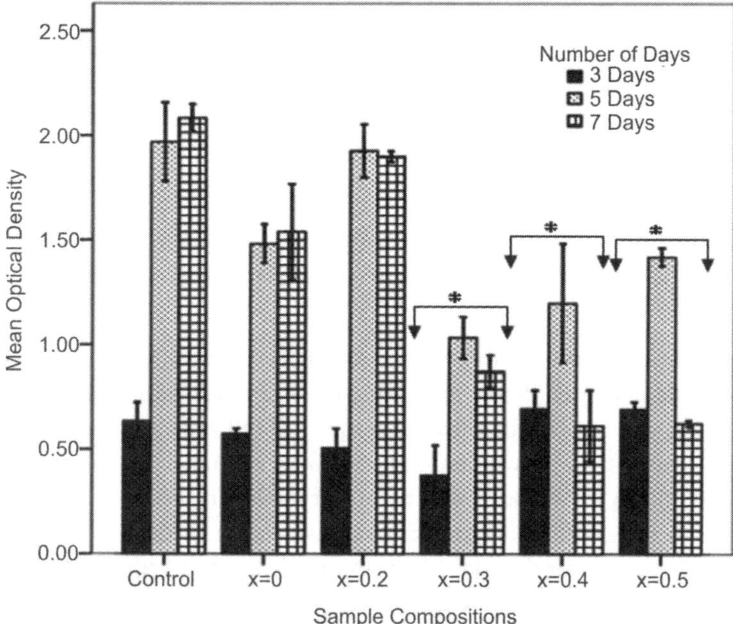

Fig. 15.8 MTT assay results showing the mouse fibroblast cells (L929) proliferation on $Ca_{10-x}Ag_x(PO_4)_6(OH)_2$ $(0.8 \leq x \leq 1.5)$ HAp samples after 3,5 and 7 days of culture[1023].

Note: *represents significant difference at $p < 0.05$ with respect to control

15.5 | *in vitro* Biocompatibility of Ball Milled and Sintered HA–Ag

It can be reiterated here that both the cell functionality and bactericidal property depend on the way antimicrobial agent (e.g. Ag) is introduced into HA. In the preceding section, such results are analysed for chemically doped HA. In this section, the results obtained with mechanically mixed Ag-HA will be reviewed. Here again, the antimicrobial property against E. coli and cytocompatibility with L929 cells for differently synthesized HA-Ag composites are discussed. Fig. 15.9 shows the bacterial adhesion on HA-Ag composites, HA and TCPS control. A clear reduction in bacterial viability without any sign of viable E. coli can be clearly recorded, while sintered HA and TCPS control expectedly exhibit significant bacterial colonisation. The mechanism of antimicrobial effect in the presence of Ag is already described in published papers[1038,1039,1040].

The effectiveness of antimicrobial effect mainly depends on the Ag^+ release kinetics, *in vitro*.

Song et al.[1041] reported that Ag causes plasmolysis, i.e., cytoplasm of bacteria separated from bacterial cell wall. Egger et al.[1042] speculated that, peptidoglycan present in the bacterial cell wall contains teichoic acids or lipoteichoic acids, which feature a strong negative charge. This phenomenon can contribute to sequestering free Ag^{+2} ions. Rannie and Barggs[1043] demonstrated that silver ions can interact with sulfydryl (-SH) groups of proteins as well as the bases of DNA, leading to the inhibition of respiratory processes. Batarseh[1044] reported DNA unwinding in presence of Ag ions. From the above mentioned literature results, it can be concluded that the use of Ag could nullify the post-operative infection after the implantation.

One of the major concerns in reference to HAp–Ag composites is that Ag addition, in an amount greater than a critical value just sufficient for anti microbial property, can reduce cell proliferation.

In order to investigate whether HA–10% Ag can support the attachment of biological cells, L929 mouse fibroblast cells were used. Some interesting features of cell adhesion were recorded, when cultured for 72 hours with L929 cells with cell density of 1×10^5/ml (see Fig. 15.10). Figure 15.10a shows extreme flattening the cells with multiple filopodia, extended from all the directions of the cell surface. Figure 15.10b shows the contact between two flattened cells on the composite surface. Figure 15.10c reveals a large cell, spreading across a cluster of Ag particles. The high magnification image in Fig. 15.10d shows that the filopodia is thinner in contact with silver particles. Also, the lamelopodia are absent in the silver contact zone. Summarizing, HA-10% Ag, mechanically mixed and sintered, does support cell adhesion and spreading.

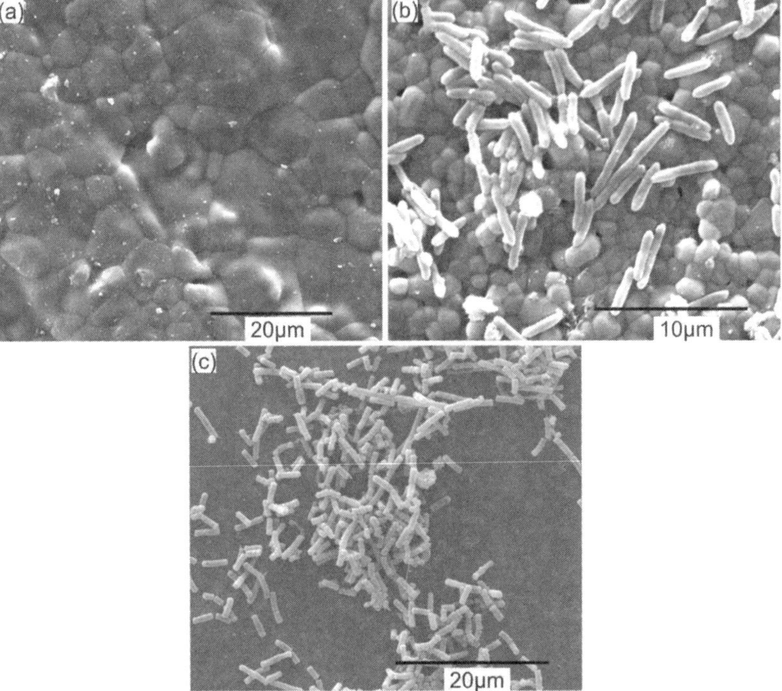

Fig. 15.9 SEM images revealing E. coli bacterial cell adhesion on (a) HAp–10 wt% composite (sintered at 1200°C), (b) pure HAp (sintered at 1200°C) and (c) negative control (tissue culture polysterene) after 4 hours of inoculation[1023].

The dependence of Ag-dopant on cell and bacterial viability can be assessed. For example, a significant cell growth inhibition in x = 0.5 compositions has been observed by SEM images. The mean optical density is decreasing in x = 0.3, x = 0.4 and x = 0.5 compositions as the culture time increases to 7 days. Such observations indicate lowering in cell growth behaviour on the silver containing compositions.

MTT reagent (Mitochondrial dehydrogenase enzyme) directly reacts with the mitochondria to form formazan crystals in metabolically active cells and MTT assay expression can be directly correlated with the number of viable cells, provided MTT reagent does not react with biomaterial substrate (e.g. carbon based materials).

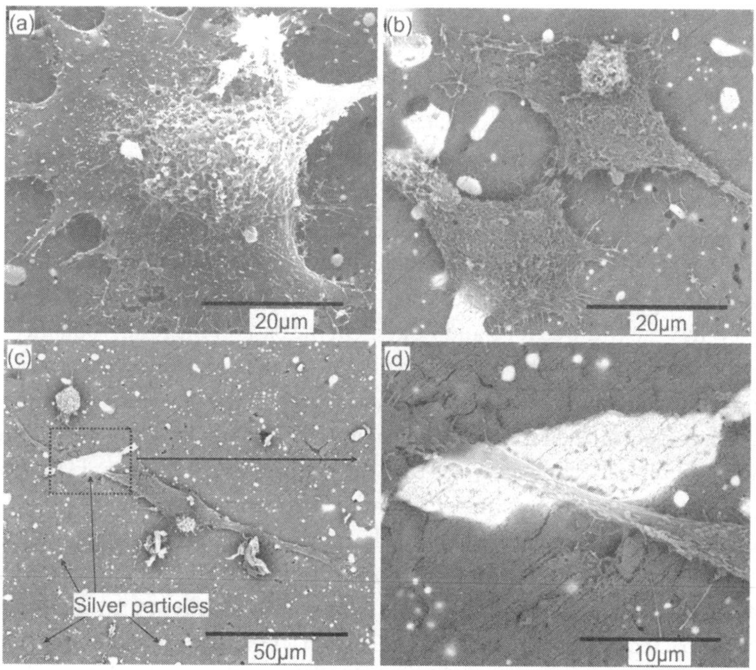

Fig. 15.10　SEM images of L929 cells adhered on HAp–10Ag (mechanically mixed and sintered) revealing the morphology of adhered cells; the white particles in the BSE images (b, c, and d) are silver particles. Results were obtained after 72 hours of culture[1023].

The origin of cytotoxicity in silver doped compositions can be traced to the presence of silver ions, as shown by optical absorption measurements for x = 0.5 composition. These measurements show the optimum value of silver doping is less than x = 0.3 in $Ca_{10-x}Ag_x(PO_4)_6(OH)_2$ ($0.0 \leq x \leq 0.5$) compositions to keep the cytotoxicity in control with the silver content.

The gram negative *E. coli* bacteria are used to explore the antimicrobial activity of the parent and silver doped hydroxyapatite samples. The antimicrobial effects of silver in powder hydroxyapatite and in thin film coatings have shown the strong inhibitory and bactericidal effects as well as a broad spectrum of activity[1021]. Similar to cell culture studies and MTT analysis, the silver doped HAp samples show the increased antimicrobial activity against gram negative bacteria compared to parent HAp (x = 0.0) irrespective of the form of the sample pellet or powder. The cytotoxicity behaviour of these chemically synthesized silver doped HAp samples is different from the mechanically mixed silver with HAp. The results also reveal that $Ca_{10-x}Ag_x(PO_4)_6(OH)_2$ $(0.0 \leq x \leq 0.5)$ compositions show cytotoxicity at x > 0.3 and such substitution level is in fact lower than HAp–10 wt% Ag. Although not shown in this chapter, both ALP and osteocalcin assay results do not reveal any statistically significant difference in terms of early and late stage osteoblast differentiation between HAp and HAp–10 wt% Ag composition, respectively. Importantly, the chemically doped HAp show the increased antimicrobial behaviour with less amount of silver, compared to mechanically mixed compositions.

15.6 | Closure

The discussion in this chapter establishes that chemically synthesized and silver doped HAp samples are antimicrobial and biochemically active due to the presence of silver ions. The presence of silver ions is confirmed by optical absorption spectroscopy in chemically doped sintered samples. MTT analysis reveals that Ag substitution in $Ca_{10-x}Ag_x(PO_4)_6(OH)_2$ $(0.0 \leq x \leq 0.5)$ needs to be optimized in order to obtain a good combination of cellular functionality and antimicrobial property. MTT results also implicate cytotoxic behaviour of $Ca_{10-x}Ag_x(PO_4)_6(OH)_2$ $(0.0 \leq x \leq 0.5)$ for x ≥ 0.3, when L929 fibroblast cells were cultured for 5 and 7 days. On the basis of the results obtained using bacteria culture and turbidimetric analysis, it is confirmed that the doping of silver $Ca_{10-x}Ag_x(PO_4)_6(OH)_2$ $(0.0 \leq x \leq 0.5)$ significantly improves the antibacterial property, when compared to undoped HAp.

In case of mechanically mixed HA - 10 % Ag composites, the results of the cell culture experiments with L929 mouse fibroblast cell line reveals that the addition of 10 % Ag to HAp matrix does not cause any significant difference in terms of overall cell attachment, cellular bridge formation and cell proliferation behaviour, in reference to pure HAp. Also, HAp–10 % Ag favourably supports cell attachment and proliferation of L929 mouse fibroblast cells, *in vitro*. An important result discussed in this chapter is that the addition of 10 % Ag can provide excellent anti-microbial property in terms of excellent bactericidal property against *E. coli* bacteria without compromising the *in vitro* cytocompatibility/cytotoxicity property of HAp. At closure, the results discussed in this chapter demonstrate the synthesis dependent antimicrobial properties of HA–Ag composites and such dependence can be rationalized to the presence of silver in ionic or metallic state. For more details on many of the results summarized in this chapter, the reader can follow a number of recent research papers from the author's group[1023].

16

Case Study: HA-CaTiO$_3$ based Multifunctional Composites

In last few decades, significant efforts are invested to develop synthetic materials with bone mimicking properties. Considering various limitations of HA (the inorganic constituent of natural bone), various groups have attempted either to increase fracture toughness or to increase strength or to increase functional properties like electrical conductivity. Another approach is to enhance multiple such properties in designed microstructures and this concept is realized in the development of multifunctional biomaterials. In this context, this chapter discusses the published results from the author's own research to highlight how to design HA–CaTiO$_3$ based multifunctional biocomposites with electrical conductivity property. The cytocompatibility study to probe into the influence of the substrate conductivity on cell fate processes is limited. In order to address this issue, HA–CaTiO$_3$ (hydroxyapatite–calcium titanate) is used as a model material system to showcase the effect of varying conductivity on cell functionality of electroactive mouse myoblast cells (C2C12). The cell culture results conclusively establish the positive impact of the substrate conductivity towards cell proliferation and differentiation as well as confirms the efficacy of HA–CaTiO$_3$ biocomposites as conductive platforms to facilitate the growth, orientation and fusion of myoblasts. In order to address the efficacy of electroactive HA–CaTiO$_3$ composites in better early stage osteogenesis, the results of pre-clinical study with HA–80% CaTiO$_3$ based cylindrical implants in femoral bone defects of rabbit model are also discussed.

16.1 | Introduction

In recent years, research in the field of biomaterials for orthopedic applications has been focused to develop new strategies for better and effective osseointegration of implant materials into the host. In its narrowest sense, the cell–substrate interaction has been proven to regulate cellular fate processes in the vicinity of the implants[1045,1046]. Some biological tissues exhibit a wide range of electrical activities for maintaining cellular homeostasis and modulating molecular events, engaged in development, adaptation, repair and regeneration of tissues[1047,1048,1049]. Specifically, cardiac, neural, bone and muscle tissues use the mechanism of electrical conductivity, i.e., accumulation and flow of charges (bioelectricity) to regulate its physiological behaviour and to propagate electrical potentials through their cellular components[1050,1051,1052].

> In the process of bone regeneration, piezoelectric property of the collagen is hypothesized to generate electric field to facilitate bone remodelling.

This, in turn, establishes the fact that electrical stimulation offers repair and healing of tissues in the clinical situation, where normal healing is impaired. In the last two decades, electrical stimulation has been investigated as a treatment option to heal the damage of the fractured bone tissue[1053,1054]. A study with a range of animal models has provided evidence that electrical stimulation can enhance the bone healing[1055]. In these models, electrical stimulation using direct current has been delivered locally to damaged bone through metal electrodes (stainless steel, platinum, and titanium). At the end of the treatment process, the implanted metal electrode was removed from the site of newly healed bone tissue via surgical procedure. The major disadvantages of this approach are the risk of complication of surgery and damage to the newly formed bone tissue. On the other hand, direct current stimulation, while adequate for *in vivo* application has shortcomings for *in vitro* culture. This arises from the accumulation of charged compound on the electrodes and this leads to a decrease in the magnitude of the electrical stimulus. This concomitantly limits the effectiveness of bone healing. In this perspective, an alternative approach to the external electric field application is the use of electroconductive implants. However, the *in vivo* biocompatibility property of electroconductive implants vis-a-vis that of the implants with poor conductivity are not widely investigated.

Hydroxyapatite (HA) is well known as a highly biocompatible and bioactive ceramic that promotes bone growth *in vivo*.[1056]

> Poor electrical and mechanical properties limit the application of monolithic HA as an implant material alone for orthopedic application.

The addition of ceramic/metallic reinforcement (mullite, Ti) has been studied extensively to improve the mechanical properties of HA.[1057,1058,1059,1060,1061] The development of the electrically active HA based materials ($HA/BaTiO_3$ and $HA/CaTiO_3$) as bone substitute implant materials has also attracted the attention of many biologists and materials scientists.[1062,1063,1064,1065,1066]

The main idea to develop such HA-based electroconductive composite has focused on the improvement of desirable mechanical and electrical properties, while retaining biocompatibility property. Such research is motivated from the fact that natural bone has significant electrical properties. For example, piezoelectricity, pyroelectricity[1067,1068] and such properties help in maintaining the bone structure and fracture healing.[1069,1070] The studies have shown that polarization of HA ceramics may modulate new bone formation and resorption. In addition to polarization of HA, the electrical properties of HA can be enhanced by the addition of conductive phases, like $CaTiO_3$. According to published literature reports, $CaTiO_3$ has better biocompatibility,[1071] increased osteoblast cell adhesion and enhanced osseointegration[1072] as compared to monolithic HA. For this purpose, the composites of HA and 80% CaTiO$_3$ have been developed as an electrically active bone substitute implant using multi-stage spark plasma sintering technique.[1073]

Given the potential of electroconductive biomaterials for orthopedic applications, this chapter showcases how CaTiO$_3$ addition can enhance toughness in microstructure dependent manner together with electrical conductivity. *in vitro* studies such as cellular adhesion, spreading, proliferation and differentiation of C2C12 myoblasts will demonstrate how myoblast cells behave on hydrophobic stiff bioceramic substrates in a conductivity dependent manner. Following ISO-10993 guidelines, the pre-clinical studies involving rabbit femoral defect models are conducted and the results are analysed to establish better early osseointegration of HA–80% CaTiO$_3$. Throughout this chapter, the materials are designated as HAxCT (where x=20,40,60,80 wt.% CaTiO$_3$; e.g. HA80CT means HA-80 wt.% CaTiO$_3$).

16.2 | CaTiO$_3$ Dependent Toughness Enhancement

First, the ceramic processing and relevant characterization is briefly mentioned. Hydroxyapatite, Ca$_{10}$(PO$_4$)$_6$(OH)$_2$ powders with finer particle sizes (D$_{50}$~1.12 μm) were synthesized using precursors (CaO and orthophosphoric acid) by adopting wet precipitation route. The crystalline calcium titanate (CT) (~5 μm) was synthesized using the mechanochemical activation of the mixture of CaO and TiO$_2$ (anatase), followed by calcination at 900°C for 2 hrs. HA–CT powder mix with varying amounts of CT (20, 40, 60 and 80 wt%) were densified using multistage spark plasma sintering. The graphite die/punch assembly was heated by pulsed direct current to a temperature of 850°C and held for 5 min. In the same heating cycle, the powder compact was subsequently heated to a temperature of 950°C with a dwell time of 5 min, followed by final stage of sintering at temperature of 1200°C for holding time of 5 min. During the entire heating cycle, the powder compact was under uniaxial pressure of 50 MPa.

The fracture toughness of HA–CT composites was reliably determined using single edge V-notched beam (SEVNB) technique with across head speed of 0.5 mm/min. As part of the sample preparation, V-notch of radius 15–20 μm was introduced in the specimen using a diamond abrasive paste on a commercial V-notch preparation machine. The sample dimensions were 3.5 mm × 4.0 mm × 20.0 mm in width, height and length, respectively. The samples were loaded under 3-point bending condition and the fracture toughness (K_{Ic}) was calculated using the following relationship,

$$K_{Ic} = \frac{3PS}{2B^2W} a^{1/2} Y \tag{16.1}$$

where,

$$Y = \frac{1.99 - \alpha(1 - \alpha)(2.15 - 3.93\alpha + 2.7\alpha^2)}{(1 + 2\alpha)(1 - \alpha)^{3/2}} \tag{16.2}$$

and

$$\alpha = \frac{a}{W}$$

Here, P is the load at fracture, S is the span length of specimen, B is the specimen thickness, W is the specimen width; a is the pre-crack length and Y, the shape factor.

The flexural strength was determined using 3-point bending using the same universal testing machine with a crosshead speed of 0.5 mm/min. The flexural tests of the sintered and polished specimens were carried out on bars with dimensions of 3.5 mm width × 4.0 mm height × 20.0 mm length. The flexural strengths (σ_f) were calculated according to the expression in equation (3) for 3-point bending configuration.

$$\sigma_f = \frac{3PL}{2WB^2} \tag{16.3}$$

where, P is the break load, L is the outer support span, B is the specimen thickness, and W is the specimen width.

To start with a discussion on toughness enhancement, the finer microstructural details are now analyzed. The finer scale microstructural analysis was carried out with bright field (BF) and dark field (DF) imaging techniques using TEM and the phases were identified with the help of Selected Area Electron Diffraction (SAED) patterns. The grain sizes of the monolithic HA, CT and HA–CT composites were estimated. It was found that the grains of HA were around 2 μm, whereas CT and HA–CT composites had finer equiaxed grains ~1 μm. Figure 16.1a shows a DF image of the twinned region of CT and the SADP analysis (not shown) indicates that the twinning observed in CT grains is 180° rotational twin around [101]. Twin lamellar microstructure is also observed in HA–CT composites as seen in Fig. 16.1b with a lamellar spacing of 200 nm.

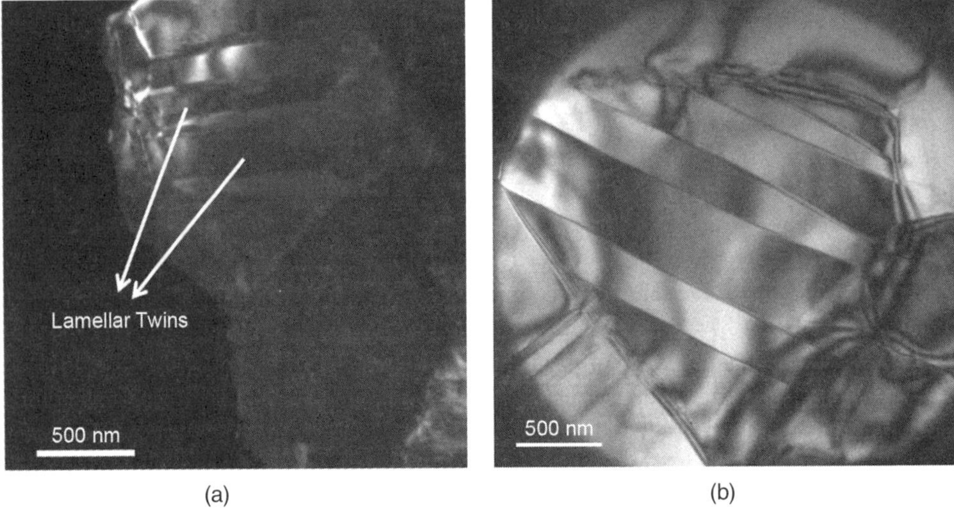

(a)   (b)

Fig. 16.1 (a) Representative dark field image revealing the lamellar twinning (characteristic to perovskite structure) with a spacing of about 200 nm in CT, (b) twins in HA–80CT with lamellar spacing of around 200 nm[1062].

As far as the mechanical strength properties are concerned, the bending strength of HA–CT composites were measured using 3-point flexural testing and the results are plotted along with SEVNB fracture toughness measurements in Fig. 16.2. Such properties are important for biomedical applications, as an implant may experience various loading conditions, *in vivo*. Another essential requirement for an implant is crack growth resistance; this was measured by evaluating the long crack fracture toughness property using single edge V-notch beam technique.

In terms of fracture toughness, HA–CT composites show a radically different behaviour. Among the composites, HA20CT with the least amount of CT shows a moderate increase in fracture toughness (1.1 MPa.m$^{1/2}$) as compared to HA (0.7 MPa.m$^{1/2}$). However, a significant increase of almost 250% in toughness when compared to monolithic HA was recorded with 40% CT (1.7 MPa.m$^{1/2}$) addition to HA. Further addition of CT resulted in a relatively insignificant decrease in toughness at higher CT content (60% and 80%) as seen from the values obtained for HA60CT and HA80CT with 1.3 MPa.m$^{1/2}$ and 1.4 MPa.m$^{1/2}$, respectively. Nevertheless, a two fold increase of toughness in HA80CT and HA40CT compared to HA was recorded. From the error bars in toughness data, the crack growth resistance of composites with CT content of 40% or more is found to be broadly comparable.

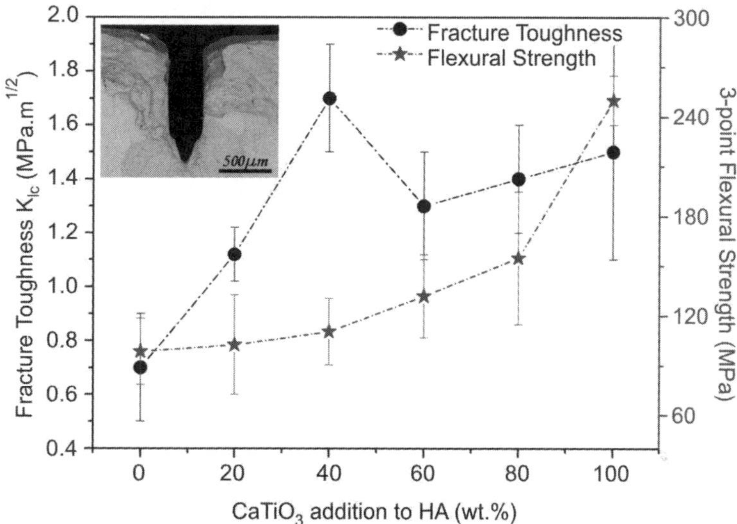

Fig. 16.2 Variation of flexural strength and fracture toughness measured with monolith HA, CT and HA–CT composites, as a function of CT content[1062].

The observed twinning in $CaTiO_3$ grains is believed to be the result of both uniaxial compression applied during spark plasma sintering and due to the loss in symmetry during the tetragonal to orthorhombic transition.

Figure 16.3 provides a representative SEM image of the fracture surface of monolithic CT. Clearly, the transgranular fracture is found to be dominant failure mode, as the cracks are found to propagate through the grain. The considerable deflection of the crack front from the twin boundaries is also observed (white arrows in Fig. 16.3). The presence of a large number of twins is observed in many of the coarse CT grains (also see Fig. 16.1). Brittle fracture is also reflected from the observation of cleavage steps on fracture surfaces. The presence of closed porosity is noticed and a few tabular grains with aspect ratio of 4–6 are also observed. From the fracture surface morphology, it is clear that the twins play an important role in the fracture behaviour of CT.

It has been earlier reported that the domain/twin structure is a consequence of a transformation from the paraelastic phase to the ferroelastic phase. In the above context, the CT grains should be considered to be ferroelastic in nature in HA–CT composites.

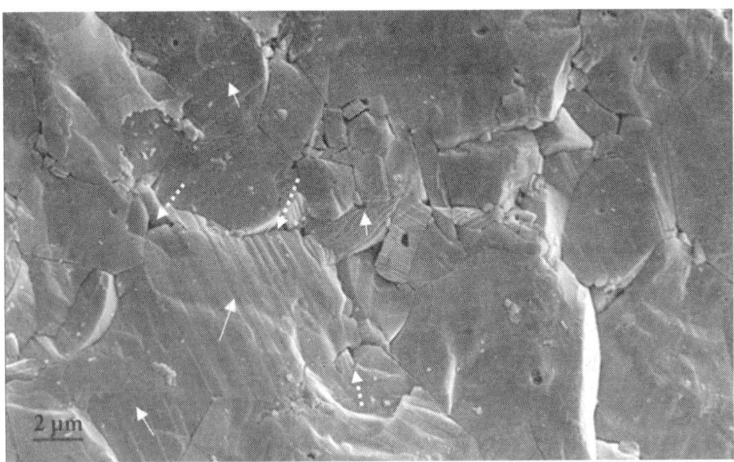

Fig. 16.3 Fracture surface of monolithic $CaTiO_3$ containing twinned grains, showing transgranular fracture; the solid arrows indicate the characteristic twins and the dotted arrows indicate the twin-crack interface[1062].

16.3 | Electrical Conductivity Property

For dielectric and electrical measurements, the polished samples of thickness 1–2 mm were coated with silver paste followed by curing, which acts as electrodes. The capacitance and dielectric loss were measured as a function of frequency (100 Hz–1 MHz) from room temperature (RT) to 200°C by precision impedance analyser. The samples were heated at a heating rate of 2°C/min. The dielectric constant (ε_r) was calculated using the following formula,

$$\varepsilon_r = \frac{C \times d}{\varepsilon_0 A} \tag{16.4}$$

where, C, d and A are the capacitance, thickness and area of the sample, respectively, and ε_0 is the dielectric constant of the free space (= 8.854×10^{-12} F m^{-1}). The dissipation factor (D) has been recorded directly from the instrument. AC conductivity (σ_{ac}) was calculated using the following relationship,

$$\sigma_{ac} = \frac{G \times d}{A} (\Omega\, cm)^{-1} \tag{16.5}$$

where, G, d and A are the conductance, thickness and area of the sample, respectively.

Figure 16.4 shows the variation of AC conductivity with temperature at frequency of 1 MHz for HA–CT composites. While the conductivity of HA linearly increases in the measured temperature range, the conductivity of HA20CT and HA40CT samples appear to be independent of temperature. The spark plasma sintered HA80CT ceramic demonstrated the highest value of conductivity due to high dissipation factor, which increases further with an increase in temperature. With an increase in temperature, the dehydration of chemically bound water causes ionic conduction due to the movement of OH⁻ ions and H⁺ proton along the columnar channels in HA crystals[1074]. The conductivity increases due to the presence of surface bound water and hydroxyl ions in HA, while conductivity of pure CT and HA80CT composite is dominated by ionic polarization. In case of monolithic CT, the mobility of O^{2-} ions, located at the face centred position in the crystal lattices of $CaTiO_3$ along c-axis contributes as the dominant conduction mechanism. It is particularly due to the presence of Ti^{3+} ions in SPSed $CaTiO_3$.

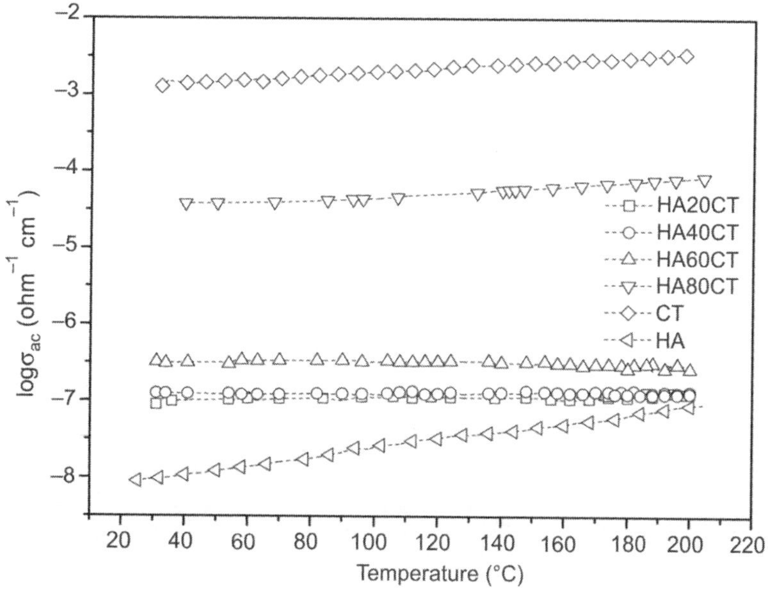

Fig. 16.4 Variation of AC conductivity with temperature for spark plasma sintered HA–CT composites at 1 MHz of frequency[1063].

16.4 | **Substrate Conductivity Dependent Muscle Cell Proliferation/Differentiation, *in vitro***

The surgical procedure during orthopedic implantation results in injury of skeletal muscles, in and around the defect site. Hence, it is necessary to ensure that the implant material that we employ as bone substitute should also provide a favourable platform for the muscle cells to grow and proliferate. With this fact in mind, we shall now analyze the behaviour of myoblast cells on stiff conductive ceramic substrates. The major emphasis is therefore to illustrate how the conductivity property of substrate influences the muscle cell functionality. For this purpose, HA–CaTiO₃ substrates by varying the $CaTiO_3$ content (0, 40, 80, and 100 wt%) were obtained with different conductivity.

Key issues are,
(a) How does the substrate conductivity influence the cell growth?
(b) How would the time in culture influence cell growth on electroconductive substrates?
(c) Can inherent substrate conductivity influence cell functionality *in vitro*, even in the absence of (electric field stimulated conditions?
(d) Can the differentiation of myoblast to myotubes be regulated on electroconductive substrates?

In addressing all the above issues in an integrated manner, a planned set of *in vitro* culture experiments were conducted using C2C12 mouse myoblast cells. Muscle tissue uses electrical conductivity for maintaining homeostasis, development and regeneration[1075] and hence myoblasts were chosen as a model cell line to assess the effect of conductivity exerted by the substrate in which the cells reside.

In order to visualize the details of cytoskeletal structures and the shape of the nuclei, myoblasts grown for 3 days were immunostained for F-actin and counterstained for nuclei. After 3 days, myoblasts reached complete confluency in all the samples. Moreover, the cells appeared more organized and strongly oriented in samples with higher $CaTiO_3$ content (Figs. 16.5c and 16.5d), whereas cells appeared to be perturbed and nonaligned on HA and HA 40CT substrates (Figs. 16.5a and 16.5b).

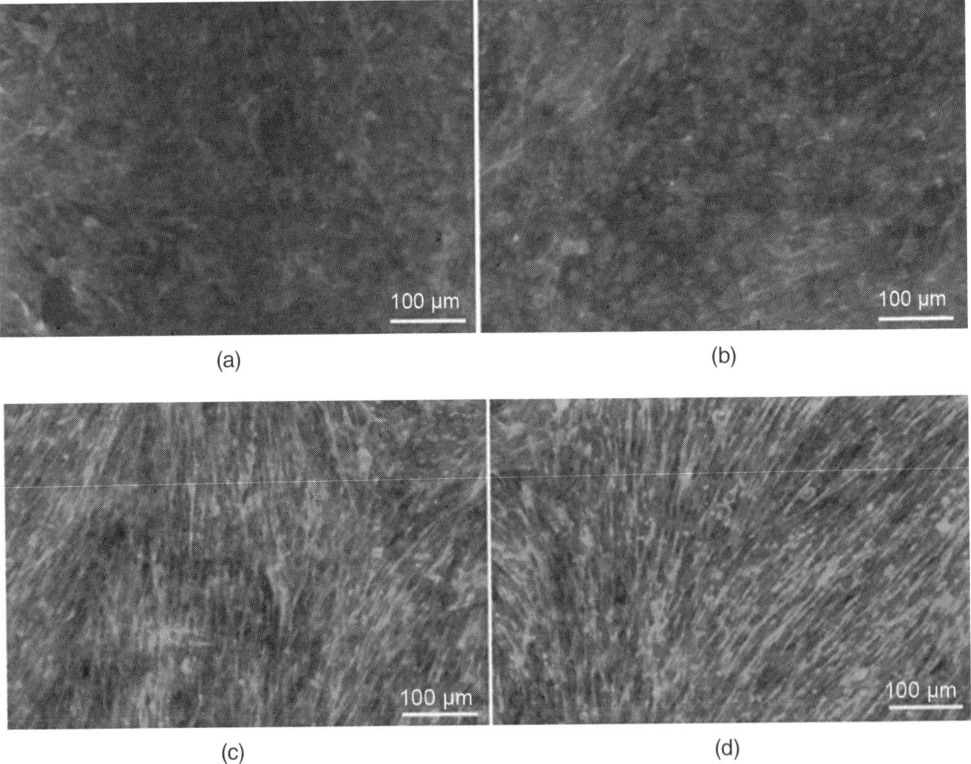

Fig. 16.5 Fluorescence image of myoblast cells proliferated (after 72 hours) of culture on (a) HA, (b) HA40CT, (c) HA 80CT, and (d) CT[1064] [see Colour Plate].

Upon reaching confluency, the cells, kept in serum starvation condition, started fusing into myotubes and the immune stained images in Fig. 16.6 allow clear visualization of long myotubes formed on HA–CT samples. HA samples, with lower conductivity showed shorter, less prominent myotubes with irregular morphology, compared to all other CaTiO₃ containing samples. As the time of culture progressed to 5 days, the differentiated cells displayed more defined and neatly aligned myotubes, with maximum number of myotubes on CT substrates (Fig. 16.6). Regardless of the detection of myogenin on all samples (Fig. 16. 6), well defined assembly of myoblasts was observed only on HA 80CT and CT substrates (Figs. 16.6c and 16.6d), demonstrating the capability of conductivity in inducing the expression of myogenic regulatory factors.

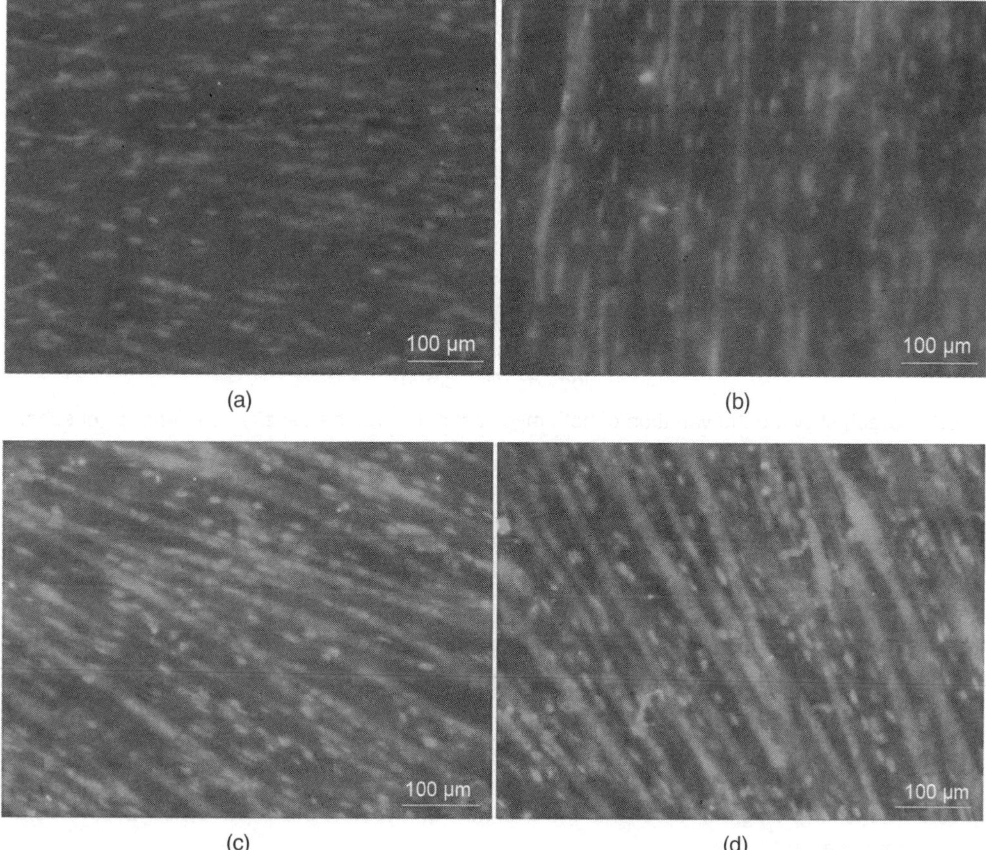

Fig. 16.6 Immunostaining of myogenin expressed on (a) HA (b) HA40CT (c) HA80CT (d) CT samples[1064] [see Colour Plate].

Also, the density of myoblast cells, obtained by counting the cell number from fluorescence images was significantly higher in HA–CaTiO₃ composites than HA. Moreover, a further investigation was carried out on these composites in order to correlate the effect of inherent electrical conductivity on the attachment and proliferation of myoblast cells. It was observed that cell density followed the same trend as that of the electrical conductivity with increasing amount of CaTiO₃ (Fig. 16.7). Similar trend is also recorded for myotube density.

Substrate conductivity positively influences the proliferation and differentiation of myoblast cells.

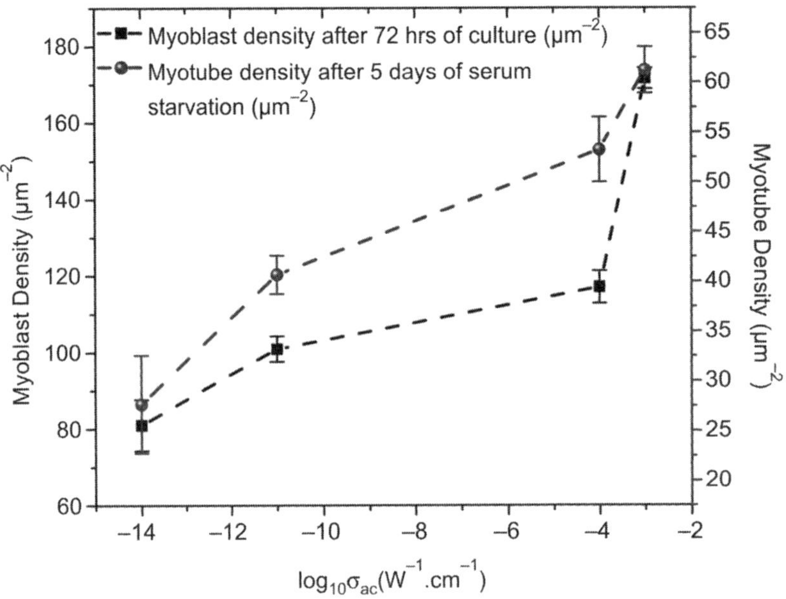

Fig.16.7 Graph showing the variation of both myoblast and myotube density as a function of substrate conductivity[1064].

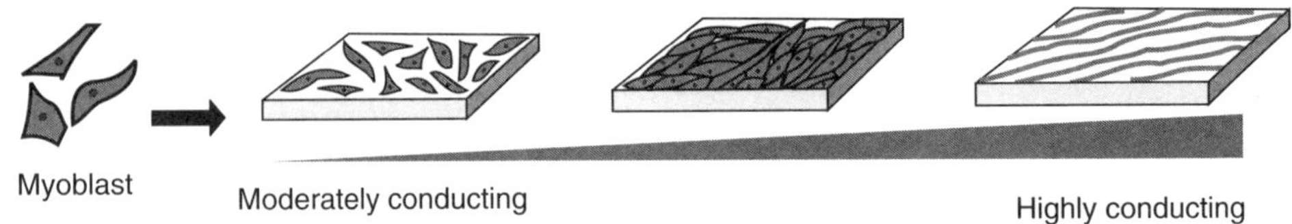

Fig. 16.8 Schematic diagram represents the overview of myoblast behaviour on conductive ceramic substrates[1064].

The assembly of myoblasts to form multinucleated myotubes is one of the prerequisite to promote muscle regeneration and function. The presence of myogenin was poorly detected on cells grown on lower conducting substrates (Figs. 16.6a and 16.6b), whereas mature myotubes with higher levels of myogenin expression was detected on substrates with higher conductivity (see schematic in Fig. 16.8). The up-regulation of myogenin induces expression of myotube specific genes, causing myofibril formation[1076,1077,1078,1079]. Such a prominent rise in the expression of myogenin on $CaTiO_3$ substrates illustrates the dependency of substrate conductivity on the entry of muscle progenitors to terminal differentiation. The orientation of myotubes in this parallel manner helps in the generation of large forces to enable muscle contraction during movement[1080]. In addition, such preferential organization and well-defined linear geometries of myotubes defines its physiological shape and governs the formation of myofibrils, which are the structural unit of skeletal muscles[1081]. While in most cases of conventional cultures, differentiation of myoblast on glass coverslips results in the development of complex and branched network of myotubes that has no physiological relevance to native muscle tissue[1082].

16.5 | Osseointegration in Rabbit Model

Inspired by the *in vitro* cytocompatibility results, the results of the pre-clinical study using femoral defect model in experimental rabbits will now be discussed. The implantation of HA–CaTiO$_3$ in the femoral defects in rabbits are shown in Fig.16.9.

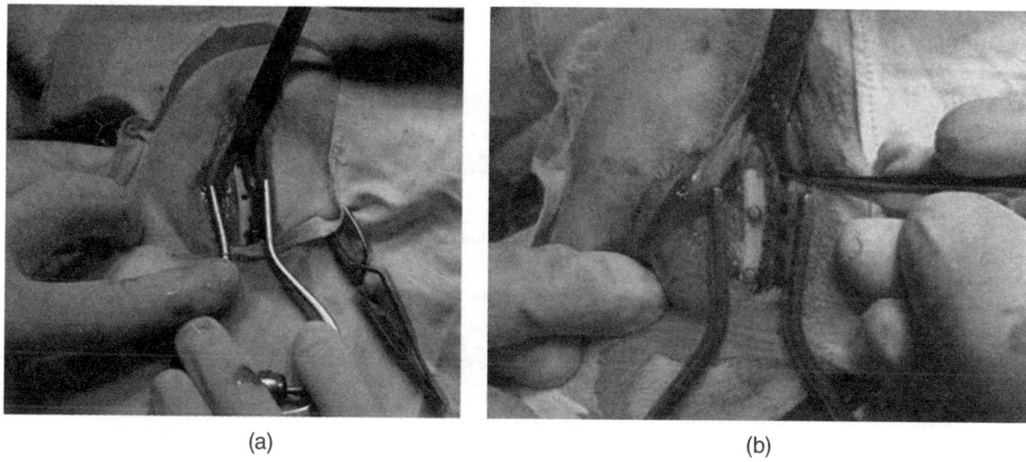

Fig. 16.9 Gross macroscopic images of (a) a bone defects on rabbit femur and (b) inserted implants immediately after the implantation[1065].

During the post implantation period, all the rabbits recovered uneventfully for post-operative periods of up to 12 weeks. Neither any observable superficial nor any deep inflammation was recorded during post-implantation. The Stevenel's blue staining technique was used to give a fairly accurate estimate of the bone area or volume. Typically, the fibrous tissue is stained by Stevenels' blue through the entire thickness of the histology sections (10–15 μm). This means that fibrous tissue was overestimated, compared to bone. In this study, neobone formation observed at both sides of the bone/implant interface in the 12 weeks post-implantation appears yellowish brown, when stained with Stevenel's blue.

As an example, new woven bone indicates the earlier bone formation, while new lamellar bone is the reflection of mature bone formation at bone/implant interface. Furthermore, the histological images, as shown in Fig. 16.10, can be analyzed from the perspective of physiological behaviour of bone/implants interface after post-implantation.

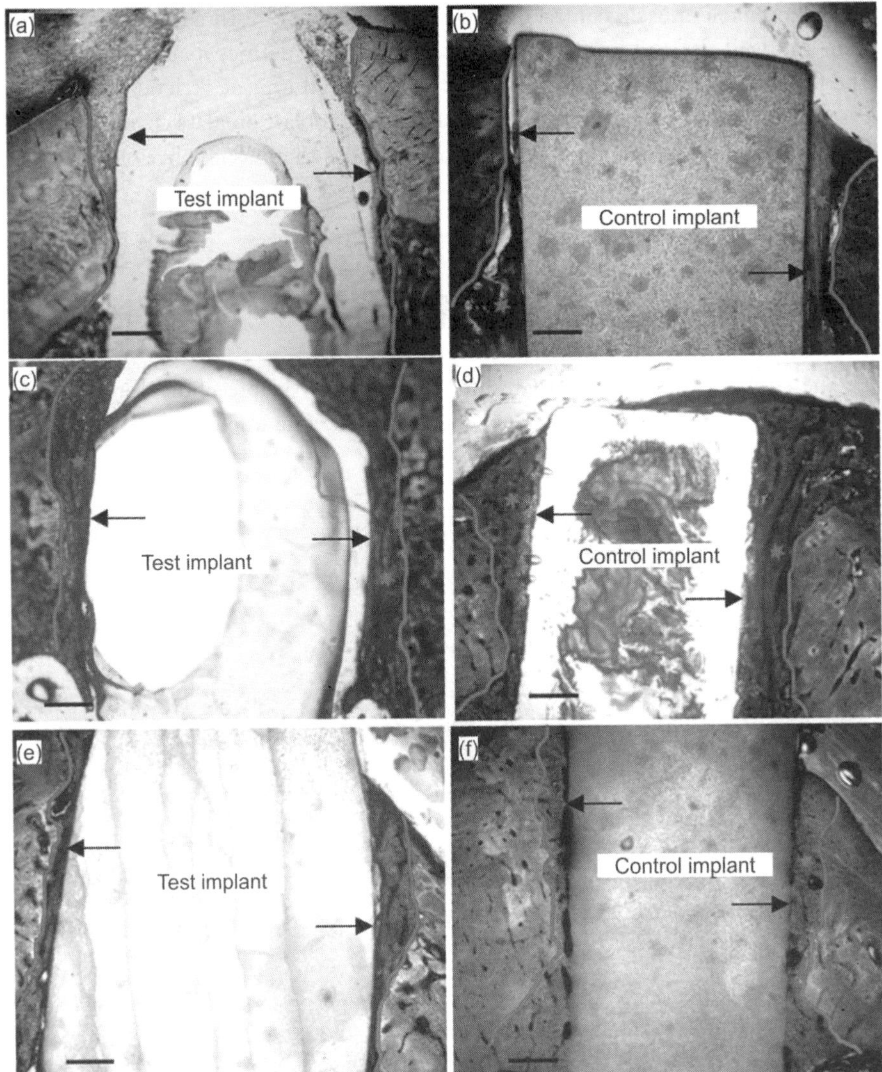

Fig. 16.10 Representative histological image of Stevenel's blue stained regions of HA80CT implant/ host bone interface after (a) 1 week, (c) 4 weeks and (e) 12 weeks as well as for HA implant (control)/host bone interface after (b) 1 week, (d) 4 weeks and (f) 12 weeks; black arrow indicates implant/new bone interface and border line indicates the new bone/host bone interface, red star indicates the osteogenesis/new lamellar bone, short red arrow indicates presence of osteoblast on the surface trabecular bone and short black arrow indicates the presence of neovascularization. image scale bar (250 µm)[1065] [see Colour Plate].

Importantly, the histological observations indicate the difference in the overall healing pattern in terms of the morphology, growth and amount of neobone formation between test (HA80CT) and control (HA) implants. After 12 weeks postoperatively, the abundant mature new lamellar bone as well as new woven bone could be seen around both HA80CT and HA materials (Fig.

16.10). The presence of mild neovascularization and lamellar bone indicates the occurrence of bone remodelling. However, the morphology of new lamellar bone formation around HA is different from the morphology of HA80CT composite. Also, the osseointegration across the interface of bone implant was significantly higher in 12 weeks compared to that after 4 weeks and this also varies along the axial direction of implants. On the other hand, bone resorption could not be observed at the endosteal aspect of cavity in both test and control samples during this period of implantation. To estimate the quantity of new bone formation at bone/implant interfaces of test and control, the histomorphometric analysis is presented to illustrate percentage of new bone formation around HA80CT and HA implant surfaces after varying periods of post-implantation.

Histomorphometry was carried out on multiple sections taken from the proximal site of each implant to determine the area percentage of neo bone available in the area between the host bone and implants. The quantification of specific area of elected tissue sections was performed using professional software packages (image pro plus, DSS image tech), which are widely used to obtain quantitative information about microscopic structures. Histomorphometry of volume fraction of neobone was performed on images taken by Zeiss microscope with attached camera at a standard magnification of 10X. The multiple sections were examined from each specimen (1 week, 4 weeks and 12 weeks). The volume fraction of neobone formation was estimated by the point intercept method on the area of neobone in between the implant and host bone. First, the entire image of the sample was covered with grid as region of interest (ROI) and the number of grid points was read. Then, the mineralized new bone area was selected using multiple selection tools and the measurement points. The volume percentage of bone formation was calculated using the following formula,

$$\% \text{ neobone volume} = \frac{\text{Total number points of bone hits}}{\text{Total number of points}}$$

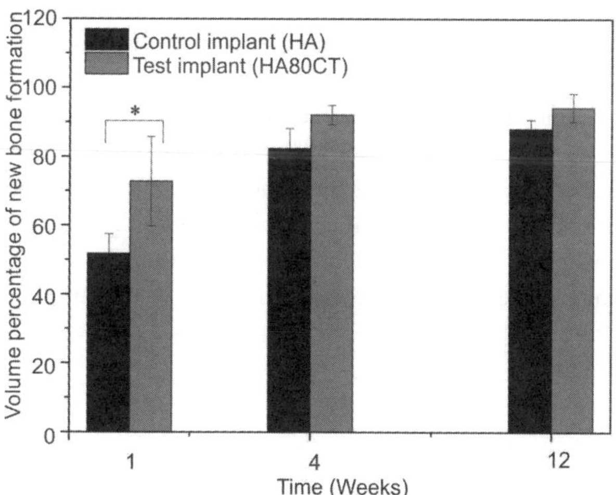

Fig. 16.11 Volume percentage of bone formation estimated from the region of interest (% ROI) between the host and implants of HA and HA80CT after 1, 4 and 12 weeks *in vivo*[1065]; error bars represent standard deviation. Error bar indicates ± standard deviation and * indicates statistically significance at p<0.05.

From Fig. 16.11, it is clear that volume percentage of bone formation of HA80CT in early stage of implantation for one week was significantly higher than that of HA. The volume percentage of bone formation around HA80CT implant after 4 and 12 weeks was 92±2.0 and 94±4.0, respectively, while volume percentage of bone formation around HA after 4 and 12 weeks was 82±5.0 and 88±3.0, respectively. Also, histomorphometry analysis of HA80CT implant reveals more volume and area percentage of bone formation after 1, 4, and 12 weeks than that of HA control with more significant difference after 1 week of implantation[1083,1084,1085,1086].

In addition, the influence of surface polarization on enhancement of osteoblastic activity of HA implanted in calvarial bones of rats is also reported[1087]. An increase in bone formation on negatively charged surfaces was attributed to the accumulation of Ca^{2+} ions on the surfaces. In contrast, the positively charged surface supported the adhesion of molecules such as osteocalcin, fibronectin, and BMPs, which altogether helps in improving the osteoblast cell migration. In addition, HA–BaTiO$_3$ based piezoelectric ceramic composite has been shown to induce the bone growth around the implant[1088]. However, none of the studies on osteogenesis ever reported the influence of substrate electrical conductivity on bone osteogenesis at implant/host bone interface. This study for the first time, attempted to compare the new bone formation of electrically active HA80CT composite with respect to synthetic HA in a rabbit animal model. The results of this study demonstrate the superiority of bone osteogenesis and neobone formation of HA80CT composites over pure HA. However, the exact mechanism of bone osteogenesis by electrically active biomaterials is not known till date. Furthermore, the results of this study significantly highlight the role of substrate electrical conductivity of HA80CT composite, which is critically responsible for neobone formation[1089,1090].

> The bone formation takes place in two directions, from implant surface towards bone and from bone towards implant surface, also known as contact osteogenesis and distance osteogenesis, respectively.

In contact osteogenesis, bone forms at a faster rate compared to distance osteogensis. This indicates that early bone formation in HA80CT implants is primarily due to contact osteogenesis. In the current study, bone formation on the implant surface has taken place during remodelling procedure, which is also known as de novo bone formation[1089]. In distance osteogenesis, new bone formation occurs at the implant surface, but implant gets surrounded by bone during cortical bone healing. In particular, the bone formation in HA implant (control) is due to the distance osteogenesis. These results highlight the possibility of using conductive substrate to realize faster new bone formation as a new strategy in bone tissue engineering towards the tissue regeneration in clinical applications.

16.6 | Closure

At the closure, the development of HA-based electroconductive biocomposites with reasonable toughness properties should have a significant impact on the potential biomedical applications. In reference to the biocompatibility property, the mouse myoblast cells were cultured on HA–CT ceramic composites. Also, HA–80% CT substrates are proven to support the modulation of myoblast proliferation and differentiation to myotube formation *in vitro* in a conductivity dependent manner[1064]. Based on the histological analysis, the progression of defect healing as well as bone regeneration on HA–80% CT implants in rabbit model has been established[1065]. Overall, spark plasma sintered HA–CT composites exhibit better multifunctional properties and even better *in vitro* and *in vivo* biocompatibility property than sintered HA. The multi-functional properties of HA–CT, such as *in vitro* cytocompatibility, *in vivo* osseointegration, electrical conductivity and better fracture toughness qualifies HA–CaTiO₃ as a new class of materials for bone tissue engineering applications (Fig. 16.12).

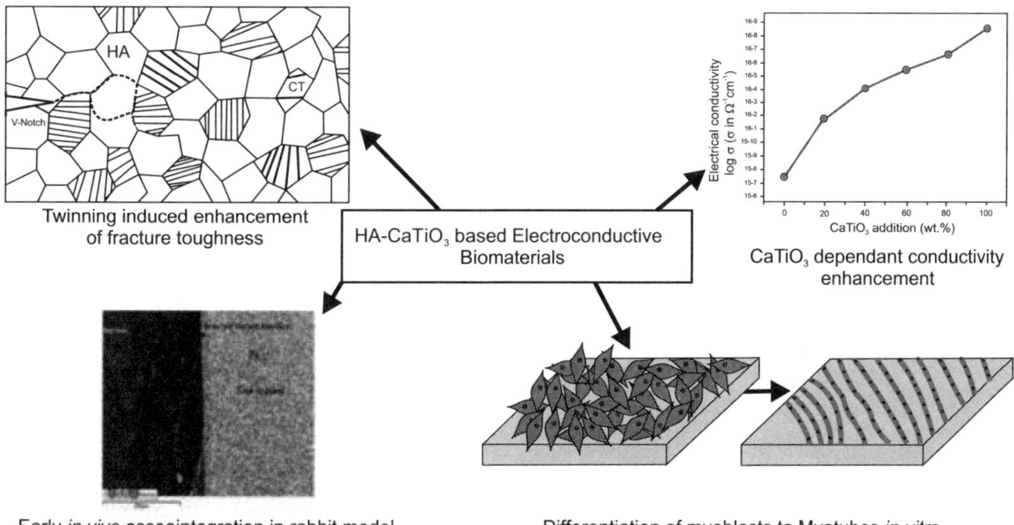

Fig. 16.12 Overview of multifunctional properties of HA–CaTiO₃ composites[1062].

Summarizing, the results discussed in this chapter establish HA–CT composites as a unique class of implantable materials with bone-mimicking electrical conductivity, mechanical and *in vitro/in vivo* biocompatibility properties. In fact, the wide spectrum of mechanical, functional and biocompatibility properties, like in HA–CT system has not yet been established in any other HA based ceramic composite. More details of the results discussed in this chapter can be found in a number of research papers published by the author's research group[1062,1063,1064,1065,1066].

Case Study: Compatibility of Neuronal/Cardiac Cells with Patterned Substrates

In the context of cell compatibility with synthetic biomaterials, the biomaterials substrate composition, surface topographical features as well as ability of a material to support the proliferation of 'difficult-to-culture' cell type constitute some of the major challenges. In addressing each of these challenges, the results obtained with carbon-based substrates and their compatibility with neural as well as cardiac cell types will be discussed in this chapter. In particular, the functionality of Neuroblastoma and Schwann cells on amorphous carbon substrates of varying micro- and nano-geometries is analyzed. Three different regular geometric patterns, long rectangular micro-stripes of varying width and channel spacing, arrays of square pillars and uniform circular quarries, with length scales similar to the cell sizes were fabricated on carbon substrates by photolithography of a negative photoresist, SU-8 followed by its curing. In contrast with flat carbon surfaces, dexterity of carbon fibres in catering short-ranged physical guidance to Schwann cells has been demonstrated. Another part of this chapter discusses cardiac tissue cell functions (specifically, cardiomyocytes and neurons) on composites of carbon nanofiber reinforced poly lactic-co-glycolic acid(50:50 wt% (PLA:PGA) PLGA) to ascertain their potential for myocardial tissue engineering applications. The results indicated that cardiomyocyte density increased with greater amounts of CNFs in PLGA (up to 25:75 wt% PLGA:CNF) for up to 5 days. For neurons, a similar trend to cardiomyocytes was observed, indicating that these conductive materials promoted the adhesion and proliferation of cardiomyocytes important for myocardial tissue engineering applications.

17.1 | Introduction

Major neuronal damage may engender loss of memory, impaired sensory or motor function, muscle dysfunctions and neuropathic pains. Neuronal dysfunctions, stemming from neurodegeneration or injury to the neuronal pathway of the central (CNS) or peripheral nervous system (PNS), are widespread[1091].

The prevalent therapeutic approaches include end-to-end surgical reconnection and nerve autografts for bridging smaller and larger nerve defect gaps respectively[1092]. A multitude of other therapies like allografts[1093], xenograft, acellular grafts[1094] and pre-degenerated nerve grafts[1095] are of limited effectiveness. The post-injury inhibitory environment in the central nervous system often poses further challenges to render these strategies less effective[1096]. A major challenge is in regeneration of nerves across gaps larger than 6 mm, as lesser injuries heal by formation of fibrin cables across the gap[1097]. Thus, approaches involving tissue engineering are becoming increasingly attractive for nerve repair[1098,1099,1100]. Scaffold materials for nerve tissue regeneration can be broadly classified under three different categories: natural (e.g., Chitosan, chitin, collagen, gelatin, alginate), synthetic [non-degradable (e.g., silicone), biodegradable (e.g., PLGA, poly-(ε-caprolactone) (PCL), poly L-lactic acid (PLLA)], and conducting polymers (polypyrrole, polyaniline)[1101] and biosynthetic materials, which integrate merits peculiar to both the aforementioned classes[1100,1101].

> Tissue engineering applications entail specific arrangement of cells into organized structures, such as a neural network, in conformation with the definite arrangement of micro- and nano-biological components, archetypal to the extracellular matrix (ECM).

Hence, while engineering artificial tissue constructs, simulation of the native micro-architecture corresponding to the concerned tissue is of importance[1102]. A variety of techniques including nanofibers self-assembly, solvent casting and particulate leaching, gas foaming, emulsification/freeze-drying, liquid–liquid phase separation, electrospinning, surface patterning and 3D printing have been adopted for the fabrication of tissue engineering scaffolds. Numerous studies have evidenced patterned neuronal cell growth via topographic cues generated by micro-patterning[1103,1104,1105,1106]. Multiple micro-patterning techniques, such as photolithography, soft-lithography, electron-beam lithography, nano-imprint lithography, dip-pen lithography, have been used extensively for fabricating micron- or nano-sized features onto various surfaces in order to emulate some aspects of the structure and length-scale of native tissue topography[1107,1108]. A variety of cell lines, like fibroblasts, osteoblasts, etc., have demonstrated oriented migration along the edges of a patterned surface. Miller et al.[1109,1110] reported neurite alignment and outgrowth along the direction of grooves fabricated on PDLA surfaces. Thompson et al.[1111] structured cell monolayers of multi-cellular aggregates on microlithographically patterned laminin surfaces. Also, the alignment of rat Schwann cells was demonstrated on laminin-conjugated micro-patterned PMMA surface[1112]. Corey et al.[1113] documented localization of neuroblastoma cells via micro patterning technology. The electrospun fibrous scaffolds have garnered much attention owing to their geometric similarity to the native protein fibres of the extracellular matrix.

Electro-spinning affords easy manufacture of fibrous mesh from a wide range of fibres including polymers, ceramics and composites. The efficiency of such fibrous scaffolds as the substrates for neural tissue engineering has been detailed by several research groups[1114,1115]. Kim et al.[1116] reported the potency of electrospun fibres of acrylo-nitrile-co-methlacrylate in fostering neuronal cell growth and development. Such studies have also been supplemented by investigations on neuronal cell development on fibrous meshes of poly-ε-caprolactone (PCL)[1117], poly-L-lactide[1118], collagen, etc. There is also evidence that the rate of axon regeneration is promoted on conducting materials in the presence of an external electric field, which is in contrast with non-conducting materials. Hence, an electrically conducting biocompatible scaffold serves as an added advantage for faster axonal growth and functional recovery[1119].

The increasing interest in carbon materials for biomedical applications (e.g. bioMEMs, biosensors), particularly in orthopedics and neural tissue engineering, is due to their attractive electrochemical and physical properties and the ease of functionalization and porosity control.

Carbon nanofibers have been widely investigated for applications in regenerative medicine[1120,1121,1122,1123,1124,1125,1126]. Carbon nanofibers can be fabricated by pyrolysis of suitable fibre-reinforced polymer precursors[1127,1128,1129] and can be functionalized with biomolecules to form scaffolds for hosting neuron hybrids[1130,1131]. Zhang et al. has evinced guided neurite growth and formation of synaptically communicative networks on functionalized vertical nanotube arrays[1132]. Long-term neural-cytocompatibility of carbon nanotubes and their dexterity in promoting cell adhesion, differentiation has been widely investigated[1132,1133,1134]. Pyrolytic carbon is being used in artificial valves, due to its ability in establishing a coherent interface with living tissue and non-thrombogenicity in contact with blood[1135]. Photolithography of polymer photoresists followed by pyrolysis in an inert atmosphere warrants an easy and robust method for fabrication of textured carbon surfaces[1136,1137,1138]. A prime advantage with such carbon scaffolds rests on easy tunability of the porosity, electrical and electrochemical properties, functionalization and mechanical properties[1139]. Biocompatibility of such carbon-microelectromechanical systems (MEMS) have been studied by several research groups[1140,1141,1142]. The results presented in the first part of this chapter will demonstrate the lack of cytotoxicity of amorphous carbon towards neural cells as well as how surface topographical features can be used to guide cell attachment on carbon substrates.

In continuation to the discussion on relevance of biomaterials science towards human healthcare, it is important to reiterate that none of the existing technology offers the perfect solution for stopping CVD from progressing as well as the instruction of CHF. To address these problems, rapid technological advancements of cell biology, genomics, and proteomics combined with discoveries in material sciences and bioengineering have created a new field of medicine, nanomedicine, utilizing materials and systems which possess at least one physical dimension between 1–100 nm to construct structures, devices, and systems that have novel properties[1143,1144,1145].

An attractive approach to the management of LV dilation and CHF, in relation to CVD, would use a novel nano-inspired conductive elastic patch to the infarcted area and to make a patch both highly elastic and biodegradable so that the material would erode over several weeks and by avoiding cellular components and a polymer, that is FDA approved, that would be more readily translatable to clinical investigations.

One of the innovative idea would be to create a cardio-patch system that would act as a 'Band-Aid' for the infarcted area and allow the regeneration of cardiomyocytes (a specialized contractile muscle cell that generates part of the myocardium tissue of the heart) over the infarcted area.

Recent research has shown when one combines poly lactic-co-glycolic acid (PLGA) and carbon nanofibers (CNFs), which are conductive and are usually grown by catalytic decomposition of certain hydrocarbons where the diameter and length of CNFs[1146], that one can increase the adhesion and proliferation of human cardiomyocytes. Stout et al. showed that using a PLGA: CNF cardio-patch improved cardiomyocyte and neuron functions[1147]. Specifically for cardiomyocyte results, cardiomyocyte density increased by more than 500% when using a patch with CNFs than without. This was also shown with cardiomyocyte proliferated on the PLGA:CNF nano-cardio-patch material; an increase in proliferation density from 530% at day 1 to 700% at day 5 resulted between the cardio-patches and no substrate system, thus indicating that not only does a patch system help in antagonizing LV remodelling due to a myocardial infarction, but the use of a nano-material promote the growth of cardiomyocytes. Furthermore, resent research has shown when electrical stimulation is used on with the PLGA:CNF cardio-patches, the composites promoted human cardiomyocyte proliferation more compared to pure PLGA.

For instance, a myocardial infarction, also known as a heart attack, usually occurs because a major blood vessel supplying the heart's left ventricle is suddenly blocked by an obstruction, such as a blood clot[1148,1149,1150]. During myocardial infarction, part of the cardiac muscle, or myocardium, is deprived of blood and therefore oxygen, which destroys cardiomyocytes and neurons leaving dead tissue[1148,1149,1150] as well as denervation of the myocardium[1151,1152]. In particular, nerve damage to cardiac tissue can result in nerve sprouting in the left ventricle[1151,1153] and development of arrhythmias[1154,1155]. Scarred cardiac muscle results in heart failure for millions of heart attack survivors worldwide.

Cardiomyocytes (a specialized contractile muscle cell that generates part of the myocardial tissue of the heart) depends on continuous conductivity to function. However, such conductivity may break down during heart disease or cardiac dysfunction.

Numerous articles suggest that using nanotechnology can specifically promote cell functions on a variety of materials, ranging from titanium to silicon[1156,1157,1158] due to optimal surface chemistry and wettability which control protein adsorption onto the surface. While some degree of nanostructured features may promote tissue growth, it is also known that certain topographies can hinder cell activity[1159,1160]. For example, using a 1718 gene microarray, Dalby et al. were able to suggest that nano-rough topographies greater than 40 nm decreased human fibroblast cell adhesion[1161].

For the above reasons, the last part of this chapter will discuss cardiomyocyte proliferation on PLGA–CNF composites. A model polymer was used consisting of poly lactic-co-glycolic acid (PLGA), since it has been approved by the Food and Drug Administration (FDA) for therapeutic devices and has desirable biodegradable and biocompatible properties[1162,1163]. More importantly, carbon nanofibers (CNFs), which are conductive and are usually grown by catalytic decomposition of certain hydrocarbons, were studied here The addition of CNFs[1164] can transform non-conductive polymers to be conductive and can be made to mimic natural proteins like collagen.[1165,1166,1167,1168,1169,1170]

17.2 | Neuronal Cell Adaptability on Textured Carbon Substrates

Resorcinol (1,3-dihydroxy benzene), denoted by R, formaldehyde, denoted by F, N, N-Dimethylformamide, denoted by DMF, Polyacrylonitrile, denoted by PAN having Mw = 1,50,000, PMMA and epoxy-based negative photoresist, SU-8, 2000 series formulation (cyclopentanone-based) solvent, as well as SU-8 201J series are used to synthesize carbon substrates. Electrospinning, a simple yet versatile method was used to produce continuous carbon fibres with a wide range of fibre diameters spanning from the micro to nano-domains. On application of appropriate electric field, the polymer solution or melt, introduced within a capillary, accelerates through a spinneret towards a collector kept at an opposite polarity, in a rapid, whipping action, whereby the polymer solvent evaporates leaving the polymer fibres on the collector[1171]. A number of parameters, like the polymer solution concentration and viscosity, rate of dispensing the polymer through the spinneret, the applied electric field strength and pattern, can be adjusted to tailor the architecture and morphology of the resulting scaffold[1172]. For fabrication of microtracks and square quarries, the standard photolithography technique was employed[1137]. SU-8 2015, the photoresist used, was spin coated at an rpm of 3000 for 30 secs, whereby, a film of thickness ~15 μm could be generated. The SU-8 patterns were obtained on bare silicon wafers as well as on PAN (spin-coating parameters similar to that of RF) coated silicon wafers. Eventually, Si wafers were used as scaffold substrate throughout the experiments. The SU-8 posts once fabricated were subjected to pyrolysis, under aforementioned conditions. The resultant thickness of SU-8 derived carbon posts was evaluated to be ~1 μm. SEM images of the porous amorphous carbon substrate are shown in Fig. 17.1, while different surface topographical patterns in Fig. 17.2.

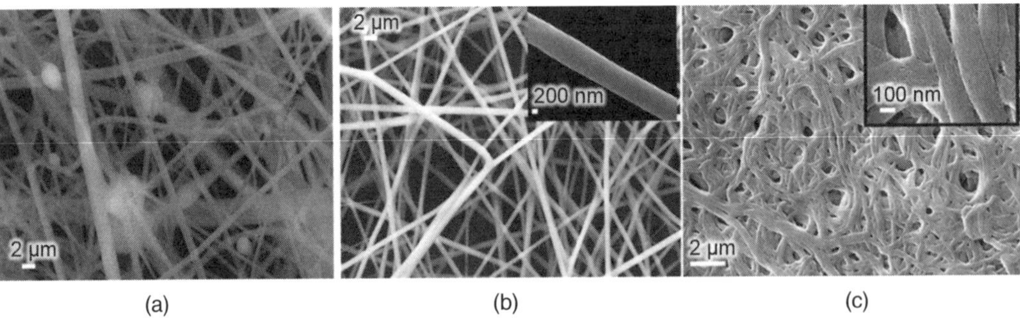

Fig. 17.1 SEM images of (a) electrospun beaded RF/PMMA fibres obtained at a feed rate of 9 µL/min and applied voltage of 1.2 kV/cm; (b) similar fibres obtained at 3 µL/min feed rate and 1.2 kV/cm (optimized conditions) [inset: magnified view (of (b)) of a single fibre]; (c) post-pyrolysis mat of carbon fibres obtained from electrospun RF/PMMA fibres obtained at 3 µL/min feed rate and 1.2 kV/cm [inset: magnified view of (c), showing pores on the fibre surface][1130].

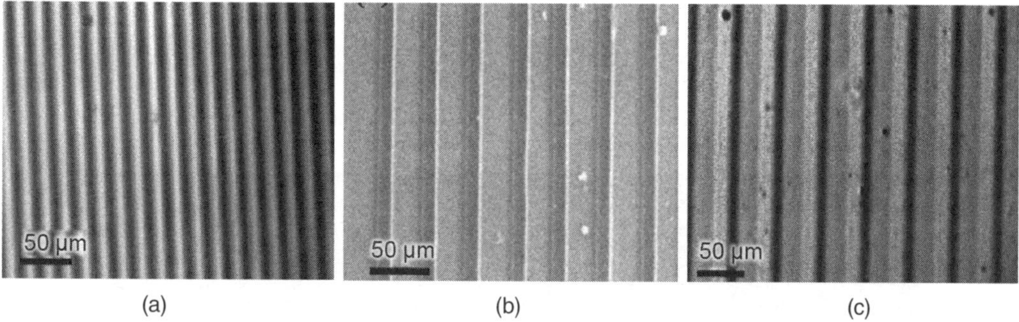

(a)   (b)   (c)

Fig. 17.2 Optical micrographs of microtracks (stripes) of SU-8, fabricated by photolithography (a) 10 µm tracks spaced 10 µm apart; (b) 15 µm tracks spaced 30 µm apart and (c) 20 µm tracks spaced 30 µm apart[1130].

17.2.1 | Neuroblastoma cell functionality on patterned carbon surfaces with microstripes

Neurite guidance, particularly axonal guidance as well as neurite outgrowth, stands as one of the pivotal processes of neural tissue engineering[1173]. Neural cells respond to the various environmental cues, signalling molecules supplied with the growth media, via an expanded terminal structure called the growth cone, at the end of the axon shaft[1174]. A growth cone consists of microtubular structures: (lamellipodia and filopodia), and their interplay with the external cues, materializes into changes in their spatial positions within the growth cone framework[1174]. The phenomenon by which a cell adheres to the substratum, which in turn generates appropriate chemical and physical cues, is commonly known as contact guidance[1175]. Chemical contact guidance is governed by the chemical nature of the substrate, embodied in its cytocompatibility, biodegradability, etc.[1175,1176]. On the other hand, physical contact guidance can be manoeuvered by tailoring the structural properties of the scaffold. For instance, during the development of forebrain, neurons, in the ventricular zone, migrate to the cortical plate via tracks laid down by the radial glial cells[1175]. Hence, the effect of track features must be understood before predicting physical contact guidance of neural cells. In same lines, for exemplifying the potentiality of patterned carbon surfaces for neural tissue regeneration, N2a cells were seeded on carbon microtracks of different dimensions and subsequently imaged under SEM, following 2 days of incubation (Fig. 17.3).

It was observed that, the cells elongated and aligned on and along the microtracks, when the track dimension (15–30 µm) was comparable to the cell dimension (~20 µm) (Figs. 17.3 a, b, and d). In between microtracks, the cells assumed a flattened morphology, with no preferential directionality, akin to flat surfaces (Figs. 17.3 a and b). Indeed, the cells oriented randomly on the groove floor, bridging two parallel tracks. Interestingly, when seeded onto patterned surfaces characterized by parallel stripes intercalated with bare silicon surfaces, (Fig. 17.3d) convalescent cells preferentially aligned on the microtracks or along their edges, in contrast to the exposed silicon surface. This finding might be denotive of superior cytocompatibility of carbon surfaces with respect to silicon surfaces. However, when the track width is smaller than that of the cell, the cells elongate across the tracks, crossing these 'topographic obstacles' to which they would have otherwise aligned (Fig. 17.3c). In a nutshell, depending on the width of the track, neuronal cell behaviour can be classified as 'on', 'off' or 'across' the stripes. Imperatively, the cells 'on' the stripes elongated along the stripes' directions. The cells 'off' the stripes spread on the planar surface between two stripes and the cells 'across' the stripes elongated randomly, unaligned with respect to the stripes. In fact, the latter attached to the stripes and/or to the planar space in between, i.e. straddled between two or more parallel tracks.

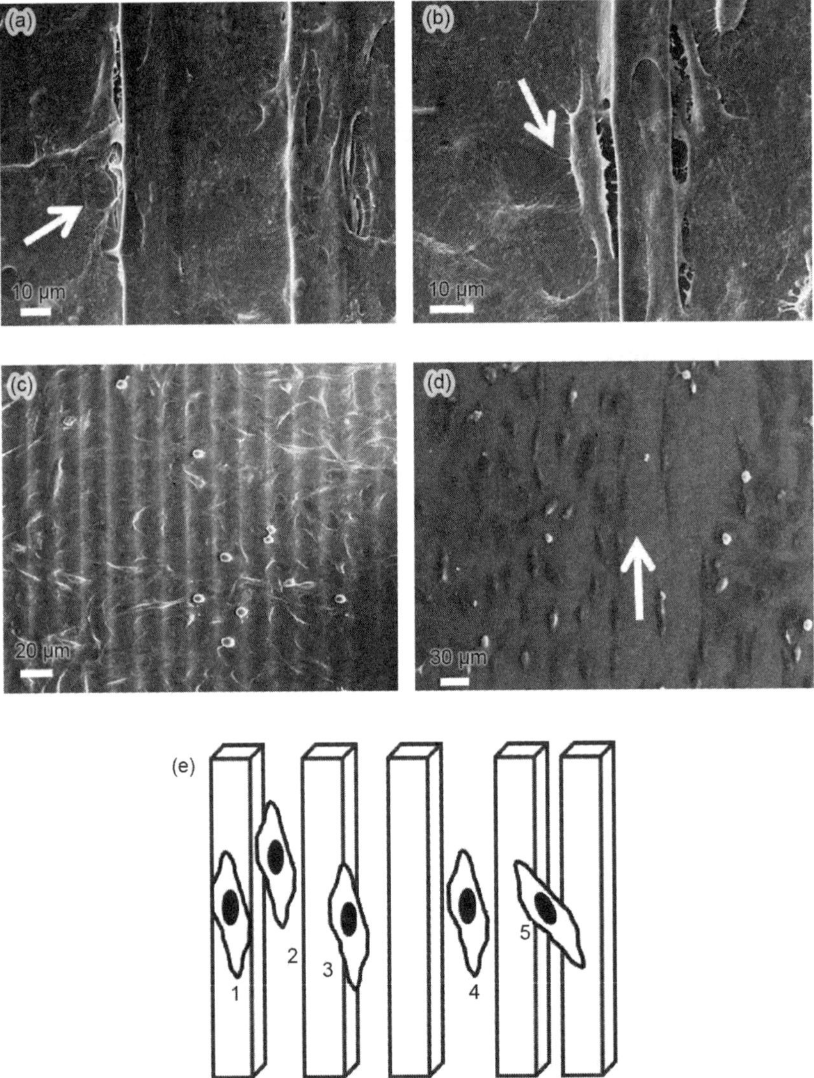

Fig. 17.3 SEM images of N2a cells, cultured for 48 hrs. on micro-patterned carbon surfaces: (a), (b) 15 μm wide stripes, 45 μm apart; (c) 10 μm wide stripes, 10 μm apart and (d) 20 μm wide stripes, 30 μm apart; [(a)–(c): PAN derived carbon film, lying beneath the SU-8 derived carbon microtracks; (d) carbon microtracks on bare silicon wafer]; arrows indicate neurite-like features (a), (b) and cell alignment (d); (e) schematic representation of alignment of N2a cells: (1) 'on', (2, 3) 'along', (4) 'off' and (5) 'across' carbon microtracks; the phenomena 1–4 can be observed when the track as well as inter-track widths are in the range of 15–30 μm; 5 is observed when the aforementioned widths are ~10 μm or less[1130].

In Figs. 17.3a and 17.3b the cells pervaded the total area of the planar groove space, rather than only hugging either of the groove walls. However, no crossing of ridges could be observed for a groove width of 45 µm. In contrast, the cells predominantly aligned along the groove walls, instead of spreading across the surface (Fig. 17.3d). The difference in cell behaviour may be attributed to the apparent chemical anisotropy among the textured surfaces and hence, differential adsorption of serum proteins[1177]. Evidently, the exposed surface chemistry dictates the conformation of adsorbed surface molecules, which in turn govern the cell adhesion mechanism[1178,1179]. The cell alignment on parallel micro-tracks can be visualized with the help of schematic shown in Fig. 17.3e.

Another interesting feature of the cultured cells is the outgrowth of neurite-like features, as shown in Fig. 17.3. Typically, neurites outgrow in the presence of differentiating medium. However, the observed features in the present case suggest neurite-like outgrowth in multiple directions, even in absence of differentiating medium, indicating the possible influence of additional physical cues provided by microtracks towards cellular responses.

The phenomenon in which the cells move along groove edges is known as 'ridge walking'[1177,1180]. A possible annotation to the mechanism of cell alignment can be the inflexibility of cytoskeleton at the ridge edge, leading to demented traction in that direction. As an alternate explanation, it was argued that the inability of the cell to form focal contact beyond the edge might prompt their alignment at the ridge edge[1181]. However, the above results clearly provide evidence towards the formation of focal contacts on the ridge top, side wall and groove floor.

17.2.2 | Schwann cell functionality on fibrous and flat amorphous carbon scaffolds

In the peripheral nervous system, the glial cells are embodied in the form of Schwann cells. In part, columns of Schwann cells direct peripheral axon growth during development and regeneration by catering appropriate geometrically and adhesively permissive substrata[1174]. Conjointly, the Schwann cells also accelerate the rate of dissemination of electrical signals and produce adhesion molecules on the surface of their plasma membranes and trophic factors, like the brain derived neurotrophic factor, glial-derived triophic factor, extracellular matrix molecules, etc., which are essential for neuron growth[1182].

> The tissue engineering of scaffolds with nanoscale features can accentuate the specificity and utility of materials for neural probes, as well as providing guidance channels for neural regeneration.

However, the effects of morphology of such nano-features on cell fate processes of Schwann cells are yet to be elaborated[1183]. Ghasemi–Mobarekh et al.[1184] investigated cell proliferation on both aligned and random nanofibrous polycaprolactone/gelatin scaffolds. Their experiments indicated better proliferation of Schwann cells on random fibres than on their aligned counterparts. It was argued that interconnected pores and surface roughness induced superior adhesion and proliferation of cells on the random fibres. Contradictorily, some researchers have upheld aligned fibres to be more efficacious in sustaining cell growth and development than randomly oriented fibres on account of stronger contact guidance provided by the former[1185,1186].

Interestingly, from the above studies, the differences in behaviour of N2a and Schwann cells, particularly in terms of morphology, when cultured on carbon fibres can be easily adduced. Contrary to the N2a cells, which adopted a polygonal spread-out morphology on carbon fibres, elongated filopodial outgrowths could be observed in Schwann cells, in response to the short-range topographic signals. Some representative fluorescence micrographs also corroborate the cytocompatibility of carbon scaffolds used in our experiments (Fig. 17.4). The formation of extensive inter-cellular junctions and actin filament extensions can be commonly observed for the cultured Schwann cells on carbon substrates. The intertwining of filopodia or cytoplasmic extensions is clearly visible on the carbon film, which is comparable with that on control. Evidently, inter-digitated filopodia protruding from opposing cells facilitated cellular interplay. In contrast, a more condensed reorganized filopodia appearing in the form of strongly aligned long unbranched thin fibres of sizes of ~300 nm are observed on RF-gel derived carbon fibres and no intertwining of such features are observed (Fig. 17.4d). It can be presaged that the fibre diameters being of the same order as filopodia diameter (100–300 nm), the Schwann cells responded to the topographic cues extensively on textured carbon substrates. Concomitantly, filopodial elongation along the length of fibre could be observed. In all probability, contact guidance provided by the fibres culminated in short-ranged alignment along the fibre length.

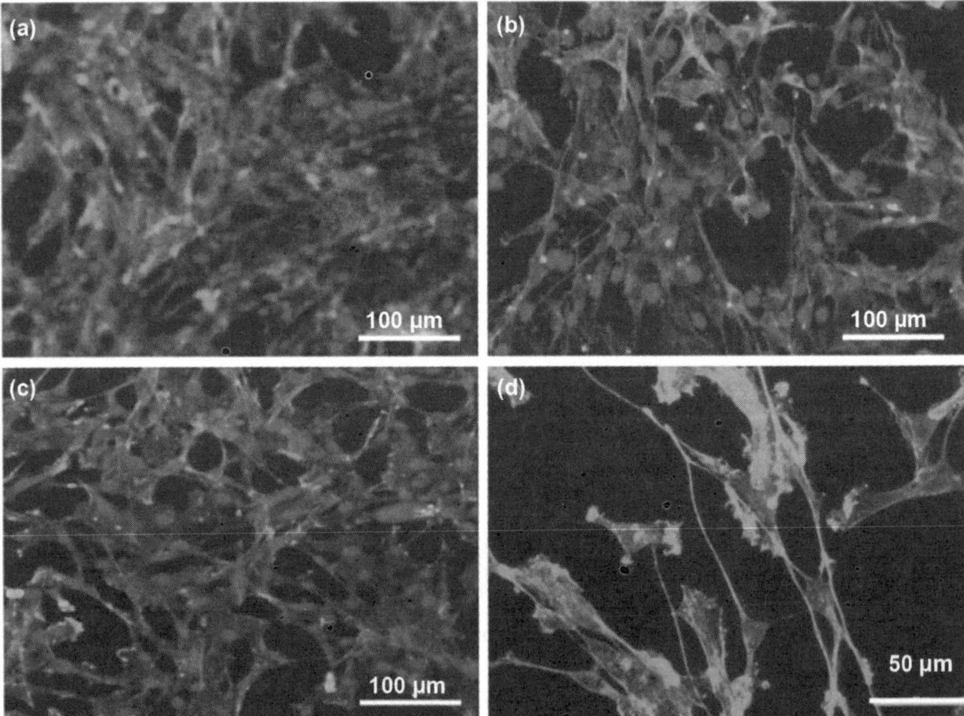

Fig. 17.4 Fluorescent images of Schwann cells, following 2 days incubation, after seeding on (a) control, (b) RF derived carbon film, (c) RF derived carbon fibres and (d) magnified view of (c) showing short-ranged alignment of cells along the fibre lengths; blue stained region represents the nucleus, while green represents the actin cytoskeleton[1130] [see Colour Plate].

Conjointly, filopodia mediates cell–cell cross-talk, whereby a cell searches and communicates with another cell in the cellular microenvironment .

Also, lamellar sheet-like lamellipodial structures aiding cell/substratum anchorage could be distinguished. A neuronal circuit can be conceived from such extensive cell/substratum and intercellular interactions.

The above results are interesting in the perspective of earlier studies reported in the literature. For example, Chew et al.[1187] have upheld that electrospun fibrous scaffolds promote maturation of Schwann cells, when compared with a flat surface. Albeit much dubiety over the ability of seeded Schwann cells in expressing neurotrophins with respect to endogenous Schwann cells, recent investigations have advocated the potency of electrospun scaffolds in emulating the extracellular matrix. The mechanisms of ECM regulated myelination, are yet to be elucidated. However, it can be suggested that integrins interact with the cytoskeleton or modulate myelin-related gene expression[1188]. Webster et al.[1189] also reported the role of specific proteins in modulation of osteoblast adhesion on nanoceramics. In fact, it has been demonstrated that topographic cues, generated from electrospun fibres can actuate myelination in human Schwann cells[1185]. Side by side, their potential in promoting Schwann cell maturation was also established[1185]. It can therefore be surmised that, electrospun fibrous scaffolds can guide the cells to maintain a more physiological phenotype and hence provide better microenvironment for Schwann cell attachment and growth[1190].

17.2.3 | Schwann cell functionality on square and circular patterns

In addition, Schwann cells were seeded on square micropatterns, and representative SEM images of which are shown in Fig. 17.5a. The cells elongated and aligned on and along the edges of the patterns as well as the groove walls, similar to the behaviour on microtracks, demonstrated previously. Simultaneously, the cells on square patches also inter-communicated with cells on the planar grooves via their filopodial extensions. Filopodial bending along the edges of the pattern could also be observed. Interestingly, although cell migration behaviour on the planar groove was somewhat non-complementary to that on a planar surface (Fig. 17.5a), oriented growth of cells, as in case of microstripes, could not be observed either. It may be suggested that, cells adhering on the inter-planar grooves encounter mutually orthogonal topographic cues, furnished by the substrate geometry as well as by the bi/multi-polar extensions, spanning the edges of the square pattern. As a consequence, the cells adopted somewhat disordered orientation on the grooves with respect to the striped substrates.

Likewise, the Schwann cells assumed different configurations in response to circular pits with horizontal and vertical pitches of 10 and 30 μm, respectively. (Fig. 17.5b) The cells not only formed focal contacts along the edges of the pits, but also across the pits and ridges. Lamellipodial extensions probing the boundaries of the pit walls can also be observed. Besides, diagonal extensions across a circular quarry, as also between quarries can be evidenced. A rough schematic exemplification of cell behaviour in terms of filopodia bending along the walls of the patterns or cell growth 'on' and 'across' the square patterns and circular pits respectively have been demonstrated in Fig. 17.5c and 17.5d.

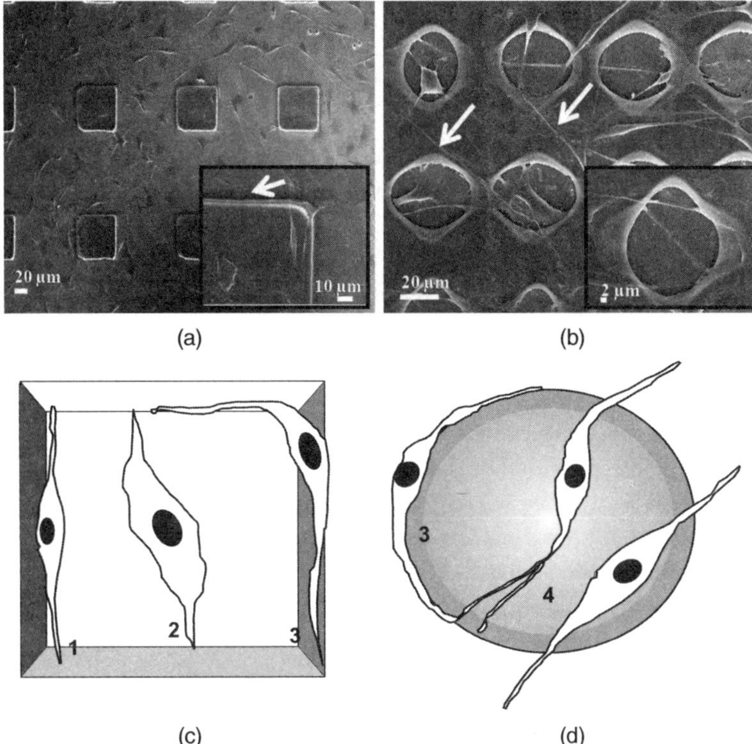

Fig. 17.5 SEM images of Schwann cells on (a)square-patterns of carbon on carbon surface 60 μm wide squares, 90 μm apart [inset: magnified view of (a), showing alignment of cell and filopodia bending along the edge of a square pattern (see arrow)]; (b) circular grooves of radii 20 μm on carbon surface, with horizontal and vertical pitches of 10 μm wide squares, 30 μm apart [inset: magnified view of (b), showing diagonal bridging the edges of a groove, by filopodia]; (c) and (d) Schematic representation of adaptability of Schwann cells on (c) square patterns and (d) circular pits: (1) 'along' the pattern wall, (2) 'on' the pattern surface, (3) 'filopodia bending' along the edges of the patterns and (4) 'across' circular pits[1130].

17.3 | Implications of Neuronal Cell Adaptability on Patterned Carbon Substrates

It is worthwhile to reiterate that despite being widely investigated for potential biomedical applications, the toxicity of carbon substrates remains a concern in the biomaterials community.

This aspect has been presently studied in case of RF-gel derived carbon substrates using two neuronal cell types. In view of their ability to emulate the natural environment in a cellular milieu, thereby arraying appropriate physical cues for manipulation of cellular functions, nanofibrous scaffolds have already garnered attention as promising substrates in the tissue regeneration processes. Electrospun fibres can direct neurite outgrowth and are promising for nerve regeneration applications. While various precursors, like PAN are extensively used to produce carbon scaffolds, the results of an *in vitro* study of cytocompatibility of neuronal cells on carbon fibrous substrates, synthesized from RF-gel, are not widely reported. In particular, textured carbon surfaces, including electrospun fibres, films and micro-topographied surfaces, were engineered and neural cell viability and proliferation were assayed individually. The results undoubtedly asseverated the cytocompatibility of the scaffolds, with respect to neuronal cells, regardless of the precursor chemistry. Nonetheless, the feedback mechanism in response to specific topographic cues is cell-type dependent. For example, although short-ranged alignment along the fibre length was evident with Schwann cells seeded on carbon nanofibers, no such behaviour could be observed in case of N2a cells. However, both the cell types could colonize the entire surface for flat films. In this case, presumably, the cellular orientation and adherence were solely guided by chemical cues, and the cell movements were not constrained by the presence of grooves and ridges. Additionally, the cells could efficiently orient along the microstripes of dimension comparable to their own (~20 µm). Based on the morphological observations of the cultured cells, Table 17.1 summarizes the structural adaptability of two types of neuronal cells.

Table 17.1 Adaptability and orientation of Neuronal cells (Neuroblastoma and Schwann cells) in response to textured carbon surfaces[1130].

Texture dimension (D)	d >> D (carbon nanofibers, ~ 200 nm in diameter)	Dmin < d < Dmax (micro-patterned carbon surfaces)		D >> d (flat carbon films with no surface patterning)
Cell type Neuroblastoma	Extensive spreading out of polygonal cells	Cell alignment on and along the micro-track length for micro-track width and inter-track distances between 15 and 30 µm. Random criss-cross arrangement, bridging two micro-tracks of pitch 10 µm.	Both the cell types demonstrate the well-known phenomenon, called, 'ridge walking', along the edges of the micro-stripes in response to the 'physical contact guidance cues' furnished by the micro-patterns.	Spread-out flattened geometry of cells irrespective of cell type.
Schwann cells	Short-ranged oriented growth and elongation of cells along the carbon fibres	Cell alignment on and along the micro-tracks for width and inter-spacing between 15 and 75 µm.		

In particular, the oriented growth of neuronal cells in response to 'physical contact guidance' manifested by the multidimensional topographic cues can be analyzed in terms of three different length scales, depending on the relative size of topographic feature (D), with respect to the cell size (d) : (a) D << d, (b) $D_{min} < d < D_{max}$, (c) d << D.

(a) *D << d:* The N2a cells assumed flattened, spread-out, polygonal shape with pincer-like edges. The Schwann cells appeared spherical or spindle-shaped with bipolar/multipolar extensions, along the length of the electrospun fibres. Carbon fibres catered topographic cues for short-ranged orientation of Schwann cells along the lengths of fibres. Absence of such behaviour for N2a suggests that multi-dimensional cues generated by random nexus of fibres induce the cells to spread-out establishing extensive filopodial contacts. In other words, the arrangement of fibres and their dimensions, fall short of actuating necessary response mechanisms sacrosanct for cellular alignment.

(b) $D_{min} < d < D_{max}$: Evidently, as suggested by the above studies, micro-tracks or micro-patterns can elicit cellular morphological adaptability, conforming to the patterns engineered, only when the dimensions of the topography features are varied within a narrow window (15–30 µm), irrespective of the cell-type. However, the upper and lower limits to D are exclusive to the cell-type. The Schwann cells demonstrated rough orientation along the pattern edge even, when micro-tracks were spaced at a distance as high as ~ 75 µm (and 90 µm as well). Contrarily, N2a cells spread out in a manner similar to that obtained with flat films, at a lower pattern spacing of 45 µm.

Similarly, although the orientation of N2a cells were lost as the micro-track size was reduced to 10 µm, short-ranged orientation of Schwann cells were preserved to some extent, on carbon fibres, an order of magnitude smaller in dimension than the corresponding cells.

(c) *d << D:* irrespective of the cell type, the cells manifested spread-out flattened geometry on flat films. Extensive filopodial branching could be observed as the cells established inter-cellular contacts in absence of topographic cues generated from the substrate.

> Schwann cells are better adept in responding to topographic cues, when compared with the neuroblastoma cells.

17.4 | Cardiac Tissue-specific Cell Proliferation on PLGA-Carbon Nanofiber Substrates

This section reviews the published results to demonstrate the efficacy of PLGA-CNF substrates to support cardiomyocyte adhesion and proliferation. After obtaining the separately sonicated PLGA and CNF solutions, various PLGA:CNF weight percent ratios were developed (100:0, 75:25, 50:50, 25:75, 0:100) by adding the appropriate amount of CNF to PLGA in 20 ml disposable scintillation vials. The CNF weight ratios were measured using a laboratory balance. When the appropriate ratios were added, each composite material was sonicated at 10W for 20 minutes each.

Human cardiomyocytes were seeded at a cell density of 3.5×10^4 cells/cm² for the cell adhesion assay and 1.5×10^4 cell/cm² for the cell proliferation assay on PLGA:CNF composites in complete growth media supplemented with 10% fatal bovine serum and 1% antibiotics. Cells were seeded onto 12-well human cardiomyocyte stem cell culture extracellular matrix plates on top of the various PLGA:CNF samples, and 22 mm diameter microscope cover slips as controls. The samples were incubated for 4 hours for the cell adhesion assay and 1, 3, and 5 days for the proliferation assay under standard incubator conditions (at 5% CO_2 95% humidified air and 37°C, changing the media every other day). This is followed by fluorescence microscopy observations for qualitative and quantitative assessment of cell morphology.

Some representative SEM images of the as-synthesized PLGA:CNF composite surfaces are shown in Fig. 17.6. CNFs were uniformly dispersed within the PLGA matrix and, as expected, more CNFs were observed for the higher CNF ratio samples. A typical SEM image showed that the CNFs had uniform diameters (nm) and the variation of the fibre diameters were within ±15 nm. Also, no fibre clustering was noticed even in the sample with the highest CNF amounts (25:75) [PLGA:CNF (wt/wt)].

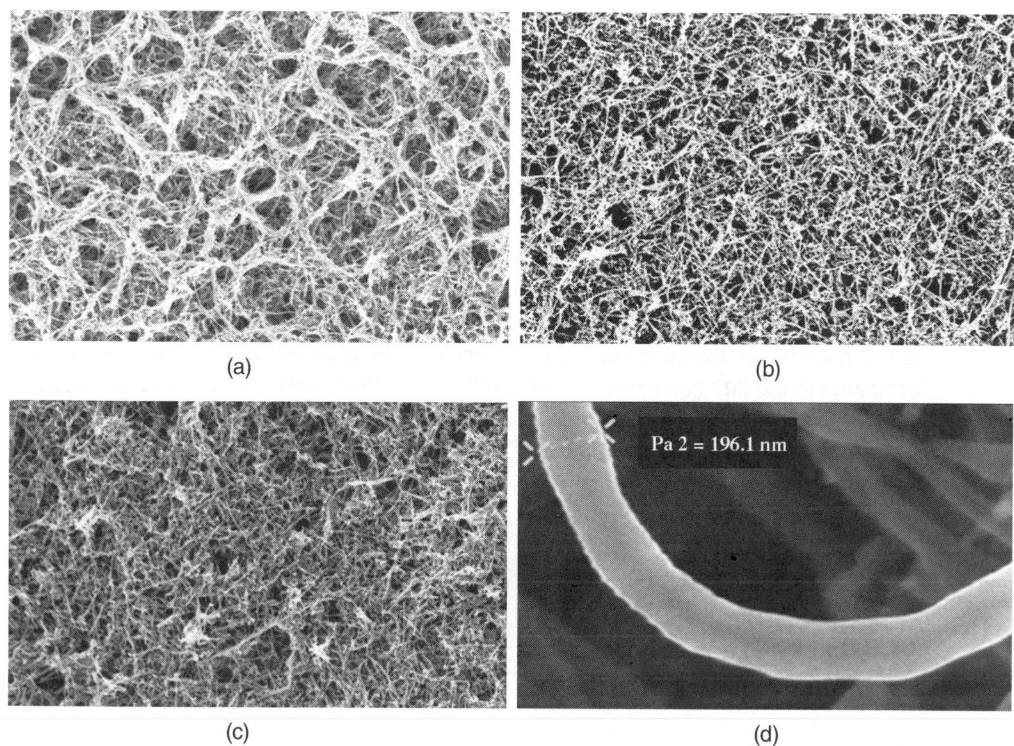

Fig. 17.6 SEM images at 5000x magnification showing the distribution of CNF fibres in the PLGA matrix at different PLGA:CNF (wt:wt) material ratios for (a) 75:25, (b) 50:50, (c) 25:75 and (d) a typical high resolution image revealing one CNF diameter is also shown[1147].

The results also provided evidence that increasing the CNF weight ratio in PLGA increased the conductivity as shown in Fig. 17.7. The pure PLGA matrix (i.e., 100:0 [PLGA:CNF (wt:wt)]) had negligible conductivity. Clearly, the increased conductivity of the composites was due to the uniform distribution of more conductive CNFs. Therefore, CNF addition increased conductivity irrespective of CNF diameter (Fig. 17.7).

Complimentary evidence of the confirmation of the crystallinity was obtained using Raman spectroscopy and the results are provided in Fig. 17.8. Two sets of Raman bands are clearly visible in the composites with one set of Raman shift at 173 and 278 cm^{-1} recorded in the presence of CNF. These Raman bands are characteristic of C-C bonds. The Raman bands at 1345 cm^{-1} and 1570 cm^{-1} arose particularly due to C=C bending or bonding.

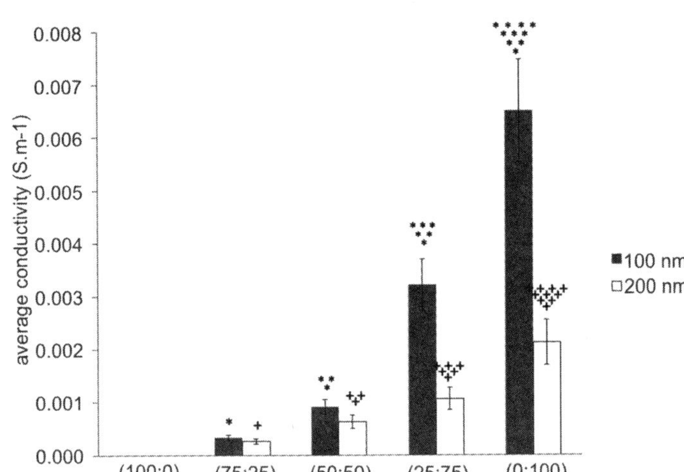

Fig. 17.7 Conductivity of PLGA-CNF substrates; data are mean conductivity values ± s.d. n=3; *p < 0.05 compared to the 100 nm diameter CNF 100:0 (PLGA:CNF) ratio; **p < 0.05 compared to the 100 nm diameter CNF 75:25 (PLGA:CNF) ratio; ***p < 0.05 compared to the 100 nm diameter CNF 50:50 (PLGA:CNF) ratio; ****p < 0.05 compared to the 100 nm diameter CNF 25:75 (PLGA:CNF) ratio + p < 0.05 compared to the 200 nm diameter CNF 100:0 (PLGA:CNF) ratio; ++ p < 0.05 compared to the 200 nm diameter CNF 75:25 (PLGA:CNF) ratio. +++ p < 0.05 compared to the 200 nm diameter CNF 50:50 (PLGA:CNF) ratio. ++++ p < 0.05 compared to the 200 nm diameter CNF 25:75 (PLGA:CNF) ratio[1147].

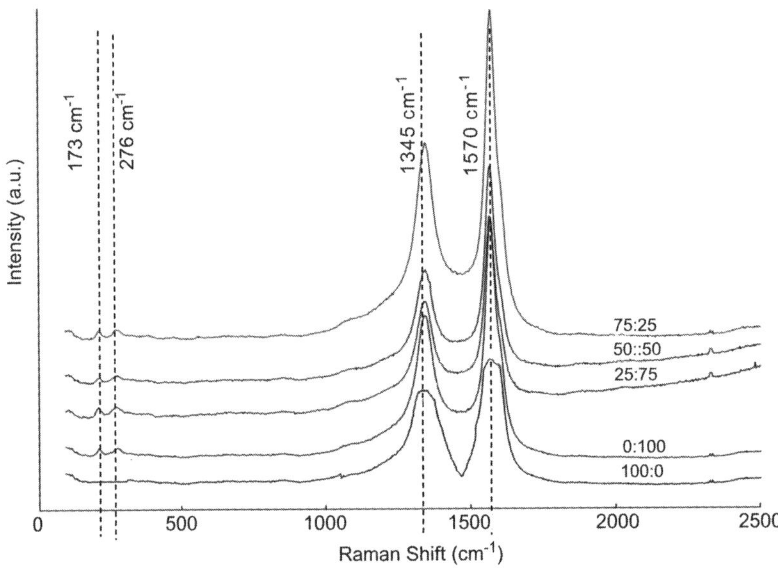

Fig. 17.8 Raman spectroscopy results obtained for the 200 nm CNF diameter composites of different PLGA:CNF ratios. As CNF addition increases, the amount of 'different' C-C bonding decreased and the 1345 and 1570 cm^{-1} peak decreased in intensity; peaks at 173 and 276 cm^{-1} were due to the carbon in CNFs; similar results were found for the 100 nm diameter CNFs[1147].

In order to assess the compatibility of cardiomyocyte cells on PLGA–CNF substrates, both cell adhesion and proliferation were quantified. For cell adhesion, the cell culture experiments were conducted for four hours and the initial attachment of cells were further quantified in terms of cell density. All the PLGA–CNF substrates exhibit statistically significantly higher cell attachment after 4 hours with respect to PLGA. As far as the influence of CNF size is concerned, no unique trend with CNF addition was observed. For example, PLGA:CNF (75:25) composite, 100 mm diameter seems to favour better cell adhesion. However, at higher CNF loading ($\geq$75%), 200 nm CNF appears to promote better cell adhesion. Nevertheless, the CNF addition clearly enhances the cardiomyocyte cell adhesion, *in vitro*.

Considering cell proliferation, the cell density is quantified after culture experiments at 1, 3 and 5 days and the results obtained with different PLGA-CNF formulations having 100 nm CNF as shown in Fig. 17.10.

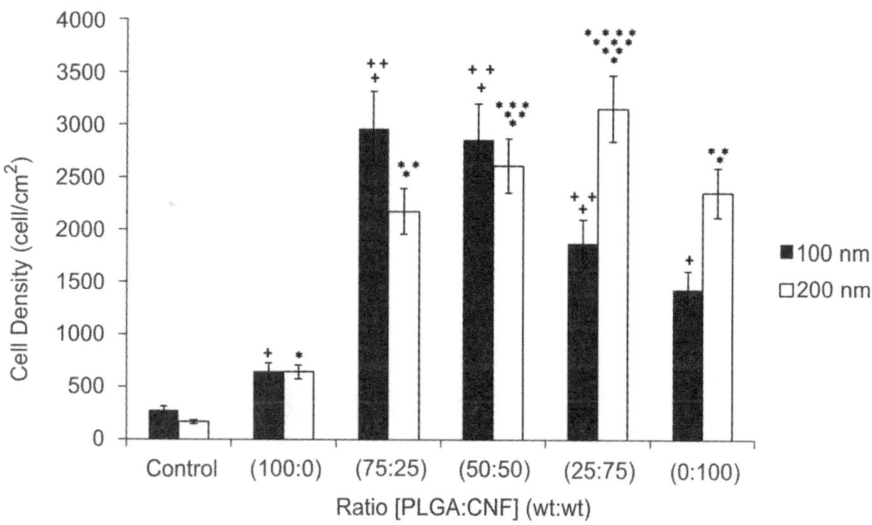

Fig. 17.9   Cardiomyocyte cell adhesion after 4 hours on different PLGA-CNF substrates; seeding density = 3500 cells/cm²; white bars indicate the 200 nm diameter CNFs and grey bars indicate the 100 nm diameter CNFs; data are mean density values ± s.d. n=3. Control was a glass substrate; data are mean density values ± s.d. n=3; *p < 0.05 compared to the 200 nm diameter CNF 100:0 (PLGA:CNF) ratio; **p < 0.05 compared to the 200 nm diameter CNF 75:25 (PLGA:CNF) ratio; ***p < 0.05 compared to the 200 nm diameter CNF 50:50 (PLGA:CNF) ratio; ****p < 0.05 compared to the 200 nm diameter CNF 0:100 (PLGA:CNF) ratio; +p < 0.05 compared to the 100 nm diameter CNF 100:0 (PLGA:CNF) ratio; ++p < 0.05 compared to the 100 nm diameter CNF 0:100 (PLGA:CNF) ratio[1147].

Cardiomyocytes are one of the important examples of slower proliferating cells with typical doubling time of around 72 hours. These cell types do NOT grow on many biomaterals.

Interestingly, a clear trend and transition in cell proliferation can be traced while analyzing the data presented in Fig. 17.10. For example, cell proliferation increases with 25% CNF addition to PLGA and thereafter systematically decreases with more CNF addition. Such a trend is observed independent of time in culture in a statistically significant manner (see Fig. 17.10). Interesting enough, the cell proliferation is equally low on both PLGA and CNF alone. Overall, the data presented in Fig. 17.10 clearly suggest that Cardiomyocyte cell proliferation is enhanced only in a narrow window of CNF addition to PLGA and clearly, there is a need to tailor CNF addition. If one compares the conductivity data of Fig. 17.7 and cell proliferation data of Fig. 17.10, it is apparent that PLGA:CNF (75:25) with moderate conductivity can support better cell proliferation than PLGA–CNF substrates with better conductivity. Representative SEM images of cardiomyocytes attached to PLGA:CNF samples are shown in Fig. 17.11, indicating close interactions between cardiomyocytes and all substrates after 48 hours of seeding. This was determined by examining three characteristics of cell morphology on the (PLGA:CNF) composites: cell filopodia and lamellipodia extension, (b) cell elongation, (c) cell-to-cell attachment. These aspects facilitate numerous cellular interactions with the substrates. Also, it can be seen that extensive cardiomyocyte spreading occurred over multiple CNFs. Another observation was the slight shift of these Raman bands in higher CNF containing PLGA composites or with 100% CNF. Collectively, such evidence provides evidence that some of the CNFs were present on the surface of the PLGA composites.

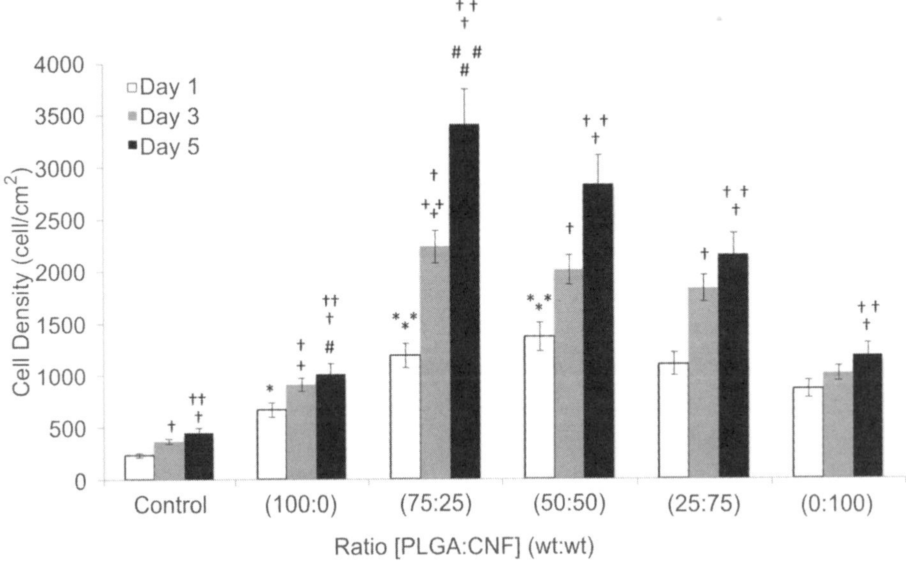

Fig. 17.10 Cardiomyocyte cell proliferation after 1, 3, and 5 days on PLGA-CNF (100 nm diameter) composites; seeding density = 1500 cells/cm^2; data are mean counts to n=3; control was glass substrate; †p < 0.05 compared to day 1 proliferation of same material ratio; ††p < 0.05 compared to day 3 proliferation of same material ratio *p < 0.05 compared to 100:0 (PLGA:CNF) on day 1; +p < 0.05 compared to control (PLGA:CNF) on day 3; ++p < 0.05 compared to control (PLGA:CNF) on day 3; the numbers along X-axis means PLGA:CNF ration in wt%[1147].

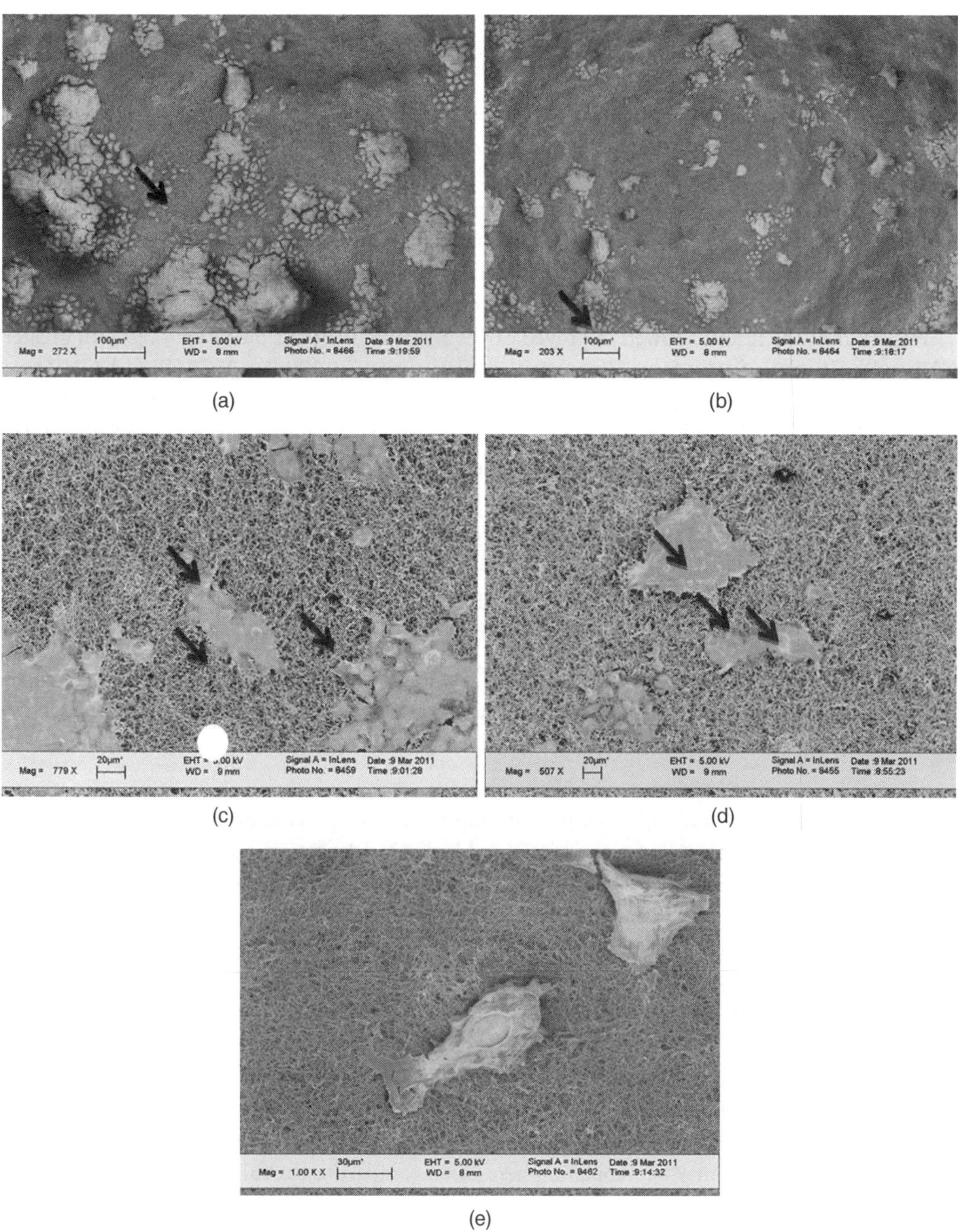

(a) (b) (c) (d) (e)

Fig. 17.11 Representative SEM images revealing specific cardiomyocyte morphology on PLGA: CNF composites with 100 nm diameter CNF: (a) 75:25 [PLGA: CNF (wt:wt)], (b) 50:50 [PLGA:CNF (wt:wt)], (c) 25:75 [PLGA:CNF (wt:wt)], and (d) 0:100 [PLGA:CNF (wt:wt)] composites; detailed close up of a single cardiomyocyte (e); arrows indicate important specific cell morphology, cell spreading and filopodia extension[1147].

17.5 | Implications of Cardiomyocyte Cell Proliferation

Recently, it has been shown that composite materials composed of different matrices (conductive vs. non-conductive and biodegradable vs. non-biodegradable) can stabilize heart tissue to decrease the consequences of a myocardial infarction and such approaches include but are not limited to culturing patient tissue, direct cell injection , biomaterial scaffolds (materials ranging from elastin to 3D PLGA fibres), 3D printing and injectable scaffolds using fibrin based materials. The results discussed in this chapter assessed a new material, PLGA:CNF composites, for enhancing conductivity and cytocompatibility properties necessary for cardiovascular applications.

Importantly, the results showed improved cardiomyocyte functions on composites with greater CNF in PLGA. Specifically, for cardiomyocyte results, comparing 100:0 to 25:75 [PLGA:CNF (wt/wt)] composites, cardiomyocyte density increased by more than 500%. As cardiomyocytes proliferated on the different PLGA:CNF ratio materials, an increase in proliferation density from 530% at day 1 to 700% at day 5 resulted between the 100:0 and 25:75 [PLGA:CNF (wt:wt)] ratio materials.

It is worth speculating why greater cardiomyocyte was observed on such composites with increasing CNF in PLGA. Researchers found that the adsorption and bioactivity of vitronectin increased on nanophase ceramics and that this correlated to enhanced osteoblast functions (including adhesion, proliferation, and differentiation), these studies reported the closer the nanometer roughness of bone was emulated in such ceramics, the more bone growth that resulted[1189].

CNFs possess nanoscale geometries which imitate the extracellular matrix of various tissues (such as the heart), potentially leading to improved cytocompatibility properties.

It can be suggested that CNFs can play an important role towards promoting cardiomyocyte and neuron density by increasing victronectin and lamimin adsorption, which in turn will induce cell adhesion and proliferation[1190]. While the mechanism of enhanced cardiomyocyte and neuron density is not thoroughly known at this time, it could have to do with the topography of PLGA:CNF composites and/or the increased presence of CNFs on PLGA surfaces which controlled initial protein adsorption events through altered surface energetics.

Importantly, it is also known that a material can be too rough and can hinder cellular activity. For example, diamond films have been created with nanometer and micron scale topographies, through microwave plasma enhanced chemical vapour deposition and hydrogen plasma treatment, and cell studies have shown decreased osteoblast adhesion and proliferation on micron-sized diamond topographies compared to nano-sized diamond topographies. Although requiring more studies, the same events may be happening here when comparing the fully dense CNF substrates to the PLGA:CNF composites.

In addition, the results discussed in this chapter showed that when adding CNFs to PLGA, the composite became conductive, whereas the PLGA matrix alone was not conductive. Pedrotty et al. showed that numerous cardiac cell functions (including adhesion, proliferation, and migration) may be modulated by electrical stimulation[1191], hence requiring the use of a conductive material in cardiac applications. Also, Mihardjo et al. demonstrated that enhanced myocardial repair following an ischemic injury could be achieved using conductive polymers, such as polypyrrole[1192]. The conductivity values measured in the present work were lower than that of the heart tissue (ranging from 0.16 S/m longitudinally to 0.005 S/m transversely)[1193], but future techniques (such as CNF alignment) may increase the anisotropic conductivity to match that of heart tissue. In addition, here, the pure CNF materials may have been too conductive and surpassed a stimulatory conductivity for the cells, thus, explaining decreased cell functions on the pure CNF compared to PLGA:CNF composites; the clearly requires more investigation.

17.6 | Closure

The *in vitro* studies using neuronal cells confirm that the length scales of amorphous carbon structures have a strong influence on the adaptability of neuroblastoma and Schwann cells. On juxtaposing the respective behavioural aspects of two cell types, it can be inferred that Schwann cells are more adaptive towards topographic cues than Neuroblastoma cells, which demonstrated aligned, oriented growth on and along micro-tracks and grooves of width 15–30 μm. However, pattern guided growth of the Schwann cells could be observed even at greater groove dimensions of 75 μm. This could be attributed to the innate ability of Schwann cells in directing end-to-end parallel alignment among themselves. Also, short-ranged alignment along fibre axes was noted for Schwann cells on electrospun carbon fibres, in direct opposition to the spread-out random arrangement for N2a cells. However, no morphological difference could be perceived, when both the cell lines were seeded independently on flat 2D carbon scaffolds. The results summarized in this chapter therefore substantiate the potential of amorphous carbon substrates towards application as 3D scaffold systems for peripheral nerve regeneration, in terms of adhesion and proliferation of Neuroblastoma and Schwann cells. More details on the results discussed in this chapter related to the influence of surface patterning on the cell functionality can be found elsewhere[1194]. It is hoped that future studies will emphasize more on the advantages offered by the amorphous carbon substrates such as easy and versatile tuning of micro/nano textures, porosity, functionalization, and electrochemical and mechanical properties.

The results shown in this chapter also illustrated that for cardiomyocytes proliferation, the best composite for 200 nm diameter CNFs was the 25:75 [PLGA:CNF (wt:wt)] composite and for the 100 nm diameter CNFs, 75:25 [PLGA:CNF (wt:wt)] composite was the best. More results are discussed elsewhere[1147].

Perspectives

This chapter highlights the need for an integrated understanding or a unified approach, while designing biomaterials and assessing their biocompatibility. In addition, the important pathways to be followed in translating lab-scale research to clinics (bench to bedside translation) are also described. Overall, the discussion in this chapter is focused on materials development, biocompatibility assessment, and product commercialization related challenges. Importantly, this chapter also summarizes the major challenges in the design and smart fabrication of patient-specific implants and translational research. The education and training of young researchers as well as the involvement of clinicians and biomedical enterprenauers are emphasized.

18.1 | Integrated Understanding of Biomaterials Development

As discussed in chapter 3, most of the conventional scaffolds developed so far lack molecular structures and surfaces that can effectively interact with biological systems in a manner, most desired to induce favourable cell/tissue response for the targeted biomedical applications Some research groups have identified new approaches that are needed, usually through the use of decellularized extracellular matrices[1195,1196] or nanostructured biopolymer hydrogels[1197,1198]. If we consider the microstructure of the majority of conventional scaffold materials, we realize that a manufactured microporous synthetic polymer or ceramic, produced for example by electrospinning or solid free-form fabrication[1199], does not represent the normal microenvironment of the natural biological system, within which cells function. In fact, we cannot expect, either stem cells or fully differentiated cells, to express new tissue when they are in such an unnatural environment[1200]. Additionally, these synthetic scaffold materials have been designed to be 'safe' through molecular structures and surfaces that do not interact with biological systems (see references under 'For Further Reading').

Biomaterials science is a rapidly growing interdisciplinary field of study of translational nature, bringing together specialists from different engineering disciplines, including materials science with those in cell/molecular biology and medical sciences with a comprehensive focus on the repair and replacement of human tissues.

To substantiate the above discussion, the recent advances in stem cell science[1201] and developmental biology[1202,1203] offer exciting possibilities of stimulating the tissue generation processes that exist in the embryo and foetus, but clinical application has proved to be elusive in adults[1204]. This process could conceivably be induced through cell therapies, wherein certain target cells could be injected into an appropriate site and given the suitable signals that cause them to express the new tissue[1205]. Whatever the merits of this approach, it is widely recognized that any tissue that is generated by the activity of these cells will require a physical structure. This implies that the regenerating tissue will need a guiding template, and there will also be a need for coordination, with spatio-temporal control of the molecular and physical signalling, again implying the use of a supporting template[1206]. The use of such a template (often referred to as a scaffold) alongside the target cells and signalling molecules constitutes the essential paradigm shift in the biomaterials science approach[1207]. The choice of the material for such templates remains an unresolved issue, that severely limits progress in promoting and fully adopting these technologies[1208].

It is worthwhile to reiterate that the adult human body has a very limited innate ability for self-repair and regeneration. Bone and skin can heal under several different circumstances[1209], and tissues within a few organs such as the liver have some powers of regeneration[1210]. However, functional regeneration of most tissues within, for example, the musculoskeletal[1211], nervous[1212], cardiovascular[1213] and urological systems[1214] is rarely achieved. Thus, some acute traumatic conditions, such as spinal cord injury, degenerative diseases, such as Parkinson's disease, Alzheimer's disease and macular degeneration, and auto-immune diseases such as Type I diabetes, have no possibilities of cure through regenerative processes.

The futuristic objective should be to determine the generic rules that govern how the functional properties of biomaterials in tissue engineering templates control their ability to facilitate tissue repair and regeneration.

18.2 | Unified Approach of Biocompatibility

Traditionally, the structure-property correlation is an important concept in the field of materials science. The development of materials follows the microstructure-property correlation, which is based on both the qualitative and quantitative relationship between microstructural parameters and mechanical or functional properties. For example, the strength of a metal with conventional microstructure is inversely proportional to the square root of grain size, as per the Hall–Petch relationship. Similarly, the magnetic or electrical or thermal transport properties depend on the elements of the material microstructure. The key microstructure elements are grain size, grain boundary phases or triple pocket phases and distribution of second phases (toughening/strengthening elements). In view of such considerations, researchers in the field of materials science tailor the processing parameters to closely controlling the microstructure, while developing a new material.

The above described concept cannot directly be extended to understand how the microstructure influences the biocompatibility property. Nevertheless, biomaterial scientists often correlate the biocompatibility property with surface properties, such as hydrophilicity/hydrophobicity and surface stiffness. This can be further substantiated by the fact that material surface properties influence protein absorption, which in turn determines the cell response. More quantitative approaches should be adopted to establish the correlation between the above-mentioned microstructural elements and biocompatibility properties.

In the case of degradable biomaterials, both scaffold composition and pore architecture have significant influence on the degradation kinetics. However, here again quantitative understanding is largely missing, limiting predictive modelling approaches. At present, degradation data are fitted to well-known kinetic equations to establish the order of degradation kinetics. This may require some thoughtful experiments and careful analytical approaches.

Despite substantial experimental evidence of the influence of substrate stiffness/surface energy/conductivity on the cell functionality modulation, the absence of any serious analytical attempt to capture quantitatively such correlation is mainly due to the involvement of multiple factors, such as the use of different cell types, the cells being in different growth stages even when a similar cell type is used in different studies, the difference in culture conditions (time in culture, medium composition) and difference in material substrate composition.

With reference to the *in vivo* biocompatibility properties, the neobone formation around the implant or the degradation *in vivo* (fibrous capsule formation kinetics) can not, at present, be predicted in a time-dependent manner on the basis of the implant composition and surface characteristics in a given animal model. Again, this is a difficult task since various defect models influence the biocompatibility of implants/scaffolds, *in vivo*.

The long-term implant stability is another important parameter to be considered in the case of orthopedic biomaterials. One of the factors having paramount influence is the elastic stiffness of the implants. Although radiological evidence is largely available to support the fact that the difference in the elastic stiffness of the implant with the neighbouring host bone is the major factor, none of the studies has ever come up with a thorough understanding of the different factors contributing to long-term implant stability in an integrated manner.

If the field of biomaterials science and tissue engineering is to grow in the next few decades, quantitative understanding has to evolve in a significant manner and this will pave the way for designing futuristic biomaterials in a predictable manner.

18.3 | Patient-specific Implants

The greatest difficulty in conceptualizing materials development in the context of biomedical applications is that one has to perceive the importance of biocompatibility. This is in contrast to mechanical or functional (optical, magnetic, electrical) properties requirements, that are of prime importance for non-biomedical or more precisely for structural or functional applications. While the former is the domain of a biomaterials researcher, the latter is the domain of a material scientist interested in developing materials for non-biomedical applications.

It may be worthwhile to mention that a given material can be suitable for both biomedical and structural applications. Ti6Al4V is one of the popular examples. In the context of biomedical applications, HA-coated Ti6Al4V has attracted wider attention, and the coating of bioactive HA on a bioinert Ti6Al4V ensures better biocompatibility or osseointegration. This demands the selection of an appropriate deposition technology to develop coatings with desired composition and adherence properties. The conventional metallurgical manufacturing methodologies can be utilized to produce implants with desired shape. Two popular biomedical applications of Ti6Al4V are as stems in total hip joint replacement and as screws in dental implants.

While the conventional processing approaches can be used as the first step in development of any new biomaterial, the fabrication of patient – specific implants require advanced manufacturing approaches, like additive manufacturing. Similarly, conventional structure characterisation tools are to be necessarily used in the first level of analysis, 3D microstructure characterisation using micro-computed tomography would provide more quantitative insight into pore architecture in scaffold-based tissue engineering research. Also, micro-CT, a non-destructive technique, can be used to image biomedical devices in 3D space. Similarly, the biomechanical properties of implantable biomaterials for load-bearing biomedical applications (articulating joints) are to be evaluated with device prototypes. In lab-scale research, one measures the strength, elastic modulus, fracture toughness of biomaterials using conventional property measurement technique as described in chapter 6 of this book. These properties are typically measured on test samples having simple geometrical shape (rectangular or cylindrical) or smaller dimensions. The biomedical device prototype, based on lab-scale tested biomaterials with unaltered composition, can have different values of material properties, as the defects or pores are expected to increase with the volume of test sample. Nonetheless, it needs to be kept in mind that the mechanical properties, measured in lab-scale research, enable researchers to compare various test compositions before designing *in vitro* experiments. However, such lab-scale results should in no way be extrapolated to predict the device performance. This indicates the importance of device-level testing.

Although additive manufacturing technologies, like laser-engineered net shaping (LENS) are being investigated to develop patient-specific Ti6Al4V implants, the combination of strength and elastic modulus of such implants needs to match conventionally (metallurgical) processed implants. This aspect needs to be addressed in future research.

With reference to medical devices, the advances in the field of materials and manufacturing enable the construction of high-quality electronic and optoelectronic based devices, which can readily integrate with the soft, curvilinear surfaces of the human body. The resulting capabilities create new opportunities for studying disease states, improving surgical procedures, monitoring health/ wellness, establishing human-machine interfaces, and performing other functions. In particular, the usefulness of such devices in terms of their integration with the brain, the heart, and the skin needs to be established further through extensive clinical trials. With the advent of new manufacturing technologies, the fabrication of custom made biomaterials is being attempted in various research groups. Typically, rapid prototyping allows implant prototypes to be produced in a wide range of materials with remarkable precision in a couple of hours.

The combined use of medical imaging tools (CT scanner) and rapid prototyping technologies enables the cost-effective and time-efficient development of personalized biomedical devices. To this end, the use of micro-CT, as described in an earlier chapter, is extremely relevant. Computed Tomography (CT) based techniques are becoming increasingly popular in the field of biomedical engineering. Fundamentally, CT is described as a 3D X-ray imaging method that involves obtaining X-ray position images at many angles of view around an axis through an object and then applying a tomographic reconstruction algorithm to produce a series of thin tomographic images of continuous trans-axial slices through the investigated object. Depending on temporal and spatial resolution, CT based techniques can be used to image objects over a wide length scale. For example, a mouse or an entire liver can be imaged using mini-CT, while nano-CT can be used to image cellular organelles (less than 1 micron). It is expected that in near future, various classes of micro-CT approaches, e.g., attenuation based scanning, fluorescence based scanning, scatter based scanning, phase contrast scanning, can be adopted to develop better bioimaging techniques in pre-clinical studies.

It should therefore be well perceived that translational research requires meaningful and close interaction among active researchers from multiple disciplines as well as the participation of clinicians and biomedical entrepreneurs.

> Considering the fact that numerous groups around the world have invested significant efforts to develop a large array of biomaterials, often within small variation in process parameters/ compositional window, an important issue remains to be explored is how to establish the conceptual and biologically relevant bases for the generic specifications for the biomaterial templates.

18.4 | Design and Smart Fabrication of Implantable Biomaterials

Disease and trauma often lead to a loss of function of tissues and organs in the human body. The supply of donor organs is rather limited, and insufficient to meet the needs of the large number of patients needing transplants. A biomaterial is defined as any natural or synthetic substance engineered to interact with biological systems for a specific targeted biomedical application, including diagnostic, screening, care and treatment. The first generation of biomaterials consisted of off-the-shelf materials, that were designed for other applications, but found use in medicine because they were deemed to be minimally toxic and "inert". In recent times, there has been an increasing interest in design and engineering of materials intended for use in biomedical applications. Polymers represent the largest and most versatile class of materials being extensively applied in a wide variety of biomedical applications. This versatility is attributed to the relative ease with which polymers can be synthesized with a wide variety of structures, and the ability to tune their physical, chemical, and surface properties to mimic the biologically relevant properties in nature. Synthesis and processing of smart biodegradable polymers is an active area of research with the aim of improving the therapeutic value of various water-soluble/insoluble drugs and bioactive molecules by enhancing bioavailability, solubility and retention time with reduced risks of toxicity.

Smart polymers used for delivery can offer various advantages, such as improved bioavailability, bioactivity, and controlled and targeted delivery. For example, polymers can be synthesized by incorporating drugs within the polymer backbone with different chemical conjugation techniques, which will allow for tuned release kinetics of the loaded drug. Such drug-loaded bioresorbable polymers can be implanted in a patient after a surgical procedure to control inflammation. Alternatively, such polymers can be placed as a patch in the gastrointestinal tract or airways of the lungs for sustained release of the drug for treatment of severe bacterial infections, for example.

The above approach also offers unprecedented opportunities to further utilize such polymers for diagnostics and treatment. A number of diseases can be potentially treated using such novel strategies. Smart nanoparticles can be designed to respond to intrinsic and extrinsic triggers for delivery of their cargo. Intrinsic triggers include various changes in the local microenvironment *in vivo*, such as temperature, pH, and metabolite concentration, etc. Alternatively, extrinsic triggers such as magnetic fields, IR radiation, etc., may be used to guide the particles and trigger their action in the body. In addition to therapy and treatment, such particles can also be used for diagnostics by augmenting the signal from the pathological site in different imaging techniques, such as fluorescence and MRI. Smart and responsive nanoparticles can be developed for "theranostics", wherein the multifunctional nanoparticles will serve a dual purpose of diagnostics and therapy. In general, the nanoparticles can be designed to deliver drugs for chemotherapy, proteins for immunotherapy or therapeutic genes for gene therapy.

> 3D printing emerges as a viable strategy for the rapid preparation of tissue scaffolds, which can recapitulate many of the architectural complexities of human tissues.

To extend the above discussion further, biodegradable polymers can be moulded into complex 3D scaffolds by 3D printing for transplantation at a target anatomical site. An additional advantage is that scaffold design can be tailored to fit the exact dimensions of the site of transplantation, which vary depending on the patient's physique, gender, age, etc. Secondly, one can also adopt the concept of printing cell-seeded scaffolds for soft tissues such as liver, intestine, cornea, etc., wherein the organ-specific cells can be encapsulated *in situ* in order to prepare the scaffold. The polymer can be bio-functionalized to maximize the desired biological outcomes.

As discussed above, 3D printing or in general, additive manufacturing routes emerge as novel approaches for fabricating patient-specific implants for orthopedic, dental and cranio-maxillofacial surgery. The biomedical advantage of such implants can only be realized using appropriate biocompatibility assessment with target application specific cell lines *in vitro*, prior to pre-clinical study and clinical trials. It is worthwhile to mention that materials with different shapes/sizes/ internal architecture (having unaltered composition) will have different biocompatibility property.

It is worthwhile to mention that conventional cytocompatibility studies using 2D culture conditions cannot be used for 3D scaffolds or implants. In order to better simulate the physiological conditions in the human body, 3D culture of cells on scaffold in dynamic condition should be adopted more extensively. This would also necessitate the optimisation of 3D culture parameters, in terms of flow parameters in a bioreactor. The cell morphological changes deep into 3D pore architecture, *in vitro* or neobone formation, *in vivo* would also require one to use, micro-CT, which provides qualitative and quantitative analysis of cell/tissue growth in 3D porous architecture.

18.5 | Adopting a Systems Biology Related Approach

In healthcare, the clinical space has often been dominated by empiricism and experience. In recent times, giant strides in biological research, biomedical technology and computational methods have however started to enhance rationality in clinical practice. The overall development has led to innovative drugs and medical devices, which have the potential to address human maladies efficiently and affordably. This positive development has been made possible for three reasons.

(a) Diagnostic, molecular and systems level knowledge and therapeutics are being considered by the clinicians, synergistically. An overall paradigm of an integrated problem-solving approach has started to emerge in healthcare.

(b) Knowledge of human biology obtained through reductionist approaches is integrated into a holistic view, where the biologists attempt to understand how novel properties of complex systems emerge through the formation of networks and interactions of smaller components. The experimental measurements made at different length scales, dimensions and timescales are integrated using mathematical models. Technology is now available to simultaneously monitor the expression of several genes or the activities of several enzymes or the emission of several neurons in response to a single stimulus and such expressions may be integrated computationally. Such approaches are essential for the complete understanding of both the normal and the pathological states of organisms. While the study of the biological components provides the building blocks, comprehensive understanding is denied, if studies are confined to the components alone.

(c) An attempt is being made to design and build novel biological functions and systems (that do not necessarily exist in the natural world) by incorporating insights gained from basic biology, chemistry, physics, materials science and engineering disciplines. On the one hand, new systems are synthesized to perform novel functions by programming cells on the basis of our knowledge of genetic regulatory network. On the other, 3D scaffolds, incubators and bioreactors are built to mimic biological functionalization in terms of signal delivery. Novel biomaterials are being developed to direct the organization, growth and differentiation of cells. These cells then go on to form functional tissue in response to physical and chemical stimuli, that are provided during the culture process.

> The above-described synthesis at the level of clinical practice, integration of component level data generated by biologists and engineers, modelling of systems while incorporating such information using computational biology approaches and fusion of these approaches to evolve affordable medical devices, drugs and treatments, through intense interaction among pharmacologists, physicists, chemists, engineers and clinicians, will be the roadmap for the future.

In the aforementioned framework, it is important to build up a thrust to address 'clinically relevant biomaterials research'. The thrust would require bringing together the existing expertise in relevant ongoing work in microbiology, biophysics, biochemistry, bioengineering, biodesign and computational biology. For the thrust to be meaningful, it is important to posit it in a clinical/innovation space. This would mean bringing the clinical practices related to some endemic diseases such as diabetes, cancer, hepatitis, genetic disorders as well as the related innovation and translational activities into one holistic effort. The integration would entail networking the components-related work in intra-institutional spaces and creating inter-institutional networks. Such networks have led to the large multi-institutional research projects in the developed nations around the world, with similar efforts still in their infancy in many developing nations. It is expected that the above–described approach will be integrated more intensely in the mainstream research program on biomaterials and tissue engineering in the near future.

18.6 | Translational Challenges and Involvement of Clinicians

With increasing life expectancy, particularly in India, there is an extremely high demand for the development of smart biomedical devices. For example, musculoskeletal disorders continue to be the most widespread problem, particularly with an increasing rate of trauma and diseases such as osteoporosis and osteoarthritis, which most often occur in an aging population. Other major clinical challenges include various cardiovascular and neurological diseases. In the context of addressing human diseases, there is an immediate need to manufacture medical devices and implants, which should help in the repair and replacement of diseased and damaged parts of the human skeleton, heart, bone and teeth, thereby restoring the function of the otherwise functionally compromised structures. In particular, the implants for load-bearing applications should possess excellent biocompatibility, superior corrosion resistance in the body environment, an excellent combination of high strength and low modulus (matching with host bone), high fatigue and wear resistance, and osseointegration with tissues. Hence, in order to have a long-term clinical performance, along with the intent to create bio-interactive devices, there is an immediate need to bring together engineers, biologists and clinicians to translate newer design strategies and manufacturing approaches to develop patient-specific biomedical implants and devices.

The field of biomaterials is therefore distinguished from other fields of science and engineering by virtue of its interdisciplinary nature. The advancement of fundamental research on biomaterials demands the effective integration of research ideas and expertise from multiple engineering disciplines (materials science, chemical engineering, mechanical engineering, electrical engineering, computer science) with that of biological science.

Research on biomedical implants has a significant societal impact. This impact is better utilized if the fundamental research pursued in the lab can be translated into products or devices for human healthcare. Clearly, such translational research requires meaningful and close interaction of the active researchers with the clinicians and biomedical entrepreneurs. The impact of translational research would be felt more if there were full-time research involvement of clinicians. Unfortunately, the number of such programmes being pursued in many of the developing nations, including India, is small. In the absence of much translational research leading to product commercialisation, the healthcare expenses remain significantly high in these countries, because of expensive imported devices, which are often beyond the reach of financially challenged citizens. Therefore, translational research programmes need to be given more importance and initiatives need to be taken by the researchers in academia.

> In many of the developed countries, the translational research in the field of biomaterials is being facilitated through the necessary 'ecosystem' (e.g. hospital within the university campus) and the availability of clinicians with full-time research involvement. Such an 'ecosystem' needs to be developed in many developing nations.

In the Indian perspective, the major translational research leading to the development of the first successful indigenous biomaterial device was led by a clinician, Dr. M. S. Valiathan, a cardiac surgeon. He led a large team of researchers at Sree Chitra Thirunal Institute of Medical Science and Technology in developing artificial heart valves, which were further commercialized in the Indian market by the TTK group. Since then, some attempts are being made in different parts of India, but the commercialization aspect has not been pursued to a significant extent.

In order to facilitate such activities, it is important that more entrepreneurs should be encouraged to establish start-ups in the biomedical devices sector. This will enable 'bench to bedside' translation into biomedically relevant products.

> The transfer of new technologies to entrepreneurs for commercialization of biomedical devices involves assessment of Technology Readiness Level (TRL) and competitive cost-structure analysis of any newly manufactured device with respect to similar products currently sold worldwide.

18.7 | Education and Training of Next Generation Researchers

The other challenges in the field of biomaterials include the lack of widespread academic programmes in this subject as well as effective training of young researchers in this interdisciplinary field. At present, the subject of biomaterials is not being taught as a core compulsory course in most of the academic institutions in many countries. As a result, many young researchers do not have any clear idea about the fundamental concepts and its significant impact on human healthcare.

Another challenge is the creation and maintenance of a large research infrastructure, spanning from 'materials' to 'biology' domain. In fact, a biomaterials lab in a traditional materials science department should have a cell culture and/or bacteria culture lab. Considering the requirement of distinctly different skillsets/expertise, the maintenance of cell lines or a culture lab by non-biologists has often been a major challenge. All these challenges make this field even more interesting, considering the opportunities to explore a new horizon at the materials/biological sciences interface.

At present, the research component of an MS/MDS thesis is largely based on patient data or clinical observations. It would therefore be prudent if clinicians, pursuing postgraduate degrees (MD/MS/MDS programs) should be encouraged to do a part of their thesis work on biomaterials related projects. It is important to mention that such exposure will orient future clinicians towards biomaterials based solutions to existing clinical problems.

An extensive network of institutes/universities/medical colleges/national labs needs to be developed for training and dissemination of knowledge in the field of biomaterials. The cumulative outcome may include globally efficient yet affordable products, and treatment modalities for the advancement of human healthcare.

In the last few decades, it has been widely observed that more and more researchers with undergraduate or core education in an engineering discipline have significantly contributed to the field of biomaterials science and tissue engineering or, in general, biomedical engineering. In view of the growing importance of the field of biomaterials science, a number of universities around the world have already started new interdisciplinary academic programmes for graduate students under the broad discipline name 'Biological Engineering' or 'Bio-engineering' or 'Biosystems Science and Engineering' or 'Medical Science and Technology' or 'Clinical Engineering' etc. Such academic programmes provide a useful platform for researchers from multiple disciplines to interact as well as jointly supervise PhD students.

In the backdrop of the above discussion, it is also recommended to include 'Biology' as a core/compulsory course for freshmen/junior undergraduate students across all the disciplines in Engineering. This practice is currently being followed in some IITs and BITS-Pilani in India as well as other universities in the world. Moreover, the course 'Biomaterials Science' or a similar course must be included as an additional core/compulsory course for undergraduate students in the discipline of Materials Science and Engineering / Metallurgy and Materials Engineering. It is recommended that active researchers can also make an effort to popularize this important and socially relevant research area through outreach activities, like public lectures in engineering colleges/high schools to stimulate young minds.

Appendix A

I. Multiple choice questions

Please select the appropriate answer from the given choices:

1. Functionality of cells grown on a biomaterial surface and those in suspension is
   (a) identical   (b) different   (c) similar   (d) opposite

2. Shape of the material is unimportant for studying its properties
   (a) Chemical   (b) Mechanical   (c) Physical   (d) all of the above

3. Technique preferred for 3D microstructural study of biomaterials is
   (a) Confocal   (b) Fluorescent   (c) XRD   (d) micro-CT

4. Compatibility of a material with the blood is referred to as
   (a) cytocompatibility   (b) biocompatibility
   (c) histocompatibility   (d) haemocompatibility

5. Histocompatibility is the term used to define the compatibility of a material with
   (a) human & animal tissues   (b) plants
   (c) bacteria   (d) fungi

6. Incompatibility that can cause implant loosening
   (a) Biomaterial toxicity   (b) Biomechanical
   (c) Biochemical reactivity   (d) All of the above

7. Critical size defect of bone means the defect is
   (a) round   (b) square
   (c) 2-3 times longer than diameter   (d) 2-3 times shorter than diameter

8. Osteons show a regular arrangement of Hydroxyapatite (HA) crystals of the size
   (a) 50-100 nm   (b) 10-20 μm   (c) 1-2 cm   (d) 1-2 μm

9. Mechanical property of cortical bone in comparison to cancellous bone is
   (a) superior   (b) inferior   (c) similar   (d) none of the above

10. Haversian and Volkmann canals channelize the blood in the
    (a) heart   (b) bones   (c) kidney   (d) skin

11. Degree of mineralization is represented by BMD which stands for
    (a) bone meta density
    (b) bone mineral density
    (c) beta material density
    (d) none of the above

12. Standard implant size for studying *in vivo* biocompatibility (osseointegration) is
    (a) 2 mm & length 6 mm
    (b) 2 μm & length 6 μm
    (c) 2 cm & length 6 cm
    (d) none of the above

13. Retrieval and microscopic evaluation of bone implant reveal information regarding
    (a) osseointegration
    (b) host immune response
    (c) fibrosis
    (d) all of the above

14. For bone mineralization the most suitable Ca and P ratio in HA is
    (a) 2.56
    (b) 1.67
    (c) 8.67
    (d) 5.00

15. Two major components of natural bone are
    (a) collagen and Ti
    (b) collagen and HA
    (c) HDPE and HA
    (d) fibrinogen and HA

16. HA is a form of
    (a) Calcium phosphate
    (b) Calcium chloride
    (c) Calcium carbonate
    (d) Calcium sulphate

17. Cytoskeleton consists of
    (a) collagen fibres
    (b) actin filaments
    (c) fibronectin filaments
    (d) elastin filaments

18. Total number of proteins in a prokaryotic cell is in the order of
    (a) 10^{16}
    (b) 10^{26}
    (c) 10^6
    (d) 10^{36}

19. Nucleus can be visualized under microscope using
    (a) Microtracer
    (b) Phalloidin
    (c) Fluorescein
    (d) DAPI

20. Continuous depolymerisation-repolymerisation of cytoskeleton filaments, leading to cell motility takes place, when cell
    (a) size increases
    (b) undergoes division
    (c) size decreases
    (d) changes shape

21. Maturation of osteoblasts to osteocytes is known as
    (a) atrophy
    (b) hypertrophy
    (c) metaplasia
    (d) differentiation

22. α-helix and β-sheet represent the following structural formats of Proteins.
    (a) primary
    (b) secondary
    (c) tertiary
    (d) quaternary

23. Transmembrane proteins are attached to
    (a) mitochondria
    (b) ribosome
    (c) nucleus
    (d) plasma membrane

24. Focal adhesion complex is formed by interacting
    (a) cell receptor and surface protein
    (b) cell membrane and surface protein
    (c) nuclear protein and surface protein
    (d) nuclear receptor and surface protein

25. Size of bacteria falls in the range of
    (a) nanometers
    (b) picometers
    (c) micrometers
    (d) millimeters

26. The peptidoglycan layer in Gram negative bacteria is
    (a) thin   (b) smooth   (c) thick   (d) wavy

27. Triple helix structure is exhibited by
    (a) Collagen   (b) Antigen   (c) ATP   (d) DNA

28. The doubling time for bacterial cells is typically
    (a) 20-40 minutes   (b) 20-40 hours   (c) 20-40 days   (d) 20-40 seconds

29. Following is not the characteristic feature of a cell undergoing apoptosis
    (a) blebbing   (b) nuclear disintegration
    (c) migration   (d) shrinking

30. LDH enzyme is released in the culture medium upon cell
    (a) division   (b) adherence   (c) migration   (d) death

31. Embryonic stem cells are
    (a) multipotent   (b) pluripotent   (c) totipotent   (d) unipotent

32. Heart muscles are formed by
    (a) Chondrocytes   (b) Keratinocytes   (c) Adipocytes   (d) Cardiomyocytes

33. Preferred model for studying the large segmental bone defects is
    (a) mouse   (b) rabbit
    (c) non-human primates   (d) rat

34. The processing of ceramic materials involves,
    (a) Melting and casting   (b) Sintering
    (c) Hot rolling   (d) Hot extrusion

35. Hardness is a measure of resistance against
    (a) fracture   (b) crack growth
    (c) permanent deformation   (d) pore growth

36. Polymer processing involves
    (a) forging   (b) hot rolling
    (c) injection molding   (d) all of the above

37. Bacteriostatic agents are the ones, which
    (a) kill the bacteria   (b) inhibit the bacterial growth
    (c) mutate the bacteria   (d) none of the above

38. The addition of Ti to HA is expected to increase
    (a) fracture toughness   (b) magnetic property
    (c) bactericidal property   (d) all of the above

39. TKR is a clinical term for
    (a) Thorough knee replacement   (b) Total knee replacement
    (c) Total knot resonance   (d) Titanium-potassium rate

40. Microscope that is generally not used to characterize grain structure of a material is,
    (a) Confocal   (b) Fluorescent   (c) Inverted   (d) all

41. For facilitating cell interaction, the surface of implant should be
    (a) smooth   (b) rough   (c) glossy   (d) all

42. Biocompatibility mainly involves surface interactions, whereas biomechanical compatibility is largely determined by the
    (a) structure
    (b) elastic modulus
    (c) chemical composition
    (d) all of the above

43. Qualitative assessment of material-tissue interaction is generally done by
    (a) hematology   (b) histology   (c) spectroscopy   (d) none of the above

44. Host bone engagement in the implant is assessed by
    (a) Convergent beam computer tomography (CBCT)
    (b) Resonance frequency analyzer (RFA)
    (c) Infrared spectroscopy
    (d) UV spectroscopy

45. Osteocytes is the mature form of
    (a) osteoblast   (b) osteons   (c) osteoclast   (d) none

46. Bone is a .... tissue type
    (a) muscular   (b) connective   (c) nervous   (d) cardiovascular

47. Haversian canals in bones contain
    (a) blood vessels & nerves
    (b) only blood vessels
    (c) only nerves
    (d) none of the above

48. Smallest structure-function unit of bone is
    (a) bone-matrix   (b) osteon   (c) Haversion canal   (d) Volkmann's canal

49. Crack in a material represents discontinuity in
    (a) structure   (b) function   (c) shape   (d) none

50. Biomedical grade Ti surfaces are etched to observe microscopic features by using ....
    reagent
    (a) Kroll's
    (b) Benedict's
    (c) phenolphthalein
    (d) none of the above

51. For *in vivo* testing, the biomaterial is implanted either subcutaneously or intramuscularly so that they can
    (a) dissolve fast   (b) be retrieved easily (c) integrate well   (d) none of the above

52. The perceived advantage of HA-barium titanate composite biomaterial is
    (a) piezoelectric property
    (b) magnetic property
    (c) increased strength
    (d) all of the above

53. Stem cells are characterized by
    (a) potency   (b) clonality   (c) stemness   (d) all of the above

54. Typical eukaryotic cell size is of the order of
    (a) 100-150 μm
    (b) 25-30 μm
    (c) 50-100 mm
    (d) 0.1-1 cm

55. Total proteins in a eukaryotic cell is in the order of
    (a) 10^{19} (b) 10^{9} (c) 10^{29} (d) 10^{39}

56. Calcium channels are sensitive to
    (a) pH change
    (b) Na⁺ ions
    (c) K⁺ ions
    (d) Voltage difference across the cell membrane

57. Hypertrophy means
    (a) increase in cell size (b) decrease in cell size
    (c) decrease in cell number (d) increase in cell number

58. Atrophy is the term used for reduction in tissue functionality due to
    (a) degeneration (b) reduction in size (c) diminished use (d) all of the above

59. Metaplasia is a phenomenon that is
    (a) reversible (b) irreversible (c) permanent (d) destructive

60. Cells communicate to neighbouring cells and environment through
    (a) signaling molecules (b) cell-junctions
    (c) cell receptors (d) all of the above

61. The example(s) of Gram positive bacteria is/are,
    (a) S. aureus (b) S. epidermidis (c) E. coli (d) both (a) and (b)

62. Osteocytes are part of
    (a) Cartilage (b) Bone (c) Muscle (d) Epithelia

63. Apoptosis is a cell fate process that represents
    (a) differentiation (b) necrosis
    (c) programmed cell death (d) motility

64. Necrosis is the accidental cell death and is generally accompanied by
    (a) migration (b) differentiation (c) adhesion (d) inflammation

65. Prior to cell division,
    (a) nucleus size increases (b) cell size decreases
    (c) cell membrane disappears (d) cytoplasm dries up

66. Chondrocytes are the constituent cells of
    (a) Bone (b) Cartilage (c) Lungs (d) none of the above

67. Which animal model is closer to humans for bone implantation study?
    (a) mouse (b) rabbit (c) pig (d) dog

68. Preferred animal models for toxicity study of drugs and biomaterials are,
    (a) mouse and rats (b) non-human primates
    (c) dog (d) pig

69. Bactericidal agents are the ones which
    (a) kill the bacteria (b) inhibit the bacteria
    (c) mutate the bacteria (d) none of the above

70. Spark plasma sintering is used to sinter HA-BaTiO₃ composites in order to
    (a) Reduce sintering time (b) Prevent sintering reactions
    (c) Prevent grain coarsening (d) All of the above

71. The fracture toughness of bone depends on
    (a) Bone density   (b) Bone length   (c) Bone thickness   (d) None of the above

72. Addition of $CaTiO_3$ to HA enhances the following property(ies)
    (a) Electrical conductivity   (b) Strength and Toughness
    (c) Dielectric property   (d) All of the above

73. Doping of Ag in HA-Ag composites can be evaluated qualitatively by
    (a) Thermo-gravimetric analysis   (b) Vibrating Sample Magnetometer
    (c) Infrared spectroscopy   (d) None of the above

74. The inhibitory effect of HA-Ag composites on bacteria is due to
    (a) Ca^{2+} ions   (b) $(PO_4)^{3-}$ ions   (c) Ag^+ ions   (d) None of the above

75. The tooth consists of the following hard material/s
    (a) Enamel   (b) Periodontal ligament
    (c) Pulp   (d) None of these

76. The component responsible for machinability of glass ceramics is
    (a) Amorphous matrix   (b) Mica crystals
    (c) SiO_2 (d) None of these

77. Following is not related to antibacterial study
    (a) minimum inhibitory concentration (MIC)
    (b) zone-inhibition assay (ZIA)
    (c) enzyme-linked immunosorbant assay (ELISA)
    (d) colony forming unit (CFU)

78. Establishing biocompatibility of a material involves
    (a) *in vitro* study   (b) *in vivo* study   (c) stem cell study   (d) both a and b

79. Ethical approvals are not required prior to
    (a) stem cell experiments   (b) clinical trials
    (c) animal experiment   (d) biomaterials processing

80. Biocompatibility study involves
    (a) cell viability   (b) cell growth
    (c) cell differentiation   (d) all of the above

81. The term biocompatibility encompasses studies related to
    (a) cytocompatibility   (b) histocompatibility
    (c) haemocompatibility   (d) all of the above

82. THR is a clinical term for
    (a) Total hip retainment   (b) Total hair replacement
    (c) Total hip joint replacement   (d) Three hour joint replacement

83. The components of a biological system, that interact first with the implant are,
    (a) cells   (b) proteins   (c) blood   (d) all of the above

84. Biocompatibility of a material is
    (a) dependent on target application   (b) independent of target application
    (c) independent of material   (d) none of the above

85. Which of the following can be a candidate material for hip joint application
    (a) Copper   (b) Aluminium (c) Titanium alloys (d) PLGA

86. Which of the following is widely used in making cardiovascular stents
    (a) Stannum or Tin (Sn)   (b) Nickel (Ni)
    (c) Nickel-Titanium alloy (Nitinol)   (d) High density polyethylene (HDPE)

87. The biocompatibility encompasses
    (a) Cell type dependent response *in vitro*
    (b) Animal strain dependent tissue response *in vivo*
    (c) Blood compatibility
    (d) All of the above

88. A biocompatible material is NOT expected to encourage
    (a) Cell growth   (b) tissue growth
    (c) bacterial growth   (d) All of the above

89. An implant is suitable for use in humans if it is
    (a) tested *in vitro*   (b) sterile
    (c) tested *in vivo*   (d) passed through clinical trials

90. Mismatch of elastic modulus of implant biomaterial with host bone may lead to
    (a) fracture   (b) aseptic loosening
    (c) infection   (d) all

91. In long bones, bone marrow is surrounded by numerous
    (a) blood vessels (b) osteons   (c) nerves   (d) lipids

92. Stochiometric formula of hydroxyapatite is
    (a) $Ca(OH)_2$ (b) $Ca_3(PO_4)_2$
    (c) $Ca_{10}(PO_4)_6(OH)_2$ (d) none of the above

93. Cortical bone is
    (a) thick   (b) compact   (c) spongy   (d) thin

94. Ratio of stress and strain that a material can tolerate without permanent deformation is defined by its
    (a) elastic modulus (b) elastic strength (c) ductility   (d) fracture strength

95. Cortical bone is not
    (a) dense   (b) flexible   (c) hard   (d) thick

96. Volkmann canals transmit blood vessels to the bone from
    (a) periosteum (b) osteons   (c) muscle   (d) none of above

97. Percentage of mineralized phase in bone is
    (a) 10-20%   (b) 20-30%   (c) 60-70%   (d) 30-40%

98. Water content in bone is in the order of about
    (a) 30 wt.%   (b) 50 wt.%   (c) 90 wt.%   (d) 10 wt.%

99. Compressive strength of a scaffold depends on
    (a) biomaterial composition   (b) processing
    (c) scaffold porosity   (d) all of the above

100. Bone density depends upon
     (a) bone size   (b) mineral content   (c) bone length   (d) none of above

101. If you are asked to choose light weight biomaterials, which one will you choose
     a) Ti   b) HDPE   c) Co-Cr-Mo   d) stainless steel

102. For steel/Ti/Co-Cr-Mo based metallic biomaterials, most biologically inert material in terms of corrosion resistance in clinically relevant potential range is
     (a) Stainless steel   (b) Ti based alloys   (c) Co-Cr-Mo   (d) all of them.

103. Number of cells per ml volume in a normal mammalian tissue is of the order of
     (a) 10^9 (b) 10^{19} (c) 10^{29} (d) 10^{39}

104. Actin filaments can be visualized under microscope using
     (a) Microtracer   (b) Phalloidin   (c) Fluorescein   (d) MTT

105. Mitochondria can be visualized under microscope using
     (a) MitoTracker   (b) Phalloidin   (c) Fluorescein   (d) MTT

106. The voltage difference across a eukaryotic cell membrane is
     (a) 70 kV   (b) 70 mV   (c) 70 V   (d) 70 μV

107. Nuclear pore-complex allows transport of
     (a) only water   (b) selective molecules
     (c) only Na^+ ions   (d) none of the choices

108. Subcellular structures that constitute the intracellular cytoskeleton are
     (a) actin   (b) intermediate filaments
     (c) microtubule   (d) all of the above

109. Proteins typically acquire structural formats
     (a) primary   (b) secondary   (c) tertiary   (d) all of the above

110. Primary structure of a protein is represented by
     (a) amino acid sequence   (b) α-Helix
     (c) β-sheet   (d) none of the above

111. Ribosome is a
     (a) membrane bound organelle   (b) RNA-Protein complex
     (c) DNA-Protein complex   (d) Lipid-Protein complex

112. Ribosomal particles are present in
     (a) eukaryotes   (b) prokaryotes   (c) both a and b   (d) none of the above

113. Luria broth is mostly used for culturing
     (a) fungi   (b) animal cells   (c) bacteria   (d) plant cells

114. The antibacterial effect of a biomaterial should be tested, when the bacterial culture is in
     (a) lag phase   (b) log phase   (c) stationary phase   (d) death phase

115. Bacterial density is commonly recorded using spectrophotometer in the
     (a) UV-Visible range   (b) Far-infrared range
     (c) Infrared or IR range   (d) IR-UV range

116. Adhesion and growth of bacteria on a material surface is faster than human cells due to their
     (a) size   (b) faster doubling time
     (c) shape   (d) all of the above

117. Changes in cell shape require reorganization of
     (a) Cell surface proteins   (b) Golgi apparatus
     (c) Endoplasmic reticulum   (d) Cytoskeleton

118. Cell differentiation study should be conducted minimum for
     (a) 3-4 hours   (b) 3-4 days   (c) 3-4 months   (d) 3-4 weeks

119. Cell migration rate is in the order of
     (a) 1-100 nm/h   (b) 1-100 m/h
     (c) 1-100 μm/h   (d) 1-100 mm/h

120. TGF is a
     (a) hormone   (b) enzyme   (c) gene   (d) growth factor

121. A paracrine signal can efficiently transfer the message to the maximum distance of
     (a) 200 mm   (b) 200 cm   (c) 200 μm   (d) 200 m

122. Internalization of signaling molecule typically takes
     (a) 15-30 days   (b) 15-30 s   (c) 15-30 hours   (d) 15-30 min

123. Transcription in eukaryotic cells takes place in
     (a) cytoplasm   (b) mitochondria   (c) nucleus   (d) ribosome

124. Translation in prokaryotic cells takes place in
     (a) endoplasmic reticulum   (b) cytoplasm
     (c) nucleus   (d) mitochondria

125. RNA that transfers the genetic information from nucleus to cytoplasm is called
     (a) transmembrane RNA   (b) messenger RNA
     (c) transfer RNA   (d) ribosomal RNA

126. The ultrastructural details of collagen fibers can be sufficiently visualized using
     (a) SEM   (b) Optical microscope
     (c) TEM   (d) All of the above

127. Function of transfer RNA or tRNA is to transport
     (a) genetic information   (b) protein
     (c) nucleic acid   (d) amino acid

128. Following is not a cell fate process
     (a) proliferation   (b) differentiation   (c) osseointegration   (d) migration

129. Lamellopodia helps the cell in
     (a) locomotion   (b) division   (c) growth   (d) apoptosis

130. Following process is not related to cell migration
     (a) adhesion   (b) cytoskeletal re-organization
     (c) lamellopodia   (d) proliferation

131. Number of stages a cell undergoes to complete one cell cycle are
     (a) Six   (b) Four   (c) Two   (d) Five

132. Cell doubling time varies with the
     (a) cell type   (b) cell age
     (c) environmental conditions   (d) all of the above

133. PCR is an abbreviated form of technique
     - (a) polymerase catalytic reaction
     - (b) polymerase chain reaction
     - (c) polycarbonate reaction
     - (d) polymer chain rate

134. Accidental cell death does not lead to
     - (a) apoptosis
     - (b) necrosis
     - (c) inflammation
     - (d) pain

135. Cell fate processes can be best analyzed quantitatively by
     - (a) densitometer
     - (b) nuclear magnetic resonance or NMR
     - (c) flow cytometry
     - (d) infrared or IR spectroscopy

136. In MTT assay, yellow compound reduces to Formazan, the colour of which is
     - (a) red
     - (b) green
     - (c) white
     - (d) purple

137. Colorimetric assays are preferred because they are
     - (a) easy to observe
     - (b) simple to perform
     - (c) simple to record/no sophisticated instrument is required
     - (d) all of the above

138. Picogreen assay is used for quantifying
     - (a) single stranded RNA
     - (b) double stranded RNA
     - (c) single stranded DNA
     - (d) double stranded DNA

139. Alkaline phosphatase assay (ALP) is performed to evaluate cell
     - (a) division
     - (b) adherence
     - (c) differentiation
     - (d) death

140. Osteocalcein is a
     - (a) lipid
     - (b) protein
     - (c) polysaccharide
     - (d) none of above

141. RUNX2 is a
     - (a) nucleic acid
     - (b) lipid
     - (c) transcription factor
     - (d) none of above

142. Quantification of live and dead cells is possible through
     - (a) electron microscopy
     - (b) inverted microscopy
     - (c) FACS analysis
     - (d) none of above

143. Totipotent cells can differentiate into
     - (a) multiple cell types
     - (b) a specific cell type
     - (c) all cell types
     - (d) none of above

144. Mesenchymal and haematpoitic stem cells are
     - (a) unipotent
     - (b) multipotent
     - (c) pluripotent
     - (d) totipotent

145. Aborted embryos are the source of
     - (a) mesenchymal stem cells
     - (b) adult stem cells
     - (c) embryonic stem cells
     - (d) none of above

146. Cells that form nails and hair are called
     - (a) chondrocyte
     - (b) osteocyte
     - (c) keratinocyte
     - (d) none of above

147. Insulin is produced by
     (a) Islet cells   (b) chondrocyte   (c) osteocyte   (d) keratinocyte

148. Ability of stem cells to divide while retaining its potency is called
     (a) differentiation   (b) self renewal   (c) stemness   (d) none of the above

149. Adult stem cells can be harvested from
     (a) all the tissues   (b) aborted embryos (c) limited tissues   (d) none of the above

150. Adult stem cells are generally
     (a) unipotent   (b) multipotent   (c) pluripotent   (d) impotent

151. The flip-flop mechanism of phosphotidylserine (PS) molecules is involved in the cell
     (a) growth   (b) apoptosis   (c) motility   (d) differentiation

152. Which cytoskeletal structures make up the mitotic spindle during cell division
     (a) Golgi apparatus (b) microtubules   (c) ribosomes   (d) centrosomes

153. In which phase does DNA replication occur during the cell cycle
     (a) G1   (b) M
     (c) S   (d) None of these answers are correct

154. Which of the following results from hypoxia to cells
     (a) apoptosis   (b) necrosis   (c) pyroptosis   (d) differentiation

155. Duration of cell culture experiments should be based on
     (a) cell type   (b) cell doubling time
     (c) cell fate to be investigated   (d) all of the above

156. Microscopic study cannot be conclusively used for evaluating
     (a) cell differentiation   (b) cell division
     (c) cell growth/proliferation   (d) cell shape

157. Under blood flow conditions *in vitro*, 'no thrombus', 'adhesion thrombus' and 'shed thrombus' are the tests for the quantification of
     (a) WBC   (b) RBC   (c) platelet adhesion (d) thrombin

158. Following represents the standards in biocompatibility testing for implants
     (a) ISO 993   (b) ISO 1993   (c) ISO 100993   (d) ISO 10993

159. Toxicity is preferably studied in mice and rats as they
     (a) are small
     (b) have short life cycle
     (c) exhibit toxicity to all synthetic materials
     (d) both (a) and (b)

160. Recommended animal models for cardiovascular study are
     (a) sheep and pig   (b) mice and rats   (c) rabbit and dog   (d) none of above

161. Experts from following streams are involved in Clinical Trials
     (a) Biomaterials science   (b) Medical science
     (c) Biostatistics   (d) all of the above

162. The first step towards cell adhesion on a biomaterial substrate involves,
     (a) Protein absorption   (b) Cells shape change
     (c) DNA replication   (d) Cell differentiation

163. Which of the following technique can be used for quantitative analysis of cell cycle
     (a) SEM
     (b) Immunofluorescence
     (c) RT-PCR
     (d) Flow cytometry

164. Micro-CT or Micro computed-tomography is used for assessing the structural features of a material
     (a) Qualitatively
     (b) Quantitatively
     (c) in destructive manner
     (d) none of above

165. General advantages of metallic biomaterials is their
     (a) reproducible properties
     (b) reproducible shape and size
     (c) better clinical reliability
     (d) all of the above

166. The manufacturing of rolled plates is done at or above the temperature
     (a) $T_m/4$
     (b) $2T_m$
     (c) $T_m/2$
     (d) T_m, where T_m is the melting temperature of the metal

167. Advanced sintering processes include
     (a) Hot isostatic pressing
     (b) Hot pressing
     (c) Spark plasma sintering
     (d) all of the above

168. Degree of polymerization depends on
     (a) number of monomers in a polymer
     (b) type of polymers
     (c) arrangement of polymers
     (d) none above

169. What should be the sintering temperature of a composite A+B if the melting point of metal A is 230 K and that of B is 540 K.
     (a) 540 K
     (b) 600 K
     (c) 810 K
     (d) 345 K

170. Crystal morphology in ceramics depends on
     (a) composition
     (b) sintering temperature
     (c) both (a) and (b)
     (d) none

171. The grain size increases, when (in reference to sintering conditions to reach fully dense ceramics)
     (a) sintering time is longer
     (b) sintering time is shorter
     (c) sintering temperature is lower
     (d) none of the choices

172. To synthesize custom made fibrous porous scaffold, you need to use
     (a) salt leaching
     (b) Electrospinning
     (c) high temperature sintering
     (d) 3D printing

173. Hydroxyapatite is sintered
     (a) above 1000°C
     (b) below 500°C
     (c) above 2000°C
     (d) above 1500°C

174. During sintering of ceramics
     (a) only pore size changes
     (b) only pore shape changes
     (c) both pore size and pore shape changes
     (d) maximum neck growth occurs by vapor phase transport

175. If the crack growth resistance increases, then
     (a) fracture toughness will decrease
     (b) fracture toughness will increase
     (c) hardness will increase
     (d) hardness will decrease

176. Strength of a material describes
     (a) load bearing capability
     (b) resistance against permanent deformation
     (c) scratch resistance
     (d) corrosion resistance

177. Stress concentration at the crack tip,
     (a) will have linear dependency on crack length
     (b) will have inverse square root dependency on crack length
     (c) is independent of crack tip radius of curvature
     (d) will be proportional to square root of ratio of crack length and crack root radius of curvature

178. For Steel/Ti alloy/Co-Cr-Mo based metallic biomaterials
     (a) the highest elastic modulus is exhibited by Ti alloys
     (b) the lowest elastic modulus is exhibited by Ti alloys
     (c) both Ti alloy and steel have comparable elastic modulus
     (d) elastic modulus of steel is closer to that of cortical bone

179. NiTi stent for cardiovascular applications can be manufactured using
     (a) rolling
     (b) forging
     (c) extrusion
     (d) wire drawing

180. Elements generally incorporated in the biomaterial for their antibacterial properties are
     (a) silver
     (b) calcium
     (c) aluminium
     (d) magnesium

181. Proteins are the natural polymer consisting of
     (a) nucleic acids
     (b) nucleotides
     (c) amino acids
     (d) phosphates

182. $BaTiO_3$ is used as an additive to HA for bone replacement applications because of
     (a) Piezoelectric property
     (b) Strength property
     (c) Toughness property
     (d) None of the above

183. Electrical properties of HA-$BaTiO_3$ composites depend on
     (a) $BaTiO_3$ content
     (b) Sintering conditions
     (c) Phase connectivity
     (d) All of the above

184. Following factors affect the cell adhesion and growth on substrates
     (a) Surface charge
     (b) Conductivity
     (c) Elastic stiffness
     (d) All of the above

185. Fracture toughness of HA can be increased by
     (a) Metal additives
     (b) Crack bridging
     (c) Crack deflection
     (d) All of the above

186. Functional proteins in general are structurally
     (a) simple
     (b) straight
     (c) globular
     (d) none of the above

187. The advantage of ZnO in HA-ZnO antimicrobial composites is
     (a) ZnO is a semiconductor
     (b) ZnO produces H_2O_2 in solution
     (c) ZnO increases the strength
     (d) None of the above

188. Bacterial colony on a biomaterial can potentially be formed from
     (a) a single bacterium   (b) 10 bacteria
     (c) 100 bacteria   (d) 1000 bacteria

189. Choose the processing steps in the order that they are followed during the fabrication of glass ceramic: (A) Quenching (B) Melting (C) Annealing
     (a) A-C-B   (b) B-A-C   (c) A-B-C   (d) B-C-A

190. The finer debris particles may generate through
     (a) friction   (b) corrosion   (c) wear   (d) all of the above

191. 45S5 is a bioglass that contains
     (a) 4.5% silica   (b) 0.45% silica   (c) 45% silica   (d) 0.045% silica

192. Glass melting during the processing of dental ceramics is done in Platinum crucible, because Pt
     (a) is inert   (b) prevents contamination
     (c) has high melting point   (d) all of the above

193. The dye usually used to observe host bone-implant interface is
     (a) MTT   (b) Hoechst   (c) Xylenol orange   (d) Phalloidin

194. The glass transition can be conclusively identified by the following technique(s)
     (a) Optical microscopy   (b) Differential scanning calorimetry
     (c) SEM   (d) None of the above

195. The crystal volume fraction in a glass ceramic enhances
     (a) machinability   (b) Cytocompatibility
     (c) optical properties   (d) None of the above

196. In an articulating joint with hard and soft materials experiencing relative motion
     (a) hard material wears soft material   (b) soft material wears hard material
     (c) strength of the interface increases   (d) none of the above

197. Short term implantation to study bone formation in animal studies is usually carried out for
     (a) 12 hours   (b) 12 days   (c) 12 weeks   (d) 12 months

198. Polymeric acetabular socket is usually manufactured by the following technique
     (a) Rolling   (b) Forging
     (c) Compression Molding   (d) Sintering

199. Annealing/heat treatment of amorphous glass is carried out to induce
     (a) Crystallization   (b) Thermal stress relief
     (c) Phase transformation   (d) All of the above

200. Macor based glass ceramics contain the following
     (a) Limestone   (b) Calcite   (c) Kaolinite   (d) Phlogopite

201. Prosthetic implants which are used in the case of bone damage sometimes result in gradual decline of bone density (osteopenia). Which of these could be possible causes for the phenomenon?
    (a) Nanoparticles from prosthesis result in cell toxicity
    (b) Implants shield stress from bone to osteoblasts
    (c) Implants transfer stress from bone to osteoblasts
    (d) None of the above

202. Based on the theory of tribology, which of the following (self-mated) will have lowest friction
    (a) HDPE   (b) Ti6Al4V   (c) SS 316L   (d) Zirconia

203. The CaP compounds, which are not suitable for biomedical applications, have
    (a) Ca/P < 1.5   (b) Ca/P = 1.67   (c) Ca/P < 2.0   (d) Ca/P = 1.5

204. 10, 000 fire ants are added to a beaker. Ants link to one another to form a bulk material. Ants can form links which can resist some stress but also have the ability to break links and form new ones. Which of the following types of material best describes the 'ant-colony' bulk material?
    (a) Brittle ceramic   (b) Fluid
    (c) Viscoelastic polymer   (d) Ductile material

205. Except few Co-Cr-Mo based alloys, most of the as cast/wrought Co-Cr-Mo in general, have elongation to fracture
    (a) less than 5%   (b) less than 0.05%
    (c) more than 50%   (d) less than 0.5%

206. Why are cross-linking proteins necessary for movement of cilia and flagella?
    (a) They assemble microtubules in cilia and flagella.
    (b) They transfer ATP to dynein. Dynein uses ATP as an energy source for muscle contraction.
    (c) They connect microtubule doublets together so that when dynein moves along one microtubule relative to the other, the cilium or flagellum bends.
    (d) They are the motor proteins responsible for the walking movement of microtubule doublets.
    (e) They anchor microtubule doublets to the basal body.

207. Why are microtubules important for the movement of neurotransmitters within neurons?
    (a) They cause cilia to beat, moving neurotransmitters in the process.
    (b) They cause cytoplasmic streaming of materials.
    (c) They guide vesicles containing neurotransmitters from the cell body down the axon.
    (d) They transport vesicles containing neurotransmitters out of the cell.
    (e) Microtubules load neurotransmitters into synaptic vesicles.

208. Which of the following is NOT the evidence, that supports the endosymbiotic theory of the origin of mitochondria and chloroplasts?
    (a) Chloroplasts and mitochondria have many cilia and are able to swim around freely
    (b) Chloroplasts and mitochondria contain prokaryote-like ribosomes
    (c) Chloroplasts and mitochondria contain prokaryote-like single circular chromosomes
    (d) Chloroplasts and mitochondria are surrounded by a double membrane
    (e) Chloroplasts and mitochondria divide independently by a process similar to binary fission

209. Which of the following is found in plant cells, but NOT in animal cells?
    (a) mitochondria
    (b) a nucleus
    (c) central vacuole/tonoplast
    (d) rough endoplasmic reticulum

210. Eukaryotic cells possess which of the following traits?
    (a) compartmentalization
    (b) chromosomes that float freely in the cytoplasm
    (c) membrane-bound organelles
    (d) both compartmentalization and membrane-bound organelles

211. Which cell structure best exemplifies the compartmentalization of a chemical reaction that requires environmental conditions that differ from those of the cytosol?
    (a) ribosome
    (b) chromosome
    (c) lysosome
    (d) cytoskeleton

212. Which of the labeled organelles would most likely be involved in calcium ion (Ca^{2+}) storage?
    (a) mitochondrion
    (b) rough endoplasmic reticulum
    (c) smooth endoplasmic reticulum
    (d) Golgi apparatus

213. Under favorable environmental conditions, how do most prokaryotes reproduce?
    (a) meiosis
    (b) mitosis
    (c) budding
    (d) binary fission

214. How does cytokinesis begin in an animal cell?
    (a) Actin forms a ring outside the cell membrane
    (b) Chromosomes line up across the middle of the cell
    (c) A cell plate forms
    (d) A cleavage furrow forms

215. What is the role of centrosomes in mitosis?
    (a) holding together the two sister chromatids
    (b) organizing the microtubules that form the mitotic spindle
    (c) acting as an attachment point between microtubules and chromosomes
    (d) producing RNA and ribosomes

216. Without the mitotic spindle, how would mitosis be different?
    (a) The cell membrane would not form around daughter nuclei.
    (b) Crossing over would not occur, so genes would more often segregate independently.
    (c) Sister chromatids would have to physically segregate by some other means.
    (d) More mistakes would occur during DNA replication.

217. How is DNA arranged in a eukaryotic cell during the G1 phase of interphase?
    (a) in pairs of sister chromatids
    (b) by microtubule length
    (c) in chromatin fibers
    (d) in condensed chromosomes

218. *in cellulo* study is conducted
    (a) using glassware
    (b) inside an organism
    (c) with cells only
    (d) none of the choices

219. Which of the following steps of mitosis occurs first in plants and animals?
    (a) Chromatids line up in the middle of the cell
    (b) The nuclear membrane breaks apart
    (c) Sister chromatids separate
    (d) The cell membrane pinches together in the middle

220. Polymerisation-depolymerisation takes place in an interactive manner for the following cell fate process,
    (a) migration   (b) differentiation   (c) apoptosis   (d) proliferation

221. Collagen molecule has dimension in the range of
    (a) 1-100 cm   (b) 1-100 μm   (c) 1-100 nm   (d) 1-100 mm

222. Which of the following is involved in blood clotting,
    (a) RBCs   (b) WBCs   (c) platelets   (d) all of the above

223. The diameter of pili/flagellum are in the order of magnitude,
    (a) nm   (b) μm   (c) mm`   (d) cm.

224. Which among the following assays can be used to measure the mitochondrial activity?
    (a) MTT assay   (b) Picogreen assay   (c) LDH assay   (d) None of the above

225. Indirect immunofluorescence technique involves
    (a) Secondary antibody
    (b) Fluorescent labelled primary antibody
    (c) Fluorescent labelled secondary antibody
    (d) All of the above

226. Taq polymerase is used in PCR because
    (a) the enzyme catalyzes 3'→5' synthesis of DNA
    (b) it has lesser half-life at 95°C
    (c) it is highly thermostable
    (d) None of the above

227. The signaling molecules that help in the movement of cells towards the inflammation
    (a) Neurotransmitters   (b) Cytokines
    (c) Hormones`   (d) None of the above

228. The following cells are recruited at the site of implantation
    (a) Neutrophils   (b) Monocytes   (c) Fibroblasts   (d) all of the above

229. In order to view and distinguish nucleus and cytoskeleton of cultured cells on a biomaterial surface, one needs to use the following technique,
    (a) Optical microscopy   (b) Fluorescence microscopy
    (c) Scanning electron microscopy   (d) Transmission electron microscopy

230. Lasers are not used in the following techniques
    (a) FACS   (b) TEM
    (c) Fluorescence microscope   (d) all of the above

231. Complex structures of proteins generally involve
    (a) hydrogen bonds   (b) Van der Waals forces
    (c) covalent bonds   (d) all of the above

232. Cells may respond to any biophysical signal by
    (a) changing shape   (b) cell division
    (c) cell differentiation   (d) all of the above

233. Annexin V-stain is used to detect
  (a) Cell cycle (b) Apoptosis (c) Proliferation (d) Differentiation

234. Typically, the cell culture with biomaterials takes place over a time period of
  (a) Less than 1 hour (b) Less than 12 hours
  (c) Less than 24 hours (d) More than 24 hours

235. The pH for all the culture medium is kept at
  (a) 5.4 (b) 3.4 (c) 7.4 (d) 9.4

236. The atmosphere in a CO_2 incubator can be described as,
  (a) 25% CO_2, 75% humidity (b) 50% CO_2, 50% humidity
  (c) 5% CO_2, 95% humidity (d) 75% CO_2, 25% humidity

237. Laminar flow hood is required during culture experiments to ensure
  (a) sterile environment (b) turbulent flow
  (c) moist environment (d) warm environment

238. Bacterial adhesion on bulk biomaterials can not be directly visualized using
  (a) SEM (b) Fluorescence microscope
  (c) TEM (d) AFM

239. The last step prior to cell culture is
  (a) Sterilization of the sample (b) Mechanical indentation
  (c) Heat treatment (d) Acid treatment

240. The culture medium for eukaryotic cells
  (a) remains the same, irrespective of cell type (stem cells or primary cells or transformed cell line)
  (b) contains antibiotics only
  (c) contains nutrients only
  (d) contains proteins, nutrients and antibiotics

241. When cells adhere on a biomaterial substrate, the following takes place,
  (a) Cells spread on the surface
  (b) Cells always shrink in size
  (c) Cell size remain unchanged
  (d) Contact between neighbouring cells established

242. For bacteria culture, the incubator atmosphere should have
  (a) CO_2 supply (b) sterile environment (37°C)
  (c) 100% humidity (d) 50% humidity

243. The temperature during all culture experiments is maintained at
  (a) 27°C (b) 37°C (c) 47°C (d) 57°C

244. Turbidometric analysis is carried out to assess the cell density in bacterial cultures by measuring
  (a) Absorbance at 620 nm (b) Scattering at 620 nm
  (c) Interference at 620 nm (d) Scattering at 280 nm

245. In fluoroscence microscope, the emission wavelength is .... than/like excitation
wavelength
    (a) longer   (b) shorter   (c) same   (d) none of the choices

246. The *in vivo* osseointegration of a bone implant depends on
    (a) animal model   (b) defect model
    (c) implantation time   (d) all of the above

247. Wear debris particles can cause
    (a) inflammatory reaction   (b) toxicity
    (c) systemic response   (d) all of the above

248. Which of the following are multipotent in nature?
    (a) Mesenchymal stem cells   (b) Hematopoietic stem cells
    (c) Neural stem cells   (d) both (a) and (b)

249. Which one among the following is a master gene for adipogenesis?
    (a) RUNX2   (b) PPARγ2   (c) SOX2   (d) both (a) & (b)

250. Which of the following is not true in the context of bacterial viability?
    (a) MIC > MBC   (b) MIC < MBC
    (c) MIC = MBC   (d) None of the above

251. The idea of encapsulating islet cells in a semi-permeable membrane is to allow the entry of
    (a) glucose/macrophages   (b) nutrients/killer T cells
    (c) both (a) and (b)   (d) none of the above

252. Acute response is normally observed within the time period of implantation
    (a) 72 hours   (b) 24 hours   (c) 30 days   (d) 90 days

253. Agarose gel electrophoresis is used to separate
    (a) DNA/RNA molecules   (b) protein molecules
    (c) polysaccharides   (d) All of the above

254. The coefficient of friction of HDPE composite/alumina couple in SBF typically lies around
    (a) 0.1   (b) 1.0   (c) 0.001   (d) 0.0001

255. As a first step towards bacterial adhesion the following takes place
    (a) Protein adsorption   (b) Direct interaction of bacteria with substrate
    (c) Bacteria changes in shape   (d) Biofilm formation

256. The knowledge on bacterial adhesion on biomaterial substrates is required to assess the
potentiality of
    (a) Tissue formation   (b) Microbial infection
    (c) Substrate for cell proliferation   (d) None of the above

257. If a material is biocompatible, but not bioactive, that means
    (a) No cell adhesion will take place at all
    (b) Limited, but to a visible extent, cells will adhere
    (c) Cells may not survive
    (d) none of the given choices

258. From the biomaterials compositional perspective, it is important to develop materials, which do not encourage
     (a) Myoblast cell growth
     (b) Eukaryotic cell growth
     (c) Stem cell growth
     (d) Bacteria growth

259. Cells, grown on biocompatible material will change its shape due
     (a) Cytokinesis
     (b) Chemokinesis
     (c) Cell apoptosis
     (d) Material-protein interactions

260. MTT assay should not be used to quantify
     (a) Number of dead cells
     (b) Number of viable cells on graphene
     (c) Number of cells at S phase
     (d) All of the above

261. The cell viability can be assessed using two complimentary assays,
     (a) MTT and LDH
     (b) ALP and osteocalcin
     (c) ALP and RUNX2
     (d) Osteocalcin and osteopontin

262. TEM is primarily used to study
     (a) Cell signaling
     (b) Cell differentiation
     (c) Cellular ultrastructure
     (d) Cell fate processes

263. Intercellular signaling pathway is activated by
     (a) Target proteins
     (b) Intracellular signaling proteins
     (c) Receptor protein
     (d) Extracellular signal molecule

264. A cell can stay for a variable length of time in
     (a) G1 phase
     (b) S phase
     (c) G2 phase
     (d) M phase

265. In DNA replication, new strand is synthesized in
     (a) $5' \rightarrow 3'$ direction
     (b) $3' \rightarrow 5'$ direction
     (c) both of above
     (d) None of the above

266. The biological assay known as marker of dead cells
     (a) LDH assay
     (b) Picogreen assay
     (c) ALP assay
     (d) RUNX2 assay

267. BAI stands for
     (a) Biomaterial associated infection
     (b) Bacteria associated infection
     (c) Biologically activated infection
     (d) none of the above

268. Experiments which are far more relevant to assess biocompatibility of a biomaterial
     (a) *in vivo*
     (b) *in vitro*
     (c) both
     (d) none of the above

269. Ability of biomaterials to be in contact with proliferating cells without producing an adverse effect *in vitro* is
     (a) biocompatibility
     (b) cytocompatibility
     (c) haemocompatibility
     (d) All of above

270. Histological unit of mammalian bone is
     (a) Haversian canal
     (b) Osteon
     (c) Chondrion
     (d) Hydroxyapatite

271. Constituent of bone matrix is
     (a) Ceramic
     (b) Polymeric
     (c) Both (a) and (b)
     (d) Metallic

272. .... is defined as a direct structural and functional connection between ordered, living bone and the surface of a load-bearing implant.
   (a) Osseoinduction
   (b) Osseointegration
   (c) Osseogeneration
   (d) Osseoconduction
   (e) All of the above

273. The doubling time for human cells is about
   (a) 12 minutes
   (b) 12 hours
   (c) 12 seconds
   (d) 12 milliseconds

274. A human cell contains long DNA of length
   (a) 2 m
   (b) 2 km
   (c) 2 cm
   (d) $2\ \mu m$

275. Bone can be described as
   (a) Vascularized tissue
   (b) Hybrid structure of hard surface with spongy core
   (c) Polymer–Ceramic biocomposite
   (d) All of the above

276. The foreign body response is indicated by the presence of the
   (a) components of granulation tissue only
   (b) foreign body giant cells, but absence of the components of granulation tissue
   (c) foreign body giant cells and components of granulation tissue
   (d) Angiogenesis as a result of chemotaxis of foreign body giant cells

277. Which of the following cannot be considered as a parameter for evaluation of tissue response
   (a) Number and distribution of inflammatory cells
   (b) Osseointegration
   (c) Thickness and quality of fibrous tissue surrounding material
   (d) Tissue calcification

278. Chondrocytes are part of
   (a) Cartilage
   (b) Blood
   (c) Muscle
   (d) Epithelial

279. LDH assay
   (a) Quantifies dead cells
   (b) is complimentary to ALP assay
   (c) Quantifies gene expression
   (d) Quantifies live cells

280. The transition from one phase to another during cell division under favorable physiological conditions is
   (a) irreversible in nature
   (b) reversible in nature
   (c) both (a) and (b)
   (d) none of the above

281. Electric field stimulation to a cell population in culture can cause
   (a) only cell proliferation
   (b) only cell apoptosis
   (c) both cell proliferation and lysis
   (d) cell necrosis

282. The broader electrochemical condition of a living human body is characterised by a range of pH values
   (a) 3-6
   (b) 7-8
   (c) 2-12
   (d) 6-8

283. Typical cell migration speed of L929 cells is
    (a) 30 µm/s   (b) 30 µm/min   (c) 30 µm/ms   (d) 30 µm/h

284. Calcine blue assay is used to assess
    (a) cell viability   (b) cell motility
    (c) cell differentiation   (d) cell apoptosis

285. For steel/Ti alloys/Co-Cr-Mo based metallic biomaterials,
    (a) Corrosion can be envisaged as a problem to be experienced in clinical applications
    (b) These metals can have a better wear resistance than ceramic articulating implants
    (c) These metals can have higher reliable strength and thus better clinical acceptability than bioceramic implants
    (d) These metals can have larger tensile deformation to failure than thermoplastic biopolymers

II. Fill in the blanks with clearly defined terms/symbols, or tick the most appropriate answer

1. The type of force that decreases the length and increases the diameter of the material is called ....

2. Transcription process takes place in the .... of prokaryotic cells.

3. PLGA is a co-polymer of PLA or Poly lactic acid and .... (abbreviation and full form).

4. .... combine through peptide bonds to form proteins.

5. A natural human bone can be best described as a composite of Hydroxyapatite and .... protein.

6. Cytoskeleton can be described by a combination of actin filaments, intermediate filaments and ....

7. Translation of genetic information into protein takes place in .... of the cells.

8. Ordered polymer chains in the context of polymer structure refer to ....

9. Ca/P ratio in stoichiometric HA is ....

10. Nitinol is essentially an alloy of Nickel and ....

11. If osteoblast cells, under certain physiological conditions, change to osteocyte cell type such transition is known as ....

12. Polydispersity index is defined as the ratio of M_w/M_n where M_w is weight average of molecular weight and M_n is ....

13. The type of force that increases the length and decreases the diameter of the material is called ....

14. Transfer of genetic information from DNA to mRNA in nucleus is called ....

15. The ceramics are best characterized by the combination of .... bonding.

16. Zipper mechanism relates the interaction between ....

17. Conformation in the context of polymer structure is defined ....

18. In order to make hydrophobic surface to hydrophilic surface, ... treatment may be adopted.

19. Instrumented indentation can be used to measure both .... and ....

20. The Hooke's law in tensorial notation is ....

21. If the fracture surfaces of a specimen under tension are flat, it can be concluded that it has undergone ....

22. The relationship between shear stress and shear strain in the elastic region is ....

23. The mode II loading involves the application of .... force at the crack faces.

24. When a crack propagates in the bone, the total surface energy of the cracked bone must ....

25. The mechanisms involving energy dissipation around a crack tip always enhance ....
    of a material.

26. Unstable configuration or conformation of a protein molecule is known as ....

27. L929 is an example of .... (type) tissue cells of .... (name the animal).

28. Cell-cell junction by actin filament is known as ....

29. Fibroblasts are contained in .... (type of tissue).

30. Primary structure of various protein molecules is distinguished by .... of amino
    acids.

31. An eukaryotic cell contains .... number of protein molecules.

32. Osteolysis is typically defined as ....

33. Lymphocytes are one of the cell type of .... tissue (name of tissue), responsible
    for ....

34. Pin-on-disc tester is used to evaluate .... properties of materials.

35. Each organ is made of a number of tissues and examples of at least four such tissues are
    ....

36. The function of connective tissue is ....

37. The fundamental structural differences between eukaryotic and prokaryotic cells are
    .... and ....

38. The sequence of .... determine the protein structure, which is essentially
    .... and flexibility of such structure is due to .... across the chains.

39. The measure of *in vivo* biocompatibility is the thickness of .... around the implant.

40. The staining used for histopathological testing of implant/bone interface is ....

41. MTT is a .... assay, used to assess the .... cells . MTT reacts with a
    particular cellular organelle, i.e. .... and after the reactions .... crystals
    form.

42. Optical density to quantify the bacterial concentration is measured using ....

43. DAPI is used as a florescent dye, the fluorescence emission is .... in colour, and
    labels .... of cell.

44. For cell fate analysis of a 3D scaffold, one needs to use .... (equipment) and for the
    quantification of porosity or pore interconnectivity, one has to use .... (equipment).

45. Short term implantation involves experiments over timescale of .... and for the *in
    vivo* biocompatibility study of bone replacement materials, typically .... animal
    model is used.

46. HAp is an electrical insulator and one of the ways to improve electrical conducting property
    is by .... to form a composite material.

47. Embryonic stem cells are .... in nature and have the ability to differentiate into
    ....

48. The idea of encapsulating islet cells in a semi-permeable membrane is to allow the entry of
.... as well as to protect the cells from ....

49. .... are cellular appendages for bacterial motility/locomotion and ....
for adhesion onto material/tissue surface.

50. .... are the bacterial ribosomal subunits .... are eukaryotic ribosomal
subunits.

51. .... is an enzyme that relieves strain during unwinding of DNA for replication.

52. .... is one of the method for direct enumeration of viable bacteria in a sample.

53. In a FACS dot plot of FSC vs SSC, the information provided by FSC is .... and
SSC is ....

54. Osteogenesis involves conversion from transitory Osteoblast to Osteocytes to Osteoclast to
....

55. Chondrogenesis involves conversion from Chondroblast to Chondrocyte to Hypertrophic
chondrocyte to ....

56. Myogenesis involves conversion from Myoblast to Myoblast fusion to Myotube formation
to ....

57. The fracture toughness of cancellous bone is .... than cortical bone ((i) less; (ii)
more; (iii) equal; (iv) none of above).

58. Blinding is a strategy in .... (*in vitro* testing, *in vivo* testing, clinical trials,
histological testing)

59. Ethical committee approval becomes more stringent when one uses .... animal
model. (choose from rat/dog/rabbit)

60. Human subjects recruited for clinical trials are also called .... (doctors, volunteers,
saints, helpers)

61. Fill in the blank with the correct choice from the parenthesis below
(Post market evaluation, extensive study in larger population, initial safety and therapeutic
dose range in human subject, Efficacy in human subject)
    (a) Phase I clinical trial involves ....
    (b) Phase II clinical trial involves ....
    (c) Phase III clinical trial involves ....
    (d) Phase IV clinical trial involves ....

62. Fill in the blank
    In the context of mechanical property of Biomaterials:
    (a) Tension refers to ....
    (b) Compression refers to ....
    (c) Shear refers to ....

63. The ease of processability in injection molding technique depends on .... or
.... (characteristic of melt).

64. The crystal morphology in glass ceramics typically depends on its .... and
....

65. HDPE is an example of .... polymer.

66. An example of polymeric bone cement is .... and its monomer unit is ....

67. Polydispersity index is defined as the ratio of ....

68. The polymer, that can be used as heart valve is ....

69. Human blood has a pH of .... and body core has a temperature of ....

70. The binding of ligands to a protein molecule occurs via .... bonds.

71. Blood is called as a tissue becaused it comprises of .... and ....

72. In order to assess whether a material without any hydroxyapatite can have good biocompatibility, ... test can be used.

73. Biocompatibility is a term, incorporating .... and .... compatibility in reference to a specific biomedical application.

74. .... microscopy observations of stained cells can provide cell morphological features.

75. Depending on cell type or substrate composition, cell can take .... hours to grow in a reasonable number and .... days for any change in cellular functionality.

76. Depending on application of biomaterial, ... and .... should be used to assess the biocompatibility of material *in vitro and in vivo*, respectively.

77. The functionality or activity of cells in culture medium, when grown in isolation is different from the cells when grown on ....

78. Assessment of biocompatible properties should be primary concern, irrespective of .... properties of any bio-mimicking material.

79. All the organelles in eukaryotes are contained in well-defined .... membrane structure.

80. DNA is dispersed in cytoplasm of prokaryotes known as ....

81. Size of eukaryotic cell is 25-35 μm, while that of prokaryotes is ....

82. Small proteins that stimulate cell growth by promoting synthesis of protein and by inhibiting their degradation are called as ....

83. .... are highly dynamic structures that are used to guide the migration and growth of a cell in a concentration field.

84. In order to assess the biocompatibility of a material for bone replacement application, ... cell line is to be used.

85. A cell activates suicide program by a process called as ....

86. M phase starts at the end of .... phase and ends at the start of next .... phase.

87. .... is the accidental cell death characterized by membrane disruption, cell swelling and rupture leading to inflammation.

88. Conversion of MTT reagent into purple formazan can be directly related to the number of .... cells.

89. Molecules used for their fluorescent property are known as ....

90. *in vitro* is a Latin word, which means ....

91. Formation of new blood vessel is called as ....

92. Chemical or biological agents, that prevent bacterial multiplication, while not necessarily killing them are called as .... agent.

93. CFU stands for ....

94. Tailoring antimicrobial agent in any implant should be in such a quantity that kills the bacterial but not ....

95. Doses of .... rays should be optimized for sterilization of specific biomaterial.

96. Growth factors as defined as ....

97. Focal contacts are assemblies of .... and facilitate binding of .... to ....

98. The process of inter-conversion of fibroblast to chondrocytes is known as ....

99. Typically, [Ca^{+2}] concentration in cytoplasm is .... than that in ECM and is around ....

100. Normally, the friction coefficient of any biomaterial in synovial fluid or *in vitro* medium is .... than that in dry/ambient environment.

101. Stem cells are defined as ....

102. qRT – PCR means ....

103. Higher the optical density measured in MTT assay, ... will be the cell viability.

104. Osteocalcin gene is responsible for ....

105. Collagen has a characteristic .... structure with a characteristic length scale of ....

106. The doubling time for adult chondrocytes is ....

107. Fluorescence is defined as ....

108. Pluripotent stems cells are defined as ....

109. Totipotent stems cells are defined as ....

110. Stemness in the context of stems cells means that ....

111. In case of necrotic cells, PI stain intercalates with ...., causing ....

112. An example of anti-apoptotic gene is ....

113. In natural bone, individual osteon have size of ....

114. The blood platelets participate in ....

115. In immunofluorescence techniques, antibodies are chemically conjugated to .... dyes.

III. Identify whether the following statements are True/False

1. The enzyme responsible for reducing the MTT is the dehydrogenase. [True/False]
2. Cell cycle analysis is possible using FACS analysis. [True/False]
3. The enzyme responsible for reducing the LDH is the lactate dehydrogenase. [True/False]
4. Stem cells are the best cell line for differentiation. [True/False]
5. Haematopoietic cells are generally resourced from embryo. [True/False]
6. When cells multiply to produce same kind of cells, it is called as differentiation. [True/False]
7. Alkaline phosphate (ALP) and RUNX2, both are differentiation markers. [True/False]
8. FACS stands for fluorescence activated cell sorter. [True/False]
9. Stem cell potency indicates their capacity to differentiate into various cell types. [True/False]
10. *Staphylococcus* is an example of Gram positive bacteria. [True/False]
11. *Escherichia coli* is an example of Gram positive bacteria. [True/False]
12. Culturing bacteria in test tube is also called planktonic culture. [True/False]
13. Planktonic culture is used for ascertaining minimum inhibitory concentration (MIC) of an antibacterial agent. [True/False]
14. The process of formation of two separate nucleus with the replicated chromosomes is known as Cytokinesis. [True/False]
15. Acute effect or response study lasts for 24 hrs but sub-acute lasts for 14-28 days. [True/False]
16. Thrombus formation on Titanium like metals can be prevented by non-metallic coating [True/False]
17. *in vivo* degradation of implant is measured by change in volume with time. [True/False]
18. Clinical trials for biomaterials is known as "First-in-man-studies". [True/False]
19. Histology is a Qualitative method of analysis. [True/False]
20. Randomization in clinical trials prevents selection and accidental bias. [True/False]
21. Multi-centric clinical trials mean more than 2 hospitals are involved in the study. [True/False]
22. The term concealment is used to keep certain specific information as secret. [True/False]
23. Micro-CT is a quantitative method of analysis. [True/False]
24. Metallic biomaterials are shaped using melt-molding technique. [True/False]
25. Convergent beam computer tomography (CBCT) is used for analyzing bone height and width around the implanted orthodontic device. [True/False]
26. Ball milling technique helps in bringing the ceramic particles from micron to sub-micron size. [True/False]
27. AABABABBAA represents a homopolymer. [True/False]
28. AABABABBAA represents a heteropolymer. [True/False]

29. The antibacterial action of silver is through disruption/alterations of bacterial membrane. [True/False]

30. Bone strength and fracture toughness decreases with age. [True/False]

31. UHMWPE is used as acetabular socket mainly due to a better combination of biocompatibility and tribological properties. [True/False]

32. Wear particulates from improper choice of materials for acetabular socket and femoral head can cause inflammation. [True/False]

33. Diamond like carbon coating prevents thrombus formation on Titanium. [True/False]

34. Dynamic host tissue response cannot be modulated *in vitro*. [True/False]

35. The antibacterial action of silver is through changes in bacterial DNA. [True/False]

36. The most reliable method of evaluating fracture toughness of ceramics is through single edge V-notch bend test. [True/False]

37. Resistance to crack growth propagation in a material is called fracture toughness. [True/False]

38. Natural bone is Piezoelectric. [True/False]

39. Osteoconduction phenomenon allows bone cells ingrowth in the implant. [True/False]

40. Bone marrow, a spongy material that fills the medullary cavity of long bone is a good source of Stem cells. [True/False]

41. Cancellous bone is mechanically inferior to cortical bone. [True/False]

42. Transfer of genetic information from DNA to mRNA in nucleus is called Transcription. [True/False]

43. Results of *in vitro* assays performed on biomaterials can be extrapolated directly to predict biocompatibity of clinically. [True/False]

44. Host response or foreign body response is the reaction of living system to presence of a foreign material *in vitro*. [True/False]

45. The functionality or activity of cells in culture medium, when grown in isolation is different from the cells when grown on biomaterial substrates. [True/False]

46. The important property to be evaluated for blood contacting devices is haemocompatibility. [True/False]

47. All the biomaterials scaffolds may be considered as implants and vice versa. [True/False]

48. Cilia help in motility or migration of the bacterium. [True/False]

49. Mesenchymal stem cells (MSCs) are relatively rare in the bone marrow. [True/False]

50. Embryonic stem cells can be received from placental cord. [True/False]

51. Stem cells are defined by self-renewal, potency and clonality. [True/False]

52. Photobleaching can be minimized by increasing the intensity or time span of light exposure. [True/False]

53. Late stage marker of osteogenesis is known as osteocalcin. [True/False]

54. Differentiation involves an expression of unique genes that are specific to a cell type with an irreversible change towards a particular cellular function phenotype. [True/False]

55. If extracellular conditions are favorable and signals to grow and divide are present, cells in early G1 progress through a commitment point near the end of G1. [True/False]

56. One of the most abundant (50% of all major organic molecules) intracellular organic molecule/biological polymer made up of smaller units known as carbohydrates. [True/False]

57. A single bacteria can form biofilm. [True/False]

58. Endocrine signaling depends on endocrine cells, which secrete hormones into the bloodstream, that are then distributed widely throughout the body. [True/False]

59. Some of the molecules involved in direct cell-cell contact (known as cell-junction molecules) allow for direct cytoplasmic communication. [True/False]

60. A complete set of information in an organism's DNA is called as gene. [True/False]

61. DNA composed of two DNA strands (long polypeptide chains) held together by covalent bonds between the paired bases. [True/False]

62. Cell differentiation is the same as cell replication. [True/False]

63. Non-linear stress strain response is observed by actin filaments. [True/False]

64. Any material to be an ideal bone replacement material, biocompatibility property is a necessary but not sufficient property. [True/False]

65. All blood cells have nucleus. [True/False]

66. Erythrocytes are responsible for blood clotting. [True/False]

67. In case of a bone implant (X), which is/are true –

    a. X should be biocompatible for all other applications, like cardiovascular applications

    b. X may not be biocompatible for neural tissue engineering applications

    c. X must be biocompatible for bone tissue engineering applications, *in vivo*

    d. X can support osteogenesis of human mesenchymal stem cells

68. Non-human primates closely resemble the human anatomical structure and physiological function. [True/False]

69. Osteopontin is a bone marker. [True/False]

70. GATA-4 is a marker for onset of cardiomyocyte differentiation. [True/False]

IV. **Match the following**

A.
1. Decrease in cell size
2. Increase in cell number
3. Increase in cell size
4. Disordered growth in cellular shape, size, and/or organization
5. Change in cell type

  (a) Metaplasia
  (b) Hyperplasia
  (c) Atrophy
  (d) Dysplasia
  (e) Hypertrophy

B.
1. Cells signal neighboring cells by diffusion
2. Cell secrets growth factor into blood stream, carried into target cell
3. Cell signals itself
4. Growth factors for cell proliferation and differentiation

  (a) Cytokines
  (b) Paracrine
  (c) Autocrine
  (d) Endocrine

C.
1. Angiogenesis
2. Haversian canal
3. Necrosis and apoptosis
4. Macrophages
5. Fibroblast

  (a) Inflammatory tissue
  (b) Bone tissue
  (c) Connective tissue
  (d) Poor prognosis
  (e) Blood vessels

V. Diagram identification

1. Complete or identify the following figures with appropriate labeling, where applicable:

(a) Prokaryotic cell

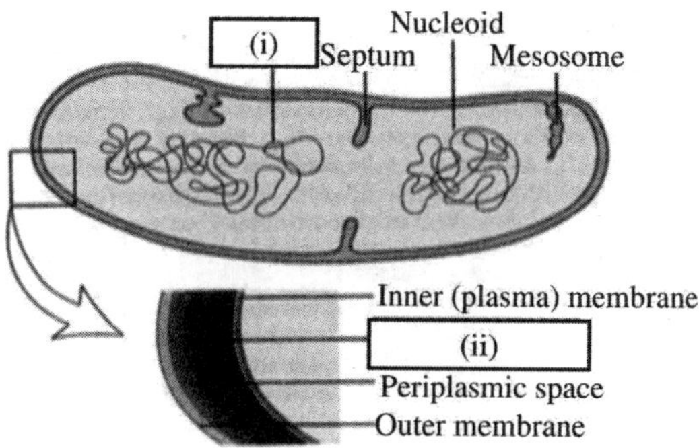

(b) Eukaryotic cell

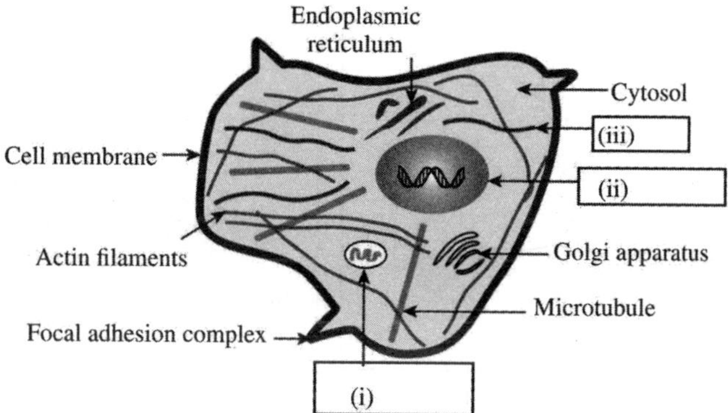

(c)

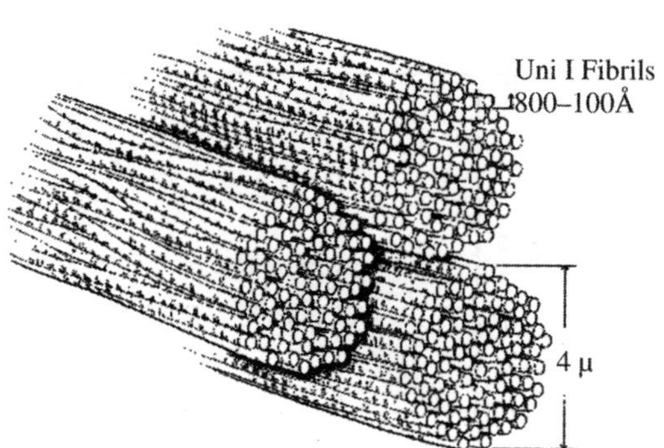

[see Colour Plate]

2.

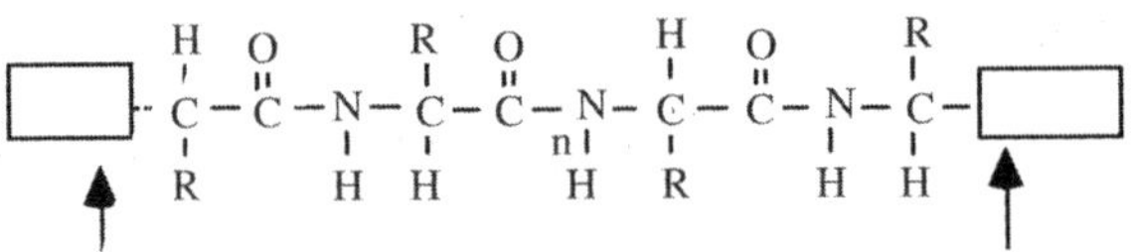

3.

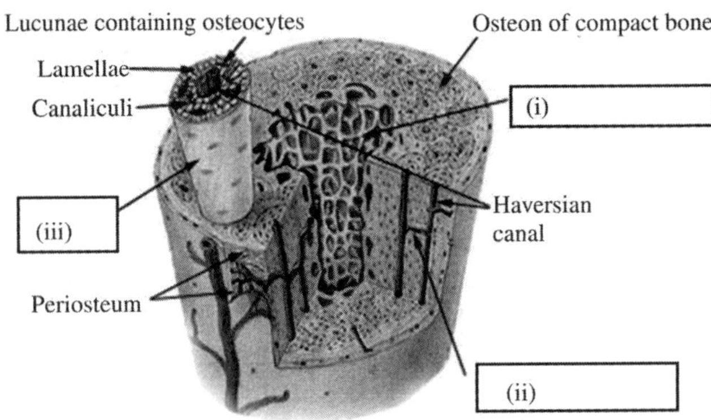

[see Colour Plate]

4. Identify the type of cell signaling:

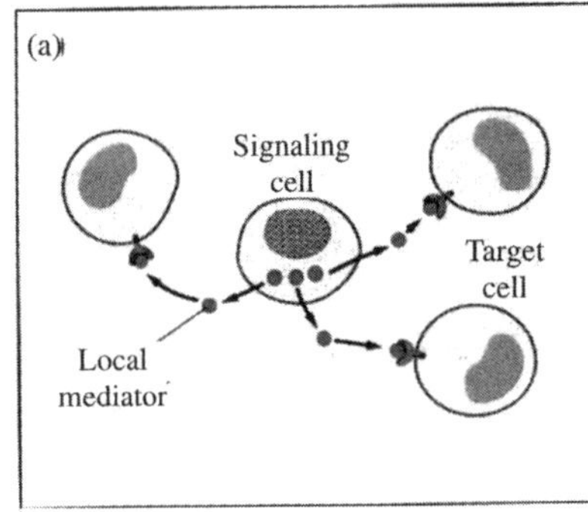

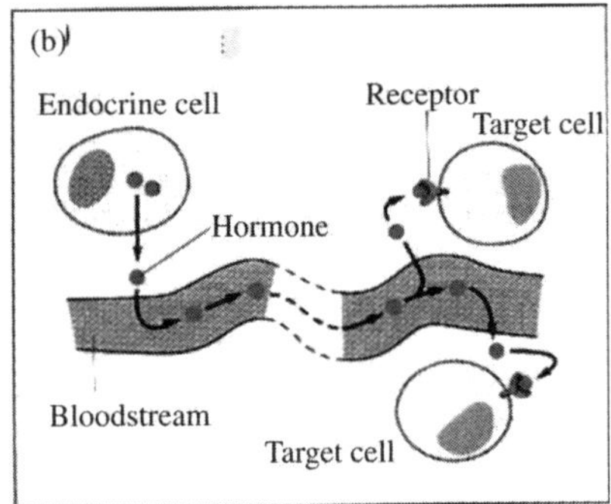

[see Colour Plate]

5. In the figure below, label the four quadrants of a FACS plot:

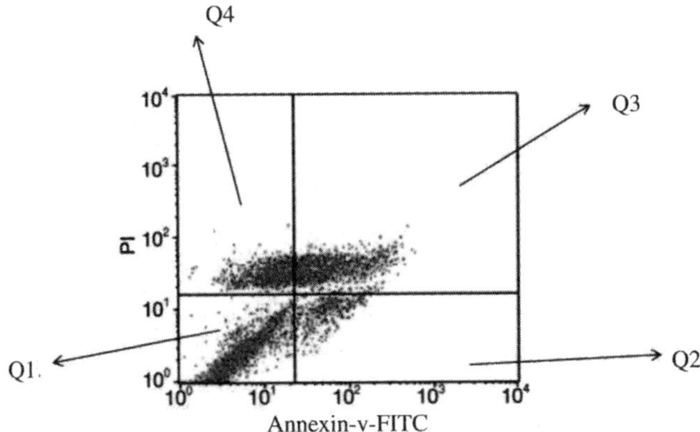

6. Identify the following molecule and then identify the components/dimensions of various parts:

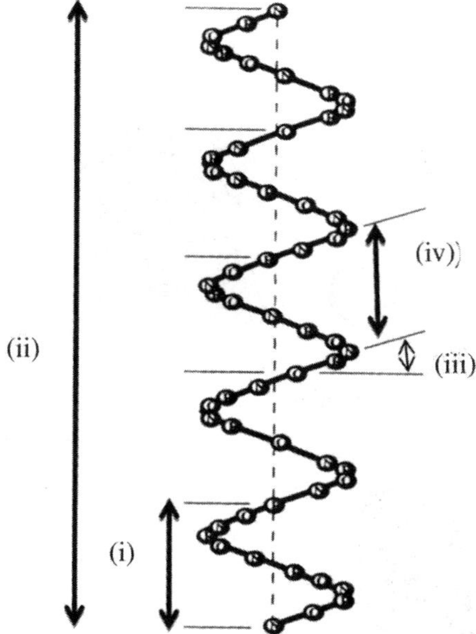

7. Below is a stress–strain curve for various biological samples. Identify from the graph, which of the following exhibit highest fracture toughness.

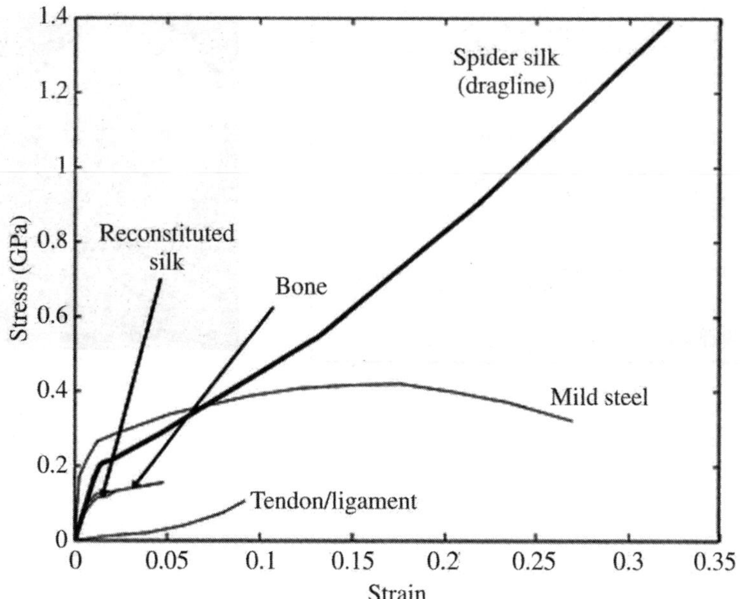

8. Identify the following in terms of either, outcome or the progress of some process or characterization technique or, state of cell population in various co-ordinates or the marked regions/steps:

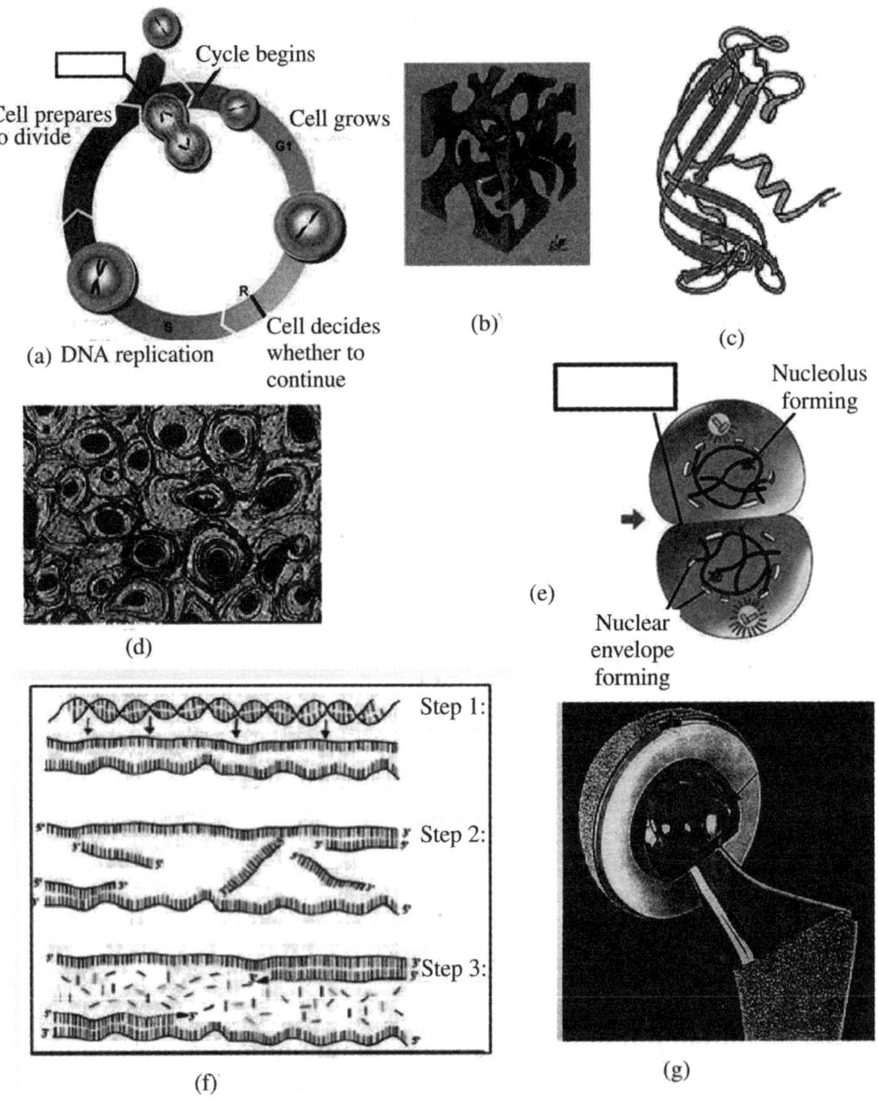

[see Colour Plate]

VI. Short answer type

1. What are the three criteria that define stem cells? Briefly define and explain each of them.

2. Define the following:
   (a) Biological cell
   (b) Biocompatibility
   (c) Host response
   (d) Cytocompatibility
   (e) Biofilm
   (f) Cellular organelle
   (g) Osseointegration
   (h) Osteoinduction
   (i) Osteoconduction
   (j) Grain and grain boundary (microstructure of a crystalline material)
   (k) Crack, in the context of fracture of ceramics
   (l) Infrared spectroscopy
   (m) Polarizability of a molecule
   (n) Elastic scattering
   (o) Structure factor, in the context of XRD
   (p) Fluorescence
   (q) Resolution and numerical aperture of a microscope
   (r) Cell shape index
   (s) Haemocompatibility

3. (i) One line description for the following
   (a) Hematopoietic stem cell
   (b) Mesenchymal stem cell
   (c) Oligodendrocyte
   (d) Islet cell
   (e) Hepatocyte
   (f) Astrocyte
   (g) Adipocyte
   (ii) Name the adult stem cells that can be isolated form the following source
   (a) Bone marrow
   (b) Brain
   (c) Liver
   (d) Blood
   (e) Heart
   (f) Umbilical cord
   (g) Eye

4. Provide an outline of the gram staining protocol and the rationale for the differential staining.

5. Schematically illustrate the following:
   (a) Two different methods of toughening mechanisms (one each from process zone and bridging zone)

    (b) Fracture toughness measurement of ceramic implants using indentation cracking and SEVNB technique

    (c) Difference between Hardness and Strength (definition and measurement technique) of ceramics

    (d) Glide of a positive edge dislocation in metal and stresses around the defect.

    (e) The differences between polymer, ceramic and ductile metals w.r.t tensile stress-strain response.

    (f) Different modes of loading and displacements at crack faces in ceramics (mode I, II and III)

    (g) Evolution of cone cracks in alumina implant under indentation loading

    (h) Difference in mechanisms of bulk and surface erosion of polymers

    (i) Difference between precipitation hardening and dispersion strengthening

    (j) Stress-strain response and the crack propagation in ceramics under uniaxial compression

6. Explain the following with suitable graphical illustration and equations, as necessary

    (a) Instrumented indentation plot of a polymer and a ceramic

    (b) Extrusion of polymers

    (c) Progressive deformation of a semi-crystalline polymer

    (d) Process to make a thin metallic wire for cardiovascular

    (e) Porosity effect on strength and elastic modulus

    (f) Graft co-polymer and block co-polymer (in terms of chain configuration)

    (g) Viscoelastic response and change in macromolecular structure of polymer during tensile deformation.

    (h) Fatigue behavior of stainless steel based implant materials

    (i) Fracture toughness measurement of ceramic implants using indentation cracking and SEVNB technique

    (j) Tensile deformation and compressive deformation of a metallic implant

    (k) Small angle grain boundary

    (l) Lack of deformation of alumina implant

    (m) Fatigue properties of Ti implants in air and SBF solution

    (n) Cell division process (stage wise)

    (o) Cell migration (various stages)

    (p) Checkpoints in cell division process

    (q) m-RNA synthesis from DNA in an eukaryotic cell

    (r) Stages in foreign body reaction towards an implant material, *in vivo*

    (s) Cell-cell adhesion mechanisms

    (t) Cell-material interaction *in vitro*

    (u) Blood-material interaction, (types and mechanisms)

    (v) Quantification of live/viable cells

    (w) Late stage differentiation assay

    (x) Fluorescence microscope (working principle)

    (y) Different cytokine based assays to study inflammatory response

    (z) Explain the expected reason for the high toughness of Zirconia based dental implants.

    (aa) Explain the origin of the strengthening of stainless steel implants.

7. Classify bacteria based on their shape and provide examples of pathogenic bacteria in each category. Briefly mention the materials related factors, that influence the adhesion of prokaryotic cells on a biomaterial.

8. Derive the rate equation for bacterial growth kinetics in the exponential phase.

9. Distinguish with suitable graphical illustrations and examples (wherever applicable):
   (a) Scaffold vs. implant
   (b) Genotoxicity vs. Haemocompatibility
   (c) Transcription vs. Translation process
   (d) Gram positive vs. Gram negative bacteria
   (e) Hyperpalasia vs. Metaplasia
   (f) Necrosis vs. Apoptosis
   (g) FGF vs. TGF
   (h) Cytotoxicity vs. Genotoxicity
   (i) Viscoelasticity of polymers vs. Plasticity of metals
   (j) DNA structure vs. RNA structure
   (k) Crystallinity in metals vs. Crystallinity in polymers
   (l) Hardness vs. strength (in case of metals)
   (m) Coupling agent (polymers) vs. alloying elements (metals)
   (n) Strengthening vs. toughening mechanism
   (o) Hot pressing vs. spark plasma sintering
   (p) 3D printing vs. 3D plotting
   (q) Depth of field vs. depth of focus
   (r) Fluorescence vs. confocal microscopy
   (s) Flow cytometry vs RT-PCR
   (t) *in vitro* vs. *in vivo* tests
   (u) Abrasive wear vs. Adhesive wear
   (v) Paracrine signaling vs. Endocrine signaling
   (w) Shear vs. compression behavior of metallic biomaterial
   (x) Cell apoptosis vs. Cell necrosis

10. What are the modes of action of antibiotics? Provide examples of the group of antibiotics under each category. Also outline the mechanisms of antimicrobial resistance.

11. Briefly describe the two phases of bacterial adhesion onto material surfaces. List the key factors governing bacterial adhesion with examples.

12. Expand and provide reaction schemes for ONPG and LDH biochemical assays.

13. Provide a reaction schematic for the generation and detoxification of ROS and RNS.

14. Provide examples of surface modification and applications for – a) Non-adhesive surfaces, b) Contact-killing surfaces, c) Tissue integrating surfaces, d) Antimicrobial releasing surfaces and e) Multi-functional coatings.

15. Explain various cardiogenic differentiation assays.

16. Carbonaceous materials (CNT, graphene) are gaining importance in biomaterial application. How can one reliably quantify cell viability on such surfaces (principles and protocols of complimentary assays)?

17. Strength is an important property for load bearing metallic implants. What rule links the microstructure and strength property. Explain it.

18. How does cell spreading on material surface occur? Explain the steps of cell material interaction with diagram?

19. Explain all the models for Protein adsorption in detail.

20. Explain process of cell migration with diagram.

21. Cytoskeleton structure plays an important role in various cellular functions. With a neat sketch, explain the morphological features of each of the structural elements of the cytoskeleton.

22. What is Photobleaching? How it can be minimized?

23. Magnetoactive biomaterials are receiving importance for biomedical applications. Using M-H plot, explain the difference between ferromagnetism and paramagnetism. How would one measure the magnetic properties (principles and guidelines of measurements)?

24. What is bactericidal agent? Give examples.

25. Define tissue. With help of a schematic, explain the formation of myotubes in the context of muscle tissue.

26. Draw a neat sketch of a gram negative bacterium (label important features) and discuss the mechanism of bacteria-material interaction leading to biomass formation. Mention key differences between such interactions with that of eukaryotic cell material interaction. Mention two additives that you would add to enhance antimicrobial property of HA.

27. If one claims X to be cytocompatible using bone cells, mention the following:
    (a) Sample preparation (just prior to culturing) and individual stages of biological culture protocol.
    (b) Clearly mention cell lines to be used, culture time, including the name of equipments.
    (c) Fundamental details of the assay to be used to quantify live cells on material
    (d) How surface roughness and porosity influences cell adhesion?
    (e) What are the differences between adhesion of eukaryotic and prokaryotic cells?
    (f) Why fibroblast cells are widely used as initial screening test for cytocompatibility?

28. (a) Mention the importance of checkpoints in the process of cell division and how many checkpoints are there for eukaryotic cell division?
    (b) What happens in metaphase?

29. (a) What do you mean by polypeptide chains have structural polarity?
    (b) What is the importance of cell membrane structure?

30. What are stem cells? Explain in detail, the types of stem cells, their classification based on potency and the types of cell division

31. Give three examples for each of the following:
    (a) Synthetic biodegradable materials
    (b) Protein based biomaterials
    (c) Polysaccharide based biomaterials
    (d) Ceramic based biomaterials

32. List down five biophysical cues that can influence stem cell fate.

33. Explain briefly the guidelines to be followed for the following property measurements in case of bioceramic implants: (a) hardness, (b) fracture toughness, (c) strength.

34. Using a schematic, explain the interaction of electron beam with a material surface, that takes place in a scanning electron microscope. What is Lorentz force? How would the probe current depend on various parameters?

35. With a neat sketch, draw the schematic of an eukaryotic cell and label at least five cellular organelles.

36. Stainless steel plates are widely used in orthopedic surgery. Explain the different stages of processing such plates, starting from steel melt. Use a process flow diagram. Also, explain the selection of a specific flow stress region where the plates can undergo bulk deformation processes.

37. Casting is one of the bulk metal fabrication route. Explain the rule that governs the casting time of a metal. What is investment casting? Which biomaterial is widely processed by this casting route?

38. With a flow diagram, explain how can one design the animal experiments with biomaterials and also assess the outcome, qualitatively and quantitatively? What are the different ethical aspects of pre-clinical animal experiments?

39. What are the general biosafety issues to be addressed before conducting any cell culture study? What are the various aspects of ethical approval that one must follow before designing any stem cell study?

40. What are the general culture conditions to assess the viability and proliferation of bone cell on biomaterial, in terms of incubator environment, culture medium composition, etc? Do such conditions vary on different cell type? How are those different for prokaryotes?

VII. Analytical problems

1. A thin wire of a shape memory alloy based cardiovascular stent with diameter of 0.75 mm experiences a tensile load of 250 N and at this load, it is found that the 100 m wire elongates by 5 cm. Calculate the engineering stress and engineering strain at this point. After the load is removed, the length of the wire is found to be 100.03 m. Calculate the yield strength, if the elastic modulus is given to be 200 GPa.

2. A plate made of an unknown oxide bioceramic has a fracture toughnessof 2 MPa.m$^{1/2}$ and a plate of alumina with a fracture toughness of 5 MPa.m$^{1/2}$. Due to difference in manufacturing techniques used to produce these two materials, the longest surface crack is found to be 3 µm in the unknown bioceramic, whereas it is 7 µm in alumina. Determine which plate is to be used, if the maximum allowable stress is 250 MPa in the direction perpendicular to the crack.

3. The plain strain fracture toughness of a hypothetical biomedical grade Ti-alloy is, K_{Ic}=24 MPa.m$^{1/2}$. The tensile yield strength is 500 MPa. Assume that Y = 1.5 for the test specimen in question. What is the smallest crack that will cause brittle failure in this sample? If the elastic modulus is 71 GPa, what is the surface energy? (Hint: Use the crack length obtained in the previous section and apply the Griffith's theory for brittle fracture).

4. What would be the allowance crack size for Al_2O_3 (K_{Ic} = 3 MPa.m$^{1/2}$) and how it will be different for ZrO_2 (K_{Ic} = 10 MPa.m$^{1/2}$), when both experience 100 MPa tensile stress? Explain each term in the equation that you use. Assume, Y=1.

5. Determine the load at which zirconia ceramic implant with a critical crack size of 1 µm (K_{Ic}=10 MPa.m$^{1/2}$, E=200 GPa) will fracture under four point loading. What will be the radius of curvature at fracture for the same implant sample (square cross-section with thickness 5 mm and span length 40 mm).

6. Determine the fracture strength of an alumina ceramic with a critical surface crack size of 1 µm (K_{Ic}=4 MPa.m$^{1/2}$, E= 400 GPa). What will be the radius of curvature at fracture for the same alumina bar (rectangular cross-section with thickness 5 mm) under flexural loading?

VIII. Descriptive type questions

1. Why are HA-based materials widely being investigated? What is the difference between HA and TCP (in terms of chemical composition and properties)? What is BCP? How can one independently produce macroporous and microporous BCP using conventional ceramic processing route? What characterization tools, would you use to identify the presence of phosphate?

2. How is the glass ceramics processed in the laboratory? What are the characteristic tools that you would use to confirm the presence of glass and crystalline phases? Can you use glass-ceramics for biomedical applications in various shapes and sizes, since the glass-based materials are widely known for its brittleness?

3. The (eukaryotic) cell-material interaction is the central theme of the Biomaterials Science.
    (a) With suitable sketches, explain the stages of cell-material interaction, *in vitro*.
    (b) Comment on how the material related specific properties would influence such interaction.
    (c) Describe how would you grow eukaryotic cells on a synthetic biomaterial substrate (general cell culture protocol) and mention various equipment and conditions at each stage?
    (d) What are the microscopic techniques that you would use to study cell morphological changes?
    (e) What are the various cell morphological parameters that one would quantify to assess the change in cell morphology at various timepoints in culture?

4. On a stress-strain plot, schematically illustrate the difference among the cytoskeletal proteins (actin, tubulin and intermediary filaments) in terms of elastic modulus and deformation behavior.

5. You are asked to develop an implantable tissue engineered scaffold for cardiovascular application (cardiac patch). Answer the following with appropriate explanation.
    (a) What physical property of scaffold would be important to promote application-specific cell functionality?
    (b) What would be the cell type that you select for cytocompatibility study?
    (c) Can you use stem cells and assess the functionality changes on such scaffold?
    (d) Explain a brief protocol for pre-clinical testing of such scaffolds for the targeted application, together with necessary ethical issues.

6. What are the differences between adhesion of eukaryotic and prokaryotic cells?

7. Why fibroblast cells are widely used as initial screening test for cytocompatibility?

8. Discuss the deformation and fracture mechanisms of human bone at different length scales. How does the fracture toughness property measured to get reliable values experimentally (sketch the sample configuration)? How does the fracture toughness of natural cortical and cancellous bone vary with age/mineral density?

9. Sketch a complete assembly of implantable biomaterials to be used in Total Hip joint Replacement (THR). Based on the criterion of resistance to Corrosion and Wear as well as high biocompatibility, what would be the ideal material combination of THR? Justify your answer. What are the differences between cemented and uncemented hip joint replacement?

10. Draw the 'mer' unit of a) PE, b) PDMS, c) Teflon and d) Sketch the tensile behaviour of the composite and show how it is changed w.r.t polymer 'P'

11. With a schematic, explain various aspects or property requirements, to be carefully considered to develop bone analogue materials.

12. A polymer 'P' (thermoplastic) is bioinert and has very low E-modulus and strength. In order to improve bioactivity and mechanical property, how and what would you do for the following:
    (a) Selection of composite composition
    (b) Processing condition and technique
    (c) Type of coupling agent

13. Bacterial adhesion is the first step in prosthetic infection. In this context, answer the following:
    (a) Draw a neat sketch of a gram negative bacterium (label important features)
    (b) Discuss the mechanism of bacteria – material interaction leading to biofilm formation.
    (c) Mention key differences between such interactions with that of eukaryotic cell-material interaction.
    (d) Briefly mention the materials related factors that influence the adhesion of prokaryotic cells on a biomaterial.
    (e) HA does not have necessary antimicrobial properties. How can you compositionally design HA-based materials with enhanced antimicrobial property of HA?
    (f) How can you confirm the presence of the antimicrobial agent in the modified HA using specific characterization tools?
    (g) Briefly explain various biochemical assays to quantitatively and qualitatively assess the bacterial viability, ROS generation, biomass and biofilm characteristics.

14. The understanding as how an implant interfaces and interacts with host bone, *in vivo* is one step further towards confirming biocompatibility. In this context, answer the following,
    (a) Describe various stages of tissue-material interaction (at various timepoints) that a bioinert material would experience when implanted in a rabbit model, with suitable illustrations.
    (b) How does such mechanism would differ in case of a biodegradable material in identical animal model?
    (c) How does the porosity of bone replacement material influence biocompatibility, *in vivo*?

15. Briefly explain the criterion to be followed while selecting cell type and culture duration, while studying cell proliferation on (a) carbon-based biomaterials and (b) Hydroxyapatite-based biomaterials. What are the complimentary biochemical assays that you would use to quantify your results?

16. What are the criteria to be followed while selecting a specific animal model to study (a) bone regeneration, (b) cardiovascular diseases, and (c) neural tissue engineering?

17. (a) Describe the various stages to be followed in designing any animal experiments to assess the biocompatibility of any material for bone tissue engineering applications.
    (b) What are the various techniques or characterization tools to qualitatively and quantitatively assess the bone regeneration?

18. If one wants to develop a NiTi wire of a novel composition for cardiovascular stent, staring from a workpiece (several times larger in diameter) of identical composition, answer the following:
    (a) What manufacturing protocol that one has to follow?
    (b) What are the process parameters that will determine the final diameter of the unfractured wire?
    (c) Do you need any intermediate heat treatment?
    (d) Can such wire have any special property?
    (e) How can you confirm the presence of various phases in the drawn wire?
    (f) What are the biocompatibility tests that one has to adapt before clinical trials?

19. You are asked to develop a patient-specific ceramic femoral head prototype. Describe the various stages of the conventional processing that you have to adopt to obtain femoral head with close dimensional similarity.

20. In order to enhance mechanical properties without compromising on biocompatibility property, one would like to develop HA-Y composite, where Y is a second phase addition (reactive with HA, Hydroxyapatite for prolonged sintering time). In this context, answer the following.
    (a) What would be the ideal sintering technique?
    (b) Discuss briefly sintering mechanism involved.
    (c) What are the mechanical properties of relevance and mention corresponding technique to evaluate them?
    (d) What would be justifiable amount of Y that you would like to add?

21. If a material 'M' is to be used for peripheral nerve tissue regeneration,
    (a) What are the cell type(s) to be used for *in vitro* experiment and why?
    (b) Sketch the morphology of such cells after they are grown on material 'M'.
    (c) What would be the appropriate animal model and why?
    (d) Describe the stages of animal study.

22. You are asked to prove that a material 'A' should not cause gene level toxicity. How would you go ahead to design *in vitro* experiments and what assay would you use? Describe the fundamentals principles of such assays.

23. If one intends to develop HA-based electroconductive bioceramic composite, answer the following :
    (a) Why do you need to develop electroconductive biomaterial, in first place? [Hint: Consider bone properties]
    (b) How would you design such composite composition and process such material?
    (c) How would you quantify electrical conductivity property?
    (d) Is it possible to enhance mechanical properties (e.g. fracture toughness) along with electrical conductivity for such composites?

24. Corrosion is a major issue for metallic biomaterials.
    (a) How would one design Ti-based alloys (in terms of alloying additions) for better corrosion resistance?
    (b) What are the experimental protocols that one would use to quantify corrosion rate?
    (c) What are the microstructural features that characterize corroded surfaces?

25. Fracture resistance of the ceramics is largely determined by the combination of fracture toughness and fracture strength. HA, like any other ceramic, is known for its brittleness. In this context, answer the following:
    (a) How can you plan to design HA-based composite composition with better fracture resistance?
    (b) What are the different experimental techniques that you will adopt to reliably evaluate fracture resistance (strength and toughness) of such composites? Remember, better is the fracture resistance, more would be the clinical acceptability.

26. Fracture is an important performance-limiting property of ceramics. Explain Griffith's model of fracture to define the critical crack size. In this context, define stress intensity factor (mode I). How can one define fracture toughness?

27. What are the material related parameters, that would influence polymer processing? Explain the role played by each of them. If one has to prepare near-net shaped polymeric acetabular socket without any machining step, what processing approach one has to adopt? Explain the process details with suitable diagram.

28. Biodegradable porous polymers are widely used in tissue engineering applications. Name one of them and write the formula of its 'mer' unit. How would one measure various properties (including porous architecture) of relevance, that would influence the biocompatibility? What characterisation tools would you use?

29. What are the different cellular adaptation processes? Describe, with sketches, various cell fate processes, that a cell is expected to experience while growing on a biomaterial.

30. Write the reaction of two amino acids to form peptide bond. Describe various levels of protein structure (use numbers to quantify some structural features).

31. If you want to enhance the strength of any metallic biomaterial, what are the strengthening mechanisms that you would adopt?

32. You are asked to develop Ti-based implants for maxillofacial reconstruction. What is the processing strategy that you would adopt to develop patient-specific implant, starting from metal powders? Remember you are not allowed to use room temperature additive manufacturing technique. What are the different process variables in your processing approach? What are the different post-processing techniques, that one has to follow.

33. What are the various approaches to enhance biocompatibility by modifying the surface characteristics? What are the various parameters that would influence the cell compatibility of a patterned biomaterial substrate?

34. What are the different joining processes used for metallic biomaterials? Describe the working principle for any two of such joining processes.

35. What are the natural biological materials? Provide a few examples. Is the natural bone a composite material? Explain. Can a synthetic biomaterial be fabricated using the best possible 'Good Manufacturing Protocols'? Do you think such a biomaterial can match the bone in terms of physicochemical, functional and biocompatibility property? Explain with sufficient reasoning.

36. What is ECM? How is it synthesized when eukaryotic cells are cultured on a biomaterial?

37. Force is a vector.
    (a) Can stress be a vector too? If not, explain why it is not.
    (b) Define stress at a point in a solid experiencing forces from different direction, using appropriate mathematical expression. Remember, an implant in a human patient also experiences biomechanical (muscle) forces from different directions.

38. (a) Can strain be related to displacement in a solid experiencing stress? If so, explain in terms of necessary sketches and mathematical expressions.
    (b) Describe the relationship between stress and strain in tensorial form for a linear elastic solid, like ceramic implant.

39. On a single stress-strain plot, distinguish the difference in deformation behavior of a brittle ceramic, ductile metal and thermoplastic polymer, when all experience independently uniaxial tension.

40. (a) Write the relationship between stress and strain in the most appropriate (from mechanics point of view) form to describe the deformation behavior of a ductile metal.
    (b) Describe the underlying mechanisms to explain the characteristic mechanical deformation of the thermoplastic polymer.

Appendix B

Key Answers

I. Multiple Choice Questions

1. (b) different
2. (d) all of the above
3. (d) micro-CT
4. (d) haemocompatibility
5. (a) human & animal tissues
6. (b) Biomechanical
7. (c) 2-3 times longer than diameter
8. (a) 50-100 nm
9. (a) superior
10. (b) bones
11. (b) bone mineral density
12. (a) 2 mm & length 6 mm
13. (d) all of the above
14. (b) 1.67
15. (b) collagen and HA
16. (a) Calcium phosphate
17. (b) actin filaments
18. (c) 10^6
19. (d) DAPI
20. (b) undergoes division
21. (d) differentiation
22. (b) secondary
23. (d) plasma membrane
24. (a) cell receptor and surface protein
25. (c) micrometers
26. (a) thin
27. (a) Collagen
28. (a) 20-40 minutes
29. (c) migration
30. (d) death
31. (b) pluripotent
32. (d) Cardiomyocytes
33. (b) rabbit
34. (b) Sintering
35. (c) permanent deformation
36. (c) injection molding
37. (b) inhibit the bacterial growth
38. (a) fracture toughness
39. (b) Total knee replacement
40. (d) all
41. (b) rough
42. (b) elastic modulus

43. (b) histology
44. (a) Convergent beam computer tomography (CBCT)
45. (a) osteoblast
46. (b) connective
47. (a) blood vessels & nerves
48. (b) osteon
49. (a) structure
50. (a) Kroll's
51. (b) be retrieved easily
52. (a) piezoelectric property
53. (d) all of the above
54. (b) 25-30 μm
55. (b) 10^9
56. (d) Voltage difference across the cell membrane
57. (a) increase in cell size
58. (d) all of the above
59. (a) reversible
60. (d) all of the above
61. (d) both (a) and (b)
62. (b) Bone
63. (c) programmed cell death
64. (d) inflammation
65. (a) nucleus size increases
66. (b) Cartilage
67. (b) rabbit
68. (a) mouse and rats
69. (a) kill the bacteria
70. (d) All of the above
71. (a) Bone density
72. (d) All of the above
73. (c) Infrared spectroscopy
74. (c) Ag^+ ions
75. (a) Enamel
76. (b) Mica crystals
77. (c) enzyme-linked immunosorbant assay (ELISA)
78. (d) both a and b
79. (d) biomaterials processing
80. (d) all of the above
81. (d) all of the above
82. (c) Total hip joint replacement
83. (d) all of the above
84. (a) dependent on target application
85. (c) Titanium alloys
86. (c) Nickel-Titanium alloy (Nitinol)
87. (d) all of the above
88. (c) bacterial growth
89. (d) passed through clinical trials
90. (b) aseptic loosening
91. (b) osteons
92. (c) $Ca_{10}(PO_4)_6(OH)_2$
93. (b) compact
94. (a) elastic modulus
95. (b) flexible
96. (a) periosteum
97. (c) 60-70%
98. (d) 10 wt.%
99. (d) all of the above
100. (b) mineral content
101. (b) HDPE
102. (b) Ti based alloys
103. (a) 10^9
104. (b) Phalloidin
105. (a) MitoTracker
106. (b) 70 mV

107. (b) selective molecules
108. (d) all of the above
109. (d) all of the above
110. (a) amino acid sequence
111. (b) RNA-Protein complex
112. (c) both a and b
113. (c) bacteria
114. (b) log phase
115. (a) UV-Visible range
116. (b) faster doubling time
117. (d) Cytoskeleton
118. (d) 3-4 weeks
119. (c) 1-100 μm/h
120. (d) growth factor
121. (d) 200 μm
122. (d) 15-30 min
123. (c) nucleus
124. (b) cytoplasm
125. (b) messenger RNA
126. (c) TEM
127. (d) amino acid
128. (c) osseointegration
129. (a) locomotion
130. (d) proliferation
131. (b) Four
132. (d) all of the above
133. (b) polymerase chain reaction
134. (a) apoptosis
135. (c) flow cytometry
136. (d) purple
137. (d) all of the above
138. (d) double stranded DNA
139. (c) differentiation
140. (b) protein
141. (c) transcription factor
142. (c) FACS analysis
143. (c) all cell types
144. (b) multipotent
145. (c) embryonic stem cells
146. (c) keratinocyte
147. (a) Islet cells
148. (c) stemness
149. (c) limited tissues
150. (b) multipotent
151. (b) apoptosis
152. (b) microtubules
153. (c) S
154. (b) necrosis
155. (d) all of the above
156. (a) cell differentiation
157. (c) platelet adhesion
158. (d) ISO 10993
159. (d) both (a) and (b)
160. (a) sheep and pig
161. (d) all of the above
162. (a) Protein absorption
163. d) Flow cytometry
164. (b) Quantitatively
165. (d) all of the above
166. (c) $T_m/2$
167. (d) all of the above
168. (a) number of monomers in a polymer
169. (d) 345 K
170. (c) both (a) and (b)
171. (a) sintering time is longer
172. (b) Electrospinning
173. (a) above 1000°C
174. (c) both pore size and pore shape changes

175. (b) fracture toughness will increase

176. (a) load bearing capability

177. (d) will be proportional to square root of ratio of crack length and crack root radius of curvature

178. (b) the lowest elastic modulus is exhibited by Ti alloys

179. (d) wire drawing

180. (a) silver

181. (c) amino acids

182. (a) Piezoelectric property

183. (d) All of the above

184. (d) All of the above

185. d) All of the above

186. (c) globular

187. (b) ZnO produces H_2O_2 in solution

188. (a) a single bacterium

189. (b) B–A–C

190. (d) all of the above

191. (c) 45% silica

192. (d) all of the above

193. (c) Xylenol orange

194. (b) Differential scanning calorimetry

195. (a) machinability

196. (a) hard material wears soft material

197. (c) 12 weeks

198. (c) Compression Molding

199. (d) All of the above

200. (d) Phlogopite

201. (b) Implants shield stress from bone to osteoblasts

202. (a) HDPE

203. (a) Ca/P < 1.5

204. (c) Viscoelastic polymer

205. (a) less than 5 %

206. (c) They connect microtubule doublets together so that when dynein moves along one microtubule relative to the other, the cilium or flagellum bends

207. (c) They guide vesicles containing neurotransmitters from the cell body down the axon

208. (a) Chloroplasts and mitochondria have many cilia and are able to swim around freely

209. (c) central vacuole/tonoplast

210. (d) both compartmentalization and membrane-bound organelles

211. (c) lysosome

212. (b) rough endoplasmic reticulum

213. (d) binary fission

214. (d) A cleavage furrow forms

215. (b) organizing the microtubules that form the mitotic spindle

216. (c) Sister chromatids would have to physically segregate by some other means.

217. (c) in chromatin fibers

218. (c) with cells only

219. (b) The nuclear membrane breaks apart

220. (a) migration

221. (c) 1-100 nm

222. (c) platelets

223. (a) nm

224. (a) MTT assay

225. (c) Fluorescent labelled secondary antibody
226. (c) it is highly thermostable
227. (b) Cytokines
228. (d) all of the above
229. (b) Fluorescence microscopy
230. (b) TEM
231. (d) all of the above
232. (d) all of the above
233. (b) Apoptosis
234. (c) Less than 24 hours
235. (c) 7.4
236. (c) 5% CO_2, 95% humidity
237. (a) sterile environment
238. (c) TEM
239. (a) Sterilization of the sample
240. (d) contains proteins, nutrients and antibiotics
241. (a) Cells spread on the surface
242. (b) sterile environment (37°C)
243. (b) 37°C
244. (b) Scattering at 620 nm
245. (a) longer
246. (a) all of the above
247. (d) all of the above
248. (e) both (a) and (b)
249. (b) PPARγ2
250. (a) MIC > MBC
251. (c) Both (a) and (b)
252. (b) 24 hours
253. (d) all of the above
254. (a) 0.1
255. (a) Protein adsorption
256. (b) Microbial infection
257. (b) Limited, but to a visible extent, cells will adhere
258. (d) Bacteria growth
259. (d) Material-protein interactions
260. (d) all of the above
261. (a) MTT and LDH
262. (c) Cellular ultrastructure
263. (d) Extracellular signal molecule
264. (a) G1 phase
265. (a) $5' \rightarrow 3'$ direction
266. (a) LDH assay
267. (a) Biomaterial associated infection
268. (a) *in vivo*
269. (b) cytocompatibility
270. (b) Osteon
271. (c) both (a) and (b)
272. (b) Osseointegration
273. (b) 12 hours
274. (a) 2 m
275. (d) all of the above
276. (c) foreign body giant cells and components of granulation tissue
277. (d) Tissue calcification
278. (a) cartilage
279. (a) Quantifies dead cells
280. (a) irreversible in nature
281. (c) both cell proliferation and lysis
282. (b) 7-8
283. (d) 30 μm/h
284. (a) cell viability
285. (a) Corrosion can be envisaged as a problem to be experienced in clinical applications

II. Fill in the blanks

1. Compression
2. Cytoplasm
3. PGA/Polyglycolic acid
4. Amino acids
5. Collagen
6. Microtubules
7. Cytoplasm
8. crystallinity/crystallization.
9. 1.67
10. Titanium
11. Differentiation
12. number average molecular weight
13. Tension
14. Transcription
15. Ionic and Covalent
16. bacteria cell protein and host cell receptors
17. as order that arises from the rotation of molecules about the single bond
18. plasma/ion sputtering
19. Hardness and Elastic modulus
20. $\sigma_{ij} = S_{ijkl}\,\epsilon_{kl}$
21. brittle facture
22. shear modulus
23. shear
24. increase
25. toughness
26. protein folding
27. connective, mouse
28. Adheren junction
29. Connective tissue
30. linear sequencing
31. $2\text{-}3 \times 10^9$
32. pathological destruction/disappearance of bone tissue
33. lymphoid , foreign body response
34. wear
35. muscular, connective, nervous and epithelial
36. To transport essential nutrients and oxygen within the body, and also to provide structural framework and support

37. compartmentalization and membrane bound organelles
38. amino acids , polypeptide chain , hydrogen bonding
39. fibrous capsule
40. hematoxylin and eosin and/or Alcian blue
41. colorimetric, number of viable, mitochondria , formazan
42. UV-visible spectrophotometer
43. Blue , nucleus
44. flow cytometer, micro-CT
45. 12 weeks , rabbit
46. adding conducting second phase
47. pluripotent, multiple cell types
48. nutrients, foreign invaders
49. Flagella, Pili
50. 30s and 50s, 40s and 60s
51. Topoisomerase
52. Colony counting
53. Cell Size, Cell granularity
54. Bone
55. Cartilage
56. Muscle
57. less
58. clinical trials
59. dog
60. volunteers
61. (a) initial safety and therapeutic dose range in human subject
    (b) Efficacy in human subject
    (c) extensive study in larger population
    (d) post market evaluation
62. (a) Increase in length and decrease in diameter
    (b) decrease in length and increase in diameter
    (c) Angular distortion
63. molecular weight or viscosity
64. composition and sintering temperature
65. Homo
66. Polymethyl methacrylate (full form) , methyl methacrylate
67. M_w/M_n where M_w is weight average of molecular weight and M_n is number average molecular weight.
68. PTFE/Teflon
69. 7.35 to 7.45, 37oC

70. Ionic/hydrogen/Van der Waals
71. Cells , fluids matrix/plasma
72. *in vitro*
73. *in vitro* and *in vivo*
74. Fluorescence
75. 12-24, 21-28
76. Cells, animals
77. Biomaterial substratum
78. Physical or mechanical
79. Double-layered
80. Central nucleoid
81. 0.1-5.0 μm
82. Growth factors
83. Lamellipodia (actin filaments)
84. Osteoblast
85. Apoptosis
86. G_2 , G_1
87. Necrosis
88. Viable
89. Fluorophores
90. "in test tubes"/"in glass"
91. Angiogenesis
92. Bacteriostatic agent
93. Colony forming unit
94. Cells
95. UV
96. Naturally occurring substance capable of stimulating cellular growth, proliferation and differentiation
97. Large number of proteins, cell, underlying substrate
98. Transdifferentiation
99. $2\text{-}10\text{X}10^4$ fold lower , 100 nm
100. Lower
101. Undifferentiated biological cells that can differentiate into specialized cells and can divide through mitosis to produce more stem cells
102. Quantitative reverse transcription polymerase chain reaction
103. Greater
104. Body's metabolic regulation and it is pre-osteoblastic or bone-building by nature
105. Triple helix, approximately 300 nm
106. 17-19 hour

107. The emission of light that occurs within nanoseconds after the absorption of light that is typically of a shorter wavelength

108. Cells that can give rise to all the cell types that makes up the body e.g., Embryonic cell

109. Cells that have potential to develop into any cell found in the human body. e,g., Placental cells

110. Cells with the capacity to create more stem cells

111. Base pairs of DNA, red fluorescence in dead cells

112. BCL-X$_L$

113. 0.2 mm

114. Blood clot/thrombogenesis

115. fluorescent

III. True/False

1. True	2. True	3. True	4. True
5. False	6. False	7. True	8. True
9. True	10. True	11. False	12. True
13. True	14. False	15. True	16. True
17. True	18. True	19. True	20. True
21. True	22. True	23. True	24. False
25. True	26. True	27. False	28. True
29. False	30. True	31. True	32. True
33. True	34. True	35. False	36. True
37. True	38. True	39. True	40. True
41. True	42. True	43. False	44. False
45. True	46. True	47. False	48. False
49. False	50. False	51. True	52. False
53. True	54. True	55. True	56. False
57. True	58. True	59. True	60. False
61. False	62. False	63. True	64. True
65. False	66. False	67. (d)	68. True
69. True	70. True		

IV. Match the following

A. 1(c), 2(b), 3(e), 4(d), 5(a)

B. 1(b), 2(d), 3(c), 4(a)

C. 1(e), 2(b), 3(d), 4(a), 5(c)

V. Diagram identification

1. (a) Prokaryotic cell

    (i) DNA

      (ii) Cell wall

  (b)   (i) Mitochondrion

      (ii) Nucleus

      (iii) Intermediate filament/actin/microtubules of cytoskeleton

  (c) Collagen fibers

2. (i) free amine group ($-NH_2$)

  (ii) free carboxyl group ($-COOH$)

3. (i) Trabeculae of spongy bone

  (ii) Volkmann's canal

  (iii) Osteon

4. (a) Paracrine signaling

  (b) Endocrine signaling

5. Q1. Livecell

  Q2. Early apoptotic cell

  Q3. Late apoptotic cell

  Q4. Necrotic cell

6. (i) single turn

    5.4 Å pitch

    3.6 residues

  (ii) 18 residues (length 27 Å)

  (iii) 26°

  (iv) 5.1 Å

7. Spider silk (dragline)

8. (a) Cell division

  (b) Cancellous/spongy/trabecular bone

  (c) Quaternary structure of protein

  (d) Microstructure of cortical bone

  (e) Cleavage furrow

  (f) polymerase chain reaction involved in PCR cycle

    (i) Step 1: Denaturation at 94° C (1 min)

    (ii) Step 2: Annealing at 54° C (45 s)

    (iii) Step 3: Extension at 72° C (2 min)

  (g) Total hip replacement (THR) biomedical device (comprising of acetabular socket, femoral ball head and femoral stem)

VII. Analytical problems

1. Answer: Engineering stress = 565.88 MPa, Engineering strain = 5×10^{-4}, Yield strength = 40 MPa
2. Answer: Alumina plate is to be used σ_c = 337 MPa
3. Answer: Surface Energy = 1.8 kJ/m^2
4. Answer: Maximum allowable surface crack: Al_2O_3 = 0.28 mm, ZrO_2 = 3.18 mm. [Hint: multiply surface crack length by 2 to get maximum allowable volume crack].
5. Answer: Load at fracture = 23.5 kN, Radius of curvature at fracture = 88.62 mm
6. Answer: Fracture strength σ_f = 713.65 MPa, R = 1.4 m

References

Chapter 1

1. Basu, B., D. Katti, and A. Kumar. 2009. *Advanced Biomaterials: Fundamentals, Processing and Applications.* (ISBN: 978-0-470-19340-2). USA: John Wiley & Sons Inc.
2. Basu, Bikramjit. 2017. *Biomaterials for Musculoskeletal Regeneration: Concepts.* Singapore: Springer Nature.
3. Basu, Bikramjit, and Sourabh Ghosh. 2017. *Biomaterials for Musculoskeletal Regeneration: Applications; Concepts.* Singapore: Springer Nature.
4. Ratner, B. D. 2004. 'A history of biomaterials.' In *Biomaterial Science: An Introduction to Material in Medicine*, edited by B. D. Ratner, A. S. Hoffman, F. J. Schoen, and J. E. Lemons, 2nd edition, 10-19. London: Elsevier Academic Press.
5. Bishagratna, K. L. 1916. *An English Translation of The Sushruta Samhita: Kaviraj Kunja Lal Bishagratna.* Calcutta.
6. Bruggeman, J. 2010. *Biodegradable Polyol-Based Polymers: A polymer platform for biomedical applications.* Erasmus MC: University Medical Center Rotterdam.
7. Scott, M. 1983. '32,000 years of sutures.' *NATNEWS.* 20. 15–17.
8. Dandona, L., R. Dandona, R. Anand, M. Srinivas, and V. Rajashekar. 2013. 'Outcome and number of cataract surgeries in India: policy issues for blindness control.' *Clinical and Experimental Ophthalmology.* 31. 23–31.
9. Lindstrom, R. 2015. 'Thoughts on cataract surgery: 2015.' *Review of Ophthalmology.* Accessed April 13, 2017. *http://www.reviewofophthalmology.com/content/t/surgical_education/c/53422.*
10. Crawford, R. W., D. W. Murray. 1997. 'Total hip replacement: indications for surgery and risk factors for failure.' *Annals of the Rheumatic Diseases.* 56. 455–457.
11. Ratner, Buddy D. Allan S. Hoffman, Frederick J. Schoen, and Jack E. Lemons. 2004. *Biomaterials Science: An Introduction of Materials in Medicine.* Academic Press.
12. Apple, David J., and John Sims. 1996. 'Harold ridley and the invention of the intraocular lens.' *Survey of Ophthalmology.* 40(4). 279–292.
13. Hench, Larry L. 2006. 'The story of bioglass.' *Journal of Materials Science: Materials in Medicine.* 17(11) 967–978.
14. MarketsandMarkets. 2016. 'Biomaterials market by type of material (metallic, ceramic, polymers, natural) & application (cardiovascular, orthopedic, dental, plastic surgery, wound healing, neurology, tissue engineering, ophthalmology) - global forecast to 2021. Accessed April 13, 2017. http://www.marketsandmarkets.com/.

15. Black, Jonathan. 1999. *Biological Performance of Materials: Fundamentals of Biocompatibility*, 3rd edition. CRC Press.

16. Williams, D. F. 2009. 'On the nature of biomaterials.' *Biomaterials*. 30: 5897–5909.

17. Biomaterials. Accessed 13 April 2017. http://www.nature.com/subjects/biomaterials

18. Hench, L. L., and J. M. Polak. 2002. 'Third-generation biomedical materials.' *Science*. 295: 1014-1017.

19. Wise, D. L., D. J. Transtolo, D. E. Altobelli, M. J. Yaszemski, J. D. Gresser, and E. R. Schwartz. 1995. *Encyclopedic Handbook of Biomaterials and Bioengineering*. New York: Marcel Dekker.

20. Williams, D. F. 2008. 'On the mechanisms of biocompatibility.' *Biomaterials*. 29. 2941–2953.

21. Dee, K. C., D. A. Puleo, and R. Bizios. 2003. *An Introduction to Tissue–Biomaterial Interactions*. John Wiley and Sons Inc.

22. Kulkarni, M. M., C. S. Sharma, A. Sharma, S. Kalmodia, and B. Basu. 2012. 'Multiscale micro-patterned polymeric and carbon substrates derived from buckled photoresist films: fabrication and cytocompatibility. *J.' Mat. Sc.* 47. 3867.

23. Saha, N., A. K. Dubey, and B. Basu. 2012. 'Cellular proliferation, cellular viability, and biocompatibility of HA–ZnO composites.' *J. Biomed. Mater. Res. Part B*. 100B. 256-264.

24. Tripathi, G., and B. Basu. 2012. 'A porous hydroxyapatite scaffold for bone tissue engineering: Physico-mechanical and biological evaluations.' *Ceramics Int*. 38. 341-349.

25. Dubey, A. K., B. Basu, K. Balani, R. Guo, and A. S. Bhalla. 2011. 'Dielectric and pyroelectric properties of HAp–BaTiO$_3$ composites.' *Ferroelectrics*. 423. 63-76.

26. Dubey, A. K., B. Basu, K. Balani, R. Guo, and A. S. Bhalla. 2011. 'Multifunctionality of perovskites BaTiO$_3$ and CaTiO$_3$ in a composite with hydroxyapatite as orthopedic implant materials.' *Integr. Ferroelectr*. 131. 119-126.

27. Dubey, A., A. Tiwari, R. Kumar, S. Gupta, and B. Basu. 2009. 'Time constant determination for electrical equivalent of biological cells.' *J. Appl. Phys*. 105. 84705.

28. Dubey, A., S. Dutta-Gupta, and B. Basu. 2011. 'Optimization of electrical stimulation conditions for enhanced fibroblast cell proliferation on biomaterial surfaces.' *J. Biomed. Mater. Res.: Part B- Appl. Biomater*. 98 B. 18-29.

29. Dubey, A. K., M. Banerjee, and B. Basu. 2011. 'Biological cell–electrical field interaction: stochastic approach.' *J. Biol. Phys*. 37. 39-50.

30. Morais, J. M., F. Papadimitrakopoulos, and D. J. Burgess. 2010. Biomaterials/tissue interactions: possible solutions to overcome foreign body response. *AAPS J*. 12 (2). 188-96.

31. Albrektsson, T., and C. Johansson. 2001. 'Osteoinduction, osteoconduction and osseointegration.' *Eur. Spine J*. 10. S96-S101.

32. Stout, D. A., B. Basu, and T. J. Webster. 2011. 'Poly(lactic–co-glycolic acid): Carbon nanofiber composites for myocardial tissue engineering applications.' *Acta Biomaterialia*. 7(8). 3101–3112.

33. Williams, David. 2014. *Essential Biomaterials Science*. Cambridge Texts in Biomedical Engineering Series. Cambridge University Press.

34. (a) Greeshma, T. 2015. 'Electric stimuli as instructive cues to guide cellular differentiation on electrically conductive biomaterial substrates *in vitro*.' PhD Dissertation. Indian Institute of Science, Bangalore. (b) Dubey, A. K., 2011. 'Electric field stimulated enhanced cell response on electrically active Hydroxyapatite–BaTiO$_3$ composite.' PhD Dissertation. Indian Institute of Technology, Kanpur. (c) Chandorkar, Y. 2015. 'Multi-functional, Anti-inflammatory and Tunable Polyesters- A Novel Polymer Platform for Tissue Engineering and Drug Delivery Applications.' PhD Dissertation. Indian Institute of Science, Bangalore, India. (d) Ravi, Kumar K. 2015. 'Toughness enhancement of multifunctional HA–CaTiO$_3$ composites and Experimental/Analytical studies on electrical field stimulated cell functionality *in vitro*.' PhD Dissertation. Indian Institute of Science, Bangalore. (e) Sunil, Kumar B. 2015. 'Developmental strategies to address prosthetic infection and magneto-responsive biomaterials for orthopedic applications.' PhD Dissertation. Indian Institute of Science, Bangalore. (f) Kumar,

Alok. 2013. 'Microstructure development, biomineralization and osteoblast cell fate processes on high toughness Hydroxyapatite–Titanium composites.' PhD Dissertation. Indian Institute of Technology, Kanpur.

35. (a) Accessed April 13, 2017. http://www.sushruta.in/the-name.html (b) Champaneria, M. C., A. D. Workman, and S. C. Gupta. 2014. 'Sushruta: Father of Plastic Surgery.' *Annals of Plastic Surgery*. 73. 2–7. (c) Accessed April 13, 2017. http://romanmedicine.co.uk/key_figures (d) Accessed April 13, 2017. http://ideas.ted.com/a-history-of-biomaterials

36. Yang, P., and J-M. Tarascon. 2012. 'Towards systems materials engineering.' *Nat. Mater.* 11. 560–563.

For Further Reading

a. Williams, David F., and R. Roaf. 1973. *Implants in Surgery*. W.B. Saunders.

b. Williams, David F. 1981. *Fundamental Aspects of Biocompatibility*. CRC Press.

c. Williams, David F. 1983. *Biocompatibility of Clinical Implant Materials*. CRC Press.

d. Williams, David F., Robert W. Cahn, Peter Haasen, and E. J. Kramer.1992. *Materials Science and Technology: A Comprehensive Treatment Medical and Dental Materials*. Wiley VCH.

e. Williams David F. 1999. *The Williams Dictionary of Biomaterials*. Liverpool, Great Britain: Liverpool University Press.

f. Silver Frederick H., and David L. Christiansen. 1999. *Biomaterials Science and Biocompatibility*. Springer: USA.

g. Grimnes, Sverre, and Orjan G. Martinsen. 2000. *Bioimpedance and Bioelectricity Basics*. Academic Press.

h. Alberts, Bruce, Alexander Johnson, Julian Lewis, Martin Raff, Keith Roberts, and Peter Walter. 2002. *Molecular Biology of the Cell*. Garland Science: New York.

i. Pelczar, Michael J., JR., E.C.S. Chan, and Noel R. Krieg. 2004. *Microbiology*. Tata McGraw-Hill Publishing Company Limited: New Delhi.

j. Hench, Larry L., and Julian R. Jones. 2005. *Biomaterials, Artificial Organs and Tissue Engineering*. Woodhead Publishing Ltd.

k. Freshney, R. Ian. 2005. *Culture of Animals Cells*. John Wiley & Sons, Inc.

l. Barnes Frank S., and Ben Greenebaum. 2007. *Biological and Medical Aspects of Electromagnetic Fields*. CRC press.

m. Waigh, Tom A. 2007. *Applied Biophysics-A molecular Approach for Physical Scientists*. John Wiley & Sons.

n. Ramalingam, Murugan. 2012. *Integrated Biomaterials for Biomedical Technology*. Wiley Publication.

o. Webster, Thomas J. 2012. *Nanomedicine Technologies and Applications*. Woodhead Publications.

p. Agrawal, C. Mauli, Joo L. Ong, Mark R Appleford, and Gopinath Mani. 2014. *Introduction to Biomaterials: Basic Theory with Engineering Applications*. Cambridge University Press.

q. Vallet-Regi, Maria. 2014. *Bio-Ceramics with Clinical Applications*. Wiley Publications.

r. Balani, K., Vivek Verma, Arvind Agarwal, and Roger Narayan. 2015. *Biosurfaces: A Materials Science and Engineering Perspective*. Wiley.

Chapter 2

37. Hench, L. L. 1980. 'Biomaterials.' *Science*. 208. 826-831.

38. Leuer, L. H., J. M. Gross, and K. M. Johnson. 1996. 'Material properties, biocompatibility, and wear resistance of the medtronic pyrolytic carbon.' *The Journal of Heart Valve Disease*. 5. S105–109.

39. Habal, Mutaz B. 1984. 'The biologic basis for the clinical application of the silicones: a correlate to their biocompatibility.' *Archives of Surgery*. 119(7). 843–848.

40. Stepan, Lenka L., Daniel S. Levi, and Gregory P. Carman. 2005. 'A thin film nitinol heart valve.' *Journal of Biomechanical Engineering.* 127(6). 915–918.

41. Jovanovic, Sascha A., Hubertus Spiekermann, and E. Jürgen Richter. 1992. 'Bone regeneration around titanium dental implants in dehisced defect sites: a clinical study.' *The International Journal of Oral and Maxillofacial Implants.* 7(2). 233–245.

42. Müller, M. E. 1970. '7 Total Hip Prostheses.' *Clinical Orthopedics and Related Research.* 72. 46–68.

43. Hench, L. Larry, and June Wilson. 1984. 'Surface-active biomaterials.' *Science.* 226(4675). 630–636.

44. Hench, Larry L., and Julia M. Polak. 2002. 'Third-generation biomedical materials.' *Science.* 295(5557). 1014–1017.

45. Groot, K. De, C. P. A. T. Klein, J. G. C. Wolke, J. M. A. de Blieck-Hogervorst, T. Yamamoto, L. L. Hench, and J. Wilson. 1990. *CRC Handbook of Bioactive Ceramics. Volume II: Calcium Phosphate and Hydroxylapatite Ceramics.* CRC Press.

46. Zhang, Ruiyun and Peter X. Ma. 1999. Poly(A-hydroxyl acids)/hydroxyapatite porous composites for bone-tissue engineering. i. preparation and morphology.

47. Chang, B.S., K.S. Hong, H.J. Youn, H.S. Ryu, S.S. Chung, and K.W. Park. 2000. 'Osteoconduction at porous hydroxyapatite with various pore configurations.' *Biomaterials.* 21(12). 1291–1298.

48. Dalton, Jeanette E., Stephen D. Cook, Kevin A. Thomas, and John F. Kay. 1995. 'The effect of operative and hydroxyapatite coating on the mechanical and biological response to implants.' *The Journal of Bone and Joint Surgery.* 77(1). 97–110.

49. Ducheyne, P., L. L. Hench, A Kagan II, M. Martens, A. Bursens, and J. C. Mulier. 1980. 'Effect of hydroxyapatite impregnation on skeletal bonding of porous coated implants.' *Journal of Biomedical Materials Research.* 14(3). 225–237.

50. Suchanek, Wojciech and Masahiro Yoshimura. 1998. 'Processing and properties of hydroxyapatite-based biomaterials for use as hard tissue replacement implants.' *Journal of Materials Research.* 13(01). 94–117.

51. Cochran, David L. 1999. 'A comparison of endosseous dental implant surfaces.' *Journal of Periodontology.* 70(12). 1523–1539.

52. Stamboulis, A., L. L. Hench, and A. R. Boccaccini. 2002. 'Mechanical Properties of Biodegradable Polymer Sutures Coated with Bioactive Glass.' *Journal of Materials Science: Materials in Medicine.* 13(9). 843–848.

53. Hoffman, Allan S. 2008. 'The origins and evolution of "controlled" drug delivery systems.' *Journal of Controlled Release.* 132(3). 153–163.

54. Jeong, Byeongmoon, You Han Bae, Doo Sung Lee, and Sung Wan Kim. 1997. 'Biodegradable block copolymers as injectable drug-delivery systems.' *Nature.* 388(6645). 860–862.

55. Becker, William, Burton E. Becker, Michael G. Newman, and Sture Nyman. 1989. 'Clinical and microbiologic findings that may contribute to dental implant failure.' *The International Journal of Oral and Maxillofacial Implants.* 5(1). 31–38.

56. Schoen, F. J. and R. J. Levy. 1984. 'Bioprosthetic Heart Valve Failure: Pathology and Pathogenesis.' *Cardiology Clinics.* 2(4). 717–739.

57. Engh, C. A., J. D. Bobyn, and A. H. Glassman. 1987. 'Porous-coated hip replacement: the factors governing bone ingrowth, stress shielding, and clinical results.' *Journal of Bone and Joint Surgery, British Volume.* 69(1). 45–55.

58. Kruger, J. 1979. 'Fundamental aspects of the corrosion of metallic implants.' *Corrosion and Degradation of Implant Materials.* Baltimore: American Society for Testing Materials. 107–127.

59. Basu, B., D. Katti, and A. Kumar. 2009. *Advanced Biomaterials: Fundamentals, Processing and Applications.* USA: John Wiley and Sons Inc.

60. Jones, Alan L., Robert W. Bucholz, Michael J. Bosse, Sohail K. Mirza, T. R. Lyon, L. X. Webb, A. N. Pollak, J. D Golden, and A. Valentinopran. 2006. 'Recombinant human bmp-2 and allograft compared

with autogenous bone graft for reconstruction of diaphysealtibial fractures with cortical defects.' *The Journal of Bone and Joint Surgery*. 88(7). 1431–1441.

61. Wilson, June and Samuel B. Low. 1992. 'Bioactive ceramics for periodontal treatment: comparative studies in the patus monkey.' *Journal of Applied Biomaterials*. 3(2). 123–129.

62. Fansa, Hisbam, Gerburg Keilhoff, Gerald Wolf, and Wolfgang Schneider. 2001. 'Tissue engineering of peripheral nerves: A comparison of venous and acellular muscle grafts with cultured schwann cells.' *Plastic and Reconstructive Surgery*. 107(2). 485–494.

63. Akhyari, Payam, Akhyari, P., P. W. Fedak, R. D. Weisel, T. Y. J. Lee, S. Verma, D. A. Mickle, and R. K. Li., 2002. 'Mechanical stretch regimen enhances the formation of bioengineered autologous cardiac muscle grafts.' *Circulation*. 106(12 suppl 1). I–137.

64. Currie, Lachlan J., Justin R. Sharpe, and Robin Martin. 2001. 'The use of fibrin glue in skin grafts and tissue-engineered skin replacements.' *Plast. Reconstr. Surg.* 108. 1713–1726.

65. Pham, Quynh P., Upma Sharma, and Antonios G. Mikos. 2006. 'Electrospinning of polymeric nanofibers for tissue engineering applications: a review.' *Tissue Engineering*. 12(5). 1197–1211.

66. Yang, Shoufeng, Kah-Fai Leong, Zhaohui Du, and Chee-Kai Chua. 2001. 'The Design of Scaffolds for Use in Tissue Engineering. Part I. Traditional Factors.' *Tissue Engineering*. 7(6). 679–689.

67. Mano, J. F., R.A. Sousa, L. F. Boesel, N. M. Neves, and R. L. Reis. 2004. 'Bioinert, biodegradable and injectable polymeric matrix composites for hard tissue replacement: state of the art and recent developments.' *Composites Science and Technology*. 64. 789–817.

68. Hakkarainen, M., A. C. Albertsson, and S. Karlsson. 1996. 'Weights losses and molecular weight changes correlated with the evolution of hydroxyacids in simulated *in vivo* degradation of homo- and copolymers of PLA and PGA.' *Polymer Degradation and Stability*. 52. 283–291.

69. (a)Andrew, S. D., G. C. Phil, and K. G. Marra. 2001. 'The Influence of Polymer Blend Composition on the Degradation of Polymer/Hydroxyapatite Biomaterials.' *J. Mater. Sci: Mater. Med.* 12. 673–677 (b) Heidemann, W., S. Jeschkeit, K. Ruffieux, J. H. Fischer, M. Wagner, G. Krüger, E. Wintermantel, and K. L. Gerlach. 2001. 'Degradation of Poly(D,L)lactide Implants with or without Addition of Calciumphosphates *in vivo*.' *Biomaterials*. 22. 2371–81

70. Sun, H., L. Mei, C. Song, X. Cui, and P. Wang. 2006. 'The *in vivo* degradation, absorption and excretion of PCL-based implant.' *Biomaterials*. 27. 1735–1740

71. Woodruff, M. A. and D. W. Hutmacher. 2010. 'The return of a forgotten polymer—Polycaprolactone in the 21st century.' *Progress in Polymer Science*. 35. 1217–1256.

72. Griffin, J., R. Delgado-Rivera, S. Meiners, and K. E. Uhrich. 2011. 'Salicylic acid-derived poly(anhydride-ester) electrospun fibers designed for regenerating the peripheral nervous system.' *Journal of Biomedical Materials Research Part A*. 97. 230–242.

73. Hubbell, J. A. 1995. Biomaterials in Tissue Engineering. *Nature Biotechnology*. 13. 565–576.

74. Supèr, H., D. W. Grijpma, and A. J. Pennings. 1994. 'Incorporation of Salicylates into Poly(l-lactide).' *Polymer Bulletin*. 32. 509–515.

75. Bergsma, J. E., W. C. De Bruijn, F. R. Rozema, R. R. Bos, and G. Boering. 1995. 'Late Degradation Tissue Response to Poly(L-lactide) Bone Plates and Screws.' *Biomaterials*. 16. 25–31.

76. Yang, Z., S. M. Best, and R. E. Cameron. 2009. 'The Influence of α-Tricalcium Phosphate Nanoparticles and Microparticles on the Degradation of Poly(D, L-lactide-co-glycolide).' *Advanced Materials*. 21. 3900–3904.

77. Ji, W., F. Yang, H. Seyednejad, Z. Chen, W. E. Hennink, J. M. Anderson, J. J. van den Beucken, and J. A. Jansen. 2012. 'Biocompatibility and Degradation Characteristics of PLGA-based Electrospun Nanofibrous Scaffolds with Nanoapatite Incorporation.' *Biomaterials*. 33. 6604–6614.

78. Gunatillake, P. A. and R. Adhikari. 2003. 'Biodegradable Synthetic Polymers for Tissue Engineering.' *European Cells and Materials*. 5. 1–16.

79. Yang, S., K. F. Leong, Z. Du, and C. K. Chua. 2001. 'The design of scaffolds for use in tissue engineering. Part I. Traditional factors.' *Tissue Eng.* 7. 679.

80. Niiranen, H., T. Pyhalto, P. Rokkanen, M. Kellomaki, and P. Tormala. 2004. '*In vitro* and *in vivo* behaviour of self-reinforced bioabsorbable polymer and self-reinforced bioabsorbable polymer/bioactive glass composites.' *J. Biomed. Mater. Res. A.* 69A. 699.

81. Gollwitzer, H., P. Thomas, P. Diehl, E. Steinhauser, B. Summer, S. Barnstorf, L. Gerdesmeyer, W. Mittelmeier, and A. Stemberger. 2005. 'Biomechanical and allergological characteristics of a biodegradable poly(D,L-lactic acid) coating for orthopedic implants.' *J. Orthop. Res.* 23. 802.

82. Wu, L. and J. Ding. 2004. '*In Vitro* Degradation of three-dimensional porous poly(DL-lactide-co-glycolide) scaffolds for tissue engineering.' *Biomaterials.* 25. 5821-5830

83. (a) Schmidmaier, G., B. Wildemann, H. Bail, M. Lucke, T. Fuchs, A. Stemberger, A. Flyvbjerg, N. P. Haas, and M. Raschke. 2001. 'Local application of growth factors (insulin like growth factor-1 and transforming growth factor-β1) from a biodegradable Ploy(D,L-lactide) coating of osteosynthetic implants accelerates fracture healing in rats.' *Bone.* 28. 341. (b) Schmidmaier, G., B. Wildemann, A. Stemberger, N. P. Haas, and M. Raschke. 2001. 'Biodegradable Poly(D,L-lactide) coating of implants for continuous release of growth factors.' *J. Biomed. Mater. Res.* 58. 449.

84. (a) Gollwitzer, H., K. Ibrahim, H. Meyer, W. Mittelmeier, R. Busch, and A. Stemberger. 2003. 'Antibacterial Poly(D,L-lactic acid) Coating of Medical Implants using a Biodegradable Drug Delivery Technology.' *J. Antimicrob. Chemother.* 51. 585. (b) Stashevskaya, K., E. Markvicheva, S. Strukova, I. Prudchenko, V. Zubov, and C. Grandfils. 2007. 'Thrombin Receptor Agonist Peptide Entrapped in Poly-(D,L)-lactide-co-glycolidemicroparticles: Preparation and Characterization.' *J. Microencapsul.* 24. 129.

85. Rai, B., S. H. Teoh, D. W. Hutmacher, T. Kao, and K. H. Ho. 2005. 'Novel PCL-based honeycomb scaffolds as drug delivery systems for rhBMP-2.' *Biomaterials.* 26. 3739.

86. Wang, S., L. Lu, and M. J. Yaszenski. 2006. 'Bone tissue-engineering material Poly(propylene fumarate): Correlation between molecular weight, chain dimensions, and physical properties.' *Biomacromolecules.* 7. 1976–1982.

87. (a) Temenoff, J. S. and A. G. Mikos. 2000. 'Injectable biodegradable materials for orthopedic tissue engineering.' *Biomaterials.* 21. 2405-2412 (b) Peter, S. J., L. Lu, D. J. Kim, G. N. Stamatas, M. J. Miller, M. J. Yaszemski, and A. G. Mikkos. 2000. 'Effects of transforming growth factor β-1 released from biodegradable polymer microparticles on marrow stromal osteoblasts cultured on poly(propylene fumarate) substrates.' *J. Biomed. Mater. Res.* 50. 452–462.

88. Doi, Y., S. Kitamura, and H. Abe. 1995. 'Microbial synthesis and characterization of poly(3-hydroxybutyrate-co-3-hydroxyhexanoate).' *Macromolecules.* 28. 4822–4838.

89. Li, H. Y., R. L. Du, and J. Chang. 2005. 'Fabrication, characterization, and *in vitro* degradation of composite scaffolds based on PHBV and bioactive glass.' *J. Biomater. Appl.* 20. 137–155.

90. Chen, G. Q. and Q. Wu. 2005. 'The application of polyhydroxyalkanoates as tissue engineering materials.' *Biomaterials.* 26. 6565–6578.

91. (a) Doyle, C., E. T. Tanner, and W. Bonfield. 1991. '*In vitro* and *in vivo* evaluation of polyhydroxybutyrate and of polyhydroxybutyrate reinforced with hydroxyapatite.' *Biomaterials.* 12. 841–847 (b) Hayati, A. N., S. M. Hosseinalipour, H. R. Rezaie, and M. A. Shokrgozar. 2012. 'Characterization of poly(2-hydroxybutyrate)/nano-hydroxyapatite composite scaffolds fabricated without the use of organic solvents for bone tissue engineering applications.' *Mater. Sci. and Eng.: C.* 32. 416–422.

92. Nair, L. S., and C. T. Laurencin. 2006. 'Polymers as biomaterials for tissue engineering and controlled drug delivery.' *Adv. Biochem. Eng. Biotechnol.* 102. 47–90.

93. Hench, L. L., and J. Wilson, eds. 1999. *An Introduction to Bioceramics*, 2nd edition. London: Word Scientific.

94. Hench, L. L., R. J. Splinter, and W. C. Allen. 1971. 'Bonding mechanisms at the interface of ceramic prosthetic materials.' *J. Biomed. Mater. Res. Symp.* 2. 117–141.

95. Hench, L. L. 1991. 'Bioceramics: from Concept to Clinic.' *J. Am. Ceram. Soc.* 74. 1487–1510.

96. Mitra, J., G. Tripathi, A. Sharma, and B. Basu. 2013. 'Scaffolds for bone tissue engineering: role of surface patterning on osteoblast response.' *RSC Advances.* 3. 11073–11094.

97. Braun, S., S. Rappopart, R. Zusman, D. Avnir, and M. Ottolenghi. 1990. 'Biochemically active sol-gel glasses: the trapping of enzymes.' *Mater. Lett.* 10. 1–5.

98. Keshaw, H., A. Forbes, and R. M. Day. 2005. 'Release of angiogenic growth factors from cells encapsulated in alginate beads with bioactive glass.' *Biomaterials.* 26. 4171–4179.

99. Rahaman, M. N., D. E. Day, B. S. Bal, Q. Fu, S. B. Jung, L. F. Bonewald, and A. P. Tomsia. 2011. ' 'Bioactive glass in tissue engineering.' *Acta Biomater.* 7. 2355–2373.

100. (a) Jell, G. and M. M. Stevens. 2006. 'Geneactivation by Bioactive Glasses.' *J. Mater. Sci: Mater. Med.* 17. 997-1002 (b) Valerio, P., M. M. Pereira, A. M. Goes, and M. F. Leite. 2004. 'The effect of ionic products from bioactive glass dissolution on osteoblast proliferation and collagen production.' *Biomaterilas.* 25. 2941–2948.

101. (a) Boccaccini, A. R., I. Notingher, V. Maquet, and R. Je´ro^me. 2003. 'Bioresorbable and bioactive composite materials based on polylactide foams filled with and coated by Bioglasss particles for tissue engineering applications.' *J. Mater. Sci: Mater. Med.* 14. 443. (b) Boccaccini, A. R. and V. Maquet. 2003. 'Bioresorbable and bioactive polymer/Bioglass® composites with tailored pore structure for tissue engineering applications.' *Compos. Sci. Technol.* 63. 2417.

102. Xynos, I. D., M. V. J. Hukkanen, L. D. K. Buttery, L. L. Hench, and J. M. Polak. 2000. 'Bioglasss 45S5 stimulates osteoblast turnover and enhances bone formation *in vitro*: implications and applications for bone tissue engineering.' *Calcif. Tissue Int.* 67. 321.

103. Peitl, O., G. P. LaTorre, and L. L. Hench. 1996. 'Effect of crystallization on apatite layer formation of bioactive glass 45S5' *J. Biomed. Mater. Res.* 30. 509.

104. Buchanan, L. A., and A. El-Ghannam. 2010. 'Effect of bioactive glass crystallization on the conformation and bioactivity of adsorbed proteins.' *J. Biomed. Mater. Res. A.* 93. 537.

105. Clupper, D. C., and L. L. Hench. 2003. 'Crystallization kinetics of tape cast bioactive glass 45S5.' *J. Non-Cryst. Solids.* 318. 43.

106. Thompson, I. D., and L. L. Hench. 1998. 'Mechanical properties of bioactive glasses, glass-ceramics and composites. *Proc. Inst. Mech. Eng.* H 212. 127.

107. Barrere, F., C. A. van Blitterswijk, and K. de Groot. 2006. 'Bone regeneration: molecular and cellular interactions with calcium phosphate ceramics.' *Int. J. Nanomedicine.* 1.

108. Barrere, F., C. M. van der Valk, R. A. Dalmaijer, G. Meijer, C. A. van Blitterswijk, K. de Groot, and P. Layrolle. 2003. 'Osteogenecity of osteocalcium phosphate coatings applied on porous metal implants.' *J. Biomater. Res. A.* 66. 779.

109. Chu, T. M. G., D. G. Orton, S. J. Hollister, S. E. Feinberg, and J. W. Halloran. 2002. 'Mechanical and *in vivo* performance of hydroxyapatite implants with controlled architectures.' *Biomaterials.* 23. 1283.

110. de Groot, K., C. P. A. T. Lein, J. G. C. Wolke, and J. M. A. de Bliek-Hogervost. 1990. 'Chemistry of Calcium Phosphate Bioceramics.' In *Handbook of Bioactive Ceramics*, Vol 2, edited by Yamamuro, T., L. L. Hench, and J. Wilson Boca Raton, FL: CRC Press. 3–16.

111. Kokubo, T., H. M. Kim, and M. Kawashita. 2003. 'Novel bioactive materials with different mechanical properties.' *Biomaterials.* 24. 2161.

112. (a) Miyazaki, T., H. M. Kim, T. Kokubo, H. Kato, and T. Nakamura. 2001. 'Induction and acceleration of bonelike apatite formation on tantalum oxide gel in simulated body fluid.' *J. Sol–Gel Sci. Technol.* 21. 83. (b) Miyazaki, T., H. M. Kim, T. Kokubo, C. Ohtsuki, and T. Nakamura. 2001. 'Apatite-forming

ability of niobium oxide gels in a simulated body fluid.' *J. Ceram. Soc. Japan.* 109. 929. (c) Kim, H. W., H. E. Kim, and J. C. Knowles. 2004. 'Hard-tissue-engineered zirconia porous scaffolds with hydroxyapatite sol-gel and slurry coatings.' *J. Biomed. Mater. Res. B. Appl. Biomater.* 70. 270. (d) Li, P. J., C. Ohtsuki, T. Kokubo, K. Nakanishi, N. Soga, and K. deGroot. 1994. 'The of hydrated silica, titania, and alumina in inducing apatite on implants.' J. Biomed. Mater. Res. 28. 7.

113. Staiger M. P., A. M. Pietak, J. Huadmai, and G. Dias. 2006. 'Magnesium and its alloys as orthopedic biomaterials: A review.' *Biomaterials.* 27(9). 1728–1734.

114. Kumar, Alok, Awadesh Kumar Mallik, Nurcan Calis Acikbas, Merve Yaygingol, Ali Celic, Ferhat Kara, Hasan Mandal, Debabrata Basu, Krishanu Biswas, and Bikramjit Basu. 2012. 'Cytocompatibility property evaluation of gas pressure sintered SiAlON-SiC composites with L929 fibroblast cells and Saos-2 osteoblast-like cells.' *Mat. Sc. Engg. C.* 32. 464–469.

115. (a) Hoffman, Allan S. 2012. 'Hydrogels for biomedical applications.' *Advanced Drug Delivery Reviews.* 64. 18–23. (b) Lutolf, M. P. 2009. 'Biomaterials: Spotlight on hydrogels.' *Nature Materials.* 8(6). 451–453.

116. Mata, Alvaro, Aaron J. Fleischman, and Shuvo Roy. 2005. 'Characterization of polydimethylsiloxane (PDMS) properties for biomedical micro/nanosystems.' *Biomedical Microdevices.* 7(4). 281–293.

117. Chawla, K. K. 2012. *Compites Materials: Science and Engineering,* 3rd edition. New York: Springer.

118. Nath, Shekhar, S. Kalmodia, and Bikramjit Basu. 2010. 'Densification, Phase stability and *in vitro* biocompatibility property of hydroxyapatite-10 wt% silver composites.' *Journal of Materials Science: Materials in Medicine.* 21. 1273–1287.

119. Saha, Naresh, Ashutosh K. Dubey and Bikramjit Basu. 2012. 'Cellular proliferation, cellular viability, and biocompatibility of HA–ZnO composites.' *J. Biomed. Mater. Res. Part B.* 100B. 256–264.

120. Dubey, Ashutosh Kumar, Kantesh Balani, and Bikramjit Basu. 2013. 'Multifunctional properties of multistage spark plasma sintered $HA–BaTiO_3$-based piezobiocomposites for bone replacement applications.' *Journal of the American Ceramic Society.* 96(12). 3753–3759.

121. Zheng, Xuebin, Minhui Huang, and Chuanxian Ding. 2000. 'Bond strength of plasma-sprayed hydroxyapatite/ti composite coatings.' *Biomaterials.* 21(8). 841–849.

122. Chenglin, Chu, Zhu Jingchuan, Yin Zhongda and Wang Shidong. 1999. 'Hydroxyapatite–Ti functionally graded biomaterial fabricated by powder metallurgy.' *Materials Science and Engineering: A.* 271(1). 95–100.

123. Salman, S., O. Gunduz, S. Yilmaz, M. L. Öveçoglu, Robert L. Snyder, S. Agathopoulos, and F. N. Oktar. 2009. 'Sintering effect on mechanical properties of composites of natural hydroxyapatites and titanium.' *Ceramics International.* 35(7). 2965–2971.

124. Göller, Gültekin, Faik N. Oktar, D. Toykan, and E. S. Kayali. 2002. 'The improvement of titanium reinforced hydroxyapatite for biomedical applications.' *Key Engineering Materials.* 240. 619–622.

125. Kumar, Alok, Krishanu Biswas, and Bikramjit Basu. 2013. 'On the toughness enhancement in hydroxyapatite-based composites.' *Acta Materialia.* 61(14). 5198–5215.

126. Gu, Y. W., K. A. Khor, and P. Cheang. 2003. '*in vitro* studies of plasma-sprayed hydroxyapatite/Ti-6Al-4V composite coatings in simulated body fluid (SBF).' *Biomaterials.* 24(9). 1603–1611.

127. Khor, K. A., Y. W. Gu, D. Pan, and P. Cheang. 2004. 'Microstructure and mechanical properties of plasma sprayed HA/YSZ/Ti-6Al-4V composite coatings.' *Biomaterials.* 25(18). 4009–4017.

128. Arifin, Amir, Abu Bakar Sulong, Norhamidi Muhamad, Junaidi Syarif, and Mohd Ikram Ramli. 2014. 'Material processing of hydroxyapatite and titanium alloy (HA/Ti) composite as implant materials using powder metallurgy: a review.' *Materials and Design.* 55: 165–175.

129. Debao, Liu Chen Minfang, and Wang Xiaowei. 2008. 'Fabrication and corrosion biodegradable properties of the HA/Mg biocomposite.' *Rare Metal Materials and Engineering.* 12: 029.

130. Ye, Xinyu, Minfang Chen, Meng Yang, Jun Wei, and Debao Liu. 2010. '*in vitro* corrosion resistance and cytocompatibility of nano-hydroxyapatite reinforced Mg-Zn-Zr composites.' *Journal of Materials Science: Materials in Medicine.* 21(4). 1321–1328.

131. Yang, Yunzhi, Kyo-Han Kim, C. Mauli Agrawal, and Joo L. Ong. 2004. 'Interaction of hydroxyapatite–titanium at elevated temperature in vacuum environment.' *Biomaterials.* 25(15). 2927–2932.

132. Nath, Shekhar, Rajesh Tripathi, and Bikramjit Basu. 2009. 'Understanding phase stability, microstructure development and biocompatibility in calcium phosphate–titania composites, synthesized from hydroxyapatite and titanium powder mix.' *Materials Science and Engineering: C.* 29(1). 97–107.

133. Priya, Ashok, Shekhar Nath, Krishanu Biswas, and Bikramjit Basu. 2010. '*in vitro* dissolution of calcium phosphate-mullite composite in simulated body fluid.' *Journal of Materials Science: Materials in Medicine.* 21(6). 1817–1828.

134. Nath, Shekhar, Bikramjit Basu, Mira Mohanty, and P. V. Mohanan. 2009. '*in vivo* response of novel calcium phosphate-mullite composites: Results up to 12 weeks of implantation.' *Journal of Biomedical Materials Research Part B: Applied Biomaterials.* 90(2). 547–557.

135. Towler, M. R., I. R. Gibson, and S. M. Best. 2000. 'Novel processing of hydroxyapatite-zirconia composites using nano-sized particles.' *Journal of Materials Science Letters.* 19(24). 2209–2211.

136. Kim, Hae-Won, Young-Hag Koh, Byung-Ho Yoon, and Hyoun-Ee Kim. 2002. 'Reaction sintering and mechanical properties of hydroxyapatite–zirconia composites with calcium fluoride additions.' *Journal of the American Ceramic Society.* 85(6). 1634–1636.

137. Quan, Renfu, Disheng Yang, Xiaochun Wu, Hongbin Wang, Xudong Miao, and Wei Li. 2008. '*in vitro* and *in vivo* biocompatibility of graded hydroxyapatite–zirconia composite bioceramic.' *Journal of Materials Science: Materials in Medicine.* 19(1). 183–187.

138. Kim, Sona, Young-Min Kong, In-Seop Lee, and Hyoun-Ee Kim. 2002. 'Effect of calcinations of starting powder on mechanical properties of hydroxyapatite–alumina bioceramic composite.' *Journal of Materials Science: Materials in Medicine.* 13(3). 307–310.

139. Balani, Kantesh, Debrupa Lahiri, Anup K. Keshri, S. R. Bakshi, Jorge E. Tercero, and Arvind Agarwal. 2009. 'The nano-scratch behaviour of biocompatible hydroxyapatite reinforced with aluminum oxide and carbon nanotubes.' *JOM Journal of the Minerals, Metals and Materials Society.* 61(9). 63–66.

140. Tripathi, Garima, Julie E. Gough, Amit Dinda, and Bikramjit Basu. 2013. '*in vitro* cytotoxicity and *in vivo* osseointergration properties of compression-moulded HDPE-HA-Al$_2$O$_3$ hybrid biocomposites.' *Journal of Biomedical Materials Research Part A.* 101(6). 1539–1549.

141. Mezahi, F. Z., Abdel Hamid Harabi, Souleiha Zouai, S. Achour, and Didier Bernache- Assollant. 2005. 'Effect of stabilised ZrO$_2$, Al$_2$O$_3$ and TiO$_2$ on sintering of hydroxyapatite.' *Materials Science Forum.* 492. 241–248.

142. Wang, M. and D. Porter. 1994. 'Processing, characterisation, and evaluation of hydroxyapatite reinforced polyethylene.' *British Ceramic Transactions.* 93(3). 91.

143. Wang, M., S. Deb, and W. Bonfield. 2000. 'Chemically coupled hydroxyapatite-polyethylene composites: processing and characterisation.' *Materials Letters.* 44(2). 119–124.

144. Fang, Liming, Yang Leng, and Ping Gao. 2006. Processing and mechanical properties of HA/UHMWPE nanocomposites. *Biomaterials.* 27(20). 3701–3707.

145. Chu, K. T., Y. Oshida, E. B. Hancock, M. J. Kowolik, T. Barco, and S. L. Zunt. 2003. 'Hydroxyapatite/PMMA composites as bone cements.' *Biomedical materials and engineering.* 14(1). 87–105.

146. Yang, Chun-Chen, Che-Tseng Lin, and Shwu-Jer Chiu. 2008. 'Preparation of the PVA/HAp composite polymer membrane for alkaline dmfc application.' *Desalination.* 233(1). 137–146.

147. Dang, Zhi-Min, Chun-Yan Tian, Jun-Wei Zha, Sheng-Hong Yao, Yu-Juan Xia, Jian-Ying Li, Chang-Yong Shi, and Jinbo Bai. 2009. 'Potential bioelectroactive bone regeneration polymer nanocomposites with high dielectric permittivity.' *Advanced Engineering Materials.* 11(10):B144–B147.

148. Balani, Kantesh Rebecca Anderson, Tapas Laha, Melanie Andara, Jorge Tercero, Eric Crumpler, and Arvind Agarwal. 2007. 'Plasma-sprayed carbon nanotube reinforced hydroxyapatite coatings and their interaction with human osteoblasts *in vitro*.' *Biomaterials.* 28(4). 618–624.

149. Wang, M., and W. Bonfield. 2001. 'Chemically coupled hydroxyapatite–polyethylene composites: structure and properties.' *Biomaterials*. 22(11). 1311–1320.

150. Fang, Liming, Ping Gao, and Yang Leng. 2007. 'High strength and bioactive hydroxyapatite nanoparticles reinforced ultrahigh molecular weight polyethylene.' *Composites Part B: Engineering*. 38(3). 345 – 351.

151. Di Silvio, L., M. Dalby, and W. Bonfield. 1998. '*In vitro* response of osteoblasts to hydroxyapatite-reinforced polyethylene composites.' *Journal of Materials Science: Materials in Medicine*. 9(12). 845–848.

152. Dalby, M. J., L. Di Silvio, E. J. Harper, and W. Bonfield. 1999. '*In vitro* evaluation of a new polymethylmethacrylate cement reinforced with hydroxyapatite.' *Journal of Materials Science: Materials in Medicine*. 10(12). 793–796.

153. Nair, L. S,. and C. T. Laurencin. 2007. 'Biodegradable polymers as biomaterials.' *Progress in Polymer Science*. 32(8–9). 762–798.

154. Kumar, Alok, Sourav Mandal, Srimanta Barui, Ramakrishna Vasireddi, Uwe Gbureck, Michael Gelinsky, and Bikramjit Basu. 2016. 'Low Temperature Additive Manufacturing of Three Dimensional Scaffolds for Bone-Tissue Engineering Applications: Processing related Challenges and Property Assessment.' *Materials Science and Engineering: R: Reports* 103: 1–39.

155. Alok Kumar, K. Biswas and B. Basu. 2005. 'Hydroxyapatite-titanium bulk composites for bone tissue engineering applications.' *J. Biomed. Mat. Res. Part A* 103: 791-806.

For Further Reading

a. Kim, Hae-Won, Young-Min Kong, Young-Hag Koh, Hyoun-Ee Kim, Hyun-Man Kim, and Jea Seung Ko. 2003. 'Pressureless Sintering and Mechanical and Biological Properties of Fluor-hydroxyapatite Composites with Zirconia.' *Journal of the American Ceramic Society* 86 (12): 2019-2026.

b. Amaral, M., Lopes, M. A., Silva, R. F., and Santos, J. D. 2002. 'Densification route and mechanical properties of Si_3N_4–bioglass biocomposites.' *Biomaterials*. 23(3): 857-862.

c. Vasconcelos, Cristina. 2012. 'New challenges in the sintering of HA/ZrO_2 composites.' *Sintering of Ceramics: New Emerging Techniques*.

d. Chiba, A., Kimura, S., Raghukandan, K., and Morizono, Y. 2003. 'Effect of alumina addition on hydroxyapatite biocomposites fabricated by underwater-shock compaction.' *Materials Science and Engineering*. A 350(1): 179-183.

Chapter 3

156. Langer, R. and J. P. Vacanti. 1993. 'Tissue engineering.' *Science*. 260. 920.

157. Mano, J. F., R. A. Sousa, L. F. Boesel, N. M. Neves, and R. L. Reis. 2004. 'Bioinert, biodegradable and injectable polymeric matrix composites for hard tissue replacement: state of the art and recent developments.' *Compos. Sci. Technol*. 64. 789.

158. Salgado, A. J., O. P. Coutinho, and R. L. Reis. 2004. 'Bone tissue engineering: State of the art and future trends.' *Macromol. Biosci*. 4. 743.

159. (a) Prockop, D. J., and K. I. Kivirikko. 1995. 'Collagens: molecular biology, diseases, and potentials for therapy.' *Annu. Rev. Biochem*. 64. 403. (b) Olsen, B. R., A. M. Reginato, and W. Wang. 2000. 'Bone development.' *Ann. Rev. Cell. Dev. Biol*. 16. 191.

160. Hartgerink, J. D. 2004. 'Covalent capture: a natural complement to self-assembly.' *Curr. Opin. Chem. Biol*. 8. 604.

161. Stevens, M. M., and J. H. George. 2005. 'Exploring and engineering the cell surface interface.' *Science.* 310. 1135.

162. Hubbell J. A. 1995. 'Biomaterials in tissue engineering.' *Nature Biotechnology.* 13. 565–576.

163. Shoichet, M. S. 2009. 'Polymer scaffolds for biomaterials applications.' *Macromolecules.* 43. 581–591.

164. Langer, R. 2009. 'Perspectives and challenges in tissue engineering and regenerative medicine.' *Advanced Materials.* 21. 3235–3236.

165. Hollister, S. J. 2005. 'Porous scaffold design for tissue engineering.' *Nature Materials.* 4. 518–524.

166. Grover, C. N., R. E. Cameron, and S. M. Best. 2012. 'Investigating the morphological, mechanical and degradation properties of scaffolds comprising collagen, gelatin and elastin for use in soft tissue engineering.' *Journal of the Mechanical Behavior of Biomedical Materials.* 10. 62–74.

167. Place, E. S., N. D. Evans, and N. M. Stevens. 'Complexity in biomaterials for tissue engineering.' *Nat. Mater.* 8. 457–470.

168. Lysaght, M. J., A. Jaklenec, and E. Deweerd. 2008. 'Great expectations: private sector activity in tissue engineering, regenerative medicine, and stem cell therapeutics.' *Tissue Engineering Part A.* 14. 305–315.

169. Hutmacher, D. W., and S. Cool. 2007. 'Concepts of scaffold-based tissue engineering the rational to use solid free-form fabrication techniques.' *J. Cell. Mol. Med.* 11. 654.

170. Gibson, L. J. and M. F. Ashby. 1997. *Cellular Solids: Structure and Properties.* Cambridge: Cambridge University Press.

171. Bose, S., M. Roy, and A. Bandopadhyay. 2012. 'Recent advances in bone tissue engineering scaffolds.' *Trends Biotechnol.* 30. 546.

172. Sa´lcedo, S., D. Arcos, and M. Vallet-Regı´. 2008. 'Upgrading calcium phosphate scaffolds for tissue engineering applications.' *Key Eng. Mater.* 377. 19.

173. Mikos, Antonios G., and J. S. Temenoff. 2000. 'Formation of highly porous biodegradable scaffolds for tissue engineering.' *Electron. J. Biotechnol.* 3. 23-24

174. (a) Levenberg, S., and R. Langer. 2004. 'Advances in tissue engineering.' *Curr. Top Dev. Biol.* 61. 113. (b) Griffith, L. G. 2002. 'Emerging design principles in biomaterials and scaffolds for tissue engineering.' *Ann. NY Acad. Sci.* 961. 83. (c) Karageorgiou, V. and D. Kaplan. 2005. 'Porosity of 3D biomaterial scaffolds and osteogenesis.' *Biomaterials.* 26. 5474.

175. (a) Sell, S. A., P. S. Wolfe, K. Garg, J. M. McCool, I. A. Rodriguez, and G. L. Bowlin. 2010. 'The use of natural polymers in tissue engineering: A focus on electrospun extracellular matrix analogues.' *Polym.* 2. 522. (b) Place, E. S., J. H. George, C. K Williams, and M. M. Stevens. 2009. 'Synthetic polymer scaffolds for tissue engineering.' *Chem. Soc. Rev.* 38. 1139.

176. Hench, L. L. 1998. 'Bioceramics.' *J. Am. Ceram. Soc.* 81. 1705.

177. Kim, H. W., J. C. Knowles, and H. E. Kim. 2004. 'Hydroxyapatite/poly(ε-caprolactone) composite coatings on hydroxyapatite porous bone scaffold for drug delivery.' *Biomaterials.* 25. 1279.

178. Lee, K., E. A. Silve, and D. J. Mooney. 2011. 'Growth factor delivery-based tissue engineering: general approaches and a review of recent developments.' *J. R. Soc. Interface.* 8. 153.

179. Sepulveda, P., F. Ortega, M. D. M. Innocentini, and V. C. Pandolfelli. 2001. 'Properties of highly porous hydroxyapatite obtained by the gel casting of foams.' *J. Am. Ceram. Soc.* 83. 3021.

180. Cyster, L. A., D. M. Grant, S. M. Howdle, F. Rose, D. J. Irvine, D. Freeman, C. A. Scotchford, and K. M. Shakesheff. 2005. 'The influence of dispersant concentration on the pore morphology of hydroxyapatite ceramics for bone tissue engineering.' *Biomaterials.* 26. 697.

181. Fu, Q., M. N. Rahaman, B. S. Bal, W. Huang, and D. E. Day. 2007. 'Preparation and bioactive characteristics of a porous 13-93 glass and fabrication into the articulating surface of a proximal tibia.' *J. Biomed. Mater. Res. A.* 82. 222.

182. Chang, B. S., C. K. Lee, K. S. Hong, H. J. Youn, H. S. Ryu, S. S. Chung, and K. W. Park. 2000. 'Osteoconduction at porous hydroxyapatite with various pore configurations.' *Biomaterials.* 21. 1291.

183. Deville, S., E. Saiz, and A. P. Tomsia. 2006. 'Freeze casting of hydroxyapatite scaffolds for bone tissue engineering.' *Biomaterials*. 27. 5480.

184. Nam, Y. S., J. J. Yoon, and T. G. Park. 2000. 'A novel fabrication method of macroporous biodegradable polymer scaffolds using gas foaming salt as porogen additive.' *J. Biomed. Mater. Res.* 53. 1.

185. Vallet-Regi, M. 2008. 'Current trend for porous inorganic materials for biomedical applications.' *Chem. Eng. J.* 137. 1.

186. Pek, Y. S., S. Gao, M. S. Arshad, K. J. Leck, and J. Y. Ying. 2008. 'Porous collagen-apatite nanocomposite foams as bone regeneration scaffolds.' *Biomaterials*. 29. 4300.

187. Geiger, M., R. H. Li, and W. Friess. 2003. 'Collagen sponges for bone regeneration with rhBMP-2.' *Adv. Drug Deliv. Rev.* 55. 1613.

188. Zhang, Y., V. J. Reddy, S. Y. Wong, X. Li, B. Su, S. Ramakrishna, and C. T. Lim. 2010. 'Enhanced biomineralization in osteoblasts on a novel electrospun biocomposite nanofibrous substrate of hydroxyapatite/collagen/chitosan.' *Tissue Eng. Part A*. 16. 1949.

189. Chen, M., D. Q. S., Le, A. Baatrup, J. V. Nygaard, S. Hein, L. Bjerre, M. Kassem, X. Zou, and C. Bünger. 2011. 'Self-assembled composite matrix in a hierarchical 3D scaffold for bone tissue engineering.' *Acta Biomater*. **7**. 2244.

190. Fini, M., A. Motta, P. Torricelli, G. Giavaresi, N. Nicoli Aldini, M. Tschon, R. Giardino, and C. Migliaresi. 2005. 'The healing of confined critical size cancellous defects I presence of silk fibroin hydrogel.' *Biomaterials*. 26. 3527.

191. Meinel, A. J., K. E. Kubow, E. Klotzsch, M. Garcia-Fuentes, M. L. Smith, V. Vogel, H. P. Merkle, and L. Mienel. 2009. 'Optimization strategies for electrospun silk fibroin tissue engineering scaffolds.' *Biomaterials*. 30. 3058.

192. Meinel, L., O. Betz, R. Fajardo, S. Hofmann, A. Nazarian, E. Cory, M. Hilbe, J. McCool, R. Langer, G. Vunjak-Novakovic, H. P. Merkle, B. Rechenberg, D. L. Kaplan, and C. Kirker-Head. 2006. 'Silk based biomaterials to heal critical sized femur defects.' *Bone*. 39. 922.

193. Kim, H. J., U. J. Kim, H. S. Kim, C. Li, M. Wada, C. G. Leisk, and D. L. Kaplan. 2008. 'Bone tissue engineering with premineralized silk scaffolds.' *Bone*. 42. 1226.

194. Li, C., C. Vepari, H. J. Jin, H. J. Kim, and D. L. Kaplan. 2006. 'Electrospun silk-BMP-2 scaffolds for tissue engineering.' *Biomaterials*. 27. 3115.

195. Kim, J. Y., J. Y. Choi, J. H. Jeong, E. S. Jang, A. S. Kim, and S. G. Kim. 2010. 'Low molecular weight silk fibroin increases alkaline phosphatase and type I collagen expression in MG63 cells.' *BMB Rep*. 43. 52.

196. Mandal, B., A. Grinberg, E. S. Gil, B. Panilaitis, and D. S. Kaplan. 'High-strength silk protein scaffolds for bone repair.' *Proc. Natl. Acad. Sci. USA*. 2012, doi: 10.1073/*Proc. Natl. Acad. Sci. USA*. 1119474109

197. Hota, M. K., M. K. Bera, B. Kundu, S. C. Kundu, and C. K. Maiti. 2012. 'A natural silk fibroin protein-based bio-memristor.' *Adv. Funct. Mat*. doi: 10.1002/adfm.201200073.

198. Lutolf, M. P. 2009. 'Integration column: artificial ECM: expanding the cell biology toolbox in 3D.' *Integr. Biol (Camb)*. 1. 235.

199. Lutolf, M. P., P. M. Gilbert., and H. M. Blau. 2009. 'Designing materials to direct stem-cell fate.' *Nature*. 462. 433.

200. Narayanan, K., K. J. Leck, S. Gao, and A. C. Wan. 2009. 'Three-dimensional reconstituted extracellular matrix scaffolds for tissue engineering.' *Biomaterials*. 30. 4309.

201. Evans, N. D., E. Gentleman, X. Chen, C. J. Roberts, J. M. Polak, and M. M. Stevens. 2010. 'Extracellular matrix-mediated osteogenic differentiation of murine embryogenic stem cells.' *Biomaterials*. 31. 3244.

202. Hidalgo-bastida, L. A. and S. H. Cartmell. 2010. 'Mesenchymal stem cells, osteoblasts and extracellular matrix proteins: enhancing cell adhesion and differentiation for bone tissue engineering.' *Tissue Eng.: Part B*. 16. 405.

203. He, X., J. Ma, and E. Jabbari. 2008. 'Effect of grating RGD and BMP-2 protein-derived peptides to a hydrogel substrate on osteogenic differentiation of marrow stromal cells.' *Langmuir.* 24. 12508.

204. Whitesides, G. M., and B. Grzybowski. 2002. 'Self-assembly at all scales.' *Science.* 295. 2418.

205. Zhang, S., F. Gelain, and X. Zhao. 2005. 'Designer self-assembling peptide nanofibers scaffolds for 3D tissue cell cultures.' *Semin. Cancer Biol.* 15. 413.

206. Dinca, V., E. Kasotakis, J. Catherine, A. Mourka, A. Ranella, A. Ovsianikov, B. N. Chichkov, M. Farsari, A. Mitraki, and C. Fotakis. 2008. 'Directed three-dimensional patterning of self-assembled peptide fibrils.' *Nano. Lett.* 8. 538.

207. (a) Ryadnov, M. G., A. Bella, S. Timson, and D. N. Woolfson. 2009. 'Molecular design of peptide fibrilar nano- to microstructures.' *J. Am. Chem. Soc.* 131. 13240. (b) Gribbon, C., K. J. Channon, W. Zhang, E. F. Banwell, E. H. Bromley, J. B. Chaudhuri, R. O. Oreffo, and D. N. Woolfson. 2008. 'Magic wand: a single designed peptide that assembles to stable orderd alpha-helical fibers.' *Biochemistry.* 47. 10365.

208. Kyle, S., A. Aggeli, E. Ingham, and M. J. McPherson. 2009. 'Production of self-assembling biomaterials for tissue engineering.' *Trends in Biotechnol.* 27. 423.

209. (a) Zhang, L., J. Rodriguez, J. Raez, A. J. Myles, H. Fenniri, and T. J. Webster. 2009. 'Biologically inspired rosette nanotubes and nanocrystalline hydroxyapatite hydrogel nanocomposites as improved bone substitutes.' *Nanotechnology.* 20. 175101. (b) Zhang, L., Y. Chen, J. Rodriguez, H. Fenniri, and T. J. Webster. 2008. 'Biomimetic helical rosette nanotubes and naocrystalline hydroxyapatite coatings on titanium for improving orthopedic implants.' *Int. J. Nanomed.* 3. 323.

210. Misawa, H., N. Kobayashi, A. Soto-Gutierrez, Y. Chen, A. Yoshida, J. D. Rivas-Carrillo, N. Navarro-Alvarez, K. Tanaka, A. Miki, J. Takei, T. Ueda, M. Tanaka, H. Endo, N. Tanaka, and T. Ozaki. 2006. 'PuraMatrix facilitates bone regeneration in bone defects of calvaria in mice.' *Cell Transpl.* 15. 903.

211. Woolfson, D. N. and M. G. Ryadnov. 2006. 'Peptide-based fibrous biomaterials: Some things old, new and borrowed.' *Curr. Opin. Chem. Biol.* 10. 559.

212. (a) Branco, M. C., and J. P. Schneider. 2009. 'Self-assembling materials for therapeutic delivery.' *Acta Biomater.* 5. 817. (b) Branco, M. C., D. J. Pochan, N. J. Wagner, and J. P. Schneider. 2009. 'Macromolecular diffusion and release from self-assembled beta-hairpin peptide hydrogels.' *Biomaterials.* 30. 1339. (c) Rudra, J. S., Y. F. Tian, J. P. Jung, and J. H. Collier. 2010. 'A self-assembling peptide acting as an immune adjuvant.' *Proc. Natl. Acad. Sci. USA.* 107. 622.

213. Gauba, V. and J. D. Hartgerink. 2008. 'Synthetic collagen heterotrimers: structural mimics of wild-type and mutant collagen type I.' *J. Am. Chem. Soc.* 130. 7509.

214. Petka, W. A., J. L. Harden, K. P. McGrath, D. Wirtz, and D. A. Tirrell. 1998. 'Reversible hydrogels from self-assembling artificial proteins.' *Science.* 281. 389.

215. (a) Liu, B., A. K. Lewis, and W. Shen. 2009. 'Physical hydrogels photo-cross-linked from self-assembled macromers for potential use in tissue engineering.' *Biomacromolecules.* 10. 3182. (b) Fischer, S. E., L. Mi, H. Q. Mao, and J. L. Harden. 2009. 'Biofunctional coatings via targeted covalent cross-linking of associating triblock proteins.' *Biomacromolecules.* 10. 2408. (c) Wheeldon, I. R., E. Campbell, and S. J. Banta. 2009. 'A chimeric fusion protein engineered with disparate functionalities-enzymatic activity and self-assembly.' *Mol. Biol.* 392. 129. (d) Wheeldon, I. R., J. W. Gallaway, S. C. Barton, and S. Banta. 2008. 'Bioelectrocatalytic hydrogels from electron-conducting metallopolypeptides coassembled with bifunctional enzymatic building blocks.' *Proc. Natl. Acad. Sci. USA.* 105. 15275. (e) Fischer, S. E., X. Liu, H. Q. Mao, and J. L. Harden. 2007. 'Controlling cell adhesion to surfaces via associating bioactive triblock proteins.' *Biomaterials.* 28. 3325. (f) Ryadnov, M. G., D. Papapostolou, and D. N. Woolfson. 2008. 'The leucine zipper as a building block for self-assembled protein fibers.' *Methods Mol. Biol.* 474. 35.

216. Gajjeraman, S., G. He, K. Narayanan, and A. George. 2008. 'Biological assemblies provide novel templates for the synthesis of biocomposites and facilitate cell adhesion.' *Adv. Funct. Mater.* 18. 3972.

217. Brun, P., F. Ghezzo, M. Roso, R. Danesin, G. Palù, A. Bagno, M. Modesti, I. Castagliuolo, and M. Dettin. 2011. 'Electrospun scaffolds of self-assembling peptides with poly(ethylene oxide) for bone tissue engineering.' *Acta Biomater.* **7**. 2526.

218. Drury, J. L., and D. J. Mooney. 2003. 'Hydrogels for tissue engineering: scaffold design variables and applications.' *Biomaterials.* **24**. 4337.

219. Burdick, J. A., and K. S. Anseth. 2002. 'Photoencapsulation of osteoblasts in injectable RGD-modified PEG hydrogels for bone tissue engineering.' *Biomaterials.* **23**. 4315.

220. Lutolf, M. P., and J. A. Hubbell. 2005. 'Synthetic biomaterials as instructive extracellular microenvironments for morphogenesis in tissue engineering.' *Nat. Biotechnol.* **23**. 47.

221. Lin, C. C. and K. S. Anseth. 2009. 'PEG hydrogels for controlled release of biomolecules in regenerative medicine.' *Pharm. Res.* **26**. 631.

222. Shah, R. N., N. A. Shah, M. M. Del Rosario Lim, C. Hsieh, G. Nuber, and S. I. Stupp. 2010. 'Supramolecular design of self-assembling nanofibers for cartilage regeneration.' *Proc. Natl. Acad. Sci. USA.* **107**. 3293.

223. Cui, H., M. J. Webber., and S. I. Stupp. 2010. 'Self-assembly of peptide amphiphiles: from molecules to nanostructures to biomaterials.' *Biopolymers.* **94**. 1.

224. Nowak, A. P., V. Breedveld, L. Pakstis, B. Ozbas, D. J. Pine, D. Pochan, and T. J. Deming. 2002. 'Rapidly recovering hydrogel scaffolds from self-assembling diblock copolypeptide amphiphiles.' *Nature.* **417**. 424.

225. (a) Hosseinkhani, H., M. Hosseinkhani, F. Tian, H. Kobayashi, and Y. Tabata. 2007. 'Bone regeneration on a collagen sponge self-assembled peptide amphiphile nanofibers scaffold.' *Tissue Eng.* **3**. 11. (b) Hosseinkhani, H., M. Hosseinkhani, F. Tian, H. Kobayashi, and Y. Tabata. 2006. 'Ectopic bone formation in collagen sponge self-assembled peptide amphiphile nanofibers hybrid scaffolds in a perfusion culture bioreactor .' *Bioamterials.* **27**. 5089.

226. Semino, C. E. 2008. 'Self-assembling peptides: from bio-inspired materials to bone regeneration.' *J. Dent. Res.* **87**. 606.

227. Lutolf, M. P., and J. Hubbell. 2005. 'Synthetic biomaterials as instructive extracellular microenvironments for morphogenesis in tissue engineering.' *Nat. Biotechnol.* **23**. 47.

228. (a) Barnes, C. P., S. A. Sell, E. D. Boland, D. G. Simpson, and G. L. Bowlin. 2007. 'Nanofiber technology: designing the next generation of tissue engineering scaffolds.' *Adv. Drug Deliv. Rev.* **59**. 1413. (b) Liang, D., B. S. Hsiao, and B. Chu. 2007. 'Functional electrospun nanofibrous scaffolds for biomedical applications.' *Adv. Drug. Deliv. Rev.* **59**. 1392. (c) Zhang, X., M. Reagan, and D. L. Kaplan. 2009. 'Electrospun silk biomaterial scaffolds for regenerative medicine.' *Adv. Drug. Deliv. Rev.* **61**. 988. (d) Lee, K. Y., L. Jeong, Y. O. Kang, S. J. Lee, and W. H. Park. 2009. 'Electrospinning of polysaccharides for regenerative medicine.' *Adv. Drug Deliv. Rev.* **61**. 1020.

229. Pham, Q. P., U. Sharma, and A. G. Mikos. 2006. 'Electrospun poly (e-caprolactone) microfiber and multilayer nanofiber/microfiber scaffolds: characterization of scaffolds and measurement of cellular infiltration.' *Biomacromolecules.* **7**. 2796.

230. (a) Sell, S. A., M. J. McClure, K. Garg, P. S. Wolfe, and G. L. Bowlin. 2009. 'Electrospinning of collagen/biopolymers for regenerative medicine and cardiovascular tissue engineering.' *Adv. Drug Deliv. Rev.* **61**. 1007. (b) Lim, S.H. and H. Q. Mao. 2009. 'Electrospun scaffolds for stem cell engineering.' *Adv. Drug Deliv. Rev.* **61**. 1084. (c) Cao, H. Q., T. Liu, and S. Y. Chew. 2009. 'The application of nanofibrous scaffolds in neural tissue engineering.' *Adv. Drug Deliv. Rev.* **61**. 1055.

231. Burg, K. J. L., S. Porter, and J. F. Kellam. 2000. 'Biomaterial development for bone tissue engineering.' *Biomaterials.* **21**. 2347.

232. Kim, H. W., H. S. Yu, and H. H. Lee. 2007. 'Nanofibrus matrices of poly(lactic acid) and gelatin polymeric blends for the improvement of cellular responses.' *J. Biomed. Mater Res. Part A.* **87A**. 25.

233. Shin, M., H. Yoshimoto, and J. P. Vacanti. 2004. '*In vivo* bone tissue engineering using mesenchymal cells on a novel electrospun nanofibrous scaffold.' *Tissue Eng.* 10. 33.

234. Badami, A. S., M. R. Kreke, M. S. Thompson, J. S. Riffle and A. S. Goldstein. 2006. 'Effect of fiber diameter on spreading, proliferation nd differentiation of osteoblastic cells on electrospun poly(lactic acid) substrates.' *Bioamterials.* 27. 596.

235. Sombatmankhong, K., N. Sanchavanakit, P. Pavasant, and P. Supaphol. 2007. 'Bone scaffolds from electrospun fiber mats of poly(3-hydroxybutyrate), poly(3-hydroxybutyrate-co-3-hydroxyvalerate) and their blend.' *Polymer.* 48. 1419.

236. Zhang, Y., H. Ouyang, C. T. Lim, S. Ramakrishna, and Z. M. Huang. 2004. 'Electrospinning of gelatin fibers and gelatin/PCL composite fibrous scaffolds.' *J. Biomed. Mater. Res. Part B: Appl. Biomater.* 72. 156.

237. Luong-Van, E., L. Grøndahl, S. J. Song, V. Nurcombe, and S. Cool. 2007. 'The *in vivo* assessment of a novel scaffold containing heparan sulfate for tissue engineering with human mesenchymal stem cells.' *J. Mol. Hist.* 38. 459.

238. (a) Ma, K., C. K. Chan, S. Liao, W. Y. K. Hwang, Q. Feng, and S. Ramakrishna. 2008. 'Electrospun nanofiber scaffolds for rapid and rich capture of bone marrow-derived hematopoietic stem cells.' *Biomaterials.* 29. 2096. (b) Ma, Z., W. He, T. Yong, and S. Ramakrishna. 2005. 'Grafting of gelatin on electrospun poly (caprolactone) nanofibers to improve endothelial cell spreading and prolifera- tion and to control cell orientation.' *Tissue Eng.* 11. 1149.

239. (a) Matthews, J. A., G. E. Wnek, D. G. Simpson, and G. L. Bowlin. 2002. 'Electrospinning of collagen nanofibers.' *Biomacromolecules.* 3. 232. (b) Shin, Y. R. V., C. N. Chen, S. W. Tsai, Y. J. Wang, and O. K. Lee. 2006. 'Growth of mesenchymal stem cells on electrospun type I collagen nanofibers.' *Stem Cells.* 24. 2391. (c) Zeugolis, D. I., S. T. Khew, E. S. Y. Yew, A. K. Ekaputra, Y. W. Tong, L. L. Yung, D. W. Hutmacher, C. Sheppard, and M. Raghunath. 2008. 'Electro-spinning of pure collagen nano- fibres- just an expensive way to make gelatin?' *Biomaterials.* 29 (15): 2293-2305.

240. Fang, X., and D. H. Renecker. 1997. 'DNA fibers by electrospinning.' *Macromol. Sci-Phys. B.* 36. 169.

241. Kim, H. W., H. E. Kim, and J. C. Knowles. 2006. 'Production and potential of bioactive glass nanofibers as a next-generation biomaterial.' *Adv. Funct. Mater.* 16. 1529.

242. (a) Kim, H. W. and H. E. Kim. 2005. 'Nanofiber generation of hydroxyapatite and fluor-hydroxyapatite bioceramics.' *J. Biomed. Mater. Res. Part B: Appl. Biomater.* 22B. 323. (b) Wu, Y., L. L. Hench, J. Du, K. L. Choy, and J. Guo. 2004. 'Preparation of hydroxyapatite fibers by electrospinning technique.' *J. Am. Ceram. Soc.* 87. 1988. (c) Dai, X. and S. Shivkumar. 2007. 'Electrospinning of PVA-calcium phosphate sol precursors for the production of fibrous hydroxyapatite.' *J. Am. Ceram. Soc.* 90. 1412.

243. Sakai, S., Y. Yamada, T. Yamaguchi, and K. Kawakami. 2006. 'Prospective use of electrospun ultra-fine silicate fibers for bone tissue engineering.' *Biotechnol J.* 1. 958.

244. Kim, H. W., H. H. Lee, and J. C. Knowles. 2008. 'Nanofibrous glass tailored with apatite- fibronectin interface for bone cell stimulation.' *J. Nanosci. Nanotechnol.* 8. 3013.

245. (a) Kim, H. W., J. H. Song, and H. E. Kim. 2006. 'Bioactive glass nanofiber-collagen nanocomposite as a novel bone regeneration matrix.' *J. Biomed. Mater. Res. Part A.* 79A. 698. (b) Kim, H. W., H. H. Lee, and G. S. Chun. 2008. 'Bioactivity and osteoblast responses of novel biomedical nanocomposites of bioactive glass nanofiber filled poly(lactic acid).' *J. Biomed. Mater. Res. Part A.* 85A. 651.

246. Linhart, W., F. Peters, W. Lehmann, K. Schwarz, A. F. Schilling, M. Amling, J. M. Reuger, and M. Epple. 2001. Biologically and chemically optimized composites of carbonated apatite and polyglycolide as bone substitution materials. *J. Biomed. Mater. Res.* 54. 162.

247. Hu, Y., C. Zhang, S. Zhang, Z. Xiong, and J. Xu. 2003. 'Development of a porous poly(L-lactic acid)/ hydroxyapatite/collagen scaffold as a BMP delivery system and its use in healing canine segmental bone defect.' *J. Biomed. Mater. Res. A.* 67A. 591.

248. Olszta, M. J., X. Cheng, S. S. Jee, R. Kumar, Y. Y. Kim, M. J. Kaufman, E. P. Douglas, and L. B. Gower. 2007. 'Bone structure and formation: a new perspective.' *Mater. Sci. Eng. R.* 58. 77.

249. Rezwan, K., Q. Z. Chen, J. J. Blaker, and A. R. Boccaccini. 2006. 'Biodegradable and bioactive porous polymer/inorganic composite scaffolds for bone tissue engineering.' *Biomaterials.* 27. 3413.

250. Kim, H. W., J. H. Song, and H. E. Kim. 2005. 'Nanofiber generation of gelatin-hydroxyapatite biomimetics for guided tissue regeneration.' *Adv. Funct. Mater.* 15. 1988.

251. Song, J. H., H. E. Kim, and H. W. Kim. 2008. 'Electrospun fibrous web of collagen-apatite precipitated nanocomposite for bone regeneration.' *J. Mater. Sci. Mater. Med.* 19. 2925.

252. Zhang, Y., J. R. Venugopal, A. El-Turki, S. Ramakrishna, B. Su, and C. T. Lim. 2008. 'Electrospun biomimetic nanocomposite nanofibers of hydroxyapatite/chitosan for bone tissue engineering.' *Biomaterials.* 29. 4314.

253. Fujihara, K., M. Kotaki, and S. Ramakrishna. 2005. 'Guided bone regeneration membrane made of polycaprolactone/calcium carbonate composite nano-fibers.' *Biomaterials.* 26. 4139.

254. Kim, H. W., H. H. Lee, and J. C. Knowles. 2006. 'Electrospinning biomedical nanocomposite fibers of hydroxyapaite/poly(lactic acid) for bone regeneration.' *J. Biomed. Mater. Res. Part A.* 79. 643.

255. Song, J. H., B. H. Yoon, H. E. Kim and H. M. Kim. 2007. 'Bioactive and degradable hybridized nanofibers of gelatin–siloxane for bone regeneration.' *J. Biomed. Mater. Res. Part A.* 84A. 875.

256. Whited, B. M., J. R. Whitney, M. C. Hofmann, Y. Xu, and M. N. Rylander. 2011. 'Pre-osteoblast infiltration and differentiation in highly porous apatite-coated PLLA electrospun scaffolds.' *Biomaterials.* 32. 2294.

257. Francis, L., J. Venugopal, M. P. Prabhakaran, V. Thavasi, E. Marsano, and S. Ramakrishna. 2010. 'Simultaneous electrospin–electrosprayed biocomposite nanofibrous scaffolds for bone tissue regeneration.' *Acta Biomater.* 6. 4100.

258. Balasundaram, G. and T. J. Webster 2006. 'A perspective on nanophase materials for orthopedic implant applications.' *J. Mater. Chem.* 16. 3737.

259. Heino, J. and J. Käpylä. 2009. 'Cellular receptors of extracellular matrix molecules.' *Curr. Pharm. Des.* 15. 1309.

260. Mammoto, T. and D. E. Ingber. 2010. 'Mechanical control of tissue and organ development. *Development.*' 137. 1407.

261. Curtis, A., and M. Riehle. 2001. 'Tissue engineering: the biophysical background.' *Phys. Med. Biol.* 46. 47.

262. Curtis, A. S., N. Gadegaard, M. J. Dalby, M. O. Riehle, C. D. Wilkinson, and G. Aitchison. 2004. 'Cells react to nanoscale order and symmetry in their surroundings.' *IEEE Trans. Nanobiosci.* 3. 61.

263. Teixeira, A. I., P. F. Nealey, and C. J. Murphy. 2004. 'Responses of human keratocytes to micro- and nano structured substrates.' *J. Biomed. Mater. Res. A.* 71. 369.

264. (a) Price, R. L., K. Ellison, K. M. Haberstroh, and T. J. Webster. 2004. 'Nanometer surface roughness increases select osteoblasts adhesion on nanofiber compacts.' *J. Biomed. Mater. Res. A.* 70. 129. (b) Price, R. L., K. M. Haberstroh, and T. J. Webster. 2003. 'Enhanced functions of osteoblasts on nanostructured surfaces of carbon and alumina.' *Med. Biol. Eng. Comput.* 41. 372. (c) Zinger, O., G. Zhao, Z. Schwatz, J. Simpson, M. Wieland, D. Landolt, and B. Boyan. 2005. 'Differential regulation of osteoblasts by substrate microstructural features.' *Biomaterials.* 26. 1837. (d) Webster, T. J., L. S. Schadler, R. W. Siegel, and R. Bizios. 2001. 'Mechanisms of enhanced osteoblast adhesion on nanophase alumina involve vitronectin.' *Tissue Eng.* 7. 291. (e) Yim, E. K., R. M. Reano, S. W. Pang, A. F. Yee, C. S. Chen, and K. W. Leong. 2005. 'Nanopattern-induced changes in morphology and motility of smooth muscle cells.' *Biomaterials.* 26. 5405.

265. Dalby, M. J., N. Gadegaard, R. Tare, A. Andar, M. O. Riehle, P. Herzyk, C. D. W. Wilkinson, and R. O. C. Oreffo. 2007. 'The control of human mesenchymal cell differentiation using nanoscale symmetry and disorder.' *Nat. Mater.* 6. 997.

266. Pattison, M. A., S. Wurster, T. J. Webster, and K. M. Haberstroh. 2005. 'Three-dimensional, nano-structured PLGA scaffolds for bladder tissue replacement applications.' *Biomaterials.* 26. 2491.

267. (a) Falconnet, D., G. Csucs, H. M. Grandin, and M. Textor. 2006. 'Surface engineering approaches to micropattern surfaces for cell-based assays.' *Biomaterials.* 27. 3044. (b) Ogaki, R., M. Alexander, and P. Kingshott. 2010. 'Chemical patterning in biointerface science.' *Mater. Today.* 13. 22. (c) Christman, K. L., V. Enriquez-Rios, and H. D. Maynard. 2006. 'Nanopatterning proteins and peptides.' *Soft Matter.* 2. 928.

268. Walker, R. A., V. T. Cunliffe, J. D. Whittle, D. A. Steele, and R. D. Short. 2009. 'Submillimeter-scale surface gradients of immobilized protein ligands.' *Langmuir.* 25. 4243.

269. Ochsner, M., M. R. Dusseiller, H. M. Grandin, S. Luna-Morris, M. Textor, V. Vogel, and M. L. Smith. 2007. 'Micro-well arrays for 3D shape control and high resolution analysis of single cells.' *Lab. Chip.* 7. 1074.

270. Liu, X., P. K. Chu, and C. Ding. 2010. 'Surface nanofunctionalization of biomaterials.' *Materials Science and Engineering R.* 70. 275.

271. Webster, T. J., R. W. Siegel, and R. Bizios. 1999. 'Osteoblast adhesion on nanophase ceramics.' *Biomaterials.* 20. 1221.

272. Sinha, N. and J. T. W. Yeow. 2005. 'Carbon nanotubes for biomedical applications.' *IEEE Trans. Nanobiosci.* 4. 180.

273. MacDonald, R. A., B. F. Laurenzi, G. Viswanathan, P. M. Ajayan, and J. P. Stegeman. 2005. 'Collagen-carbon nanotube composite materials as scaffolds in tissue engineering.' *J. Biomed. Mater. Res.* 74A. 489.

274. Boccaccini, A. R., F. Chicatun, J. Cho, O. Bretcanu, J. A. Roether, S. Novak, and Q. Z. Chen. 2007. 'Carbon nanotube coatings on bioglass-based tissue engineering scaffolds.' *Adv. Funct. Mater.* 17. 2815.

275. Ward, B. C. and T. J. Webster. 2007. 'Increased functions of osteoblasts on nanophase metals.' *Mater. Sci. Eng. C.* 27. 575.

276. (a) Variola, F., J. H. Yi, L. Richert, J. D. Wuest, F. Rosei, and A. Nanci. 2008. 'Tailoring the surface properties of Ti6Al4V by controlled chemical oxidation.' *Biomaterials.* 29. 1285. (b) Shi, G. S., L. F. Ren, L. Z. Wang, H. S. Lin, S. B. Wang, and Y. Q. Tong. 2009. 'H_2O_2/HCl and heat-treated Ti-6Al-4V stimulates pre-osteoblast proliferation and differentiation.' *Oral Surg. Oral Med. Oral Pathol. Oral Radiol. Endod.* 108. 368.

277. (a) Wang, X. X., S. Hayakawa, K. Tsuru, and A. Osaka. 2002. 'Bioactive titania gel layers formed by chemical treatment of Ti substrate with a H_2O_2/HCl solution.' *Biomaterials.* 23. 1353. (b) Sun, T. and M. Wang. 2008. 'Low-temperature biomimetic formation of apatite/TiO_2 composite coatings on Ti and NiTi shape memory alloy and their characterization.' *Appl. Surf. Sci.* 255. 396.

278. Guo, J. L., R. J. Padilla, W. Ambrose, I. J. D. Kok, and L. F. Cooper. 2007. 'The effect of hydrofluoric acid treatment on TiO_2 grit blasted titanium implants on adherent osteoblasts gene expression *in vitro in vivo.*' *Biomaterials.* 28. 5418.

279. Lamolle, S. F., M. Monjo, M. Rubert, H. J. Haugen, S. P. Lyngstadaas, and J. E. Ellingsen. 2009. 'The effect of hydrofluoric acid treatment of titanium surface on nanostructural and chemical changes and the growth of MC3T3 cells –E1 cells.' *Biomaterials.* 30. 736.

280. Uchida, M., H. M. Kim, F. Miyaji, T. Kokubo, and T. Nakamura. 2002. 'Apatite formation on zirconium metal treated with aqueous NaOH.' *Biomaterials.* 23. 313.

281. Miyazaki, T., H. M. Kim, F. Miyaji, T. Kokubo, H. Kato, and T. Nakamura. 2000. 'Bioactive tantalum metal prepared by NaOH treatment.' *J. Biomed. Mater. Res.* 50. 35.

282. Ummethala, R., F. Despang, M. Gelinsky, and B. Basu. 2011. '*In vitro* corrosion and mineralization of novel Ti-Si-C alloy.' *Electrochimica Acta.* 56. 3809.

283. (a) Macak, J. M., H. Tsuchiya, and P. Schmuki. 2005. 'High aspect-ratio TiO_2 nanotubes by anodization of titanium.' *Angew. Chem. Int. Ed.* 44. 2100. (b) Macak, J. M., H. Tsuchiya, L. Taveira, S. Aldabergerova,

and P. Schmuki, P. 2005. 'Smooth anodic TiO$_2$ nanotubes.' *Angew. Chem. Int. Ed.* 44. 7463. (c) Paulose, M., K. Shankar, S. Yoriya, H. E. Prakasam, O. K. Varghese, G. K. Mor, T. J. LaTempa, A. Fitzgerald, and C. A. Grimes. 2006. 'Anodic growth of highly ordered TiO$_2$ nanotube arrays to 134 microm in length.' *J. Phys. Chem. B.* 110. 16179. (d) Albu, S. P., A. Ghicov, S. Aldabergenova, P. Drechsel, D. LeClere, G. E. Thompson, J. M. Macak, and P. Schmuki. 2008. 'Formation of TiO$_2$ nanotubes and robust anatase membranes.' *Adv. Mater.* 20. 4135. (e) Albu, S. P., D. Kim, and P. Schmuki. 2008. 'Growth of aligned TiO$_2$ bamboo-type nanotubes and highly ordered nanolace.' *Angew. Chem. Int. Ed.* 47. 1916. (f) Wang, D. A., Y. Liu, B. Yu, F. Zhou, and W. M. Liu. 2009. 'TiO$_2$ nanotubeswith tunable morphology, diameter and length: synthesis and photo-electrical/catalytic performance.' *Chem. Mater.* 21. 1198.

284. (a) Park, J., S. Bauer, K. Van der Mark, and P. Schmuki. 2007. 'Nanosize and vitality: TiO$_2$ nanotube diameter directs cell fate.' *Nano. Lett.* **7**. 1686,. (b) Park, J., S. Bauer, K. A. Schlegel, F. W. Neukam, K. Von der Mark, and P. Schmuki. 2009. 'TiO$_2$ nanotube surfaces: 15 nm-an optimal length scale of surface topography for cell adhesion and differentiation.' *Small.* 5. 666.

285. Oh, S., C. Daraio, L. H. Chen, T. R. Pisanic, R. R. Finönes, and S. J. Jin. 2006. 'Significantly accelerated osteoblasts cell growth on aligned TiO$_2$ nanotubes.' *Biomed. Mater. Res. A.* 78. 97.

286. Yerokhin, A. L., X. Nie, A. Leyland, A. Matthews, and S. J. Dowey. 1999. 'Plasma electrolysis for surface engineering.' *Surf. Coat. Technol.* 122. 73.

287. Han, Y., S. H. Hong, and K. W. Xu. 2002. 'Porous nanocrystalline titania coatings by plasma electrolytic oxidation.' *Surf. Coat. Technol.* 154. 314.

288. Arrabal, R., E. Matykina, F. Viejo, P. Skeldon, and G. E.Thompson. 2008. 'Corrosion resitance of WE43 and AZ91D magnesium alloys with phosphate PEO coatings.' *Corrosion Science.* 50 (6): 1744-1752.

289. Jin, F. Y., P. K. Chu, K. Wang, J. Zhao, A. P. Huang, and H. H. Tong. 2008. 'Thermal stability of titamia films prepared on titanium by micro-arc oxidation.' *Mater. Sci. Eng. A.* 476. 78.

290. Shin, Y. K., W. S. Chae, Y. W. Song, and Y. M. Sung. 2006. 'Formation of titania photocatalyst films by by micro-arc oxidation of Ti and Ti-6Al-4V alloys.' *Electrochem. Commun.* 8. 465.

291. Ceschini, L., E. Lanzoni, C. Martini, D. Prandstraller, and G. Sambogna. 2008. 'Comparison of dry sliding friction and wear of Ti6Al4V alloy treated by plasma electrolytic oxidation and PVD coating.' *Wear.* 264. 86.

292. Jun, Y. K., H. S. Kim, J. H. Lee, and S. H. Hong. 2006. 'Co sensing performance in micro-arc oxidized TiO$_2$ films for air quality control.' *Sensor Actuator B.* 120. 69.

293. Yang, B., M. Uchida, H. M. Kim, X. Zhang, and T. Kobuko. 2004. 'Preparation of bioactive titanium metal via anodic oxidation treatment.' *Bioamterials.* 25. 1003.

294. Song, W. H., Y. K. Jun, Y. Han, and S. H. Hong. 2004. 'Biomimetic apatite coatings on micro-arc oxidized titania.' *Biomaterials.* 25. 3341.

295. Sul, Y. T. 2003. 'The significance of the surface properties of oxidized titanium to the bone response: special emphasis on potential biochemical bonding of oxidized titanium implant.' *Biomaterials.* 24. 3893.

296. Li, L. H., Y. M. Kong, H. W. Kim, Y. W. Kim, H. E. Kim, S. J. Heo. and J.Y Koak 2004. 'Improved biological performance of Ti implants due to surface modification by micro-arc oxidation.' *Biomaterials.* 25. 2867.

297. (a) Li, Y., I. S. Lee, F. Z. Cui, and S. H. Choi. 2008. 'The biocompatibility of nanostructured calcium phosphate coated on micro-arc oxidized titanium.' *Biomaterials.* 29. 2025. (b) Wei, D. Q., Y. Zhou, D. C. Jia, Y. M. Wang. 2007. 'Characteristic and *in vitro* bioactivity of microarc-oxidized TiO$_2$-based coating after chemical treatment.' *Acta Biomater.* 3. 817. (c) Han, Y., G. H. Chen, J. F. Sun, Y. M. Zhang, and K. W. Xu. 2008. 'UV-enhanced bioactivity and cell response of micro-arc oxidized titania coatings.' *Acta Biomater.* 4. 1518.

298. (a) Yan, Y. Y., and Y. Han. 2007. 'Structure and activity of micro-arc oxidized zirconia films.' *Surf. Coat. Technol.* 201. 5692. (b) Bai, A., and Z. J. Chen. 'Effect of electrolyte additives on anti-corrosion

ability of micro-arc oxide coatings formed on magnesium alloy.' *Surf. Coat. Technol.* 203. 1956. (c) Yan, Y. Y., Y. Han, and C. G. Lu. 2008. 'The effect of chemical treatment on apatite- forming ability of macroporous zirconia films formed by micro-arc oxidation.' *Appl. Surf. Sci.* 254. 4833.

299. Piveteau, L. D. 2001. 'Sol-gel coatings.' In *Titanium in Medicine*, edited by D. M. Brunette, P. Tengvall, M. Textor and P. Thomson, 267-282. Berlin: Springer.

300. Brinker, C. J., A. J. Hurd, P. R. Schunk, G. C. Frye, and C. S. Ashley. 1992. 'Review of sol-gel thin-film formation.' *J. Non-Cryst. Solids.* 147. 424.

301. Acros, D., and M. Vallet-Regi. 2010. 'Sol-gel silica based biomaterials and bone tissue regeneration.' *Acta Biomater.* 6. 2874.

302. Advincula, M. C., F. G. Rahemtulla, R. C. Advincula, E. T. Ada, J. E. Lemons, and S. L. Bellis. 2006. 'Osteoblast adhesion and matrix mineralization on sol-gel-derived titanium oxide.' *Biomaterials.* 27. 2201.

303. He, G., J. Hu, S. C. Wei, J. H. Li, X. H. Liang, and E. Luo. 2008. 'Surface modification of titanium by nano-TiO_2/HA bioceramic coating.' *Appl. Sur. Sci.* 255. 442.

304. Jokinen, M., M. Patsi, H. Rahiala, T. Peltola, M. Ritala, and J. B. Rosenholm. 1998. 'Influence of sol and surface properties on *in vitro* bioactivity of sol–gel-derived TiO_2 and TiO_2–SiO_2 films deposited by dipcoating method.' *J. Biomed. Mater. Res.* 42. 295.

305. Kommireddy, D. S., S. M. Sriram, Y. M. Lvov, and D. K. Mills. 2006. 'Stem cell attachment to layer-by-layer assembled TiO2 nanoparticle thin films.' *Biomaterials.* 27. 4296.

306. Liu, J. X., D. Z. Yang, F. Shi, and Y. J. Cai. 2003. 'Sol-gel deposited TiO2 film on NiTi surgical alloy for biocompatibility improvement.' *Thin Solid Films.* 429. 225.

307. Eisenbarth, E., D. Velten, and J. Breme. 2007. 'Biomedical Implant coatings.' *Biomol. Eng.* 24. 27.

308. (a) Liang, B., C. Ding, H. Liao, and C. Coddet. 2006. 'Phase composition and stability of nanostructured 4.7 wt% yttria stabilized zirconia coatings deposited by atmospheric plasma spraying.' *Surf. Coat. Tech.* 200. 4549. (b) Liang, B. and C. Ding. 2005. 'Phase composition of nanostructured zirconia coatings deposited by air plasma spraying.' *Surf. Coat. Tech.* 191. 267. (c) Wang, G. C., X. Y. Liu, J. H. Gao, and C. X. Ding. 2009. '*in vitro* bioactivity and phase stability of plasma-sprayed nanostructured 3Y-TZP coatings.' *Acta Biomater.* 5. 2270.

309. Mattox, D. M. 1998. *Handbook of Physical Vapor Deposition (PVD) Processing: Film Formation, Adhesion, Surface Preparation and Contamination Control.* New Jersey: Noyes Publications.

310. (a) Grill, D. 2003. 'Diamond-like carbon coatings as biocompatible materials- an overview.' *Diam. Rel. Mater.* 12. 166. (b) Kwok, S. C. H., W. Zhang, G. J. Wan, D. R. McKenzie, M. M. M. Bilek, and P. K. Chu. 2007. 'Hemocompatibility and antibacterial properties of silver doped diamond-like carbon prepared by pulsed filtered cathodic arc deposition.' *Diam. Relat. Mater.* 16. 1353. (c) Xie, Y. T., X. B. Zheng, C. X. Ding, X. Y. Liu, and P. K. Chu. 2008. 'Mechanism of apatite formation on silicon suboxide film prepared by pulsed metal vacuum arc deposition.' *Mater. Chem. Phys.* 109. 342. (d) Liu, X. Y., A. P. Huang, C. X. Ding, and P. K. Chu. 2006. 'Bioactivity and cytocompatibility of zirconia (ZrO_2) films fabricated by cathodic arc deposition.' *Biomaterials.* 27. 3904. (e) Liu, X. Y., P. K. Chu, and C. X. Ding. 2007. 'Nanofilm and coating for biomedical application prepared by plasma-based technologies.' In: Ila D, J. Baglin, N. Kishimoto, and P. K. Chu, eds. *'Ion-Beam-Based-Nanofabrication'* *Materials Research Society Symposium Proceeding.* 81–89. (f) Cohen, A., P. Liu-Synder, D. Storey, and T. J. Webster. 2007. 'Decreased fibroblast and increased osteoblasts functions on on ionic plasma deposited nanostructured Ti coatings.' *Nanoscale Res. Lett.* 2. 385.

311. Huang, N., P. Yang, Y. X. Leng, J. Wang, H. Sun, J. Y. Chen, and G. J. Wan. 2004. 'Surface modification of biomaterials by plasma immersion ion implantation.' *Surf. Coat. Tech.* 186. 218.

312. Wan, Y. Z., S. Raman, F. He, and Y. Huang. 2007. 'Surface modification of medical metals by ion implantation of silver and copper.' *Vacuum.* 81. 1114.

313. Wang, J., J. X. Li, L. R. Shen, L. A. Ren, Z. J. Xu, A. S. Zhao, L. Yongxiang, and N. Huang. 2007. 'The biomedical properties of polyethylene terepthalate surface modified by silver ion implantation.' *Nucl. Instrum. Meth. Phys. Res. B.* 257. 141.

314. Zhang, W., Y. J. Luo, H. Y. Wang, S. H. Pu, and P. K. Chu. 2009. 'Biocompatibility of silver and copper plasma doped polyethylene.' *Surf. Coat. Tech.* 203. 2550.

315. Williams, O. A., M. Nesladek, M. Daenen, S. Michaelson, A. Hoffman, E. Osawa, K. Haenen., and R. B. Jackman. 2008. 'Growth, electronic properties and applications of nanodiamond.' *Diam. Relat. Mater.* 17. 1080.

316. Braga, N. A., C. A. A. Cairo, E. C. Almeida, M. R. Baldan, and N. G. Ferreiraa. 2008. 'From micro to nanocrystalline transition in the diamond formation on porous pure titanium.' *Diam. Relat. Mater.* 17. 1891.

317. Lee, S. W., B. K. Oh, R. G. Sanedrin, K. Salaita, T. Fujigaya, and C. A. Mirkin. 2006. 'Biologically active protein nanoarrays generated using parallel dip-pen nanolithography.' *Adv. Mater.* 18. 1133.

318. Yang, J. Y., Y, C. Ting, J. Y. Lai, H. L. Liu, H. W. Fang, and W. B. Tsai. 2008. 'Quantitative analysis of osteoblasts-like cells (MG63) morphology on nanogrooved substrata with various groove and ridge dimensions.' *J. Biomed. Mater. Res. A.* 90. 629.

319. Tsai, W. B., Y. C. Ting, J. Y. Yang, J. Y. Lai, and H. L. Liu. 2009. 'Fibronectin modulates the morphology of osteoblast-like cells (MG63) on nano-grooved substrates.' *J. Mater. Sci–Mater. M.* 20. 1367.

320. Khang, D., G. E. Park, and T. J. Webster. 2008. 'Enhanced chondrocyte densities on carbon nanotube composites: the combined role of annosurface roughness and electrical simulation.' *J. Biomed. Mater. Res. A.* 86. 253.

321. Reising, A., C. Yao, D. Storey, and T. J. Webster. 2008. 'Greater osteoblast long-term functions on ionic plasma deposited nanostructured orthopedic implant coatings.' *J. Biomed. Mater. Res.* 87A. 78.

322. Park, G. E., M. A. Pattison, K. Park, and T. J. Webster. 2005. 'Accelerated chondrocyte functions on NaOH-treated PLGA scaffolds.' *Biomaterials.* 26. 3075.

323. (a) Deeg, J. A., I. Louban, D. Aydin, C. Selhuber-Unkel, H. Kessler, and J. P. Spatz. 2011. 'Impact of local versus global ligand density on cellular adhesion.' *Nano. Lett.* 11. 1469. (b) Huang, J. H., S. V. Grater, F. Corbellinl, S. Rinck, E. Bock, R. Kemkemer, H. Kessler, J. Ding, and J. P. Spatz. 2009. 'Impact of order and disorder in RGD nanopatterns on cell adhesion.' *Nano. Lett.* 9. 1111. (c) Valsesia, A., I. Mannelli, P. Colpo, F. Bretagnol, and F. Rossi. 2008. 'Protein nanopatterns for improved immunodetection sensitivity.' *Anal. Chem.* 80. 7336.

324. Malmstrom, J., B. Christensen, H. P. Jakobse, J. Lovmand, R. Foldbjerg, E. S. Sorensen, and D. S. Sutherland. 2010. 'Large area protein patterning reveals nanoscale control of focal adhesion development.' *Nano. Lett.* 10. 686.

325. Zhang, J., Y. Li, X. Zhang, and B. Yang. 2010. 'Colloidal self-assembly meets nanofabrication: from two-dimensional colloidal crystals to nanostructure arrays.' *Adv. Mater.* 22. 4249.

326. Gonuguntla, M., A. Sharma, and S. A. Subramanian. 2006. 'Elastic Contact Induced Self-Organized Patterning of Hydrogel Films.' *Macromolecules.* 39. 3365.

327. Singh, G., S. Pillai, A. Arpanaei, and P. Kingshott. 2011. 'Highly-ordered mixed protein patterns over large areas from self-assembly of binary colloids.' *Adv. Mater.* 23. 1519.

328. Kosiorek, A., W. Kandulski, H. Glaczynska, and M. Giersig. 2005. 'Fabrication of nanoscale rings, dots, and rods by combining shadow nanosphere lithography and annealed polystyrene nanosphere masks.' *Small.* 1. 439.

329. Ogaki, R., F. Lyckegaard, and P. Kingshott. 2010. 'High resolution surface chemical analysis of a trifunctional pattern made by sequential colloidal shadowing.' *Chem. Phys. Chem.* 11. 3609.

330. Singh, G., H. J. Griesser, K. Bremmell, and P. Kingshott. 2011. 'Highly-ordered nanopatterns by plasma polymerisation through masks of self-assembled binary colloid crystals.' *Adv. Funct. Mater.* 21. 540.

331. Wright, J. P., O. Worsfold, C. Whitehouse, and M. Himmelhaus. 2006. 'Ultraflat ternary nanopatterns fabricated using colloidal lithography.' *Adv. Mater.* 18. 421.

332. Ogaki, R., M. A. Cole, D. S. Sutherland, and P. Kingshott. 2011. 'Micro-cup array patterns of four chemical regions with nanoscale precision.' *Adv. Mater.* 23. 1876.

333. (a) Feng, X. F., J. Zhang, H. K. Xie, Q. H. Hu, Q. Huang, and W. W. Liu. 2003. 'The RF plasma polymer of lysine and the growth of human nerve cells on its surface.' *Surf. Coat. Tech.* 171. 96. (b) Yuan, Y., C. S. Liu, Y. Zhang, M. Yin, and C. O. Shi. 2007. 'Radio frequency plasma polymers of n-butyl methacrylate and their controlled drug release characteristics: I. 'Effect of the oxygen gas.' *Surf. Coat. Tech.* 201. 6861.

334. Zhang, W., Y. J. Luo, H. Y. Wang, J. Jiang, S. H. Pu, P. K. Chu. 2008. 'Ag and Ag/N2 plasma modification of polyethylene for enhancement of antimicrobial properties and cell growth/proliferation.' *Acta Biomater.* 4. 2028.

335. Siow, K. S., L. Britcher, S. Kumar, and H. J. Griesser. 2006. 'Plasma methods for generation of chemically reactive surfaces for biomolecule immobilization and cell colonization- A review.' *Plasma Process Polym.* 3. 392.

336. Teixidor, G. T., R. A. Gorkin III, P. P. Tripathi, G. S. Bisht, M. Kulkarni, T. K. Maiti, T. K. Bhattacharyya, J. R. Subramanium, A. Sharma, B. Y. Park, and M. Madou. 2008. 'Carbon microelectromechanical systems as a substratum for cell growth.' *Biomed. Mater.* 3. 34116.

337. (a) Shin, M., H. Yoshimoto, and J. P. Vacanti. 2004. '*In vivo* Bone Tissue Engineering Using Mesenchymal Stem Cells on a Novel Electrospun Nanofibrous Scaffold.' *Tissue. Eng.* 10. 33. (b) Murugan, R. and S. Ramakrishna. 2006. 'Nano-featured scaffolds for tissue engineering: a review of spinning methodologies.' *Tissue Eng.* 12. 436.

338. Jayawarna, V., M. Ali, T. A. Jowitt, A. F. Miller, A. Saiani, J. E. Gough, and R. V. Ulijn. 2006. 'Nanostructured hydrogels for three-dimensional cell culture through self-assembly of fluorenylmethoxycarbonyl-dipeptides.' *Adv. Mater.* 18. 611.

339. (a) Woo, K. M., V. J. Chen, and P. X. Ma. 2003. 'Nano-fibrous scaffolding architecture selectively enhances protein adsorption contributing to cell attachment.' *J. Biomed. Mater. Res. A.* 67. 531. (b) Mart, R. J., R. D. Osborne, M. M. Stevens, and R. V. Ulijn. 2006. 'Peptide-based stimuli-responsive biomaterials.' *Soft Matter.* 2. 822. (c) Stevens, M. M., S. Allen, M. C. Davies, C. J. Roberts, J. K. Sakata, S. J. B. Tendler, D. A. Tirrell, and P. M. Williams. 2005. 'Molecular level investigations of the inter- and intramolecular interactions of pH responsive artificial triblock proteins.' *Biomacromolecules.* 6. 1266. (d) Yang, Z. M. and B. Xu. 2007. 'Conjugates of naphthalene and dipeptides confer molecular hydrogelators with high efficiency of hydrogelation and superhelical nanofibers.' *J. Mater. Chem.* 17. 850. (e) Zhang, S. 2003. 'Fabrication of novel biomaterials through molecular self-assembly.' *Nat. Biotechnol.* 21. 1171. (f) Ryadnov, M. G. and D. N. Woolfson. 2003. 'Engineering the morphology of a self-assembling protein fiber.' *Nat. Mater.* 2. 329.

340. Li, W. J., R. Tuli, X. Huang, P. Laquerriere, and R. S. Tuan. 2005. 'Multilineage differentiation of human mesenchymal stem cells in a three-dimensional nanofibrous scaffold.' *Biomaterials.* 26. 5158.

341. Li, W. J., C. T. Laurencin, E. J. Caterson, R. S. Tuan, and F. K. Ko. 2002. 'Electrospun nanofibrous structure: a novel scaffold for tissue engineering.' *J. Biomed. Mater. Res.* 60. 613.

342. Bonzani, I. C., J. H. George, and M. M. Stevens. 2006. 'Novel materials for bone and cartilage regeneration.' *Curr. Opin. Chem. Biol.* 10. 568.

343. (a) Benoit, D. S. and K. S. Anseth. 2005. 'The effect on osteoblast function of colocalized RGD and PHSRN epitopes on PEG surfaces.' *Biomaterials.* 26. 5209. (b) Keselowsky, B. G., D. M. Collard, and A. J. Garcia. 2005. 'Integrin binding specificity regulates biomaterial surface chemistry effects on cell differentiation.' *Proc. Natl. Acad. Sci. USA.* 102. 5953. (c) Maheshwari, G., G. Brown, D. A. Lauffenburger, A. Wells, and L. G. Griffith. 2000. 'Cell adhesion and motility depend on RGD motif

clustering.' *J. Cell. Sci.* 113. 1677. (d) Palecek, S. P. 1997. 'Integrin-ligand bonding properties govern cell migration speed.' *Nature.* 385. 537.

344. Bökel, C. and N. Brown. 2002. 'Integrins in development: moving on, responding to, and sticking to the extracellular matrix.' *Dev. Cell.* 3. 311.

345. Lutolf, M. P., J. L. Lauer-Fields, H. G. Schmoekel, A. T. Metters, F. E. Weber, G. B. Fields, and J. A. Hubbell. 2003. 'Synthetic matrix metalloproteinase-sensitive hydrogels for the conduction of tissue regeneration: engineering cell-invasion characteristics.' *Proc. Natl. Acad. Sci. USA.* 100. 5413.

346. (a) Benoit, D. S. W., A. R. Durney, and K. S. Anseth. 2007. 'The effect of heparin-functionalized PEG hydrogels on three-dimensional human mesenchymal stem cell osteogenic differentiation.' *Biomaterials.* 28. 66. (b) Casper, C. L., N. Yamaguchi, K. L. Kiick, and J. F. Rabolt. 2005. 'Functionalizing electrospun fibers with biologically relevant macromolecules.' *Biomacromolecules.* 6. 1998. (c) Yamaguchi, N. and K. L. Kiick. 2005. 'Polysaccharide-poly (ethylene glycol) star copolymer as a scaffold for the production of bioactive hydrogels.' *Biomacromolecules.* 6. 1921.

347. Jiang, T., Y. Khan., L. S. Nair, W. I. Abdel-Fattah, and C. T. J. Laurencin. 2010. 'Functionalization of chitosan/poly(lacticacid-glycolic acid) sintered microsphere scaffolds via surface heparinization for bone tissue engineering.' *J. Biomed. Mater. Res. A.* 93. 1193.

348. Singh, S., B. M. Wu, and J. C. Y. Dunn. 2011. 'The enhancement of VEGF-mediated angiogenesis by polycaprolactone scaffolds with surface cross-linked heparin.' *Biomaterials.* 32. 2059.

349. Sun, X. Y., R. Shankar, H. G. Börner, T. K. Ghosh, and R. J. Spontak. 2007. 'Field driven biofunctionalization of polymer fiber surfaces during electrospinning.' *Adv. Mater.* 19. 87.

350. Korzhikov, V. A., E. G. Vlakh, and T. B. Tennikova. 2012. 'Polymers in orthopedic surgery and tissue engineering: from engineering materials to smart biofunctionalization of a surface.' *Polym. Sci. Series A.* 54. 585.

351. (a) Anselme, K. 2000. 'Osteoblast adhesion on biomaterials.' *Biomaterials.* 21. 667. (b) Ito, Y. 1999. 'Surface micropatterning to regulate cell functions.' *Biomaterials.* 20. 2333. (c) Tan, J., H. Shen, and W. M. Saltzman. 2001. 'Micron-scale positioning of features influences the rate of polymorphonuclear leukocyte migration.' *Biophys. J.* 81. 2569.

352. (a) Wilkinson, C. D. W., M. Riehle, M. Wood, J. Gallagher, and A. S. G. Curtis. 2002. 'The use of materials patterned on a nanoand micro-metric scale in cellular engineering.' *Mater. Sci. Eng. C.* 19. 263. (b) Curtis, A., and C. Wilkinson. 1997. 'Topographical control of cells.' *Biomaterials.* 18. 1573.

353. Anselme, K., P. Davidson, A. M. Popa, M. Giazzon, M. Liley, and L. Ploux. 2010. 'The interaction of cells and bacteria with surfaces structured at the nanometre scale.' *Acta Biomater.* 6. 3824.

354. Yang-Tse, C. and D. E. Rodak. 2005. 'Is the lotus leaf superhydrophobic.' *Appl. Phys. Lett.* 86. 144101.

355. Martines, E., K. Seunarine, H. Morgan, N. Gadegaard, and C. D. W. Wilkinson. 2005. 'Superhydrophobicity and superhydrophilicity of regular nanopatterns.' *Nano. Lett.* 5. 2091.

356. Bhushan, B., D. R. Tokachichu, M. T. Keener, and S. L. Lee. 2005. 'Nanoscale adhesion, friction and wear suited of biomolecules on silicon based surfaces.' *Acta Biomater.* 2. 39.

357. Selhuber, C., J. Blümmel, F. Czerwinski, and J. P. Spatz. 2006. 'Tuning surface energies with nanopatterned substrates.' *Nano. Lett.* 6. 267.

358. Decuzzi, P. and M. Ferrari. 2010. 'Modulating cellular adhesion through nanotopography.' *Biomaterials.* 31. 173.

359. Jain, S., A. Sharma, and B. Basu. '*In vitro* cytocompatibility assessment of amorphous carbon structures using neuroblastoma and Schwann cells.' *J. Biomed. Mater. Res. Part B : Applied Biomaterials.* 104. 520-531. (in Press, 2012).

360. Kemkemer, R., S. Jungbauer, D. Kaufmann, and H. Gruler. 2006. 'Cell orientation by a microgrooved substrate can be predicted by automatic control theory.' *Biophys. J.* 90. 4701.

361. Risken H. 1984. *The Fokker-Planck Equation.* Heidelberg, Germany: Springer. 1–110.

362. Meyer, U., A. Büchter, H. P. Wiesmann, U. Joos, and D. B. Jones. 2005. 'Basic reactions of osteoblasts on structured material surfaces.' *Eur. Cells. Mater.* 9. 39.

363. Elwing, H. 1998. 'Protein absorption and ellipsometry in biomaterial research.' *Biomaterials.* 19. 397.

364. Vroman, L. 1987. 'The importance of surfaces in contact phase reactions.' *Semin. Thromb. Hemost.* 13. 79.

365. (a) Rezania, A. and K. E. Healy. 2006. 'The effect of peptide surface density on mineralization of a matrix deposited by osteogenic cells.' *J. Biomed. Mater. Res.* 52. 595. (b) Globus, R. K., S. B. Doty, J. C. Lull, E. Homulhamedov, M. J. Humphries, and C. H. Dmasky. 1998. 'Fibronectin is a survival factor for differentiated osteoblasts.' *J. Cell. Sci.* 111. 1385. (c) Bentmann, A. K., N. Kawelke, D. Moss, H. Zentgraf, Y. Bala, R. Berger, J. A. Gasser, and I. A. Nakchbandi. 2009. 'Circulating fibronectin affects bone matrix, whereas osteoblasts fibronectin modulates osteoblast function.' *J. Bone. Miner. Res.* 25. 706.

366. (a) Kantlehner, M., P. Scahffner, D. Finsinger, J. Meyer, A. Jonczyk, B. Diefenbach, B. Nies, G. Hölzemann, S. L. Goodman, and H. Kessler. 2000. 'Surface coating with cyclic RGD peptides stimulates osteoblasts adhesion and proliferation as well as bone formation.' *Chem. BioChem.* 1. 107. (b) Siebers, M. C., P. J. Ter Brugge, X. F. Walboomers, and J. A. Jansen. 2005. 'Integrins as linker proteins between osteoblasts and bone replacing materials. A critical review.' *Biomaterials.* 26. 137. (c) Horii, A., A. Wang, F. Gelain, and S. Zhang. 2007. 'Biological designer self-assembling peptide nanofibers scaffolds significantly enhance osteoblast proliferation, differentiation and 3D migration.' *PLoS One.* 2. e190.

367. Dalby, M. J., L. Di Silvio, G. W. Davies, and W. Bonfield. 2000. 'Surface topography and HA filler volume effect on primary human osteoblasts *in vitro*.' *J. Mat. Sci.: Mater. Med.* 11. 805.

368. Cooper, L. F. 2000. 'A role for surface topography in creating and maintaining bone at titanium endosseous implants.' *J. Prosthet. Dent.* 84. 522.

369. Webster, T. J., C. Ergun, R. H. Doremus, R. W. Siegel, and R. Bizios. 2000. 'Enhanced functions of osteoblasts on nanophase ceramics.' *Biomaterials.* 21. 1803.

370. Bettinger, C. J., R. Langer, and J. T. Borenstein. 2009. 'Engineering substrate topography at the micro- and nano scale to control cell function.' *Angew. Chem. Int. Ed.* 48. 5406.

371. Lenhert, S., M. B. Meier, U. Meyer, L. Chi, and H. P. Wiesmann. 2004. 'Osteoblast alignment, elongation and migration on grooved polystyrene patterned by Langmuir-Blodgett lithography.' *Biomaterials.* 26. 563.

372. Petit, V. and J. P. Thiery. 2000. 'Focal adhesion: Structure and dynamics.' *Biol. Cell.* 92. 477.

373. Cavalcanti-Adam, E. A., A. Micoulet, J. Blümmel, J. Auernheimer, H. Kessler, and J. P. Spatz. 2006. 'Lateral spacing of integrin ligands influences cell spreading and focal adhesion assembly.' *Eur. J. Cell. Biol.* 85. 219.

374. Dalby, M. J., N. Gadegaard, A. S. G. Curtis, and R. O. C. Oreffo. 2007. 'Nanotopographical control of human osteoprogenitor differentiation.' *Curr. Stem. Cell. Res. Ther.* 2. 129.

375. Fujita, S., M. Ohshima, and H. Iwata. 2009. 'Time-lapse observation of cell alignment on nanogrooved patterns.' *J. R. Soc. Interface.* 6. S269.

376. Wang, J. H., B. P. Thampatty, J. S. Lin, and H. J. Im. 2007. 'Mechanoregulation of gene expression in fibroblasts.' *Gene.* 391. 1.

377. Forgacs, G. 1995. 'On the possible role of cytoskeletal filamentous networks in intracellular signaling: an approach based on percolation.' *J. Cell. Sci.* 108. 2131.

378. Ingber, D. E. 1997. 'Tensegrity: the architectural basis of cellular mechanotransduction.' *Annu. Rev. Physiol.* 59. 575.

379. (a) Dalby, M. J. 2005. 'Topographically induced direct cell mechanotransduction.' *Med. Eng. Phys.* 27. 730. (b) Dalby, M. J., M. O. Riehle, D. S. Sutherland, H. Agheli, and A. S. Curtis. 2004. 'Use of nanotopography to study mechanotransduction in fibroblasts – methods and perspectives.' *Eur. J. Cell.*

Biol. 83. 159. (c) Dalby, M. J., M. J. Biggs, N. Gadegaard, G. Kalna, C. D. Wilkinson, and A. S. Curtis. 2007. 'Nanotopographical stimulation of mechanotransduction and changes in interphase centromere positioning.' *J. Cell. Biochem.* 100. 326.

380. Frey, M. T., I. Y. Tsai, T. P. Russell, S. K. Hanks, and Y. L. Wang. 2006. 'Cellular responses to substrate topography: role of myosin II and focal adhesion kinase.' *Biophys. J.* 90. 3774.

381. Wojciak-Stothard, B., A. Curtis, W. Monaghan, K. MacDonald, and C. Wilkinson. 1996. 'Guidance and activation of murine macrophages by nanometric scale topography.' *Exp. Cell. Res.* 223. 426.

382. Dalby, M. J., D. Pasqui, and S. Affrossman. 2004. 'Cell response to nano-islands produced by polymer demixing: a brief review.' *IEEE Proc. Nanobiotechnol.* 151. 53.

383. Dalby, M. J., M. O. Riehle, H. Johnstone, S. Affrossman, and A. S. Curtis. 2004. 'Investigating the limits of filopodial sensing: a brief report using SEM to image the interaction between 10 nm high nano-topography and fibroblast filopodia.' *Cell. Biol. Int.* 28. 229.

384. Lim, J. Y., A. D. Dreiss, Z. Zhou, J. C. Hansen, C. A. Siedlecki, R. W. Hengstebeck, J. Cheng, N. Winograd, and H. J. Donahue. 2007. 'The regulation of integrin-mediated osteoblast focal adhesion and focal adhesion kinase expression by nanoscale topography.' *Biomaterials.* 28. 1787.

385. Washburn, N. R., K. M. Yamada, C. G. Simon Jr, S. B. Kennedy, and E. J. Amis. 2004. 'High throughput investigation of osteoblast response to polymer crystallinity: influence of nanometer-scale roughness on proliferation.' *Biomaterials.* 25. 1215.

386. (a) de Oliveira, P. T., S. F. Zalzal, M. M. Beloti, A. L. Rosa, and A. Nanci. 2007. 'Enhancement of *in vitro* osteogenesis on titanium by chemically produced nanotopography.' *J. Biomed. Mater. Res. A.* 80. 554. (b) Swan, E. E. L., K. C. Popat, C. A. Grimes, and T. A. Desai. 2005. 'Fabrication and evaluation of nanoporous alumina membranes for osteoblast culture.' *J. Biomed. Eng.* 72. 288. (c) Peng, L., M. L. Eltgroth, T. J. LaTempa, C. A. Grimes, and T. A. Desai. 2009. 'The effect of TiO2 nanotubes on endothelial function and smooth muscle proliferation.' *Biomaterials.* 30. 1268. (d) Richert, L., F. Vetrone, J. H. Yi, S. F. Zalzal, J. D. Wuest, F. Rosei, and A. Nanci. 2008. 'Surface nanopatterning to control cell growth.' *Adv. Mater.* 20. 1488. (e) Porter, J. R., A. Henson, and K. C. Popat. 2009. 'Biodegradable poly(epsilon-caprolactone) nanowires for bone tissue engineering applications.' *Biomaterials.* 30. 780. (f) Oh, S., C. Daraio, L. H. Chen, T. R. Pisanic, R. R. Finones, and S. Jin. 2006. 'Significantly accelerated osteoblast cell growth on aligned TiO2 nanotubes.' *J. Biomed. Mater. Res. A.* 78A. 97.

387. Webster, T. J., L. Schadler, R. W. Siegel, and R. Bizios. 2001. 'Mechanisms of enhaced osteoblast aadhesion on nanophase alumina involve vitronectin.' *Tissue Eng.* 7. 291.

388. Brammer, K. S., S. Oh, C. J. Cobb, L. M. Bjursten, J. van der Heyde, and S. Jin. 2009. 'Improved bone-forming functionality on diameter-controlled TiO2 nanotube surface.' *Acta Biomater.* 5. 3215.

389. Oh, S., K. S. Brammer, Y. S. J. Li, D. Teng, A. J. Engler, S. Chien, and S. Jin. 2009. 'Stem cell fate dictated solely by altered nanotube dimension.' *Proc. Natl. Acad. Sci. USA.* 106. 2130.

390. Andersson, A. S., J. Brink, U. Lidberg, and D. S. Sutherland. 2003. 'Influence of systematically varied nanoscale topography on the morphology of epithelial cells.' *IEEE Trans. Nanobiosci.* 2. 49.

391. Biggs, M. J. P. S., R. G. Richards, and M. J. Dalby. 2010. 'Nanotopographical modification: a regulator of cellular function through focal adhesions.' *Nanomedicine.* 6. 619.

392. (a) Matsuzaka, K., X. F. Walboomers, M. Yoshinari, T. Inoue, and J. A. Jnasen. 2003. 'The attachment and growth behavior of osteoblasts-like cells on microtextured surfaces.' *Bioamterials.* 24. 2711. (b) Wang, J. H., E. S. Grood, J. Floorer, and R. Wenstrup. 2000. 'Alignment and proliferation of MC3T3-E1 osteoblasts in microgrooved silicone substrata subjected to cyclic stretching.' *J. Biomech.* 33. 729.

393. Teixeira, A. I., G. A. Abrams, P. J. Bertics, C. J. Murphy, and P. F. Nealey. 2003. 'Epithelial contact guidance on well-defined micro- and nanostructured substrates.' *J. Cell. Sci.* 116. 1881.

394. Andersson, A. S., P. Olsson, U. Lidberg, and D. Sutherland. 2003. 'The effects of continuous and discontinuous groove edges on cell shape and alignment.' *Exp. Cell. Res.* 288. 177.

395. Dalby, M. J., D. McCloy, M. Robertson, H. Agheli, D. Sutherland, S. Affrossman, and R. O. Oreffo. 2006. 'Osteoprogenitor response to semi-ordered and random nanotopographies.' *Biomaterials.* 27. 2980.

396. Lim, J. Y., J. C. Hansen, C. A. Siedlecki, R. W. Hengstebeck, J. Cheng, N. Winograd, and H. J. Donahue. 2005. 'Osteoblast adhesion on poly(L-lactic acid)/polystyrene demixed thin film blends: effect of nanotopography, surface chemistry and wettability.' *Biomacromolecules.* 6. 3319.

397. Biggs, M. J. P., R. G. Richards, N. Gadegaard, R. J. McMurray, S. Affrossman, C. D. W. Wilkinson, R. O. Oreffo, and M. J. Dalby. 2009. 'Interactions with nanoscale topography: adhesion quantification and signal transduction in cells of osteogenic and multipotent lineage.' *J. Biomed. Mater. Res. A.* 91A. 195.

398. Biggs, M. J., R. G. Richards, N. Gadegaard, C. D. Wilkinson, R. O. Oreffo, and M. J. Dalby. 2009. 'The use of nanoscale topography to modulate the dynamics of adhesion formation in primary osteoblasts and ERK/MAPK signalling in STRO-1+ enriched skeletal stem cells.' *Biomaterials.* 30. 5094.

399. Biggs, M. J. P., R. G. Richards, S. McFarlane, C. D. W. WilkinsonR. O. C. Oreffo, and M. J. Dalby. 2008. 'Adhesion formation of primary human osteoblasts and the functional response of mesenchymal cells to 330 nm deep microgrooves.' *J. R. Soc. Interface.* 5. 1231.

400. Rice, J. M., J. A. Hunt, J. A. Gallagher, P. Hanarp, D. S. Sutherland, and J. Gold. 2003. 'Quantitative assessment of the response of primary derived human osteoblasts and macrophages to a range of nanotopography surfaces in a single culture model *in vitro*.' *Biomaterials.* 24. 4799.

401. Kunzler, T. P., C. Huwiler, T. Drobek, J. Voros, and N. D. Spencer. 2007. Systematic 'study of osteoblast response to nanotopography by means of nanoparticle-density gradients.' *Biomaterials.* 28. 5000.

402. Walter, N., C. Selhuber, H. Kessler, and J. P. Spatz. 2006.' Cellular unbinding forces of initial adhesion processes on nanopatterned surfaces probed with magnetic tweezers.' *Nano. Lett.* 6. 398.

403. Arnold, M., E. A. Cavalcanti-Adam, R. Glass, J. Blümmel, W. Eck, M. Kantlehner, H. Kessler, and J. P. Spatz. 2004. 'Activation of integrin function by nanopatterned adhesive interfaces.' *Chem. Phys. Chem.* 5. 383.

404. Arnold, M., V. C. Hirschfeld-Warneken, T. Lohmuller, P. Heil, J. Blummel, E. A. Cavalcanti-Adam, M. Lopez-garcia, P. Weather, H. Kessler, B. Geiger, and J. P. Spatz. 2008. 'Induction of cell polarization and migration by a gradient of nanoscale variations in adhesive ligand spacing.' *Nano. Lett.* 8. 2063.

405. Bigerelle, M., K. Anselme, B. Noël, I. Ruderman, P. Hardouin, and A. Iost. 2002. 'Improvement in the morphology of surfaces for cell adhesion: a new process to double human osteoblast adhesion on Ti-based substrates.' *Biomaterials.* 23. 1563.

406. (a) Hart, A., N. Gadegaard, C. D. W. Wilkinson, R. O. C. Oreffo, and M. J. Dalby. 2007. 'Osteoprogenitor response to low-adhesion nanotopographies originally fabricated by electron beam lithography.' *J. Mater. Sci. Mater. Med.* 18. 1211. (b) Biggs, M. J. P., R. G. Richards, N. Gadegaard, C. D. W. Wilkinson, and M. J. Dalby. 2007. 'The effects of nanoscale pits on primary human osteoblast adhesion formation and cellular spreading.' *J. Mater. Sci. Mater. Med.* 18. 399.

407. Möller, K., U. Meyer, D. H. Szulczewski, H. Heide, B. Priessnitz, and D. B. Jones. 1994. 'The influence of zeta potential and and interfacial tension on osteoblast-like cells.' *Cell. Mater.* 4. 263.

408. Redey, S. A., M. Nardin, D. Bernache-Assolant, C. Rey, P. Delannoy, L. Sedel, and P. J. Marie. 2000. 'Behavior of human osteoblastic cells on stoichiometric hydroxyapatite and type A carbonate apatite: role of surface energy.' *J. Biomed. Mater. Res.* 50. 353.

409. Ishikawa, K., T. Akasaka, S. Abe, Y. Yawaka, M. Suzuki, and F. Watari. 2011. 'Application of imogolite, almino-silicate nanotube as scaffold for the mineralization of osteoblasts.' *Bioceramics Development and Applications.* 1. 1.

410. dos Santos E. A., M. Farina, G. A. Soares, and K. Anselme. 2009. 'Chemical and topographical influence of hydroxyapatite and beta-tricalcium phosphate surfaces on human osteoblastic cell behavior.' *J. Biomed. Mater. Res. A.* 89. 510.

411. di Silvio, L., M. J. Dalby, and W. Bonfield. 2002. 'Osteoblast behaviour on HA/PE composite surfaces with different HA volumes.' *Biomaterials.* 23. 101.

412. Lehnert, D., B. Wehrle-Haller, C. David, U. Weiland, C. Ballestrem, B. A. Imhof, and M. Bastmeyer. 2004. 'Cell behaviour on micropatterned substrata: limits of extracellular matrix geometry for spreading and adhesion.' *J. Cell. Sci.* 117. 41.

413. Slater, J. H., and W. Frey. 2008. 'Nanopatterning of fibronectin and the influence of integrin clustering on endothelial cell spreading and proliferation.' *J. Biomed. Mater. Res. A.* 87A. 176.

414. Mitchell, E. A., B. T. Chaffey, A. W. McCaskie, J. H. Lakey, and M. A. Birch. 2010. 'Controlled spatial and conformational display of immobilized bone morphogenetic protein-2 and osteopontin signaling motifs regulates osteoblast adhesion and differentiation *in vivo*.' *BMC Biol.* 8. 57.

415. Berry, C. C., A. S. G. Curtis, R. O. C. Oreffo, H. Agheli, and D. S. Sutherland. 2007. 'Human fibroblast and human bone marrow cell response to lithographically nanopatterned adhesive domains on protein rejecting substrates.' *IEEE Trans. Nanobiosci.* 6. 201.

416. Arnold, M., M. Schwieder, J. Blummel, E. A. Cavalcanti-Adam, M. Lopez-Garcia, H. Kessler, B. Geiger, and J. P. Spatz. 2009. 'Cell interactions with hierarchically structured nanopatterned adhesive surfaces.' *Soft. Matter.* 5. 72.

417. Lord, M. S., M. Foss, and F. Besenbacher. 2010. 'Influence of nanoscale surface topography on protein adsorption and cellular response.' *Nano. Today.* 5. 66.

418. Elmengaard, B., J. E. Bechtlod, and K. Soballe. 2005. '*In vivo* study of the effect of RGD treatment on bone ongrowth on press-fit titanium alloy implants.' *Biomaterials.* 26. 3521.

419. Kane, R. S., and A. D. Stroock. 2007. 'Nanobiotechnology: protein–nanomaterial interaction.' *Biotechnol. Prog.* 23. 316.

420. Yap, F. L., and Y. Zhang. 2007. 'Protein and cell micropatterning and its integration with micro/nanoparticles assembly.' *Biosens. Bioelectron.* 22. 775.

421. Denis, F. A., P. Hanarp, D. S. Sutherland, J. Gold, C. Mustin, P. G. Rouxhet, and Y. F. Dufrene. 2002. 'Protein adsorption on model surfaces with controlled nanotopography and chemistry.' *Langmuir.* 18. 819.

422. Wilson, C. J., R. E. Clegg, D. I. Leavesley, and M. J. Pearcy. 2005. 'Mediation of biomaterial–cell interactions by adsorbed proteins: a review.' *Tissue Eng.* 11. 1.

423. Meyer, U., H. P. Wiesmann, T. Fillies, and U. Joos. 2003. 'Early tissue reaction at the interface of immediate loaded dental implants.' *Int. J. Oral Maxillofac. Impl.* 18. 489.

424. Büchter, A., J. Kleinheinz, H. P. Wiesmann, J. Kersken, U. Joos, and U. Meyer. 2005. 'Biological and biomechanical evaluation of bone remodelling and implant stability after using an osteotome technique.' *Clin. Oral. Impl. Res.* 16. 1.

425. Frost, H. M. 2000. 'The Utah paradigm of skeletal physiology: An overview of its insights for bone, cartilage and collagenous tissue organs.' *J. Bone. Miner. Metab.* 18. 305.

426. Spadaro, J. A. 1997. 'Mechanical and electrical interactions in bone remodeling.' *Bioelectromagnetics.* 18. 193.

427. (a) Piekarski K and M. Munro. 1977. 'Transport mechanism operating between blood supply and osteocytes in long bones.' *Nature.* 269. 80. (b) Rubin, J., C. Rubin, and C. R. Jacobs. 2006. 'Molecular pathways mediating mechanical signaling in bones.' *Gene.* 367. 1.

428. (a) Taylor, A. F., M. M. Saunders, D. M. Singhle, J. M. Cimbala, Z. Zhou, and H. J. Donahue. 2007. 'Mechanically stimulated osteocytes regulate osteoblastic activity via gap junctions.' *Am. J. Physiol. Cell. Physiol.* 292. C545. (b) Robling, A. G., K. M. Duijvelaar, J. V. Geevers, N. Ohashi, and C. H. Turner. 2001. 'Modulation of appositional and longitudinal bone growth in the rat ulna by applied static and dynamic force.' *Bone.* 29. 105. (c) Saunders, M. M., L. A. Simmerman, G. L. Reed, N. A. Sharpley, and A. F. Taylor. 2010. 'Biomimetic bone mechanotransduction modeling in neonatal rat femur organ cultures: structural verification of proof of concept.' *Biomech. Model Mechanobiol.* 9. 539.

429. Büchter, A., D. Wiechmann, S. Koerdt, H. P. Wiesmann, J. Piffko, and U. Meyer. 2005. 'Load related implant reaction of mini-implants used for orthodontic anchorage.' *Clin. Oral Impl. Res.* 16. 473.

430. Jacobs, C. R., S. Temiyasathit, and A. B. Castillo. 2010. 'Osteocyte mechanobiology and pericellular mechanics.' *Annu. Rev. Biomed. Eng.* 12. 369.

431. Jones, D. B., E. Broeckmann, T. Pohl, and E. L. Smith. 2003. 'Development of mechanical testing and loading system for trabercular bone studies for long term cultures.' *Eur. Cells. Mater.* 5. 48.

432. Knothe Tate, M. L., R. Steck, and E. J. Anderson. 2009. 'Bone as an inspiration for a novel class of mechanoactive materials.' *Biomaterials.* 30. 133.

433. Yang, Y., J. L. Magnay, and L. Cooling. 2002. 'Development of a "mechano-active" scaffold for tissue engineering.' *Biomaterials.* 23. 2119.

434. Büchter, A., D. Wiechmann, S. Koerdt, H. P. Wiesmann, J. Piffko, and U. Meyer. 2005. 'Load related implant reaction of mini-implants used for orthodontic anchorage.' *Clin. Oral. Impl. Res.* 16. 473.

435. (a) Hartig, M., U. Joos, and H. P. Wiesmann. 2000. 'Capacitively coupled electric field accelerate proliferation of osteoblasts-like primary cells and increase bone extracellular matrix formation *in vitro.*' *Eur. Biophys. J.* 7. 499. (b) Markx, G. H. 2008. 'The use of electric fields in tissue engineering.' *Organogenesis.* 4. 11.

436. Hartig, M., U. Joos, and H. P. Wiesmann. 2000. 'Capacitively coupled electric fields accelerate proliferation of osteoblast-like primary cells and increase bone extracellular matrix formation *in vitro.*' *Eur. Biophys. J.* 29. 499.

437. Wiesmann, H., M. Hartig, U. Stratmann, U. Meyer, and U. Joos. 2001. 'Electrical stimulation influences mineral formation of osteoblast-like cells *in vitro.*' *Biochim. Biophys. Acta.* 1538. 28.

438. Ba, X., M. Hadjiargyrou, E. di Masi, Y. Meng, and M. Simon. 2011. 'The role of moderate static magnetic fields on biomineralization of osteoblasts on sulfonated polystyrene fields.' *Biomaterials.* 32. 7831.

439. Polte. T. R., M. Shen, J. Karavitis, M. Montoya, J. Pendse, S. Xia, E. Mazur, and D. E. Ingber. 2007. 'Nanostructured magnetizable materials that switch cells between life and death.' *Biomaterials.* 28. 2783.

440. (a) Sharma, S., and T. A. Desai. 2005. 'Nanostructured antifouling poly(ethylene glycol) films for silicon-based microsystems.' *J. Nanosci. Nanotechnol.* 5, 235. (b) Dusseiller, M. R., D. Schlaepfer, M. Koch, R. Kroschewski, and M. Textor. 2005. 'An inverted microcontact printing method on topographically structured polystyrene chips for arrayed micro-3D culturing of single cells.' *Biomaterials.* 26. 5917. (c) Lensen, M. C., P. Mela, A. Mourran, J. Groll, J. Heuts, H. T. Rong, and M. Möller. 2007. 'Micro- and nanopatterned star poly(ethylene glycol) (PEG) materials prepared by UVbased imprint lithography.' *Langmuir.* 23. 7841. (d) Blummel, J., N. Perschmann, D. Aydin, J. Drinjakovic, T. Surrey, M. Lopez-Garcia, H. Kessler, and J. P. Spatz. 2007. 'Protein repellent properties of covalently attached PEG coatings on nanostructured SiO_2-based interfaces.' *Biomaterials.* 28. 4739. (e) Epstein, A. K., T. S. Wong, R. A. Belisle, E. M. Boggs, and J. Aizenber. 2012. 'Liquid infused structured surfaces with exceptional anti-biofouling perfoemance.' *Proc. Natl. Acad. Sci. USA.* 2012, doi: 10.1073/pnas.1201973109.

441. Linneweber, J., P. M. Dohmen, U. Kertzscher, K. Affeld, Y. Nose, and W. Konertz. 2007. 'The effect of surface roughness on activation of the coagulation system and platelet adhesion in rotary blood pumps.' *Artif. Organs.* 31. 345.

442. (a) Velve-Casquillas, G., M. Le Berre, M. Piel, and P. T. Tran. 2010. 'Microfluidic tools for cell biological research.' *Nano. Today.* 5. 28. (b) Didar, T. F., and M. Tabrizian. 2010. 'Adhesion based detection, sorting and enrichment of cellsin microfluidic Lab-on-Chip devices.' *Lab. Chip.* 10. 3043. (c) Kim, S. M., S. H. Lee, and K. H. Suh. 2008. 'Cell research with physically modified microfluidic channels: A review.' *Lab. Chip.* 8. 1015. (d) Kwon, K. W., S. S. Choi, S. H. Lee, B. Kim, S. N. Lee, M. C. Park., P. Kim, S.Y. Hwang, and K. Y. Suh. 2007. 'Label-free, microfluidic separation and enrichment of human breast cancer cells by adhesion difference.' *Lab. Chip.* 7. 1461.

443. Hench, L. L, and J. Wilson. 1984. 'Surface-active biomaterials.' *Science.* 226. 630.

444. Mitra, J., G. Tripathi, A. Sharma, and B. Basu. 2013. 'Scaffolds for bone tissue engineering: role of surface patterning on osteoblast response.' *RSC Advances.* 3. 11073.

445. Stevens, M. M. 2008. 'Biomaterials for bone tissue engineering.' *Mater. Today.* 11. 18.

446. Tripathi, G. and B. Basu. 2012. 'B. A porous hydroxyapatite scaffold for bone tissue engineering: Physico-mechanical and biological evaluations.' *Ceramics Int.* 38. 341.

447. Chen, Q. Z., and A. R. Boccaccini. 2006. 'Poly(D,L-lactic acid) coated 45S5 Bioglass-based scaffolds for bone engineering.' *J. Biomed. Mater. Res. A.* 77. 445.

448. George, A. and S. Ravindran. 2010. 'Protein templates in hard tissue engineering.' *Nano. Today.* 5. 254.

449. Wu, S. L., X. M. Liu, T. Hu, P. K. Chu, J. P. Y. Ho, Y. L. Chan, K. W. Yeung, C. L. Chu, T. F. Hung, K. F. Huo, C. Y. Chung, W. W. Lu, K. M. Cheung, and K. D. Luk. 2008. 'A biomimetic hierarchical scaffold: Natural growth of nanotitanates on three-dimensional microporous Ti-based metals.' *Nano. Lett.* 8. 3803.

450. Zhao, B. H., I. S. Lee, I. H. Han, J. C. Park, and S. M. Chung. 2007. 'Effects of surface morphology on human osteosarcoma cell response.' *Curr. Appl. Phys.* 7S1. e6.

451. Yang, L., B. W. Sheldon, and T. J. Webster. 2009. 'The impact of diamond nanocrystallinity on osteoblasts functions.' *Biomaterials.* 30. 3458.

452. (a) Janson, I. A. and A. J. Putnam. 2015. 'Extracellular matrix elasticity and topography: Material-based cues that affect cell function via conserved mechanisms.' *J. Biomed. Mater. Res. Part A.* 103A. 1246–1258. (b) Woodard J. R., A. J. Hilldore, S. K. Lan, C. J. Park, A. W. Morgan, J. A. C, Eurell, S. G. Clark, M. B. Wheeler, R. D. Jamison, and A. J. W. Johnson. 2007. 'The mechanical properties and osteoconductivity of hydroxyapatite bone scaffolds with multi-scale porosity.' *Biomaterials.* 28. 45–54.

Chapter 4

453. Basu, Bikramjit. 2017. *Biomaterials for Musculoskeletal regeneration: Concepts.* Springer Nature. Germany

454. Petrofes N. F., and A. M. Gadalla. 1988. 'Processing aspects of shaping advanced materials by electrical discharge machining.' *Adv. Mater. Manuf. Process.* 3(1). 127–153.

455. Lauwers, B., J. P. Kruth, W. Liu, W. Eeraerts, B. Schcht, and P. Bleys. 2004. 'Investigation of material removal mechanisms in EDM of ceramic composite materials.' *J. Mater. Proc. Tech.* 149. 347–352.

456. Sanchez, J. A., I. Cabanes, L. N. Lopez de Lacalle, and A. Lamikiz. 2001. 'Development of optimum electro discharge machining technology for advanced ceramics.' *Int. J. Adv. Manuf. Tech.* 18(12). 897–905.

457. Zhao, C., Vleugels, J., Groffils, C., Luypaert, P.J. and Van der Biest, O., 2000. Hybrid sintering with a tubular susceptor in a cylindrical single-mode microwave furnace. Acta materialia, 48(14), 3795-3801.

458. Langer, J., M. J. Hoffmann, and O. Guillon. 2009. 'Direct comparison between hot pressing and electric field-assisted sintering of submicron alumina.' *Acta. Mater.* 57. 5454–5465.

459. Bernard-Granger, G., A. Addad, G. Fantozzi, G. Bonnefont, C. Guizard, and D. Vernat. 2010. 'Spark plasma sintering of a commercially available granulated zirconia powder: Comparison with hot-pressing.' *Acta. Mater.* 58. 3390–3399.

460. Shen, Z., Z. Zhao, H. Peng, and M. Nygren. 2002. *Nature.* 417. 266–268.

461. Kumar, R., K. H. Prakash, P. Cheang, and K. A. Khor. 2005. 'Microstructure and mechanical properties of spark plasma sintered zirconia-hydroxyapatite nano-composite powders.' *Acta. Mater.* 53. 2327–2335.

462. Orru, R., R. Licheri, A. M. Locci, A. Cincotti, and G. Cao. 2009. 'Consolidation/synthesis of materials by electric current activated/assisted sintering.' *Mater. Sci. Eng. R.* 63. 127–287.

463. Aman, Y., V. Garnier, and E. Djurado. 2009. 'Influence of green state processes on the sintering behaviour and the subsequent optical properties of spark plasma sintered alumina.' *J. Eur. Ceram. Soc.* 29[16]. 3363–3370.

464. Venkateswaran, T., B. Basu, G. B. Raju, and D. Y. Kim 2006. 'Densification and properties of transition metal borides-based cermets via spark plasma sintering.' *J. Eur. Ceram. Soc.* 26. 2331–2340.

465. Booy, M. L. 1978. *Polymer Engineering and Science*. 18(12).

466. Erdmenger, R. June 7, 1966. US Patent 3,254,367.

467. Padmanabhan, B. 'Melting and Mixing Capability Enhancement: Technology Revealed.' Aug 31, 2004. US Patent 6.783,270.

468. Tadmor, Z., and C. G. Gogos. 1979. Chapter 9, Section 9.1 'Modeling of Processing Machines Using Elementary Steps.' In *Principles of Polymer Processing*. New York: Wiley-Interscience.

469. Vasireddi, R. and B. Basu, 2015. 'Conceptual design of three-dimensional scaffolds of powder-based materials for bone tissue engineering applications.' *Rapid Prototyp. J.* 21. 716–724.

470. Coopers, P.W. 2014. '3D printing and the new shape of industrial manufacturing.'

471. Lantada, A. D., and P. L. Morgado. 2012. *Annu. Rev. Biomed. Eng.* 14. 73–96.

472. Yeong, W.-Y., C.-K. Chua, K.-F. Leong, and M. Chandrasekaran. 2004. 'Rapid prototyping in tissue engineering: challenges and potential.' *Trends Biotechnol.* 22. 643–652.

473. US Food and Drug Administration. 2013. 'Paving the way for personalized medicine: FDA's role in a new era of medical product development.' Silver Spring, MD: US Food and Drug Administration.

474. Zhao, S., M. Zhu, J. Zhang, Y. Zhang, Z. Liu, Y. Zhu, and C. Zhang. 2014. 'Three dimensionally printed mesoporous bioactive glass and poly (3-hydroxybutyrate-co-3-hydroxyhexanoate) composite scaffolds for bone regeneration.' *J. Mater. Chem. B.* 2. 6106–6118.

475. Klammert, U., U. Gbureck, E. Vorndran, J. Rödiger, P. Meyer-Marcotty, and A. C. Kübler. 2010. '3D powder printed calcium phosphate implants for reconstruction of cranial and maxillofacial defects.' *J. Cranio-Maxillofacial Surg.* 38. 565–570.

476. Wang, J., M. Yang, Y. Zhu, L. Wang, A.P. Tomsia, and C. Mao. 2014. 'Phage nanofibers induce vascularized osteogenesis in 3D printed bone scaffolds.' *Adv. Mater.* 26. 4961–4966.

477. Kloxin, A. M., A. M. Kasko, C. N. Salinas, and K. S. Anseth. 2009. 'Photodegradable hydrogels for dynamic tuning of physical and chemical properties.' *Science.* 324. 59–63.

478. M .A. Markets. 2013. 'Additive Manufacturing Market, by Application, Forecast 2012–2017, Markets and Markets, 2013.' Accessed 8 March 2017. Available at http://www.marketsandmarkets.com/PressReleases/ additive-manufacturing.asp

479. Peltola, S. M., F. P. W. Melchels, D. W. Grijpma, and M. Kellomäki. 2008. 'A review of rapid prototyping techniques for tissue engineering purposes.' *Ann. Med.* 40. 268–280.

480. Webb, P. A. 2000. 'A review of rapid prototyping (RP) techniques in the medical and biomedical sector.' *J. Med. Eng. Technol.* 24. 149–153.

481. Kolesky, D. B. R. L. Truby, A. S. Gladman, T. A. Busbee, K. A. Homan, and J. A. Lewis 2014. '3D bioprinting of vascularized, heterogeneous cell-laden tissue constructs.' *Adv. Mater.* 26. 3124–3130.

482. Liu, F.-H. 2014. 'Fabrication of bioceramic bone scaffolds for tissue engineering.' *J. Mater. Eng. Perform.* 23. 3762–3769.

483. Landers, R., A. Pfister, U. Hübner, H. John, R. Schmelzeisen, and R. Mülhaupt, 2002. 'Fabrication of soft tissue engineering scaffolds by means of rapid prototyping techniques.' *J. Mater. Sci.* 37. 3107–3116.

484. Huang, T., X. Qu, J. Liu, and S. Chen. 2014. '3D printing of biomimetic microstructures for cancer cell migration.' *Biomed. Microdevices.* 16. 127–132.

485. Hutmacher, D. 2000. 'Scaffolds in tissue engineering bone and cartilage.' *Biomaterials.* 21. 2529.

486. Sachlos, E. and J. Czernuszka. 2003. 'Making tissue engineering scaffolds work. Review: the application of solid freeform fabrication technology to the production of tissue engineering scaffolds.' *Eur. Cells Mater.* 5. 29–39.

487. Abdullah, J. and A. Hassan. 2005. *Bone Grafts Bone Subst.* 34. 547–561.

488. Garrett, B. 2014. '3D printing: new economic paradigms and strategic shifts.' *Global Policy.* 5. 70–75.

489. Mironov, V., T. Boland, T. Trusk, G. Forgacs, and R. Markwald. 2003. 'Organ printing: computer-aided jet-based 3D tissue engineering.' *Trends Biotechnol.* 21. 157–161.

490. Peltola, S. M., F. P. W. Melchels, D. W. Grijpma, and M. Kellomäki. 2008. 'A review of rapid prototyping techniques for tissue engineering purposes.' *Ann. Med.* 40. 268–280.

491. Derby, B. 2012. 'Printing and prototyping of tissues and scaffolds.' *Science.* 338. 921–926.

492. Sun, F. G. W. 2013. *Biofabrication, Micro- and Nano-fabrication, Printing, Patterning and Assemblies (Micro and Nano Technologies).* Amsterdam: Elsevier. 288.

493. Gaytan, S. M., L. E. Murr, E. Martinez, J. L. Martinez, B. I. Machado, D. A. Ramirez, F. Medina, S. Collins, and R. B. Wicker. 2010. 'Comparison of microstructures and mechanical properties for solid and mesh cobalt-base alloy prototypes fabricated by electron beam melting.' *Metall. Mater. Trans. A.* 41. 3216–3227.

494. Murr, L. E. S. A. Quinones, S. M. Gaytan, M. I. Lopez, A. Rodela, E. Y. Martinez, D. H. Hernandez, E. Martinez, F. Medina, and R. B. Wicker. 2009. 'Microstructure and mechanical behavior of Ti–6Al–4V produced by rapid-layer manufacturing, for biomedical applications.' *J. Mech. Behav. Biomed. Mater.* 2. 20–32.

495. Palmquist, A., A. Snis, L. Emanuelsson, M. Browne, and P. Thomsen. 2011. 'Long-term biocompatibility and osseointegration of electron beam melted, free-form–fabricated solid and porous titanium alloy: Experimental studies in sheep.' *J. Biomater. Appl.* 27(8). 1003-1016.

496. Hernandez, J., S. J. Li, E. Martinez, L. E. Murr, X. M. Pan, K. N. Amato, X. Y. Cheng, F. Yang, C. A. Terrazas, and S. M. Gaytan. 2013. 'Microstructures and hardness properties for β-phase Ti–24Nb–4Zr–7.9 Sn alloy fabricated by electron beam melting.' *J. Mater. Sci. Technol.* 29. 1011–1017.

497. Luca, F., M. Emanuele, R. Pierfrancesco, M. Alberto, H. Simon, and W. Konrad. 2010. 'Ductility of a Ti-6Al-4V alloy produced by selective laser melting of prealloyed powders.' *Rapid Prototyping J.* 16. 450–459.

498. Thijs, L., F. Verhaeghe, T. Craeghs, J. V. Humbeeck, and J. P. Kruth. 2010. 'A study of the microstructural evolution during selective laser melting of Ti–6Al–4V.' *Acta Mater.* 58. 3303–3312.

499. Butscher, A., M. Bohner, C. Roth, A. Ernstberger, R. Heuberger, N. Doebelin, P. Rudolf von Rohr, and R. Müller. 2012. 'Printability of calcium phosphate powders for three-dimensional printing of tissue engineering scaffolds.' *Acta Biomater.* 8. 373–385.

500. Calvert, P. 2001. 'Inkjet printing for materials and devices.' *Chem. Mater.* 13. 3299–3305.

501. Hutmacher, D. 2000. 'Scaffolds in tissue engineering bone and cartilage.' *Biomaterials.* 21. 2529.

502. Derby, B. 2010. 'Inkjet printing of functional and structural materials: fluid property requirements, feature stability, and resolution.' *Annu. Rev. Mater. Res.* 40. 395–414.

503. Lanzetta, M. and E. Sachs. 2003. 'Improved surface finish in 3D printing using bimodal powder distribution.' *Rapid Prototyping J.* 9. 157–166.

504. Butscher, A., M. Bohner, S. Hofmann, L. Gauckler, and R. Müller. 2011. 'Structural and material approaches to bone tissue engineering in powder-based three-dimensional printing.' *Acta Biomater.* 7. 907–920.

505. Lam, C. X. F., X. Mo, S.-H. Teoh, and D. Hutmacher. 2002. *Mater. Sci. Eng.: C.* 20. 49–56.

506. Khalyfa, A., S. Vogt, J. Weisser, G. Grimm, A. Rechtenbach, W. Meyer, and M. Schnabelrauch. 2007. 'Development of a new calcium phosphate powder-binder system for the 3D printing of patient specific implants.' *J. Mater. Sci. Mater. Med.* 18. 909–916.

507. Lanzetta, M., and E. Sachs. 2003. 'Improved surface finish in 3D printing using bimodal powder distribution.' *Rapid Prototyping J.* 9. 157–166.

508. Landers, R. and R. Mülhaupt. 2000. 'Desktop manufacturing of complex objects, prototypes and biomedical scaffolds by means of computer-assisted design combined with computer-guided 3D plotting of polymers and reactive oligomers.' *Macromol. Mater. Eng.* 282. 17–21.

509. Derby, B. 2010. 'Inkjet printing of functional and structural materials: fluid property requirements, feature stability, and resolution.' *Annu. Rev. Mater. Res.* 40. 395–414.

510. Stringer, J. and B. Derby. 2009. 'Limits to feature size and resolution in ink jet printing.' *J. Eur. Ceram. Soc.* 29. 913–918.

511. Derby, B. 2008. 'Bioprinting: inkjet printing proteins and hybrid cell-containing materials and structures.' *J. Mater. Chem.* 18. 5717–5721.

512. Setti, L., A. Fraleoni-Morgera, B. Ballarin, A. Filippini, D. Frascaro, and C. Piana. 2005. 'An amperometric glucose biosensor prototype fabricated by thermal inkjet printing.' *Biosens. Bioelectron.* 20. 2019–2026.

513. Sen, A. K., and J. Darabi. 2007. 'Droplet ejection performance of a monolithic thermal inkjet print head.' *J. Micromech. Microeng.* 17. 1420–1427.

514. Saunders, R. E. and B. Derby. 2014. 'Inkjet printing biomaterials for tissue engineering: bioprinting.' *Int. Mater. Rev.* 59 (2014) 430–448.

515. Zhou, Z., F. Buchanan, C. Mitchell, and N. Dunne. 2014. 'Printability of calcium phosphate: calcium sulfate powders for the application of tissue engineered bone scaffolds using the 3D printing technique.' *Mater. Sci. Eng.: C.* 38. 1–10.

516. Bredt, J. F. 1998. ' Binder Composition for Use in Three Dimensional Printing.' Accessed April 19, 2017. https://www.google.com/patents/US5851465.

517. Gbureck, U., T. Holzel, I. Biermann, J. E. Barralet, and L. M. Grover. 2008. 'Preparation of tricalcium phosphate/calcium pyrophosphate structures via rapid prototyping.' *J. Mater. Sci. Mater. Med.* 19. 1559–1563.

518. Moseke, C., and U. Gbureck. 2010. 'Tetracalcium phosphate: Synthesis, properties and biomedical applications.' *Acta Biomater.* 6. 3815–3823.

519. Moseke, C., M. Gelinsky, J. Groll, and U. Gbureck. 2013. 'Chemical characterization of hydroxyapatite obtained by wet chemistry in the presence of V, Co, and Cu ions.' *Mater. Sci. Eng.: C.* 33. 1654–1661.

520. Vorndran, E., U. Klammert, M. Klarner, L. M. Grover, J. E. Barralet, and U. Gbureck. 2009. *Dent. Mater.* 25. e18–e19.

521. Lowmunkong, R., T. Sohmura, Y. Suzuki, S. Matsuya, and K. Ishikawa. 2009. 'Fabrication of freeform bone-filling calcium phosphate ceramics by gypsum 3D printing method.' *J. Biomed. Mater. Res. B: Appl. Biomater.* 90B. 531–539.

522. Suwanprateeb, J., F. Thammarakcharoen, K. Wasoontararat, and W. Suvannapruk. 2012. 'Influence of printing parameters on the transformation efficiency of 3D-printed plaster of paris to hydroxyapatite and its properties.' *Rapid Prototyping J.* 18. 490–499.

523. Pelham, R. J. and Y. 1. Wang. 1997. 'Cell locomotion and focal adhesions are regulated by substrate flexibility.' *Proc. Natl. Acad. Sci. U.S.A.* 94. 13661–13665.

524. Butscher A., M. Bohner, S. Hofmann, L. Gauckler, and R. Müller. 2011. 'Structural and material approaches to bone tissue engineering in powder-based three-dimensional printing.' *Acta Biomater.* 7. 907–920.

525. Luo, Y., A. Lode, F. Sonntag, B. Nies, and M. Gelinsky. 2013. 'Well-ordered biphasic calcium phosphate–alginate scaffolds fabricated by multi-channel 3D plotting under mild conditions.' *J. Mater. Chem. B.* 1. 4088–4098.

526. Landers, R., U. Hübner, R. Schmelzeisen, and R. Mülhaupt. 2002. 'Rapid prototyping of scaffolds derived from thermoreversible hydrogels and tailored for applications in tissue engineering.' *Biomaterials.* 23. 4437–4447.

527. Pfister, A., R. Landers, A. Laib, U. Huebner, R. Schmelzeisen, and R. Muelhaupt. 2004. 'Biofunctional rapid prototyping for tissue-engineering applications: 3D bioplotting versus 3D printing.' *J. Polym. Sci., A: Polym. Chem.* 42. 624–638.

528. Sun, L., S. T. Parker, D. Syoji, X. Wang, J. A. Lewis, and D. L. Kaplan. 2012. 'Direct-Write Assembly of 3D Silk/Hydroxyapatite Scaffolds for Bone Co-Cultures.' *Adv. Healthcare Mater.* 1. 729–735.

529. Franco, J., P. Hunger, M. Launey, A. Tomsia, and E. Saiz. 2010. *Acta Biomater.* 6. 218– 228.

530. De Clerck, J. P., and J. Prosthet. 1987. 'Microwave polymerization of acrylic resins used in dental prostheses.' Dentistry. 57. 650–658.

531. Tarafder, S., V. K. Balla, N. M. Davies, A. Bandyopadhyay, and S. Bose. 2013. 'Microwave-sintered 3D printed tricalcium phosphate scaffolds for bone tissue engineering.' *J. Tissue Eng. Regener. Med.* 7. 631–641.

532. Kang, S. J. L. 2004. *Sintering: Densification, Grain Growth and Microstructure,* 1st edition. Butterworth-Heinemann- Burlington.

533. Basu, B., D. S. Katti, and A. Kumar. September, 2009. 'Advanced Biomaterials: Fundamentals, Processing, and Applications.' Hokoben, NJ, USA: Wiley. 776.

534. Suk, M., S. Choi, J. Kim, Y. Kim, and Y. K. YS. 2003. 'Fabrication of a porous material with a porosity gradient by a pulsed electric current sintering process.' *Met. Mater. Int.* 9. 599–603.

535. Warnke, P. H., H. Seitz, F. Warnke, S. T. Becker, S. Sivananthan, E. Sherry, Q. Liu, J. Wiltfang, and T. Douglas. 2010. 'Ceramic scaffolds produced by computer-assisted 3D printing and sintering: Characterization and biocompatibility investigations.' *J. Biomed. Mater. Res.* 93B. 212–217.

536. Yadoji, P., R. Peelamedu, D. Agrawal, and R. Roy. 2003. 'Microwave sintering of Ni–Zn ferrites: comparison with conventional sintering.' *Mater. Sci. Eng.: B.* 98. 269– 278.

537. Cornelsen, M., S. Petersen, K. Dietsch, A. Rudolph, K. Schmitz, K. Sternberg, and H. Seitz. 2013. 'Infiltration of 3D printed tricalciumphosphate scaffolds with biodegradable polymers and biomolecules for local drug delivery.' *Biomed. Technol.* 58. 4090–4091.

538. Brooke, L. F., D. B. Toby, U. Zee, W. H. Dietmar, D. D. Paul, and R. D. Tim. 2013. 'Dermal fibroblast infiltration of poly (ε-caprolactone) scaffolds fabricated by melt electrospinning in a direct writing mode.' *Biofabrication.* 5. 025001.

539. Gbureck, U., O. Grolms, J. E. Barralet, L. M. Grover, and R. Thull. 2003. 'Mechanical activation and cement formation of β-tricalcium phosphate.' *Biomaterials.* 24. 4123–4131.

540. Gbureck, U., T. Hölzel, U. Klammert, K. Würzler, F. A. Müller, and J. E. Barralet. 2007. 'Resorbable dicalcium phosphate bone substitutes prepared by 3D powder printing.' *Adv. Funct. Mater.* 17. 3940–3945.

541. Meininger, S., S. Mandal, A. Kumar, J. Groll, B. Basu, and U. Gbureck. 2015. *Acta Biomater.* 31. 401–411.

542. Tiwari, D., B. Basu, and K. Biswas. 2009. 'Simulation of thermal and electric field evolution during spark plasma sintering.' *Ceramics International.* 35. 699–708.

543. Kumar, A., S. Mandal, S. Barui, R. Vasireddi, U. Gbureck, M. Gelinsky, and B. Basu. 2016. 'Low temperature additive manufacturing of three dimensional scaffolds for bone-tissue engineering applications: Processing related challenges and property assessment.' *Materials Science and Engineering R.* 103. 1–39.

Chapter 5

544. Chu, T. M. G., D. G. Orton, S. J. Hollister, S. E. Feinberg, and J. W. Halloran. 2003. *Biomaterials.* 23. 1283–1293.

545. Kujala, S., J. Ryhanen, A. Danilov, and J.Tuukkanen. 2003. 'Effect of porosity on the osseointegration and bone ingrowth of a weight-bearing nickel–titanium bone graft substitute.' *Biomaterials.* 24. 4691–4697.

546. Van Lenthe, G. H., H. Hagenmuller, M. Bohner, S. J. Hollister, L. Meinel, and R. Muller. 2007. 'Nondestructive micro-computed tomography for biological imaging and quantification of scaffold–bone interaction *in vivo.*' *Biomaterials.* 28. 2479–2490.

547. Ho, S., and D. Hutmacher. 2006. 'A comparison of micro CT with other techniques used in the characterization of scaffolds.' *Biomaterials.* 27. 1362–1376.

548. Van Lenthe, G. H., H. Hagenmüller, M. Bohner, S. J. Hollister, L. Meinel, and R. Müller. 2007. 'Nondestructive micro-computed tomography for biological imaging and quantification of scaffold–bone interaction *in vivo*.' *Biomaterials*. 28. 2479–2490.

549. Basu, B., A. Sabareeswaran, and S. J. Shenoy. 2015. *Journal of Biomedical Materials Research Part B: Applied Biomaterials*. 103. 1168–1179.

550. Butscher, A., M. Bohner, C. Roth, A. Ernstberger, R. Heuberger, N. Doebelin, P. Rudolf von Rohr, and R. Müller. 2012. 'Printability of calcium phosphate powders for three-dimensional printing of tissue engineering scaffolds.' *Acta Biomaterialia*. 8. 373–385.

551. Jin, Q. M., H. Takita, T. Kohgo, K. Atsumi, H. Itoh, and Y. Kuboki. 2000. 'Effects of geometry of hydroxyapatite as a cell substratum in BMP-induced ectopic bone formation.' *Journal of Biomedical Materials Research*. 52. 841–851.

552. Wang, F., L. Shor, A. Darling, S. Khalil, W. Sun, and S. Guceri. 2004. 'Multi-nozzle deposition for construction of 3D biopolymer tissue scaffolds.' *Rapid Prototyping Journal*. 10. 42–44.

553. Hanlon, E. B., R. Manoharan, T-W. Koo, K. E. Shafer, J. T. Motz, M. Fitzmaurice, J. R. Kramer, I. Itzkan, R. R. Dasari, and M. S. Feld. 2000. 'Prospects for *in vivo* Raman spectroscopy.' *Phys. Med. Biol.* 45. R1–R59.

554. David, B. Williams, and C. Barry Carter. 2009. *Transmission Electron Microscopy: A Textbook for Materials Science*. Springer.

555. Harry van Lenthe, G., Henri Hagenmüller, Marc Bohner, Scott J. Hollister, Lorenz Meinel, and Ralph Müller. 2007. 'Nondestructive micro-computed tomography for biological imaging and quantification of scaffold–bone interaction *in vivo*.' *Biomaterials*. 28. 2479–2490.

556. Wertheim, Gunther K. 1964. *Mossbauer effect: principle and applications*. Academic Press.

557. Royal Society of Chemistry. 2017. 'Introduction to Mössbauer Spectroscopy: Part 2.' Accessed 13 April 2017. Available at http://www.rsc.org/Membership/Networking/InterestGroups/MossbauerSpect/part2.asp.

For Further Reading

a. Van der Heide, Paul. 2011. *X-Ray Photoelectron Spectroscopy: An Introduction to Principles and Practices*. John Wiley & Sons Inc.

b. Williams, David B., and C. Barry Carter. 2009. *Transmission Electron Microscopy: A Textbook for Materials Science*. Springer.

c. Cullity, B. D. 1978. *Elements of X-ray Diffraction*. Addison-Wesley.

d. (i) Reimer, Ludwig. 1998. *Scanning Electron Microscopy: Physics of Image Formation and Microanalysis*. Springer.

(ii) Goldstein, J., Dale E. Newbury, D. C. Joy, C.E. Lyman, P. Echlin, E. Lifshin, L. Sawyer, and J.R. Michael. 2003. *Scanning Electron Microscopy and X-Ray Microanalysis*. Springer.

e. Haugstad, G. 2012. *Atomic Force Microscopy: Understanding Basic Modes and Advanced Applications*. John Wiley & Sons Inc.

f. Pawley, J. 2006. *Handbook of Biological Confocal Microscopy*, Springer.

g. Kubitscheck, U. 2013. *Fluorescence Microscopy: From Principles to Biological Applications*. Wiley-Blackwell.

h. Sharma, V. K., G. Klingelhofer, and T. Nishida. 2013. *Mössbauer Spectroscopy: Applications in Chemistry, Biology, and Nanotechnology*. John Wiley & Sons Inc.

i. Smith, E., and Dent G. 2005. *Modern Raman Spectroscopy: A Practical Approach*. John Wiley & Sons, Ltd.

j. Keeler, J. 2005. *Understanding NMR Spectroscopy, 2nd edition*. John Wiley & Sons Ltd.

k. Brian C. Smith. 2011. *Fundamentals of Fourier Transform Infrared Spectroscopy*, CRC press (Taylor & Francis).

l. Barsoukov, E., and J. R. Macdonald. 2005. *Impedance Spectroscopy: Theory, Experiment, and Applications.* John Wiley & Sons.

Chapter 6

558. Basu, Bikramjit. 2017. *Biomaterials for Musculoskeletal regeneration: Concepts.* Germany: Springer Nature.

559. Basu, Bikramjit, and Kantesh Balani. 2011. *Advanced Structural Ceramics.* John Wiley & Sons.

560. Basu, Bikramjit, and Mitjan Kalin. 2011. *Tribology of Ceramics and composites: Materials Science Perspective.* John Wiley & Sons.

561. Inglis, C. E. 1913. 'Stresses in a plate due to the presence of cracks and sharp corners.' *Trans. Inst. Nav. Archit.* 55. 219–241.

562. Orowan, E. 1952. In 'Fatigue and Fracture of Metals', Symposium at Massachusetts Institute of Technology, USA, John Wiley & Sons, Inc., NewYork.

563. Wiederhorn, S. M. 1968. 'Moisture assisted crack growth in ceramics.' *The International Journal of Fracture Mechanics.* 4(2). 171–177.

564. Ovri, J., and T. Davies. 1987. *Materials Science and Engineering.* 96. 109–116.

565. Purwar, A., T. Venkateshwaran, and Bikramjit Basu. 2017. 'Experimental and computational analysis of thermo-structural stability of ZrB_2-SiC-Ti composite during Arc-jet testing.' *Journal of American Ceramic Society* (In press, DOI:10.1111/jace.15001)

566. Swarnakar, A. K., O. Van Der Biest, and B. Baufeld. 2011. 'Young's modulus and damping in dependence on temperature of Ti–6Al–4V components fabricated by shaped metal deposition.' *J. Mater. Sci.* 46. 3802.

567. Swarnakar, A. K., A. Stamboulis, D. Holland, and O. Van Der Biest. 2013. 'Improved Prediction of Young's Modulus of Fluorine-Containing Glasses Using MAS-NMR Structural Data.' *J. Am. Ceram. Soc.* 96(4). 1271–1277.

568. Roebben, G., B. Basu, J. Vleugels, J. Van Humbeeckand, and O. Van Der Biest, 2000. 'The effect of the electrical properties on the pulsed electric current sintering behavior of ZrO_2 based ceramic composites.' *J. Alloys and Compounds.* 310(1–2). 284–287.

569. Guicciardi, S., A. K. Swarnakar, O. Van Der Biest, and D. Sciti. 2010. 'Temperature dependence of the dynamic Young's modulus of ZrB_2–$MoSi_2$ ultra-refractory ceramic composites.' *Scripta Materialia.* 62. 831–834.

570. Swarnakar, A. K., L. Donzel, J. Vleugels, and O. Van Der Biest. 2009. 'High temperature properties of ZnO ceramics studied by the impulse excitation technique.' *Journal of European Ceramic Society.* 29. 2991–2998.

571. (a) Standard Test Method for Dynamic Young's Modulus, Shear Modulus and Poisson's Ratio for Advanced Ceramics by Impulse Excitation of Vibration ASTM Designation. June, 1995. C1259–95. 375. (b) Carr, J., Milhet, X., Gadaud, P., Boyer, S.A.E., Thompson, G.E., and Lee, P., 2015. 'Quantitative characterization of porosity and determination of elastic modulus for sintered micro-silver joints.' *Journal of Materials Processing Technology.* 225. 19-23.

572. Choi, S. R., and N. P. Bansal. 2005. 'Mechanical behavior of zirconia/alumina composites.' *Ceram. Inter.* 31. 39–46.

573. Mizuno, M., and H. Okuda. 1995. 'VAMAS round robin on fracture toughness of silicon nitride.' *J. Am. Ceram. Soc.* 78. 1795–1801.

574. Anstis, G. R., P. Chantikul, B. R. Lawn, and D. B. Marshall. 1981. 'A critical evaluation of indentation techniques for measuring fracture toughness: I, direct crack measurements.' *J. Am. Cer. Soc.* 64. 533–538.

575. Kaliszewski, M. S., G. Behrens, A. H. Heuer, M. C. Shaw, D. B. Marshall, G. W. Dransmann, and R. W. Steinbrech. 1994. 'Indentation Studies on Y_2O_2-Stabilized ZrO_2: I, Development of Indentation-Induced Cracks.' *J. Am. Cer. Soc.* 77(5). 1185–1193.

576. Evans, A. G., and T. R. Wilshaw. 1976. 'Quasi-static solid particle damage in brittle solids—I. Observations analysis and implications.' *Acta Metallurgica.* 24(10). 939–956.

577. Niihara, K., R. Morena, and D. P. H. Hasselman. 1982. 'Evaluation of K Ic of brittle solids by the indentation method with low crack-to-indent ratios.' *J. Mater. Sci. Lett.* 1. 13–16.

578. (a) Shetty, D. K., I. G. Wright, P. N. Mincer, and A. H. Heuer. 1985. 'Indentation fracture of WC-Co cermets.' *J. Mat. Sc.* 20. 1873–1882. (b) Purwar, A., Ragini Mukherjee, Krishnamurthy Ravikumar, S. Ariharan, Nagarajan Kirupakaran Gopinath, and Bikramjit Basu. 2016. 'Development of ZrB_2-SiC-Ti by Multi Stage Spark Plasma Sintering at 1600°C, *Journal of the Ceramic Society of Japan.* 124 (4). 393-402.

579. Dieter, George E. 1986. *Mechanical Metallurgy, 3rd edition.* Mcgraw-Hill Series in Mechanical Engineering, McGraw-Hill Education.

580. (a) Khvorova, I. A. 2012. 'Materials Science.' Tomsk Polytechnic University- Tomsk: TPU Publishing House. 137. (b) NDT Resource Center. 'Linear Defects - Dislocations.' Accessed 13 April 2017. https://www.nde-ed.org/EducationResources/CommunityCollege/Materials/Structure/linear_defects. htm

581. Lawn, Brian. 1993. *Fracture of Brittle Solids, 2nd edition.* Cambridge Solid State Science Series. Cambridge University Press.

Chapter 7

582. Basu, Bikramjit. 2017. *Biomaterials for Musculoskeletal Regeneration: Concepts.* Singapore: Springer Nature.

583. 'Types of cytoskeleton and their molecular structure.' Accessed April 24, 2017. http://csls-text.c.u-tokyo.ac.jp/active/06_01.html

584. Takahashi, Kazutoshi, and Shinya Yamanaka 2006. 'Induction of pluripotent stem cells from mouse embryonic and adult fibroblast cultures by defined factors.' *Cell.* 126(4). 663–676.

585. Stem Cell Information. 'Stem cell basics: What are the potential uses of human stem cells and the obstacles that must be overcome before these potential uses will be realized?' World Wide Web site. Bethesda, MD: National Institutes of Health, U.S. Department of Health and Human Services.

586. Bianco, Paolo, and Pamela Gehron Robey. 2001. 'Stem cells in tissue engineering.' *Nature.* 414(6859). 118–121.

587. Detsch, R, F. Uhl, U. Deisinger, and G. Ziegler. 2008. '3d-cultivation of bone marrow stromal cells on hydroxyapatite scaffolds fabricated by dispense-plotting and negative mould technique.' *Journal of Materials Science: Materials in Medicine.* 19(4). 1491–1496.

588. Ghasemi-Mobarakeh, Laleh, Molamma P. Prabhakaran, Mohammad Morshed, Mohammad-Hossein Nasr-Esfahani, and Seeram Ramakrishna. 2008. 'Electrospun poly (-caprolactone)/gelatin nanofibrous scaffolds for nerve tissue engineering.' *Biomaterials.* 29(34). 4532–4539.

589. Prabhakaran, Molamma P., Jayarama Reddy Venugopal, and Seeram Ramakrishna. 2009. 'Mesenchymal stem cell differentiation to neuronal cells on electrospun nanofibrous substrates for nerve tissue engineering.' *Biomaterials,* 30(28). 4996–5003.

590. Ghasemi-Mobarakeh, Laleh, Molamma P. Prabhakaran, Mohammad Morshed, Mohammad Hossein Nasr-Esfahani, and Seeram Ramakrishna. 2009. 'Electrical stimulation of nerve cells using conductive nanofibrous scaffolds for nerve tissue engineering.' *Tissue Engineering Part A.* 15(11). 3605–3619.

591. Segers, Vincent F. M., and Richard T. Lee. 2008. 'Stem-cell therapy for cardiac disease.' *Nature.* 451(7181). 937–942.

592. Premkumar, K. 2004. *The Massage Connection Anatomy and Physiology.* Baltimore: Lippincott, Williams and Wilkins.

593. Alberts, Bruce, Alexander Johnson, Julian Lewis, Martin Raff, Keith Roberts, and Peter Walter. 2002. *Molecular Biology of the Cell, 4th edition.* New York: Garland Science.

594. Alberts, Bruce, Dennis Bray, Karen Hopkin, Alexander Johnson, Julian Lewis, Martin Raff, Keith Roberts, and Peter Walter. 2010. *Essential Cell Biology, 3rd edition* Garland Science.

595. Sunil Kumar B, Developmental strategies to address prosthetic infection and magneto-responsive biomaterials for orthopedic applications, PhD Dissertation, Indian Institute of Science, Bangalore, 2015.

596. Cooper, G. M. 2000. *The Cell: A Molecular Approach, 2nd edition.* Sunderland (MA): Sinauer Associates.

597. Joanne Willey, Linda Sherwood, Chris Woolverton, Prescott/Harley/Klein's Microbiology, 7th edition, McGraw Hill Education, 2007.

598. Conde, Cecilia, and Alfredo Cácere. 2009. 'Microtubule assembly, organization and dynamics in axons and dendrites.' *Nature Reviews Neuroscience.* 10. 319–332.

599. Klug, William S., Michael R. Cummings, Charlotte A. Spencer, and Michael A. Palladino. 2012. *Concepts of Genetics, 10th edition.* Pearson Education, Inc.

600. Meregalli, M., A. Farini, and Y. Torrente. 2011. 'Mesenchymal Stem Cells as Muscle Reservoir.' *J. Stem Cell Res. Ther.* 1. 105.

Chapter 8

601. Williams, D. F. 1987. *Definitions in Biomaterials, Progress in Biomedical Engineering.* Amsterdam. The Netherlands: Elsevier Publishers.

602. Gonzalez-Simon, A., and O. Eniola-Adefeso. 2012. Host Response to Biomaterials. In *Engineering Biomaterials for Regenerative Medicine*, edited by S. K. Bhatia. 143-159. New York: Springer.

603. Young, V. R., William P. Steffee, Paul B. Pencharz, Joerg C. Winterer, and Nevin S. Scrimshaw. 1975. 'Total human body protein synthesis in relation to protein requirements at various ages.' *Nature.* 253. 192–194.

604. Schmidt, D. R., H. Waldeck, and W. J. Kao. 2009. 'Protein adsorption to biomaterials.' *Biological Interactions on Materials Surfaces: Understanding and Controlling protien, cell and tissue responses.* , edited by R. Bizios and David A Puleo. New York: Springer.

605. Kasemo, B. 2002. 'Biological surface science.' *Surface Science.* 500. 656–677.

606. Ward, W. K. 2008. 'A review of the foreign-body response to subcutaneously-implanted devices: the role of macrophages and cytokines in biofouling and fibrosis.' *Journal of Diabetes Science and Technology.* 2. 768–777.

607. Ratner B. D., A. S. Hoffman, F. J. Schoen, and J. E. Lemons. 2004. *Biomaterials Science: An Introduction to Materials in Medicine, 2nd edition.* Academic Press.

608. Anderson, J. M., A. Rodriguez, and D. T. Chang. 2008. 'Foreign body reaction to biomaterials.' *Semin Immunol.* 20. 86–100.

609. Ingber, D. E. 2003. 'Mechanosensation through integrins: Cells act locally but think globally.' *Proceedings of the National Academy of Sciences.* 100. 1472–1474.

610. Frederick, S., and R. Mitchell. 2004. Tissues, the Extracellular Matrix, and Cell–Biomaterial Interactions. *Biomaterials Science: An Introduction to Materials in Medicine, 2nd edition.* Elsevier Academic Press. 260–281.

611. Engler, A. J., S. Sen, H. L. Sweeney, and D. E. Discher. 2006. 'Matrix elasticity directs stem cell lineage specification.' *Cell*. .126. 677–689.

612. Fletcher, D. A., and R. D. Mullins. 2010. 'Cell mechanics and the cytoskeleton.' *Nature*. 463. 485–492.

613. Kholodenko, B. N. 2006. 'Cell-signalling dynamics in time and space.' *Nat. Rev. Mol. Cell. Biol.* 7(3). 165–176.

614. Lodish, H, A. Berk, S. L. Zipursky, P. Matsudaira, D. Baltimore, and J. Darne 2000. 'Overview of Extracellular Signaling.' *Molecular Cell Biology, 4th edition*. New York: W. H. Freeman.

615. Schuldiner, Maya, Ofra Yanuka, Joseph Itskovitz-Eldor, Douglas A Melton, and Nissim Benvenisty. 2000. 'Effects of eight growth factors on the differentiation of cells derived from human embryonic stem cells.' *Proceedings of the National Academy of Sciences*. 97(21). 11307–11312.

616. Kim, Jungju, In Sook Kim, Tae Hyung Cho, Kyu Back Lee, Soon Jung Hwang, Giyoong Tae, Insup Noh, Sang Hoon Lee, Yongdoo Park, and Kyung Sun. 2007. 'Bone regeneration using hyaluronic acid-based hydrogel with bone morphogenic protein-2 and human mesenchymal stem cells.' *Biomaterials*. 28(10). 1830–1837.

617. Bravo-Cordero, Jose Javier, Marco A. O. Magalhaes, Robert J. Eddy, Louis Hodgson, and John Condeelis. 2013. 'Functions of cofilin in cell locomotion and invasion.' *Nature Reviews Molecular Cell Biology*. 14. 405–415.

618. Lichtman, J., and J. A. Conchello. 2005. 'Fluorescence microscopy.' *Nature Methods*. 2. 910–920.

619. Terasaki, M., and M. E. Dailey. 1995. 'Confocal microscopy of living cells.' *Handbook of Biological Confocal Microscopy*. 327–346. US: Springer.

620. Vielreicher, M., S. Schürmann, R. Detsch, M. A. Schmidt, A. Buttgereit, A. Boccaccini, O. Friedrich 2013. 'Taking a deep look: modern microscopy technologies to optimize the design and functionality of biocompatible scaffolds for tissue engineering in regenerative medicine.' *Journal of the Royal Society Interface*. 10(86). 20130263.

621. Booth, M. J., D. Débarre, and A. Jesach. 2012. 'Adaptive optics for biomedical microscopy.' *Opt. Photon. News*. 23. 22–29.

622. 'Explaining the Point-Spread Function (PSF), and how it can be used for deconvolution.' Accessed 24 April, 2017. Available at https://microscopysolutions.ca/

623. Tata, B. V. R., and B. Raj. 1998. 'Confocal laser scanning microscopy: applications in material science and technology.' *Bull. Mater. Sci.* 21. 263–278.

624. Vilches, J., J. I. Vilches-Perez, and M. Salido. 2007. 'Cell-Surface interaction in biomedical implants assessed by simultaneous fluorescence and reflection confocal microscopy.' *Modern Research and Educational Topics in Microscopy*. Spain: Formatex 60–67.

625. Ettinger, A., and T. Wittmann. 2014. 'Fluorescence Live Cell Imaging.' *Methods in Cell Biology*. 123. 77–94. doi:10.1016/B978-0-12-420138-5.00005-7.

626. 'Education in Microscopy and Digital Imaging.' Accessed 24 April, 2017. http://zeiss-campus.magnet.fsu.edu/articles/livecellimaging/techniques.html

627. Waters, J.C., 2009. 'Accuracy and precision in quantitative fluorescence microscopy.' *The Journal of Cell Biology*. 185(7). 1135–1148.

628. Xu, F., T. Beyazoglu, E. Hefner, U. A. Gurkan, and U. Demirci. 2011. 'Automated and Adaptable Quantification of Cellular Alignment from Microscopic Images for Tissue Engineering Applications.' *Tissue Engineering Part C, Methods*. 17(6). 641–649.

629. Willerth, Stephanie M., and Shelly E. Sakiyama-Elbert. 2008. 'Combining stem cells and biomaterial scaffolds for constructing tissues and cell delivery.' *StemBook*. Cambridge (MA): Harvard Stem Cell Institute.

630. Engler, A. J., S. Sen, H. L. Sweeney, and D. E. Discher. 2006. 'Matrix elasticity directs stem cell lineage specification.' *Cell*. 126(4). 677–689.

631. Rowlands, A. S., P. A. George, and J. J. Cooper-White. 2008. 'Directing osteogenic and myogenic differentiation of MSCs: interplay of stiffness and adhesive ligand presentation.' *American Journal of Physiology- Cell Physiology*. 295(4). C1037–C1044.

632. Park. J. S., J. S. Chu, A. D. Tsou, R. Diop, Z. Tang, A. Wang, and S. Li. 2011. 'The effect of matrix stiffness on the differentiation of mesenchymal stem cells in response to TGF-beta.' *Biomaterials*. 32(16). 3921–3930.

633. Wang, P. Y., W. B. Tsai, and N. H. Voelcker. 2012. 'Screening of rat mesenchymal stem cell behaviour on polydimethylsiloxane stiffness gradients.' *Acta Biomater*. 8(2). 519–530.

634. Pek, Y. S., A. C. A. Wan, and J. Y. Ying. 2010. 'The effect of matrix stiffness on mesenchymal stem cell differentiation in a 3D thixotropic gel'. *Biomaterials*. 31(3). 385–391.

635. Wei, Z. C. Wang, H. Liu, S. Zou, and Z. Tong. 2012. 'Facile fabrication of biocompatible PLGA drug-carrying microspheres by O/W pickering emulsions.' *Colloids Surf. B Biointerfaces*. 91(0). 97–105.

636. Justin, R. Tse, and Adam J. Engler. 2011. 'Stiffness gradients mimicking *in vivo* tissue variation regulate mesenchymal stem cell fate.' *PloS one*. 6 (1): e15978.

637. Meinel, L, R. Fajardo, S. Hofmann, R. Langer, J. Chen, B. Snyder, G. Vunjak-Novakovic, and D. Kaplan. 2005. 'Silk implants for the healing of critical size bone defects.' *Bone*. 37(5). 688–698.

638. Fortier, L A. 2005. 'Stem cells: classifications, controversies, and clinical applications.' *Vet. Surg.* 34(5). 415–423.

639. Hu, X., S-H. Park, E. S. Gil, X-X. Xia, A. S. Weiss, and D. L. Kaplan. 2011. 'The influence of elasticity and surface roughness on myogenic and osteogenic-differentiation of cells on silk-elastin biomaterials.' *Biomaterials*. 32(34). 8979–8989.

640. Lee, S., J. Kim, T. J. Park, Y. Shin, S. Y. Lee, Y. M. Han, S. Kang, and H. S. Park. 2011. 'The effects of the physical properties of culture substrates on the growth and differentiation of human embryonic stem cells.' *Biomaterials*. 32(34). 8816–8829.

641. Buxboim, A., K. Rajagopal, A. E. Brown, and D. E. Discher. 2010. 'How deeply cells feel: methods for thin gels.' *J. Phys. Condens. Matter*. 22(19). 194116.

642. Guilak, F., D. M. Cohen, B. T. Estes, J. M. Gimble, W. Liedtke, and C. S. Chen. 2009. 'Control of stem cell fate by physical interactions with the extracellular matrix.' *Cell Stem Cell*. 5(1). 17–26.

643. Yim, E. K. F., E.M. Darling, K. Kulangara, F. Guilak, and K. W. Leong. 2010. 'Nanotopography-induced changes in focal adhesions, cytoskeletal organization, and mechanical properties of human mesenchymal stem cells.' *Biomaterials*. 31(6). 1299–1306.

644. Ferreira, L. J. M. Karp, L. Nobre, and R. Langer. 2008. 'New opportunities: the use of nanotechnologies to manipulate and track stem cells.' *Cell Stem Cell*. 3(2). 136–146.

645. Brammer, K. S., C. Choi, C. J. Frandsen, S. Oh, and S. Jin. 2011. 'Hydrophobic nanopillars initiate mesenchymal stem cell aggregation and osteo-differentiation.' *Acta Biomater*. 7(2). 683–690.

646. Jager, E. W. H., M. H. Bolin, K. Svennersten, X. Wang, A. Richter-Dahlfors, and M. Berggren. 2009. 'Electroactive surfaces based on conducting polymers for controlling cell adhesion, signaling, and proliferation.' *IEEE*. 1778–1781.

647. Oh, S., K. S. Brammer, Y. S. J. Li, D. Teng, A. J. Engler, S. Chien, and S. Jin. 2009. 'Stem cell fate dictated solely by altered nanotube dimension.' *Proceedings of the National Academy of Sciences*. 106(7):. 2130–2135.

648. Engel, E., E. Martínez, C. A. Mills, M. Funes, J. A. Planell, and J. Samitier. 2009. 'Mesenchymal stem cell differentiation on microstructured poly (methyl methacrylate) substrates.' *Annals of Anatomy- Anatomischer Anzeiger*. 191(1). 136–144.

649. Olivares-Navarrete, R., S. L. Hyzy, D. L. Hutton, C. P. Erdman, M. Wieland, B. D. Boyan, and Z. Schwartz. 2010. 'Direct and indirect effects of microstructured titanium substrates on the induction of mesenchymal stem cell differentiation towards the osteoblast lineage.' *Biomaterials*. 31(10). 2728–2735.

650. Phillips, J. E., T. A. Petrie, F. P. Creighton, and A. J. García. 2010. 'Human mesenchymal stem cell differentiation on self-assembled monolayers presenting different surface chemistries.' *Acta Biomaterialia.* 6(1). 12–20.

651. Guimard, N. K., N. Gomez, and C. E. Schmidt. 2007. 'Conducting polymers in biomedical engineering.' *Prog. Polym. Sci.* 32(8–9). 876–921.

652. Kotwal, A., and C. E. Schmidt. 2001. 'Electrical stimulation alters protein adsorption and nerve cell interactions with electrically conducting biomaterials.' *Biomaterials.* 22(10). 1055–1064.

653. Collazos-Castro, J. E., J. L. Polo, G. R. Hernández-Labrado, V. Padial-Cañete, and C. García-Rama. 2010. 'Bioelectrochemical control of neural cell development on conducting polymers.' *Biomaterials.* 31(35). 9244–9255.

654. Ravichandran, R. S. Sundarrajan, J. R. Venugopal, S. Mukherjee, and S. Ramakrishna. 2010. 'Applications of conducting polymers and their issues in biomedical engineering.' *J. R. Soc. Interface.* 6(7). 7.

655. Yow, S. Z., T. H. Lim, E. K. F. Yim, C. T. Lim, and K. W. Leong. 2011. 'A 3D Electroactive Polypyrrole-Collagen Fibrous Scaffold for Tissue Engineering.' *Polymers.* 3(1). 527–544.

656. Jun, I., S. Jeong, and H. Shin. 2009. 'The stimulation of myoblast differentiation by electrically conductive sub-micron fibers.' *Biomaterials.* 30(11). 2038–2047.

657. Babensee, J. E., J. M. Anderson, L. V. McIntire, and A. G. Mikos. 1998. 'Host response to tissue engineered devices.' *Advanced Drug Delivery Reviews.* 33.111–139.

658. Ratner, B. D., A. S. Hoffman, F. J. Schoen, and J. Lemons. 2004. *Biomaterials Science: A Multidisciplinary Endeavor.* California: Elsevier Inc.

659. Sharma, C. P. 2004. 'Biomaterials and artificial organs- current status, Biomaterials, Biomedical Technology & Quality System.' Published by IIPC, SCTIMST.

660. Ratner, B. D. 2002. 'Reducing capsular thickness and enhancing angiogenesis around implant drug release systems.' *Journal of Controlled Release.* 78. 211–218.

661. Anderson, J. M. 2001. 'Biological responses to materials.' *Annual Review of Materials Research.* 31. 81–110.

662. Zhang, Lei, Zhiqiang Cao, Tao Bai, Louisa Carr, Jean-Rene Ella-Menye, Colleen Irvin, Buddy D. Ratner, and Shaoyi Jiang. 2013. "Zwitterionic hydrogels implanted in mice resist the foreign-body reaction." *Nature Biotechnology.* 31(6). 553.

663. Ratner, B. D. S. F., J. J. Lemons, A. S. Hoffman. 2004. *Biomaterials Science- An Introduction to Materials in Medicine, 2nd edition.* Elsevier Academic press.

664. Anderson, J. M., A. Rodriguez, and D. T. Chang. 2008. 'Foreign body reaction to biomaterials.' *Seminars in Immunology.* 20. 86–100.

665. Anderson, J. M. 2001. 'Biological responses to materials.' *Annual Review of Materials Research.* 31. 81–110.

666. Horbett, T. 2004. 'The role of adsorbed proteins in tissue response to biomaterials.' *Biomaterials Science- An Introduction to Materials in Medicine.* Second ed. San Diego, CA: Elsevier Academic press. 237–246.

667. Anderson, J. M., A. Rodriguez, and D. T. Chang. 2008. 'Foreign body reaction to biomaterials.' *Seminars in Immunology.* Elsevier. 86–100.

668. Kyriakides, T. R., M. J. Foster, G. E. Keeney, A. Tsai, C, M. Giachelli, I. Clark-Lewis, B.J. Rollins, and P. Bornstein. 2004. 'The CC chemokine ligand, CCL2/MCP1, participates in macrophage fusion and foreign body giant cell formation.' *The American Journal of Pathology.* 165. 2157–2166.

669. Ratner, B. D. 2002. 'Reducing capsular thickness and enhancing angiogenesis around implant drug release systems.' *Journal of Controlled Release.* 78. 211–218.

670. Sharkawy, A. A., B. Klitzman, G. A. Truskey, and W. M. Reichert. 1998. 'Engineering the tissue which encapsulates subcutaneous implants. II. Plasma–tissue exchange properties.' *Journal of Biomedical Materials Research.* 40. 586–597.

671. Morais, J. M., F. Papadimitrakopoulos, and D. J. Burgess. 2010. 'Biomaterials/tissue interactions: possible solutions to overcome foreign body response.' *The AAPS Journal*. 12. 188–196.

672. Scuderi, G. J., A. R. Vaccaro, L. N. Fitzhenry, S. Greenberg, and F. Eismont. 2004. 'Long-term clinical manifestations of retained bullet fragments within the intervertebral disk space.' *Journal of Spinal Disorders & Techniques*. 17.108–111.

673. Zhang, L, Z. Cao, T. Bai, L. Carr, J. -R. Ella-Menye, C. Irvin, B. D. Ratner, and S. Jiang. 2013. 'Zwitterionic hydrogels implanted in mice resist the foreign-body reaction.' *Nature Biotechnology*. 31(6). 553–556.

674. Hetrick, E. M., H. L. Prichard, B. Klitzman, and M. H. Schoenfisch. 2007. 'Reduced foreign body response at nitric oxide-releasing subcutaneous implants.' *Biomaterials*. 28. 4571–4580.

675. Wang, Y., S. Vaddiraju, L. Qiang, X. Xu, F. Papadimitrakopoulos, and D. J. Burgess. 2012. 'Effect of dexamethasone-loaded poly (lactic-co-glycolic acid) microsphere/poly (vinyl alcohol) hydrogel composite coatings on the basic characteristics of implantable glucose sensors.' *Journal of Diabetes Science and Technology*. 6. 1445–1453.

676. Patil, S. D., F. Papadmitrakopoulos, and D. J. Burgess. 2007. 'Concurrent delivery of dexamethasone and VEGF for localized inflammation control and angiogenesis.' *Journal of Controlled Release*. 117. 68–79.

677. Gilligan, B. C., M. Shults, R. K. Rhodes, P. G. Jacobs, J. H. Brauker, T. J. Pintar, and S. J. Updike. 2004. 'Feasibility of continuous long-term glucose monitoring from a subcutaneous glucose sensor in humans.' *Diabetes Technology & Therapeutics*. 6. 378–386.

678. Langer, R. 2009. 'Perspectives and challenges in tissue engineering and regenerative medicine.' *Advanced Materials*. 21. 3235–3236.

679. Williams, D. F. 2008. 'On the mechanisms of biocompatibility.' *Biomaterials* 29. 2941–2953.

680. Sharkawy, A. A., B. Klitzman, G. A. Truskey, and W. M. Reichert. 1997. 'Engineering the tissue which encapsulates subcutaneous implants. I. Diffusion properties.' *Journal of Biomedical Materials Research*. 37. 401–412.

681. Woodward, S. C. 1982. 'How fibroblasts and giant cells encapsulate implants: considerations in design of glucose sensors.' *Diabetes Care*. 5. 278–281.

682. He, Y., J. Hower, S. Chen, M. T. Bernards, Y. Chang, and S. Jiang. 2008. 'Molecular simulation studies of protein interactions with zwitterionic phosphorylcholine self-assembled monolayers in the presence of water.' *Langmuir*. 24. 10358–10364.

683. Basu, Bikramjit. 2017. *Biomaterials for Musculoskeletal Regeneration: Concepts*. Singapore: Springer Nature.

684. Alberts, Bruce, Alexander Johnson, Julian Lewis, Martin Raff, Keith Roberts, and Peter Walter. 2002. *Molecular Biology of the Cell, 4th edition*. New York: Garland Science.

685. Suresh, Subra. 2007. *Acta Biomaterialia*. 3. 413–438.

686. Versaevel, M., T. Grevesse, and S. Gabriele. 2012. 'Spatial coordination between cell and nuclear shape within micropatterned endothelial cells.' *Nat. Commun.* 3. 671.

687. Discher, D. E., P. Janmey, and Y. 1. Wang. 2005. 'Tissue cells feel and respond to the stiffness of their substrate.' *Science*. 310(5751). 1139–1143.

688. Grainger, D. W. 2013. 'All charged up about implanted biomaterials.' *Nature Biotechnology*. 31. 507–509.

Chapter 9

689. Mosmann, T. 1983. 'Rapid colorimetric assay for cellular growth and survival: Application to proliferation and cytotoxicity assays.' *Journal of Immunological Methods*. 65(1–2). 55-–63.

690. Fields, R. D., and M. V. Lancaster. 1993. *Am. Biotechnol. Lab.* 11. 48–50.

691. Larson, E. M., D. J. Doughman, D. S. Gregerson, and W. F. Obritsch. 1997. 'A new, simple, nonradioactive, nontoxic *in vitro* assay to monitor corneal endothelial cell viability.' *Invest. Ophthalmol. Vis. Sci.* 38. 1929–1933.

692. Ishiyama, M., Y. Miyazono, K. Sasamoto, Y. Ohkura, and K. Ueno. 1997. 'A new, simple, nonradioactive, nontoxic *in vitro* assay to monitor corneal endothelial cell viability.' *Talanta.* 44. 1299–1305.

693. TaKaRa. Accessed 24 April, 2017. Available at http://www.clontech.com/

694. 'ELISA Principle Basis and Extension.' Accessed 24 April, 2017. http://www.elisa-antibody.com/ELISA-Introduction/ELISA-Principle.

695. J. Paul Robinson PhD, Jennifer Sturgis BS and George L. Kumar. 2009. 'Immunofluorescence.' *Immunohistochemical Staining Methods.* California. 61–65.

696. Odell, I. D., and D. Cook,. 2013. 'Immunofluorescence techniques.' *Journal of Investigative Dermatology.* 133. 1–4.

697. Kumar, Alok ,Thomas J. Webster, K. Biswas, and Bikramjit Basu. 2013. 'Flow cytometry analysis of human foetal osteoblast fate processes on spark plasma sintered Hydroxyapatite-Titanium biocomposites.' *J. Biomed. Mater. Res. Part A.* 101 (10). 2925–2938.

698. Cusella-De Angelis, M. G., G. Lyons, C. Sonnino, L. De Angelis, E. Vivarelli, K. Farmer, W.E. Wright, M. Molinaro, M. Bouche, and M. Buckingham. 1992. 'MyoD, myogenin independent differentiation of primordial myoblasts in mouse somites.' *J. Cell Biol.* 116(5). 1243–1255.

699. Miller, Jeffrey Boone. 1990. 'Myogenic programs of mouse muscle cell lines: expression of myosin heavy chain isoforms, MyoD1, and myogenin.' *The Journal of Cell Biology.* 111(3). 1149–1159.

700. Young, D., J. DeQuach, and K. Christman. 2011. 'Human cardiomyogenesis and the need for systems biology analysis.' *Wiley Interdisciplinary Reviews Systems Biology and Medicine.* 3(6). 666–680.

701. Zhang, F., and K. B. S. Pasumarthi. 2007. 'Ultrastructural and immunocharacterization of undifferentiated myocardial cells in the developing mouse heart.' *Journal of Cellular and Molecular Medicine.* 11(3). 552–560.

702. Ueyama, T., H. Kasahara, T. Ishiwata, Q. Nie, and S. Izumo. 2003. 'Myocardin expression is regulated by Nkx2.5, and its function is required for cardiomyogenesis.' *Molecular and Cellular Biology.* 23(24). 9222–9232.

703. Layland, J., R. J. Solaro, and A. M. Shah. 2005. 'Regulation of cardiac contractile function by troponin I phosphorylation.' *Cardiovascular Research.* 66(1). 12–21.

704. Voronova, A., A. Al Madhoun, A. Fischer, M. Shelton, C. Karamboulas, and I. S. Skerjanc. 2012. 'Gli2 and MEF2C activate each other's expression and function synergistically during cardiomyogenesis *in vitro*.' *Nucleic Acids Research.* 40(8). 3329–3347.

705. Carden, M. J., J. Q. Trojanowski, W. W. Schlaepfer, and V. M. Lee. 1987. 'Two-stage expression of neurofilament polypeptides during rat neurogenesis with early establishment of adult phosphorylation patterns.' *J. Neurosci.* 7(11). 3489–3504.

706. Tice, R. R. E. Agurell, D. Anderson, B. Burlinson, A. Hartmann, H. Kobayashi, Y. Miyamae, E. Rojas, J. C. Ryu, and Y. F. Sasaki. 2000. 'Single cell gel/comet assay: guidelines for *in vitro* and *in vivo* genetic toxicology testing.' *Environ. Mol. Mutagen.* 35(3). 206.

707. Kalmodia, S., V. Sharma, A. K. Pandey, A. Dhawan, and B. Basu. 2011. 'Cytotoxicity and genotoxicity property of hydroxyapatite-mullite eluates.' *J. Biomed. Nanotechnol.* 7. 74–75.

708. Basu, Bikramjit. 2017. *Biomaterials for Musculoskeletal Regeneration: Concepts.* Singapore: Springer Nature.

709. Hulsart-Billström, G., J. I. Dawson, S. Hofmann, R. Müller, M. J. Stoddart, M. Alini, H. Redl, A. El Haj, R. Brown, V. Salih, J. Hilborn, S. Larsson, and R. O. Oreffo. 2016. 'A surprisingly poor correlation between *in vitro* and *in vivo* testing of biomaterials for bone regeneration: results of a multicentre analysis.' *Eur. Cell Mater.* 31. 312–322.

710. Brey, P. 2009. 'Biomedical Engineering Ethics.' In *A Companion to Philosophy of Technology.* edited by Jan Kyrre Berg Olsen, Stig Andur Pedersen, and Vincent F. Hendricks. NJ: John Wiley & Sons.

711. Horch, Raymund E., L. M. Popescu, Charles Vacanti, and Giovanni Maio. 2008. 'Ethical issues in cellular and molecular medicine and tissue engineering.' *Journal of Cellular and Molecular Medicine.* 12 (5b). 1785–1793.

712. 2009. 'Consensus guidance for banking and supply of human embryonic stem cell lines for research purposes.' *Stem Cell Rev.* 5(4). 301–314.

713. Bubela, Tania, Jenilee Guebert, and Amrita Mishra. 2015. 'Use and misuse of material transfer agreements: lessons in proportionality from research, repositories, and litigation.' *PLoS. Biology.* 13(2). e1002060.

714. Bennett, A. B., W. D. Streitz, and R. A. Gracel. 2007. Specific Issues with Material Transfer Agreements. In: *Intellectual property management in health and agricultural innovation: A handbook of best practices.* Oxford, edited by Krattiger A, R. T. Mahooney, and L. Nelson.697-716 MIHR and PIPRA.

715. Master, J. R., J. A. Thompson, B. Daly-Burns, Y. A. Reid, W. G. Dirks, P. Packer, L. H. Toji, T. Ohno, H. Tanabe, C. F. Arlett, L. R. Kelland, M. Harrison, A. Virmani, T. H. Ward, K. L. Ayres, and P. G. Debenham. 2001. 'Short tandem repeat profiling provides an international reference standard for human cell lines.' *Proc. Natl. Acad. Sci. USA.* 98. 8012–8017.

716. (a) Kraehenbuehl, T. P., R. Langer, and L. S. Ferreira. 2011. 'Three-dimensional biomaterials for the study of human pluripotent stem cells.' *Nature Methods.* 8(9). 731–736. (b) Lutolf, M. P., P. M. Gilbert, and H. M. Blau. 2009. 'Designing materials to direct stem-cell fate.' *Nature.* 462(7272). 433–441.

717. Abbas Abul K., Andrew H. Lichtman, and Jordan S. Pober. 2000. *Cellular and Molecular Immunology.*, 4th edition. Philadelphia: Saunders.

718. Fritschy, J. M. and W. Härtig. 2001. 'Immunofluorescence.' Willey Online Publication.

719. Pfaffl, Michael W. 2004. 'Quantification Strategies in Real-time PCR.' In: *A–Z of quantitative PCR edited by S. A. Bustin.* International University Line, La Jolla, CA, USA.

720. Primrose, S. B., and R. M. Twyman. 2006. *Principles of Gene Manipulation and Genomics, 7th edition.* Blackwell Publishing.

721. Santos, C. F. D., V.T. Sakai, C.F. Santos, V.T. Sakai , M.A. Machado, D.N. Schippers, and A. S. Greene. 2004. 'Reverse transcription and polymerase chain reaction: principles and applications in dentistry.' *Journal of Applied Oral Science.* 12. 1–11.

722. Bentzinger, C. F., Y. X. Wang, and M. A. Rudnicki. 2012. 'Building muscle: molecular regulation of myogenesis.' *Cold Spring Harb Perspect Biol.* 4(2). a008342.

For Further Reading

a. Liang, Peng, and Arthur B. Pardee. 1992. 'Differential Display of Eukaryotic Messenger RNA by Means of the Polymerase Chain Reaction.' *Science.* 257. 967.

b. Nolan,John P., and Larry A. Sklar. 1998. 'The emergence of flow cytometry for sensitive real-time measurements of molecular interactions.' *Nature.* 16. 633-638.

c. Lippincott-Schwartz, Jennifer, and Suliana Manley. 2009. 'Putting super-resolution fluorescence microscopy to work.' *Nature.* 6. 21-23.

d. Rust, Micheal J., Mark Bates, and XiaoweiZhuang. 2006. 'Sub-diffraction-limit imaging by stochastic optical reconstruction microscopy (STORM).' *Nature.* 3. 793-795.

e. Perfetto,Stephen P., Pratip K. Chattopadhyay, and Mario Roederer. 2004. 'Seventeen colour flow cytometry: unravelling the immune system.' *Nature.* 4. 648-655.

f. Schmittgen,Thomas D., and Kenneth J Livak. 2008. 'Analyzing real-time PCR data by the comparative CT method. *Nature.* 3. 1101-1108.

g. Subach, Fedor V., George H Patterson, Suliana Manley, Jennifer M Gillette, Jennifer Lippincott-Schwartz, and Vladislav V Verkhusha. 'PhotoactivablemCherry for high-resolution two-color fluorescence microscopy.' *Nature.* 6. 153-159.

h. Cheng Suzanne, Sheng-Yung Chang, Patti Gravitt, and Richard Respess. 1994. 'Long PCR.' *Nature.* 369. 684-685.

i. Inoue, Haruhiro, Shin-eiKudo, and Akira Shiokawa. 2005. 'Technology Insight: laser-scanning confocal microscopy and endocytoscopy for cellular observation of the gastrointestinal tract.' *Nature.* 2. 31-37.

j. Hell Stefan W. 'Toward Fluorescence Nanoscopy.' *Nature.* 21. 1347-1355.

k. Lichtman, Jeff W., and Jose-Angel-Conchello. 2005. 'Fluorescence Microscopy.' *Nature.* 2. 910-919.

l. Yuste, Rafael. 2005. 'Fluorescence Microscopy Today. *Nature.* 2. 902-904.

m. Muirhead K. A., P. K. Horan, and G. Poste. Flow Cytometry: Presenr and Future.' *Nature.* 3. 337-356.

n. Bailey Brent, Daniel L. Farkas, D. Lansing Taylor, and Frederick Lanni. 1993. 'Enhancement of axial resolution in fluorescence microscopy by standing-wave excitation.' *Nature.* 366. 44-48.

o. Brakenhoff,G. J., H. T. M. van der Voort, E. A. van Spronsen, W. A. M. Linnemans, and N. Nanninga. 'Three-dimensional chromatin distribution in neuroblastoma nuclei shown by confocal scanning laser miscroscopy.' *Nature.* 317. 748-749.

p. White J.G., and W. B. Amos. 1987. 'Confocal microscopy comes of age.' Nature 328. 183-184.

q. Siebert, Paul D., and James W. Larrick. 1992. 'Competitive PCR.' *Nature.* 359. 557-558.

r. Saiki, Randall K., Teodorica L. Bugawan, Glenn T. Horn, Kary B. Mullis, and Henry A.1986. 'Erlich, Analysisi of enzymatically amplified β-globin and HLA-DQα DNA with allele-specific oligonucleotide probes.' *Nature.* 324. 163-166.

s. Juette, Manuel F., Travis J Gould, Mark D Lessard, Michael J Mlodzianoski, Bhupendra S Nagpure, Brain T Bennett, Samuel T Hess, and Joerg Bewersdorf. 2008. 'Three-dimensional sub-100 nm resolution fluorescence microscopy of thick samples. *Nature.* 5. 527-529.

t. Vermes, István, Clemens Haanen, Helga Steffens-Nakken, and Chris Reutelingsperger.1995. 'A novel assay for apoptosis flow cytometric detection of phosphatidylserine expression on early apoptotic cells using fluorescein labeled Annexin V.' *Journal of Immunological Methods.* 184. 39-51.

Chapter 10

723. Basu, Bikramjit, and Kantesh Balani. 2011. 'Advanced Structural Ceramics.' USA: John Wiley & Sons, Inc., and American Ceramic Society.

724. Basu, Bikramjit. 2017. *Biomaterials for Musculoskeletal Regeneration: Concepts.* Singapore: Springer Nature.

725. Katsikogianni, M., and Y. F. Missirlis. 2004. 'Concise review of mechanisms of bacterial adhesion to Biomaterials and of techniques used in estimating bacteria material interactions.' Eur Cell Mater. 8. 37–57.

726. Pulverer, G., P. G. Quie, and G. Peters, Eds. 1987. *Pathogenesis and Clinical Significance of Coagulase-Negative Staphylococci.* Stuttgart: Fisher Verlag.

727. Walsh, Evelyn J., Helen Miajlovic, Oleg V. Gorkun, and Timothy J. Foster. 2008. 'Identification of the staphylococcus aureus MSCRAMM clumping factor B (ClfB) binding site in the *αC*-domain of human fibrinogen.' *Microbiology.* 154. 550–558.

728. Dickinson, R. B., J. A. Nagel, D. McDevitt, T. J. Foster, R. A. Proctor, and S. L. Cooper. 1995. 'Quantitative comparison of clumping factor- and coagulase-mediated Staphylococcus aureus adhesion to surface-bound fibrinogen under flow.' *Infect. Immun.* 63(3). 143–3150

729. Kojima, S. and D. Blair. 2004. 'The bacterial flagellar motor: structure and function of a complex molecular machine.' *Int. Rev. Cytol.* 233. 93–134.

730. Beachey, E. 1981. 'Bacterial adherence: adhesin-receptor interactions mediating the attachment of bacteria to mucosal surface.' *J. Infect. Dis.* 143(3). 325–345.

731. Meroueh, Samy O., Krisztina Z. Bencze, Dusan Hesek, Mijoon Lee, Jed F. Fisher, Timothy L. Stemmler, and Shahriar Mobashery. 2006. 'Three-dimensional structure of the bacterial cell wall peptidoglycan.' *Proc Natl Acad Sci U S A*. 103. 4404–4409.

732. Ong, Y. L., A. Razatos, G. Georgiou, and M. M. Sharma. 1999. 'Adhesion forces between *E. coli* bacteria and biomaterial surfaces.' *Langmuir*. 15. 2719–2725.

733. Schmidt, L. M., J. J. Delfino, J. F. Preston, and St Laurent. 1999. 'Biodegradation of low aqueous concentration pentachlorophenol (PCP) contaminated groundwater.' *Chemosphere*. 38. 2897–2912.

734. Hasty, D. L., I. Ofek, H. S. Courtney, and R. J. Doyle. 1992. 'Multiple adhesins of streptococci.' *Infect. Immun*. 60. 2147–2152.

735. Li, B., and B. E. Logan. 2004. 'Bacterial adhesion to glass and metal-oxide surfaces.' *Colloids and Surfaces B: Biointerfaces*. 36. 81–90.

736. Van Oss, C. J. 1994. *Interfacial Forces in Aqueous Media*. New York: Marcel Dekker.

737. Israelachvili, J. 1992. *Intermolecular and Surface Forces, 2nd edition*. London: Academic Press Limited.

738. Emerson IV, R. J., and T. A. Camesano. 2004. 'Nanoscale investigation of pathogenic microbial adhesion to a biomaterial.' *Applied Environmental Microbiology*. 70. 6012–6022.

739. Van Loosdrecht, M. C., J. Lyklema, W. Norde, G. Schraa, and A. J. Zehnder. 1987. 'The role of bacterial cell wall hydrophobicity in adhesion.' *Applied Environmental Microbiology*. 53. 1893–1897.

740. An, Y. H., R. J. Friedman, R. A. Draughn, E. Smith, C. Qi, and J. F John. 1993. 'Staphylococci adhesion to orthopedic biomaterials.' *Trans. Soc. Biomater*. 16. 148.

741. Sugarman, B., and D. Musher, 1981. 'Adherence of bacteria to suture materials.' *Proc. Soc. Exp. Biol. Med*. 167. 156–160.

742. Gristina, A. G., C. D. Hobgood, and E. Barth. 1987. 'Biomaterial specificity, molecular mechanisms, and clinical relevance of S. epidermidis and S.aureus infections in surgery.' In: *Pathogenesis and Clinical Significance of Coagulase-Negative Staphylococci.*, edited by G. Pulverer, P. G. Quie, and G. Peters. Stuttgart: Fisher Verlag. 143–157.

743. Duran, L. W., J. A. Pietig, and J. E. Driemeyer. 1993. 'Prevention of microbial colonization on medical devices by photochemical immobilization of antimicrobial peptides.' *Trans. Soc. Biomater*. 16. 35.

744. Merritt, K., J. Shafer, and S. A. W. Brown. 1979. 'Implant site infection rates with porous and dense materials.' *J. Biomed. Mater. Res*. 13. 101–108.

745. McAllister, E. W., L. C. Carey, P. G. Brady, R. Heller, and S. G. Kovacs. 1993. 'The role of polymeric surface smoothness of biliary stents in bacterial adhesion, biofilm deposition, and stent occlusion.' *Gastrointest. Endosc*. 39. 422–425.

746. Baker, A. S., and L. W. Greenham. 1998. 'Release of gentamicin from acrylic bone cement: elution and diffusion studies.' *J. Bone Joint Surg*. 70. 1551–1557.

747. Hogt, A. H., J. Dankert, J. A. de Vries, and J. Feijen. 1983. 'Adhesion of coagulase-negative staphylococci to biomaterials.' *J. Gen. Microbiol*. 129. 1959–1968.

748. Fletcher, M., and G. I. Loeb. 1979. 'Influence of substratum characteristics on the attachment of a marine pseudomonad to solid surfaces.' *Appl. Environ. Microbiol*. 37. 67–72.

749. Satou, N., J. Satou, H. Shintani, and K. Okuda. 1988. 'Adherence of streptococci to surface-modified glass.' *J. Gen. Microbiol*. 134. 1299–1305.

750. Busscher, H. J., A. H. Weerkamp, H. C. van der Mei, A. W. van Pelt, H. P. de Jong, and J. Arends. 1984. 'Measurement of the surface free energy of bacterial cell surfaces and its relevance for adhesion.' *Appl. Environ. Microbiol*. 48. 980–983.

751. Krekeler, C., H. Ziehr, and J. Klein. 1989. 'Physical methods for characterization of microbial cell surfaces. *Experientia*.' 45. 1047–1054.

752. Dankert, J., A. H. Hogt, and J. Feijen. 1986. 'Biomedical polymers: bacterial adhesion, colonization, and infection.' *CRC Crit. Rev. Biocompat*. 2. 219–301.

753. Harden, V. P., and J. O. Harris. 1953. 'The isoelectric point of bacterial cells.' *J. Bacteriol*. 65. 269–271.

754. Kuusela, P. 1978. 'Fibronectin binds to *Staphylococcus aureus.*' *Nature.* 276. 718–720.

755. Kuusela, P., T. Vartio, M. Vuento, and E. B. Myhre. 1985. 'Attachment of staphylococci and streptococci on fibronectin, fibronectin fragments, and fibrinogen bound to a solid phase.' *Infect. Immunol.* 50. 77–85.

756. Flock, J. I.; G. Fröman, and K. Jönsson. 1987. 'Cloning and expression of the gene for a fibronectin-binding protein from *Staphylococcus aureus.*' *EMBO J.* 6. 2351–2357.

757. Mosher, D. F., and R. A. Proctor. 1980. 'Binding and factor XIIIa-mediated cross-linking of a 27 kilodalton fragment of fibronectin to *Staphylococcus aureus.*' *Science.* 209. 927–929.

758. Gibbons, R. J., and I. Etherden. 1983. 'Comparative hydrophobicities of oral bacteria and their adherence to salivary pellicles.' *Infect. Immunol.* 41. 1190–1196.

759. Reynolds, E. C. and A. Wong. 1983. 'Effect of adsorbed protein on hydroxyapatite zeta potential and *Streptococcus mutans* adherence.' *Infect.Immunol.* 39. 1285–1290.

760. Herrmann, M., P. E. Vaudaux, and D. Pittit. 1988. 'Fibronectin, fibrinogen, and laminin act as mediators of adherence of clinical staphylococci isolates to foreign material.' *J. Infect. Dis.* 158. 693–701.

761. Muller, E., S. Takeda, D. Goldmann, and G. B. Pier. 1991. 'Blood proteins do not promote adherence of coagulase-negative staphylococci to biomaterials.' *Infect. Immunol.* 59. 3323–3326.

762. Brokke, P., J. Dankert, J. Carballo, and J. Feijen. 1991. 'Adherence of coagulase-negative staphylococci onto polyethylene catheters *in vitro* and *in vivo:* a study on the influence of various plasma proteins.' *J. Biomater. Appl.* 5. 204–226.

763. Pelczar, M. J. 1993. *Microbiology.* TATA-McGraw Hill.

764. Christensen, G. D., L. M. Baddour, B. M. Madison., J.T.Parisi, S.N. Abraham, D.L. Hasty, J.H. Lowrance, J. A. Josephs, and W. A Simpson. 1990. 'Colony morphology of staphylococci on memphis agar phase variation of slime production resistance to betalactam antibiotics in virulence.' *J. Infect. Dis.* 161. 1153–1169.

765. Cooper, M., S. M. Batchelor, and J. I. Prosser. 1995. 'Is cell density signalling applicable to biofilms?' In: *The Life and Death of Biofilm,* edited by Wimpenny J., P. Handley, P. Gilbert, H. Lappin-Scott Cardiff: Bioline Press. 93–97.

766. Schierholz, J. M., J. Beuth, and G. Pulverer. 1993. 'Adherent bacteria and activity of antibiotics.' *J. Antimicrob. Chemother.* 43. 158–160.

767. Costerton , J. W., P. S. Stewart, and E. P. Greenberg. 1999. 'Bacterial biofilms: a common cause of persistent infections.' *Science.* 284. 1318–1322.

768. Wang, H., H. Cheng, D. Wei, and F. Wang. 2011. 'Comparison of methods for measuring viable E. coli cells during cultivation: great differences in the early and late exponential growth phases.' *J. Microbiol. Methods.* 84(1). 140–143.

769. Netuschil, L., T. M. Auschill, A. Sculean, and N. B. Arweiler. 2014. 'Confusion over live/dead stainings for the detection of vital microorganisms in oral biofilms--which stain is suitable?' *BMC Oral Health.* 14(2). 2.

770. Boda, S. K., J. Broda, F. Schiefer, J. Weber-Heynemann, M. Hoss, U. Simon, B. Basu, and W. Jahnen-Dechent. 2015. 'Cytotoxicity of ultrasmall gold nanoparticles on planktonic and biofilm encapsulated Gram-positive Staphylococci.' *Small.* 11(26). 3183–3193.

771. Chakraborty, S. P., S. K. Sahu, P. Pramanik, and S. Roy. 2012. '*in vitro* antimicrobial activity of nanoconjugated vancomycin against drug resistant *Staphylococcus aureus.*' *Int. J. Pharm.* 436(1–2). 659–676.

772. Furtado, G. L. and A. A. Medeiros. 1980. 'Single-disk diffusion testing (Kirby-Bauer) of susceptibility of Proteus mirabilis to chloramphenicol: significance of the intermediate category.' *Journal of Clinical Microbiology.* 12(4). 550–553.

773. Boda, S. K., K. Ravikumar, D. K. Saini, and B. Basu. 2015. 'Differential viability response of prokaryotes and eukaryotes to high strength pulsed magnetic stimuli.' *Bioelectrochemistry.* 106. Part B: 276–289.

774. Liu, H., Y. Du, X. Wang, and L. Sun. 2004. 'Chitosan kills bacteria through cell membrane damage.' *International Journal of Food Microbiology.* 95(2). 147–155.

775. Bajpai, I., N. Saha, and B. Basu. 2012. 'Moderate intensity static magnetic field has bactericidal effect on *E. coli* and *S. epidermidis* on sintered hydroxyapatite.' *J. Biomed. Mater. Res. B. Appl. Biomater.* 100(5). 1206–1217.

776. Dalai, S., S. Pakrashi, R. S. S. Kumar, N. Chandrasekaran, and A. Mukherjee. 2012. 'A comparative cytotoxicity study of TiO2 nanoparticles under light and dark conditions at low exposure concentrations.' *Toxicology Research.* 1(2). 116–130.

777. Nebe-von-Caron, G., P. J. Stephens, C. J. Hewitt, J. R. Powell, and R. A. Badley. 2000. 'Analysis of bacterial function by multi-colour fluorescence flow cytometry and single cell sorting.' *Journal of Microbiological Methods.* 42. 97–114.

778. Joyce, E., A. Al-Hashimi, and T. J. Mason. 2011. 'Assessing the effect of different ultrasonic frequencies on bacterial viability using flow cytometry.' *J. Appl. Microbiol.* 110(4). 862–870.

779. Vatansever, F., W. C. de Melo, P. Avci, D. Vecchio, M. Sadasivam, A. Gupta, R. Chandran, M. Karimi, N. A. Parizotto, R. Yin, and G. P. Tegos. 2013. 'Antimicrobial strategies centered around reactive oxygen species-bactericidal antibiotics, photodynamic therapy, and beyond.' *FEMS Microbiol. Rev.* 37(6). 955–989.

780. Novo, D., N. G. Perlmutter, R. H. Hunt, and H. M. Shapiro. Jan, 1999. 'Accurate flow cytometric membrane potential measurement in bacteria using diethyloxacarbocyanine and a ratio metric technique.' *Cytometry.* 35(1). 55–63.

781. Zhang, L., P. Dhillon, H. Yan, S. Farmer, and R. E. W. Hancock. 2000. 'Interactions of bacterial cationic peptide antibiotics with outer and cytoplasmic membranes of *Pseudomonas aeruginosa.*' *Antimicrobial Agents and Chemotherapy.* 44(12). 3317–3321.

782. Tote, K., D. Vanden Berghe, L. Maes, and P. Cos. 2008. 'A new colorimetric microtitre model for the detection of *Staphylococcus aureus* biofilms.' *Lett. Appl. Microbiol.* 46(2). 249–254.

783. Merritt, J. H., D. E. Kadouri, and G. A. O'Toole. 2005. 'Growing and analyzing static biofilms.' *Curr. Protoc. Microbiol.* Chapter 1 (Unit 1B).

784. Zotta, T., A. Guidone, P. Tremonte, E. Parente, and A. Ricciardi. 2012. 'A comparison of fluorescent stains for the assessment of viability and metabolic activity of lactic acid bacteria.' *World J. Microbiol. Biotechnol.* 28(3). 919–927.

785. Tawakoli, P. N., A. Al-Ahmad, W. Hoth-Hannig, M. Hannig, and C. Hannig. 2012. 'Comparison of different live/dead stainings for detection and quantification of adherent microorganisms in the initial oral biofilm.' *Clinical Oral Investigations.* 17(3). 841–850.

786. Bakke R., and P. Q. Olsson. 1986. 'Biofilm thickness measurements by light microscopy.' *Journal of Microbiological Methods.* 5(2). 93–98.

787. Willey, Joanne, Linda Sherwood, and Chris Woolverton. 2007. *Prescott, Harley, and Klein's Microbiology, 7th edition.* McGraw-Hill Education.

788. Alberts, Bruce, Alexander Johnson, Julian Lewis, Martin Raff, Keith Roberts, and Peter Walter. 2002. *Molecular Biology of the Cell, 4th edition.* New York: Garland Science.

789. Black, Jacquelyn G. 2005. *Microbiology: Principles and Explorations, 6th Edition.* John Wiley & Sons.

790. Bayoudh, Sonia, Ali Othmane, Laurence Mora, and Hafedh Ben Ouada. October 2009. 'Assessing bacterial adhesion using DLVO and XDLVO theories and the jet impingement technique.' *Colloids and Surfaces B: Biointerfaces.* 73(1). 1–9.

Chapter 11

791. Chandorkar Y, Multi-functional, Anti-inflammatory and Tunable Polyesters – A Novel Polymer Platform for Tissue Engineering and Drug Delivery Applications, PhD Dissertation, Indian Institute of Science, Bangalore, India, 2015.

792. Garcia, Y., A. Breen, k. Burugapalli, P. Dockery, and A. Pandit. 2007. 'Stereological methods to assess tissue response for tissue-engineered scaffolds.' *Biomaterials.* 28(2). 175–186.

793. Accessed 24 April, 2017. https://www.aaalac.org/resources/SOP_CPCSEA.pdf

794. Saraf, S. K., and V. Kumaraswamy. 2013. 'Basic research: Issues with animal experimentations.' *Indian J. Orthop.* 47(1). 6–9.

795. Ghasemi, M., and A. R. Dehpour. 2009. 'Ethical considerations in animal studies.' *J. Med. Ethics. Hist. Med.* 2. 12.

796. Nordgren, A. 2004. 'Moral imagination in tissue engineering research on animal models.' *Biomaterials.* 25(9). 1723–1734.

797. Benazzo, F., F. Falez, M. Dietrich, S. Affatato, W. Leardini, and M. Zavalloni. 2006. *Bioceramics and Alternative Bearings in Joint Arthroplasty.* Steinkopff. 171.

798. Olofsson, J. 2011. 'Friction and Wear Mechanisms of Ceramic Surfaces: With Applications to Micro Motors and Hip Joint Replacements.' Phd Thesis, Uppsala: Acta Universitatis Upsaliensis. Digital Comprehensive Summaries of Uppsala Dissertations from the Faculty of Science and Technology. 841.

799. Nevelos, J. E. E. Ingham, C. Doyle, A. B. Nevelos, and J. Fisher. 2001. 'The influence of acetabular cup angle on the wear of "BIOLOX Forte" alumina ceramic bearing couples in a hip joint simulator.' *Journal of Materials Science: Materials in Medicine.* 12. 141.

800. Totten, G. E., and H. Liang. 2004. *Mechanical Tribology: Materials, Characterization, and Applications.* Taylor & Francis.

801. Douglas, C. Hansen. 2008. 'Metal Corrosion in the Human Body: The Ultimate Bio-Corrosion Scenario.' *The Electrochemical Society Interface.* 17.

802. Kasper, C., F. Witte, R. Portner, D. Das, Z. Zhang, T. Winkler, M. Mour, C. Gunter, M. Morlock, H.-G. Machens, and A. Schilling. 2012. In *Tissue Engineering III: Cell - Surface Interactions for Tissue Culture.* Berlin Heidelberg: Springer.

803. Fruh, H. J., G. Willmann, and H. G. Pfaff. 1997. 'Wear characteristics of ceramic-on-ceramic for hip endoprostheses.' *Biomaterials.* 18. 873.

804. Willmann, G., H. J. Fruh, and H. G. Pfaff. 1996. 'Wear characteristics of sliding pairs of zirconia (Y-TZP) for hip endoprostheses.' *Biomaterials.* 17. 2157.

805. Catelas, I., A. Petit, R. Marchand, D. J. Zukor, L. Yahia, and O. L. Huk. 1999. 'Cytotoxicity and macrophage cytokine release induced by ceramic and polyethylene particles *in vitro*.' *J. Bone Joint Surg. Br.* 81(3). 516–521.

806. Hatton, A., J. E. Nevelos, J. B. Matthews, J. Fisher, and E. Ingham. 2003. 'Effects of clinically relevant alumina ceramic wear particles on TNF-α production by human peripheral blood mononuclear phagocytes.' *Biomaterials.* 24. 1193.

807. Imbert, L. 2011. *Working principle of the dual mobility (total hip replacement): wear mechanisms and design optimization.* Göteborg: Chalmers University of Technology.

808. Katzer, A., S. Hockertz, G. H. Buchhorn, and J. F. Loehr. 2003. Toxicology 190. 145–154.

809. Dorlot, J. M., P. Christel, and A. Meunier. 1989. 'Wear analysis of retrieved alumina heads and sockets of hip prostheses.' *J. Biomed. Mater. Res.* 23. 299–310.

810. Liagre, B., S. Moalic, P. Vergne, J. L. Charissoux, D. Bernache-Assollant, and J. L. Beneytout. 2002. 'Effects of alumina and zirconium dioxide particles on arachidonic acid metabolism and proinflammatory interleukin production in osteoarthritis and rheumatoid synovial cells.' *J. Bone. Joint. Surg. Br.* 84. 920–930.

811. Li, J., Y. Liu, L. Hermansson, and R. Soremark. 1993. 'Evaluation o biocompatibility of various ceramic powders with human fibroblasts *in vitro*.' *Clin. Mater.* 12. 197–201.

812. Wada, K., W. Yu, M. Elazizi, S. Barakat, M. A. Ouimet, R. Rosario-Meléndez, J. P. Fiorellini, D. T. Graves, and K. E. Uhrich. 2013. 'Locally delivered salicylic acid from poly(anhydride-ester): impact of diabetic bone regeneration.' *Journal of Controlled Release.* 171. 33–37.

813. Erdmann, L., B. Macedo, and K. E. Uhrich. 2000. 'Degradable poly(anhydride ester) implants: effects of localized salicylic acid release on bone.' *Biomaterials.* 21. 2507–2512.

814. Bruggeman, J. P. B. J. de Bruin, C. J. Bettinger, and R. Langer. 2008. 'Biodegradable poly (polyol sebacate) polymers.' *Biomaterials.* 29. 4726.

815. Zhang, L., Z. Cao, T. Bai, L. Carr, J-R. Ella-Menye, C. Irvin, B. D. Ratner, and S. Jiang. 2013. 'Zwitterionic hydrogels implanted in mice resist the foreign-body reaction.' *Nature Biotechnology.*

816. Shikanov, A., S. Shikanov, B. Vaisman, J. Golenser, and A. J. Domb. 2008. 'Paclitaxel tumor biodistribution and efficacy after intratumoral injection of a biodegradable extended release implant.' *International Journal of Pharmaceutics.* 358. 114–120.

817. Ranganath, S. H., Y. Fu, D. Y. Arifin, I. Kee, L. Zheng, H. S. Lee, P. K. Chow, and C. H. Wang. 2010. 'The use of submicron/nanoscale PLGA implants to deliver paclitaxel with enhanced pharmacokinetics and therapeutic efficacy in intracranial glioblastoma in mice.' *Biomaterials.* 31. 5199–5207.

818. Jaiswal, M., F. Naz, A. K. Dinda, and V. Koul. 2013. '*in vitro* and *in vivo* efficacy of doxorubicin loaded biodegradable semi-interpenetrating hydrogel implants of poly (acrylic acid)/gelatin for post surgical tumor treatment.' *Biomedical Materials.* 8. 045004.

819. Naraharisetti, Kumar P., B. Yung Sheng Ong, J. Wei Xie, T. Kam Yiu Lee, C. H. Wang, and N. V. Sahinidis. 2007. '*in vivo* performance of implantable biodegradable preparations delivering Paclitaxel and Etanidazole for the treatment of glioma.' *Biomaterials.* 28. 886–894.

820. Gou, M., K. Men, H. Shi, M. Xiang, J. Zhang, J. Song, J. Long, Y. Wan, F. Luo, X. Zhao, and Z. Qian. 2011. 'Curcumin-loaded biodegradable polymeric micelles for colon cancer therapy *in vitro* and *in vivo.*' *Nanoscale.* 3. 1558–1567.

821. Seil, J. T. and T. J. Webster. 2010. 'Electrically active nanomaterials as improved neural tissue regeneration scaffolds.' *WIREs Nanonedicine and Nanobiotechnology.* 2. 635–647.

822. Gupta, D. J. Venugopal, M. P.Prabhakaran, V. G. Dev, S. Low, A. T. Choon, and S. Ramakrishna. 2009. 'Aligned and random nanofibrous substrate for the *in vitro* culture of Schwann cells for neural tissue engineering.' *Acta Biomaterialia.* 5. 2560–2569.

823. Yaghoub, M. B., R. Tremblay, R. Voicu, G. Mealing, R. Monette, C. Py, K. Faid, and M. Sikorska. 2005. 'Neurogenesis and neuronal communication on micropatterned neurochips.' *Biotecnology and Bioengineering.* 92. 336–345.

824. Chiono, V, C. Tonda-Turo, and G. Ciardelli. 2009. 'Artificial scaffolds for peripheral nerve reconstruction.' *International Review of Neurobiology.* 87. 173–198.

825. Kotwal, A., and C. E. Schmidt. 2001. 'Electrical stimulation alters protein adsorption and nerve cell interactions with electrically conducting biomaterials.' *Biomaterials.* 22. 1055–1064.

826. Schmidt, C. E., V. R. Shastri, J. P. Vacanti, and R. Langer. 1997. 'Stimulation of neurite outgrowth using an electrically conducting polymer.' *Proc. Natl. Acad. Sci. USA.* 94. 8948–8953.

827. Archibald, S. J., J. Shefner, C. Krarup, and R. D. Madison. 1995. 'Monkey median nerve repaired by nerve graft or collagen nerve guide tube.' *The Journal of Neuroscience.* 15. 4109–4123.

828. Chiono, V., C. Tonda-Turo, and G. Ciardelli. 2009. 'Artificial scaffolds for peripheral nerve reconstruction.' *International Review of Neurobiology.* 87. 173–198.

829. Kotwal, A., and C. E. Schmidt. 2001. 'Electrical stimulation alters protein adsorption and nerve cell interactions with electrically conducting biomaterials.' *Biomaterials.* 22. 1055–1064.

830. Matsumoto, K., K. Ohnishi, T. Kiyotani, T. Sekine, H. Ueda, T. Nakamura, K. Endo, and Y. Shimizu. 2000. 'Peripheral nerve regeneration across an 80-mm gap bridged by a polyglycolic acid (PGA)–collagen tube filled with laminin-coated collagen fibers: a histological and electrophysiological evaluation of regenerated nerves.' *Brain Research.* 868. 315–328.

831. Evans, G. R. D. K. Brandt, M. S. Widmer, L. Lu, R. K. Meszlenyi, P. K. Gupta, A. G. Mikos, J. Hodges, J. Williams, A. Gürlek, and A. Nabawi. 1999. '*in vivo* evaluation of poly(L-lactic acid) porous conduits for peripheral nerve regeneration. *Biomaterials.*' 20. 1109–1115.

832. (a) Basu, Bikramjit, and Sourabh Ghosh. 2017. *Biomaterials for Musculoskeletal regeneration: Applications, Concepts*. Singapore: Springer Nature. (b) Thrivikraman, Greeshma, Giridhar Madras, and Bikramjit Basu. 2014. '*in vitro/in vivo* assessment and mechanisms of toxicity of bioceramic materials and its wear particulates.' *RSC Advances*. 4. 12763–12781. (c) Chandorkar, Yashoda, Nitu Bhaskar, Giridhar Madras, and Bikramjit Basu. 2015. 'Long term, sustained release of salicylic acid from crosslinked, biodegradable polyesters induces reduced foreign body response in mice.' *Biomacromolecules*. 16(2). 636–649.

833. LaVan, D., T. McGuire, and R. Langer. 2003. 'Small-scale systems for *in vivo* drug delivery.' *Nature Biotechnology*. 21(10). 1184–1191.

834. Lee, T. T., J. R. García, J. I. Paez, A. Singh, E. A. Phelps, S. Weis, Z. Shafiq, A. Shekaran, A. Del Campo, and A. J. García. 2015. 'Light-triggered *in vivo* activation of adhesive peptides regulates cell adhesion, inflammation and vascularization of biomaterials.' *Nature Materials*. 14(3). 352–360.

835. Van Oosten, M., T. Schäfer, J. A. Gazendam, K. Ohlsen, E. Tsompanidou, M. C. De Goffau, H. J. Harmsen, L. M. Crane, E. Lim, K. P. Francis, and L. Cheung. 2013. 'Real-time *in vivo* imaging of invasive- and biomaterial-associated bacterial infections using fluorescently labelled vancomycin.' *Nature Communication*. 4. 2584.

836. Pearce, A.I., R. G. Richards, S. Milz, E. Schneider, and S. G. Pearce. 2007. 'Animal models for implant biomaterial research in bone: a review.' *Eur. Cell Mater*. 13. 1–10.

837. Beauchamp, Tom L., and James F. Childress. 2009. *Principles of Biomedical Ethics*. Oxford University Press.

838. Johnson, Paula D., and David G. Besselsen. 2002. 'Practical aspects of experimental design in animal research.' *Institute for Animal Research Journal*. 43 (4). 202–206.

839. Nath, Shekhar. 2008. 'Development of Novel Calcium Phosphate–Mullite Composites for Orthopedic Applications', PhD Dis. Indian Institute of Technology, Kanpur, India.

840. Peckham, M. 'Histology at a Glance.' Accessed 24 April, 2017. http://wiley-vch.e-bookshelf.de.

841. Anderson, J. M. 2001. 'Biological responses to materials.' *Annual Review of Materials Research*. 31(1). 81–110.

842. Robinson, Rebecca. 2011. '3Rs of animal testing for regenerative medicine products.' *Science Translational Medicine*. 3(112). 112fs11.

843. Chandorkar, Yashoda. 2015. 'Multi-functional, Anti-inflammatory and Tunable Polyesters—A Novel Polymer Platform for Tissue Engineering and Drug Delivery Applications', PhD Dis. Indian Institute of Science, Bangalore, India.

Chapter 12

844. The Titanium Information Group, 'Titanium Alloys in Medical Applications' *AZo Journal of Materials Online*. Accessed 24 April, 2017. http://www.azom.com/details.asp?ArticleID=1794.

845. J. J. Jacobs, J. L. Gilbert, R. M. Urban, J. Bone Joint Surg. 80. 'Corrosion of metal orthopedic implants.' *National Center for Biotechnology Information, U.S.A.* 268.

846. Katti, K. S. 1998. 'Biomaterials in total joint replacement.' *National Center for Biotechnology Information, U.S.A.* 39(3). 133.

847. J. A. Disegi. 'Titanium alloys for fracture fixation implant, injury.' Int. J. Care Injured 31: Supplement 4 (200) D14.

848. G. He, M. and Hagiwara,. 2006. 'Ti alloy design strategy for biomedical applications.' *Mater. Sci. Engin.* C. 26 (2006), 14.

849. Lütjering, Gerd, and J. C. Williams. 2007. *Titanium, Engineering materials, processes*. Springer-Verlag, Berlin.

850. Long, M., and H.J. Rack. 1998. 'Titanium alloys in total joint replacement--a materials science perspective.' *Biomaterials*. 19. 1621.

851. Leventhal, G. C. 1951. Titanium: a metal for surgery. *J Bone Joint Surg*. 33 . 473.

852. William, D. F. 1981. *Biocompatibility of Clinical Implant Materials, Volume 1*. CRC Press.

853. McMaster, J. A. 1970. 'Titanium for Prosthetic Devices.' Presented at IMD Dental-Medical Committee, American Institute of Mechanical Engineers, Cleveland. October 21.

854. Hille, Guenther H. 1966. 'Titanium for surgical implants.' *J MATER*. 2. 373-383.

855. Solar, R. J., S. R. Pollack and E. Korostoff. 1979. '*In vitro* corrosion testing of titanium surgical implant alloys: An approach to understanding titanium release from implants.' *J. Biomed Mater Res*. 13 . 217.

856. Niinomi, M. 2002. 'Recent metallic materials for biomedical applications.' *Metallurgical and Materials Transactions*. Springer:Verlag. 33. 477.

857. De Groot, k. and R. le Geros. 1988. 'Position Papers.' Biocermanics: Material Characteristics Versus *in Vivo* Behavior. Ann New York Acad. Sci. 523, 227, 268,272.

858. Geetha, M., A. K. Singh, R. Asokamani, and A. K. Gogia. 2009. 'Ti based biomaterials the ultimate choice of orthopedic implants.' A review. *Prog Mater Sci*. 54. 397.

859. Jackson, J., W. Ahmed (Ed.) 2007. *Titanium and Titanium Alloy Applications in Medicine: Surface Engineered Surgical Tools and Medical Devices*. Springer.

860. Oliveira, V. R. R. Chaves, R. Bertazzoli, and R. Caram. 1998. 'Preparation and characterization of Ti-Al-Nb orthopedic implants.' *Brazilian Journal of Chemical Engineering*. 17. 326.

861. Boyer, R. R. 1996. 'An Overview of the Use of Titanium in the Aerospace Industry.' *Mater. Sci. Engin*. 213. 103.

862. Ferrero, J. G. 2005. 'Candidate materials for high strength faster applications in both the aerospace and automotive industry.' *Journal of Materials Engineering and Performance* 14. 691.

863. Luckey H. A., and Kubli Jr F. (Eds.). *Titanium Alloys in Surgical Implants*. ASTM Publication.

864. Donachie, M. J. 2000. *Titanium a Technical Guide, 2nd edition;* ASM International.

865. Atkinson, J. R., and B. Jobbins. 1981. 'Properties of engineering materials for use in body.' In *Introduction to Biomechanics of Joint and Joint Replacement*. London: Mechanical Engineering Publications.

866. Solar R. J. 'Corrosion Resistance of Titanium surgical implant Alloys, A Review, Corrosion and Degradation of Implant Materials.' *American Society for Testing of Materials,*. 259-273.

867. Wang, K. 1996. 'The use and Properties of Titanium and Titanium Alloys for Medical Applications in USA.' Mater. Sci. Engin. 213. 134.

868. Semlitsch, M., F. Staub, and H. Weber,. 1985., Titanium- Aluminum- Niobium alloy, development for biocompatible, high- strength surgical implants. '*Biomedizinische Technik* 30. 334.

869. Steinemann SG. 1980. ' Corrosion of surgical implants-*in vivo* and *in vitro* tests.' In: *Evaluations of Biomaterials*, edited by G. D. Winter, J. L. Leray, K. de Groot. New York: John Wiley & Sons.

870. Okazaki, Y., Y. Ito, K. Kyo, and T. Tateishi. 1996. 'Corrosion resistance and corrosion fatigue strength of new titanium alloys for medical implants without V and Al.' *Material Science Engineering*. 213. 138-147.

871. Niinomi M. 'Mechanical properties of biomedical alloys.' *Materials Science and Engineering*. 243. 231-236.

872. Scales J. T, and Black J. 'Staining around Titanium alloy Prosthesis--an orthopedic enigma.' *Journal of Bone and Joint Surgery*. 73. 534-536.

873. Khan, M. A., R. L. William, and D. E. Williams. 1996. '*in vitro* corrosion and wear of titanium Alloys in the biological environment. *Biomaterials*. 17. 212-217.

874. Williams, D. F. 1973. 'The deterioration of materials in use.' In *Implants in Surgery*, edited by D. F. Williams, R. Roaf, and W. B. Saunders. London. 137-201.

875. Atkinson, J. R., and B. Jobbins. 1981. 'Properties of engineering materials for use in body.' In *Introduction to Biomechanics of Joint and Joint Replacement*. London: Mechanical Engineering Publications.

876. Parks, J. B., and R. S. Lakes. 'Metallic Implant Materials.' *Biomaterials- an Introduction*. New York: Plenum Press.

877. Brown, S. A., and K. Merritt. 1981. 'Fretting Corrosion in Saline and Serum.' *Journal of Biomedical Materials Research*. 15. 867.

878. Urban, R. M., J.J. Jacobs, J.L. Gilbert, and J.O. Galante. 1994. 'Migration of corrosion products, from modular hip prostheses, Particle microanalysis and histopathological findings. *Journal of Bone and Joint Surgery*. 76. 1345.

879. Black, J. 1985. 'Metallic Ion Release and its relationship to Oncogenesis.' *Proceedings of the thirteenth Open Scientific Meeting of the Hip Society*. St. Louis: Mosby.

880. Torgerson, S., and N.R. Gjerdet. 'Retrieval study of Stainless Steel and Titanium miniplates and screws used in Maxillofacial Surgery.' *Journal of Material Science: Materials Medicine* 5. 256.

881. Bundinski, K. G. 1991. 'Tribological Properties of Titanium alloys.' *Wear*. 151. 203.

882. Hutchings, L. M., and A.P. Mercer, The influence of atmosphere composition on the abrasive wear of titanium and Ti-6Al-4V. Wear. 124. 165.

883. Donachie, J. Mathew. 2000. *Titanium :a Technical Guide,* 2nd edition. ASM International.

884. (a) Choubey, A., B. Basu, and R. Balasubramaniam. 2004. 'Electrochemical Behavior of Intermetallic Ti$_3$Al based Alloys in Simulated Human Body Fluid Environment.' *Intermetallics*. 12. 679-682, (b) Choubey, A., B. Basu, and R. Balasubramaniam. 2004. 'Effect of Replacement of V by Nb and Fe on the Electrochemical and Corrosion Behaviour of Ti-6Al-4V in Simulated Physiological Environment. *Journal of Alloys and Compounds*. 381. 288-294, (c) Choubey, Animesh, Bikramjit Basu, and R. Balasubramaniam. 2004. 'Tribological behaviour of Ti-based Alloys in simulated body fluid solution at fretting contacts.' *Materials Science and Engineering*. 379. 234-239.

885. Handtrack, D., C. Sauer, and B. Kieback. 2008. 'Microstructure and properties of ultrafine-grained and dispersion -strengthened titanium materials of implants.' *J. Mater. Sci.* 43. 671.

886. Handtrack, D., F. Despang, C. Sauer, B. Kieback, N. Reinfried, and Y. Grin. 2006. 'Fabrication of the ultrafine grained and dispersion - strengthened titanium material by spark plasma sintering.' *Mater. Sci. Engin.* 437. 423.

887. Jones D. A. 1992. *Principles and Prevention of Corrosion*. Maxwell Macmillan: New York. 75-76.

888. Despang, E., A. Bernhardt, A. Lode, T. Hanke, D. Handtrack, B. Kieback, and M. Gelinsky. 2010. 'Response of human bone marrow stromal cells to a novel ultra-fine-grained and dispersion-strengthened titanium-based material.' *Acta Biomater.* 6. 1006.

889. Aragon, P. J., and S. F. Hulbert. 1972. 'Corrosion of Ti-6A1-4V in simulated body fluids and bovine plasma.' *J. Biomed. Mater.* Res. 6. 155.

890. Choubey, A. 2003. 'Electrochemical and tribological behaviour of novel titanium based alloys for human body implants.' Masters Thesis, Indian Institute of Technology Kanpur, India 2003.

891. Davidson, J. A. A. K. Mishra, P. Kovacs, and R. A. Poggie, .1994. 'New surface-hardened, low-modulus, corrosion-resistant Ti-13Nb-13Zr alloy for total hip arthroplasty.' Biomed. Mater. Eng., 4. 231.

892. Jaffee, R. L., and H. M. Brute. Eds. 1973. *Titanium Science and Technology, Volume4*. Plenum Press: London.

893. Speck, K. M., and A. C. Fraker,. 1980. 'Anedic polarization behavior of Ti- Ni and Ti-6A1-4Vin simulated physiological solutions.' *J Dent. Res.* 59. 1590.

894. Covino Jr, S. Bernard, and E. Alman David. 2002. 'Corrosion of Titanium Matrix Composites.' No. DOE/ARC-2002-011. Albany Research Center (ARC): Albany.

895. Verkhoturov, A. D., H. A. Kuzenkova, N. V. Lebukhova, and I. A. Podchernyaeva. 1988. *Poroshkovaya Metallurgiya* 302(2) 81.

896. Cheng, Y., and Y. F. Zheng,. 2007. 'Characterization of TiN, TiC and TiCN coatings on Ti–50.6 at.% Ni alloy deposited by PIII and deposition technique.' *Surf. Coatings Technol.* 201. 4909.

897. Turgoose, S., and R.A. Cottis. 1999. 'Corrosion Testing Made Easy: Electrochemical Impedance and Noise,' NACE International, Houston, TX.

898. Assis, S. L., S. Wolynec, and I. Costa,. 2008. 'The electrochemical behaviour of Ti-13Nb-13Zr alloy in various solutions.' *Materials and Corrosion* 59. 739.

899. Tamilselvi, S., and N. Rajendran,. 2006. 'Electrochemical Studies on the Stability and Corrosion Resistance of Ti-5Al-2Nb-1Ta Alloy for Biomedical Applications.' *Trends Biomater. Artif. Organs*, 20. 49.

900. (a) Raghunandan, U. 2010. 'Corrosion and *in vitro* mineralization of Ti-Si-C alloy,' Masters thesis, Indian Institute of Technology Kanpur, India.
(b) Ummethala, Raghunandan. Florian Despang, Michael Gelinsky, and Bikramjit Basu,. 2011. '*in vitro* corrosion and mineralization of novel Ti–Si–C alloy.' Electrochimica Acta, Volume 56. 3809-3820.

901. Feng, B., Y. Chen, and X. D. Zhang. 2002. 'Effect of water vapor treatment on apatite formation on precalcified titanium and bond strength of coatings to substrates.' *J Biomed. Mater. Res.* 59 (2002) 12.

902. Takadama, H., H. M. Kim, T. Kokubo, and T. Nakamura,. 2001. 'TEM-EDX study of mechanism of bonelike apatite formation on bioactive titanium metal in simulated body fluid.' *J Biomed. Mater. Res.* 57. 441.

903. Habibovic, P., F. Barrère, C. A. van Blitterswijk, K. de Groot, and P. Layrolle,. 2002. 'Biomimetic hydroxyapatite coating on metal implants.' *J. Am. Ceram.* Soc. 85. 517.

904. Fatehi, K., F. Moztarzadeh, M. S. Hashjin, M. Tahriri, M. Rezvannia, and R. Ravarian,. 2008. '*In vitro* biomimetic deposition of apatite on alkaline and heat treated Ti6Al4V alloy surface.' *Bull. Mater. Sci.* 31. 101.

905. Wang, X., Y. Li, J. Lin, P. D. Hodgson, and C. Wen, . 2008. 'Effect of heat-treatment atmosphere on the bond strength of apatite layer on Ti substrate.' *Dental materials.* 24. 1549.

906. Athanasiou, k., C. F. Zhu, D. R. Lanctot, D. R.; C.M. Agrawal, and X. Wang,. 2000. 'Fundamentals of biomechanics in tissue engineering of bone.' *Tissue Engineering.* 6. 361.

907. Datye, A. V., M. Jaramillo, and K. H. Wu. 2004. 'Corrosion behaviour of cardiovascular stents.' *LACCEI 2004*, Challenges and Opportunities for Engineering Education, Research and Development,Miami, Florida, USA.

908. Caia, Z., T. Shafer, I. Watanabe, M. E. Nunn, and T. Okabe. 2003.'Electrochemical characterization of cast titanium alloys.' *Biomaterials* 24. 213.

Chapter 13

909. Gautier, S., E. Champion, and D. B. Assollant. 1997. 'Processing, microstructure and toughness of Al_2O_3 platelet-reinforced hydroxyapatite.' *J. Euro. Ceram. Soc.* 17. 1361–1369.

910. Li, J., B. Fartash, and L. Hermansson. 1995. 'Hydroxyapatite–alumina composites and bone-bonding.' *Biomaterials.* 16. 417–422.

911. Rao, R. R., and T. S. Kannan. 2002. 'Synthesis and sintering of hydroxyapatite–zirconia composites.' *Mat. Sci. Engg. C.* 20. 187–193.

912. Silva, V. V., and F. S. Lameiras. 2001. 'Domínguez RZ. Microstructural and mechanical study of zirconia-hydroxyapatite (ZH) composite ceramics for biomedical applications.' *Compo. Sci. Tech.* 61. 301–310.

913. Goller, G., H. Demirkiran, F. N. Oktar, and E. Demirkesen. 2003. 'Processing and characterization of bioglass reinforced hydroxyapatite composites.' *Ceram. Inter.* 29. 721–724.

914. Suchanek, W. M. Yashima, M. Kakihana, and M. Yoshimura. 1997. 'Hydroxyapatite/hydroxyapatite–whisker composites without sintering additives: mechanical properties and microstructural evolution.' *J. Am. Ceram. Soc.* 80(11). 2805–2813.

915. Schneider, H., K. Okada, and J. Pask. 1994. *Mullite and Mullite Ceramics.* John Wiley & Sons. 260.

916. (a) Nath, S., K. Biswas, and B. Basu. 2008. 'Phase stability and microstructure development in hydroxyapatite–mullite system.' *Scrip. Mat.* 58(12). 1054–1057. (b) Nath, Shekhar. 2008. 'Development of Novel Calcium Phosphate–Mullite Composites for Orthopedic Applications.' PhD Dissertation. Indian Institute of Technology, Kanpur, India. (c) Dubey, Ashutosh Kumar, Geet Sitesh, Shekhar Nath, and Bikramjit Basu. 2011. 'Spark plasma sintering to restrict sintering reactions and enhance properties of hydroxyapatite–mullite biocomposites.' *Ceramics International.* 37. 2755–2761. (d) Nath, Shekhar, Bikramjit Basu, Mira Mohanty, and P. V. Mohanan. 2009. '*in vivo* response of novel hydroxyapatite–mullite composites: results up to 12 weeks of implantation.' *Journal of Biomedical Materials Research: Part B.* 90B. 547–557. (e) Nath, Shekhar, A. Dey, A. K. Mukhopadhyay, and Bikramjit Basu. 2009. 'Nanoindentation response of novel hydroxyapatite–mullite composites.' *Mat. Sc. Engg.* A513–514. 197–201.

917. Barnes, C. A. A. Elsaesser, J. Arkusz, A. Smok, J. Palus, A. Lesniak, A. Salvati, J. P. Hanrahan, W. H. Jong, E. Dziubaltowska, M. Stepnik, K. Rydzynski, G. McKerr, I. Lynch, K. A. Dawson, and C. V. Howard. 2008. 'Reproducible comet assay of amorphous silica nanoparticles detects no genotoxicity.' *Nano. Lett.* 8(9). 3069.

918. Katzer, A., H. Marquardt, J. Westendorf, J. V. Wening, and G. von Foerster. 2002. 'Polyetheretherketone—cytotoxicity and mutagenicity *in vitro*.' *Biomaterials.* 23(8). 1749.

919. Muller, B. P., A. Eisentrager, W. Jahnen-Dechent, W. Dott, and J. Hollender. 2003. 'Effect of sample preparation on the *in vitro* genotoxicity of a light curable glass ionomer cement.' *Biomaterials.* 24(4). 611.

920. Muller, B. P., S. Ensslen, W. Dott, and J. Hollender. 2002. 'Improved sample preparation of biomaterials for *in vitro* genotoxicity testing using reference materials.' *J. Biomed. Mater. Res.* 61(1). 83.

921. Brown, C., S. Williams, J. L. Tipper, J. Fisher, and E. Ingham. 2007. 'Characterisation of wear particles produced by metal on metal and ceramic on metal hip prostheses under standard and microseparation simulation.' *J. Mater. Sci. Mater. Med.* 18(5). 819.

922. Hatton, A., J. E. Nevelos, A. A. Nevelos, R. E. Banks, J. Fisher, and E. Ingham. 2002. 'Alumina-alumina artificial hip joints. Part I: a histological analysis and characterisation of wear debris by laser capture microdissection of tissues retrieved at revision.' *Biomaterials.* 23(16). 3429.

923. Kleinsasser, N. H., K. Schmid, A. W. Sassen, U. A. Harréus R. Staudenmaier, M. Folwaczny, J. Glas, and F. X. Reichl. 2006. 'Cytotoxic and genotoxic effects of resin monomers in human salivary gland tissue and lymphocytes as assessed by the single cell microgel electrophoresis (Comet) assay.' *Biomaterials.* 27(9). 1762.

924. Papageorgiou, I., C. Brown, R. Schins, S. Singh, R. Newson, S. Davis, J. Fisher, E. Ingham, and C. P. Case. 2007. 'The effect of nano-and micron-sized particles of cobalt–chromium alloy on human fibroblasts *in vitro*.' *Biomaterials.* 28(19). 2946.

925. Jagota, A., E. D. Boyes, and R. K. Bordia. 1992. 'Sintering of glass powder packings with metal Inclusions.' *Proc. Mat. Res. Soc. Symp.* 249. 475–480.

926. Yin, X. and M. J. Stott. 2003. 'α- and β-tricalcium phosphate: a density functional study.' *Phy. Rev. B.* 68. 205205.

927. Pierre, P. D. S. S. 1954. 'Constitution of bone china: I, high temperature phase equilibrium studies in the system tricalcium phosphate-alumina-silica.' *J. Am. Ceram. Soc.* 36(6) 243–258.

928. Mackay, A. L. 1953. 'A preliminary examination of the structure of α-Ca$_3$(PO$_4$)$_2$.' *Acta. Cryst.* 6. 743.

929. Arends, J., J. Christoffersen, M. R. Christoffersen, H. Eckert, B. O. Fowler, J. C. Heughebaert, G. H. Nancollas, J. P. Yesinowski, and S. J. Zawacki. 1987. 'Hydroxyapatite precipitated from an aqueous solution: an international multimethod analysis.' *J. Cryst. Gr.* 84. 515–532.

930. Awaji, H., and Y. Sakaida. 1990. 'V-notch technique for single-edge notched beam and chevron notch methods.' *J. Am. Ceram. Soc.* 73. 3522–3523.

931. Zhang, X., L. Xu, S. Du, J. Han, P. Hu, and W. Han. 2008. 'Fabrication and mechanical properties of ZrB_2–SiC_w ceramic matrix composite.' *Mater. Lett.* 62. 1058–1060.

932. Suchanek, W., M. Yashima, M. Kakihana, and M. Yoshimura. 1996. 'Processing and mechanical properties of hydroxyapatite reinforced with hydroxyapatite whiskers.' *Biomaterials.* 17. 1715–1723.

933. Metcalfe, B. L., and J. H. Sant. 1975. 'The synthesis, microstructure and physical properties of high purity mullite.' *Trans. Br. Ceram. Soc.* 74. 193–201.

934. Mrudiyasni, K. S., and L. M. Brown. 1972. 'Synthesis and mechanical properties of stoichiometric aluminum silicate (mullite).' *J. Am. Ceram. Soc.* 55. 548–552.

935. Taya, M., S. Hayashi, A. S. Kobayashi, and H. S. Yoon. 1990. 'Toughening of a particulate-reinforced ceramic-matrix composite by thermal residual stress.' *J. Am. Ceram. Soc.* 73. 1382–1391.

936. Reis, R. L. and T. S. Román. 2005. *Biodegradable Systems in Tissue Engineering and Regenerative Medicine.* United States of America: CRC Press.

937. Ivanov, B., A. M. Hakimian, G. M. Peavy, and. R. F. Jr. Haglund. 2003. 'Mid-infrared laser ablation of a hard biocomposite material: mechanistic studies of pulse duration and interface effects.' *Appl. Surf. Sc.* 208–209. 77–84.

938. Declercq, H. A., R. M. H. Verbeeck, L. I. F. J. M. D. Ridder, E. H. Schacht, and M. J. Cornelissen. 2005. 'Calcification as an indicator of osteoinductive capacity of biomaterials in osteoblastic cell cultures.' *Biomaterials.* 26. 4964–4974.

939. Arinzeh, T. L., T. Tran, J. Mcalary, and G. Daculsi. 2005. 'A comparative study of biphasic calcium phosphate ceramics for human mesenchymal stem-cell-induced bone formation.' *Biomaterials.* 26. 3631–3638.

940. Ku, C. H., D. P. Pioletti, M. Browne, and P. J. Gregson. 2002. 'Effect of different Ti–6Al–4V surface treatments on osteoblasts behavior.' *Biomaterials.* 23. 1447–1454.

941. Es-Souni, M., and H. F. Brandies. 2005. 'Assessing the biocompatibility of NiTi shape memory alloys used for medical applications.' *Anal. Bioanal. Chem.* 381. 557–567.

942. Koshihara, M., M. Masuyama, M. Uehara, and K. Suzuki. 2004. 'Effect of dietary calcium: Phosphorus ratio on bone mineralization and intestinal calcium absorption in ovariectomized rats.' *Bio Factors.* 22.39–42.

943. Thian, E. S., J. Huang, S. M. Best, Z. H. Barber. , R. A. Brooks, N. Rushton, and W. Bonfield. 2006. 'The response of osteoblasts to nanocrystalline silicon-substituted hydroxyapatite thin films.' *Biomaterials.* 27. 2692–2698.

944. Kue, R., A. Sohrabi, D. Nagle, C. Frondoza, and D. Hungerford. 1999. 'Enhanced proliferation and osteocalcin production by human osteoblast-like MG63 cells on silicon nitride ceramic discs.' *Biomaterials.* 20. 1195–1201.

945. Misra, A., R. P. Z. Lim, Z. Wu, and T. Thanabalu.2007. 'N-WASP. plays a critical role in fibroblast adhesion and spreading.' *Biochem. Biophy. Res. Comm.* 364. 908–912.

946. West, K. A., H. Zhang, and M. C. Brown. 2001. 'The LD4 motif of paxillin regulates cell spreading and motility through an interaction with paxillin kinase linker (PKL).' *J. Cell. Biol.* 154. 161–176.

947. Chang, X., Z. M. Hou, S. Yasuaki, T. Tomoo, and Y. Akira. 2005. 'Quantitative study of alkaline phosphatase and osteocalcin in the process of new bone.' *West China J. Stomat.* 23. 424–426.

948. Shiratori, K., K. Matsuzaka, Y. Koike, S. Murakami, M. Shimono, and T. Inoue. 2005. 'Bone formation in β-tricalcium phosphate-filled bone defects of the rat femur: Morphometric analysis and expression of bone related protein mRNA.' *Biomed. Res.* 26. 51–59.

949. Matsuno, T., Y. Hashimoto, T. Nakahara, K. Kuremoto, T. Nakamura, and T. Satoh. 2005. 'β-TCP/Collagen Sponge Composite Enhances the Osteogenic Differentiation of Mesenchymal Stem Cells.' *J. Hard Tissue Bio.* 14. 189–191.

950. Piattelli, A., A. Scarano, and C. Mangano. 1996. 'Clinical and histologic aspects of biphasic calcium phosphate ceramic (BCP) used in connection with implant placement.' *Biomaterials.* 17. 1767–1770.

951. Uoshima, M. H., I. Ishikawa, A. Kinoshita, H. T. Weng, and S. Oda. 1995. 'Clinical and histological observation of replacement of biphasic calcium phosphate by bone tissue in monkeys.' *Int. J. Periodont. Rest. Dent.* 15. 204–213.

952. Tice, R. R., E. Agurell, D. Anderson, B. Burlinson, A. Hartmann, H. Kobayashi, Y. Miyamae, E. Rojas, J. C. Ryu, and Y. F. Sasaki. 2000. 'Single cell gel/comet assay: guidelines for *in vitro* and *in vivo* genetic toxicology testing.' *Environ. Mol. Mutagen.* 35(3). 206.

953. Singh, N. P., M. T. McCoy, R. R. Tice, and E. L. Schneider. 1988. 'A simple technique for quantitation of low levels of DNA damage in individual cells.' *Exp. Cell. Res.* 175(1). 184.

954. Bajpayee, M., A. K. Pandey, D. Parmar, N. Mathur, P. K. Seth, and A. Dhawan. 2005. 'Comet assay responses in human lymphocytes are not influenced by the menstrual cycle: a study in healthy Indian females.' *Mutat. Res.* 565(2). 163.

955. Kadoya, Y., A. Kobayashi, and H. Ohashi. 1998. 'Wear and osteolysis in total joint replacements.' *Acta. Orthop. Scand. Suppl.* 278. 1.

956. Jiranek, W. A., M. Machado, M. Jasty, D. Jevsevar, H. J. Wolfe, S. R. Goldring, M. J. Goldberg, and W. H. Harris. 1993. 'Production of cytokines around loosened cemented acetabular components. Analysis with immunohistochemical techniques and *in situ* hybridization.' *J. Bone Joint Surg. Am.* 75(6). 863.

957. Neale, S. D., and N. A. Athanasou. 1999. *Acta. Orthop. Scand.* 70(5). 452.

958. Hirayama, T., Y. Fujikawa, I. Itonaga, and T. Torisu. 2001. 'Effect of particle size on macrophage-osteoclast differentiation *in vitro.*' *J. Orthop. Sci.* 6(1). 53.

959. Nath, S., S. Kalmodia, and B. Basu. 2010. 'Densification, phase stability and *in vitro* biocompatibility property of hydroxyapatite-10 wt% silver composites.' *J. Mater. Sci. Mater. Med.* 21(4). 1273.

960. Heddle, J. A. 1973. 'A rapid *in vivo* test for chromosomal damage.' *Mutat. Res.* 18(2). 187.

961. Schmid, W. 1975. 'The micronucleus test.' *Mutat. Res.* 31(1). 9.

962. Clark, P. A., A. Rodriguez, D. R. Sumner, M. A. Hussain, and J. J. Mao. 2005. 'Modulation of bone ingrowth of rabbit femur titanium implants by *in vivo* axial microechanical laoding.' *J. App. Physio.* 98. 1922–1929.

963. Ratner, B. D., A. S. Hoffman, F. J. Schoen, and J. E. Lemons. 2004. *Biomaterials Science–An Introduction to Materials in Medicine, 2nd edition.* New York: Academic Press.

964. Barros, V. M. R., L. A. Salata, C. E. Sverzut, S. P. Xavier, R. V. Noort, A. Johnson, and P. V. Hatton. 2002. '*in vivo* bone tissue response to a canasite glass–ceramic.' *Biomaterials.* 23. 2895–2900.

965. Basu, Bikramjit. 2017. *Biomaterials for Musculoskeletal Regeneration: Concepts.* Singapore: Springer Nature.

Chapter 14

966. Basu, B., D. Katti, and A. Kumar. 2009. *Advanced Biomaterials: Fundamentals, Processing and Applications, 1st edition.* USA: Wiley.

967. Sepulveda, P., F. S. Ortega, M. D. M. Innocentini, and V. C. Pandolfelli. 2001. 'Properties of highly porous hydroxyapatite obtained by the gelcasting of foams.' *J. Am. Chem. Soc.* 83. 3021–3024.

968. Bruder S. P. and B. S. Fox. 1999. 'Tissue engineering of bone: cell based strategies.' *Clin. Orthop.* 367S. 68–83.

969. Kokubo, T., H. M. Kim, and M. Kawashita. 2003. 'Novel bioactive materials with different mechanical properties.' *Biomaterials.* 24. 2161–2175.

970. Wang, M. 2003. 'Developing bioactive composite materials for tissue replacement.' *Biomaterials.* 24. 2133–2151.

971. Le Guehennec, L., P. Layrolle, and G. Daculsi. 2004. 'A review of bioceramics and fibrin sealant.' *Eur. Cells Mater.* 8. 1–11.

972. Hopp, S. G., J. G. Dahners, and L. A. Gilbert. 1989. 'A study of the mechanical strength of long bone defects treated with various bone autograft substitutes: an experimental investigation in the rabbit.' *J. Orthop. Res.* 7. 579–584.

973. Perry. C. R. 1999. 'Bone repair techniques, bone graft and bone graft substitutes.' *Clin. Orthop.* 360. 71–86.

974. Peter, S. Y., M. J. Miller, A. W. Yasko, M. J. Yaszemski, and A. G. Mikos. 1998. 'Polymer concepts in tissue engineering.' *J. Biomed. Mater. Res.* 43. 422–427.

975. Korkusuz, P., and F. Korkusuz. 2004. 'Hard Tissue–Biomaterial Interactions.' In: *Biomaterials in Orthopedics.*, edited by Yaszemski, M. J., D. J. Trantolo, K. U. Lewandrowski, V. Hasırcı, D. E. Altobelli, and D. L. Wise. New York: Marcel Dekker Inc. 1–40.

976. Ignner, K. E., C. Doyle, and W. Bonfield. Eds. Heimke, G., U. Soltesz, and A. J. C. Lee. 1990. 'The Strength of the Interface Developed Between Biomaterials and Bone.' *Advances in Biomaterials.* Amsterdam: Elsevier Science Publishers 9. 149-154.

977. Doyle, C., K. E. Tanner, and W. Bonfield. 1991. '*in vitro* and *in vivo* evaluation of polyhydroxybutyrate and polyhydroxybutyrate reinforced with hydroxyapatite.' *Biomaterials.* 12. 841–847.

978. Wang, M., D. Porter, and W. Bonfield. 1994. 'Processing, characterization and evaluation of hydroxyapatite reinforced polyethylene composites.' *British Ceramics Transactions.* 93. 91–95.

979. Huang, J., L. Di Silvio, M. Wang, I. Rehman, C. Ohtsuki, and W. Bonfield. 1997. 'Evaluation of *in vitro* bioactivity and biocompatibility of Bioglass-reinforced polyethylene composite.' *J. Mater. Sci. Mater. Med.* 8(12). 809.

980. Hench, L. L. 1998. 'Bioceramics.' *J. Am. Cer. Soc.* 81. 1705.

981. Downes, R. N., S. Vardy, K. E. Tanner, and W. Bonfield. 1991. 'Hydroxyapatite-Polyethylene Composite in Orbital Surgery.' In: *Bioceramics 4*, edited by Bonfield, W., K. E. Tanner, and G. W. Hastings. Oxford: Butterworth Heinemann.

982. Tanner, K. E., R. N. Downes, and W. Bonfield. 1994. 'Clinical applications of hydroxyapatite reinforced materials.' *Brit. Ceram. Trans.* 93. 104.

983. Deb, S., M. Wang, K. E. Tanner, and W. Bonfield. 1996. 'Hydroxyapatite-polyethylene composites: effect of grafting and surface treatment of hydroxyapatite.' *J. Mater. Sci. Mater. Med.* 7. 191.

984. Takuma, M., S. Harada, F. Kurokawa, F. Takashima, S. Miyauchi, and T. Maruyama. 1987. 'Experimental study on the functional adaptation to aluminium oxide,hydroxyapatite and titanium implants.' *J. Osaka. Univ. Dent. Sch.* 27. 111–121.

985. Zetterqvist, L., G. Anneroth, and A. Nordenram. 1991. 'Tissue integration of AlzO,-ceramic dental implants: an experimental study in monkeys.' *Int. J. Oral. Maxillofac. Implants.* 6. 285–293.

986. Davies, J. E., N. Nagai, N. Takeshita, and D. C. Smith. 1991. 'Deposition of Cement Like Matrix on Implant Materials.' In: *The Bone Biomaterial Interface.*, edited by Davies, J. E. Toronto, Canada: University of Toronto Press. 285–294.

987. (a) Bodhak, Subhadip, Shekhar Nath, and Bikramjit Basu. 2009. 'Friction and wear properties of Novel HDPE-HAp-Al$_2$O$_3$ biocomposites against alumina counterface.' *J. Biomaterials Applications.* 23. 407–433. (b)Nath, Shekhar, Subhadip Bodhak, and Bikramjit Basu. 2009. 'HDPE-Al$_2$O$_3$-HAp composites for biomedical applications: processing and characterization.' *Journal of Biomedical Materials Research: Part B, Applied Biomaterials.* 88B. 1–11. (c) Bodhak, S., S. Nath, and B. Basu. 2008. 'Fretting wear properties of hydroxyapatite, alumina containing high density polyethylene biocomposites against zirconia.' *Journal of Biomedical Materials Research: Part A.* 85. 83–98. (d) Nath, S., S. Bodhak, and B. Basu. 2007. 'Tribological investigation of Novel HDPE-HAp-Al$_2$O$_3$ hybrid biocomposites against steel under dry and simulated body fluid condition.' *Journal of Biomedical Materials Research: Part A.* 83A(1). 191–208. (e) Tripathi, Garima, and Bikramjit Basu. 2014. '*in vitro* osteogenic cell proliferation, mineralization and *in vivo* osseointegration of injection molded HDPE-based hybrid composites in rabbit animal model.' *J. Biomaterials Applications.* 29(1). 142–157. (f) Tripathi, G., J. E. Dough, A. Dinda, and B. Basu. 2013. '*in vitro* cytotoxicity and *in vivo* osseointergration property of compression molded HDPE-HA-Al$_2$O$_3$ hybrid biocomposites.' *Journal of Biomedical Materials Research Part A.* 101(6). 1539–1549. (g) Tripathi, Garima, Ashutosh Dubey, and Bikramjit Basu. 2012. 'Evaluation of physico-mechanical properties and *in vitro* biocompatibility of compression molded HDPE based biocomposites with HA/Al$_2$O$_3$ ceramic fillers and titanate coupling agents.' *Journal of Applied Polymer Science.* 124. 3051–3063.

988. Xavier, S. F., J. M. Schultz, and K. J. Friedrich. 1990. 'Fracture propagation in particulate filled polypropylene composites.' *Mater. Sci.* 25. 2428.

989. Rothon, R. N. 1995. *Particulate-Filled Polymer Composites.*, 2nd edition, U.K.: Longman Scientific & Technical.

990. Iwashita, N., E. Psomiadou, and Y. Sawada. 1998. 'Effect of coupling treatment of carbon fiber surface on mechanical properties of carbon fiber reinforced carbon composites.' *Composites Part A: Appl. Sc. Manufact.* 29. 965.

991. Shiguo, D., C. Jianfei, X. Yinghong, J. Faxiang, and J. Xiaoying. 1997. 'Effect of coupling agents on mechanical property of AN/PU energy composite.' *J. Appl. Polym. Sci.* 63. 1259.

992. Menon, N., F. D. Blum, and I. R. Dharani. 1994. 'Use of titanate coupling agents in kevlar–phenolic composites.' *J. Appl. Poly. Sci.* 54. 113.

993. Singh, B., A. Verma, and M. Gupta. 1998. 'Studies on adsorptive interaction between natural fiber and coupling agents.' *J. Appl. Poly. Sci.* 70. 1847.

994. Hunt, K. N., J. R. G. Evans, and K. Woodthorpe. 1998. 'Processing character of MgO-partially stabilized zirconia (PSZ) in size grading prepared by injection molding.' *J. Polym. Engg. Sci.* 28. 1572.

995. Monte, S. J. and G. Sugerman. 1984. 'Processing of composites with titanate coupling agents.' A review. *Polym. Engg. Sci.* 24. 1369.

996. Kharade, A. Y. and D. D. Kale. 1999. 'Lignin-filled polyolefins.' *J. Appl. Polym. Sci.* 72. 1321.

997. Sousa, R. A., R. L. Reis, A. Cunha, and M. J. Bevis. 2002. 'Structure development and interfacial interactions in high-density polyethylene/hydroxyapatite (HDPE/HA) composites molded with preferred orientation.' *J. Appl. Polym. Sci.* 86. 2873.

998. Pramanik, N., S. Mohapatra, P. Bhargava, and P. Pramanik. 2009. *Mater. Sci. Engg.* C29. 228.

999. ASTM, E., 1876. 'Standard test method for dynamic young's modulus, shear modulus, and poisson's ratio by impulse excitation of vibration.' American Society for Testing and Materials (ASTM), New York.

1000. Saldana, L., V. Barranco, J. L. Gonzalez-Carrasco, M. Rodriguez, L. Munuera, and N. Vilaboa. 2007. 'Thermal oxidation enhances early interactions between human osteoblasts and alumina blasted Ti6Al4V alloy.' *J. Biomed. Mater. Res.* 81A. 334–346.

1001. Basle, M. F., A. Rebel, F. Grizon, G. Daculsi, N. Passuti, and R. Filmon. 1993. 'Cellular response to calcium phosphate ceramics implanted in rabbit bone.' *J. Mater. Sci. Mater. Med.* 4. 273–280.

1002. Boss, J. H. 1999. 'Osseointegration.' *J. Long Term Eff. Med. Impl.* 18. 1–10.

1003. Shors, E. C., and R. E. Holmes. 1993. *An Introduction to Bioceramics, Advanced Series in Ceramics.* Eds. Hench, L and J. Wilson. Singapore: World Scientific.

1004. Ratner, B. D., A. S. Hoffman, F. J. Schoen, and J. E. Lemons. 2004. *Biomaterials Science: An Introduction to Materials in Medicine, 2nd edition.* New York: Academic Press.

1005. Clark, P. A., A. Rodriguez, D. R. Sumner, M. A. Hussain, and J. J. Mao. 2005. 'Modulation of bone ingrowth of rabbit femur titanium implants by *in vivo* axial microechanical laoding.' *J. Appl. Physiol.* 98. 1922–1929.

1006. Hokrgozar, M. A., M. Farokhi, F. Rajaei, M. H. A. Bagheri, S. Azari, I. Ghasemi, F. Mottaghitalab, K. Azadmanesh, and J. Radfar. 2010. 'Biocompatibility evaluation of HDPE-UHMWPE reinforced β-TCP nanocomposites using highly purified human osteoblast cells.' *J. Biomed. Mater. Res.* 95A. 1074–1083.

1007. Bélanger, M. and Y. Marois. 2001. 'Hemocompatibility, biocompatibility, inflammatory and *in vivo* studies of primary reference materials low-density polyethylene and polydimethylsiloxane: A review.' *J. Biomed. Mater. Res.* 58. 467–477.

Chapter 15

1008. Katsikogianni, M. and Y. F. Missirilis. 2004. 'Concise review of mechanisms of bacterial adhesion to biomaterials and of techniques used in estimating bacteria-material interactions.' *Eur. Cells. Mater.* 8. 37.

1009. Osborn J. F. and H Newesely. 1980. 'The material science of calcium phosphate ceramics.' *Biomaterials.* 1. 108.

1010. Lee, I. S., D. H. Kim, H. E. Kim, Y. C. Jung, and C. H. Han. 2002. 'Biological performance of calcium phosphate films formed on commercially pure Ti by electron-beam evaporation.' *Biomaterials.* 23. 609.

1011. Nath, S., R. Tripathi, and B. Basu. 2009. 'Understanding phase stability, microstructure development and biocompatibility in calcium phosphate–titania composites, synthesized from hydroxyapatite and titanium powder mix.' *Mater. Sci. Engg. C.* 29. 97.

1012. Nath, S., K. Biswas, and B. Basu. 2008. 'Phase stability and microstructure development in hydroxyapatite–mullite system.' *Scripta Materialia.* 58. 1054.

1013. Oonishi, H. 1991. 'Orthopedic applications of hydroxyapatite.' *Biomaterials.* 12. 171.

1014. Guha, A. K., S. Singh, R. Kumaresan, S. Nayar, and A. Sinha. 2009. 'Mesenchymal cell response to oversized biphasic calcium phosphate composites.' *Colloids and Surfaces B: Biointerfaces.* 73. 156.

1015. Ning, C. Q. and Y. Zhou. 2002. '*In vitro* bioactivity of a biocomposite fabricated from HA and Ti powders by powder metallurgy method.' *Biomaterials.* 23. 2909.

1016. Jarcho, M. 1981. 'Calcium phosphate ceramics as hard tissue prosthetics.' *Clin. Orthop. Rel. Res.* 157. 259.

1017. Alt, V., T. Bechert, P. Steinrucke, M. Wagner, P. Seidel, E. Dingeldein, E. Domann, and R. Schnettler. 2004. 'An *in vitro* assessment of the antibacterial properties and cytotoxicity of nanoparticulate silver bone cement.' *Biomaterials.* 25. 4383.

1018. Kim, T. N., Q. L. Feng, J. O. Kim, J. Wu, H. Wang, G. C. Chen, and F. Z. Cui. 1998. 'Antimicrobial effects of metal ions (Ag+, Cu2+, Zn2+) in hydroxyapatite.' *J. Mater. Sci.: Mater. Med.* 9. 129.

1019. Yingguang, L., Y. Zhuoru, and C. Jiang. 2007. 'Preparation and Antibacterial Properties of Zinc-Doped Hydroxyapatite Nanoparticles [J].' *J. Rare Earths.* 25. 452.

1020. Kim, J. S., E. Kuk, K. N. Yu, J. H. Kim, S. J. Park, H. J. Lee, S. H. Kim, Y. K. Park, Y. H. Park, C. Y. Hwang, Y. K. Kim, Y. S. Lee, D. H. Jeong, and M. H. Cho. 2007. 'Antimicrobial effects of silver nanoparticles.' *Nanomedicine: Nanotechnology, Biology and Medicine.* 3. 95.

1021. Asmus, S. M. F., S. Sakakuoorand, and G. Pezzotti. 2003. 'Hydroxyapatite toughened by silver inclusions.' *J. Comp. Mater.* 37. 2117.

1022. Lee, B. T., N. Y. Shin, J. K. Han, and H. Y. Song. 2006. 'Microstructures and fracture characteristics of spark plasma-sintered HAp–5vol.% Ag composites.' *Mater. Sci. Engg. A.* 429. 348.

1023. (a) Singh, Brajendra, S. Kumar, Bikramjit Basu, and Rajeev Gupta. 2015. 'Phase stability of silver particles embedded calcium phosphate bioceramics.' *Bull. Mat. Sc.* 38. 525–529. (b) Singh, Brajendra, S. Kumar, Bikramjit Basu, and Rajeev Gupta. 2015. 'Conductivity studies of Silver, Potassium and Magnesium doped Hydroxyapatites.' *International Journal of Applied Ceramic Technology.* 12(2). 319–328. (c) Singh, Brajendra S. Kumar, Bikramjit Basu, and Rajeev Gupta. 2013. 'Enhanced ionic conduction in hydroxyapatites.' *Materials Letters.* 95. 100–102. (d) Singh, Brajendra, Ashutosh Dubey, S. Kumar, Naresh Saha, Bikramjit Basu, and Rajeev Gupta. 2011. '*in vitro* biocompatibility and antimicrobial activity of wet chemically prepared $Ca_{10-x}Ag_x(PO_4)_6(OH)_2$ (0.0 ≤ x ≤ 0.5) hydroxyapatites.' *Mat. Sc. Engg. C.* 31. 1320–1329. (e) Nath, Shekhar, S. Kalmodia, and Bikramjit Basu. 2010. 'Densification, phase stability and *in vitro* biocompatibility property of hydroxyapatite-10 wt% silver composites.' *Journal of Materials Science: Materials in Medicine.* 21. 1273–1287.

1024. Cusco, R., F. Guitian, S. d. Aza, and L. Artus. 1998. 'Differentiation between hydroxyapatite and β-tricalcium phosphate by means of μ-Raman spectroscopy.' *J. Euro. Ceramic Soc.* 18. 1301–1305.

1025. Ray Jr., W. J., J. W. Burgner II, H. Deng, and R. Callender. 1993. 'Internal chemical bonding in solutions of simple phosphates and vanadates.' *Biochemistry.* 32. 12977–12983.

1026. de Aza, P. N., F. Guitin, C. Santos, S. de Aza, R. Cusc, and L. Arts. 1997. 'Vibrational properties of calcium phosphate compounds, comparison between hydroxyapatite and β-tricalcium phosphate.' *Chem. Mater.* 9. 916-922.

1027. Tsuda, H. and J. Arends. 1997. 'Raman spectroscopy in dental research: a short review of recent studies.' *Adv. Dent. Res.* 11. 539–547.

1028. Geoffrey, I. N., A. B. Graham, and B. Metson James. 2001. 'The thermal decomposition of silver (I, III) oxide: a combined XRD, FT-IR and Raman spectroscopic study.' *Phys. Chem. Chem. Phys.* 3. 3838–3845.

1029. Pettinger, B., X. Bao, I. Wilcock, M. Muhler, R. Schlogl, and G. Ertl. 1994. 'Thermal decomposition of silver oxide monitored by Raman spectroscopy: from AgO units to oxygen atoms chemisorbed on the silver surface.' *Angew. Chem. Inf. Ed. End.* 33. 85–86.

1030. Kreibig, U., and M. Vollmer. 1995. 'Optical Properties of Metal Clusters.' Springer: Berlin.

1031. Hughes, A. E., and S. C. Jain. 1979. 'Metal colloids in ionic crystals.' *Adv. Phys.* 28. 717–828.

1032. Mulvaney, P. 1996. 'Surface plasmon spectroscopy of nanosized metal particles.' *Langmuir.* 12. 788–800.

1033. Mie, G. 1908. 'Contributions to the optics of turbid media, especially colloidal metal solutions.' *Ann. Phys.* 25. 377–445.

1034. Morones, J. R., J. L. Elechiguerra, A. Camacho, K. Holt, J. B. Kouri, J. T. Ramírez, and M. J. Yacaman. 2005. 'The bactericidal effect of silver nanoparticles.' *Nanotechnology.* 16. 2346–2253.

1035. Jiménez, J. A., S. Lysenko, and H. Liu. 2008. Photoluminescence via plasmon resonance energy transfer in silver nanocomposite glasses.' *J. Appl. Phys.* 104. 054313.

1036. Briant, J. L., and G. C. Farrington. 1980. 'Ionic conductivity in Na^+, K^+, and Ag^+ β″-alumina.' *J. Solid State Chem.* 33. 385–390.

1037. Jonscher, A. K. 1977. 'Dielectric relaxation in solids.' *Nature (London).* 267. 673–679.

1038. Matsumura, Y., K. Yoshikata, S. I. Kunisaki, and T. Tsuschido. 2003. 'Mode of bactericidal action of silver zeolite and its comparison with that of silver nitrate.' *App. Env. Micro.* 69. 4278–4281.

1039. Richards, R. M. E., R. B. Taylor, and D. K. L. Xing. 1984. 'Effect of silver on whole cells and speroplasts of a silver resistant pseudomonas aeruginosa.' *Microbios.* 39. 151–158.

1040. Russell, A. D., and W. B. Hugo. 1994. 'Antimicrobial activity and action of silver.' *Prog. Med. Chem.* 31. 351–371.

1041. Song, H. Y., K. K. Ko, I. H. Oh, and B. T. Lee. 2006. 'Fabrication of silver nanoparticles and their antimicrobial mechanisms.' *Europ. Cells. Mat.* 11. 58.

1042. Egger, S., R. P. Lehmann, M. J. Height, M. J. Loessner, and M. Schuppler. 2009. 'Antimicrobial properties of a novel silver-silica nanocomposite material.' *Appl. Environ. Microbiol.* 75. 2973–2976.

1043. Bragg, P. D., and D. J. Rannie. 1974. 'The effect of silver ions on the respiratory chain of *Escherichia coli*.' *Can. J. Microbiol.* 20. 883–889.

1044. Batarseh, K. I. 2004. 'Anomaly and correlation of killing in the therapeutic properties of silver (I) chelation with glutamic and tartaric acids.' *J. Antimicrobial Chemotherapy.* 54. 546–548.

Chapter 16

1045. Basu, B., D. Katti, and Ashok Kumar. 2009. *Advanced Biomaterials: Fundamentals, Processing and Applications.* USA: John Wiley & Sons, Inc.

1046. Magnani, A., A. Priamo, D. Pasqui, and R. Barbucci. 2003. 'Cell behaviour on chemically microstructured surfaces.' *Mater. Sci. Eng. C.* 23(3). 315–328.

1047. Kitchen, S., and S. Bazin. 2002. *Electrotherapy: Evidence-Based Practice.* Churchill Livingstone.

1048. Martino, S., F. D'Angelo, I. Armentano, J. M. Kenny, and A. Orlacchio. 2012. 'Stem cell-biomaterial interactions for regenerative medicine.' *Biotechnol. Adv.* 30(1). 338–351.

1049. Poltawski, L., and T. Watson. 2009. 'Bioelectricity and microcurrent therapy for tissue healing; a narrative review.' *Phys. Ther. Rev.* 14(2). 104–114.

1050. McCaig, C. D., A. M. Rajnicek, B. Song, and M. Zhao. 2005. 'Controlling cell behavior electrically: current views and future potential.' *Physiol. Rev.* 85(3). 943–978.

1051. McCaig, C. D., and M. Zhao. 1997. 'Physiological electrical fields modify cell behaviour.' *Bioessays.* 19(9). 819–826.

1052. Nuccitelli, R. 2003. 'Endogenous electric fields in embryos during development, regeneration and wound healing.' *Radiat. Prot. Dosim.* 106(4). 375–383.

1053. Steinberg, M. E., A. Bosch, J R A. Schwan, and R. Glazer. 1968. 'Electrical Potentials in Stressed Bone.' *Clinical Orthopedics and Related Research.* 61. 294.

1054. Bassett, C., and R. O. Becker. 1962. 'Generation of electric potentials by bone in response to mechanical stress.' *Science (New York, NY).* 137. 1063.

1055. Friedenberg, Z., E. Andrews, B. Sinolenski, B. Pearl, and C. Brighton. 1971. 'Bone reaction to varying amounts of direct current.' *Plastic and Reconstructive Surgery.* 48. 95.

1056. (a) Basu, Bikramjit. 2016. *Biomaterials for Musculoskeletal regeneration: Concepts.* Springer. (b) Basu, Bikramjit, and Sourabh Ghosh. 2016. *Biomaterials for Musculoskeletal regeneration: Applications.* Springer.

1057. Young-M, K., B. Chang-J, L. Su-Hee, K. Hae-W, and K. Hyoun-Ee. 2005. 'Improvement In biocompatibility of ZrO_2-Al_2O_3 nano-composite by addition of HA.' *Biomaterials.* 26. 509–517.

1058. Volceanov, A., E. Volceanov, and S. Stoleriu. 2006. 'Hydroxiapatite–zirconia composites for biomedical applications.' *J. Optoelctron. Adv. M.* 8(2). 585–588.

1059. Rapacz-Kmita, A., A. Ślósarczyk, and Z. Paszkiewicz. 2006. 'Mechanical properties of HAp-ZrO_2 composites.' *J. Eur. Cer. Soc.* 26(8). 1481–1488.

1060. Suchanek, W., M. Yashima, M. Kakihana, and M. Yoshimura. 1997. 'Hydroxyapatite/hydroxyapatite–whisker composites without sintering additives: mechanical properties and microstructural evolution.' *J. Am. Ceram. Soc.* 80(11). 2805–2813.

1061. Nath, S., Dubey, A. K., and B. Basu. 2011. 'Mechanical properties of novel calcium phosphate-mullite biocomposites.' *J. Biomater. Appl.* 27(1). 67–78.

1062. Ravikumar, K., Prafulla Kumar Mallik, and Bikramjit Basu. 2016. 'Twinning induced enhancement of fracture toughness in ultrafine grained hydroxyapatite–calcium titanate composites.' *Journal of the European Ceramic Society.* 36(3). 805–815.

1063. Dubey, Ashutosh K., P. K. Mallik, S. Kundu, and B. Basu. 2013. 'Dielectric and electrical conductivity properties of multi-stage spark plasma sintered $HA–CaTiO_3$ composites and comparison with conventionally sintered materials.' *Journal of the European Ceramic Society.* 33(15–16). 3445–3453.

1064. Thrivikraman, Greeshma, Prafulla K. Mallik, and Bikramjit Basu. 2013. 'Substrate conductivity dependent modulation of cell proliferation and differentiation *in vitro*.' *Biomaterials.* 34(29). 7073–7085.

1065. Mallik, Prafulla Kumar, and Bikramjit Basu. 2014. 'Better early osteogenesis of electroconductive hydroxyapatite–calcium titanate composites in a rabbit animal model.' *Journal of Biomedical Materials Research Part A.* 102(3). 842–851.

1066. Mallik, Prafulla K. June, 2013. 'Spark plasma sintered multifunctional $HA-CaTiO_3$ composite for bone tissue engineering application.' PhD Dissertation. Indian Institute of Technology. Kanpur, India.

1067. Hastings, G., and F. Mahmud. 1988. 'Electrical effects in bone.' *Journal of Biomedical Engineering.* 10. 515–521.

1068. Singh, S., and S. Saha. 1984. 'Electrical properties of bone: a review.' *Clinical Orthopedics and Related Research.* 186. 249.

1069. Ravalioli, A., and A. Kraiewski. Ed. 1992. *Bioceramics: Materials-Properties-Applications.* London: Chapman & Hall. 50–51.

1070. Yang, W., and Z. Liu. Ed. 1993. *Biomedical Engineering.* TJST, Tianjing. 29–43.

1071. Dubey, A. K., G. Tripathi, and B. Basu. 2010. 'Characterization of hydroxyapatite-perovskite (CaTiO3) composites: phase evaluation and cellular response.' *Journal of Biomedical Materials Research Part B: Applied Biomaterials.* 95. 320–329.

1072. Webster, T. J., C. Ergun, R. H. Doremus, and W. A. Lanford. 2003. 'Increased osteoblast adhesion on titanium coated hydroxylapatite that forms $CaTiO_3$.' *Journal of Biomedical Materials Research Part A.* 67. 975–980.

1073. Reddy, Madhav K., N. Kumar, and B. Basu. 2010. 'Innovative multi-stage spark plasma sintering to obtain strong and tough ultrafine-grained ceramics.' *Scripta Materialia.* 62. 435–438.

1074. Nakamura, S., H. Takeda, and K. Yamashita. 2001. 'Proton transport polarization and depolarization of hydroxyapatite ceramics.' *J. Appl. Phys.* 89. 5386–5393.

1075. Sastre, A., and T. R. Podleski. 1976. 'Pharmacologic characterization of the Na+ ionophores in L6 myotubes.' *Proc. Natl. Acad. Sci. USA.* 73(4). 1355–1359.

1076. Huang, N. F., S. Patel, R. G. Thakar, J. Wu, B. S. Hsiao, B. Chu, R. J. Lee, and S. Li. 2006. 'Myotube assembly on nanofibrous and micropatterned polymers.' *Nano. Lett.* 6(3). 537–542.

1077. Wright, W. E., D. A. Sassoon, and V. K. Lin. 1989. 'Myogenin, a factor regulating myogenesis, has a domain homologous to MyoD.' *Cell.* 56(4). 607–617.

1078. Andres, V., and K. Walsh. 1996. 'Myogenin expression, cell cycle withdrawal, and phenotypic differentiation are temporally separable events that precede cell fusion upon myogenesis.' *J. Cell Biol.* 132(4). 657–666

1079. Kang, J. S., and R. S. Krauss. 2010. 'Muscle stem cells in developmental and regenerative myogenesis.' *Curr. Opin. Clin. Nutr. Metab. Care.* 13(3). 243–248.

1080. McKeon-Fischer, K. D., and J. W. Freeman. 2012. 'Addition of conductive elements to polymeric scaffolds for muscle tissue engineering.' *Nano. LIFE.* 02(03). 1230011.

1081. Ricotti, L., A. Polini, G. G. Genchi, G. Ciofani, D. Iandolo, H. Vazao, V. Mattoli, L. Ferreira, A. Menciassi, and D. Pisignano. 2012. 'Proliferation and skeletal myotube formation capability of C2C12 and H9c2 cells on isotropic and anisotropic electrospun nanofibrous PHB scaffolds.' *Biomed. Mater.* 7(3). 035010.

1082. Griffin, M. A., S. Sen, H. L. Sweeney, and D. E. Discher. 2004. 'Adhesion-contractile balance in myocyte differentiation.' *J. Cell. Sci.* 117(24). 5855–5863.

1083. Zhang, X. 1991. 'A Study of Porous Block HA Ceramics and its Osteogenesis.' In: *Bioceramics and the Human Body.*, edited by Ravaglioli, A. and A. Krajewski Amsterdam, The Netherlands: Elsevier Science. 692–703.

1084. Yamasaki, H., and H. Sakai. 1992. 'Osteogenic response to porous hydroxyapatite ceramics under the skin of dogs.' *Biomaterials.* 13. 308–312.

1085. Yuan, H., M. Van Den Doel, S. Li, C. Van Blitterswijk, K. De Groot, and J. De Bruijn. 2002. 'A comparison of the osteoinductive potential of two calcium phosphate ceramics implanted intramuscularly in goats.' *Journal of Materials Science: Materials in Medicine.* 13. 1271–1275.

1086. Yuan, H., Y. Li, J. D. de Bruijn, K. de Groot, and X. Zhang. 2000. 'Tissue responses of calcium phosphate cement: a study in dogs.' *Biomaterials.* 21(12). 1283–1290.

1087. Itoh, S., S. Nakamura, M. Nakamura, K. Shinomiya, and K. Yamashita. 2006. 'Enhanced bone ingrowth into hydroxyapatite with interconnected pores by electrical polarization.' *Biomaterials.* 27. 5572–5579.

1088. Jianqing, F., Y. Huipin, and Z. Xingdong. 1997. 'Promotion of osteogenesis by a piezoelectric biological ceramic.' *Biomaterials.* 18. 1531–1534.

1089. Davies, J. E. 2003. 'Understanding peri-implant endosseous healing.' *Journal of Dental Education.* 67. 932–949.

1090. Puleo, D., and A. Nanci. 1999. 'Understanding and controlling the bone–implant interface.' *Biomaterials.* 20. 2311–2321.

Chapter 17

1091. Basu, B., D. S. Katti, and A. Kumar. 2009. *Advanced Biomaterials: Fundamentals, Processing and Applications.* USA: John Wiley & Sons, Inc. and American Ceramic Society.

1092. Schmidt, C. E., and J. B. Leach. 2003. 'Neural tissue engineering: strategies for repair and regeneration.' *Annu. Rev. Biomed. Eng.* 5. 293–347.

1093. Ravi, V. B. 2006. 'Peripheral nerve regeneration: an opinion on channels, scaffolds and anisotropy.' *Biomaterials.* 27(19). 3515–3518.

1094. Evans, P. J., R. Midha, and S. E. Mackinnon. 1994. 'The peripheral nerve allograft: a comprehensive review of regeneration and neuroimmunology.' *Prog. Neurobiol.* 43(3). 187–233.

1095. Kim, B. S., J. J. Yoo, and A. Atala. 2004. 'Peripheral nerve regeneration using acellular nerve grafts.' *J. Biomed. Mater. Res.* 68A(2). 201–209.

1096. Krekoski, C. A., D. Neubauer, J. B. Graham, and D. Muir. 2002. 'Metalloproteinase-dependent predegenration *in vitro* enhances axonal regeneration within acellular peripheral nerve grafts.' *J. Neurosci.* 22(23). 10408–10415.

1097. Fawcett, J. W., and R. A. Asher. 1999. 'The glial scar and central nervous system repair.' *Brain Res. Bull.* 49(6). 377–391.

1098. Daly, W., L. Yao, D. Zeugolis, A. Windebank, and A. Pandit. 2012. 'A biomaterials approach to peripheral nerve regeneration: bridging the peripheral nerve gap and enhancing functional recovery.' *J. R. Soc. Interface.* 9(67). 202–221.

1099. Nectow, A. R., K. G. Marra, and D. L. Kaplan. 2012. 'Biometarials for the development of peripheral nerve guidance conduits.' *Tissue Eng. Part B: Rev.* 18(23). 40–50.

1100. Subramanian, A., U. Maheshwari-Krishnan, and S. Sethuraman. 2009. 'Development of biomaterial scaffolds for nerve tissue engineering: Biomaterial mediated neural regeneration.' *J. Biomed. Sci.* 16. 108–118.

1101. Chiono, V., C. Tonda-Turo, and G. Ciardelli. 2009. 'Artificial scaffolds for peripheral nerve reconstruction.' *Int. Rev. Neurobiol.* 87.173–198.

1102. Barnes, C. P., S. A. Sell, E. D. Boland, D. G. Simpson, and G. L. Bowlin. 2007. 'Nanofiber technology: designing the next generation of tissue engineering scaffolds.' *Adv. Drug. Deliv. Rev.* 59(14). 1413–1433.

1103. Thompson, D. M., and H. M. Buettner. 2006. 'Neurite outgrowth is directed by Schwann cell alignment in the absence of other guidance cues.' *Ann. Biomed. Eng.* 34(4). 669–676.

1104. Hanson, J. N., M. J. Motala, M. L. Heien, M. Gillette, J. Sweedler, and R. G. Nuzzo. 2009. 'Textural guidance cues for controlling process outgrowth of mammalian neurons.' *Lab. Chip.* 9. 122–131.

1105. Johansson, F., P. Carlberg, N. Danielsen, L. Montelius, and M. Kanje. 2006. 'Axonal outgrowth on nano-imprinted patterns.' *Biomaterials.* 27(8). 1251–1258.

1106. Murugan, R., and A. Tiwari. 2010. 'Spatially controlled cell growth using patterned biomaterials.' *Adv. Mater. Lett.* 1(3). 179–187.

1107. Kim, D. H., J. A. Mitchel, and R. V. Bellamkonda. 2010. 'Topography, cell response, and nerve regeneration.' *Annu. Rev. Biomed. Eng.* 12. 203–231.

1108. Falconnet, D., G. Csucs, H. M. Grandin, and M. Textor. 2006. 'Surface engineering approaches to micropattern surfaces for cell-based assays.' *Biomaterials.* 27(16). 3044–3063.

1109. Thery, M. 2010. 'Micropatterning as a tool to decipher cell morphogenesis and functions.' *J. Cell. Sci.* 123. 4102–4113.

1110. Miller, C., H. Shanks, A. Witt, G. Rutkowski, and S. Mallapragada. 2001. 'Oriented Schwann cell growth on micropatterned biodegradable polymer surfaces.' *Biomaterials.* 22(11). 1263–1269.

1111. Miller, C., S. Jeftinija, and S. Mallapragada. 2001. 'Micropatterned Schwann cell-seeded biodegradable polymer substrates significantly enhance neurite alignment and outgrowth.' *Tissue Eng. Part B: Rev.* 7(6). 705–715.

1112. Thompson, D.M., and H. M. Buettner. 2004. 'Oriented Schwann cell monolayers for directed neurite outgrowth.' *Ann. Biomed. Eng.* 32(8). 1120–1130.

1113. Schmalenberg, K. E., and K. E. Uhrich. 2005. 'Micropatterned polymer substrates control alignment of proliferating Schwann cells to direct neuronal regeneration.' *Biomaterials.* 26(12). 1423–1430.

1114. Corey, J. M., C. C. Gertz, T. J. Sutton, Q. Chen, K. B. Mycek, B. S. Wang, A. A. Martin, S. L. Johnson, and E. L. Feldman. 2010. 'Patterning N-type and S-type neuroblastoma cells with Pluronic F108 and ECM proteins.' *J. Biomed. Mater. Res. A.* 93(2). 673–686.

1115. Nisbet, D. R., and J. S. Forsythe. 2009. 'A review of the cellular response on electrospun nanofibers for tissue engineering.' *Journal of Biomater. Appl.* 24. 7–29.

1116. Corey, J. M., C. C. Gertz, B. S. Wang, L. K. Birrell, S. L. Johnson, D. C. Martin, and E. L. Feldman. 2008. 'The design of electrospun PLLA nanofibers scaffolds compatible with serum-free growth of primary motor and sensory neurons.' *Acta Biomater.* 4(4). 863–875.

1117. Kim, Y. T., V. K. Haftel, S. Kumar, and R. V. Bellamkonda. 2008. 'The role of aligned polymer fiber-based constructs in the bridging of long peripheral nerve gaps.' *Biomaterials.* 29(21). 3117–3127.

1118. Xie, J., M. R. MacEwan, X. Li, S. E. Sakiyama-Elbert, and Y. Xia. 2009. 'Neurite outgrowth on nanofiber scaffolds with different orders, structures and surface properties.' *ACS Nano.* 3(5). 1151–1159.

1119. Corey, J. M., D. Y. Lin, K. B. Mycek, Q. Chen, S. Samuel, E. L. Feldman, and D. C. Martin. 2007. 'Aligned electrospun nanofibers specify the direction of dorsal root ganglia neurite growth.' *J. Biomed. Mater. Res. A.* 83(3). 636–645.

1120. Jenkins, G. M. 1980. 'Biomedical applications of carbons and graphites.' *Clin. Phys. Physiol. Meas.* 1(3). 171–194.

1121. Kelly, S., E. M. Regan, J. B. Uney, A. D. Dick, J. P. McGeehan, E. J. Mayer, and F. Claeyssens. 2008. 'Patterned growth of neuronal cells on modified diamond-like carbon substrates.' *Biomaterials.* 29(17). 2573–2580.

1122. Rodil, S. E., R. Olivares, and H. Arzate. 2005. '*in vitro* cytotoxicity of amorphous carbon films.' *J. Biomed. Mater. Eng.* 15. 101–112.

1123. Bokros, J. C. 1977. 'Carbon biomedical devices.' *Carbon.* 15(6). 355–371.

1124. Schmidt, C. E., V. R. Shastri, J. P. Vacanti, and R. Langer. 1997. 'Stimulation of neurite outgrowth using an electrically conducting polymer.' *Proc. Natl. Acad. Sci. USA.* 94(17). 8948–8953.

1125. Czarnecki, J.S., K. Lafdi, R. M. Joseph, and P. A. Tsonis. 2012. 'Hybrid carbon-based scaffolds for applications in soft tissue reconstruction.' *Tissue Eng. Part A.* 18(9–10). 945–956.

1126. Tran, P. A., L. Zhang, and T. J. Webster. 2009. 'Carbon nanofibers and carbon nanotubes in regenerative medicine.' *Adv. Drug. Deliv. Rev.* 61(12). 1097–1114.

1127. Rajzer, I., E. Menaszek, L. Bacakova, M. Rom, and M. Blazewickz. 2010. '*in vitro* and *in vivo* studies on biocompatibility of carbon nanofibers.' *J. Mater. Sci.: Mater. Med.* 21(9). 2611–2622.

1128. Sharma, C. S., R. Vasita, D. K. Upadhyay, A. Sharma, D. S. Katti, and R. Venkataraghavan. 2010. 'Photoresist derived electrospun carbon nanofibers with tunable morphology and surface properties.' *Ind. Eng. Chem. Res.* 49(6). 2731–2739.

1129. Maitra, T., S. Sharma, A. Srivastava, Y. K. Cho, M. Madou, and A. Sharma. 2012. 'Improved graphitization and electrical conductivity of suspended carbon nanofibers derived from carbon nanotube/polyacrylonitrile composites by directed electrospinning.' *Carbon.* 50(5). 1753–1761.

1130. Mitra, J., S. Jain, A. Sharma, and B. Basu. 2013. 'Patterned growth and differentiation of neural cells on polymer derived carbon substrates with micro/nano structures *in vitro*.' *Carbon.* 65. 140–155.

1131. Webster, T. J., M. C. Waid, J. L. McKenzie, R. L. Price, and J. U. Ejiofor. 2004. 'Nano-biotechnology: carbon nanofibres as improved neural and orthopedic implants.' *Nanotechnology.* 15(1). 48–54.

1132. Zhang, X., S. Prasad, S. Niyogi, A. Morgan, M. Ozkan, and C. S. Ozkan. 2005. 'Guided neurite growth on patterned carbon nanotube substrates.' *Sensor Actuat. B.* 106. 843–850.

1133. Wu, H. C., X. Chang, L. Liu, F. Zhao, and Y. Zhao. 2010. 'Chemistry of carbon nanotubes in biomedical applications.' *J. Mater. Chem.* 20. 1036–1052.

1134. Sinha, N. 2005. 'Carbon nanotubes for biomedical applications.' *IEEE Trans. Nanobiosci.* 4(2). 180–195.

1135. Li, X., Y. Fan, and F. Watari. 2010. 'Current investigations into carbon nanotubes for biomedical applications.' *Biomed. Mater.* 5(2). 022001.

1136. Hench, L.L. 1991. 'Bioceramics: from concept to clinic.' *J. Am. Chem. Soc.* 74(7). 1487–1510.

1137. Ranganathan, S., R. McCreery, S. M. Majji, and M. Madou. 2000. 'Photoresist-derived carbon for microelectromechanical systems and electrochemical applications.' *J. Electrochem. Soc.* 147(1). 277–282.

1138. Malladi, K., C. L. Wang, and M. Madou. 2006. 'Fabrication of suspended carbon microstructures by e-beam writer and pyrolysis.' *Carbon.* 44(13). 2602–2607.

1139. Zhou, H., A. Gupta, T. Zhu, and J. Zhou. 2007. 'Photoresist derived carbon for growth and differentiation of neuronal cells.' *Int. J. Mol. Sci.* 8(8). 884–893.

1140. Kostecki, R., X. Song, and K. Kinoshita. 1999. 'Electrochemical analysis of interdigitated carbon microelectrodes.' *Electrochem. Solid-State Lett.* 2(9). 465–467.

1141. Teixidor, G. T., R. A. G. III, P. P. Tripathi, G. S. Bisht, M. Kulkarni, T. K. Maiti, T. K. Battacharyya, J. R. Subramaniam, A. Sharma, B. Y. Park and M. Madou. 2008. 'Carbon microelectromechanical systems as a substratum for cell growth.' *Biomed. Mater.* 3(3). 034116–034123.

1142. Kulkarni, M. M., C. S. Sharma, A. Sharma, S. Kalmodia, and B. Basu. 2012. 'Multiscale micro-patterned polymeric and carbon substrates derived from buckled photoresist films: fabrication and cytocompatibility.' *J. Mater. Sci.: Mater. Med.* 47(8). 3867–3875.

1143. Silva, G. A. 2006. Neuroscience nanotechnology: progress, opportunities and challenges. *Nature Reviews Neuroscience.* 7(1). 65–74.

1144. Omenetto, F. G., and D. L. Kaplan. 2010. 'New opportunities for an ancient material.' *Science.* 329(5991). 528-531.

1145. Peter, X. M. 2008. 'Biomimetic materials for tissue engineering.' *Advanced Drug Delivery Reviews.* 60(2). 184–198.

1146. Baker, R. T. K. 1989. 'Catalytic growth of carbon filaments.' *Carbon.* 27(3). 315–323.

1147. Stout, D. A., B. Basu, and T. J. Webster. 2011. 'Poly(lactic–co-glycolic acid): carbon nanofiber composites for myocardial tissue engineering applications.' *Acta Biomaterialia.* 7(8). 3101–3112.

1148. Locca, D., C. Bucciarelli-Ducci, G. Ferrante, A. La Manna, N. G. Keenan, A. Grasso, P. Barlis, F. Del Furia, S. K. Prasad, J. C. Kaski, D. J. Pennell, and C. Di Mario. 2010. 'New universal definition of myocardial infarction: applicable after complex percutaneous coronary interventions?' *JACC: Cardiovascular Interventions.* 3. 950–958.

1149. Thygesen, K, J. S. Alpert, and H. D. White. 2007. 'Universal definition of myocardial infarction.' *Journal of the American College of Cardiology.* 50. 2173–2195.

1150. Jain A. K., and C. Knight. 2010. Myocardial infarction—new definitions and implications.' *Medicine.* 38. 421–423.

1151. Barber, M,, T. Mueller, B. Davies, R. Gill, and D. Zipes. 1985. 'Interruption of sympathetic and vagal-mediated afferent responses by transmural myocardial infarction.' *Circulation.* 72. 623–631.

1152. Oh, Y., A. Y. Jong, D. T. Kim, H. Li, C. Wang, A. Zemljic-Harpf, R. S. Ross, M. C. Fishbein, P. Chen, and L. S. Chen. 2006. 'Spatial distribution of nerve sprouting after myocardial infarction in mice.' *Heart Rhythm.* 3. 728–736.

1153. Vracko, R., D. Thorning, and R. Frederickson. 1990. 'Fate of nerve fibers in necrotic, healing, and healed rat marycardium.' *Laboratory Investigation.* 63. 490–501.

1154. Randall, W. C., M. P. Kaye, G. R. Hageman, H. K. Jacobs, D. E. Euler, and W. Wehrmacher. 1976. 'Cardiac dysrhythmias in the conscious dog after surgically induced autonomic imbalance.' *Am. J. Cardiol.* 38. 178–183.

1155. Flegal K., E. Ford, K. Furie, A. Go, K. Greenlund, and N. Haase. Lloyd-Jones. 2009. 'Heart Disease and Stroke Statistics Update: A Report From the American Heart Association Statistics Committee and Stroke Statistics Subcommittee.' 119. e21.

1156. Khang, D., J. Lu., C. Yao, K. M. Haberstroh, and T. J. Webster. 2008. 'The role of nanometer and sub-micron surface features on vascular and bone cell adhesion on titanium.' *Biomaterials* 29. 970–983.

1157. Price, R. L., M. C. Waid, K. M. Haberstroh, and T. J. Webster. 2003. 'Selective bone cell adhesion on formulations containing carbon nanofibers.' *Biomaterials* 24. 1877–1887.

1158. Ranella, A., M. Barberoglou, S. Bakogianni, C. Fotakis, and E. Stratakis. 2010. 'Tuning cell adhesion by controlling the roughness and wettability of 3D micro/nano silicon structures.' *Acta Biomaterialia* 6. 2711–2720.

1159. Yang, L., B. W. Sheldon, and T. J. Webster. 2009. 'The impact of diamond nanocrystallinity on osteoblast functions.' *Biomaterials* 30. 3458–3465.

1160. Dalby, M. J., D. Pasqui, and S. Affrossman. 2004. 'Cell response to nano-islands produced by polymer demixing: a brief review.' *Nanobiotechnology, IEE Proceedings.* 151. 53–61.

1161. Dalby, M. J., D. Giannaras, M. O. Riehle, N. Gadegaard, S. Affrossman, and A. S. G. Curtis. 2004. 'Rapid fibroblast adhesion to 27 nm high polymer demixed nano-topography.' *Biomaterials.* 25. 77–83.

1162. Joachim, Loo S. C., W. L. Jason Tan, S. M. Khoa, N. K. Chia, S. Venkatraman, and F. Boey. 2008. 'Hydrolytic degradation characteristics of irradiated multi-layered PLGA films.' *Int. J. Pharm.* 360. 228–230.

1163. Koh, L. B., I. Rodriguez, and J. Zhou. 2008. 'Platelet adhesion studies on nanostructured poly(lactic-co-glycolic-acid)-carbon nanotube composite.' *Journal of Biomedical Materials Research Part A.* 86A. 394–401.

1164. Baker, R. T. K. 1989. 'Catalytic growth of carbon filaments.' *Carbon.* 27. 315–323.

1165. Beachley, V., and X. Wen. 2010. 'Polymer nanofibrous structures: fabrication, biofunctionalization, and cell interactions.' *Progress in Polymer Science* 35. 868–892.

1166. Edwards, S. L., J. S. Church, J. A. Werkmeister, and J. A. M. Ramshaw. 2009. 'Tubular micro-scale multiwalled carbon nanotube-based scaffolds for tissue engineering.' *Biomaterials* 30. 1725–1731.

1167. Liang, D., B. S. Hsiao, and B. Chu. 2007. 'Functional electrospun nanofibrous scaffolds for biomedical applications.' *Advanced Drug Delivery Reviews* 59. 1392–1412.

1168. MacDonald, R. A., C. M. Voge, M. Kariolis, and J. P. Stegemann. 2008. 'Carbon nanotubes increase the electrical conductivity of fibroblast-seeded collagen hydrogels.' Acta biomaterialia 4. 1583–1592.

1169. Bal, S. 2010. 'Experimental study of mechanical and electrical properties of carbon nanofiber/epoxy composites.' *Materials & Design* 31. 2406–2413.

1170. Aslan, S., C. Z. Loebick, S. Kang, M. Elimelech, L. D. Pfefferle, and P. R. Van Tassel. 2010. 'Antimicrobial biomaterials based on carbon nanotubes dispersed in poly(lactic-co-glycolic acid).' *Nanoscale.* 2(9). 1789.

1171. Yaghoub, M. B., R. Tremblay, R. Voicu, G. Mealing, R. Monette, C. Py, K. Faid, and M. Sikorska. 2005. 'Neurogenesis and neuronal communication on micropatterned neurochips.' *Biotechnol. Bioeng.* 92(3). 336–345.

1172. Bhardwaj, N., and S. C. Kundu. 2010. 'Electrospinning: A fascinating fiber fabrication technique.' *Biotechnol. Adv.* 28(3). 325–347.

1173. Khan, S. P., G. G. Auner, and G. M. Newaz. 2005. 'Influence of nanoscale surface roughness on neural cell attachment on silicon.' *Nanomedicine.* 1(2). 125–129.

1174. Dent, E. W., and F. B. Gertler. 2003. 'Cytoskeletal dynamics and transport in growth cone motility and axonal guidance.' *Neuron.* 40(2). 209–227.

1175. Guan, K. L., and Y. Rao. 2003. 'Signalling mechanisms mediating neuronal responses to guidance cues.' *Nat. Rev. Neurosci.* 4(12). 941–956.

1176. Curtis, A., and C. Wilkinson. 1997. 'Topographical control of cells.' *Biomaterials.* 18(24). 1573–1583.

1177. Lee, J. W., K. S. Lee, N. Cho, B. K. Ju, K. B. Lee, and S. H. Lee. 2007. 'Topographical guidance of mouse neuronal cell on SiO2 microtracks.' *Sens and Actuators B.* 128(1). 252–257.

1178. Britland, S., H. Morgan, B. Wojiak-Stodart, M. Riehle, A. Curtis, and C. Wilkinson. 1996. 'Synergistic and hierarchical adhesive and topographic guidance of BHK cells.' *Exp. Cell. Res.* 228. 313–325.

1179. Lewandowska, K., N. Balanchander, C. N. Sukenik, and L. A. Cupl. 1989. 'Modulation of fibronectin adhesive functions for fibroblasts and neural cells by chemically derivatized substrata.' *J. Cell. Physiol.* 141(2). 334–345.

1180. Wilson, C. J., R. E. Clegg, D. I. Leavesley, and M. J. Pearcy. 2005. 'Mediation of biomaterial-cell interactions by adsorbed proteins: A review.' *Tissue Eng.* 11(1–2). 1–18.

1181. Dunn, G. A., and J. P. Heath. 1976. 'A new hypothesis of contact guidance in tissue cells.' *Exp. Cell. Res.* 101(1). 1–14.

1182. O'hara, P. T., and R. C. Buck. 1979. Contact guidance *in vitro*: a light, transmission, and scanning electron microscopic study. *Exp. Cell. Res.* 121(2). 235–249.

1183. Jessen, K. R., and R. Mirsky. 1999. 'Schwann cells and their precursors emerge as major regulators of nerve development.' *Trends Neurosci.* 22(9). 402–410.

1184. Seidlits, S. K., J. Y. Lee, and C. E. Schmidt. 2008. 'Nanostructured scaffolds for neural applications.' *Nanomedicine (Lond).* 3(2). 183–199.

1185. Ghasemi-Mobarakeh, L., M. P. Prabhakarana, M. Morshed, M. H. Nasr-Esfahani, and S. Ramakrishna. 2010. 'Biofunctionalized PCL nanofibers scaffolds for nerve tissue engineering.' *Mater. Sci. Eng. C.* 30(8). 1129–1136.

1186. Chew, S. Y., R. Mi, A. Hoke, and K. W. Leong. 2008. 'The effect of alignment of electrospun fibrous scaffolds on Schwann cell maturation.' *Biomaterials.* 29(6). 653–661.

1187. Schmalenberg, K. E., and K. E. Uhrich. 2005. 'Micropatterned polymer substrates control alignment of proliferation Schwann cells to direct neuronal regeneration.' *Biomaterials*. 26. 1423–1430.

1188. Chew, S. Y., R. Mi, A. Hoke, and K. W. Leong. 2007. 'Aligned protein-polymer compiste fibers enhance nerve regeneration: a potential tissue-engineering platform.' *Adv. Funct. Mater.* 17. 1288–1296.

1189. Webster, T. J., C. Ergun, R. H. Doremus, R. W. Siegel, and R. Bizios. 2000. 'Specific proteins mediate enhanced osteoblast adhesion on nanophase ceramics.' *Journal of Biomedical Materials Research*. 51. 475–483.

1190. Tran, P. A., L. Zhang, and T. J. Webster. 2009. 'Carbon nanofibers and carbon nanotubes in regenerative medicine.' *Advanced Drug Delivery Reviews* 61. 1097–1114.

1191. Pedrotty, D. M., J. Koh, B. H. Davis, D. A. Taylor, P. Wolf, and L. E. Niklason. 2005. Engineering skeletal myoblasts: roles of three-dimensional culture and electrical stimulation.' *Am J Physiol Heart Circ Physiol* 288. H1620–1626.

1192. Mihardja, S. S., R. E. Sievers, and R. J. Lee. 2008. 'The effect of polypyrrole on arteriogenesis in an acute rat infarct model.' *Biomaterials*.29. 4205–4210.

1193. Roberts-Thomson, K. C., P. M. Kistler, P. Sanders, J. B. Morton, H. M. Haqqani, I. Stevenson, J. K. Vohra, P. B. Sparks, and J. M. Kalman. 2009. 'Fractionated atrial electrograms during sinus rhythm: Relationship to age, voltage, and conduction velocity.' *Heart Rhythm* 6. 587–591.

1194. Shilpee, Jain. 2013. 'Investigation of neural cell fate processes on amorphous carbon nanofibers, film and patterned structures.' PhD dissertation. Indian Institute of Technology, Kanpur, India.

Chapter 18

1195. Bonandrini, B., F. Marina, P. Evangelia, M. Marina, P. Norberto, C. Federica, S. Fabio, C. Sara, B. Ariela, G. Remuzzi, and G. Remuzzi . 2014. 'Recellularization of well-preserved acellular kidney scaffold using embryonic stem cells.' *Tiss. Eng. Part A*. 20(9–10), 1486–1498.

1196. Weber, B., P. E. Dijkman, J. Scherman, B. Sanders, M. Y. Emmert, J. Grunenfelder, R. Verbeek, M. Bracher, M. Black, T. Franz, and J. Kortsmit., 2013. 'Off-the-shelf human decellularized tissue engineered heart valves in a non-human primate model.' *Biomaterials*. 34(11). 7269–7280.

1197. Li, X., L. Wang, Y. Fan, Q. Feng, F. –Z. Cui, and F. Watari. 2013. 'Nanostructured scaffolds for bone tissue engineering.' *J. Biomed. Mater. Res. Part A*. 101A. 2424–2435.

1198. Berns, E. J., S. Sur, L. Pan, J. E. Goldberger, S. Suresh, S. Zhang, J. A. Kessler, and S. I. Stupp. 2014. 'Aligned neurite outgrowth and directed cell migration in self-assembled monodomain gels.' *Biomaterials*. 35(1). 185–195.

1199. Seol, Y. J., T. Y. Kang , and D. W. Cho. 2012. 'Solid freeform fabrication technology applied to tissue engineering with various biomaterials.' *Soft Matter*. 8. 1730–1735.

1200. Farooque, T. M., C. H. Camp, C. K. Tison, G. Kumar, S. H. Parekh, and C. G. Simon. 2014. 'Measuring stem cell dimensionality in tissue scaffolds.' *Biomaterials*. 35(9). 2558–2567.

1201. Chen, F. M., L. A. Wu, M. Zhang, R. Zhang, and H. H. Sun. 2011. 'Homing of endogenous stem/progenitor cells for *in situ* tissue regeneration; promises, strategies and translation perspectives.' *Biomaterials*. 32(12). 3189–3209.

1202. Williams, D. F. 2014. *Essential Biomaterials Science*. Cambridge University Press. 640.

1203. Levin, M. 2011. 'The wisdom of the body: future techniques and approaches to morphogenetic fields in regenerative medicine, developmental biology and cancer.' *Regen. Med*. 6(6). 667–673.

1204. Hoover-Plow, J., and Y. Gong. 2012. 'Challenges for heart disease stem cell therapy.' *Vasc. Health Risk Manage*. 8(1). 99–113.

1205. Ramsden, C. M., M. B. Powner, A. J. F. Carr, M. J. K. Smart, L. de Cruz, and P. J. Coffey. 2013. 'Stem cells in retinal regeneration: past, present and future.' *Development.* 140. 2576–2585.

1206. Williams, D. F. 2006. 'To engineer is to create.' *Trends in Biotech.* 24. 4–8.

1207. Williams, D. F. 2014. 'The biomaterials conundrum in tissue engineering.' *Tiss. Eng. Part A.* 20 (7–8). 1129–1131.

1208. Williams, D. F. 2009. 'On the nature of biomaterials.' *Biomaterials.* 30(30). 5897–5909.

1209. Solanas, G., and S. A. Benitah. 2013. 'Regenerating the skin; a task for the heterogeneous stem cell pool and surrounding niche.' *Nature Rev. Mol. Cell. Biol.* 14(10). 737–748.

1210. Si-Tayeb, K., F. P. Lemaigre, and S. A. Duncan. 2010. 'Organogenesis and development of the liver.' *Dev. Cell.* 18(2). 175–189.

1211. Milner, D. J., and J. A. Cameron. 2013. 'Muscle repair and regeneration: stem cells, scaffolds, and the contributions of skeletal muscle to amphibian limb regeneration.' *Curr. Topics Microbiol. Immunol.* 367. 133–159.

1212. Gu, X., F. Ding, and D. F. Williams. 2014. 'Neural tissue engineering options for peripheral nerve regeneration.' *Biomaterials.* 35. 6143–6156

1213. Lin, Z., and W. T. Pu. 2014. 'Strategies for cardiac regeneration and repair.' *Sci. Transl. Med.* 6(239). 236rv1.

1214. Ludlow, J. W., R. W. Kelley, and T. A. Bertram. 2013. 'The future of regenerative medicine; urinary system.' *Tiss. Eng. Part B.* 18(3). 218–224.

For Further Reading

a. Langer R., and Tirrell D. 'Designing materials for biology and medicine.' *Nature.* 004;428(6982):487–92.

b. Mitragotri S, Lahann J. Physical approaches to biomaterial design. *Nature Materials* 2009; 8(1):15–23.

c. Orive G, Anitua E, Pedraz JL, Emerich DF. 2009. 'Biomaterials for promoting brain protection, repair and regeneration.' *Nature Reviews Neuroscience.* 10(9):682–92.

d. Place E. S., Evans N. D., Stevens MM. 'Complexity in biomaterials for tissue engineering.' *Nature Materials* 2009 Jun;8(6):457–70.

e. Dong-An, W., Varghese, S., Sharma, B., Strehin, I., Fermanian, S., Gorham, J., Fairbrother, D.H., Cascio, B. and Elisseeff, J.H., 2007. 'Multifunctional chondroitin sulphate for cartilage tissue-biomaterial integration.' *Nature materials* 6(5), p.385.

f. Zhang, Shuguang. 'Fabrication of novel biomaterials through molecular self-assembly.' *Nature Biotechnology* 21, no. 10 (2003): 1171.

Index

Colour Plates

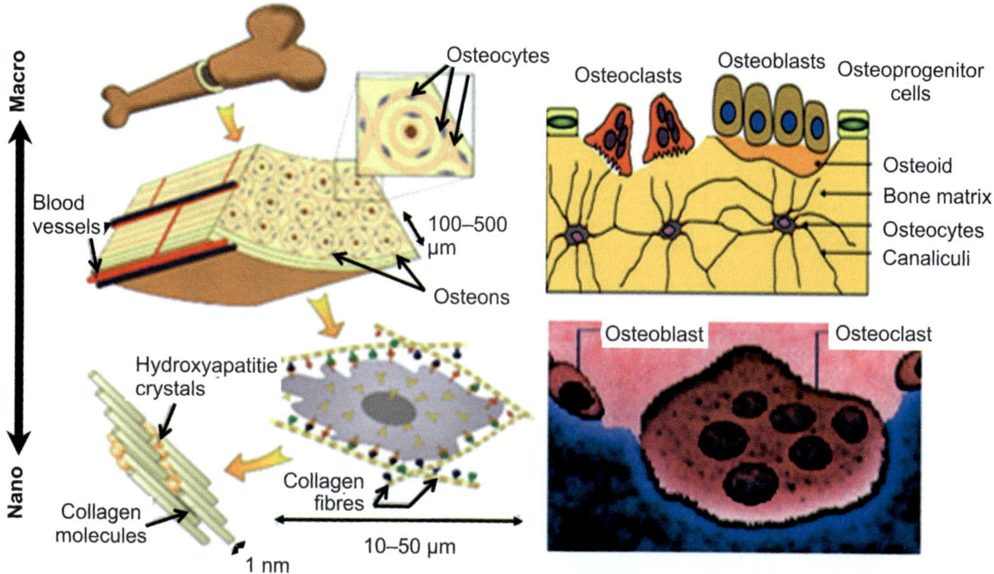

Fig. 3.1 Hierarchical structure of natural bone, including that at cellular level[444].

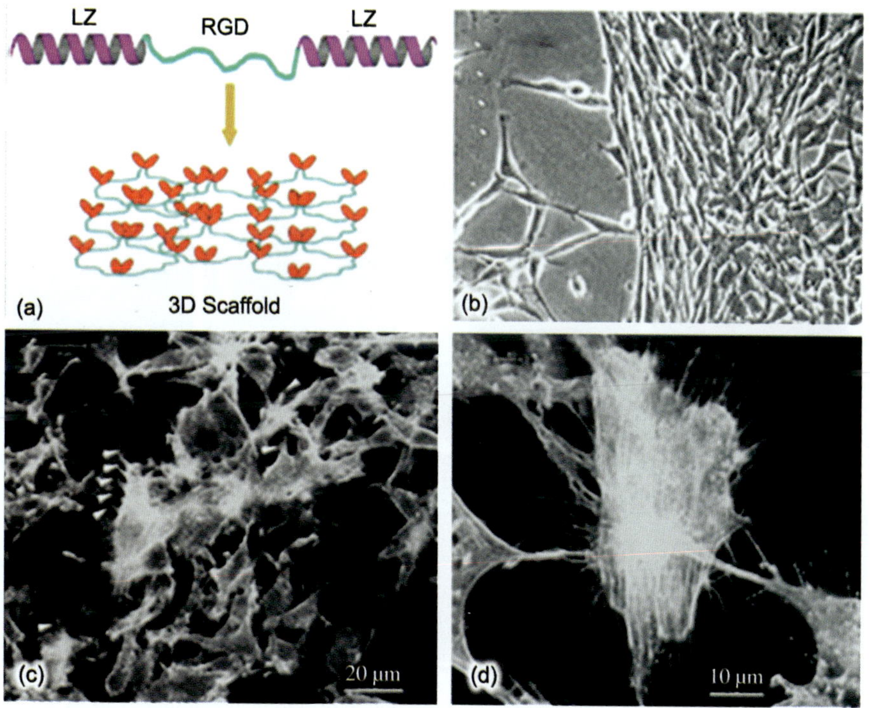

Fig. 3.8 Leucine zipper-DMP1 hydrogel facilitates cell–matrix interplay: (a) schematic representation of the leucine zipper-DMP1 chimeric protein. (b) a comparative study of cell attachment on a LZ-DMP1 substrate and glass palte (control) by light microscopy. (c) and (d) actin stress fibres and focal adhesion complexes promoting cell—cell communications, when coated on LZ-DMP1 substrate[291].

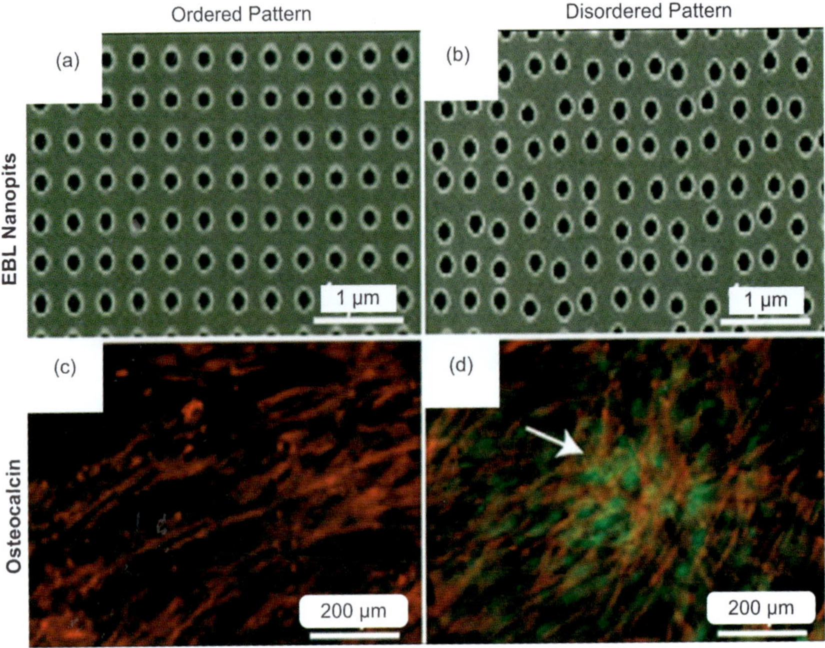

Fig. 3.12 Nanotopography and cell differentiation. (a,b) The nanopits (120 nm in diameter and 100 nm deep) were generated by electron beam lithography (EBL) (a) in a square arrangement and (b) with increasing disorder (displaced square ±50 nm from true center). The disordered topography is more proficient in stimulating human mesenchymal stem cells to increase the expression of bone-specific ECM protein osteopontin (arrow, (d)), compared to that in case of the orderly square arrangement (c) [Adapted from Ref. 263].

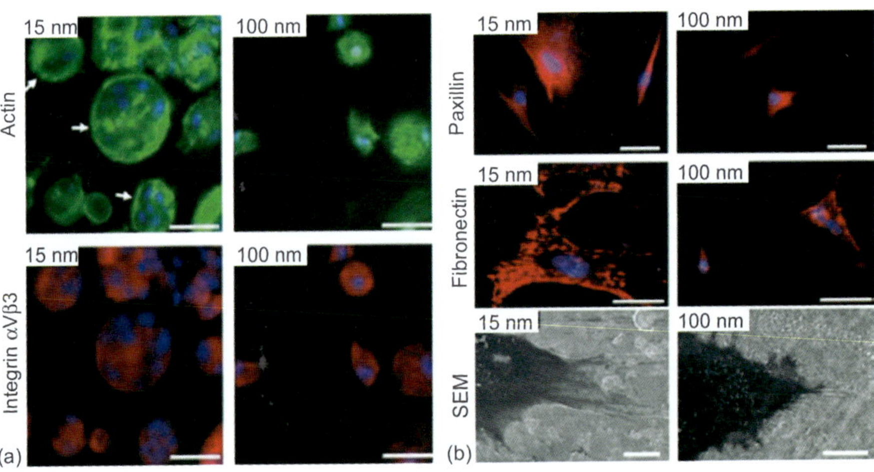

Fig. 3.15 Some illustrative results showing the clear evidence of osteoclast adhesion on 15 nm nanotubes (a). Scale bars: 50, 200, and 200 μm; (b) Morphological features of primary human osteoblasts to TiO$_2$ nanotubes. SEM images also reveal focal contacts by paxillin staining (upper panel) (scale bars: 100 μm), extracellular deposition of fibronection matrix (middle panel) (scale bars: 50 μm), and filopodia formation (lower panel) are higher on the 15 nm nanotubes (scale bars: 10 μm) [Adapted from Ref. 270].

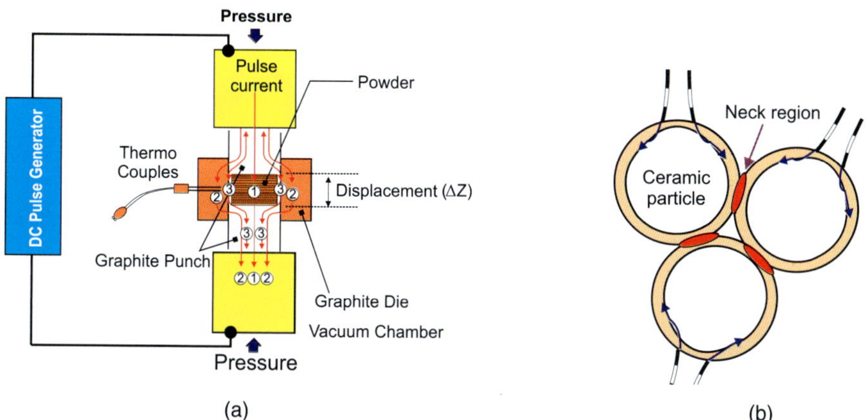

Fig. 4.17 Spark plasma sintering (SPS): (a) schematic showing the DC current flow during the sintering process, and (b) mechanism of temperature rise at the interface of particles due to Joule heating effect.

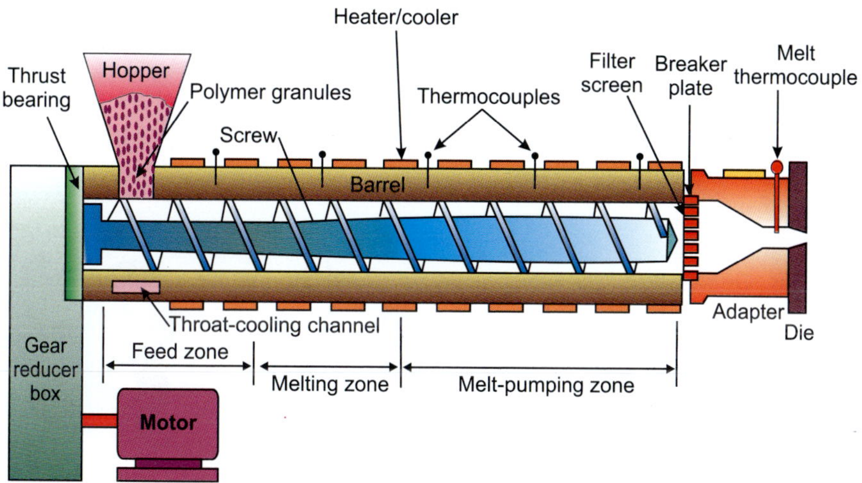

Fig. 4.18 A schematic of extrusion process to extrude polymeric rods.

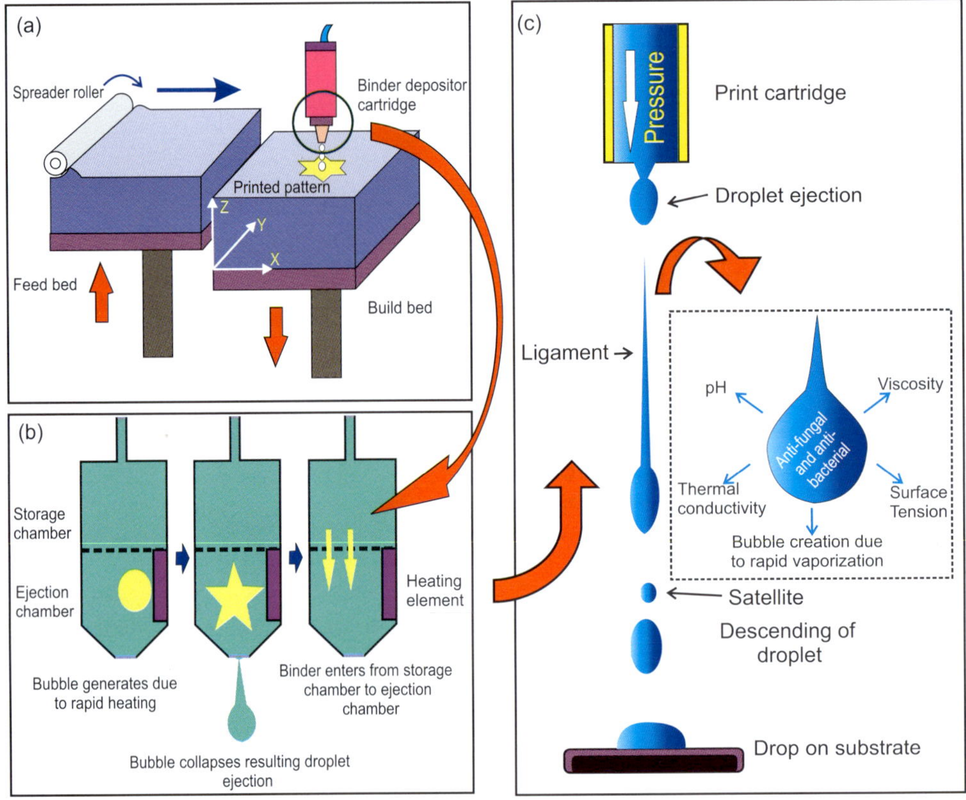

Fig. 4.21 Schematic illustration of 3D printing process: (a) the sequential stages in bubble formation inside printhead, (b) and the binder droplet ejection and subsequent interaction with powder bed, and (c) the various physical parameters of relevance of the binder are shown in inset[543].

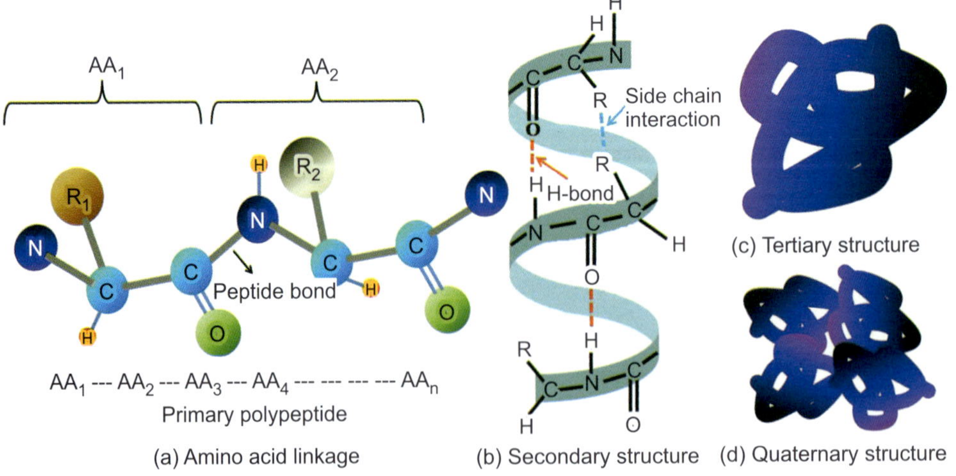

Fig. 7.3 Pictorial representation of different levels of protein structure [Adapted from Ref. 593].

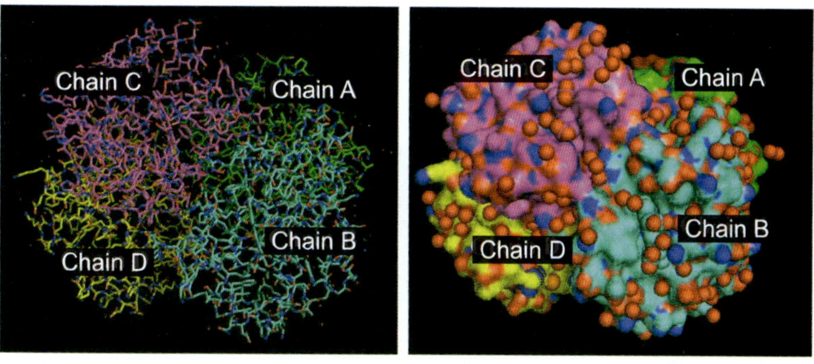

Fig. 7.5 Illustration of the quaternary structure of protein, human deoxyhemoglobin at 1.74Å using PyMOL (TM) 1.7.4.5 Edu - Educational Product.

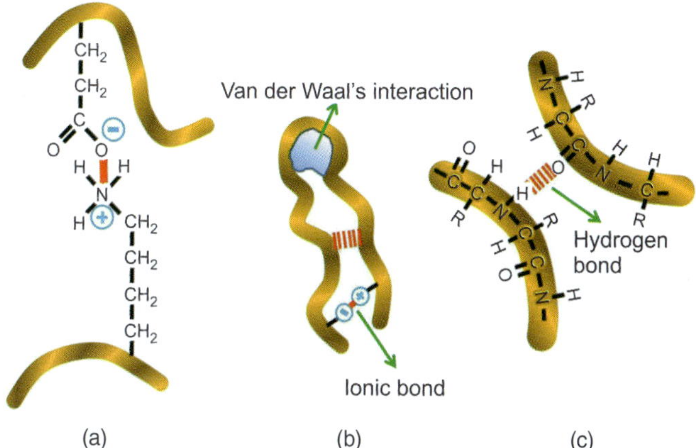

Fig. 7.6 Pictorial representation of protein-protein interaction mediated through different bond formations [Adapted from Ref. 593].

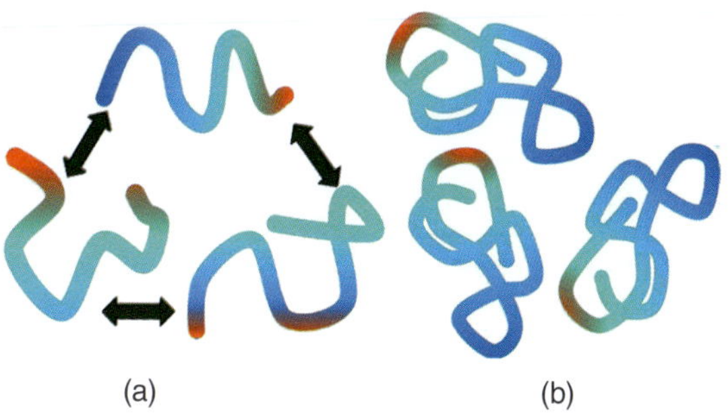

Fig. 7.7 Schematic illustration of protein conformations: (a) uncoiled protein due to denaturation and (b) coiled structure of the protein [Adapted from Ref. 594].

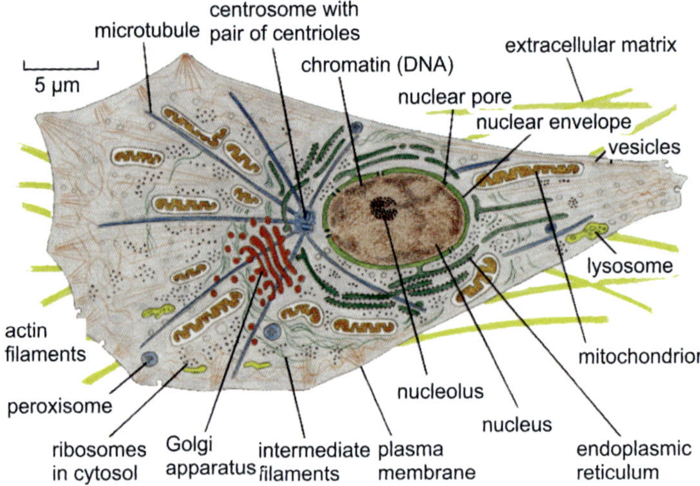

Fig. 7.9 Structure of a typical eukaryotic cell showing various cellular organelles[593].

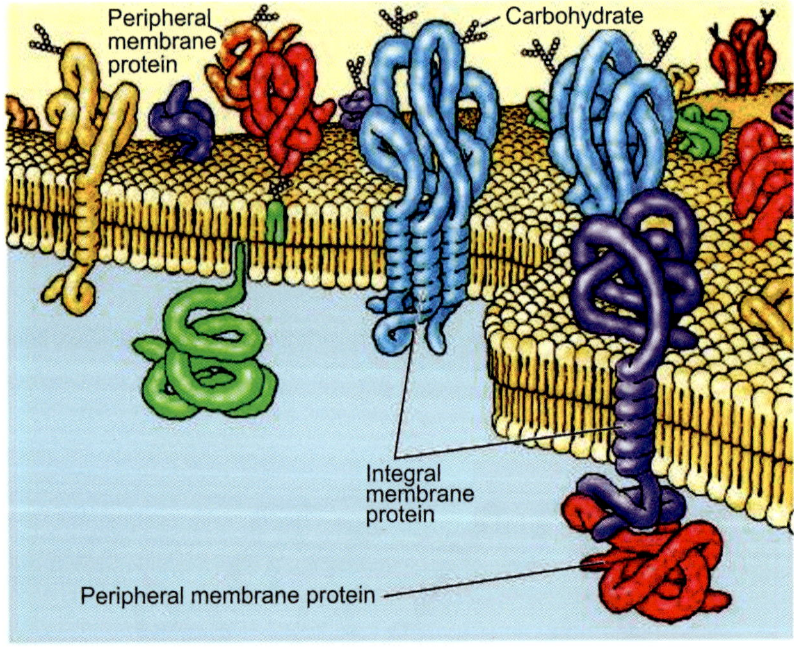

Fig. 7.12 The eukaryotic cell plasma membrane, characterised by a phospholipid bilayer embedded with proteins and cholesterol[596].

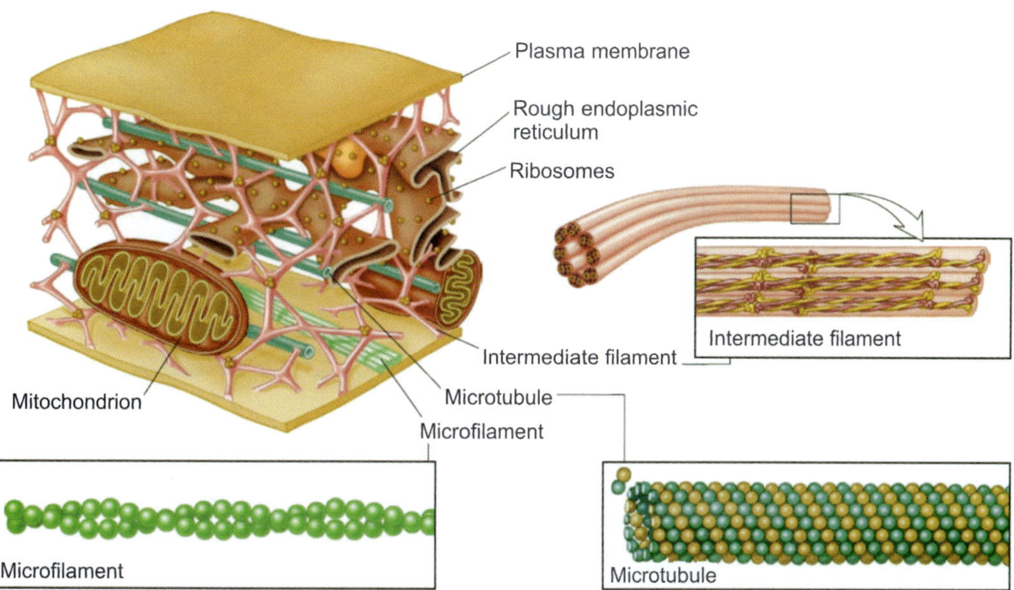

Fig. 7.13 Typical morphological features of different structural elements of cytoskeleton[597].

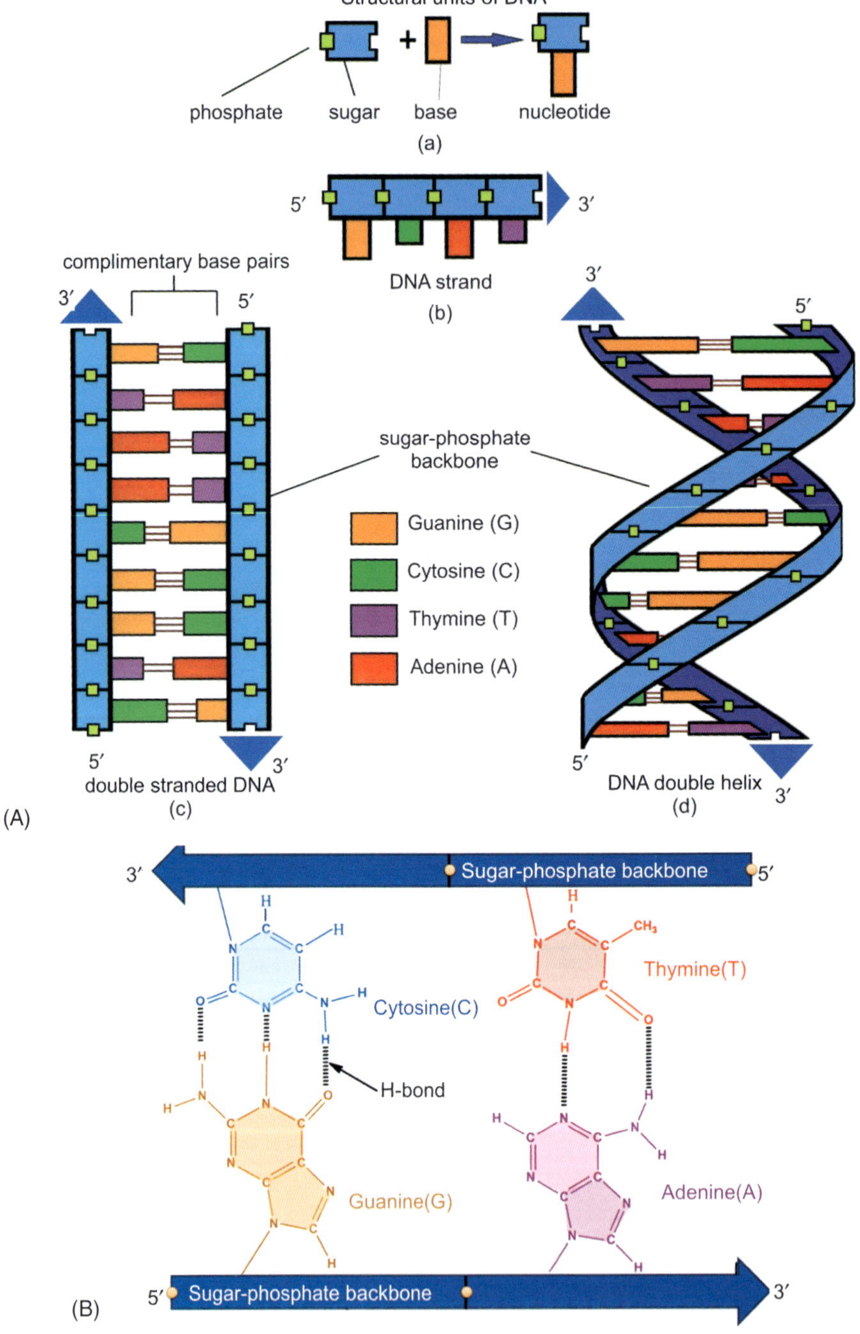

Fig. 7.20 (A) Schematic representation of the characteristic structure of DNA: (a) covalent bonding between the structural units to form nucleotide, (b) typical DNA strand, (c) complimentary base pairing in double stranded DNA, and (d) double helix structure of DNA; (B) bonding of nucleotide between two parallel backbone chains as per base pairing rule [Adapted from Ref.593].

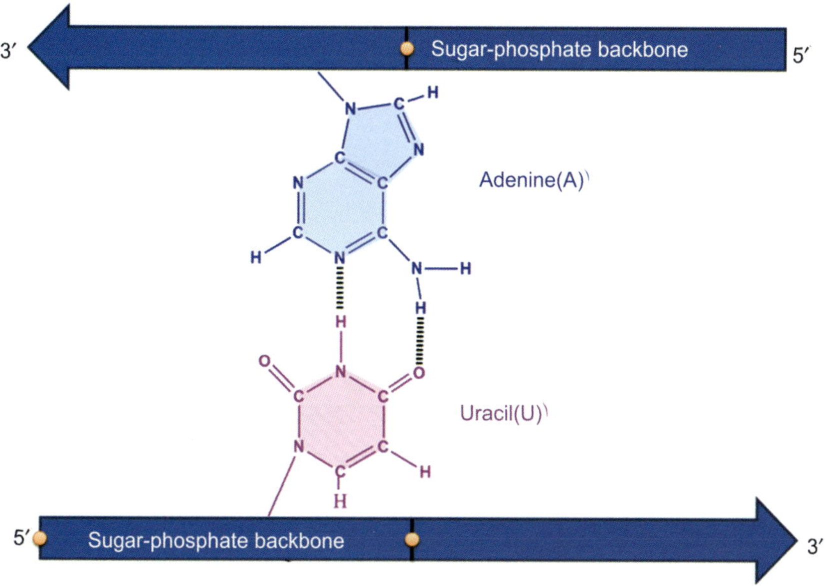

Fig. 7.21 Base pairing in RNA: A-U instead of A-T (as in case of DNA), while G-C pairing remains identical in both DNA and RNA [Adapted from Ref. 593].

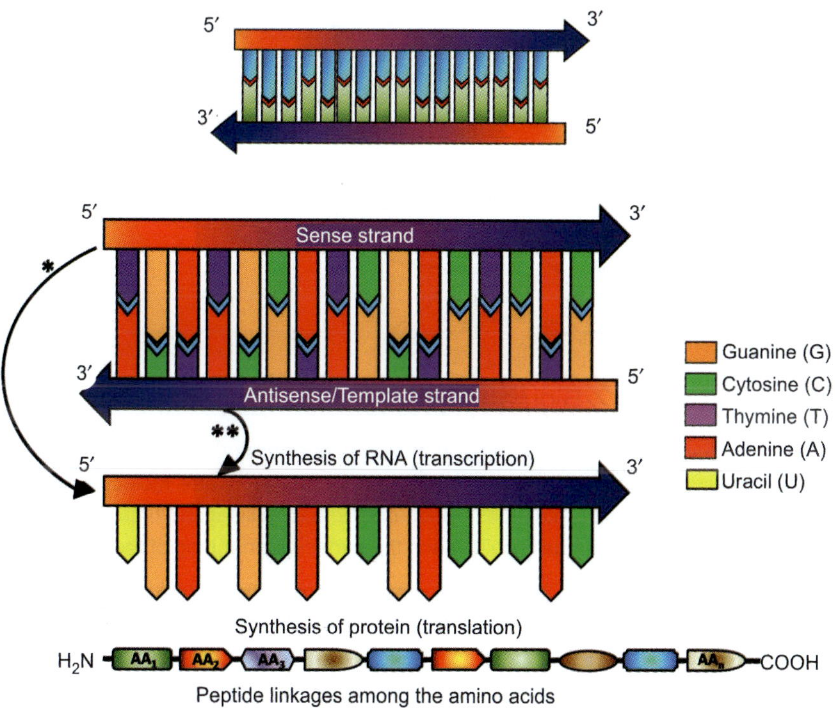

Fig. 7.23 Schematic representation of central dogma of molecular biology involving replication, transcription and translation. AA$_1$, AA$_2$,…AA$_n$ are different amino acid sequences. [Adapted from Ref. 594] (Acknowledgement: Deepak Saini).

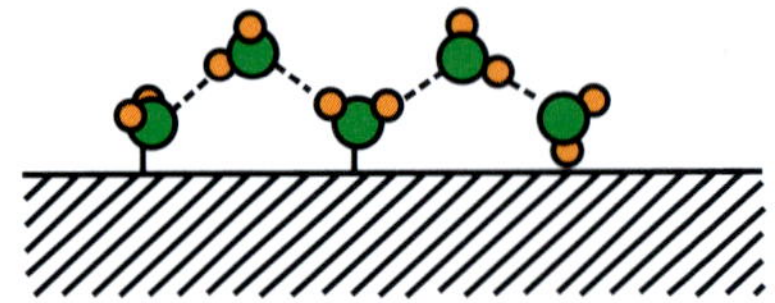

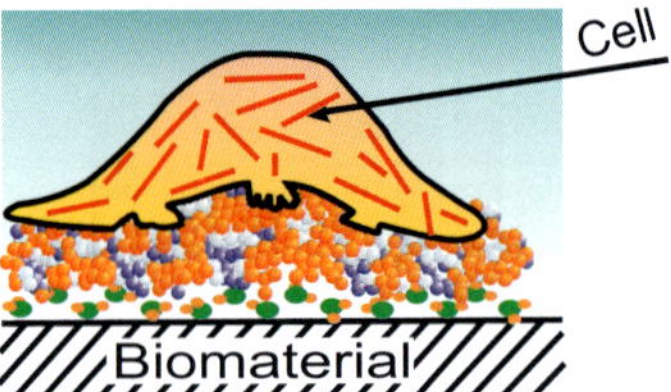

Fig. 8.1 A schematic depiction of the succession of events prior or during the interaction of a cell with a biomaterial; the primary moieties to cover the surface are water molecules (ns time scale); the water layer regulates protein adsorption on micro to millisecond time scale, and proceeds for even longer times; eventually cells reach the surface; denaturing of the proteins lowers their activity [Adapted from Ref. 605].

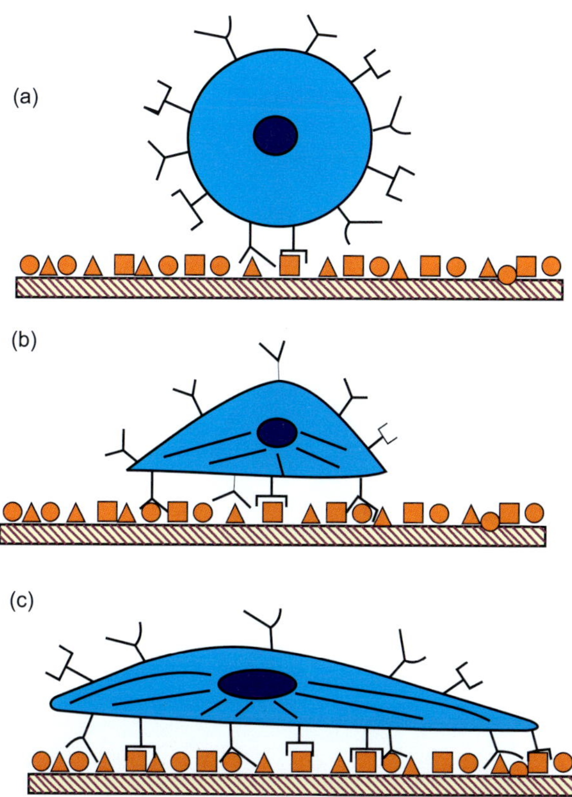

Fig. 8.3 A schematic depicting the steps in the progression of anchorage-dependent mammalian cell adhesion; (a) the initial contact of cells with the adsorbed protein layer, (b) the interactions between cell surface receptors and cell adhesion ligands that are adsorbed on the biomaterial, (c) extensive cell spreading on the biomaterial substrate with enhanced ligand–receptor interactions [Adapted from Ref. 610].

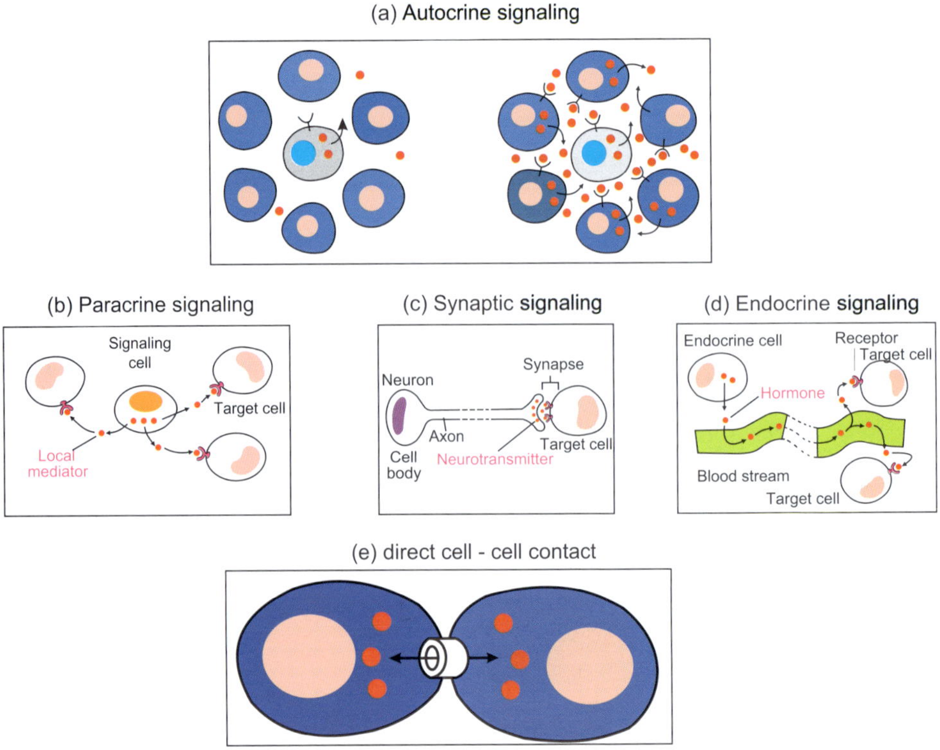

(a) Autocrine signaling

(b) Paracrine signaling

(c) Synaptic **signaling**

(d) Endocrine **signaling**

(e) direct cell - cell contact

Fig. 8.5 Types of soluble signalling. (a) autocrine signalling, (b) paracrine signalling (c) synaptic signalling (d) endocrine signalling, and (e) direct cell–cell contact [Adapted from Ref. 684].

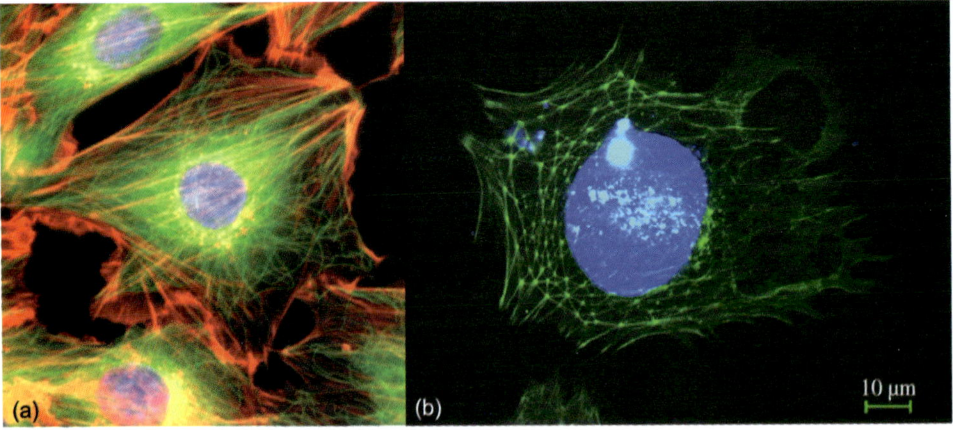

Fig. 8.18 (a) Fluorescence image of a biological cell with nucleus appearing in blue, cytoskeletal actin filaments in red and alpha-tubulin in green[685]; (b) Fluorescence image of a cell revealing the binding sites on actin filamentous structure of a cell.

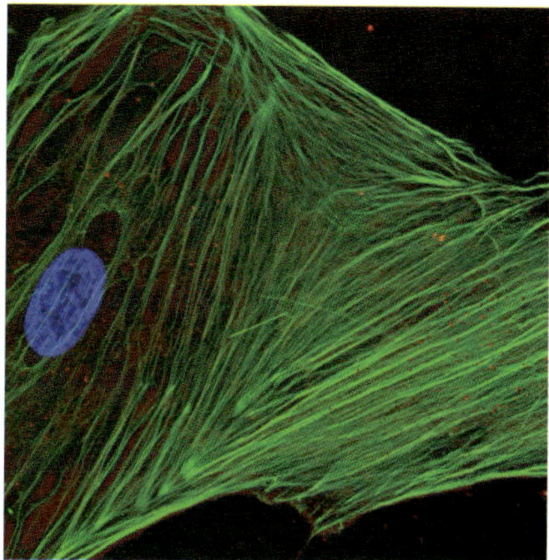

Fig. 8.21 Laser confocal fluorescence microscopic image of mesenchymal stem cells treated for 24 hours with citrate capped gold nanoparticles; cells were labelled with Alexa fluor-488 phalloidin for F-actin (green) and the nucleus was counterstained with DAPI (blue); co-localization of gold nanoparticles was observed in cells as red dots.

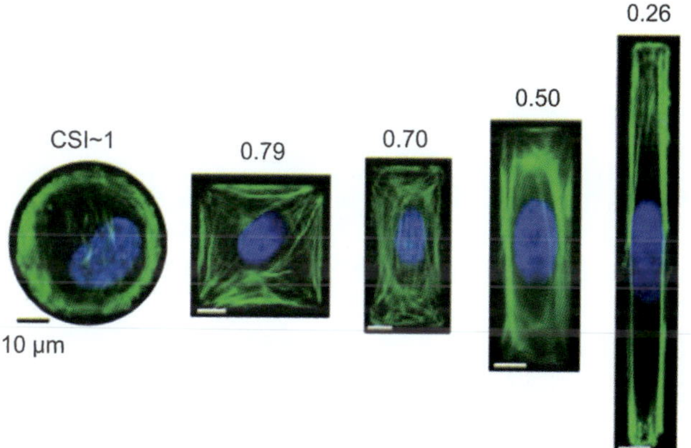

Fig. 8.22 CLSM image of cells, showing CSI ranging from 1 in circular-shaped cells to 0.26 in highly elongated cells; the nucleus is stained in blue with DAPI and actin is labelled with phalloidin Alexa 488 after 24 hours in culture[686].

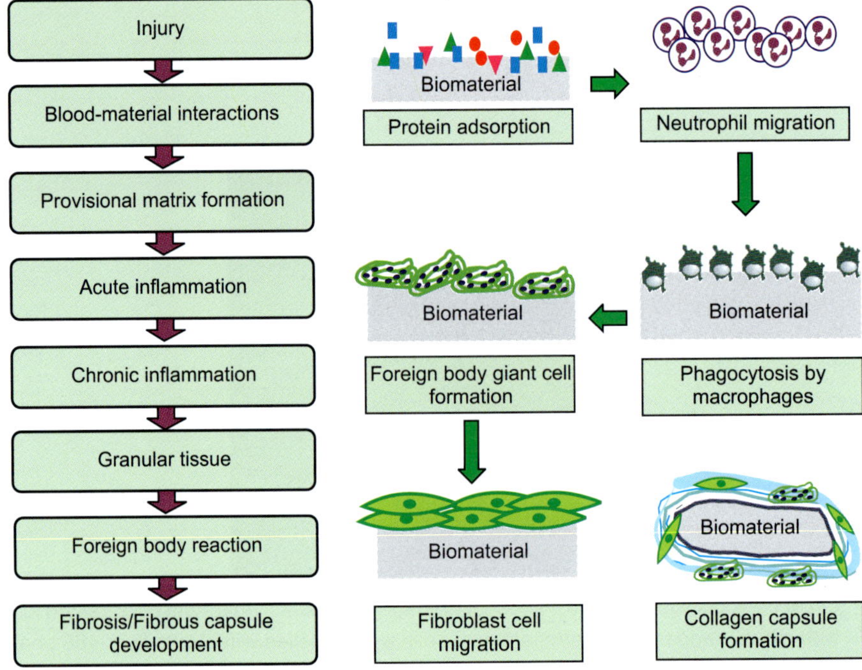

Fig. 8.28 A flowchart showing the sequence of host reactions following the implantation of a biomaterial/ medical device, along with a schematic representation of some of the stages involved in acute inflammation and chronic inflammation that ultimately lead to the formation of the fibrous capsule around the implant.

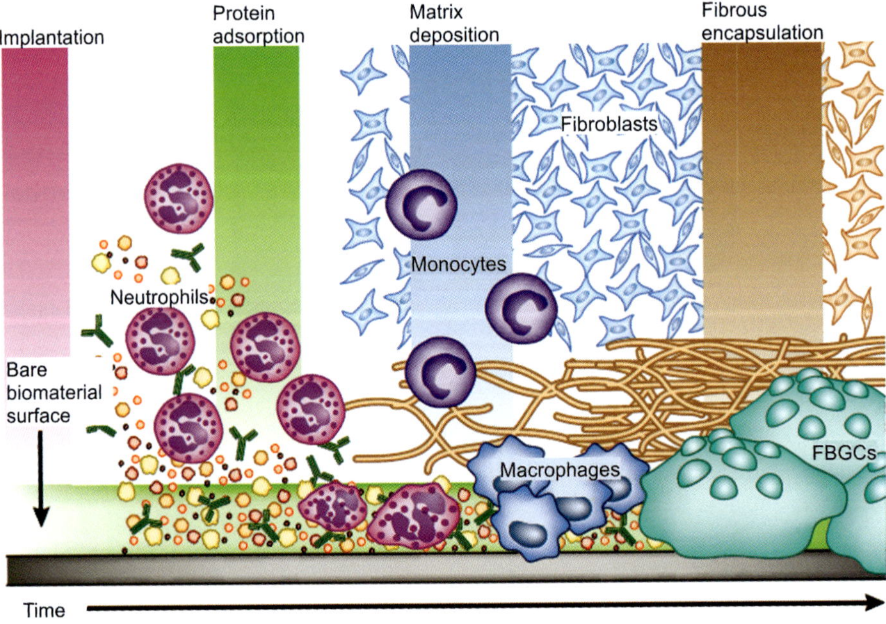

Fig. 8.30 A schematic representation of the host response upon implantation of a biomaterial (reproduced from Ref. 688].

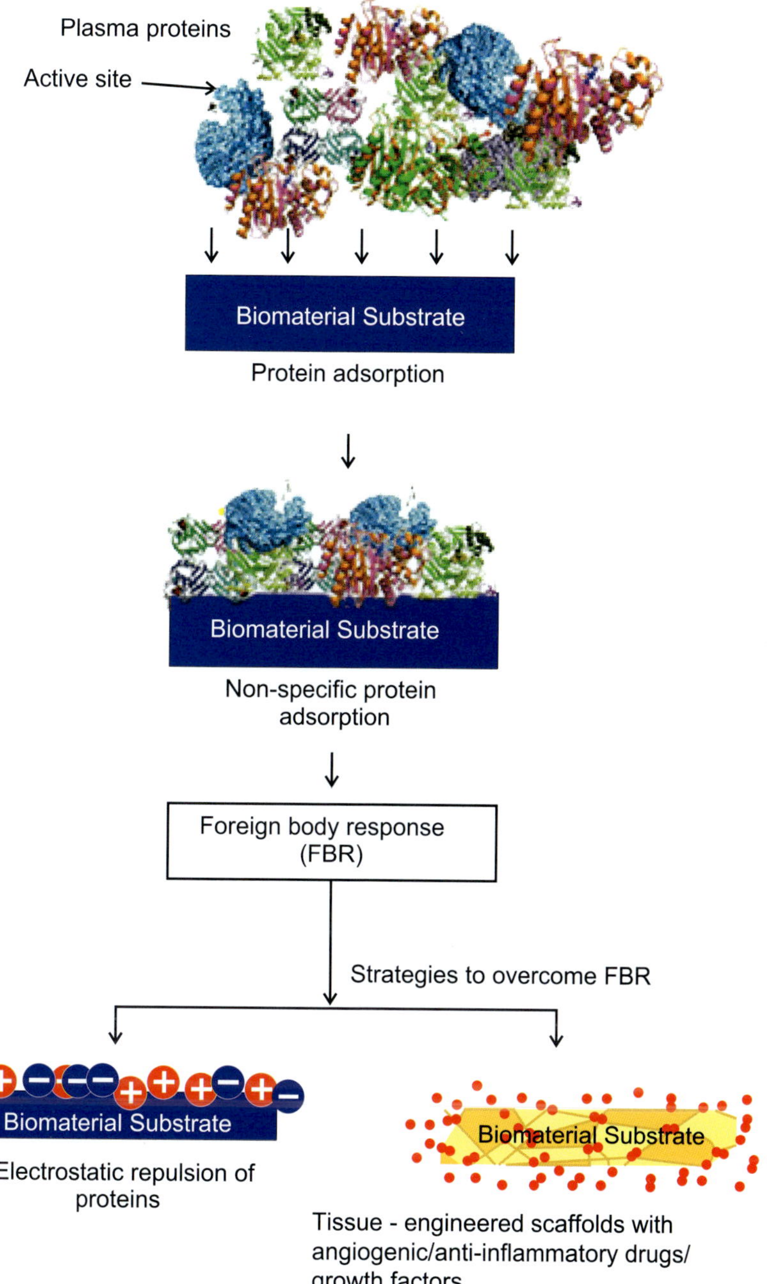

Fig. 8.31 A schematic showing the hypothesis that non-specific protein adsorption causes a foreign body response; the conformation and orientation of proteins are also distorted during non-specific protein adsorption; the foreign body response can be overcome by preventing protein adsorption or by modifying the biomaterial to release bioactive agents, such as anti-inflammatory drugs/angiogenic drugs/Vascular Endothelial Growth Factor (VEGF).

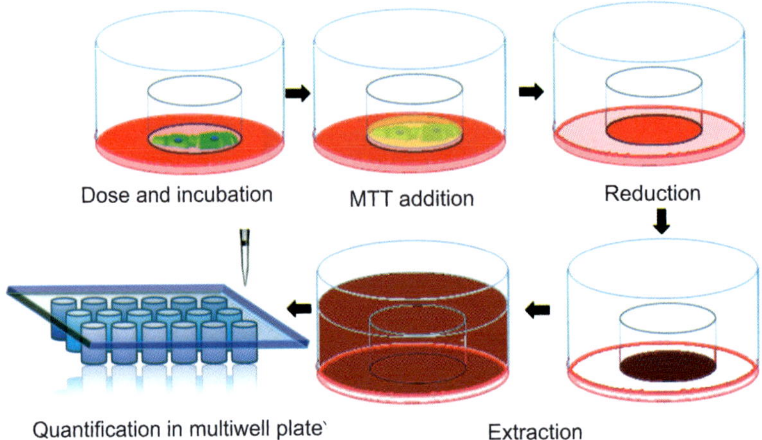

Fig. 9.3 Schematic diagram showing sequential steps of MTT assay.

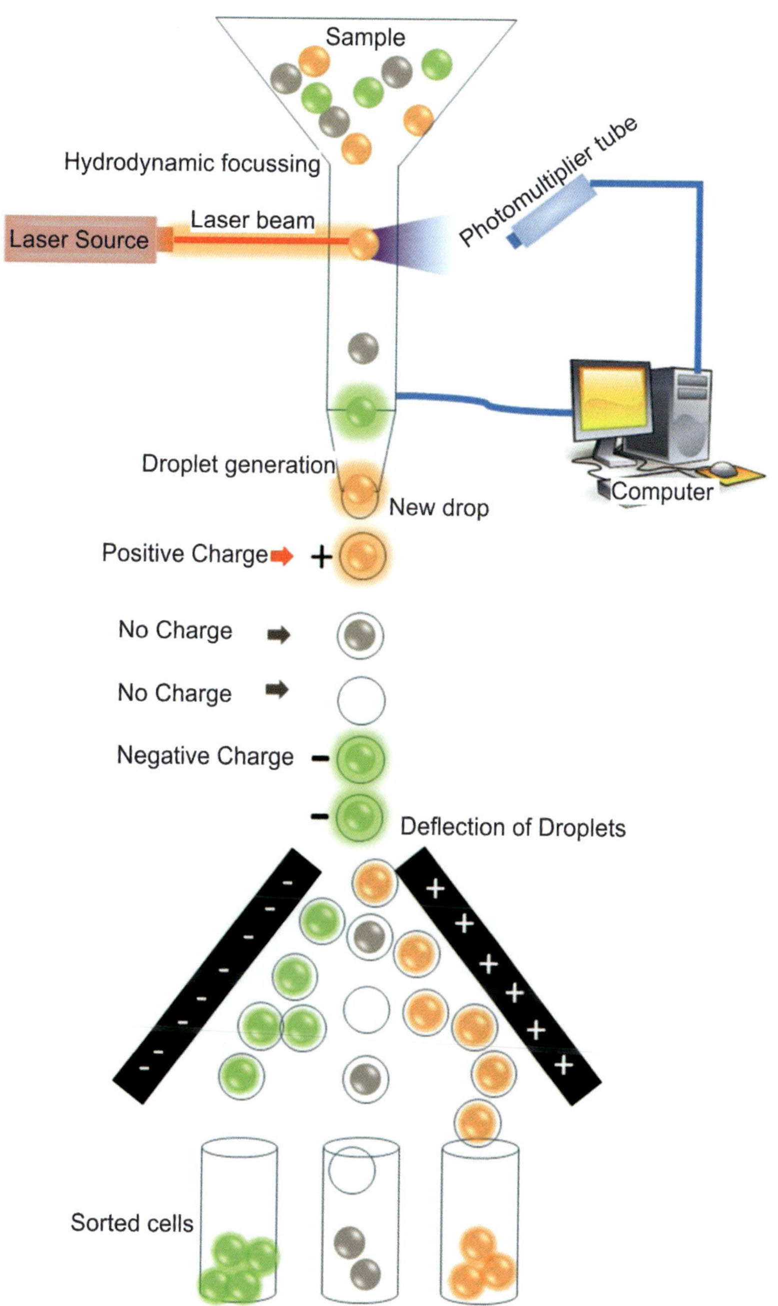

Fig. 9.7 Schematic showing the working principle of flow cytometry.

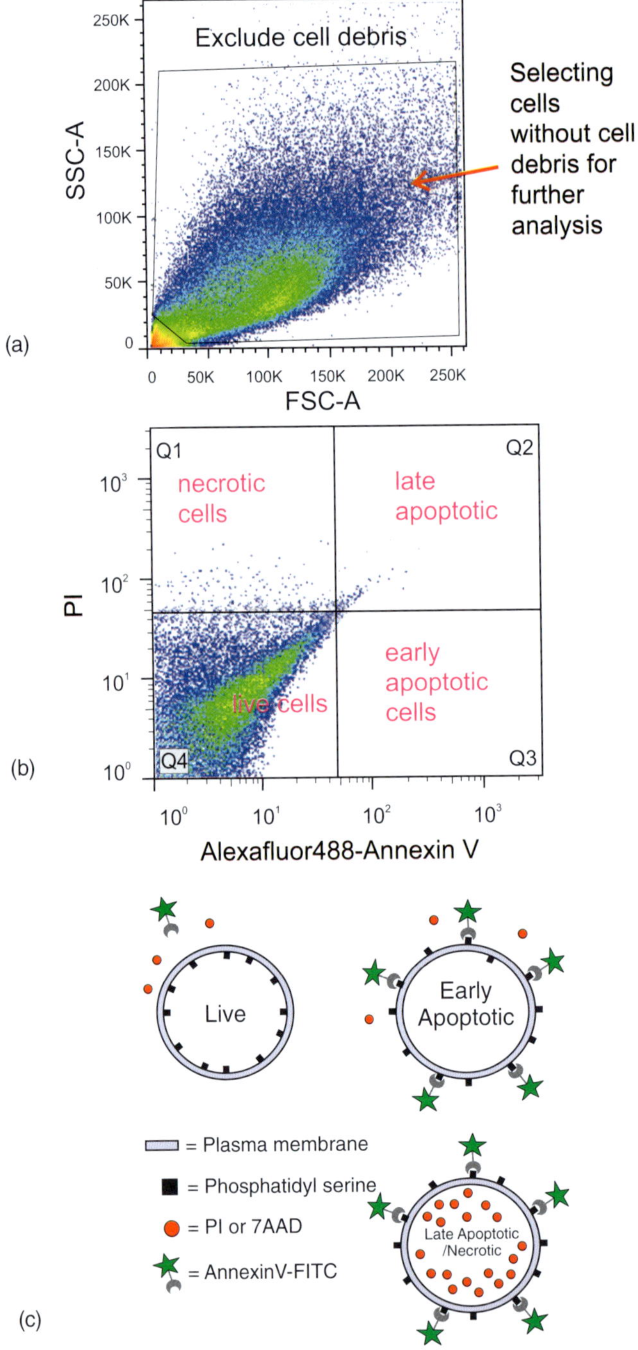

Fig. 9.12 Schematic illustration of the flow cytometry based quantification of cell apoptosis; the apoptotic cells expose phosphatidyl serine (PS), which can bind to annexin V; PI is impermeant to live cells and apoptotic cells, but stains dead cells with red fluorescence, binding tightly to the nucleic acids in the cell (c), the typical FACS plot to show how to exclude cell debris (a) and typical FACS plot to show the distinction of the apoptotic and necrotic cells (b).

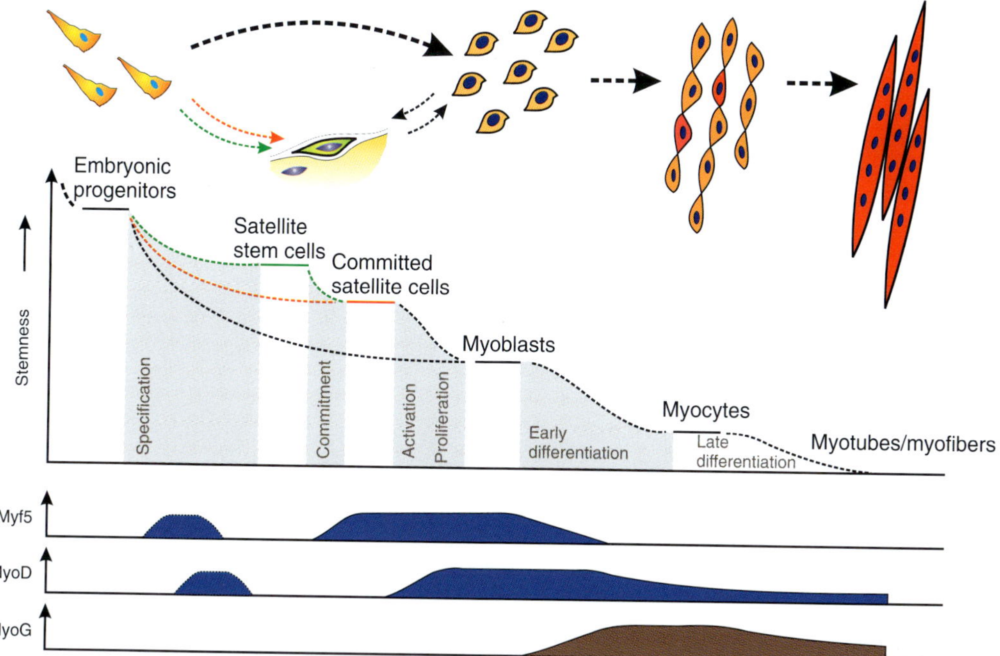

Fig. 9.27 Schematic showing myogenesis, a muti-step process involving the transcriptional activation and repression of several genes including MyoD, MyoG, Myf5, etc. [Adapted from Ref. 722].

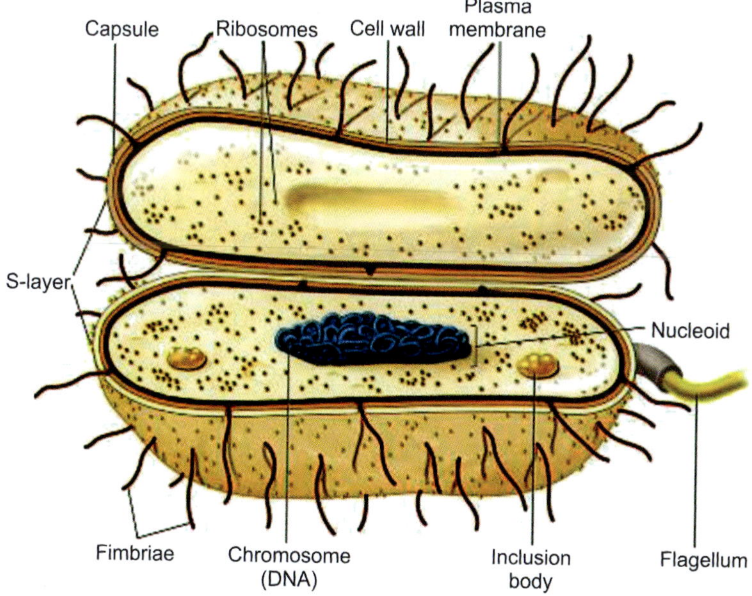

Fig. 10.1 Schematic illustration showing the anatomy of a bacteria cell[787].

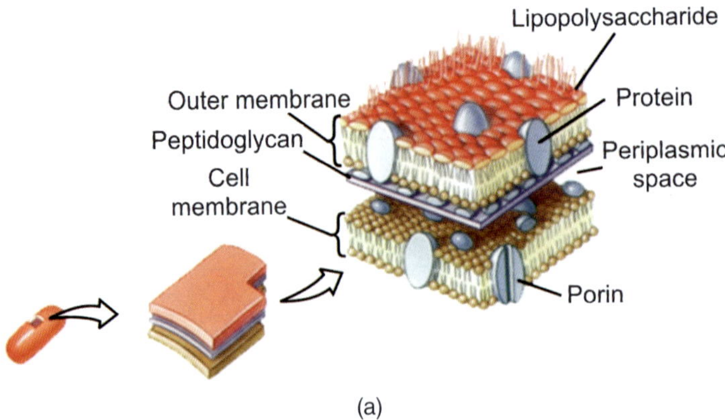

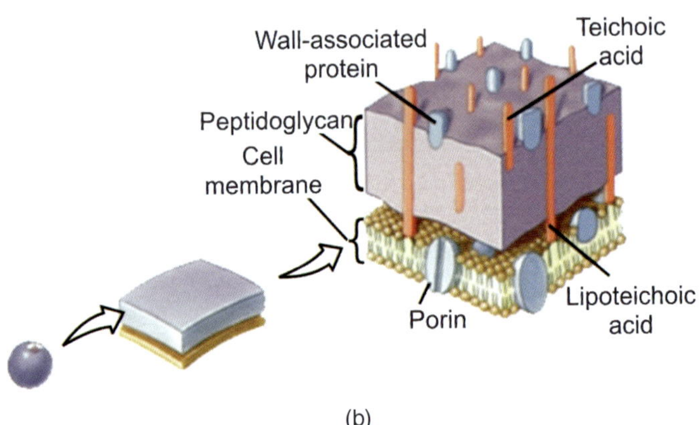

Fig. 10.3 Cell wall structure of (a) Gram-negative and (b) Gram-positive bacteria; adapted from[789].

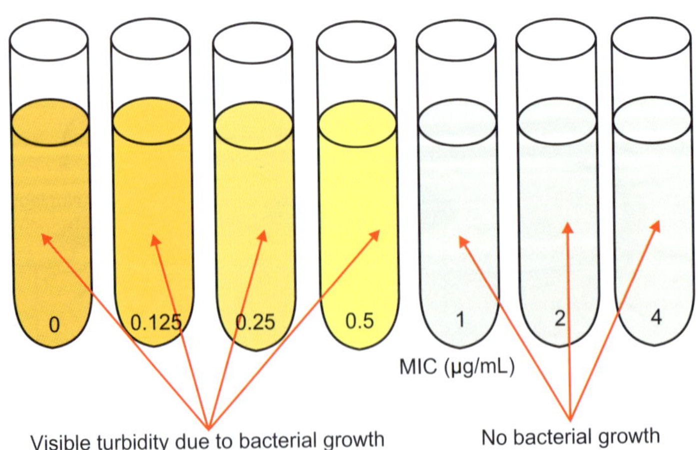

Fig. 10.7 Pictorial illustration of MIC indicated by the lowest dosage at which no bacterial growth occurs.

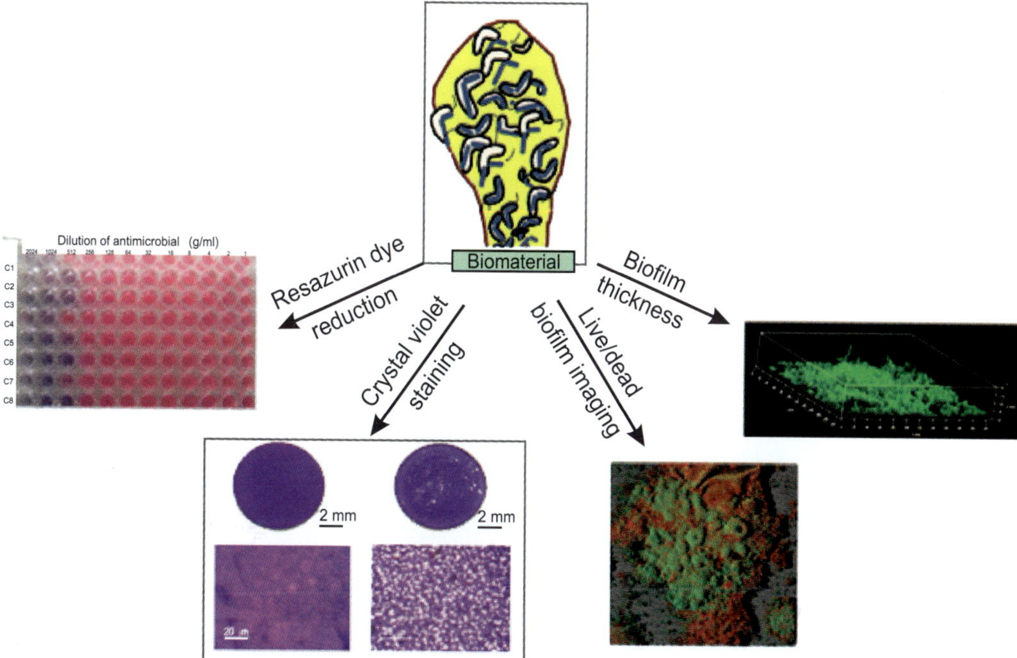

Fig. 10.15 Summary of the commonly employed methods for the evaluation of bacterial biofilms grown on a biomaterial resazurin and live/dead imaging are indicators of biofilm viability, while crystal violet staining and biofilm thickness reflect on the biomass and biofilm phenotype, respectively.

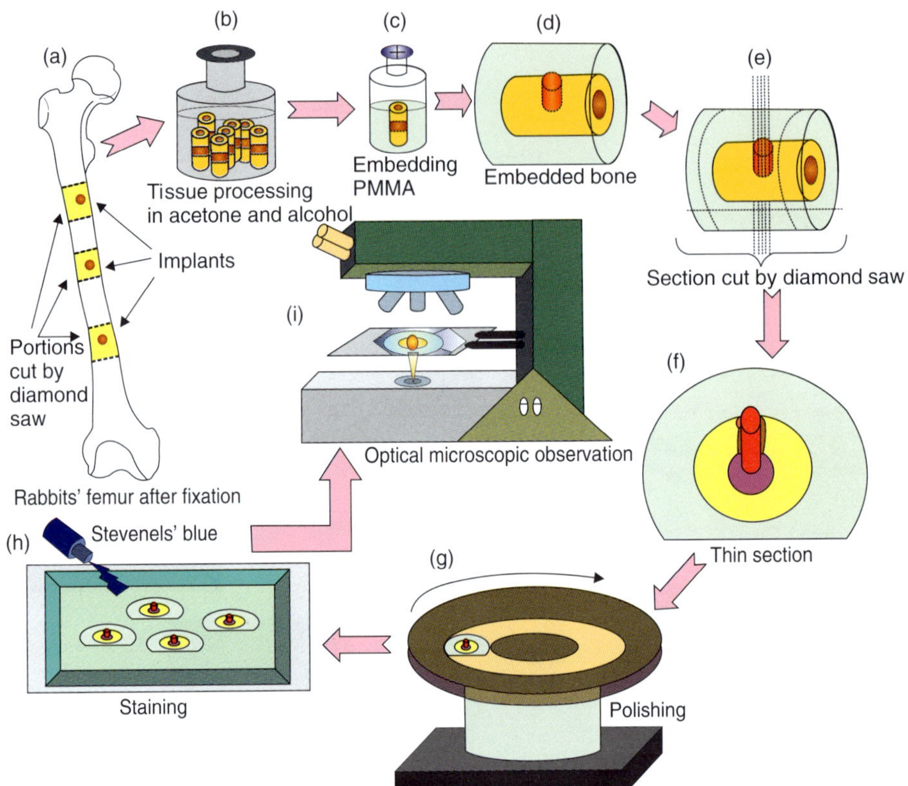

Fig. 11.5 Schematic showing step-by-step protocol of the sample preparation for histopathology analysis of the explants for the bone replacement applications. It is worthwhile to note that the basic tissue harvesting and sectioning protocol varies from tissue to tissue;. (a) showing the femur of the rabbit after fixation, the red circles indicate implants, (b) dehydration of bone pieces along with implants, (c) embedding in PMMA polymer, (d) embedded bone piece in PMMA, after removing from the bottle, (e) cutting of thin sections by diamond saw, thin dotted lines indicate the cutting path, (f) thin section showing the bone (yellow), bone marrow(pink) embedded in PMMA matrix, (g) polishing of thin section using diamond paste, (h) staining in Stevenels' blue, (i) optical microscopy observation; Note: Depending on the histological features to be examined in the thin tissue sections, one can use alternative stains instead of Stevenels' blue stain[839].

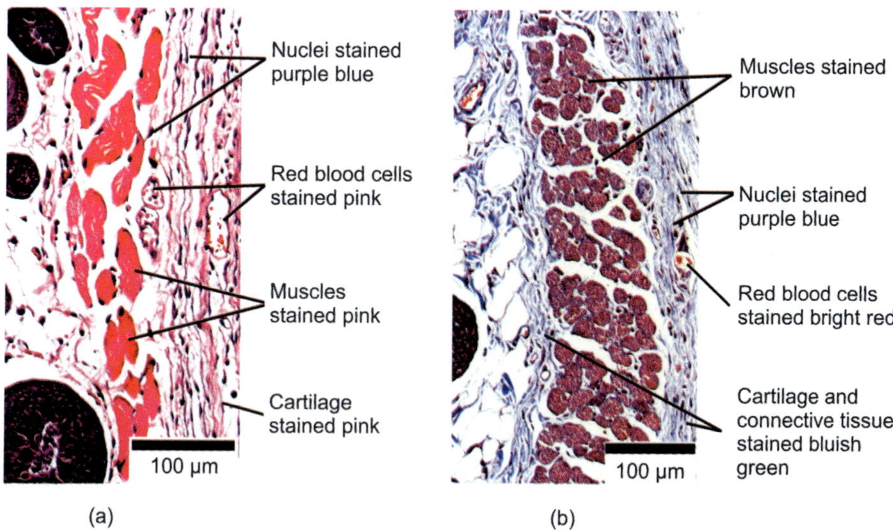

Fig. 11.6 Subcutaneous tissue sections stained with (a) Hematoxylin and Eosin stain, and (b) Masson's trichrome stain.

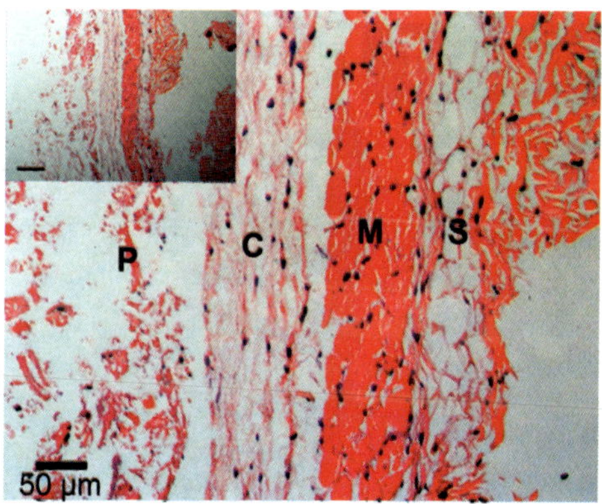

Fig. 11.7 Representative H&E stained histology image of the polymer-tissue interface in subcutaneously implanted PLGA at 16 weeks after implantation. The figure shows the presence of some tissue in the region where the polymer was present earlier. As PLGA degrades, its place is taken up by connective tissue. Collagen is stained pink while cell nuclei are stained blue. P-Polymer site; C- fibrous capsule; M- muscle; S- skin. Inset magnification - 10 X, scale bar = 150 µm.[791]

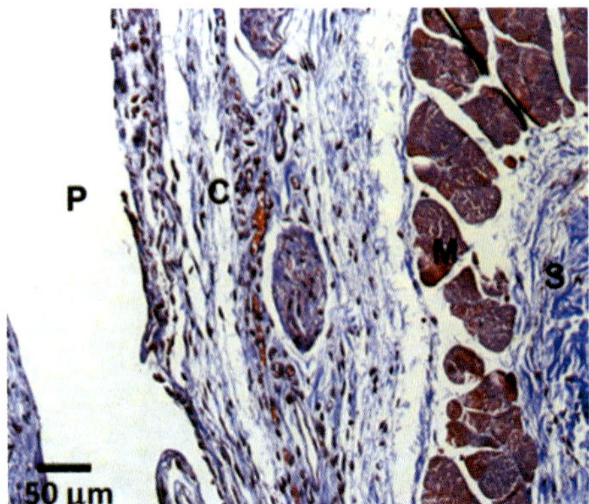

Fig. 11.8 Mason's Trichrome stained images [20 X magnification] of the tissue/implant interface of PLGA at 2 weeks post implantation. Numerous inflammatory cells are seen at the tissue-PLGA interface, while fewer cells were visible at the tissue-SAP interface.[791].

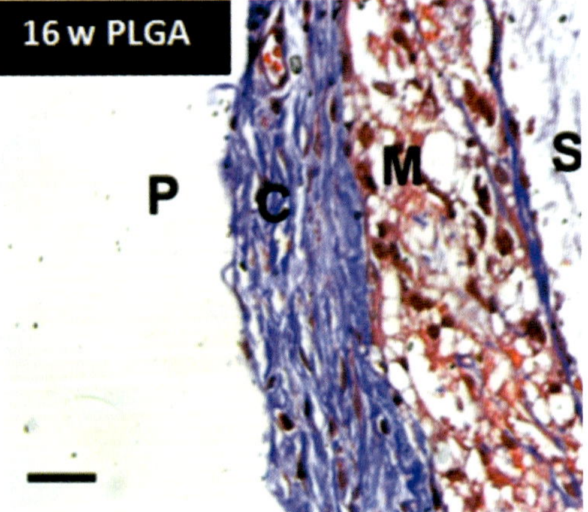

Fig. 11.9 Representative Mason's Trichrome stained images [50 X magnification] showing the implant/tissue interface for PLGA at 16 weeks post-implantation. Muscles, collagen and cell nuclei are stained red, blue and black, respectively. P-Polymer, C-collagen, M-muscle, S-skin. Scale bar = 25 µm.[791].

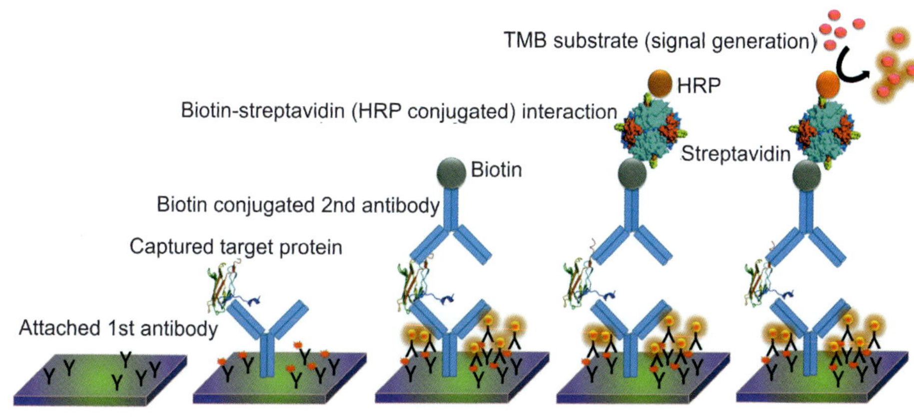

Fig. 11.11 Schematic illustration depicting the components of an ELISA used for tumour necrosis factor (TNF-α) analysis; the target protein in this assay is TNF-α.

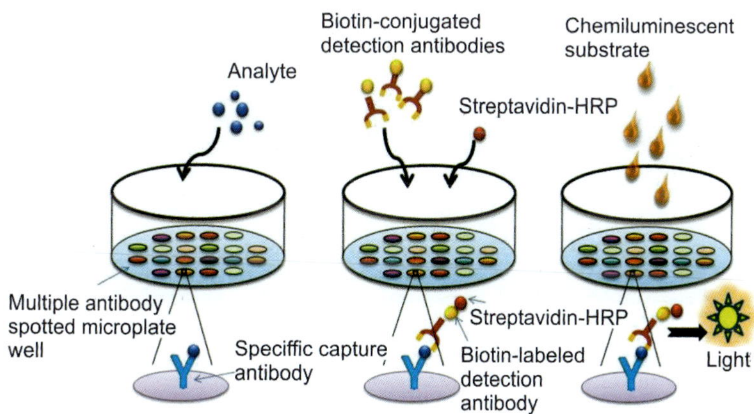

Fig. 11.12 Illustration depicting the components of an ELISA used for IL-1β analysis; the setup shown is for a multiplex system where capture and detection antibodies for multiple antigens are used and detected using an imager (ELISPOT assay).

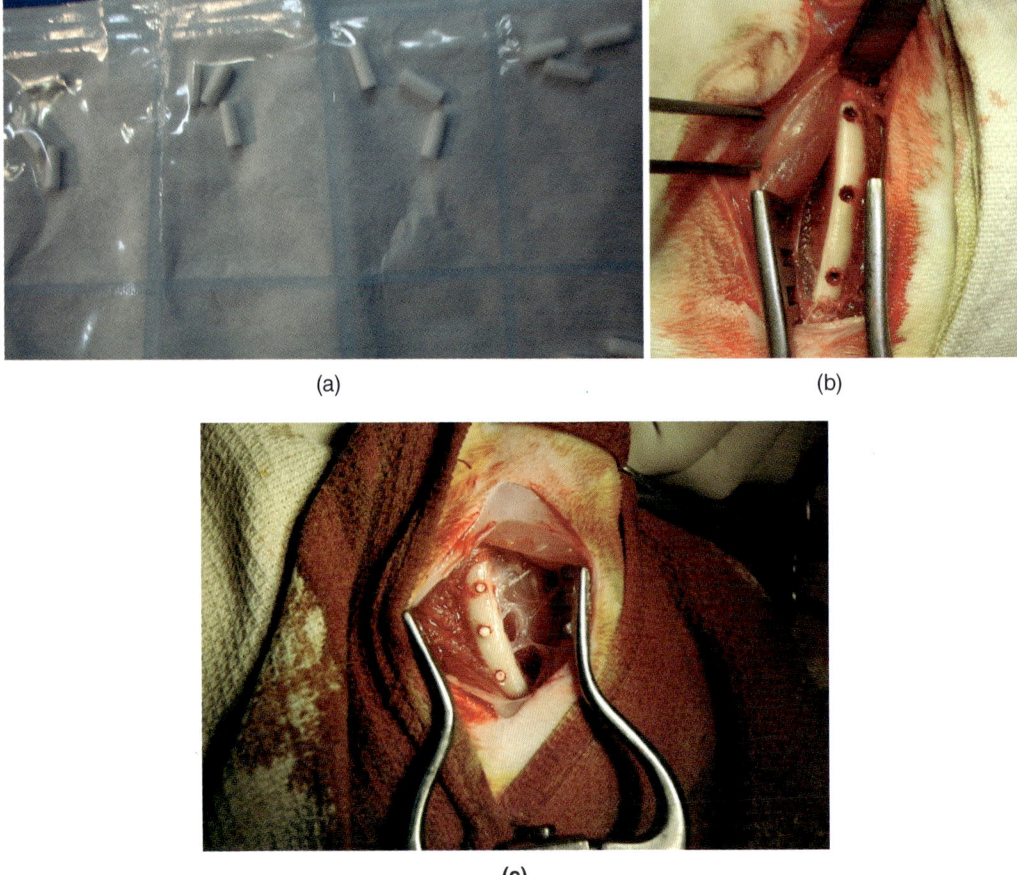

Fig. 11.15 Photograph showing (a) the sterilized cylindrical shaped implants prior to implantation, (b) creating cylindrical bone defects of a limb in an experimental rabbit during surgery and (c) three implants placed in the cylindrical bone defects, post-surgery[839].

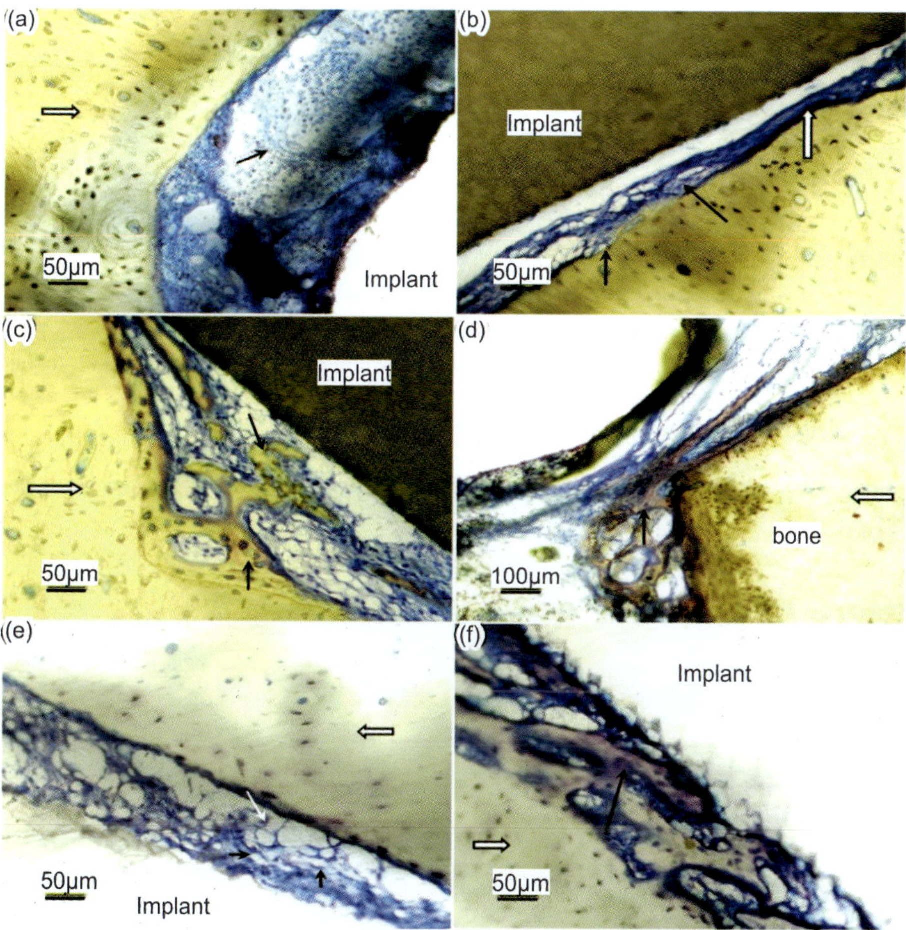

Fig. 13.13 Representative histology images of host bone-implant interface; (a–d) HAp 20 wt% mullite (test implant) and (e and f) UHMWPE (control) after one week of post-implantation[916].

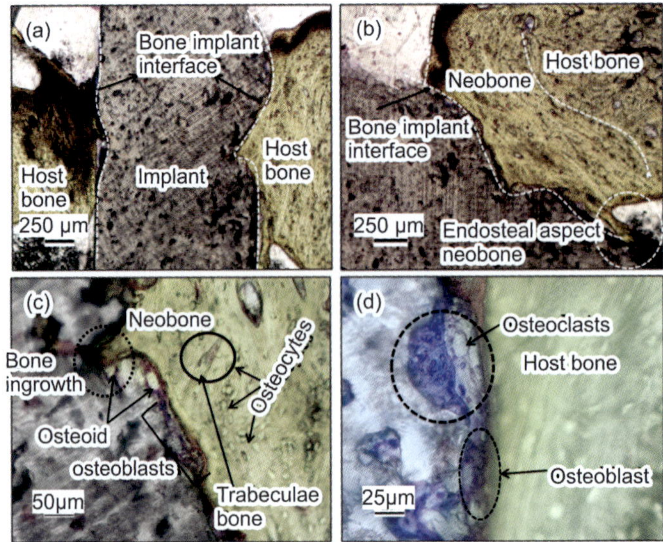

Fig. 14.7 Histology image of (a) host bone/implant interface (HDPE-20 wt% HA-20 wt% Al_2O_3), 14 weeks post-implantation, (b) neobone/host bone interface at endosteal aspect, (c) neobone and trabeculae bone, along with osteoids and osteoblasts, and (d) osteoblasts and osteoclasts together at bone/implant interface[987].

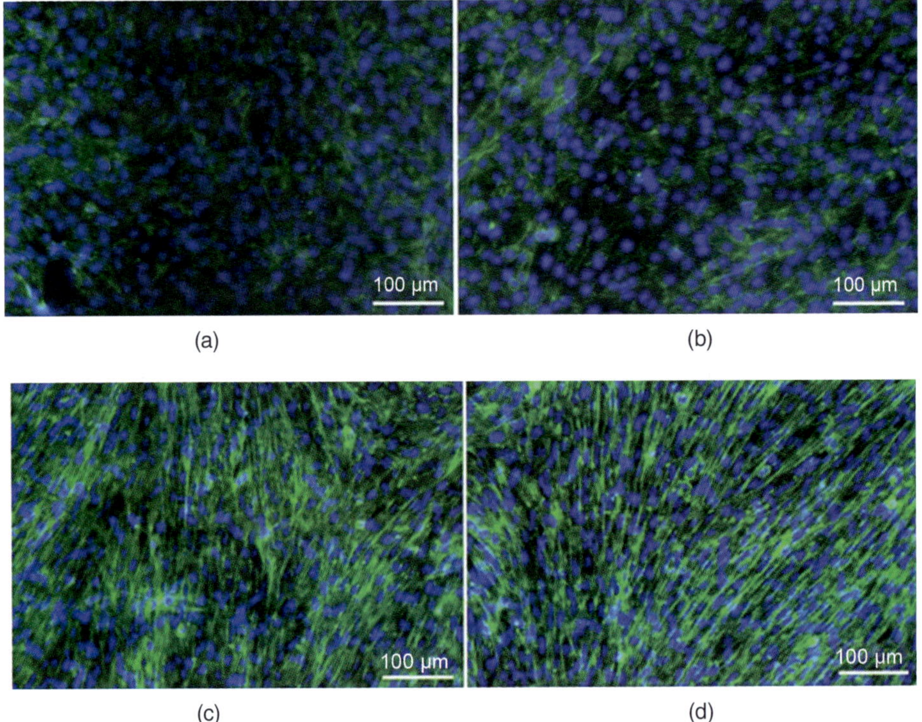

Fig. 16.5 Fluorescence image of myoblast cells proliferated (after 72 hours) of culture on (a) HA, (b) HA40CT, (c) HA 80CT, and (d) CT[1064].

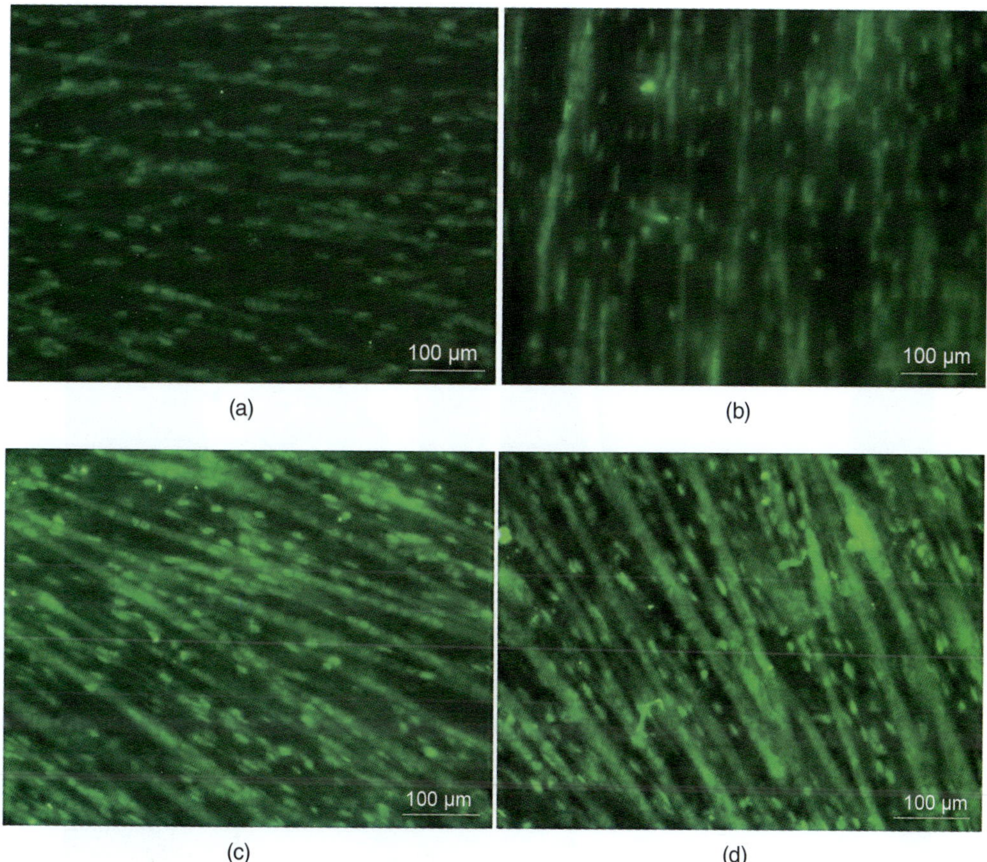

(a)

(b)

(c)

(d)

Fig. 16.6 Immunostaining of myogenin expressed on (a) HA (b) HA40CT (c) HA80CT (d) CT samples[1064].

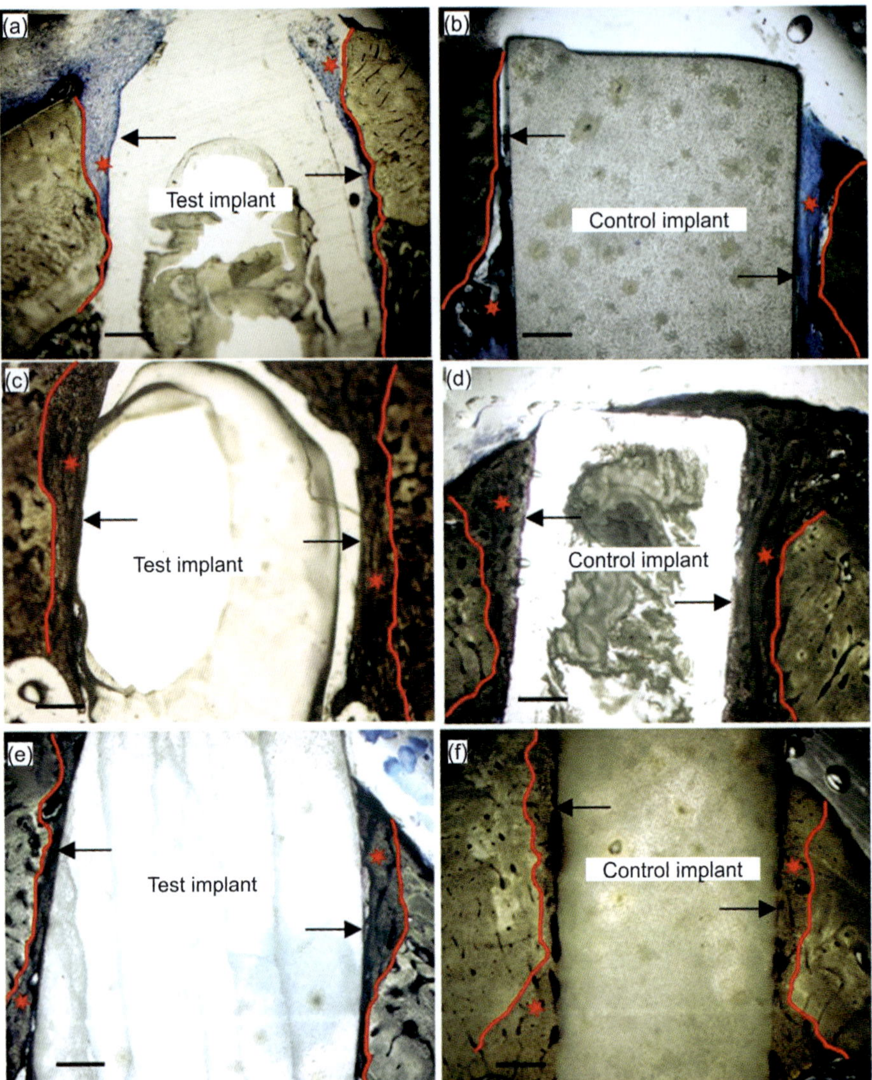

Fig. 16.10 Representative histological image of Stevenel's blue stained regions of HA80CT implant/host bone interface after (a) 1 week, (c) 4 weeks and (e) 12 weeks as well as for HA implant (control)/host bone interface after (b) 1 week, (d) 4 weeks and (f) 12 weeks; black arrow indicates implant/new bone interface and border line indicates the new bone/host bone interface, red star indicates the osteogenesis/new lamellar bone, short red arrow indicates presence of osteoblast on the surface trabecular bone and short black arrow indicates the presence of neovascularization. image scale bar (250 μm)[1065].

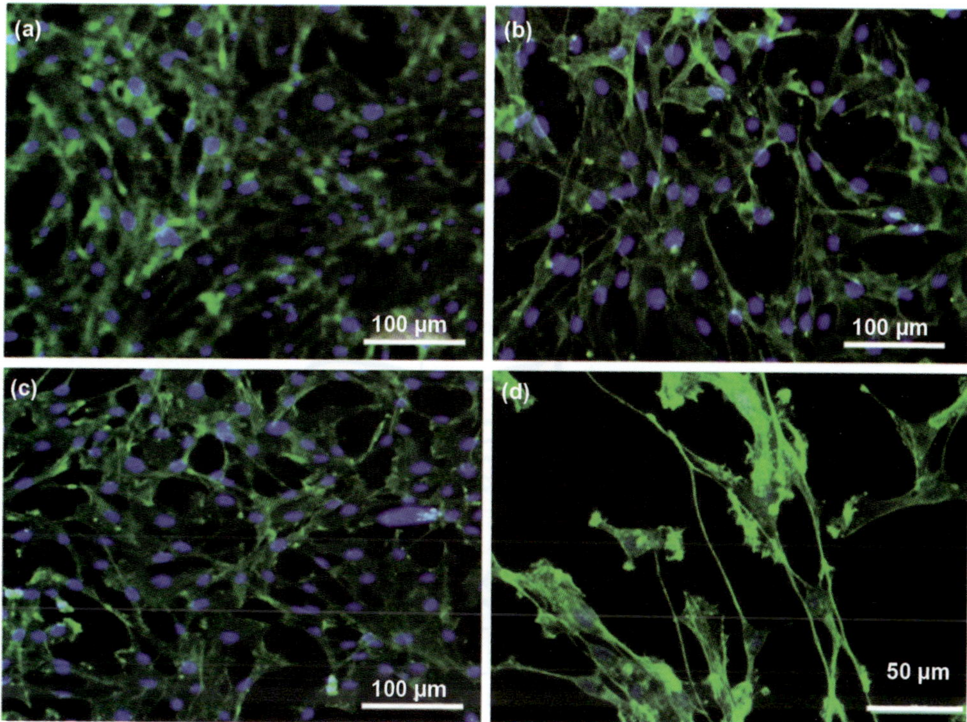

Fig. 17.4 Fluorescent images of Schwann cells, following 2 days incubation, after seeding on (a) control, (b) RF derived carbon film, (c) RF derived carbon fibres and (d) magnified view of (c) showing short-ranged alignment of cells along the fibre lengths; blue stained region represents the nucleus, while green represents the actin cytoskeleton[1130].